GEOLOGY OF PETROLEUM

A Series of Books in Geology
EDITORS: James A. Gilluly and A. O. Woodford

SECOND EDITION

GEOLOGY OF PETROLEUM

A. I. LEVORSEN

Sections on Hydrodynamics and Capillary Pressure revised and edited by
FREDERICK A. F. BERRY
University of California, Berkeley

W. H. FREEMAN AND COMPANY
New York

COPYRIGHT © 1954, 1967 BY W. H. FREEMAN AND COMPANY

The publisher reserves all rights to reproduce this book, in whole or in part, with the exception of the right to use short quotations for review of the book

Printed in the United States of America

LIBRARY OF CONGRESS CATALOGUE CARD NUMBER: 65-25242

ISBN 0-7167-0230-4

14 15 16 17 18 19 20 VB 6 5 4 3 2 1 0 8 9 8

FOREWORD

A. I. LEVORSEN died on July 16, 1965. Shortly before his death he had spent several days with his publisher considering editorial changes of his manuscript for the second edition of *Geology of Petroleum*. Because he had not been able to complete his work, I was asked by his family and by his publisher to write a foreword, read proofs, and make an index. As A. I. Levorsen's friend and former student, I was more than glad not only to undertake these tasks but also to assume the role of editor of the new material on hydrodynamics and capillary pressure. Having been asked earlier by the author for my suggestions on the revision of this book, I felt a special obligation to assist in whatever way I could in the publication of the second edition.

Many important advances have been made in the geology of petroleum since the first edition was published. The vigorous application of modern concepts of physics and chemistry to geologic problems has led to the resolution of many of the problems related to the origin, migration, and accumulation of petroleum—problems that were not well understood when the manuscript for the first edition was completed in 1953.

The principal advances that have been made in the interval between editions are the result of detailed studies of the fluids found within rocks and of the pore space in which these fluids are contained. As A. I. Levorsen pointed out in the preface to the first edition, the geology of petroleum is essentially the geology of fluids. And it is here that the real increases in understanding have been made. The work of geologists and engineers, chemists and physicists, has significantly increased our knowledge of the behavior of fluids within various types of rocks. Practical investigations aimed at determining the flow systems in various geologic provinces have enabled us to understand the diverse origins of various fluid potentials and to predict the probable effects of a given flow system upon the accumulation of petroleum. Today we have a far better understanding of the relative transmissive properties of rocks and of the geometry of fluid flow through various rocks under different hydrodynamic conditions. Much, too has been learned about the origin of various waters and their dissolved mineral content.

At the time the first edition was published, hydrologic investigations of the deep sediments of geologic basins were just being initiated as a result of investigations that demonstrated conclusively that flowing water had a significant effect on the accumulation of petroleum. The variety of ways in which fluid potential gradients alter the geometry and the size of petroleum traps has since come to be clearly understood. Similarly, the mechanics of petroleum movement as a discrete phase within reservoir rocks (secondary migration) under both hydrostatic and hydrodynamic conditions have now been described quite thoroughly.

Important contributions to our knowledge of the organic geochemistry of petroleum formation have been made since 1954. Many of the problems related to the transformation of organic matter into petroleum have been resolved. Petroleum substances have been found in Recent sediments in both marine and lacustrine environments, and petroleum has been discovered to be accumulating on a modest scale within a sand lens in Recent sediments of the Orinoco Delta, in eastern Venezuela. Chemical investigations of many kinds have supplied much useful information. For example, knowledge gained from isotopic analyses of the carbon atom, as well as studies of the distribution and behavioral differences between various hydrocarbon molecules, have been particularly important.

Although many, if not most, of the important problems related to the mechanisms involved in the origin, migration, and accumulation of petroleum have been resolved, no acceptable mechanism for primary migration has been proposed. We still cannot explain how petroleum moves from the shales—the source beds—into the sandstones, limestones, and other porous and permeable rocks.

The major changes in this edition reflect the emphasis that has recently been placed on applying the principles of hydrodynamics to the geology of petroleum. Some parts of the first edition have been eliminated, numerous new sections—both major and minor—have been added, and the chapter on the origin of petroleum has been completely rewritten. The new material deals primarily with the importance of fluid mechanics in the migration and accumulation of petroleum. Twenty-eight new illustrations have been prepared, more than half of which relate to the influence of hydrodynamics upon petroleum traps. The solid core of the first edition has been retained; the features that made the first edition unique remain.

In my revision of the new material—particularly in the parts dealing with the mechanics of fluids—I have incorporated changes that, I feel, are of the sort that the author would have made if he had lived. None of these changes, however, alter the substance of what A. I. Levorsen wrote; this is his book and his revision.

The first edition of this book was noted for its broad and thorough treat-

FOREWORD

ment of the many separate facets of geology and engineering that bear upon the geology of petroleum, and for its extensive treatment of petroleum traps. No other book has ever covered as thoroughly the various types and geometries of traps. A. I. Levorsen was a man of great intellectual honesty and intuition. Thus, he included in the first edition thorough reviews of subjects that were applicable to the problems encountered in the geology of petroleum even though the particular relevance of such subjects to these problems was not then well understood. An example of this is his treatment of temperature anomalies. The effect of temperature upon the physical properties of fluids has long been obvious, but the importance of temperature anomalies to the discernment of fluid movement was not even the subject of speculation in 1953. Nonetheless, A. I. Levorsen included a careful treatment of the temperature anomalies around various petroleum accumulations. Only recently have a few workers begun to suspect that these anomalies probably represent various fluid flow patterns.

A mark of good scholarship is that a man leaves firm blocks on which others can build. *Geology of Petroleum* has been and remains the most outstanding scholarly work of its kind. A mark of a good textbook and reference work is the quality and quantity of new ideas that it stimulates the reader to develop. The development of new ideas always interested A. I. Levorsen. No serious student or professional geologist who reads this book can fail to develop some fresh idea of his own. Despite my familiarity with the book since its inception, my work on this second edition gave me a new—and seemingly viable—idea.

Many of A. I. Levorsen's friends advised him on specific problems that arose in the course of revision of this book. He would, of course, have acknowledged each of them individually had he lived to write a preface for the second edition. I know only that he would have liked to thank Sherman A. Wengerd of the University of New Mexico, who contributed many particularly helpful suggestions, and Louis Renné, who prepared the twenty-eight new illustrations. As for the many other individuals who helped him in various ways, all I can do is acknowledge them collectively and extend Mr. Levorsen's thanks for their generous help.

My personal thanks go to Roger L. Hoeger, who read critically the new material on hydrodynamics; to John M. Hunt for reviewing Chapter 11, Origin of Petroleum; to Jon Galehouse, who prepared the index for this edition; and to Richard L. Beasley and Mary Grauberger for their help in checking the bibliographies and the cross references.

FREDERICK A. F. BERRY

Berkeley, California
October 1966

PREFACE TO THE FIRST EDITION

THIS BOOK is intended primarily for students who have had the basic courses in geology, and also for petroleum geologists who are actively exploring for oil and gas pools. Geology enters into many of the problems of the geophysicist, the reservoir and production engineer, and the wildcatter, and it is hoped there is also something of interest here for each of these. The order of presentation is what to me seems most logical: first, and of most importance, the reservoir, with particular emphasis on the trap; next the reservoir conditions of temperature and pressure and the different reservoir fluids; then the speculative ideas on origin, migration, and accumulation; and finally some of the ways of applying what has been considered in the search for new pools and provinces. In the search for new pools, we must remember, most of the ideas that concern the geology of petroleum are translated sooner or later into economic values, for they enter into one's judgment on many questions in exploration, such as whether to lease, to drill, to test, or to abandon. The book therefore includes some of the practical applications, along with the theoretical analyses, of the geological elements involved in petroleum exploration.

The work and the interest of the petroleum geologist have gradually expanded. Less than fifty years ago he merely mapped the geologic structure at the surface of the ground. Now he is concerned with all phases of exploration, from his first faint suspicion of the presence of a trap, through the drilling of the discovery well, to the final development of the pool. We are learning that the geology of petroleum is essentially the geology of fluids. Geologic concepts enter into each phase of the operation—more during the prospecting, it is true, than during the development; but even in the later period many operators continue to use geologic concepts until the pool is completely developed—some, indeed, into the period of secondary recovery. The modern petroleum geologist must know something, therefore, about the reservoir fluids, the forces involved, and the manner in which petroleum is developed and produced. In the engineering phases of exploration, as in reservoir and production mechanics, new data requiring quantitative study are being uncovered, and many new concepts of the interrelations between the rocks and their fluids are forming. From these the geologist is learning much about how petroleum acts under varying underground conditions, and he is thereby better able to search for it.

Several books would be necessary if I were to cover completely the different fields of activity now assigned to the petroleum geologist. In attempting to present the most important topics in a single volume, I have given first attention to what, in my opinion, aids most in the discovery of new oil and gas pools or in the extension of old ones: an understanding of the geologic history of an area—its stratigraphy, its sedimentation, its deformation, and especially its fluid phenomena; for it is from a wide knowledge of many principles concerning underground conditions that predictions that lead to discovery may best be made.

Two important procedures that are particularly adapted to petroleum exploration—electrical logging and geophysical surveying—are merely touched on, and their principles are discussed only enough to give the reader wholly unfamiliar with them some knowledge of their technology and application. The reason for this seeming slight is that both procedures involve complex technology, and a reasonably complete discussion of them would require much more space than can be allotted. Both have been adequately treated in a number of books and articles. To a lesser degree, other topics that are of great importance to the petroleum geologist have also been passed by— for example, aerial surveying and mapping, laboratory techniques of sample and core analysis, principles of sedimentation and paleontology, and planetable mapping. All require geological interpretation to be of use in petroleum exploration, and the emphasis in this book is on what to do with geological data *after* they are available, by whatever means they are obtained. The techniques used now in obtaining the necessary data will be supplanted— some probably even before the book is printed—by new techniques, but the principles utilizing the data should remain valid for a long time. The chief purpose of this book, then, is to discuss and analyze the principles that are employed in the discovery process rather than to describe oil pools or techniques of obtaining the data that may be used.

Probably some explanation should be made of the large number of references that are cited. They are intended to give credit to authors and publications for the ideas expressed, to be sure, but they are primarily intended to be used as a guide to study for expanding one's knowledge of the subject being considered. For that reason I have generally referred to at least one of the recent articles dealing with the subject—if possible, to a recent article containing a bibliography or a longer-than-usual list of references. The predominance of articles on the geology of petroleum in American geologic literature explains the larger number of American references used, most of which are available in college and oil-field area libraries. The specific references and also the general references at the end of each chapter may therefore be used as sources for seminar and special subject studies, and it is hoped that in this they will prove useful.

The maps and sections have nearly all been redrawn, with many inessential details omitted, and were selected to illustrate the principles discussed; they are not to be taken as descriptions of oil pools. There is no point in cataloguing producing pools except as they illustrate specific principles that apply to exploration, for our first interest is in finding new producing areas.

Many people have assisted me in preparing this book, and to each I am most grateful. Especial thanks go to Frederick A. F. Berry, who critically read the manuscript and made many useful suggestions. Gilman A. Hill, William J. McPherson, Walter Rose, and Lyndon L. Foley also read parts of the manuscript and made helpful suggestions. The figures were drafted by William A. Adent, O. T. Hayward, Jack Redmond, Clifford Bird, and Louis Renné. Most of the manuscript was typed by Bette Cornett Binford, Doris Slogar, Genevieve Scott, Jean Freeman, and Cleta Walker.

A. I. LEVORSEN

Tulsa, Oklahoma

March 1, 1954

CONTENTS

	PART ONE: INTRODUCTION	1
1	Introduction and Summary	3
2	The Occurrence of Petroleum	14

	PART TWO: THE RESERVOIR	47
3	The Reservoir Rock	52
4	The Reservoir Pore Space	97
5	Reservoir Fluids—Water, Oil, Gas	144
6	Reservoir Traps—General and Structural	232
7	Reservoir Traps (continued)—Stratigraphic and Fluid	286
8	Reservoir Traps (continued)—Combination and Salt Domes	349

	PART THREE: RESERVOIR DYNAMICS	385
9	Reservoir Conditions—Pressure and Temperature	389
10	Reservoir Mechanics	433

	PART FOUR: THE GEOLOGIC HISTORY OF PETROLEUM	495
11	The Origin of Petroleum	499
12	Migration and Accumulation of Petroleum	538

PART FIVE: APPLICATIONS 585

13 Subsurface Geology 588
14 The Petroleum Province 627
15 The Petroleum Prospect 658

APPENDIX 674

Glossary 674
Abbreviations 680
List of Bibliographies 684
Tables 685

INDEX 690

PART ONE

Introduction

1. *Introduction and Summary*
2. *The Occurrence of Petroleum*

CHAPTER 1

Introduction and Summary

Petroleum: history – origin, migration, concentration – pool – reservoir – resources and reserves – discovery – prospect.

PETROLEUM (*rock-oil,* from the Latin *petra,* rock or stone, and *oleum,* oil) occurs widely in the earth as gas, liquid, semisolid, or solid, or in more than one of these states at a single place. Chemically any petroleum is an extremely complex mixture of hydrocarbon (hydrogen and carbon) compounds, with minor amounts of nitrogen, oxygen, and sulfur as impurities. Liquid petroleum, which is called *crude oil* to distinguish it from refined oil, is the most important commercially. It consists chiefly of the liquid hydrocarbons, with varying amounts of dissolved gases, bitumens, and impurities. It has an oily appearance and feel; in fact, it resembles the ordinary lubricating oil sold at filling stations, is immiscible with water and floats on it, but is soluble in naphtha, carbon disulfide, ether, and benzene. Petroleum gas, commonly called *natural gas* to distinguish it from manufactured gas, consists of the lighter paraffin hydrocarbons, of which the most abundant is methane gas (CH_4). The semisolid and solid forms of petroleum consist of the heavy hydrocarbons and bitumens. They are called *asphalt, tar, pitch, albertite, gilsonite,* or *grahamite,* or by any one of many other terms, depending on their individual characteristics and local usage. The general term "bitumen" has long been used interchangeably with "petroleum" for both the liquid and the solid forms. *Hydrocarbon* is a term often used interchangeably with "petroleum" for any of its forms. This is not strictly correct, since hydrocarbons consist of only hydrogen and carbon, whereas petroleum contains many impurities. Definitions of some of the common forms of petroleum are given in the Appendix.

The nomenclature and scientific classification of petroleum are in a state of uncertainty and confusion. Geologists, chemists, lawyers, refiners, and highway engineers have all made attempts to define the naturally occurring forms, but, for one reason or another, few of their definitions have gained widespread acceptance. Chester, in his dictionary of minerals,[1] defines petroleum and many of the liquid and solid forms as "mineral hydrocarbons." Legally, petroleum has been called a mineral,[2] but this usage does not satisfy the common geologic definition of a mineral as an inorganic substance with chemical and physical properties either uniform or varying within narrow ranges. It has also been called a *mineraloid*,[3] a term also applied to chalcedony and amber, on the ground that it is not definite enough in chemical composition to be called a mineral. Perhaps a compromise term, such as "mineral substance" or "organic mineral," would be most useful, even though mineralogically inacceptable. Because of its association with rocks, petroleum is included among "mineral resources" and is frequently called *mineral fuel*, along with peat and coal; this term raises no fine points of definition.*

Because of its wide occurrence and its unique appearance and character, petroleum has always been readily observed by man, and is repeatedly mentioned in the earliest writings of nearly every region of the earth.[4] Oil and gas seepages and springs, and tar, asphalt, or bitumen deposits of various kinds exposed at the surface of the ground, were regarded as local curiosities and attracted visitors from great distances. From the earliest times recorded by man, petroleum is frequently mentioned as having an important part in the religious, the medical, and even the economic life of many regions. Not until after the middle of the nineteenth century, however, when it was first discovered in large quantities underground, did its potential commercial importance become apparent.

The use of petroleum spread slowly in what has been called the "kerosene age" (1859–1900), but the development of the internal-combustion engine, near the beginning of the twentieth century, set off a phenomenal growth of the petroleum industry, a growth that has not yet shown any sign of slackening. We are now in what might be called the "gasoline age," for gasoline is the chief product now being derived from petroleum. More than half of the national supply of energy in the United States is furnished by gasoline, natural gas, and other petroleum products, and the use of petroleum as a source of energy is increasing rapidly in other parts of the world as well. In addition, thousands of chemical compounds, known as *petrochemicals,* are made from petroleum. Petroleum has, in short, become one of the most important natural resources of modern civilization.

Ever since E. L. Drake drilled the first well for oil in Pennsylvania in 1859, and especially, of course, since 1900, the geology of petroleum has assumed growing importance as a special economic application of geology. From the

---

[1] The superior figures in the text refer to the reference notes at the end of each chapter.

* The United States Geological Survey, before 1924, and subsequently the United States Bureau of Mines, included petroleum in the annual reports *Mineral Resources of the United States,* and a number of bulletins of the Survey (for example, 786 and 796) include reports on petroleum under the heading Minerals Fuels.

first, geologists attempted to explain the occurrence of oil and gas in terms of geologic phenomena. Then, as the petroleum industry grew and developed, they were called in more and more to guide the programs of exploration for the raw materials upon which the industry depended. New geologic concepts relating to petroleum were thus developed, and at the same time enormous volumes of new data were made available with which to test and prove or disprove many established principles of geology. As a result, not only the petroleum industry, but the science of geology as a whole, has benefited greatly.

The term "petroleum geology" has come into use to describe the area of common interest between petroleum producers and geologists. It is doubtful, however, whether this is a proper usage. Rather, it is more accurate to say "geology of petroleum" just as we say "geology of iron" or "geology of clay," although the shorter term, "petroleum geology," is commonly used in writing or speaking informally. The geologic concepts applied to petroleum are all established and recognized geologic principles, which are merely put to practical use in finding and exploiting petroleum deposits. A person who applies these principles to finding petroleum, however, may properly be called a *petroleum geologist*.

When a petroleum pool has been discovered, we know (1) that a supply of petroleum originated in some manner, (2) that it became concentrated into a pool,* and (3) that it has been preserved against loss and destruction. The evidence for the speculative theories about the geologic history of the petroleum before it was discovered—its origin, migration, accumulation, and preservation—can come only from a study of the pool. For that reason the logical sequence for study is: (1) to examine the evidence as we find it—that is, the occurrence of petroleum, both at the surface and underground, the geological, physical, and chemical environment of the reservoir and its fluid content, and the phenomena observed and the principles involved during production; (2) to use this knowledge as the basis for speculation on the theoretical phases concerned with the history of the reservoir before discovery. We must say bluntly, at this point, that we do not know just how oil and gas originated, nor how they have moved and accumulated into pools. These problems, if solved, would aid greatly in the *main job* of the petroleum geologist—*the search for new pools*—and we shall discuss possible solutions later. A full discussion of the various elements that enter into the problems of the origin, migration, and concentration of petroleum into pools will be delayed until the reader has studied the evidence concerning reservoirs, their fluid content, and their fluid mechanics.

The fundamental geologic requirements for oil and gas pools are, of course, the same the world over. Whether one is exploring in the Americas, along the

---

* The oil or gas content of a single deposit is called an *oil pool* or a *gas pool;* if several pools are located on a single geologic feature, or are otherwise closely related, the group of pools is called a *field*.

continental shelf, the Middle East, or the Far East, the essential elements of a pool are simple. *A porous and permeable body of rock, called the reservoir rock, which is overlain by an impervious rock, called the roof rock, contains oil or gas or both, and is deformed or obstructed in such a manner that the oil and gas are trapped.*

Commercial deposits of crude oil and natural gas are always found underground, where they nearly always occur in the water-coated pore spaces of sedimentary rocks. Being lighter than water, the gas and oil rise and are concentrated in the highest part of the container; in order to prevent their escape, the upper contact of the porous rock with an impervious cover must be concave, as viewed from below. Such a container is called a *trap*, and the portion of the trap that holds the pool of oil or gas is called the *reservoir*. The significant thing is that reservoirs can be of various shapes, sizes, origins, and rock compositions.

Any rock that is porous and permeable may become a reservoir, but those properties are most commonly found in sedimentary rocks, especially sandstones and carbonates. A trap may be formed, either wholly or partly, by the deformation of the reservoir rock, which may be accomplished by folding, faulting, or both, and in either a single episode or in several episodes. Or a trap may be formed, either wholly or partly, by stratigraphic variations in the reservoir rock. These may be primary, such as original facies changes, irregular distribution of mineral particles, or diagenetic solution and cementation. Or they may result from secondary causes, such as fracturing, solution and cementation associated with erosion surfaces, or truncation and overlap along unconformities. Likewise, the direction and rate of flow of the fluids within the reservoir rock may influence, or even dominate, the position of the pool within the structural or stratigraphic trapping feature. Many traps are the result of complex combinations of structural, stratigraphic, and fluid variations that are difficult to unravel and evaluate from the data available before the pool is developed. The geologic principles involved in the formation of traps are fairly simple, but the variations in their application are almost infinite in number and complexity.

We have yet no direct method of locating a pool of petroleum. We know no physical property of underground petroleum that we can measure at the surface of the ground. The petroleum geologist's approach to the discovery problem must therefore be indirect. Each pool is unique—we may think of a pool as the end result of twenty or twenty-five variables of which only a few can be ascertained in advance. Test wells are located where an underground trap capable of trapping a pool of oil or gas is inferred from the available geologic data, and where it is believed that a pool of petroleum, if present, can be produced at a profit. Since new geologic conditions are discovered during the drilling of the test well, the petroleum geologist's interest persists until a discovery has been made and until the well begins to produce oil or gas or both. During the drilling of the test well the interests of the geologist merge with those of the petroleum engineer.

Both the amount and the location of undiscovered petroleum are, of course, unknown. The petroleum must be discovered before it can be of any use to society. For broad geologic reasons we are fairly certain that petroleum deposits will eventually be found in some regions that are not now known to contain them. We cannot say in advance, however, at what depth or at what exact location these deposits will be discovered. The actual location and size of a deposit in the earth are determined only by drilling test wells into the deposit and by producing the content of the reservoir. A test well drilled in the hope of discovering a new pool is called a *wildcat well*. The person or organization drilling a wildcat well is called a *wildcatter*.* If the well taps a deposit of petroleum, it is called a *discovery oil well* or a *discovery gas well*, according to which kind of petroleum is found. If it produces neither oil nor gas but only water, it is called a *dry hole*, a *duster*, or a *wet well*. Wells drilled into the same reservoir after a discovery has been made are called *development wells*. The next well drilled after a discovery confirms the discovery and is called a *confirmation well*.

The fundamental need of the petroleum industry is an adequate supply of its raw materials—crude oil and natural gas. Each year the "crop" of oil and gas is completely destroyed by consumption; no "seed" is left with which to start a new supply. Renewal of the domestic supply for any country depends almost entirely on the continuing discovery of new deposits. At any time the recoverable petroleum in sight, which is known as the *producible reserve,* or the *proved reserve,* is only that which has been discovered and developed but has not been consumed. For a number of years the known reserve of oil in the United States has ranged between eleven and fifteen times the annual consumption; for the rest of the world it has ranged from thirty-five to forty times. The gas reserves of the United States are eighteen times the annual consumption, but a rapid increase in the use of gas may reduce this ratio in the years to come. In the absence of a new source of energy to replace petroleum, the past steady increase in the consumption of petroleum products is creating a growing need for new deposits, which, in turn, means a continuous increase in the need for petroleum discovery.

The petroleum *reserves* of any region should be distinguished from its petroleum *resources*. The reserves consist of the oil and gas that are now available for use. The resources, which are always far in excess of the reserves, include the reserves, the prospective undiscovered reserves, and any substances

---

* In the early days of the oil industry in the United States, drillers working on wells back in the hills said they were drilling "out among the wildcats"—hence were called "wildcatters." The term "wildcatter" is an honored title in the petroleum industry; it corresponds to "prospector" in mining or "inventor" in manufacturing, and carries none of the distasteful connotation of "wildcat strike," "wildcat land boom," or "wildcat stock."

from which petroleum could be derived, either by one or by both of (1) present or improved technology and (2) present or more favorable economic conditions. Technology implies ideas, "know-how," concepts, machines, methods, and principles; economic conditions imply availability of adequate capital, incentive, profit, skilled labor, and political climate. Here is a comparison between the petroleum resources of any region and their relation to the petroleum reserves:

RESOURCES	HOW TRANSFORMED INTO RESERVES
1. Known and recoverable oil and gas deposits.	Now available.
2. Oil and gas known to have been left behind in pools but not recoverable at present.	In part by secondary recovery methods, but chiefly by new technology and more favorable economic conditions.
3. Undiscovered and undeveloped petroleum pools.	By discovery and development through present and improved technology, with present or more favorable economic conditions.
4. "Tar" and asphalt deposits, inspissated deposits, outcropping oil pools.	By present and improved technology, together with more favorable economic conditions.
5. "Oil," or kerogen, shales, torbanites, and coals.	By present and improved technology, together with more favorable economic conditions.

This book is concerned with the first three of the above resources, but our chief interest will be in the third—undiscovered petroleum deposits. This is the resource that has supplied the oil reserves of the past, and it may be expected to supply much of the reserves for a long time to come. Improvements in technology are steadily bringing the day closer when large reserves now locked up in the nonrecoverable oil, inspissated deposits and the "tar" and asphalt deposits will enter the supply picture. These resources are large and the question as to when they will become available in large quantities is primarily a matter of cost.

When the science of geology is applied to the petroleum industry, an economic element is inevitably introduced at some point. The skillful petroleum geologist translates an idea or concept into barrels of oil or cubic feet of gas at the surface of the ground. And these products must be worth more in money than they cost to produce. Some geologists may be working in a laboratory far removed from the oil fields, doing work that seems to be completely academic, but someone in the chain of discovery, sooner or later, must translate their work into terms that can be entered on the record as producing lands and finally as profits. Every geologist working in the petroleum industry should bear in mind, therefore, that the ultimate objective is *to find oil and gas that are profitable to produce*. He may find such petroleum through the discovery of new pools, or he may find it through the better development of known pools, either by extending them or by increasing production from them.

A petroleum geologist does not physically see an oil or gas pool any more than a meteorologist, for example, sees a low- or high-pressure area, though both commonly use contour lines and maps to describe the ideas they intend to convey. Both are presenting mental concepts of the conditions they believe exist. Any undiscovered oil or gas field can be mapped only as an idea or a concept in the mind of the petroleum geologist. He may, by careful mapping, judge that the rocks and the structure are favorable to the trapping of petroleum under a certain area, but until a discovery well has been drilled he does not *know* whether that area is an oil pool. Imagination, then, is an indispensable quality of the petroleum geologist. The world's future supply of petroleum is as dependent upon the imaginative powers of the petroleum geologist as on the presence of favorable rocks—of which there appears to be an abundant volume.

The actual discovery of a pool is made by the drill, but the proper location of the wildcat well to test a trap, the depth to which it should be drilled, and the detection and outlining of the oil or gas pool from what is revealed by that well and others, are wholly geologic problems. They constitute the essence of the geology of petroleum and are the most important work of the petroleum geologist. He may need to consider only a simple combination of stratigraphy and structural geology, or he may have to take account of a complex combination of data, involving such various fields as stratigraphy, sedimentation, paleontology, geologic history, fluid flow, structural geology, petrography, geophysics, geochemistry, and metamorphism. In addition to all this, he may have to draw on his own and other people's knowledge of many related sciences, such as physics, chemistry, biology, and engineering. He must do his best to work out the geology of an area from what is visible or what can be mapped at the surface, and from all available well and geophysical data for depths ranging up to three miles or more below the surface. His prediction, however, may often be based on the most fragmentary data, some of which are obtained by specialists or experts who may or may not have a working knowledge of geology, or by geologists who have worked with no thought of the petroleum possibilities of the region. This information is assembled on maps and cross sections, and fitted together in the mind of the petroleum geologist, where it is interpreted and translated into the best place to drill a well that will penetrate a trap below the surface of the ground and thereby enable the well to test the trap's content.

As the search for petroleum gets deeper below the surface, the geology becomes more complex and uncertain, and the data upon which the geologist must base his conclusions become progressively fewer. As drilling is costly, there are never as many test wells as the petroleum geologist would wish.

Every scrap of information must therefore be squeezed out of the record and put to use, and the data from each record must be projected outward in all directions. The geologic, geophysical, and engineering data must be assembled on various maps—structural, stratigraphic, facies, thickness, paleogeologic, potentiometric, productivity, isopotential, and geothermal. The aim of academic geology may be said to be the accurate correlation of formations, the detailed working out of the geologic history, or the making of a carefully contoured structural map. That of exploration geophysics, on the other hand, is to measure various physical properties of the rocks underground, such as their reflecting power, their magnetism, their electrical properties, and their relative densities. The job of the petroleum engineer is to determine the reservoir data—the pressures and pressure changes, the fluid mechanics—and to produce the oil and gas efficiently. Yet all these maps and data do not, of themselves, tell the whole story. If they are to be fully used in the discovery of petroleum, they must be interpreted, correlated, and integrated. This interpretation of the combined basic geological, geophysical, and engineering data, for the purpose of finding new oil and gas pools, is what constitutes the special province of the petroleum geologist. It results, first of all, in locating an oil and gas *prospect,* which is the set of circumstances, both geologic and economic, that will justify the drilling of a wildcat well. The petroleum geologist's work does not stop, however, when he has located a prospect; it continues during the drilling of the wildcat well. He must relate the new facts encountered in the drilling to the problem of identifying and testing the potential producing formations and of completing the well in the producing formation if the well becomes a discovery well. The petroleum geologist thus spans the gap between geology and the related sciences, on the one hand, and the oil and gas prospect and the pool, on the other. This relationship is graphically shown in Figure 1-1.

The chapters that follow attempt to show how the petroleum geologist uses the evidence of the rocks to help him discover petroleum. He obtains his evidence directly by observing outcrops, well cuttings, and well cores, and indirectly through geophysical measurements, logs, core analyses, and fluid data. He obtains additional evidence from the producing history of pools that have been discovered in the past. The techniques and methods for obtaining this evidence are many and varied, and they are constantly being improved and made more accurate. Since they have been described in many books and articles,[5] they will be only briefly mentioned in this book. It is the purpose here, rather, to consider how the evidence may be interpreted *after it has been obtained,* so that we can better predict the location of new petroleum deposits or extend old ones. What evidence is significant? And how is it to be translated into discovery?

Petroleum prospecting is an art.[6] It requires combining and blending many geologic variables in varying proportions, since each pool, field, or province is characterized by a unique combination of many different geologic

INTRODUCTION AND SUMMARY [CHAPTER 1]

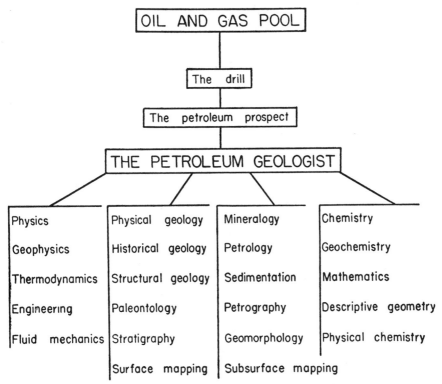

FIGURE 1-1 *Relations among the various sciences and specialized fields that are utilized by the petroleum geologist. He stands between these sciences and the oil and gas pool; his chief job is to interpret them so as to locate a prospect that, when drilled, will yield commercial oil and gas.*

conditions. Some of these conditions can be known in advance, but most cannot, and the most successful geologist is the one who can visualize the pool or locate the extension with the least advance information. He may be likened to the artist who can draw the picture with the fewest lines, or to the paleontologist who can identify a fossil vertebrate from the least number of bones. The chief objective of this book, then, is to point out the kinds of evidence, the principles, and the ideas that should prove to be the most helpful in the discovery of petroleum.

Selected General Readings

E. H. Cunningham-Craig, *Oil Finding*, Arnold, London, and Longmans, Green & Co., New York (1920), 324 pages. Chiefly of historical interest.

A. Beeby Thompson, *Oil-Field Exploration and Development*, 2nd ed., 2 vols.; Vol. 1, *Oil-Field Principles;* Vol. 2, *Oil-Field Practice;* Technical Press, London (1950). Chiefly of historical interest.

David White, "Outstanding Features of Petroleum Development in America," Bull. Amer. Assoc. Petrol. Geol., Vol. 19 (April 1935), pp. 469–502. Also printed as Sections 1–6 of Part II of the hearings in pursuance of House Resolution 441, 73rd Congress. Chiefly of historical interest.

E. DeGolyer, "Future Position of Petroleum Geology in the Oil Industry," Bull. Amer. Assoc. Petrol. Geol., Vol. 24 (August 1940), pp. 1389–1399. An informal discussion of exploration philosophy.

Max W. Ball, *This Fascinating Oil Business,* Bobbs-Merrill Co., Indianapolis (1940). A popular discussion of the oil business and its many ramifications and facets by an authority of long experience. 445 pages.

William B. Heroy, "Petroleum Geology," in *Geology 1888–1938,* Fiftieth Anniversay Volume, Geol. Soc. Amer., New York (June 1941), pp. 511–548. 101 references cited. Also published in the Report for 1943 of the Smithsonian Institution, Washington, D.C., pp. 161–198. A review of the development of the geology of petroleum, the changing concepts and applications.

Samuel W. Tait, Jr., *The Wildcatters: An Informal History of Oil-Hunting in America,* Princeton University Press, Princeton, N.J. (1946), 218 pages. A popular account of petroleum prospecting.

Wallace E. Pratt, *Oil in the Earth,* Univ. of Kansas Press (1942, 1944), 110 pages. Four lectures covering an elementary discussion of oil, where it is found, who finds it and how. Should be required reading for every petroleum geologist.

Paul H. Price, "Evolution of Geologic Thought in Prospecting for Oil and Natural Gas," Bull. Amer. Assoc. Petrol. Geol., Vol. 31 (April 1947), pp. 673–697. Bibliog. 63 items. A review of the history prior to drilling of the Drake well in 1859 and a summary of the more important geological ideas concerning oil and gas.

Parke A. Dickey, *The First Oil Well,* Jour. Petrol. Technol., Amer. Inst. Min. Met. Engrs. (January 1959), pp. 14–26. The story of the beginning of the oil industry in the United States.

Reference Notes

1. Albert Huntington Chester, *A Dictionary of the Names of Minerals,* John Wiley & Sons, New York (1896).

2. Burke v. So. Pac. R.R. Co., 234 U.S. 669, 58 L. Ed. 1527, 34 Sup. Ct. Rep. 907. Universal Oil Company, re Land Decisions, Glossary, Bull. 95, U.S. Bur. Mines (1920), p. 229: "Petroleum is a mineral, and the same may be said of salts and phosphates, and of clay containing alumina and other substances in the earth."

3. Austin F. Rogers, *Introduction to the Study of Minerals,* 3rd ed., McGraw-Hill Book Co., New York (1937), p. 262.

4. Sir Boverton Redwood, *A Treatise on Petroleum,* 4th ed., Charles Griffin & Co., London (1922), pp. 1–160. Contains descriptions of and references to occurrences throughout the world.

R. J. Forbes, *Bitumen and Petroleum in Antiquity,* E. J. Brill, Leiden (1936), 109 pages. Bibliog. 160 items. Early occurrences, especially in the Middle East.

C. R. Owens, *Histoire et archéologie du pétrole* (extract from Vol. 4, Second World Petroleum Congress, June 1937), Congrès Mondial du Pétrole, Paris.

Carey Croneis, "Early History of Petroleum in North America," Scientific Monthly, Vol. 37 (August 1933), pp. 124–133.

5. E. DeGolyer (ed.), *Elements of the Petroleum Industry,* Amer. Inst. Min. Met. Engrs., New York (1940), 519 pages. Contains 20 authoritative articles by 20 authors on various phases of the petroleum industry, including exploration and production methods.

L. W. LeRoy (ed.), *Subsurface Geologic Methods: A Symposium,* 2nd ed., Colorado School of Mines, Golden, Colorado (June 1950), 1156 pages. 59 articles by authorities on various phases of oil-field technology.

6. E. DeGolyer, *The Development of the Art of Prospecting,* Princeton University Press (1940).

CHAPTER 2

The Occurrence of Petroleum

Surface occurrences: seepages – mud volcanoes – disseminated deposits – vein deposits – kerogen shale. Subsurface occurrences: showings – pools – fields – provinces. Geographic location. Geologic age of reservoir rock.

THE OCCURRENCE of petroleum is widespread but very uneven. In some rocks it occurs only in infinitesimal amounts, measured in parts per million or even in parts per billion, whereas rocks of other areas contain enormous accumulations measured in billions of barrels. Petroleum occurs on all the continents of the world, although some continents are much richer in petroleum than others. And it occurs in all the geologic systems from Precambrian to Recent, though some systems are notably more prolific than others. The unevenness of the occurrence of petroleum, it should be remembered, is due in part to the unevenness of the exploration effort. This, in turn, depends on such variables as current geologic thought about the occurrence of petroleum, and on economic and political factors that either aid or hinder exploration.

Some petroleum occurrences are visible as outcrops at the surface of the ground. More important, however, from the standpoint of the petroleum geologist and the industry, are the underground or *subsurface* occurrences, exploited only as the result of drilling. Almost all of the world's commercial supply of oil and gas is produced from subsurface deposits.

The petroleum deposits of the world may be classified under several different categories. The most useful are these:

    1. Mode of Occurrence

  a. *Surface occurrences,* such as seepages, springs, exudates of bitumen, mud volcanoes, inspissated deposits, vug and vein fillings, and various kinds of "oil," kerogen, and bituminous shales.

b. *Subsurface occurrences,* including minor showings of oil and gas, pools, fields, and provinces.

   2. Geographic Location

The distribution by countries, continents, and other geographic units.

   3. Geologic Age of Reservoir Rock

MODE OF OCCURRENCE

The simplest classification of petroleum deposits is based on mode of occurrence; and on this basis the main division is into surface and subsurface occurrences. A deposit of either type may be of small magnitude and of only scientific interest, or may constitute a commercial deposit.*

Surface Occurrences

Petroleum occurs at the surface of the ground in a variety of ways. Some surface occurrences may be thought of as currently active, or "live," such as (1) those that form seepages and exudations of bitumen, and (2) those associated with springs and mud volcanoes and mud flows. Others may be considered as fossil, or "dead," occurrences, such as (3) bitumen-impregnated sediments, inspissated deposits, and dikes and vein fillings of solid bitumen; (4) vug and cavity fillings. Many surface deposits combine more than one of these types, and are therefore difficult to classify accurately. Another kind of surface occurrence takes the form of kerogen shales, or "oil shales." The substances that make up kerogen are a border class of hydrocarbon materials occupying a place in the chemical classification of hydrocarbons between the petroleum hydrocarbons and the coals. Although solid in the natural state, they decompose into gaseous and liquid petroleum hydrocarbons when heated to 350°C or more. They are called *pyrobitumens.* Free hydrocarbons are frequently closely associated with kerogen, but the association is thought to be due to the oil-wetting characteristics of the kerogen rather than to the kerogen as a source of the oil.

Seepages, springs, and bitumen exudates. Petroleum—whether gas, oil, or liquid asphalt—that exudes in the form of springs and seepages may reach the surface along fractures, joints, fault planes, unconformities, or bedding planes, or through any of the connected porous openings of the rocks. Some

---

* "Commercial deposit" is taken to mean an occurrence of a size and grade that warrant exploitation and sale of the product. The resulting operation may or may not be profitable. If it does not prove profitable, it probably will be abandoned, or its efficiency will be improved, or it will be continued in the hope that the economic situation improves and permits a profit.

of the more common types of seepage are shown diagrammatically in Figure 2-1. Most seepages are formed, presumably, by the slow escape of petroleum from fairly large accumulations that have been brought close to the surface and into the zone of fracturing by erosion, or that have been tapped by faults and fractures. Petroleum seepages are common in the sedimentary regions of the world, and many pools and producing regions have been discovered by drilling near seepages.[1] Occasionally a seepage and a pool are connected. More often there is no relationship that can actually be observed; the seepage merely furnishes direct evidence of the presence of petroleum in the area.

Surface seepages, either as oil or as gas, are frequently associated with water springs. The oil floats to the surface of the water, and the gas bubbles out and escapes into the atmosphere. If the water is stagnant, the oil may accumulate as a viscous-to-solid mass that remains after the water evaporates. If the oil-bearing water enters streams, the oil is carried along until it is either destroyed by oxidation or bacterial action or redeposited on the ocean floor. Oil films floating on water have a characteristic iridescent luster and are somewhat similar in appearance to films of iron oxide; but, unlike oxide films, they do not break when stirred. Gas seepages are more readily observed when they occur in swamps or bubble through water. Dry desert areas, for example, may have gas seepages that are not recognized.

Surface oil seepages may be of large dimensions. Some surface deposits of oil and asphalt, such as those in southern California, Venezuela, Trinidad, and the Baku region of the USSR, cover hundreds or even thousands of acres, and must be the residue from large oil pools. These deposits generally assume a viscous and semi-solid state, but they may become liquid and flow when warmed by the daily and seasonal rise in temperature. Earthquake activity

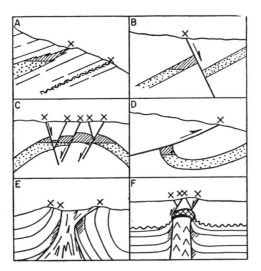

FIGURE 2-1

Sections showing the position of typical seepages with relation to the underlying structure. Seepages are marked x, and oil and gas pools are cross-hatched. The seepages in A are at the outcrop of the pool and at the outcrop of an unconformity; in B the seepage is along the outcrop of a normal fault; in C the seepages overlie a faulted anticline; in D the seepage is along the outcrop of a thrust fault; in E the seepages are associated with diapir folding; in F the seepages overlie a salt plug and are associated with the faults that occur above this intrusion.

may also account for unusually large flows from some seepages. Large seepages may consist of nearly pure oil, asphalt, or semisolid bitumens; more commonly these are mixed with varying amounts of sand, sticks, clay, leaves, peat, animal bones, and other debris.

Many of the larger seepages were once of considerable economic importance. Pitch Lake, in southwestern Trinidad near the shore of the Gulf of Paria, is one of the larger asphalt seepages and probably the most famous.[2] It lies in a nearly circular depression about 2,000 feet in diameter, and is over 135 feet deep near the center. It overlies a low structural dome containing heavy oil entrapped in Pliocene and Upper Miocene rocks. As the asphalt is mined near the center of the depression, the asphalt removed is continually replaced by what slowly wells up; so there is a slow motion outward toward the edges of the deposit. Gas accompanies the asphalt and bubbles up through the water that accumulates in the minor depressions at the surface. Many of these depressions are small synclines between folds in the asphalt surface, which has hardened as a result of exposure to the air. The gas causes the asphaltic material to assume a porous or honeycomb structure. It is believed that the lake will ultimately yield over 25 million tons of asphalt, of which more than half has been produced to date. The material mined is remarkably uniform in composition, and consists of asphalt, gas, water, sand, and clay.

Many other seepages occur in Trinidad [3] but are generally so mixed with detritus as to be of no commercial value. The seepages of the Trinity Forest Reserve, where it is possible to walk for miles without ever being out of sight of asphalt,[4] are especially noteworthy.

Another example of the petroleum seepage is the near-by Bermudez pitch lake, in eastern Venezuela. This is one of the largest deposits of pure asphalt yet discovered in the world.[5] It covers an area of over 1,100 acres and contains liquid asphalt to a depth of twenty feet, its average depth being five feet. The asphalt exudes from springs and remains soft and semiliquid under the hard crust that forms over the top. The top is covered with vegetation and pools of water, as the deposit has not been mined for many years.

Numerous tar and asphalt seepages, many of them large, are found throughout the Coast Ranges from Los Angeles to Coalinga, a distance of 140 miles. Probably the largest and most notable are those of the Ojai Valley–Sulphur Mountain district, which were described long ago by Benjamin Silliman, Jr., a professor of chemistry at Yale College. While visiting in California in 1864, he wrote back a description of the Rancho Ojai seepages: "The oil is struggling to the surface at every available point and is running down the rivers for miles and miles." [6]

The numerous surface occurrences of the Middle East are among the most famous in the world. They consist of oil and gas seepages, pitch lakes, bitumen deposits, and asphalt floating on the seas. The location of the best-known examples, as well as of the oil fields, is shown in Figure 2-2. The seepages are mentioned in many of the earliest writings, and from the beginning of recorded history down to the present time petroleum has exerted a profound influence on the lives of the peoples living in this area. Many descriptions of these ancient seepages have been recorded,[7] but only a few of the more striking examples will be mentioned here.

Probably the most famous seepage is at Hit,[8] in Iraq. (See Fig. 2-2.) Long before

FIGURE 2-2 *Map of the Middle East showing the location of some of the well-known oil and gas seepages and of the oil pools.*

the Christian era a bitumen industry had developed around this seepage, which, like many other seepages of the ancient world, acquired local importance from its reputed medicinal properties.[9]

The chief producing formation of the great Iranian oil fields is the Asmari limestone (Lower Miocene and Upper Oligocene). Where it crops out along the foothills to the northeast, many oil and gas seepages are found, either in the limestone or closely associated with it.[10] Active seepages of oil and gas overlie the Quaiyarah, Kirkuk, Naft Khaneh, Naft-i-Shah, and Masjid-i-Sulaiman oil fields.

Many references to petroleum are found in the Old Testament. "The vale of Siddon was full of slimepits," or asphalt pits (Genesis 14: 10). The builders of the tower of Babel used "oil out of the flinty rock" (Deuteronomy 32: 13). When Moses was three months old, and his mother "could no longer hide him, she took for him an ark of bulrushes, and daubed it with slime and with pitch, and put the child therein" (Exodus 2: 3). At one time the Dead Sea, around which many saturated sands, seepages, and bitumen deposits occur, was called "Lake Asphaltites." [11] Asphalt chunks are still found floating on it, but their source is unknown.

The numerous seepages of the region around the Caucasus Mountains, especially on the Apsheron Peninsula of the USSR, have also had a long history.[12] Many of them are associated with mud volcanoes (see pp. 21–23). Marco Polo, writing at the end of the thirteenth century, says of one seepage in the Baku area: "On the confines toward Geirgine there is a fountain from which oil springs in great abundance, inasmuch as a hundred shiploads might be taken from it at one time." [13] One of the gas seepages near Baku, known as the "Eternal Fires," was visited annually by thousands of fire-worshipers, many coming long distances.[14]

Marine seepages—oil and gas that escape under the ocean—occur at a number of places on the earth. The best-known are those near Santa Barbara, California;[15] off the south, east, and west coasts of Trinidad; in Consets Bay, Barbados, BWI;[16] off Yucatan in the Gulf of Mexico; southeast of Ancón, Ecuador,[17] where oil-stained spume is blown inshore from the ocean; and in the Caspian Sea off the Apsheron Peninsula. Such occurrences on the floor of the ocean are presumably associated, like land seepages, with fractures and openings leading to buried pools.

Mud Volcanoes and Mud Flows. Most mud volcanoes are caused by the diapiric intrusion of plastic clay. High-pressure gas-water seepages that often occur with them carry mud, sand, fragments of rock, and occasionally oil. Mud volcanoes are generally confined to regions underlain by incompetent softer shales, boulder and submarine landslide deposits, clays, sands, and unconsolidated sediments, such as are common in Tertiary formations.

The surface expression of a mud volcano is commonly a cone of mud through which gas escapes either continuously or intermittently. Single cones or groups of cones may cover an area of several square miles and extend more than a thousand feet in height, although they are more often measured in tens and hundreds of feet. Some mud volcanoes, however, may show at the surface either as basin-like depressions or as level stretches of ground strewn with erratic blocks of rock carried up from below. Mud volcanoes of these kinds generally occur where the rainfall is heavy, or at places on the seacoast where the tides and waves wash the soft muds away as fast as they are extruded but leave behind the erratic pebbles and boulders, many of them very large, that have been carried up from below by the mud stream. Erratic boulders have at times been encountered in drilling through subsurface unconformities. These deposits may be ancient mud volcanoes or mud flows, or they may be buried submarine landslide debris such as is thought to be the source material for many of the modern mud flows. Unless their occurrence is understood as erratic, they may cause a false interpretation of the stratigraphy and geologic history.

Flows of mud and breccia with little or no accompanying gas are often associated with mud volcanoes. The material may either come up through fractures and fissures or be squeezed out along bedding planes and faults by some diapiric folding mechanism. (See also pp. 250, 251.) The similarity of mud-volcano and mud-flow phenomena to igneous volcanism led Kugler[18] to use the term "sedimentary volcanism." The crusts of sticky mud frequently associated with the extrusions are sometimes explosively ruptured when the pressure of the gas accumulating underneath has become sufficiently high.

Many mud volcanoes, especially the larger ones, are associated with anticlines, faults, or diapiric folds. A mud volcano is especially likely to form on an anticline overlain by stiff, thick clay. During dry weather the clay becomes desiccated and cracked, and, if the cracks cut deep enough, a little of the gas manages to escape. Once an exit channel is formed, it is kept open by the escaping gas. As the gas rises, it mixes with the clay and ground water to form a mud, which erupts either steadily or spasmodically, depending on the local pressure, on the available amounts of gas, water, and mud, and on the size and shape of the opening. Mud volcanoes are most active after a long drought, probably because the desiccation cracks are then wider and penetrate deeper underground.[19]

As in igneous volcanism, auxiliary vents, from which mud flows, may open along the side of the cone when the main channel becomes clogged or when the cone grows large. In some mud volcanoes, from which large quantities of gas escape along with the mud, spectacular eruptions may occur.[20] The cone may consist in part of breccias, blocks, and fragments of rock, which are mixed with the mud. The breccias and erratic blocks that occur in some mud flows may have been picked up along the walls of the vent as the mud and gas rose to the surface. They are more likely, however, to be of shallow origin and to have come from "Wildflysch"—a buried layer of ancient submarine landslide material, such as erratic pebbles and boulders of all kinds, embedded in soft muds.* Such a mixture is incompetent when wet and charged with gas, and is easily squeezed out as folding occurs. The ascending gas and water readily pick up the soft mud with its erratic pebbles and boulders and carry it to the surface, where it is extruded as a mud flow.

The chief significance of mud volcanoes and mud-breccia flows to petroleum geologists is that they generally indicate the presence of gas. They also suggest that geologists need to be cautious in interpreting occurrences of unusual and anomalous rocks and fossils in subsurface stratigraphy, since these may come from erratic boulders contained in ancient mud flows; they may indicate

---

* H. G. Kugler in a personal communication.

the presence of near-surface, buried submarine landslip deposits. Since mud volcanoes are striking in appearance, the chances of finding new areas of mud-volcano and mud-flow activity are small, but new individual flows will undoubtedly continue to occur within the previously known areas.

Among the largest and most spectacular mud volcanoes in the world are those of the Baku region, USSR, which is part of the Apsheron Peninsula. (See Fig. 2-3.) These have been described by many writers[21] since the time of the first recorded history of the area. Here the mud volcanoes are generally associated with diapiric folding in the soft Tertiary rocks. In some places mud and mud breccias ooze without the aid of free gas from cracks in the rocks, as toothpaste oozes from a tube. The cones of the Touragai, Kinzi Dag, and Kalmes mud volcanoes rise to heights of over 1,200 feet (400 meters) above the adjacent Caspian Sea, and many lesser cones are found throughout the region. Eruptions containing great quantities of gas are frequent, and these are very spectacular when the gas catches fire. When an eruption on Opman-Box-Dag caught fire in 1922, the smoke rose to a height of 14 kilometers; and the flames from the Touragai eruption could be seen from a distance of 700 kilometers. The heat from the burning gas bakes and fuses the clays and muds into dark porous slags and porcellanites. Intraformational breccias that may be ancient mud flows or submarine landslide debris are found by drilling, and these often make subsurface

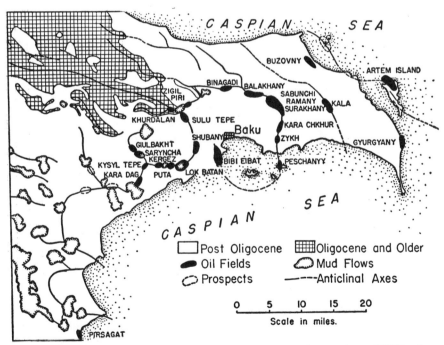

FIGURE 2-3 *Map of the Apsheron peninsula and the Baku province, USSR, showing the location of the oil fields and of the mudflows and mud volcanoes* (*See also Fig. 2-2.*)

correlations difficult. A section through a fossil mud flow or mud volcano of the Bibi-Eibat field is shown in Figure 2.4.

The mud volcanoes in Burma have long been known.[22] The most famous are on the Arakan Coast.[23] Those on the island of Cheduba and at Mimbu have at times erupted with startling violence, and the eruptions have been accompanied by burning gas. An eyewitness to the eruption in Cheduba says: "I saw what at first I took for a black cloud, but which was no doubt mud, shoot far above the trees, followed a moment afterward by very dark red flames and dense black smoke, which looked to me to shoot right up to the clouds." The Mimbu volcanoes lie along an important fault and are about two miles south of a group of oil-producing wells at the north end of the Mimbu field. There are seven or eight vents. Although some of them are mounds of pale-gray mud that reach 100 feet in height, others

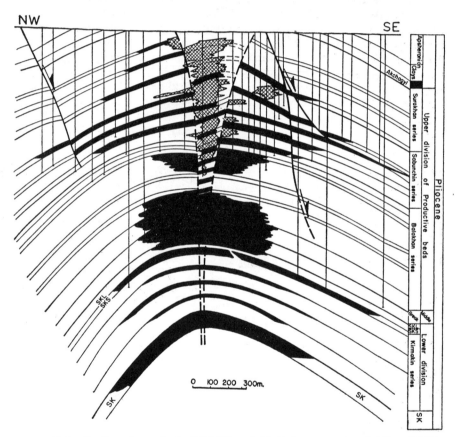

FIGURE 2-4 *Section through the Bibi-Eibat field, Baku area, Azerbaidjan, USSR. (For location see Fig. 2-3.) The top of the field is cut by an ancient or fossil mud volcano (cross-hatched). This is one of many multiple-reservoir fields in the Baku region. The structure is a large elongated dome with minor faulting. Oil pools are in black.* [Redrawn from Trust "Stalinneft," XVIIth Int. Geol. Cong., Moscow, USSR, Vol. 4 (1940), p. 120, Fig. 12.]

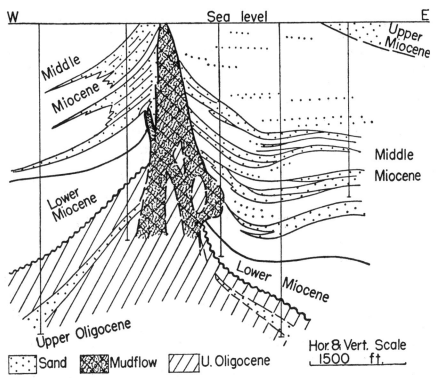

FIGURE 2-5   *Section through the Barrackpore oil field, Trinidad, showing a diapiric anticline with a mudflow core (shaded).* [Redrawn from Suter, Colonial Geology and Mineral Resources, *Vol. 3, No. 1 (1952), p. 18, Fig. 13.*]

are simply broad pools of fluid mud, through which huge bubbles of gas rise and burst every ten or fifteen seconds.

Many mud volcanoes and mud flows occur south of the Central Range on the island of Trinidad.[24] Here the volcanoes are mostly small and cone-shaped, and the eruptions are of the typical mud-and-gas variety. Along the southern coast of Trinidad, between Palo Seco and Erin, the beach is strewn with erratic blocks of calcareous sandstone. One block is about ninety cubic feet in volume, and many measure several cubic feet. These erratic blocks are left behind on the beaches, where the waves erode the mud cones and wash the mud away. Weeks[25] described one of these cones erupting nine or ten feet above sea level and throwing water six or eight feet into the air. A mud flow associated with an oil field is shown in Figure 2-5, a section through the Barrackpore field in the Digity region of Trinidad.

Occurrences of Solid Petroleum.   Petroleum is also found in forms popularly regarded as solid, although, strictly speaking, some of them are highly viscous liquids. These include "tar," * asphalt, wax, and pyrobitumen.[26] Outcrops

---

* "Tar" is a distillation product, and not a naturally occurring substance, although it is frequently spoken of as such.

of solid petroleum are found in two general forms: (1) disseminated deposits and (2) veins or dike-like deposits filling cracks and fissures.

Disseminated Occurrences. Sediments containing petroleum in the form of asphalts, bitumen, pitch, or thick, heavy oil, disseminated through the pore spaces of the rock either as a matrix or as the bonding material, are common throughout the world. They are generally called *bituminous sands* or *bituminous limestones,* depending on the nature of the host rock. Both bituminous limestones and bituminous sandstones are often quarried for direct use as road metal and paving. Their asphalt content may be as much as 25 percent but is usually between 8 and 12 percent.

Two different types of disseminated occurrences are found: (1) *inspissated deposits* and (2) *primary mixtures* of rock and bitumen.

Inspissation means "drying up." An inspissated deposit is *in situ* (in place), was probably once a pool in liquid and gaseous form, and now consists of only the more resistant and heavier residues, the lighter fractions having been lost. An inspissated deposit, then, may be thought of as a fossil oil field. As erosion gradually removes the overburden and brings the surface closer to the petroleum pool, the pressure on the fluids in the rocks is reduced. The lowering of pressure causes the gases and lighter oil fractions to come out of solution and expand, leaving the heavier hydrocarbon fractions behind. As the zone of weathering approaches the pool, the opening of incipient fractures allows the gases to escape more readily. Oxidizing agencies probably aid in solidifying the heavier oils that remain behind. Such a deposit, typical of many in California and Utah, is shown in Figure 2-6. An inspissated deposit in a fossil stream channel is described on pages 303 and 304.

Inspissated deposits occur in southern Oklahoma, in and around the Ouachita and Arbuckle Mountains. These are among the richest deposits of disseminated asphalt in the United States.[27] In this area the asphalt is generally viscous or

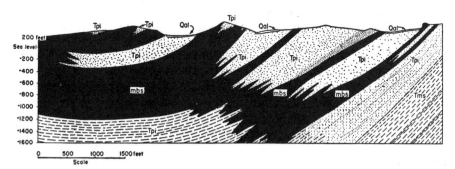

FIGURE 2-6 *Section through a bituminous black sandstone deposit* (mbs) *in the Pismo formation* (Tpi), *Pliocene-Miocene, near Edna, San Luis Obispo County, California. The asphaltic sands are underlain by the Monterey shales* (Tms). [Redrawn from Page, Williams, Hendrickson, Holmes, and Mapel, U.S. Geol. Surv., O. & G. Investig., Prelim. Map 16 (1944).]

semisolid, and occurs in rocks ranging from the Ordovician to the Cretaceous in age. The numerous buried oil pools in the region suggest that many of the surface deposits are inspissated pools that have been exposed by erosion. Some of these deposits consist entirely of sand grains held together by asphalt, which commonly constitutes 5–17 percent of the quarried rock.[28] Estimates of the total amount of asphalt in this region have run as high as 13 million tons.

The second type of disseminated deposit is one in which the sediments were mixed with the oil, asphalt, or bitumen during their deposition, the whole deposit having later been buried by younger sediments and then exposed by erosion. Such a deposit might be thought of as primary. It is hard to distinguish from an inspissated deposit, as is clearly shown by the long controversy over the origin of the Athabaska oil sands (Cretaceous) of Alberta, Canada.[29] These sands constitute the largest known single deposit of oil in the world, being estimated to contain over 600 billion barrels of oil in place, of which more than half is estimated to be recoverable. This deposit, therefore, has received a great deal of attention from geologists and engineers, and theories of its origin have passed through many cycles. Some believe in a Cretaceous oil source; others believe that a pre-Cretaceous oil came in along the unconformity separating the Cretaceous sediments (including the reservoir rock) from the truncated edges of the Paleozoic sediments.[30] It is possible that these oil sands might be considered a primary disseminated deposit in which the sand was deposited in or with the oil.

Another example of a deposit believed to be contemporaneous with the deposition of the rock is the Anacacho limestone (Cretaceous) of Uvalde County, Texas. This material, which underlies an area of many square miles, consists of limestone grains loosely cemented with asphalt. When it is treated with carbon tetrachloride, only the loose limestone grains remain.[31] The asphalt is hard and has a brilliant luster and conchoidal fracture. The asphalt content ranges from 10 to 20, and averages 15, percent.

Oil from offshore seepages, or oil carried into the ocean from land seepages, may be blown against the shore and be mixed with the sand to form bituminous sands. Some of the Pliocene to Recent oil and "tar" sands in southwest Trinidad appear to have been formed in some such manner.

Bituminous Dikes, Solid or Semisolid. A group of related bitumens occur as solid vein fillings in many places throughout the world. Although these substances are almost all very similar to one another, they are variously called asphalt, grahamite, uintaite, gilsonite, manjak, albertite, wurtzlite etc. Most of the names are derived from geographic or personal names. (See the Appendix for definitions of many of the solid bitumens.) Nearly all these solid bitumens are characterized by conchoidal-to-hackly fracture, a black or dark streak, and an appearance that somewhat resembles that of cannel coal. Some are soluble in carbon disulfide, and all require heating to free their oil content. These hydrocarbons occur as vein fillings that vary from a few inches to twenty-five feet in width. Sections through several vein deposits of solid bitumen are shown in Figure 2-7.

Occurrences of solid petroleum may be regarded as fossil or dead seepages from which the gaseous and liquid fractions have been removed, leaving only the solid residues behind. In the inspissated deposits the separation of the lighter constituents occurred in place in the rock. In the deposition of sedimentary petroleum the separation of the gas from the liquid took place before

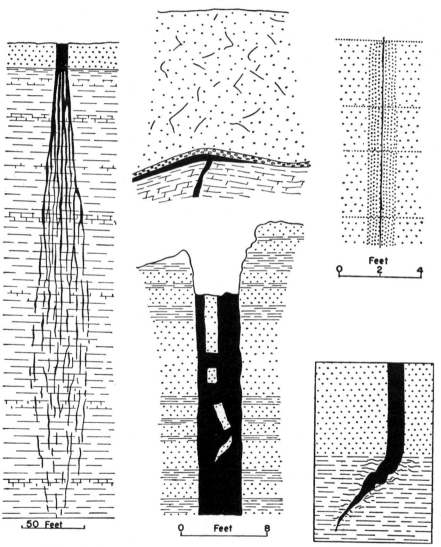

FIGURE 2-7 *Sketches of solid petroleum dikes and vein fillings.* [*Redrawn from Eldridge, U.S. Geol. Surv., 22nd Ann. Rept. (1901)*.]

the contemporaneous deposition of the oil and asphalt with the enclosing sediments. In the solid vein and dike fillings the loss of the gaseous and liquid fractions probably occurred while the petroleum was filling the opening. Most geologists would not consider occurrences of solid petroleum in a prospective region quite as favorable an indication of oil and gas pools as a live and active seepage. Solid petroleums do have significance, however, and should not be overlooked as evidence of the presence of source rocks somewhere within the region.

Miscellaneous Surface Occurrences. Liquid and solid petroleums are frequently found in fossil casts, in vug openings, in the central cavities of geodes,* and in the nuclei of concretions.[32] The oil sometimes completely fills the cavity, but generally it occupies only a portion of it. Commonly only a few of the concretions or fossil casts present contain bitumen. Since the surrounding rocks generally do not contain any visible petroleum hydrocarbons, these occurrences present a problem. How did the material get into the cavity without leaving some trace in the surrounding rock? If the cavity is lined with crystals of calcite or quartz, the petroleum presumably came in with the solutions that deposited these minerals. Other concretions may have formed around organic nuclei, which later became transformed into petroleum hydrocarbons that remain inside the cavity. More probably the concretion furnishes a small low-pressure space within an environment in which the fluid pressure would steadily increase with depth (such as would be expected with continued burial) and the few parts per million of hydrocarbons within the surrounding shales found their way into the lower-pressure space.

Indirect evidence of petroleum is sometimes given by areas of burnt clay found at various places where exuding oil or gas has caught fire and been burned up. Generally the clay burns to a red color, and it often fuses into a porcellanite or lava-like rock. Such occurrences have been found in California,[33] in Northwest Territory, Canada,[34] in Trinidad, at Burnt Hill on Barbados Island, and along the Yorkshire and Dorsetshire coast of England.[35]

Petroleum is occasionally associated with metal ores. In many quicksilver mines, for example, and some vanadium, lead, and zinc mines, asphalt and bitumen are mixed or associated with the vein material,[36] and small but measurable amounts of mercury are being brought up along with the oil in the Cymric pool of California.[37] The Black Band shale, which overlies the Freeport coal (Ohio No. 7) of the Conemaugh (Pennsylvanian) in Ohio, contains both iron ore and enough petroleum distillate to make it potentially an important commercial oil shale. The iron ore is partly a kidney ore and partly a ferruginous limestone called "mountain ore," and is thought to be a sedimentary bog-iron deposit laid down in a swamp.[38] Disseminated solid bitumen has also been found in the copper-bearing Nonesuch formation of the Keweenawan series (Precambrian) of

---

* Willard D. Pye, in a personal communication, describes oil-filled geodes in the Kaibab limestone (Permian) of the San Rafael Swell, Utah, and in the Mississippian chert of the Tri-State mining district of Oklahoma, Missouri, and Kansas.

northern Michigan.[39] Recent analyses of the shales of the Nonesuch formation show the presence of several hydrocarbons typical of those found in crude oil.* More than a hundred barrels of oil have been recovered in the Minerva fluorite mine in southern Illinois, from the fractured fluorite-bearing rocks of Chester age (Upper Mississippian). Another example is the famous Kolm black shale (Middle Cambrian) of Sweden,[40] which is mined for both oil and radioactive mineral content. Uranium is the radioactive element, and is believed to have been deposited with the shales. The Kolm shale is a low-grade "oil shale" with approximately 35 percent ash.

"Oil Shale," or Kerogen Shale. "Oil shales" are widely distributed throughout the world and throughout the geologic column.[41] They become of considerable economic interest from time to time—especially during periods when there is a fear of oil shortage—because large amounts of oil may be derived from them and may become a substitute for naturally occurring liquid petroleums.

The term "oil shale" is applied to several kinds of organic and bituminous shales, most of which consist of varying mixtures of organic matter with shale and clay. The organic matter is chiefly in the form of a mineraloid, called *kerogen,* which is of indefinite composition, insoluble in petroleum solvents, and of uncertain origin. For this reason these shales are better called *kerogen shales*. A small part of the oil recovered from kerogen shales occurs as oil in the shale; most of it is formed from the kerogen by heating. The distillation of the kerogen vapors begins at temperatures around 350°C (662°F); the kerogen in the shale is a *pyrobitumen*. Yields of up to 150 gallons of oil per short ton of shale have been encountered, but most commercial grades are on the order of 25–50 gallons per ton. Kerogen shales are neither petroleum nor coal, but rather an intermediate bitumen material with some of the properties of each. Geologists have long believed that kerogen is the primary source material of crude oil and natural gas. Modern analytical methods have shown, however, a widespread presence of small amounts of petroleum hydrocarbons; consequently, most geologists now look upon kerogen as merely a pyrobitumen with little or no genetic relation to petroleum.

Most oil shales contain free petroleum, which is recoverable by ordinary oil solvents, such as petroleum naphthas, ether, chloroform, and carbon tetrachloride. The petroleum in these shales usually occurs in fractures, fissures, bedding planes, and microscopic openings in the rock. Shales containing free oil, generally a few parts per million, may grade into fractured siltstone reservoir rocks, such as the Spraberry formation of western Texas (described on p. 121).

Kerogen, under a microscope, is seen to consist of masses of almost completely macerated organic debris, chiefly plant remains, algae, spores, spore cases, pollen, resins, waxes, and the like. In some of the richer layers it makes up 50 percent

---

* John M. Hunt, personal communication.

or more of the shale. Kerogen also contains yellow or reddish-yellow subspheroidal bodies and irregular streaks of reddish-yellow, dark-brown, and black material. This latter material may be the source of the oil obtained when the rock is distilled, but before distillation the oil is probably not any natural form of petroleum, for only small amounts can be extracted from the shale by ordinary solvents of petroleum.[42] As the kerogen content increases, the kerogen shales grade imperceptibly through torbanites and boghead coals into cannel coal, which might be termed "kerogen coal" to distinguish it from coals of the peat-to-anthracite series.[43] The frequent and often complete transitions from kerogen shales to cannel coal suggests a common mode of origin.[44]

The kerogen of the average rich kerogen shale is extremely fine-grained and is intimately mixed with inorganic clays, sands, and carbonates. The layers rich in kerogen break with a smooth conchoidal fracture and have a dull or satiny luster. The material in them burns readily and resembles cannel coal. Kerogen varies in chemical composition, by weight percentage, within the following ranges:

Carbon	69–80
Hydrogen	7–11
Nitrogen	1.25–2.5
Sulfur	1–8
Oxygen	9–17

It differs chemically from crude oil in its high content of oxygen and nitrogen, both of which must be removed in some manner before kerogen can become petroleum.

Subsurface Occurrences

Underground, or subsurface, petroleum occurrences may be broadly divided according to their size as (1) minor showings of oil and gas, (2) oil and gas pools, fields, and provinces.

Minor Showings. Some natural gas and crude oil are found in most wells drilled in sedimentary rocks, especially within the known producing regions, where nearly every exploratory well finds some indication of gas or oil, even though it may be so slight that the well is abandoned as a dry hole. Two questions are raised by any minor showing in a well: (1) "Is this a large enough showing to indicate a commercial well?" (2) "Is it at the edge of a pool?" Naturally there is no problem in deciding that a well is commercial when it "blows in" and starts to produce oil and gas at the surface, or when the core is found by analysis to be well saturated with gas and oil and capable of giving them off. A core that bubbles oil and gas at its surface as it is removed from the well is called a "live" showing, in contrast to the dull, asphaltic staining of the "dead" showing.

Many well showings, however, are on the borderline between commercial and noncommercial, and how to distinguish between commercial and noncommercial showings is one of the problems confronting every petroleum geologist and engineer concerned with drilling exploratory wells.

Oil and gas showings may be observed directly and also, by means of tests for various physical properties during the drilling of a test well, indirectly. These properties are discussed under Logging in Chapter 3. In fact, field and laboratory techniques have been developed to such a point that well data can yield fairly reliable information as to a reservoir's gas, oil, and water content, its thickness, and its porosity and permeability—its capacity, in short, to produce oil and gas. Whether the evidence obtained from any particular well will justify testing the well further, or justify an attempt to produce oil from it, immediately becomes a matter for individual judgment.

Pools, Fields, and Provinces. Commercial petroleum deposits are classified as *pools, fields,* and *provinces.*[45] Terms such as "pool," "field," "province," and "subprovince" are useful in describing and locating the various oil and gas accumulations and occurrences. They combine both geographic and geologic factors that are commonly understood by the geologists, geophysicists, and engineers of the petroleum industry. But these terms, like many others in geology, grade into one another, which makes it difficult, at times, to define their exact meaning. Local usage generally prevails eventually, even though it may not reflect the best or most accurate scientific classification and terminology.

Pool. The simplest unit of commercial occurrence is the pool.* It is defined as the body of oil or gas or both occurring in a separate reservoir and under a single pressure system. A pool may be small, underlying only a few acres, or it may extend over many square miles. Its content may be entirely gas, or it may be entirely or mainly oil. The term *major pool* is arbitrarily taken to mean a pool that will ultimately produce 50 million barrels or more of oil.†

Field. When several pools are related to a single geologic feature, either structural or stratigraphic, the group of pools is termed a *field.* The individual pools comprised in a field may occur at various depths, one above another, or they may be distributed laterally throughout the geologic feature. Geologic features that are likely to form fields are salt plugs, anticlinally folded multiple sands, and complex combinations of faulting, folding, and stratigraphic varia-

---

* Each year maps, production figures, and geologic data for all oil and gas pools in the world are published in July by World Oil, Houston 1, Texas, and in December by the Oil and Gas Journal, Tulsa, Oklahoma. In addition, both journals publish a Review and Forecast issue in January or February of each year, giving a detailed statistical summary of exploration and production throughout the world during the previous year. The Petroleum Division of the American Institute of Mining and Metallurgical Engineers, New York 18, publishes an annual list of the pools in the world, with many production and geologic data for each pool.

† The size of an oil pool is generally given as the number of barrels of crude oil that may be produced and recovered at the surface of the ground. This is but a fraction of the crude oil in place underground, usually ranging from one-quarter to three-quarters of the total amount. The oil left behind is called the *nonrecoverable* oil; the oil produced, the *recoverable* oil. The total, original amount of oil in the pool underground is called the *in-place* oil.

tion. Many examples may be seen in the maps and sections in Chapters 6, 7, and 8, on traps. The amount of oil that a pool or a field will produce is not a distinguishing characteristic. In the East Texas pool and in many of the Middle East pools, for example, the oil is obtained from a single reservoir; yet the ultimate production of each of these pools will be greater than that of many fields or even provinces. Since a field may contain several closely related pools, the terms "pool" and "field" are often confused, especially during the early development stages.

Province. A petroleum province is a region in which a number of oil and gas pools and fields occur in a similar or related geologic environment. Since the term is loosely used to indicate the larger producing regions of the world, the boundaries of a so-called province are often indistinct. The Mid-Continent province of the south-central United States, for example, has definite regional characteristics of stratigraphy, structure, and oil and gas occurrence. Consequently, the term has a specific meaning for geologists and the petroleum industry. Subprovinces may occur within provinces; within the Mid-Continent province, for example, we find the Cherokee sand subprovince of southeastern Kansas and northeastern Oklahoma, the Anadarko Basin subprovince of western Oklahoma and northwestern Texas, the Reef subprovince of west-central Texas, the Panhandle subprovince of northwestern Texas, and many others.

The Importance of Minor Occurrences. Although an understanding of the geology of petroleum must be based on a study of the large, commercial deposits, the minor, or noncommercial, occurrences are frequently of great importance to the exploration geologist. The significance of the minor occurrences is twofold:

1. Minor occurrences often furnish clues that lead to the discovery of commercial deposits. Nearly every producing region (*petroleum province*) was discovered as the result of drilling prompted by the recognition of a nearby surface or subsurface showing of gas, oil, or asphalt. Visible surface evidences, or outcrops of petroleum, have often furnished the only reason for drilling an exploration well, especially in the early days of the industry.

Minor subsurface showings have also proved valuable exploration guides. As more drilling was done and as more well logs and records became available, the geologist's explorations were guided more and more by subsurface evidences and showings of petroleum in wells. Confronted by a minor subsurface showing of oil or gas, no matter how small, the petroleum geologist must always ask himself, "Is this showing located at the edge of a pool?" The evaluation of minor subsurface showings, in fact, constitutes a large part of the work of the modern petroleum geologist.

2. Minor occurrences indicate the presence of a "source rock." Although the origin of petroleum is not known, most geologists believe that it comes from some kind of source rock or source environment. They do not agree, however, on just what a source rock is. But when an area contains petroleum in any natural form, either at the surface or underground, and in either commercial or noncommercial quantities, it is evident that petroleum was formed in some way within that par-

ticular area, or tributary to it, and that a source rock, whatever its character, is or has been present. As we shall see later, this evidence becomes important when the geologist is considering the oil and gas possibilities of a new and partially explored region. When the presence of a source rock is demonstrated only by a minor occurrence, the smallness of the occurrence may reflect the size of the trap, a lack of favorable local conditions of concentration, or a location at the edge of a pool. Even a minor accumulation, therefore, generally encourages the prospector to keep searching for larger traps and more favorable conditions.

GEOGRAPHIC LOCATION

Petroleum deposits are unevenly distributed throughout the world. Large areas in Asia, Australia, and Africa, for example, are nonproductive or only slightly productive, so far. This seeming scarcity may be due only to the lack of drilling and exploration in these areas. Other areas in each of these continents are productive and as the nonproductive areas contain large volumes of untested sediments, they too must be considered as potentially favorable for petroleum accumulation. Some large areas, on the other hand, have been found to be exceptionally rich in petroleum. The two outstanding areas, or regions, which have been termed the "oil axis" or the "oil poles" because they are on opposite sides of the earth,[46] are the Middle East region and the Gulf of Mexico–Caribbean region. The Gulf of Mexico–Caribbean pole includes the Gulf Coast province of the United States and the provinces in Mexico, Colombia, Venezuela, and Trinidad. The Middle East pole includes the provinces in Iran, Iraq, Kuwait, Saudi Arabia, and the Trans-Caucasus–Apsheron of the USSR. Two-thirds of the known reserves of the world have been found in these two regions, and future discoveries there give promise of being proportionately large. Lesser but important major productive regions in North America include the Mid-Continent province, the Appalachian province, the Illinois basin province, the Rocky Mountain province, California, the prairie provinces of Canada (especially Alberta), and the Tampico province of Mexico. In Europe the principal petroleum-bearing regions are the Polish province; the North German plains province; the Carpathian plains–Rumania–Ploesti province; the North Sea province; the Transylvanian basin province; and the Emba salt-dome province, the Perm basin province, the Fergana Valley, and the Ural-Volga province of the USSR. The Far East includes major producing areas in Ceram, Java, Borneo, Sumatra, New Guinea, Burma, and the Japanese islands. The reserves and current rates of production of each of the countries of the world are shown in Table 2-1. The growth of the world's crude-oil production may be seen in Figure 2-8.

Because of the varying political climates, the geographic situation of any petroleum prospect is extremely important in considering an exploration program.[47] In strongly nationalistic countries, such as the USSR, Mexico, Brazil,

THE OCCURRENCE OF PETROLEUM [CHAPTER 2] 33

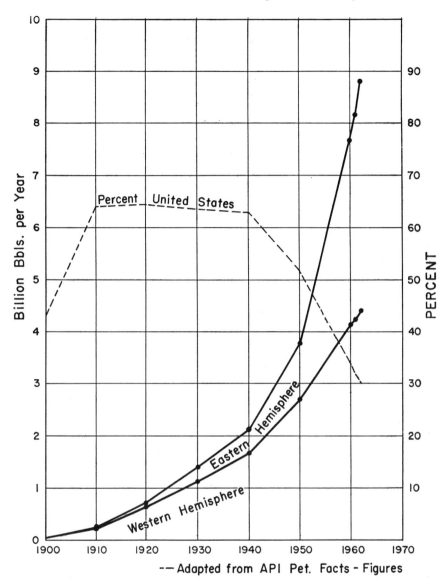

FIGURE 2-8 *World's crude oil production since 1900 in billions of barrels per year.*

and Argentina, the petroleum industry and exploration activities are closely controlled by the government itself or by government-owned companies. In less nationalistic countries, such as most European countries, Venezuela, Colombia, Peru, and the United States, the government owns some or all of the mineral rights but gives concessions or leases to companies or individuals to explore certain areas under rigid rules and regulations. This situation also

TABLE 2-1 Production and Reserves of Crude Oil

	Daily Production Est. 1963 1,000 bbl/day	Reserves Est. 1963 1,000,000 bbl
NORTH AMERICA		
Canada	717	5,675
United States	7,537	34,272
Mexico	320	2,500
Total	8,574	42,447
SOUTH AMERICA		
Argentina	262	2,300
Bolivia	9	200
Brazil	92	300
Chile	36	200
Colombia	167	900
Ecuador	7	25
Peru	5	380
Trinidad	134	500
Venezuela	3,246	17,000
Total	4,010	21,805
EUROPE		
Austria	49	240
France	50	235
Germany (West)	142	650
Italy-Sicily	34	300
Netherlands	42	250
United Kingdom	2+	1 1/2
Yugoslavia	30	250
Total	349	1,926
MIDDLE EAST		
Abu Dhabi	49	7,500
Bahrein	45	243
Iran	1,470	37,000
Iraq	1,120	25,500
Israel	3	25
Kuwait	1,930	63,500
Neutral Zone	306	10,000
Qatar	193	2,950
Saudi Arabia	1,618	60,000
Turkey	14	350
Total	6,748	207,068

	Daily Production Est. 1963 1,000 bbl/day	Reserves Est. 1963 1,000,000 bbl
AFRICA		
Algeria	502	7,000
Angola	14	200
Congo	2	10
Egypt	113	1,500
Gabon	17	150
Libya	470	7,000
Morocco	3	15
Nigeria	71	500
Total	1,192	16,375
ASIA–PACIFIC		
Australia	—	50
Burma	11	35
Formosa	—	.3
India	33	750
Indonesia	452	10,000
Japan	16	50
Malaysia	67	600
New Guinea	3	10
Pakistan	8	25
Philippines	—	.5
Total	590	11,520.8
USSR–CHINA		
USSR and controlled areas	4,121	29,500
China (estimated)	125	300
Total	4,246	29,800

SUMMARY

	Daily Production Est. 1963 1,000 bbl/day	Reserves Est. 1963 1,000,000 bbl	Reserves (Percent)
North America	8,574	42,447	12.6
South America	4,010	21,805	7.0
Europe	349	1,926	0.6
Middle East	6,748	207,068	62.5
Africa	1,192	16,375	5.0
Asia–Pacific	590	11,521	3.5
USSR–China	4,246	29,800	9.0
World total	25,709	330,942	100+%

Source: From O. & G. Jour., December 30, 1963. Most trade journals give production and reserve figures annually in December or January issues. Nearest whole numbers used.

prevails in the prairie provinces of Canada—Alberta, Saskatchewan, and Manitoba—except that the provincial government rather than the Canadian government controls most of the mineral rights, and each province sets up its own rules and regulations on exploration. The extreme opposite of the nationalistic system prevails in most of the oil-producing states of the United States, where the underground minerals are owned by the surface or fee owner of the land. There the wildcatter deals directly with the individual landowner, and both have a share in the oil and gas production.

The most advanced exploration practices have been developed in areas where the individual owns the minerals and the land. In these areas the tracts of land are generally small, and exploration becomes extremely competitive. Strong competition gives rise to new ideas and new methods, which increase production and profit. In those countries, on the other hand, in which great tracts of land are held by the government and competition is nonexistent, there is less incentive to undertake expensive and time-consuming test drilling, which is the best source of information on the many ever-present geologic variables, and is what ultimately results in the discovery of petroleum. In such countries, therefore, fewer wells are drilled, more variables remain unknown, fewer prospects are located, and less oil is discovered. This condition is less obvious, of course, in such prolific regions as the Middle East and northern Venezuela, where great pools are discovered with a minimum of exploration drilling. Yet even in these regions the proportion of oil that will ultimately be discovered is probably much less than it would be if there were a more competitive incentive for drilling.

The occurrence of petroleum is not confined to the land areas of the world. Discoveries of large deposits along the coast of southern California, along the

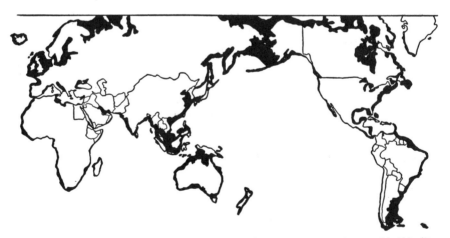

FIGURE 2-9 *Map showing the areas of continental shelf in the world (black). [Redrawn from Petroleum Press Service, London (January 1951).]*

shores of Louisiana and Texas in the Gulf of Mexico (see Figs. 6-33 and 6-35), in the North Sea, and in the Persian Gulf (see Fig. 2-2) point to the vast potential source of petroleum underlying the shallow waters that border the continents. These submerged lands, sloping down to a depth of 100 fathoms (600 feet) below sea level, are known as the *continental shelves*. Their distribution is shown in Figure 2-9. The area of the continental shelves, which are mostly under water less than 300 feet deep, has been estimated as 14 million square miles.[48] The geologic conditions prevailing on these continental shelves should be similar to those that prevail on the neighboring land areas. Exploration and development problems, however, will be quite different, because of the varying depths of water, soft bottoms, storms, uncertain ownership of the minerals, higher costs, and underwater pipe-line transportation.[49] Consequently, exploration will be much more expensive than on nearby land, and therefore slower and more cautious.

GEOLOGIC AGE OF RESERVOIR ROCK

The geologic age of the reservoir rock is a useful means of classifying a group of pools and fields or a province or subprovince, since it fits into a pre-existing classification.[50] It also serves in some cases as a descriptive classification, for the reservoir rocks of different ages frequently have different petroleum characteritics and productivity. The producing characteristics of the Permian reef limestone reservoirs of western Texas, for example, are different from those of the Ordovician limestones of the same area. The term "Miocene," applied to the Asmari limestone of the Middle East, quickly classifies the formation and brings to mind the differences between it and the Devonian limestone of the Ural-Volga region of the USSR. It is frequently desirable, therefore, to state the geologic age of a reservoir rock, for that is one means of identifying it and indicating something of its character. The age of the reservoir rock, however, does not necessarily indicate the time at which the petroleum accumulated in it. As will be pointed out later, the petroleum may have accumulated into a pool at any time subsequent to the original formation of the reservoir rock.

Two parallel classifications are used to identify (1) the time necessary to form the rock and (2) the rock units of the earth, termed geologic-time units and time-stratigraphic units, respectively. The time-stratigraphic units differ fundamentally from the descriptive classifications in that the rock boundaries are based on geologic time. That is, they are bounded by isochronous surfaces. The most commonly recognized ranks of geologic-time units and their corresponding time-stratigraphic units, in the order of descending magnitude, are:

GEOLOGIC-TIME UNITS	TIME-STRATIGRAPHIC UNITS
Era	
Period	System
Epoch	Series
Age	Stage

Thus we say "Devonian period" when considering the geologic time, but "Devonian system" when discussing the Devonian rocks. Similarly, we may speak of the Canadian series of the Ordovician system when referring to the rocks, but of the Canadian epoch of the Ordovician period when referring to the time.

The geologic time scale based on the geologic-time classification is given in Table 2-2. If the rocks of these units were being classified, the headings would be "system," "series," and "stage" instead of "period," "epoch," and "age." An estimate of the age in years is given for each of the major time boundaries.

Available evidence indicates that rocks of certain geologic ages are much richer in petroleum than those of other ages. Precambrian, Cambrian, and Triassic rocks, for example, have each produced less than 1 percent of the world's oil, and Pleistocene rocks have produced practically none. Rocks of Tertiary age, on the other hand, account for 58 percent, Cretaceous rocks for 18 percent, and Paleozoic rocks for 15 percent, of the estimated recoverable oil content of the earth. Details of the past production and the known reserves of oil, as related to geologic age, are shown in Table 2-3.

Only minor amounts of commercial oil have been found in rocks of definite Pleistocene age. The reason for this has been the subject of considerable speculation. The apparent lack of petroleum in the youngest rocks of the geologic column may be attributed, in whole or in part, to the general lack of drilling and exploration in these rocks. But it may be due in greater part to several other factors: the short time that the exposed sediments have had to form and to accumulate petroleum, the general lack of an impervious cover to create trap conditions, and the general nonmarine character of the sediments.

Petroleum is also rarely found in Precambrian rocks. Some commercial occurrences have been described,[51] but the general metamorphism and lack of permeability of most Precambrian rocks preclude them from being considered as potential reservoirs. Most of the Precambrian occurrences are found in fractures and secondary openings resulting from weathering and deformation. Significantly, these occurrences are usually associated with nearby petroleum-bearing younger sediments—which suggests a handy source of the petroleum. Oil and gas in the Precambrian rocks must be considered abnormal, and the upper surface of the Precambrian generally marks the lower limit of exploratory drilling.

The presence of organic carbon, however, has been demonstrated in rocks of Precambrian age. Rankama[52] found that the ratio of the carbon isotopes C^{12} and C^{13} in ten samples of Precambrian rocks from Finland, as measured by a

THE OCCURRENCE OF PETROLEUM [CHAPTER 2]   39

TABLE 2-2   The Geologic Time Scale

Era	North America			Europe			Age, millions of years (approx.)
	Period	Epoch	Age	Period	Epoch	Age	
Cenozoic	Quaternary	Recent Pleistocene		Neogene	Holocene Pleistocene		1
		Pliocene Miocene			Pliocene Miocene		25
	Tertiary	Oligocene		Paleogene	Oligocene		
		Eocene	Jacksonian Claibornian Wilcoxian		Eocene		
		Paleocene	Midwayan		Paleocene		63
Mesozoic	Cretaceous	Upper	Montanan Coloradoan	Cretaceous	Upper	Senonian Turonian Cenomanian	
		Lower	Comanchean Coahuilian		Lower	Albian Aptian Barremian Neocomian	135
	Jurassic	Upper Middle Lower		Jurassic	Upper Middle Lower	Malm Dogger Lias	181
	Triassic	Upper Middle Lower		Triassic		Keuper Muschelkalk Bunter	
Paleozoic	Permian	Upper Middle Lower	Ochoan Guadalupian Leonardian Wolfcampian	Permian		Tartarian Kasanian Kungurian Artinskian Sakmarian	280
	Pennsylvanian	Upper Middle Lower	Virgillian Missourian Desmoinesian Morrowan	Carboniferous	Upper	Stephanian Westphalian Dinanthian	
	Mississippian	Upper Middle Lower	Chesterian Osagean Kinderhookian		Lower	Visean Tournasian	345
	Devonian	Upper	Chautauquan Senecan	Devonian	Upper	Famennian Frasnian	
		Middle	Erian Ulsterian		Middle	Givetian Eifelian	
		Lower	Oriskanian Helderbergian		Lower	Coblenzian Gedinnian	405
	Silurian	Upper Middle Lower	Cayugan Niagaran Alexandrian	Silurian	Gotlandian	Ludlovian Wenlockian Llandoverian	
	Ordovician	Upper Middle Lower	Cincinnatian Mohawkian Chazyan Canadian		Ordovician		500
	Cambrian	Upper Middle Lower	St. Croixan Albertan Waucoban	Cambrian	Upper Middle Lower		600?
Precambrian	Proterozoic Archeozoic						

TABLE 2-3 World Production and Reserves (December 31, 1947)

Geologic Age	Total Cumulative Production to Date		Total Reserves		Total Ultimate Production	
	Billion Bbl	%	Billion Bbl	%	Billion Bbl	%
Tertiary	32.0	58.1	40.2	58.0	72.2	58.0
Cretaceous	10.8	19.6	11.7	17.0	22.5	18.3
Jurassic-Triassic	2.4	4.3	9.2	13.0	11.6	8.6
Paleozoic	9.9	18.0	8.3	12.0	18.2	15.0
Totals	55.1	100.0	69.4	100.0	124.5	100.0

Source: Adapted from G. C. Gester, *World Oil,* November 1948.

mass-spectroscopic assay, varied between 90.2 and 92.9. This ratio is nearly the same as that found in bituminous sediments, petroleum, and vegetable carbon material, including coal, where it ranges between 90.1 and 94.1. The ratio in inorganic material, such as graphite, diamond, calcite, and chalk, ranges between 87.9 and 90.2. Algae as old as Archean have been described by Gruner,[53] who also found evidence of Upper Huronian algae, bacilli, and iron bacteria.[54] He concluded that the iron and silica of the great Minnesota iron deposits were carried to the sea by rivers rich in organic matter. The disseminated bitumen found in the late Precambrian Nonesuch formation of northern Michigan[39] also suggests the possible presence of organic hydrocarbons in Precambrian time.

About the only regions of the earth in which there are virtually no prospects of future commercial petroleum discovery are those where Precambrian igneous rocks are exposed at the surface or are under the shallow cover of alluvium, glacial debris, or continental sands and gravels. The largest of these regions occur in the great shield areas of the earth—northeastern North America, Siberia, the Brazilian shield, much of central and southern Africa, and the Scandinavian countries.

As petroleum exploration expands over the world, it is becoming apparent that rocks of all geologic ages are potentially productive; the most important factor is the presence of connected porosity (*permeability*) in a manner that will trap and retain a pool of oil or gas. Lack of drilling and testing is probably the chief reason why geologists have a low opinion of rocks of some geologic ages, notably the Cambrian and the Triassic. Now that rocks of these ages have been found to be productive in various parts of the world, it is difficult to predict which rocks will eventually prove to be the most productive per unit of volume.

Nevertheless, rocks of Tertiary age continue to dominate in total productivity, and several reasons may be suggested to account for this:

1. The Tertiary contains thick sequences of unmetamorphosed marine sediments, characterized by lateral gradation, permeable reservoir rocks, adequate impervious cover, numerous traps (both local and regional), and an adequate supply of petroleum hydrocarbons.

2. Since the Tertiary system is late in the geologic time scale, only a minor part of it has been removed by erosion. Erosion attacks the topographically high rocks first and, as these normally are structurally high as well, destroys many traps filled with petroleum. A high percentage of pools, consequently, is preserved in Tertiary rocks.

3. Tertiary rocks consist largely of material eroded from pre-Tertiary anticlines, and this material may have included some of the oil that seeped out from the larger oil pools in the eroded rocks. Much of this eroded oil must, of course, have been evaporated, oxidized, destroyed by bacteria, or otherwise lost, but some of it, at least, may have been carried along with the sediments and redeposited with them.

Since the age classification of a reservoir rock offers a rough index, historically, of its potential productivity, it should be taken account of in any estimate of the chances of finding oil in a prospective region. Tertiary rocks, for example, are more likely, by and large, to contain oil pools than Triassic rocks are, because much more oil has been found in the Tertiary than in the Triassic. (It must not be forgotten, however, that some good pools have been found in Triassic rocks.) This, of course, is merely an indication of what the mathematical chances are of finding oil in a rock of a certain age at a specific place. Circumstances may occasionally justify hopes for oil in rocks belonging to systems that have yielded little oil in the past; whenever such hopes are fulfilled, there is renewed reason for believing that other areas, hitherto unproductive, have not, perhaps, been adequately tested.

CONCLUSION

Petroleum occurs in a variety of ways; some deposits are quite obvious at the surface of the earth, and others are buried at varying depths. While we are most concerned with the buried deposits, some of the large occurrences at the surface may become of first importance as our technology continues to develop. The smaller occurrences at the surface are chiefly important for the direct evidence they give of the presence of petroleum accumulations in the rocks of the area, and this is used to evaluate the prospects for deep-lying pools. If we are to include the lesser petroleum deposits, petroleum occurs in most of the countries of the earth and also in strata of all geologic ages. This is not to say that petroleum originated in each age, for the origin and the accumulation may be in separate episodes, as we shall see later. The chief interest of the petroleum geologist, however, is in knowing where petroleum is now; and one of the best places to look for new pools is the vicinity of known occurrences, irrespective of their size.

Selected General Readings

S. F. Peckham, 10th Census Report, Vol. 10 (1880), 317 pages. A comprehensive report on the occurrence of petroleum throughout the world as it was known in 1880. Extensive bibliography.

Sir Boverton Redwood, *Petroleum,* 4th ed., 3 vols., Charles Griffin & Sons, London (1922). Vol. 1 contains a detailed historical review and a discussion of petroleum occurrences throughout the world as of 1922.

A. E. Dunstan (managing editor) and many contributors, *The Science of Petroleum,* 6 vols., Oxford University Press, London and New York (1938–1953). This is a standard work of reference. The titles of the six volumes give an indication of the scope of the subject matter:
    I. Origin and Production of Crude Petroleum
    II. Chemical and Physical Principles of the Refining of Mineral Oils
    III. Refining
    IV. Utilization, Detonation and Combustion, and Bituminous Materials
    V. Crude Oils
    VI. The World's Oil-Fields

Wallace E. Pratt, *Oil In the Earth,* University of Kansas Press (1942, 1944), 110 pages. Four lectures that have become classic and are fundamental guides to petroleum geologists everywhere.

A. Beeby Thompson, *Oil-Field Exploration and Development,* 2nd ed., 2 vols., Technical Press, London (1950). Chapters 6–10 of Vol. 1 describe the occurrence of oil throughout the world, chiefly as it was before 1925.

Wallace E. Pratt and Dorothy Good (editors) and twenty contributors, *World Geography of Petroleum,* Spec. Pub. 31, Amer. Geog. Soc., Princeton University Press (1950), 464 pages. An authoritative discussion of the distribution and nature of the occurrence of petroleum in the crust of the earth. The bibliography has 346 items.

Walter K. Link, "Significance of Oil and Gas Seeps in World Oil Exploration," Bull. Amer. Assoc. Petrol. Geol., Vol. 36 (August 1952), pp. 1505–1540. Much new information on the importance of oil and gas seepages to discovery.

The American Association of Petroleum Geologists, P.O. Box 979, Tulsa, Okla., publishes numerous Special Volumes concerned with the occurrence of oil and gas.

Reference Notes

1. A. Beeby Thompson, "The Significance of Surface Oil, Indications," Jour. Inst. Petrol. Technol., Vol. 12 (1926), pp. 603–622. Discussion to p. 634.

Donald C. Barton, "The Indications of the Oilfield in the Mid-Continent and Gulf Coastal Plain of the United States," Jour. Inst. Petrol. Technol., Vol. 13 (1927), pp. 333–339.

S. E. Coomber, "Surface Indications of Oil," in *The Science of Petroleum*, Oxford University Press, London and New York (1938), Vol. 1, pp. 291–293.

V. C. Illing, "The Significance of Surface Indications of Oil," in *The Science of Petroleum*, Oxford University Press, London and New York (1938), Vol. 1, pp. 294–296.

Everett DeGolyer, "Direct Indications of the Occurrence of Oil and Gas," in *Elements of the Petroleum Industry*, Amer. Inst. Min. Met. Engrs. (1940), pp. 21–25.

Walter K. Link, "Significance of Oil and Gas Seeps in World Oil Exploration," Bull. Amer. Assoc. Petrol. Geol., Vol. 36 (August 1952), pp. 1505–1540. Includes numerous maps and sections showing seepages throughout the world.

2. A. Beeby Thompson, *Oil-Field Exploration and Development*, 2nd ed., 2 vols., Technical Press, London (1950), Vol. 1, pp. 238–342.

3. H. G. Kugler, "Summary Digest of Geology of Trinidad," Bull. Amer. Assoc. Petrol. Geol., Vol. 20 (November 1936), pp. 1439–1453.

4. E. H. Cunningham Craig, *Oil Findings*, 2nd ed., Edward Arnold, London (1921), pp. 144 and 159.

5. Ralph Alexander Liddle, *The Geology of Venezuela and Trinidad*, 2nd ed., Paleontological Research Institution, Ithaca, New York (1946), pp. 593–595.

6. Dept. of Information, Amer. Petrol. Inst., *California's Oil*, New York (1948), 28 pages.

7. Sir Boverton Redwood, *A Treatise on Petroleum*, 4th ed., Charles Griffin & Sons, London (1922), Vol. 1, pp. 208–212.

A. B. Cook and Clare Despard, "Historical Records Relating to Oil," Jour. Inst. Petrol. Technol., Vol. 13 (1927), pp. 124–134.

R. J. Forbes, *Bitumen and Petroleum in Antiquity*, E. J. Brill, Leiden (1936), 109 pages.

Laurence Lockhart, "Iranian Petroleum in Ancient and Medieval Times," Jour. Inst. Petrol. Technol., Vol. 25 (January 1939), pp. 1–18.

8. W. F. Foran, Oil from the Garden of Eden," Petrol. Engr., Vol. 14 (October 1942), p. 85.

9. R. J. Forbes, "Petroleum and Bitumen in Antiquity," Jour. Inst. Petrol. Technol., Vol. 25 (1939), pp. 19–23.

10. E. H. Cunningham Craig, *op. cit.* (note 4), p. 144.

G. M. Lees, "Pliocene Oil Seepages in Persia," Jour. Inst. Petrol. Technol., Vol. 13 (1927), pp. 321–324. Contains map showing seepages.

G. M. Lees, "The Geology of the Oilfield Belt of Iran and Iraq," in *The Science of Petroleum*, Oxford University Press, London and New York (1938), Vol. 1, pp. 140–148. Bibliog. 17 items.

11. Frederick G. Clapp, "Geology and Bitumens of the Dead Sea Area, Palestine and Transjordan," Bull. Amer. Assoc. Petrol. Geol., Vol. 20 (July 1936), pp. 881–909.

12. Sir Boverton Redwood, *op. cit.* (note 7), Vol. 1, pp. 1–31 and 201–208.

13. Col. Yule (ed.), *The Book of Ser Marco Polo the Venetian*, London (1871), i, 4.

14. Laurence Lockhart, *op. cit.* (note 7), pp. 3–14.

A. Beeby Thompson, *op. cit.* in note 2, Vol. 2, p. 547.

15. U.S. Coast and Geodetic Survey, Map. No. 5116, San Miguel Passage.

16. A. Beeby Thompson, *op. cit.* in note 2, Vol. 1, pp. 222–223 and 235–236.

17. George Shepard, "Observations on the Geology of the Santa Elena Peninsula, Ecuador, S.A.," Jour. Inst. Petrol. Technol., Vol. 13 (1927), pp. 424–461. Bibliog. 31 items.

18. H. G. Kugler, "Nature and Significance of Sedimentary Volcanism," in *The Science of Petroleum,* Oxford University Press, London and New York (1938), Vol. 1, pp. 297–299.

H. G. Kugler, "Contribution to the Knowledge of Sedimentary Volcanism in Trinidad," Jour. Inst. Petrol. Technol., Vol. 19 (1943), pp. 743–759. Discussion to p. 772. Bibliog. 26 items.

19. E. H. Cunningham Craig, *op. cit.* (note 4), p. 160.

20. A. Beeby Thompson, *op. cit.* in note 2, Vol. 1, pp. 228–229.

21. V. A. Gorin, "The Bibi-Eibat Tectonics and the Prospect of Development of the Lower Division," Azerbaijan Petroleum Industry (Baku), No. 11 and No. 12 (November and December 1933).

I. M. Goubkin, "Tectonics of Southeastern Caucasus and Its Relation to the Productive Oil Fields," Bull. Amer. Assoc. Petrol. Geol., Vol. 18 (May 1934), pp. 603–671. Bibliog. 41 items (all in Russian). Mud volcanoes, pp. 663–670.

I. M. Goubkin and S. F. Federov, "Mud Volcanoes of the Soviet Union and Their Connection with the Oil Deposits," XVIIth Int. Geol. Cong., Moscow, USSR, Vol. 4 (1937), pp. 29–59.

H. G. Kugler, "A visit to Russian Oil Districts," Jour. Inst. Petrol. Technol., Vol. 25 (1939), pp. 81–82.

22. H. L. Chhibber, *The Geology of Burma,* Macmillan & Co., London (1934), Chap. 6 ("Mud Volcanoes"), pp. 79–86. 17 references.

23. L. Dudley Stamp, "Natural Gas Field of Burma," Bull. Amer. Assoc. Petrol. Geol., Vol. 18 (March 1934), pp. 315–326; mud volcanoes, pp. 323–325.

24. H. G. Kugler, "Contribution" etc. (*loc. cit.* in note 18).

25. W. G. Weeks, "Notes on a New Mud Volcano in the Sea off the Coast of Trinidad," Jour. Inst. Petrol. Technol., Vol. 15 (1929), pp. 385–391.

26. George H. Eldridge, "The Asphalt and Bituminous Rock Deposits of the United States," 22nd Ann. Rept. U.S. Geol. Surv., Govt. Printing Office, Washington, D.C., Part I (1901), pp. 209–464.

Herbert Abraham, *Asphalts and Allied Substances,* 6th ed., 5 vols., D. Van Nostrand, Princeton (1960–1963). Extensive bibliography and references. The most complete discussion of asphalts and solid bitumens—worldwide occurrence, historical background, testing, uses.

27. L. L. Hutchison, "Rock, Asphalt, and Asphaltite in Oklahoma," Bull. 2, Okla. Geol. Surv. (March 1911), pp. 1–93.

28. L. C. Snider, "Rock Asphalts of Oklahoma and Their Use in Paving," Petroleum, Vol. 9 (1914), p. 974.

29. M. A. Carrigy, "Geology of the McMurray Formation," Memoir No. 1, Research Council of Alberta, Geological Division (1959). Contains extensive list of references.

M. A. Carrigy, "Effect of Texture on the Distribution of Oil in the Athabaska Oil Sands, Alberta, Canada," Jour. Sed. Petrol., Vol. 32 (June 1962), pp. 312–325.

S. C. Ells, *Bituminous Sands of Northern Alberta,* Canada Dept. Mines, Branch No. 632 (1926), 244 pages.

J. C. Sproule, "Origin of McMurray Oil Sands, Alberta," Bull. Amer. Assoc. Petrol. Geol., Vol. 22 (September 1938), pp. 1133–1149. Discussion to p. 1152. 43 references cited.

Alberta Society of Petroleum Geologists, "Northern Alberta Oil Sands," Bull. Amer. Assoc. Petrol. Geol., Vol. 35 (February 1951), pp. 181–184. 6 references.

Theo. A. Link, "Source of Oil in 'Tar Sands' of Athabaska River, Alberta, Canada," Bull. Amer. Assoc. Petrol. Geol., Vol. 35 (April 1951), pp. 854–862. 7 references.

30. Research Council of Alberta, Edmonton, Canada, "Athabaska Oil Sands," in K. A. Clark Volume, Inf. Series No. 5 (1963), M. A. Carrigy (ed.), 18 articles, 241 pages.
Theo. A. Link, *op. cit.* (note 29), p. 854.

31. W. V. Howard, "Lithification Processes and Early Oil Formation in Sediment," O. & G. Jour., Vol. 42 (June 17, 1943); reviewed in Jour. Inst. Petrol. Technol., Vol. 29 (1943), p. 324A.

32. T. Sterry Hunt, "Contributions to the Chemical and Geological History of Bitumen, and of Pyroschists or Bituminous Shales," Amer. Jour. Sci. & Arts, 2nd series, Vol. 35 (1863), pp. 157–171.
S. F. Peckham, *Petroleum and Its Products,* Dept. of Interior, Census Office, U.S. Geol. Ptg. Office (1885), p. 63, quoting James M. Stafford.
F. C. Phillips, "On the Occurrence of Petroleum in the Cavities of Fossils," Proc. Amer. Phil. Soc., Vol. 36 (1897), pp. 121–126.
John R. Reeves, "An Inclusion of Petroleum in a Fossil Cast" (Bloomington, Indiana), Bull. Amer. Assoc. Petrol. Geol., Vol. 9 (May-June 1925), p. 667.

33. George Homans Eldrige, "The Santa Clara Valley Oil District, Southern California," Bull. 309, U.S. Geol. Surv. (1907), pp. 22 and 77.

34. F. G. Clapp et al., *Petroleum and Natural Gas Resources of Canada,* Dept. of Mines (1915), Vol. 2, p. 262.

35. A. Beeby Thompson, *op. cit.* (note 2), Vol. 1, pp. 248–249.

36. Josiah Edward Spurr, *The Ore Magmas,* McGraw-Hill Book Co., New York (1923), Vol. 2, pp. 655–663.

37. L. P. Stockman, "Mercury in Three Wells at Cymric," Petrol. World, February 1947, p. 37.

38. H. Andrew Ireland, "Petroliferous Iron Ore of Pennsylvanian Age in Eastern Ohio," Bull. Amer. Assoc. Petrol. Geol., Vol. 28 (1944), pp. 1051–1056.

39. Charles G. Carlson, "Bitumen in Nonesuch Formation of Keweenawan Series of Northern Michigan," Bull. Amer. Assoc. Petrol. Geol., Vol. 16 (August 1932), pp. 737–740.

40. Nils Sundius, "Om oljeskiffer och skifferolje-industrien i vart land" (an account of the oil shales and the shale-oil industry of Sweden), Ymer (Svenska Sallskapet Antrop. och Geography), Argang 63 (1943), kep 1, pp. 1–16.

41. Dean E. Winchester, "Oil Shale of the Rocky Mountain Region," Bull. 729, U.S. Geol. Surv. (1923), 204 pages. Extensive bibliography, pp. 143–202.
M. J. Gavin, "Oil Shale: Historical, Technical and Economics Study," Bull. 210, U.S. Bur. Mines (1924).
Ralph H. McKee, *Shale Oil* (Mon. 25, Amer. Chem. Soc.), Reinhold Publishing Co., New York (1925), 326 pages.
The Science of Petroleum, Oxford University Press, London and New York (1938), Vol. 4, Part V. Section 43 ("Oil Shales, Torbanites, Cannels, etc.") and Section 44 ("Shale Oils and Tar Oils") contain several authoritative articles on "oil shales," the Scottish shale-oil industry, the Estonian shale-oil industry, and the occurrence and geology of "oil shales."
K. C. Heald and Eugene Ayres, "Our Reserves of Coal and Shale," in Leonard M. Fanning (ed.), *Our Oil Resources,* McGraw-Hill Book Co., New York (1945), pp. 157–209. Bibliog. 17 items.

42. U.S. Bur. Mines, "Composition of Oil Shales" (Geological Note), Bull. Amer. Assoc. Petrol. Geol., Vol. 7 (1923), pp. 296–297.

43. A. L. Down and G. W. Himus, "The Classification of Oil Shales and Cannel Coals," Jour. Inst. Petrol. Technol., Vol. 26 (July 1940), pp. 329–333.

44. James M. Schoph, "Cannel, Boghead, Torbanite, Oil Shale," Econ. Geol., Vol. 44 (January-February 1949), pp. 68–71.

45. E. G. Woodruff, "Petroliferous Provinces," Bull. 150, Amer. Inst. Min. Met. Engrs. (1919), pp. 907–912; Trans., Vol. 65 (1921), pp. 122–204. Discussion by Charles Schuchert and others, pp. 204–216.

Frederic H. Lahee, "Classification of Exploratory Drilling and Statistics for 1943," Bull. Amer. Assoc. Petrol. Geol., Vol. 28 (June 1944); Part II, definitions, pp. 703–711.

46. Wallace Pratt, *Oil in the Earth,* University of Kansas Press, Lawrence, Kansas (3rd ptg. 1944), 110 pages, pp. 31–33.

47. Western Hemisphere Oil Study Committee, Independent Petroleum Association of America, Washington 6, D.C., Report of the Subcommittee on Government Policies and Laws (October 1952). Contains a summary of the laws of the countries in the Western Hemisphere pertaining to petroleum and petroleum exploration.

48. William H. Twenhofel, *Treatise on Sedimentation,* 2nd ed., Williams & Wilkins Co., Baltimore (1932), 926 pages, p. 858.

49. National Petroleum Council, Washington, D.C., *Submerged Lands Productivity Capacity* (May 28, 1953), 43 pages, maps, and geologic sections.

50. American Commission on Stratigraphic Nomenclature, "Report No. 2—Nature, Usage, and Nomenclature of Time-Stratigraphic and Geologic-Time Units," Bull. Amer. Assoc. Petrol. Geol., Vol. 36 (August 1952), pp. 1627–1638. Contains the principles of modern rock classification.

J. Laurence Kulp, "Geologic Time Scale," Science, Vol. 133 (April 14, 1961), pp. 1105–1114.

51. Robert F. Walters, "Oil Production from Fractured Pre-Cambrian Basement Rocks in Central Kansas," Bull. Amer. Assoc. Petrol. Geol., Vol. 37 (February 1953), pp. 300–313.

52. Kalervo Rankama, "New Evidence of the Origin of Pre-Cambrian Carbon," Bull. Geol. Soc. Amer., Vol. 59 (May 1948), pp. 389–416.

53. John W. Gruner, "Algae, Believed to be Archean," Jour. Geol., Vol., 31 (1923), pp. 146–148.

54. John W. Gruner, "Contributions to the Geology of the Mesabi Range," Bull. 19, Minn. Geol. Surv. (1924), pp. 59–64.

PART TWO

The Reservoir

3. *The Reservoir Rock*
4. *The Reservoir Pore Space (Porosity and Permeability)*
5. *Reservoir Fluids—Water, Oil, Gas*
6. *Reservoir Traps—General and Structural*
7. *Reservoir Traps (continued)—Stratigraphic and Fluid*
8. *Reservoir Traps (continued)—Combination and Salt Domes*

INTRODUCTION TO PART TWO

THE PETROLEUM RESERVOIR is that portion of the rock that contains the pool of petroleum. The location of every oil and gas pool may be said to be the result of a complex of interrelated geologic conditions. Each reservoir is unique in its details, but general relations may be seen that permit broad classifications of the major elements that control a reservoir.

The petroleum reservoir consists of four essential elements, each of widely variable development, each with many gradations, and each of varying importance in the location and size of the pool of petroleum. They are:

1. The *reservoir rock,* or containing material. The composition and texture of the reservoir rock, and its continuity or lack of continuity, are of prime interest in the geology of petroleum. The edges of the reservoir rock may coincide with the edges of the petroleum pool, as where a lens is filled with oil and gas; or the reservoir rock, though extending through a large region, may become a petroleum reservoir only at locally favorable areas.

2. The *pore space,* or *void space,* sometimes called the *reservoir space,* is expressed as a fraction or percentage of the total volume of the rock (for example, 0.23 or 23%) and is called its *porosity.* The effective pore space is that portion of the reservoir rock that is available for the migration, accumulation, and storage of petroleum. The measure of the ease with which fluids may move through the interconnected pores of the rock is called its *permeability.* Porosity and permeability are properties that depend on the presence of pore space. They are of special interest because they determine the capacity of the reservoir rock both to hold and to yield petroleum.

3. The *fluid content* consists of the water, oil, and gas that occupy the effective pore space within the reservoir rock. Under favorable conditions the oil and gas are concentrated into pools, but most of the reservoir pore space outside the pools contains only water or water with petroleum measurable

in parts per million. The petroleum, then, occurs within an aquifer—within a water environment. The fluids may be in a state of either static or dynamic equilibrium; that is, at rest or in motion. During their geologic life they undoubtedly have been in motion at some time or even continuously because of changes brought about by erosion, deposition, and deformation, and any other changes that upset the equilibria of the fluid pressure, temperature, density, volume, and chemical characteristics. These changes cause the fluids to move along gradients from areas of higher energy potential toward areas of lower energy potential. Although the movements of the fluids cannot be observed directly, the concentrations of oil and gas into pools and the widespread evidence of fluid pressure gradients is evidence of such movement.

4. The *reservoir trap,* or the *trap,* is the element that holds the oil and gas in place in a pool. Most geologists think of the trap as the shape of the reservoir rock element that permits a petroleum pool to accumulate underground. As we shall see later (pp. 340–343) the trap may actually be due in part to the fluid pressure gradients that exist in the reservoir fluids. As considered here, the trap is the shape of the reservoir rock together with its pore space.

Rock traps are formed from a wide variety of combinations of structural and stratigraphic features of the reservoir rocks. A trap generally consists of an impervious cover—the *roof rock*—overlying and sealing a porous and permeable rock that contains the oil and gas. The upper boundary, as viewed from below, is concave; the top is generally arched, but it may form an angle or peak. In practice the term "trap" usually means any combination of rock structure and of permeable and impermeable rocks that will keep oil and gas from escaping, either vertically or laterally, because of differences in pressure or in specific gravity. Some petroleum reservoirs completely fill the trap, so that if any additional oil or gas were added it would spill out around the lower edges. Other reservoirs occupy merely a part of the apparent capacity of the trap.

The lower boundary of the reservoir is, either wholly or partly, the plane of contact of the oil and gas with the underlying body of water upon which the pool rests. It is known as the *oil-water contact* or *oil-water table.** The water fills all of the pore space of the reservoir rock below the oil-water contact, and that portion of the reservoir pore space that is not filled with oil and gas. If the water is at rest, the contact plane is level or approximately level. But, if the water is in motion, because of a hydrodynamic fluid potential gradient parallel to the bedding and across the pool, the lower boundary of the reservoir may be an inclined plane, and the pool is said to have an *inclined* or *tilted oil-water table*. Occasionally the tilt of the oil-water table is enough to flush the oil and

---

* The oil-water table should be distinguished from the *water table,* which has had a long priority of usage in the geology of ground water to denote the surface below which all pores are filled with water. The ground-water table is, in effect, the *air-water table,* in contrast to the oil-water table of the petroleum geologist.

INTRODUCTION TO PART TWO 51

gas out of a potential trap, in which case the rock trap is not effective, and there is neither reservoir nor pool.

These broad generalizations about the nature of a petroleum reservoir will apply to most of the known pools of petroleum in the world. The next five chapters are concerned with details of these elements as they combine in varying proportions to trap a pool of petroleum.

CHAPTER 3

The Reservoir Rock

Reservoir rock: classification – nomenclature – fragmental – chemical – miscellaneous. Well logs. Marine and nonmarine reservoir rocks.

BROADLY SPEAKING, any rock that contains connected pores may become a reservoir rock. As a matter of fact, however, nearly all reservoirs are in unmetamorphosed sedimentary rocks, and most of them in sandstones, limestones, and dolomites. Shales, slates, and igneous rocks are known to be reservoir rocks under exceptional conditions, but these conditions are rare and anomalous. A reservoir rock may be limited to the area of the pool of petroleum, or it may persist, with uniform lithological and physical characteristics, far beyond the pool.

CLASSIFICATION

Since nearly all petroleum reservoir rocks are of sedimentary origin, any classification of reservoir rocks is essentially a classification of sedimentary rocks. A number of classifications have been proposed,[1] some descriptive and some genetic, but most of them designed chiefly for the use of the specialist in sedimentary petrography.

Classifications of petroleum reservoir rocks for practical use should be as simple and broad as possible, for the petroleum geologist must keep his terminology understandable to the operator, driller, and engineer, who supply many of his basic data and to whom he has to convey his own ideas. Many terms that are perfectly good scientific descriptions, and that have clear and exact meanings for geologists, do not find much favor in the petroleum industry—such terms, for example, as "arenaceous" for "sandy," "argillaceous" for "shaly," and "rudaceous" for "conglomeratic." We need terms that are generally understood yet sufficiently definite.

A simple, broad, primary classification of reservoir rocks, based largely on the origin of the rock, divides them into three groups: (1) fragmental (clastic); (2) chemical and biochemical (precipitated); (3) miscellaneous. This may oversimplify a complex and difficult problem, but such a rough classification is useful in the geology of petroleum and is readily understandable. It is the system used here. The chief difficulty in applying any rock classification is that there are many gradational types that are hard to classify. Reservoir rocks, like all sediments, commonly grade into one another. Complex reservoir rocks are named according to their dominant constituent or rock characteristic, with an adjective to indicate the minor constituent, as in "limy sand" and "sandy lime."

It is sometimes useful to class a reservoir rock as of marine or nonmarine origin. This genetic classification may be combined with a lithologic classification, as in the terms "marine limestone," "continental sandstone," and "nonmarine conglomerate."

It is often useful, also, to place the rock in the standard geologic time scale and thereby classify it according to its geologic age. This can be done by combining an age term with other terms; we may speak, for example, of a "Permian dolomite" or a "nonmarine Oligocene sandstone" or "Devonian grit."

Nomenclature of Reservoir Rocks

A reservoir rock formation from which petroleum is produced is commonly given a specific proper name. Such names frequently begin at a well or pool during the early development stages, and, once started, they are difficult to change. A name that some driller or operator may casually give to a producing formation is first used in conversations, then in newspapers and trade journals, then in engineering and geological reports, and finally in legal contracts and scientific journals. One reservoir formation may come in this way to be called by many different names; but, as development proceeds and the continuity of the formation throughout the area becomes obvious to everyone, some names are dropped, and one name may come into general use.

The commonest names for reservoir rocks are "pay," "pay sand," "oil sand," "gas sand," "sand," and "lime," which are generally used by drillers and field men without much thought given to whether they are accurate rock names. Examples of names based on a more or less petrologic character are "Simpson sand," "McCloskey oolite," and "Welch chert." But many names have other bases. The depth relative to other pay formations is shown in such names as "First sand," "Third stray," "D-3 pay," and "First 'Wilcox.'" The relation to another reservoir is shown by "Squaw sand" for a lens detached from the "Big Injun" sand. The geologic name of the formation is used where it can be identified underground, as in "Berea sand," "Madison limestone," "Oriskany sand," and "Asmara limestone." A common method of naming a producing formation in the United States is to give it the name of the owner of the land on which the discovery well was drilled. Examples of such names are "Hoover sand" and "Jones sand."

Sometimes the name of the producing formation is taken from the name of the pool in which it was first found, as "Mirando sand" from "Mirando pool." Another method is to name the producing formation after some characteristic fossil or fossil assemblage, as in "Nodosaria sand" and "Heterostegina sands." Sometimes the sand is named after the individual responsible for its discovery, or after the operator or the geologist. Examples are the "Wilcox" sand of Oklahoma, named after Homer F. Wilcox, who drilled the discovery well; the Dibblee sand of the Cuyama Valley of California, named after Tom Dibblee, a geologist who had much to do with the discovery; the Slick sand of central Oklahoma, named after Tom Slick, a famous wildcatter; and the Vedder sand of the San Joaquin Valley, named after the geologist Dwight D. Vedder, who also owned the land on which the discovery well was drilled.

Let us now return to the main kinds of reservoir rocks. These, as was said earlier, are fragmental (clastic), chemical (and biochemical), and miscellaneous.

FRAGMENTAL RESERVOIR ROCKS

Fragmental reservoir rocks are aggregates of particles, fragments of minerals, or fragments of older rocks. They are also called *clastic* or *detrital* rocks because they consist of mineral and rock particles washed from areas that have been eroded. Their character varies with such factors as the nature of the eroded material, the distance it is transported, the climate, the steepness of the gradients, the transporting agency (whether streams, waves, currents, or winds), the biochemical conditions of the area of deposition, the distance from shore, the agencies by which the particles are sorted, and the depth of the water.

The constituent particles of fragmental reservoir rocks range in size from colloidal particles up to pebbles and boulders.* The Wentworth grade scale[2] is commonly used to identify the texture of various clastic sediments and is also used to describe the texture of reservoir rocks. It is made up of a continuous series of size grades of particles, the median size in each grade being half as great as the median size in the next coarser grade. It is shown in Table 3-1. A useful but somewhat different classification is that of the Bureau of Soils (U.S. Department of Agriculture), shown in Table 3-2. The size distribution of clastic sediments may also be indicated by giving the logarithms of the diameters of the particles, as in Figure 3-1. Most typical sediments

---

* A chart correlating the various grain-size definitions of sedimentary materials, prepared by Page E. Truesdell and David J. Varnes, and published in 1950, may be obtained from the United States Geological Survey, Washington 25, D.C.

TABLE 3-1 Particle Size Classification (Wentworth)

Classification	Grade Limits (Diameters in mm)	Microns	Retained on Mesh
Boulder	Above 256 mm		
Large cobble	256–128		
Small cobble	128–64		
Pebble			
Very large pebble	64–32		
Large pebble	32–16		
Medium pebble	16–8		
Small pebble	8–4		5
Granule	4–2		6
Sand			
Very coarse sand	2–1		12
Coarse sand	1–1/2		20
Medium sand	1/2–1/4		40
Fine sand	1/4–1/8		70
Very find sand	1/8–1/16	125–62.5	140
Silt			
Coarse silt	1/16–1/32	62.5–31.2	270
Medium silt	1/32–1/64	31.2–15.6	
Fine silt	1/64–1/128	15.6–7.8	
Very fine silt	1/128–1/256	7.8–3.9	
Clay			
Coarse clay	1/256–1/512	3.9–1.95	
Medium clay	1/512–1/1,024	1.95–0.975	
Fine clay	1/1,024–1/2,048	0.975–0.487	
Very fine clay	1/2,048–1/4,096	0.487–0.243	

are mixtures of sand, silt, and clay in varying proportions. A graphic method of showing the relations is given in Figure 3-2.[3]

Grain-size analyses of fragmental rocks offer valuable clues to the potential character of the fluid content. For example, most reservoir rocks of many analyzed from southern Trinidad, BWI, were found to be *graywackes,** which vary in grain size between 0.250 and 0.0625 mm.[4] The finer-grained rocks generally contained water, and the coarser-grained rocks carried gas or tar.

The *matrix* of a clastic rock consists of particles, distinctly smaller than the average, that partially or entirely fill the interstices between the larger grains.[5] Such particles may consist of the same minerals as the larger grains, or of other minerals, or (as is usual) of both. The matrix differs from the *cementing*

---

*Graywacke is a dark fragmental rock in which 20% or more of the constituent grains consist of other rock fragments.—Krynine.

TABLE 3-2 Particle Size Classification (U.S. Bureau of Soils)

Colloids	less than 0.001 mm	Fine sand	0.1 to 0.25 mm
Clay	0.001 to 0.005 mm	Medium sand	0.25 to 0.5 mm
Silt	0.005 to 0.05 mm	Coarse sand	0.5 to 1.0 mm
Very fine sand	0.05 to 0.1 mm	Fine gravel	1.0 to 2.0 mm

material, which is chiefly a chemical or biochemical deposit or clay that has infiltrated around both the larger particles and the matrix particles. The matrix and the cement may both be present in varying amounts, and either or both may be virtually absent from a sand. The relations of grains, matrix, and cement are graphically shown in Figure 3-3.

Among the fragmental rocks, sandstones, conglomerates, arkoses, graywackes, and siltstones are by far the most common reservoir rocks, and these sediments constitute nearly half of all reservoir rocks. Most of the fragmental reservoir rocks are siliceous, but many are fragmental carbonate rocks, such as oolites and coquinas, made up respectively of ooids and shell fragments that have been only slightly cemented or recrystallized. Shales are of only minor importance as reservoir rocks.

In some sandstones the grains are virtually all of quartz, with little or no cement. From this extreme there are all gradations, into sandstones that contain grains of other minerals in greater or lesser abundance, and into those that contain various kinds of matrix or cement, or both, in various quantities. Next to the pure-quartz type come sandstones that are highly siliceous but in which some of the grains consist of chert, fine-grained quartzite, and silicates,

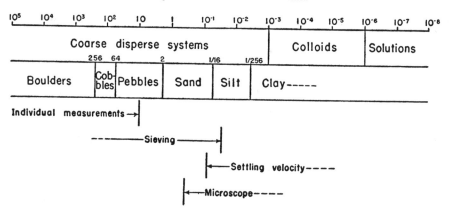

FIGURE 3-1 *Range of particle size in clastic sediments.* [*Redrawn from Krumbein and Sloss,* Stratigraphy and Sedimentation, *W. H. Freeman and Company (1963), p. 97, Fig. 4-2.*]

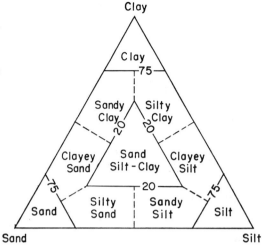

FIGURE 3-2

System used to classify sediments in terms of their content of sand, silt, and clay. [From Shepard, Jour. Sed. Petrol., Vol. 24 (1954), p. 157, Fig. 7.]

chiefly feldspar. Some sandy rocks contain feldspar, mica, and quartz in nearly the same proportion as the granites from which they are derived; these are the arkoses, or "granite washes." * Sandstones in which grains of dark, fine-grained igneous rocks are abundant are called graywackes. In all these rocks there is nearly always some matrix or cement, and usually both, between the larger grains. A reservoir rock may vary locally in texture and composition, either vertically because of bedding or laterally because of facies changes, or it may be essentially uniform throughout an entire region.

Sandstones vary from clean, well-washed quartz sandstone through all combinations produced by additions of silts and clays ("shaly," "muddy," "trashy," and "dirty" sandstones), of carbonates (limy, calcareous, and dolomitic sandstones), of silica (siliceous or quartzitic sands, volcanic ash), and of feldspars, micas, and rock fragments (arkose, granite wash, graywacke). The additions may come as primary constituents, as matrix, or as cement, and the variations in composition may be local or regionally uniform. Solution, redeposition, and recrystallization may become important in rocks containing soluble materials, especially in sandstones containing much calcite or dolomite, or rocks consisting chiefly of those minerals, as some of them do.

Clean, uniform, and continuous quartz sandstones may have been formed by the erosion of other sandstones, or they may consist of material that has been transported a long distance from the source, or been subjected to strong wave and current action during deposition, all of which make for thorough sorting and uniform texture. Where the sandstones contain abundant feldspars, micas, and other silicates, or clay or chert, they were probably derived from igneous and metamorphic rocks, from shales that had not been deeply weath-

---

* An arkose is "a sandstone containing 25 or more percent of feldspar derived from the disintegration of acid igneous rock of granitoid texture" (Bull. 98, Committee on Sedimentation, Nat. Research Council, 1935).

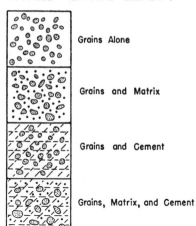

FIGURE 3-3

Possible textural elements of clastic rocks consisting of grains, cement, and matrix in varying proportions. [Redrawn from Krynine, Jour. Geol., Vol. 56 (March 1948), p. 139, Fig. 5.]

ered, or from shales and clays; the material may have been transported only short distances, or may have been deposited under such variable conditions as might prevail in deltas and flood plains. Clean, uniform sands, when continuous over wide areas, are sometimes called *blanket sands;* those containing much clay, shale, or other impurities are called *muddy* or *dirty sands* or, if they also contain fragments of other rocks, graywackes. A diagram showing the variable nature of a graywacke is shown in Figure 3-4. Graywacke reservoir rocks are found in the Bradford region of Pennsylvania and in the Gulf Coast region of Texas and Louisiana.[6]

There are more pools producing from sandstone reservoirs than from any other single rock type. The total production and ultimate reserves, however, are probably less than from carbonate reservoirs. Thousands of different pools might be mentioned, but only a few general references, where sandstone reservoir rocks are described, are listed below.

Tertiary sandstones contain most of the oil in California[7] and in the Gulf Coast region of Texas and Louisiana[8] and also many of the pools in Venezuela.[9]

Cretaceous sands are highly productive in the Burgan field of Kuwait, probably the largest single field in the world. The productive Burgan sands make up 800 feet of a total of 1,100 feet of section, and the pool is trapped in a large anticlinal fold.[10] Pools in the Baku, Maikop, and Grozny districts of the Caucasus region of the USSR—comprising some of the richest producing areas of the world—are nearly all in sandstone reservoirs.[11] The oil of the East Texas pool, the largest single pool in the United States, is found in the Woodbine sand (basal Upper Cretaceous). A structural map and a section of the East Texas pool are shown in Figure 8-3, page 351, and the location of the pool is shown in Figure 10-17, page 467.

Sandstones of Paleozoic age are the reservoir rocks in a great many pools through the eastern Mid-Continent and Rocky Mountain regions of the United States. The Devonian sands of the Bradford pool in Pennsylvania, because of the highly successful secondary recovery operations carried out there,[12] have probably been the most intensively studied reservoir rock in the world. A map of the region is shown in Figure 7-17, page 300. The "Wilcox" and Simpson sands (Middle Ordovician) of the Mid-Continent states of Kansas and Oklahoma are among the cleanest of the sandstone reservoir rocks and contain a large percentage of round, frosted sand grains. Most of the oil in the Oklahoma City pool is produced from Simpson sands. A map of the field and surrounding region is shown in Figure 14-6, page 643, and Figure 14-7, page 644.

Conglomerates, grits, and coarse sediments, chiefly composed of silica and silicates, are common reservoir rocks. Most are lenticular bodies, enclosed in finer-grained and more uniform sediments of similar composition—for example, the quartzose grits and conglomerates interbedded with the quartz sands and graywackes of the Devonian of northwestern Pennsylvania.[13] In the coarse and poorly cemented lenses of conglomerate (with pebbles up to two inches in diameter) and of grit, in the Haynesville field of Louisiana,[14] there are, because of the high permeability of these materials, many exceptionally

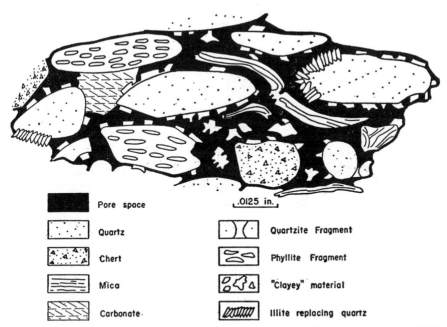

FIGURE 3-4 *Schematic section through a graywacke ("exploded" and ×125) showing the complex nature of the porosity and the mineral composition of a "dirty" sand. [Redrawn from Krynine, Jour. Geol., Vol. 56 (March 1948), p. 153, Fig. 13.]*

productive wells. The Big Injun sand (Lower Mississippian) of eastern Ohio and western West Virginia is a highly productive, massive, coarse, conglomeratic, and cross-bedded sandstone; where the same formation is thinner and consists of a more normal fine-grained sandstone, it produces only water.[15]

Basal Pennsylvanian conglomerates, composed of rounded pebbles and boulders of Mississippian chert, constitute the reservoir rock in a number of pools in Kansas;[16] and a basal conglomerate consisting mainly of schist rubble is the reservoir rock in the Playa del Rey field of California.[17] In a large area of the northern end of the East Texas field, a part of the pay zone is a conglomerate and gravel bed in the Woodbine producing formation (Upper Cretaceous), which consists mainly of normal sands, volcanic ash, and shales.[18]

Most sandstone reservoir rocks vary in texture and mineral composition, both vertically and laterally. For example, sandstone lenses and irregular patches of sand, interbedded with shale in siltstone, may thicken and thin, become shaly or clean, with no apparent reason. These sand lenses may be small and completely filled with oil or gas, and thus constitute both the reservoir rock and the trap, or they may extend continuously over larger areas and contain several pools at places where favorable trap conditions exist. When a reservoir formation that varies laterally in permeability becomes impermeable, it restrains or prevents the escape of any contained petroleum. Lateral variation in texture and mineral composition is consequently a favorable characteristic of a reservoir rock, for such a rock may trap petroleum whether deformed or not.

Many examples of rapid lateral variation in reservoir sands might be cited. Tertiary formations, especially, are characterized by numerous unpredictable lateral changes in almost every place where they are productive. Sands come and go, and are commonly cross-bedded; their areal patterns, too, are irregular: they may be long and sinuous or broad and irregular. Nearly all of the Tertiary reservoir rocks of Rumania and of the Caucasus-Apsheron province of the USSR, for example, are characterized by rapid lateral change, gradations from clean sands to silts and shales, scouring and channeling, and wide variations in grain size. Many of the Pennsylvanian sands of the Appalachian–Illinois–Mid-Continent region also are characterized by rapid and unpredictable lateral changes. Many of the Upper Cretaceous productive sands of the Rocky Mountain region grade out in short distances and are replaced at some higher or lower horizon. In the Ventura Basin of California, a highly productive province, the Tertiary rocks thicken in a few miles to 40,000 feet along a trough or belt fifty miles long, and their rapid lateral and vertical changes have been intensively studied.[19] One effect of rapid lateral variation is to make correlations difficult or even impossible. Even where abundant evidence from samples, electric logs, pressure and production data, and fossil control is available, many wells must be drilled to establish the reservoir relations with any degree of confidence. And even when this has been done, the correlations may apply only to indefinite zones rather than to individual sands or to formations.

The variable deposits of the geologic past can be explained, in part at least, by a study of deposits that are being laid down in the seas and oceans at the present time.[20] The ancient sediments, like those of modern times, were probably zoned or concentrically arranged, around the ocean deeps. The character of the deposits in each depositional zone depends on such factors as the depth of water, the distance from shore, the direction and strength of the ocean currents, the biological and biochemical environment, and the nature of the material being brought down the rivers. They were, in fact, the sum of the source materials, together with the energy impressed upon them through such forces as weathering, transportation, deposition, and diagenesis.

The depositional zone nearest the shore—the strand, or littoral zone, known as the *neritic* environment (0–600 feet of water)—is characterized by wide variations in type of deposit; it contains coarse and fine sands, silts, coquinas, clays, and shales, all subject to rapid changes in texture and composition, both vertically and laterally. Chemical and biochemical activity is there at its highest so that shale and sand deposits are likely to be interbedded with fragmental rocks composed of organic remains. Most sandstone reservoir rocks were probably formed in this variable near-shore zone. Alternate advances and retreats of the sea caused sand formations to merge and various sediments to become re-sorted and redistributed, and continental conditions therefore interfinger with marine conditions. The Tertiary reservoir rocks bordering the Caribbean and the Gulf of Mexico, those occurring throughout the East Indies, and those of the USSR in the Middle East are especially characterized by conditions such as these.

Distributary channels of deltas and tidal-flat channels may fill with sand and clastic material and thus form branching channel sand deposits enclosed in shale. These may coalesce to form patchy sands or even sand formations that are continuous over considerable areas.[21] A sand deposit in which deltaic distributary channels have been recognized is shown in Figure 3-5, a map of the Booch producing sand (Pennsylvanian) in the Hawkins field, Oklahoma. The contours show the distribution of the high-potential wells in the pool (isopotential map), and, because of the wells' higher initial yields, they are also interpreted as giving a rough measure of the areas of high permeability. (See also Fig. 13-5, p. 597.)

Seaward the variable near-shore zone grades into or interfingers with the zone of muds, where conditions are more stable—the *bathyal environment* (600–6,000 feet of water); and this zone, in turn, grades into the pelagic zone of calcareous and siliceous oozes and red clays occupying the deep basins of the oceans—the *abyssal environment* (6,000–30,000 feet of water). The boundaries between the different zones are sharp in places, but more often they are gradational.

Layers of coarse, unsorted clastic deposits alternating with uniform fine-

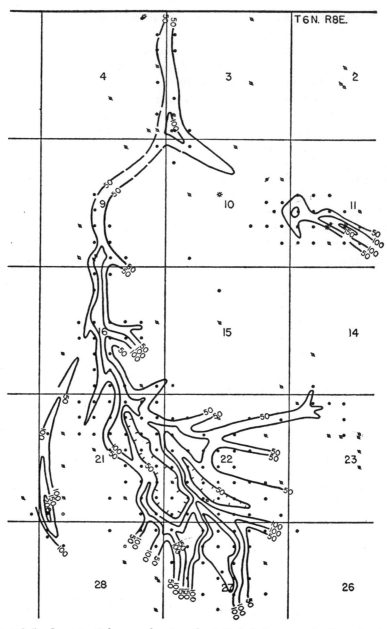

FIGURE 3-5 *Isopotential map showing the initial daily rate of oil production in the Hawkins field, Hughes County, Oklahoma. The contour interval is 50 barrels per day. Compare with Figure 13-5, an isopach map of the producing sand of the area.* [By the courtesy of Daniel A. Busch.]

grained deposits, and some graded bedding, are thought to be caused by high-density, or turbidity, currents that carry near-shore deposits far out into the ocean. They are called *turbidites*. Near-shore slumping or submarine landslides might start the movement, and the coarse material would then pass over older deposits without disturbing them, finally coming to rest in a sedimentary environment that was totally different from, and out of harmony with, the original sedimentary environment from which they began their journey.[22] That these phenomena may occur on a large scale was seen in the earthquake of 1929 off the Grand Banks of the Atlantic Ocean,[23] when material slumped off the continental slope traveled over 450 miles beyond the continental shelf and into the abyssal ocean plain, where many telegraph cables were cut. Kuenen[24] estimated the mass of moving mud to have contained 100 cubic kilometers of material, to have been about 270 meters (875 feet) thick and 350 kilometers (217 miles) wide, and to have moved over 1,000 miles from its point of origin. Coarse-grained, poorly sorted deposits, therefore, may not always indicate a near-shore environment.

Most cherts are composed of chalcedony (quartz fibers separated by films of opal), with minor impurities, such as quartz grains. Typical pure cherts have a massive, dense texture, a dull waxy luster, and conchoidal fracture. Terms such as *cherty shale, cherty limestone, cherty dolomite,* and *cherty sandstone* are applied to rocks that contain combinations of chert with other materials. All of these may form reservoir rocks. Some of them grade into rocks called *novaculite* or *porcellanite,* in which silica is the predominant material. Fragmental cherts, either produced by weathering in place, as a residual material, or transported and redeposited as clastic detritus, form reservoir rocks in some places. It is sometimes difficult to distinguish beds of a detrital chert, especially when recemented with silica, from the underlying, chemically precipitated chert, or from chert that has formed in place by weathering away of the original surrounding material. The relations are shown in Figure 3-6. A detrital chert, called "chat," forms a reservoir rock in the basal Pennsylvanian Sooy conglomerate[25] of Kansas; it is composed of boulders and smaller fragments of chert embedded in a red shale. This formation rests unconformably on the underlying Mississippian "lime," which is largely chert in its upper part and from which the chat presumably was derived. The upper part of the Mississippian limestone is weathered, fractured, and cherty, and closely resembles the overlying reworked material, especially when it is drilled up as well cuttings.

Varying amounts of feldspars are present in many sandstone reservoir rocks. As the feldspar content increases, the rock grades into an arkose, or granite wash. As with a chert reservoir rock, it is sometimes difficult to distinguish arkose from the underlying granite from which it is derived. (See Fig. 3-7.) Fresh granite may be distinguished from weathered granite by measuring the magnetic susceptibility of the well cuttings. The susceptibility of unweathered granite is generally several hundred to several thousand times

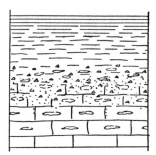

Shale

Redeposited chert.

Weathered chert.

Cherty limestone.

FIGURE 3-6

Sketch showing nature of gradational contact between chert, weathered chert, and redeposited chert, any one or all of which may be the reservoir rock.

that of weathered granite.[26] The residual, or remnant, magnetism of unweathered granite is likewise so much greater than that of weathered granite that the distinction between the two may be readily made. A striking example of arkose, or granite-wash, production is seen in the Panhandle fields of Texas. The lower part, of granite wash, contains an increasing amount of fresh granite boulders and fragments, until in places it becomes almost indistinguishable from the granite bedrock.[27] The reservoir rock is of Pennsylvanian-Permian age, and the granite basement is probably Precambrian. A section across the Texas Panhandle area is shown in Figure 3-8.

An arkosic reservoir rock, with its fresh or slightly weathered feldspars, suggests near-by mountain-building, block-faulting, or other strong deformation. It shows that basement granites have been exposed, and wedged-out sediments, including arkosic material, are likely to be found around the exposed cores. The arkosic material probably eroded rapidly, was not carried far, and was buried without extensive weathering. Of even greater importance is the fact that near-by diastrophism implies large-scale unconformities, overlaps, folding, and faulting—conditions that may form traps capable of holding large oil and gas pools. The Pennsylvanian and Permian arkoses of southern and western Oklahoma and the Panhandle of Texas are good examples. Arkoses that are free from clays and weathered debris, and that consist of permeable mixtures of quartz, feldspars, and mica, are especially good reser-

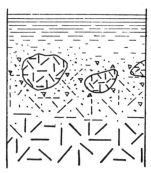

Shale

Arkose, with fresh granite boulders.

Weathered granite

Fresh granite.

FIGURE 3-7

Sketch showing the nature of the gradational contact between fresh granite, weathered granite, and reworked granite debris (arkose). Drilling into fresh granite boulders in the arkose may cause the premature abandonment of the well if they are thought to represent basement rocks.

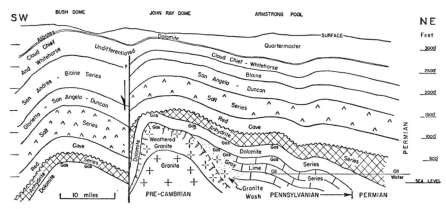

FIGURE 3-8 *Section across the buried Amarillo Arch in the Panhandle of Texas. The debris from a narrow mountain uplift several hundred miles long was deposited in the Pennsylvanian and early Permian seas along the north flank of the range. The granite core was finally buried, and the area was again folded, faulted, and eroded, with the final result that an interconnected zone of porosity, extending a distance of 125 miles, formed a trap in which over 31 trillion cubic feet of natural gas and over one billion barrels of recoverable oil accumulated. The reservoir rocks include granite wash, arkose, sand, limestone, and dolomite. An oil-water contact of nearly uniform level and a subnormal reservoir pressure characterize the field. This is but one of a number of oil and gas pools associated with the buried Wichita Mountains and Amarillo Arch.* [Redrawn from Cotner and Crum, Geology of Natural Gas, *Amer. Assoc. Petrol. Geol.* (1935), p. 388, Fig. 2.]

voir rocks. Such arkoses are said to have been "cleaned up" or "winnowed." Many arkoses, however, are so dirty with weathered fragments as to be impermeable.

Clay

Clay is of great importance in the geology of petroleum. It is present in most reservoir rocks in varying amounts, it enters into many problems connected with the porosity and permeability of the reservoir, it influences greatly the production of oil and gas in many pools, and it is especially important in the water-injection programs of secondary recovery. Most of the compaction and compressibility of sediments is due to the squeezing of water out of clay minerals; the high mineral content of oil-field waters is probably largely due to the freeing of salts adsorbed onto the clay minerals; and clays form the bulk of most drilling muds. Because of the many later references to clays, it is worthwhile to give some attention to them now. The reader is referred to the many articles on the role of clays and clay minerals in the geology of petroleum and the production of oil and gas.[28]

The clay minerals* present in nearly all reservoir rocks may be dispersed through the sandy rocks as individual grains, may fill interstices between sand grains and thus serve as a cement, or may be in thin laminae interbedded with layers of sand or carbonate. Since many of the clay minerals are platy, small amounts may be plastered over the surface of sand grains in thin films, and thus a very small amount of clay may have a surprisingly large effect on such phenomena as adhesion, adsorption, interfacial tensions, capillarity, and wettability. Some clay minerals are oleophyllic and some are hydrophyllic.

Clay minerals, when present, have much the same effect on chemical reservoir rocks as on fragmental reservoir rocks. Clay minerals occur in carbonate rocks as partings at the bedding planes, and elsewhere as thin shale laminae.[29] Such partings may consist of flocculated colloidal clay material carried to the areas of deposition, for such material will generally be flocculated (collected into small soft lumps or flakes) when it comes in contact with sea water. Styolites are common in limestones, and the material along their boundaries may consist of colloidal clay. The major Paleozoic limestone and dolomite formations of Illinois, for example, all contain clay, which is chiefly in the form of illite, but partly in that of kaolinite. The illite is considered to be authigenic (to have originated in place), since illite weathers readily into other constituents and is unstable under weathering conditions. The kaolinite is probably detrital.

Analyses of clays by x-rays and optical and electron microscopes show them to consist of aggregates of extremely minute crystalline particles of clay minerals. These clay minerals tend to be platy or lath-shaped. The smallest particles consist of a single crystal; the larger particles may contain connected groups of crystals. Individual crystals are composed of what are known as building units, which in turn make up the atomic lattices or sheets of the molecule, and are nearly identical.

Two reasons largely account for the importance of clay in reservoir studies: (1) the smallness of the individual crystal particles, many being less than two microns (8×10^{-5} inch) in diameter and some of the most active being less than 8×10^{-6} inch; (2) the chemical and physical activity of the clay minerals, especially of the montmorillonite group. The small size means a proportionately large surface area, which means a larger than normal effect on the surface phenomena of the reservoir. The chemical activity is chiefly due to the presence of loosely attached exchangeable cations, or the power of *ion-exchange*—that is, an exchange of ions of a solution for those of a solid when they are in contact. The character of both the solution and the solid undergoes a change as a result of the base-exchange. Most of the ion-exchange properties of the sediments, including reservoir rocks, are due to their clay content.[31] The physical activity of the clay minerals is due to their

---

* The chief clay minerals are kaolinite [$(OH)_8Al_4Si_4O_{10}$], illite [$(OH)_4K_y(Al_4 \cdot Fe_4 \cdot Mg_4 \cdot Mg_6)(Si_{8-y} \cdot Al_y)O_{20}$] ($y = 1$–$5$), and montmorillonite [$(OH)_4Al_4Si_8O_{20}$]. [From R. E. Grim, in *Recent Marine Sediments* (reference note 28), pp. 475–477.]

lattice, or accordion-like molecular structure, which permits the entry of water between the lattices, thereby greatly changing their volume. This permits fluid continuity, even across thick, fine-grained shales, and allows the shales to act as semipermeable membranes. Substantially large fluid pressure gradients may thus form across the shale bedding planes because of osmotic phenomena coupled with differences in the salinity of waters above and below the shale layers.

Clays that have developed fissility are called shales. Shales are not ordinarily considered reservoir rocks, but in a few places they have produced considerable oil and gas, probably contained in fractures and in films along bedding planes. The Florence pool in Colorado[32] occurs in shale of Cretaceous age, and there are gas pools in the Cherokee (Pennsylvanian) shales of eastern Kansas[33] and the Chattanooga (Mississippian-Devonian) shales of eastern Kentucky.[34] Oil production has also been found in the shales above the main pay sands in the Salt Creek and Tow Creek fields of Wyoming and in the shales above the pay formation at Rangeley, Colorado.* Production has been obtained from the shales as well as from the limestones in the shale-limestone section west of Lake Maracaibo in Venezuela.* Large quantities of oil have also been produced in the Santa Maria field, California,[35] where the reservoirs are chiefly siliceous shales, from the sandy siliceous shales of the Stevens zone in the Elk Hills field, California,[36] from the shales, siltstones, and sandy shales of the Spraberry field, Texas.[37]

Volcanic ash is sometimes an important though generally a minor constituent of producing sands. Its effect, like that of shales and the clay minerals, is to reduce permeability. The Woodbine sand of the East Texas field contains large amounts of volcanic ash:[18] the content of shale and volcanic ash together in the producing formation ranges from 30 percent in the northern part of the pool to 70 percent in parts of the southern end of the pool. The interbedding and lateral variations of the sand, shale, and volcanic ash make detailed correlations from one well to the next extremely difficult.

Cementation of Fragmental Reservoir Rocks

Some sandstone reservoir rocks consist either entirely or partly of loose, uncemented sand grains, and at times loose sand grains come up in large quantities along with the oil. The sand grains in most reservoir sandstones, however, are held together by various kinds of cementing material, chiefly carbonates, silica, or clays. Some of the cementing material may be primary, having been deposited along with the sand grains, and then precipitated chemically around and between them by a diagenetic† process. Sandstones

---

* Herman Davies in a personal communication.

† "Diagenesis denotes the processes leading to the lithification of a rock, or the conversion of newly deposited sediments into an indurated rock." (F. J. Pettijohn, *op cit.* in reference note 1, p. 476.)

cemented by primary silica are called *orthoquartzites,* as distinguished from metaquartzites, or quartzite of metamorphic origin. Other cementing material may be secondary, having been precipitated from water solutions that entered the formation after it was deposited (see also pp. 128–130).

As the proportion of cement increases, a clastic rock may be gradually changed to a chemical rock. By increase in amount of dolomitic cementing material, for example, a sandstone—within a few miles, or, even more abruptly, within a few hundred feet—may change from a clean quartz sand to a dolomitic sand, and finally to a sandy dolomite. By increase in the silica cement, a friable sand may grade into a quartzitic sand and finally into a sandy quartzite.

Different lateral aspects of equivalent sedimentary rocks that are essentially contemporaneous give rise to what are termed *facies.* These may be either local or regional in extent. Different lithologic aspects are called *lithofacies,* and different biologic aspects are called *biofacies.* Lithofacies maps are discussed on pages 600–602. The most significant lithofacies changes in reservoir rocks occur where a permeable rock grades into a less permeable rock. Such changes are of great importance in relation to the production of oil and gas from pools, the localization of traps, and the regional movement of fluids through the reservoir rock.

Clastic limestones and dolomites—the former are sometimes called calcarenites—consist of grains of calcite and dolomite that have been transported and redeposited just as grains of quartz are. The carbonate grains may consist largely of shells, shell fragments, coquina, and oolites. Rocks thus formed are always more or less recemented with recrystallized calcite and, when the process has gone far, may resemble a chemically deposited limestone or dolomite. Cross bedding may often be seen, however, on the weathered surfaces of clastic carbonate rocks, and the outlines of clastic grains may occasionally be recognized under the microscope. Carbonate rocks formed in this way are likely to be porous and therefore to be good reservoir rocks, and they probably form a larger proportion of the carbonate reservoir rocks than is generally realized.

CHEMICAL RESERVOIR ROCKS

Chemical reservoir rocks are those that are predominantly composed of chemical or biochemical precipitates. They consist of mineral matter that was precipitated at the place where the rocks first formed, and not transported as clastic grains, although clastic grains may have originated as chemically precipitated rock before being formed into grains. The dominant chemical reservoir rocks are carbonate sediments, mostly limestones and dolomites.[38] The relations may be graphically shown, as in Figure 3-9.

It is hard to tell how large a proportion of the carbonate reservoir rocks

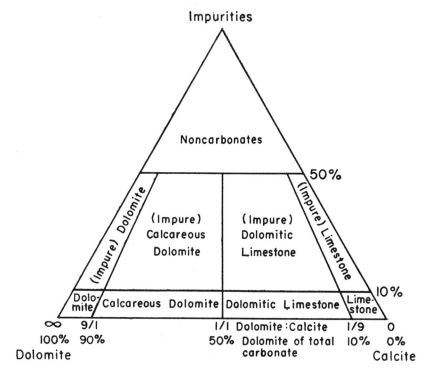

FIGURE 3-9 *Compositional graph showing the relations of the carbonate rocks. The impurities commonly consist of sands, clays, and shales.* [*After Leighton and Pendexter.*]

are truly chemical or biochemical precipitates except in thin section with the petrographic microscope;[39] a clastic limestone or dolomite may be so thoroughly recemented and recrystallized as to be readily mistaken for a limestone or dolomite formed wholly by precipitation in place. Similarly, the nomenclature of the carbonate and related rocks is a difficult problem, largely because of the gradational boundaries of the various types of carbonates. Usage is generally in the direction of compositional and structural characteristics rather than genetic.[40] Recementation and recrystallization may be considered as chemical processes in place, and a rock formed by these processes would still properly be classified as a chemical reservoir rock if it appeared as predominantly chemical in origin even if a part of it was originally clastic grains. Some chemically precipitated rocks consist wholly, or almost wholly, of silica, in the form of chert, novaculite, or orthoquartzite, but in some of these there has been a certain amount of secondary cementation with silica. Such rocks are quite common, but, compared with the carbonate rocks, they contain few reservoirs. In some reservoir rocks carbonate and siliceous precipitates may be mixed to form cherty or siliceous limestones and dolomites.

Chemically Precipitated Carbonate Rocks

The chemically precipitated carbonate reservoir rocks are usually crystalline limestones and dolomites, but sometimes consist of marl and chalk. When they consist predominantly of carbonate, their textures are generally crystalline, and they may be coarse-, medium-, or fine-grained. The relatively pure carbonate rocks may grade into more or less siliceous rocks. The siliceous components may be precipitated chert, siliceous fossils, clastic grains of quartz or chert, or shaly material more or less intimately mixed with the carbonate. As the siliceous components increase, we get such things as sandy, cherty, or shaly limestones and dolomites.

The carbonates in these rocks are almost wholly calcite ($CaCO_3$) and dolomite [$CaMg(CO_3)_2$], and the carbonate in a particular rock is likely to be almost wholly of one species or the other. The two are often intermingled, however, in various ways. Dolomitic and calcitic rocks may be interbedded, or dolomite may form irregular bodies that stand out in slight relief on weathered surfaces and give them a patchy appearance. Or the rock may have reached a certain uniform degree of dolomitization throughout, owing to replacement of calcium by magnesium, and it is then classed as a magnesian or dolomitic limestone—or in the final stage, as a dolomite in which more than 50 percent, by weight, is the mineral dolomite,[38, 41] Both patchy and uniform dolomitization are most commonly found in the older Paleozoic formations. The extensive recrystallization that accompanies dolomitization destroys many of the primary textures, fossils, and fossil casts of the original limestone. Limestones and dolomites are approximately equal in importance as reservoir rocks, differing chiefly in their permeability; the dolomites are generally more permeable.

Some carbonate rocks have closely spaced bedding planes, which may grade into massive, virtually unbedded rocks. It is rather difficult to identify bedding planes in well cuttings, but modern continuous-coring devices obtain nearly complete core recoveries, and these generally show the bedding. The massive, nonbedded rocks are usually found in organic reef deposits, and a lack of bedding is sometimes indicative of such an origin.

Limestone and dolomite reservoir rocks of biochemical origin have been identified in all terranes from Pliocene to Cambrian. Rocks of biochemical origin contain significant quantities of biologic remains along with the normal chemically precipitated material. Biochemical processes have been especially active in the organic reefs (bioherms, biostromes), which have become increasingly important as reservoir rocks; but lime-secreting organisms have contributed variable amounts of material to most carbonate rocks. (See pp. 315–316.) The main biochemical agents in forming limestones are the algae, bacteria, foraminifera, corals, bryozoa, brachiopods, and mollusks. Of these the algae are the most important as rock-builders; they have, in fact, been considered by some geologists as among the chief agents of lime secretion and deposition.[42] The carbonate secreted by living organisms is mostly $CaCO_3$, in the form of either aragonite or calcite. Magnesium carbonate as well as calcite is secreted by some organisms, but is thought to be of minor importance. Some rocks of biochemical origin were formed where the organisms lived and died, the shell fragments having later been consolidated more or less in place; others consist of shells and skeletal remains that were transported by winds and currents and concentrated into clastic deposits. The change from shell fragments to a rock results partly from the growth of algae, such as *Lithothamnium,* which filled in and solidified the loose fragments, and partly from cementation, packing, and precipitation of carbonate. After the material has been solidified and buried, the entire mass may be again dolomitized and recrystallized, leaving little or no trace of the parent material. The resultant rock is therefore a variable combination of fragmental shell debris and biochemical and chemical precipitates, localized by an environment that permitted the lime-secreting organisms to flourish and to be deposited and then modified after burial by various processes of solution and redeposition.

By increase in the proportion of insoluble clastic materials, limestones and dolomites may grade imperceptibly into shaly limestones and sandy dolomites, and from them into calcareous shales and dolomitic sandstones. The clastic sand grains, chert fragments, and other insoluble materials were probably enclosed in the carbonate material while it was being deposited. Some fragments are set in the carbonate rock like plums in a pudding, without touching one another; others make up thin laminae, lenses, or patches.

Limestones and dolomites are by far the most important of the chemical reservoir rocks; in fact, they contain nearly half of the world's petroleum reserves. The shift in importance from predominantly sandstone reservoirs to carbonate reservoirs has been a development of recent years, brought about by the discovery and development of the great oil fields of the Middle East, western Texas, and western Canada, in much of which oil is produced from limestone and dolomite reservoir rocks. Some of the more important oil-bearing carbonate formations of the world are briefly described below.

The greatest concentration of large oil fields in the world is in the Middle East, chiefly in Iran, Iraq, Kuwait, and Saudi Arabia. (See Fig. 2-2, p. 18.) Much of the production is obtained from limestone reservoir rocks, which are folded into large anticlines. Several pools contain 5 billion barrels or more each. The reservoir rock in Saudi Arabia is chiefly of the Arab zone (Upper Jurassic), consisting of several limestone and dolomite formations, each 20–200 feet thick, separated by anhydrite layers.[43] The chief producing limestone formation of Iran is the Asmari limestone (Oligocene and Miocene), 700–1,500 feet thick, which underlies a large area and has reef characteristics in places.

Prolific oil production is obtained from several pools in thick, dense, Cretaceous limestones in western Venezuela.[44] (See Fig. 14-12.) The oil is found on large, faulted anticlines; a section across one of them, the Mara field, is shown in Figure 6-31 on page 263. Cretaceous limestones also form the reservoir rock in most of the great Mexican oil fields. Many of these discoveries were made as a result of drilling near surface seepages. A structural map showing one such pool in Mexico, the Poza Rica pool, is shown in Figure 8-10, page 356.

Devonian limestone and dolomite reef deposits form the reservoir rocks of most of the large oil pools of western Canada. (See also pp. 327–329 and Fig. 7-48.) Niagaran (Silurian) limestone reef rocks form the reservoir rock of many pools, mostly small, in Ontario, Indiana, Kentucky, and Illinois.[45] (See Fig. 7-35, p. 319.)

The Kansas City–Lansing group (Pennsylvanian) of limestones, with interbedded thin shale members, ranges from 200 to 400 feet in thickness and constitutes one of the important producing sections over a large area in central Kansas.[46] The pools are localized by variations in the porosity and permeability of the limestones, together with local structural anomalies. The Kansas City–Lansing group truncates and rests unconformably on all the older formations down to the Cambro-Ordovician Arbuckle limestone on the broad, low Central Kansas Uplift.

The Arbuckle-Ellenburger group (Cambro-Ordovician)[47] contains many important pools in the Mid-Continent region of Kansas, Oklahoma, and Texas. It is also the oldest, geologically, among the major producing formations of the world. It consists predominantly of dolomite and limestone, with minor amounts of chert, sand, and shale, and ranges in thickness from about 1,000 feet near El Paso to more than 7,000 feet in the Arbuckle Mountains of Oklahoma. The limestone and dolomite members are alternately thin- and thick-bedded, and their texture varies from coarsely granular to microgranular. Many siliceous and sandy layers occur locally in the upper part in northeastern Oklahoma, where the formation is called the "siliceous lime." Most of the pools in the Arbuckle-Ellenburger rocks are trapped in folds and faulted anticlines. As the oil production declines, they produce water freely, suggesting a more than local permeability. The oil is generally of high quality, and many wells have an initial producing capacity of several thousand barrels a day.

The region of western Texas and southeastern New Mexico, also known as the Permian Basin, is one of the richest producing areas in North America. (See Fig. 3-10.) It accounts for about 17 percent of the production of the United States.[48] With the exception of the sand-trend pools (AA' in Fig. 3-10), practically all of the pools of the region are in Paleozoic limestone and dolomite reservoir rocks. In fact, all ages from Permian to Cambro-Ordovician are productive in a wide variety of traps. (See also Fig. 6-16, p. 25.)

Siliceous Reservoir Rocks

Chemically precipitated silica is common as an accessory constituent of many carbonate reservoir rocks,[49] and in places it becomes the predominant or sole constituent of the rock. The siliceous material is most often found as chert,[50] in the form of nodules, interbedded layers, and minute fragments embedded in the limestone or dolomite. It is difficult in wells to distinguish

RESERVOIR ROCK [CHAPTER 3] 73

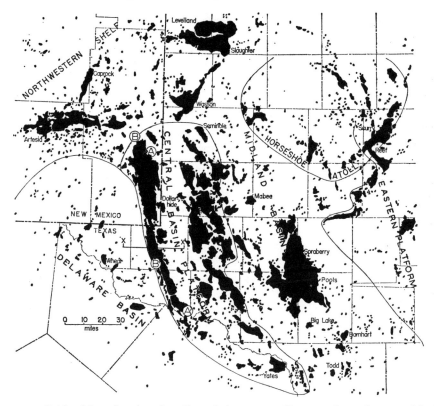

FIGURE 3-10 *Map showing the oil pools in western Texas and southeastern New Mexico.*

precipitated primary chert from eroded chert fragments later cemented by silica.

The Monterey formation (Miocene), a reservoir rock in California, consists of several kinds of siliceous rocks, classified as porcellanite, porcellanous shale, cherty shale, and chert. Bramlette[51] believes they were formed by alteration of diatomaceous rocks through solution of the delicate opaline diatom shells and reprecipitation of the silica as a cementing material. Other theories suggest that the Monterey cherts were derived from a silica gel material. Their permeability is due to cracks and fractures. (See also pp. 119–121.)

The reservoir rock of some pools in western Texas consists of a dense, hard, brittle, white-to-buff, opaque, and very fine-grained sedimentary siliceous rock that closely resembles the Caballos novaculite (Devonian), which crops out in the nearby Marathon region. As with most dense and brittle rocks, the porosity is thought to be caused by fractures, which have been enlarged in places into solution cavities and cavernous voids.[52]

MISCELLANEOUS RESERVOIR ROCKS

Miscellaneous reservoir rocks include the igneous and metamorphic rocks, mixtures of both frequently forming the "basement complexes."[53] These are interesting geologically but rarely important commercially. Where commercial production is obtained from igneous and metamorphic rocks, the reservoir is generally located up-dip from overlapping or buttressing sediments, from which the petroleum is thought to have migrated. Bedding and unconformity planes in sediments appear to have provided permeable paths for migration, and the reservoir space is usually in fractures in the brittle basement rocks.

The igneous rocks of the great volcanic fields of the earth present some rather special exploration problems, and their effect on the accumulation of oil and gas has not been fully evaluated. These volcanic fields include the Columbia Plateau of Washington and Oregon; the volcanic deposits extending up the Rocky Mountains, through British Columbia, and into Alaska; the Mexico-Arizona volcanic field; the Deccan traps of India; the volcanic deposits of the Pacific; the Paraná basin of South America; and the region comprising Iceland, North Ireland, and the Hebrides. The presence of artesian ground water in many igneous flows, and in weathered intrusive igneous rocks, shows permeability through interconnected pore spaces, which suggests that oil and gas might be found in both extrusive and intrusive igneous rocks under special conditions. The presence of lava flows does not preclude the presence of petroleum in the underlying and interbedded sediments. These sediments may well produce like any other sediment, once the hard, brittle, overlying volcanic series has been penetrated.

Igneous and metamorphic rocks from which commercial oil production has been obtained include the basalt flows, pyroclastics, and intrusive basalt and andesite dikes, known as the "Conejo volcanics" (Miocene), in the Conejo oil field of Ventura County,[54] the granitic and metamorphic basement complex (Jurassic?) in the Edison field of San Joaquin Valley,[55] and the Franciscan (Cretaceous-Jurassic?) or older schist in the El Segundo field south of Los Angeles,[56] all in California. Production is obtained from fifty or more widely scattered wells in the fractured, Precambrian, quartzite hills or knobs of central Kansas[57] and from the granite basement cores of the Mara and La Paz oil fields of western Venezuela. (See p. 125.) Commercial gas has been found in the Rattlesnake Hills field of Washington from one or two porous zones of basalt, overlain by lake clays intercalated between the igneous flows.[58] "Serpentines" are the reservoir rocks in a number of pools of central Texas.[59] (See pp. 305–306.)

The main producing formation of the Tupungato field, in Mendoza Province, Argentina, was formerly thought to be a sandstone but is now known to consist of a series of hard, volcanic tuffs, interbedded with shales through a thickness of 300 meters. The porosity is high, but the permeability is low; there is much pore space, but the pores are not well interconnected. The formation has been intensely fractured, and cores show that the openings that have not been filled with calcite or zeolites contain oil.[60]

WELL LOGS

Well logs are used to identify and correlate underground rocks and to determine the porosity of potential reservoir rocks and the nature of the fluids they contain. Since the petroleum geologist's chief area of interest is below the surface, we will refer to well logs and well data in every chapter that follows; it is desirable, therefore, that we now discuss briefly some of the different kinds of well logs, with their uses and limitations. A complete discussion of all of them is beyond the scope of this book, but the reader is referred to the many articles about and descriptions of various logging methods, a few of which are listed in the reference notes.[61]

The common types of well logging are (1) drillers' logs; (2) sample logs: (a) lithologic, and (b) paleontologic; (3) electric logs; (4) radiation logs: (a) gamma ray, and (b) neutron; (5) drilling-time logs; (6) core and mud analyses; (7) caliper logs; (8) temperature logs; (9) sonic logs; (10) dipmeters. A brief discussion of the manner in which they are made and of their uses follows.

Drillers' Logs. Most logs of wells drilled before 1930 were prepared by the drillers of the wells; the geologist has no other well data than drillers' logs in many large areas, and is forced to use and interpret them as best he can. Cable-tool drillers' logs proved relatively satisfactory, for the driller is fairly sure of the depth to the tops of the formations, of the character of the rocks, and of their content of oil, gas, or water at all times. He determines sand by the wear on his bit, and shale and limestone by the jerk on the drilling line. He knows how much water is placed in the hole and how much comes out; therefore he knows the water content of every permeable formation. In fact, the rate at which water is produced is a good measure of the permeability of a rock. He also knows immediately when small amounts of gas and oil show in the well. He describes the rocks as hard, soft, or sticky, and as red, blue, black, gray, or brown. His measurements are accurately checked each time a string of casing is run in the hole, and generally a steel-line measurement is taken at the top of the producing formation. Errors in depth, which are common, are absorbed in the last steel-line measurement. If such errors are large, they throw the log off, and a geologist generally distributes the correction up to the next check point above.

Drillers' logs of rotary-drilled holes are not nearly as dependable or usable as the cable-tool drillers' logs; in fact, rotary drillers seldom make logs any more. Rotary drillers are able to tell change in color from the cuttings that come up in the mud, and they are able to tell hard and soft formations. Since most of the early drilling, at least, was in the areas of soft formations, many of their shales and clays were described as "gumbo" or "sticky clay." "Rock" is any hard formation; "boulders" are alternating hard and soft formations; "heaving sand" is sand that is forced into the hole from the bottom; "quick sand" is sand that caves and settles rapidly; "water sand" is sand in which the cuttings come clean and bright, or sand that dilutes the drilling mud.

Sample Logs. These logs, prepared by geologists and based on an examination of the well cuttings and cores, began to be made in the period from 1920 to 1925, and have increased steadily ever since. Sample logs are made now of practically every wildcat well and of a great many production wells. The sample logs may be made from the surface of the ground to the bottom of the hole, or they may be prepared for only particularly important portions of the section.

Samples of drill cuttings from cable-tool holes are collected each time the hole is bailed, or approximately every five or ten feet, and are dried and placed in cloth sacks. As they come from the well they are generally clean and require no additional treatment. The drill cuttings are examined under a binocular microscope, at powers ranging between 12 and 24 times, and the log is then compiled, either at the well or in a laboratory.

Samples from rotary holes, called "ditch samples," are obtained from the return mud stream that comes from the well and carries the cuttings from the bottom of the hole to the surface. Well samples are collected at every five, ten, or twenty feet of drilling, and they represent the material drilled during this penetration. They are washed in water until the fine, colloidal mud material is removed, then dried and placed in sacks. The samples, together with rock fragments and cores, are examined under a binocular microscope either at the well or at the laboratory, and a lithologic log of the well is prepared. The geologist at the well is called a *well-site geologist,* and while he is there he is said to be *sitting on the well.*

Lithologic logs prepared from well cuttings, either from cable-tool holes or from rotary holes, are logs that describe the physical properties of each formation penetrated by the well. Characteristics commonly noted for each sample include the character of the material, whether it is limestone, dolomite, sandstone, silt, clay, conglomerate, anhydrite, salt, or chert; its color, its luster, whether greasy, dull, or shiny; its content of fossils; its porosity, if any, and whether the porosity is intergranular, vuggy, pin-point, or characterized by primary or secondary crystallinity; evidences of oil, gas, or solid bitumen; evidence of fracturing or slickensiding; whether the sand is loose and friable or cemented together, and the character of the cement; whether the material breaks across individual grains or whether the cuttings are clusters of sand grains cemented together; and any other physical property that may help to identify it. The time it takes for the mud stream to carry the samples from the bottom of the hole to the surface* must be considered in

---

* This time may be roughly estimated by calculating the number of strokes the mud pumps require to displace the mud in the hole and dividing by the strokes per minute. A better way is to chart the time for each well, using as control points the time it takes for a sudden change in rock material to reach the surface after recording a sudden change in drilling time rate. Rice or corn may also be added to the mud stream at the top of the drill pipe; the time it takes to make the round trip to the shale shaker is approximately twice the lag time. A rough measure is that for seven- or eight-inch holes: it will take cuttings about ten minutes per 1,000 feet to return from the bottom. [See Hiestand and Nichols, Bull. Amer. Assoc. Petrol. Geol., Vol. 23 (1939), p. 1824, and John M. Hills, *loc. cit.* in reference note 60, pp. 348–349.]

RESERVOIR ROCK [CHAPTER 3]   77

making a lithologic log from well cuttings from a rotary hole; it may take several hours for the fragments to rise from the bottom of deep holes to the surface.

Paleontologic logs, as the name implies, are logs in which the emphasis is on stratigraphy and fossil content, the fossils being chiefly the smaller forms such as foraminifera and fusilinids.

Electric Logs. An electric log is a continuous record of the electrical properties of the formations and the fluids penetrated in a well. (See Fig. 3-11.) The measurements are made in the uncased portion of the well, and are commonly made of formations in holes drilled with rotary tools while they are still full of drilling mud. Electric logs are made by passing electrodes, encased in an insulated tube called a *sonde,* down the well hole. A generator at the surface sends electric energy down

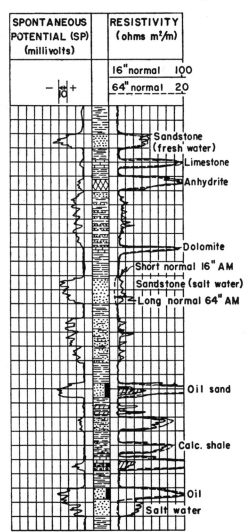

FIGURE 3-11

Typical electric log, showing the SP curve at the left, a lithologic sample log in the center, and the resistivity log at the right. [Redrawn from Stratton and Ford, "Electric Logging," in L. W. LeRoy (ed.), Subsurface Geologic Methods, *Colo. Sch. Mines,* p. 365, Fig. 152.]

one cable and out into the rock through its electrode, while other electrodes attached to other cables pick up the charge and carry it to the surface, where it is recorded on sensitized paper synchronized with the movement of the electrode along the hole. The spacing of the receiving electrodes along the sonde varies for different areas and different stratigraphic conditions.[62]

Electric logs[63] were first studied in the small French oil fields of Pechelbronn. The method was applied in Venezuela in 1929 and later in the USSR, where it rapidly became widely used. It was introduced into Romania in 1931 and since then has spread to all oil-producing regions of the world. The present standard practice is to make an electric log of every rotary well drilled. Some wells are logged at different stages in the drilling, and others are logged after the completion of the hole, depending on the immediate needs of the situation. Logging is commonly done by commercial service companies, which charge a fee for the work done.

Electric logs have become a most effective geologic tool; they are so widely understood by petroleum geologists and engineers that cross sections and correlation charts are commonly prepared with only data from electric logs to indicate the stratigraphy and structure. (See Fig. 3-12.) An electric log does not displace a lithologic log or a paleontologic log, but it furnishes added information on the rocks penetrated by the drill and their fluid content, and each kind of log supplements the data of the other. Electric logs are used chiefly for correlation purposes and for identifying and measuring porosity and reservoir fluids; their interpretation has progressed from an art to a technique to a science in the short time since they were first used.

Two values of the electrical properties of the formations and fluids drilled are determined, the electric potential and the resistivity.

Electric Potential. The log of the electric potential of a formation in a well hole is variously called the *spontaneous potential log,* the *self-potential log,* or the *S.P. log.* The S.P. log is commonly placed on the left track of the printed record. The measurement is expressed in millivolts, starting from a base line near the

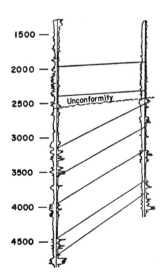

FIGURE 3-12

Section showing how electric logs may be used in correlating formations between wells. [*Redrawn from Stratton and Hamilton, in L. W. LeRoy (ed.),* Subsurface Geologic Methods, *Colo. Sch. of Mines (1951), p. 638, Fig. 336.*]

RESERVOIR ROCK [CHAPTER 3]

center of the record. Formations of higher electric potentials are shown as curves extending varying distances to the left on the millivolt scale. The electromotive forces which generate the S.P. current that is measured in the S.P. log are thought to be the result of two types of phenomena, electrofiltration and electro-osmosis. The potentials produced by both of these phenomena are cumulative, for the most part, and consequently accentuate the amplitude of the S.P. curve. (See Fig. 3-11.)

The electromotive forces caused by electrofiltration have been considered electrokinetic in nature. They are directly proportional to the pressure and electrical resistivity of the liquid, and are inversely proportional to its viscosity. The liquid is the water of the mud in the drill hole, which is an electrolyte, and is caused to flow through a pervious solid dialectric into the porous formations of the wall rock. The hydrostatic pressure of the mud is generally greater than the pressure in the permeable formations; therefore, some water filters through the mud cake into the permeable bed. The electromotive forces appear primarily where the pressure difference is the maximum, or opposite the permeable formation. The flow from the well into the formation produces a negative potential, whereas flow from the formation into the well produces a positive potential; the greater the rate of flow, the greater the potential difference. Where the pressure in the formation is the same as the pressure in the well, no current due to electrofiltration will be observed, and the recording will be zero even opposite a porous formation. Flow from the well into the formation is commonly found in practice, and the electric potential is therefore commonly negative.

The second cause of spontaneous electric potential, concentration cell potential, is thought to be electrochemical. When two electrolytes come in contact with each other, an electromotive force is generated. The electrolytes in a drill hole are the drilling mud of the well and the salt water of the formation. Below varying depths, most formation waters contain salts in variable amounts. The solution-concentration difference between the drilling mud and the formation water generates the electromotive force. If the concentration of salts, or the salinity of the water, in the formation is higher than that of the drilling mud, the electric current enters the formation, and a negative reading with respect to shale is observed opposite the porous zone. This is the common relationship found in practice. If the salinity of the mud is equal to that of the water in the formation, no potential due to electro-osmosis is generated; if the mud is more saline than the formation water, as after drilling through a salt layer, the current enters the hole, and a positive reading with respect to shale is observed opposite the porous zone.

A lesser cause of the electric potential of the formations may be the selective polar adsorption of charged ions by certain minerals within the formations, particularly clays.

The self-potential diagram of an electrical survey of a well is the resultant of the combined phenomena of electrofiltration, electro-osmosis, and ion adsorption. The common practice is to maintain the hydrostatic pressure of the fluid in the well higher than in the formation; since the concentration of salts in the formation waters is commonly higher than the concentration or salinity of the drilling-mud water, the two causes are cumulative, and the resultant is the algebraic sum of the two components. Exceptions occur occasionally, as when drilling through a salt section, where the mud becomes salty and of higher concentration than the formation water. The self-potential effect may then become positive. Shallow fresh-water sands may give a similar or negligible self-potential measurement. It may also happen that a high-pressure formation, ready to blow out from a deep sand, will give little or no self-potential effect, for the fluid from the sand enters the hole and reduces the electrofiltration effect. Occasionally, in shallow depths where there is an artesian sand, water flows into the well, and the electrofiltration effect may be reversed.

Resistivity. Rocks differ greatly in their electric conductivity and, conversely, their resistivity. The differences depend chiefly on the fluids, such as water, oil, and gas, that are in the porous and permeable portions of the rocks, for dry rocks are nonconductors. The fluids contained within the rocks are: (1) adsorbed or connate water that is present in the minute interstitial spaces of shales and clays and is incapable of circulation because the enclosing formation is not permeable; (2) fresh or salt water, that is present in permeable formations, and is free to circulate (dense rocks, such as granites, quartzites, gneisses, marbles, gypsum, anhydrite, rock salt, and coal, have so little interstitial space, and therefore so little moisture, that they are very poor conductors of electricity; they have a high resistivity); (3) oil and gas that occupy varying portions of the porosity.

Water containing one or more soluble salts is an electrolyte and electrically conductive. Oil and gas are not conductive, are highly resistive. Electric resistivity, as logged, is dependent on the relative saturations by gas, oil, and water, the concentration of the salt in the water, and the character of the rock, especially its porosity. (See also pp. 157–160 for a discussion of the formation factor.) In former years several separate resistivity logs were recorded simultaneously for different electrode spacings. These logs generally included a "short-normal" type with an electrode spacing of 10 to 20 inches, a "long-normal" type with an electrode spacing of 20 inches to 7 feet, and a "lateral" type with an electrode spacing of 15 to 20 feet. The wider spacing sent the current deeper into the surrounding formation so that it would pass through rock uncontaminated by drilling-mud filtrates. The induction log (see below) has been introduced in recent years and has supplanted the long-normal and lateral resistivity logs. The standard electric log today is an "induction-electric" log that consists of (1) a plotted log of conductivity and its reciprocal, resistivity, as determined by the induction measurement; (2) a short-normal resistivity log (16-inch electrode spacing); and (3) a self-potential (S. P.) log.

Induction Logging. Induction logs measure the resistivity (or its reciprocal, conductivity) of the strata traversed by energizing them with an induction current sent out from coils set in a sonde; no contact is made with the drilling mud. The alternating magnetic fields generated by the coils create a secondary magnetic field in a receiver coil located in the sonde. If the transmitter current is maintained at constant strength, the variations in the receiver coil are proportional to the conductivity of the strata.[64] Induction logs can be run in any uncased hole regardless of the type of fluid medium that is present. This log was first developed to measure the conductivity in wells drilled with oil-base mud where conventional resistivity determinations could not be obtained. Induction logging has since demonstrated its general superiority over conventional deep-penetration resistivity logs in holes drilled with water-base mud. Induction logs have a greater radius of investigation than either long-normal or lateral resistivity logs and, because of their superior focusing capability, determine more accurately the resistivity of thin beds.

MicroLog. Where the resistivity of the formations is much greater than that of the drilling mud (as in limestone formations, for example), the S.P. currents are short-circuited along the mud cake that forms on the well wall, and the details of the permeability are missed. The MicroLog[65] is a resistivity log in which the electrodes are spaced only one or two inches from each other and are encased in an insulated pad that is pressed against the walls of the hole. The close spacing permits the current to enter the rocks for only a short distance. Microresistivities are high opposite impermeable beds because the resistivity of such beds is approximately fifty times that of mud, and the mud cake is thin; microresistivities are low opposite permeable beds because the mud enters the formation to varying depths and forms a thick mud cake. (See Fig. 3-13.) Two different electrode spacings commonly are run. In porous and permeable zones the resistivity determined by the wide spacing is commonly higher than that determined by the short spacing. This difference is due to the relative depth of investigation of the two spacings. The short spacing measures principally the resistivity of the drilling mud; the wide spacing measures to a greater degree the resistivity of the rock formations and their fluids. Opposite permeable zones the wide spacing records the presence of high-resistivity drilling-mud filtrate that has invaded the zone; opposite zones of low permeability the resistivity measurement for the same wide spacing is commonly low because of the presence of low-resistivity formational waters with their high ion content.

Laterolog. Where the drilling mud becomes very salty, its high conductivity overshadows the effects of the conductivity of the formations. In a method known as Laterolog[66] the drilling mud is first highly charged with electricity, and this permits focusing the current laterally so that it enters the rocks. (See Fig. 3-14.)

Microlaterolog is a combination that uses the focusing feature of the Laterolog and the close spacing of the MicroLog and in certain areas gives more detailed information on the character of the reservoir than any other electric log.

Radiation Logs. Radiation logging is of two general types, one that measures the natural radioactivity of the formations, known as the *gamma-ray log,* and one that measures the effect of the bombardment of the formations by neutrons from an artificial source, known as the *neutron log.*[67] Both gamma-ray and neutron logs are

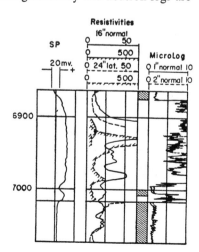

FIGURE 3-13

Typical MicroLog, in comparison with electric log, showing the differences between permeable formations (cross-hatched) and impermeable formations (blank). [Redrawn from Doll, Trans. Amer. Inst. Min. Met. Engrs., Vol. 189 (1950), p. 159, Fig. 9.]

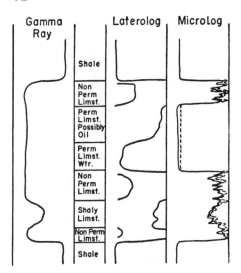

FIGURE 3-14

Gamma-ray log, Laterolog, and Micro-Log of same section of rocks. [Redrawn from Ford, Tulsa Geol. Soc. Digest (1952), p. 98.]

run simultaneously, or the gamma-ray log may be run simultaneously with the resistivity log. A sonde containing the measuring equipment is lowered into the hole at the end of an electrically conducting cable fastened to a drum at the surface and synchronized with a recording drum. The radioactivity of the strata is measured by the variations in conductivity produced by gamma rays emitted by the

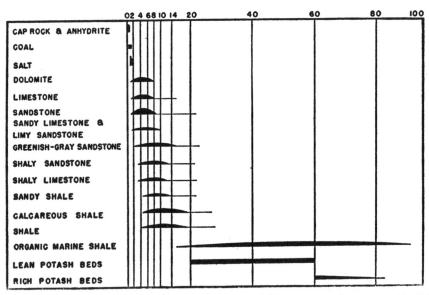

FIGURE 3-15 Relative radioactivities of various sedimentary rocks associated with petroleum pools. Radioactivity increases to the right in units of measurement of 10^{-12} gram of radium equivalent per gram of rock. [Redrawn from Russell, Bull. Amer. Assoc. Petrol. Geol., Vol. 25 (1941), p. 1775.]

rocks and penetrating the gas in an ionization chamber contained in the sonde as it is passed down the well bore. Some rocks radiate more gamma rays than others, as is seen in Figure 3-15, and these differences form the basis of the gamma-ray log. (See Fig. 3-16 and also Fig. 3-14.) Its chief use is for the correlation of formations, especially in wells where the casing has been set.

The neutron survey is obtained when a capsule containing a radium-beryllium mixture, which acts as a source of neutrons, is added to the sonde. The ionization chamber in neutron logging is affected by the gamma rays induced by the action of the neutrons on the formation, the gamma radiation emitted by the source capsule, and the natural gamma radiation of the formations, the latter being small compared with the other two radiations.

Hydrogen has a greater effect on the neutron logs than any other element, the effect being proportional to the number of hydrogen atoms per unit volume. Since hydrogen is a constituent of water, oil, and gas, the chief use of the neutron log is to locate porous zones in the formations, the interpretation being based on the fact that all porosity will be filled with one or more of the hydrogen-bearing fluids. The determinations are more accurate in limestones and dolomites than in shales and sandstones, probably because the clastic rocks generally contain elements other than hydrogen that may affect the log.[68] By calibrating neutron logs with known petrophysical data from the locality of the survey, it is possible to make reasonably accurate determinations of porosity in carbonate formations.[69]

Drilling-time Logs. This standard procedure on most rotary wells consists of measuring the time required to drill a unit depth, as the time in minutes for each foot, five feet, or ten feet of penetration, or the number of feet drilled per unit of time, as feet per hour.[70] When plotted on log strips, the change in the rate of penetration in the drilling-time logs is interpreted as a change in the lithology of the formations drilled. A geologist at the well can tell before seeing the samples the exact depth at which a change occurs, and this offers a check on his sample log and on the electric log. (See Fig. 3-17.)

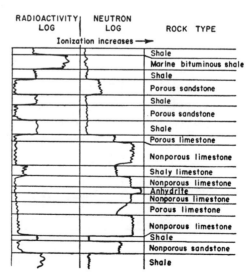

FIGURE 3-16

Typical expressions of the various rock types on gamma-ray and neutron logs. [Redrawn from Russell, Bull. Amer. Assoc. Petrol. Geol., Vol. 36 (1952), p. 327, Fig. 3.]

Core and Mud Analyses. These are made during drilling, especially of wildcat or exploratory wells. The chief purpose is to detect minute quantities of gas or oil entrained in the mud stream as it comes from the hole. Well cuttings are also tested for gas and oil.[71] The method is to pass a part of the mud stream or some of the well cuttings to a trap, where they are mixed with air and the gas separated from the mud. The air-gas mixture is passed over a "hot-wire" gas-detector instrument, where the percentage of combustible gas is measured, the temperature at which ignition occurs giving the amount of methane and the total of all hydrocarbon gases present. One improvement is a steam distillation method, which extracts the hydrocarbon gases quantitatively, and another is the gas chromatograph, which permits analysis for individual hydrocarbons. Minute showings of oil are seen when the drilling fluid is viewed under ultraviolet rays, the amount of fluorescence giving the magnitude of the showing. Logs showing drilling-mud and cuttings analyses are generally made in truck-mounted laboratories driven to the well. A part of a typical log is shown in Figure 3-17.

Caliper Logs. These show a continuous record of the diameter of the well bore. A caliper consisting of four spring-actuated arms, which when open contact the sides of the hole, is pulled through the hole. The arms are connected to a rubber, oil-filled chamber, which in turn is connected to a rheostat that measures the changing resistance. As the pressure from the arms on the chamber changes with

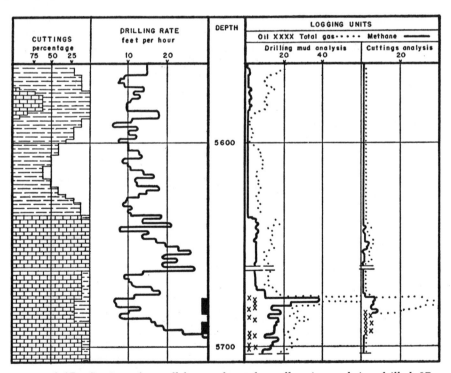

FIGURE 3-17 *Section of a well log made at the well as it was being drilled. [Redrawn from a Baroid Well Log Service log.]*

the varying diameter of the hole, the potential drop across the rheostat is measured and recorded as the caliper log.[72] Caliper logs are chiefly used to calculate the amount of cement necessary to fill up the annular space between the casing and the well, to select packer seats, and to determine accurately hole diameters for quantitative use in interpreting various electrical and radioactivity logs. These logs are also useful in locating porous zones and sometimes are helpful in identifying and correlating various lithologic types.

Temperature Logs. These are made by passing a temperature electrode down the hole. The electrode contains a twenty-inch length of platinum wire that quickly assumes the temperature of the fluids in the hole. Changes in temperature produce changes in resistance; these changes are detected at a bridge circuit in the electrode and recorded at the surface.[73] The temperature electrode measures the temperature of the drilling fluid. If the measurement is made as soon as drilling stops, the fluid will show little change from top to bottom of the hole. Most anomalies show up between twenty-four and thirty-six hours after circulation has ceased, the cooling or warming of the drilling fluid depending on the thermal conductivity of the formations penetrated and the size of the hole. Temperature logs are used chiefly to determine the height of cement in the hole by measuring the heat generated by the cement as it sets. They are also used to locate gas entering the hole, for, as it enters, it expands and cools. Formation water leaking through the casing and the position of zones where mud circulation is lost may also be determined by temperature surveys.

Sonic Logs.[74] Sonic logs are a continuous record of the time required for a sound wave to traverse a definite thickness of subsurface formation. The reciprocal of the speed of sound through the differing sediments is measured. Sound velocities vary from around 5,000 feet per second for clay to around 25,000 feet per second for dense dolomite. This corresponds to 200 microseconds per foot of clay to 40 microseconds per foot of dolomite. These logs are useful for porosity determinations, for detecting hydrocarbon-bearing zones, for determining various lithologic and stratigraphic correlations, for locating fractures, and for obtaining more accurate travel-time determinations for seismological interpretations.

Nuclear Magnetism Logs.[75] This type of log offers a direct measurement of the hydrogen in the formation and not of the rock matrix. This permits the interpretation of the results in terms of the fluids in the porous and permeable sections of the rocks and permits one to determine the water and the hydrocarbon content. It is the only log that responds solely to the formation fluids.

Dipmeters. The dip of a formation across a well bore may be measured electrically by passing three MicroLogs down the hole set 120° apart. The difference in the time at which each MicroLog passes the same formation boundary is recorded at the surface, thereby giving the angle of dip. Orientation of the instrument and its deflection from vertical are also measured at the same time so that the correct formation dip may be plotted at its exact position. Dipmeters are used, or have been used, where other types of logs are used, such as the Microlaterolog, self-potential logs, and short-normal resistivity logs.

MARINE AND NONMARINE RESERVOIR ROCK

Sedimentary reservoir rocks may be subdivided into those of marine origin and those of nonmarine, or continental, origin, but between these classes there are many gradations and intermixtures. Most petroleum is found in rocks believed to have been deposited under marine conditions, a belief that greatly influences the current ideas on its origin. Substantial deposits of petroleum have been found, however, in rocks of undoubted nonmarine origin, and it is likely that a good many more will be discovered in the future. Perhaps one reason why so few have been discovered hitherto is that many petroleum geologists are prejudiced against continental deposits as source rocks of petroleum and are consequently preoccupied with exploring in the marine sediments.

It is often difficult to determine whether a particular sediment is marine or nonmarine. The distinction is much easier at the surface, however, where depositional features such as bedding, cross-bedding, and lateral gradation can be seen in outcrops, than it is below the surface, where the evidence is all contained in well cuttings and cores. Some rocks, moreover, may be of mixed origin; the rock particles, such as those from wind-blown sands or terrace deposits, may have been originally distributed by nonmarine agencies, and the final burial may have been under marine conditions. Criteria that have been applied include:[76]

1. Marine or nonmarine fossil content.
2. Well-formed euhedral crystals of feldspar (marine).
3. New growth of secondary feldspar around nuclei of clastic feldspar (marine).
4. Aggregates of feldspar and quartz grains cemented by secondary feldspar (marine).
5. Widespread, uniformly bedded "blanket" sands (marine).
6. Thick sequences of interbedded, nonfossiliferous, unsorted fragmental rocks commonly forming lenses (nonmarine).
7. Tillites, coarse grits, and erratics (nonmarine, possibly glacial, although some may be submarine landslip debris).
8. Coal beds, formations with bone fragments, and lenticular sands (nonmarine).
9. Channel-deposited shoestring sands (probably nonmarine).

The Capitan, South Mountain, Shiells Canyon, and Bardsdale pools in the Ventura region of California[77] all produce from the continental red-bed facies of the Sespe formation, a group of sediments up to 7,500 feet thick that are widely distributed in the Ventura basin of southern California. These sediments range in age from Upper Eocene to Lower Miocene, and the proportion of marine deposits in them increases westward. They consist chiefly of red or maroon, fine-grained, arkosic sandstones, more or less silty shales and mudstones, and poorly sorted sandy conglomerates.

In northwestern Colorado the highly lenticular sand bodies of the Hiawatha member of the continental Wasatch formation (Eocene) produce oil and gas. The reservoir rocks are extremely lenticular sandstone beds, highly crossbedded and ripple-marked, interbedded with variegated mudstones and clay shales and with thin beds of coal, all containing bone fragments. The Wasatch formation is mostly of fluviatile origin, but includes occasional deposits laid down in shallow lakes. The producing formations are of fresh-water origin.[78] The oil and gas are believed to be indigenous to the Wasatch formation, for there are no marine formations within a reasonable distance that might be the source material. Farther southwest, in Utah, the interbedded sands of the Wasatch formation have also been found to contain petroleum.

Several oil pools have been found in the Uinta basin of northeastern Utah in the Green River formation (Middle Eocene), which consists of lacustrine marls, limestones, and siltstones reaching a thickness of 7,000 feet.[79] Oil-saturated sandstones occur around the margins of the basin, and numerous gilsonite veins and fracture fillings occur throughout the eastern part, where the rocks are chiefly shales. In fact, the Uinta basin has long been noted for the wide variety and abundance of hydrocarbons at the surface.

Nonmarine clastic rocks of the continental Quirequire formation (Plio-Pleistocene) form the reservoir rock of the highly productive Quirequire oil field in eastern Venezuela. The Quirequire reservoir pinches out within the field from a maximum thickness of 1,600 feet. It consists of "sandy clays interbedded with unsorted clastics ranging in size from silty sands to conglomerates to boulder beds. Thin beds of lignite, lignitic clay, tufa, travertine, and asphalt are present from the base of the formation to the surface." [80]

The reservoir rock in the San Pedro oil field of Argentina[81] is the Tupambi formation, consisting of fine sandstones, siltstones, coarse grits, and tillites, all chiefly of glacial origin. It is the basal member of the Permo-Carboniferous group. In this pool, it is believed but unproved, the source beds for the oil are the underlying Devonian shales, and the first folding to form the trap occurred in early Tertiary time.

In Romania, oil is produced in a number of pools in the Meotian (Pliocene) sediments, which consist of alternating sands, marls, sandstones, and limestones. These sediments are generally considered to be of continental, fresh-water origin.[82]

The Productive series (Miocene-Pliocene) of the rich Apsheron Peninsula (Baku area), USSR, consists of 9,500 feet or more of alternating lenticular sands, intercalated clays, and silt, which grade out into a fanglomerate. It contains many fresh-water fossils. The sediments are believed to have been deposited by the ancient Kura and Volga rivers at the places where they entered estuaries and muddy marine deltas.[83] The rich "red-bed series" of the Turkmen Republic, USSR, is equivalent in age, lithology, depositional environment, thickness, and productivity to the Productive series of the Baku area. In the South Kahetia region of Georgia, USSR,[84] numerous oil showings and some oil production have been found in the Pliocene and Miocene clays, sands, and conglomerates of continental origin.

The two large oil fields in China, Laochunmaie, near Yumen in Kansu Province, and Tsupinkai, near Wusu in Sinkiang Province,[85] both produce from Tertiary continental sandstones. Minor oil occurrences in nonmarine sediments have been described from a number of areas. One of these is in the Shensi series (Jura-Triassic), in North Shensi, China. The Shensi series is about 2,000 meters (6,600 feet) thick and of continental origin, being partly fluvial and partly lacustrine. A number of oil seepages and one pool, called the Yenang pool, have been found in it. In the Szechuan region of China some oil seepages occur in the Tzuliuching limestone (Lower Cretaceous), which is of lacustrine origin.

On the western side of the island of Madagascar, numerous oil seepages and tar springs occur in the continental Karro formation (from Upper Carboniferous to Middle Lias) and in the shallow-water fluviatile deposits of the Ankavandra Beds (Lower Lias and Triassic). The sediments are fine-to-coarse sands, conglomerates, clay shales, and micaceous sands, with a total thickness of 2,400 feet.[86]

In Pakistan, formerly northwestern India, numerous seepages occur in the lower part of the fresh-water Nimadric series (from Oligocene to Pleistocene), along and above the unconformable contact with the underlying marine nummulitic sediments (Eocene). The Nimadric series consists of over 20,000 feet of alternating sandy and silty beds, with some conglomerates, and appears to be partly of fluviatile and partly of eolian origin.[87] A somewhat similar relationship exists in Assam, where the thick nonmarine Tipam sands and clays (Miocene) unconformably overlie marine sediments (from Oligocene to Eocene) and all contain numerous oil seepages.[88]

In summary, we see that oil does occur commercially in rocks of continental, or nonmarine, origin.[89] We may conclude, then, that nonmarine sediments with porosity, permeability, adequate impermeable cover, and favorable trap conditions should not be overlooked as potentially favorable reservoir rocks. Geologists generally try to explain away the occurrence of oil in nonmarine sediments as being due to migration along fractures, faults, or bedding planes from adjacent marine sediments, even though there may be no direct evidence of such movement. The lack of exploration of nonmarine sediments has been largely due to the prevalent belief that all petroleum is of marine origin. Whether the oil originated within the nonmarine sediments or migrated into them from some neighboring marine sediment is not of practical importance; the main problem is to locate the place where it has accumulated. The evidence of existing oil and gas fields in nonmarine sediments in different parts of the world indicates that petroleum does migrate through nonmarine rocks and will accumulate into a pool where there is a trap. With a more aggressive exploration of nonmarine sediments, there seems to be every reason to expect that many more pools will be discovered in them.

CONCLUSION

The reservoir rock is the material in which oil and gas are found; it consists chiefly of sandstones, limestones, and dolomites. No one of these rocks seems to be favored ahead of the others, for large pools are found in each and in all combinations of them. Not only is the reservoir rock one of the essential elements of the single reservoir, but, as we shall see later (Chap. 14), the volume, character, and variableness of the sediments in a prospective producing region form an essential element in the judging of its petroleum possibilities. The presumption is that, if large volumes of sediments are present, they will somewhere contain potential reservoir rocks. Reservoir rocks, in fact, include so many sedimentary rock types that it is doubtful if any sedimentary basin will prove to be without some kind of rock that could become a reservoir rock.

Selected General Readings

Studies for Students: The Classification of Sedimentary Rocks, Jour. Geol., Vol. 56 (March 1948), pp. 112–165, contains the following articles: F. J. Pettijohn, "A Preface to the Classification of the Sedimentary Rocks," pp. 112–117; Robert R. Shrock, "A Classification of Sedimentary Rocks," pp. 118–129, with 55 references; Paul D. Krynine, "The Megascopic Study and Field Classification of Sedimentary Rocks," pp. 130–165. These three articles summarize many fundamental ideas on the classification and nomenclature of the sediments.

Gordon Rittenhouse, "Interpretive Petrology of Sedimentary Rocks," World Oil, October 1949, pp. 61–66. A lecture showing the deductions that may be made from studies of the composition, texture, and structure of sedimentary rocks.

F. J. Pettijohn, *Sedimentary Rocks,* 2nd ed., Harper & Brothers, New York (1957), 718 pages. A standard reference book.

Parker D. Trask, "Dynamics of Sedimentation," in *Applied Sedimentation,* John Wiley & Sons, New York (1950), pp. 3–40. 100 references. A summary of sedimentary processes.

Ph. H. Kuenen, *Marine Geology,* John Wiley & Sons, New York (1950), 568 pages, Chap. 5, "Formation of Marine Sediment," pp. 302–413. 101 references listed. Discusses phenomena related to marine sedimentation.

W. C. Krumbein and L. L. Sloss, *Stratigraphy and Sedimentation,* 2nd ed., W. H. Freeman and Company (1963), 660 pages. A standard reference book.

Howel Williams, Francis J. Turner, and Charles M. Gilbert, *Petrography,* W. H. Freeman and Company, San Francisco (1954), Part III, "Sedimentary Rocks," pp. 251–384. Special emphasis on the texture of sediments as revealed by thin sections.

Classification of Carbonate Rocks, a symposium, Amer. Assoc. Petrol. Geol., Tulsa, Okla., Memoir 1, William E. Ham, Editor (1962), 279 pages.

Geometry of Sandstone Bodies, a symposium, Amer. Assoc. Petrol. Geol., Tulsa, Okla., James E. Peterson and John C. Osmond, Editors (1961), 240 pages.

Reference Notes

1. A. W. Grabau, "On the Classification of Sedimentary Rocks," Amer. Geol., Vol. 33 (1904), pp. 228–247. See also Grabau's *Principles of Stratigraphy,* A. G. Seiler & Co., New York (1913), pp. 269–298. A genetic classification.

Paul D. Krynine, "The Megascopic Study and Field Classification of Sedimentary Rocks," Prod. Monthly, Vol. 9 (1945), pp. 12–32; also Jour. Geol., Vol. 56 (1948), pp. 130–165; also Tech. Paper 130, School of Mineral Industries, State College, Pennsylvania (1948).

Robert R. Shrock, "A Classification of Sedimentary Rocks," Jour. Geol., Vol. 56 (1948), pp. 118–129. 55 references cited. A simple classification based on composition and texture.

F. J. Pettijohn, *Sedimentary Rocks,* Harper & Brothers, New York (1949), pp. 191–194. Based on time and manner of origin and tectonism.

W. C. Krumbein and L. L. Sloss, *Stratigraphy and Sedimentation,* W. H. Freeman and Company, San Francisco (1963), pp. 150–189. Based on end-member groupings.

2. Chester K. Wentworth, "A Scale of Grade and Class Terms for Clastic Sediments," Jour. Geol., Vol. 30 (July-August 1922), pp. 377–392.

3. Francis P. Shepard, "Nomenclature Based on Sand-silt-clay Ratios," Jour. Sed. Petrol., Vol. 24 (September 1954), pp. 151–158.

4. John C. Griffiths, "Grain-Size Distribution and Reservoir-Rock Characteristics," Bull. Amer. Assoc. Petrol. Geol., Vol. 36 (February 1952), pp. 205–229. 40 references.

5. Paul D. Krynine, *op. cit.* (note 1).

6. P. D. Krynine, "Sediments and the Search for Oil," Prod. Monthly, Vol. 9 (1945), p. 22.

7. Olaf P. Jenkins (Dir.), *Geologic Formations and Economic Development of the Oil and Gas Fields of California,* Bull. 118, Calif. Div. Mines (1943), 773 pages. 126 authors. Nearly all pools in California are described.

8. Donald C. Barton and George Sawtelle (ed.), *Gulf Coast Oil Fields,* Amer. Assoc. Petrol. Geol., Tulsa, Oklahoma (1936), 1970 pages. Numerous sandstone reservoir rocks are described.

9. Caribbean Petroleum Company, "Oil Fields of Royal Dutch–Shell Group in Western Venezuela," Bull. Amer. Assoc. Petrol. Geol., Vol. 32 (1948), pp. 517–628. Many sandstone reservoir rocks are described.

E. Mencher *et al.,* "Geology of Venezuela and Its Oil Fields," Bull. Amer. Assoc. Petrol. Geol., Vol. 37 (April 1953), pp. 690–777. Many sandstone reservoir rocks are described.

10. G. M. Lees, "The Middle East," in *The Science of Petroleum*, Oxford University Press, London and New York, Vol. 6 (1953), Part I, pp. 67–72. 6 references.

11. XVIIth Int. Geol. Cong., Moscow (1937). Vol. 4 contains many articles on these areas.

12. Charles R. Fettke, *The Bradford Oil Field, Pennsylvania and New York*, Bull. M 21, Fourth Series, Penn. Geol. Surv. (1938), 454 pages.
Paul D. Krynine, *Petrology and Genesis of the Third Bradford Sand*, Bull. 29, Penn. State College (1940), 134 pages.

13. R. E. Sherrill et al., "Types of Stratigraphic Oil Pools in Venango Sands of Northwestern Pennsylvania," in *Stratigraphic Type Oil Fields*, Amer. Assoc. Petrol. Geol., Tulsa, Okla. (1941), pp. 507–558.

14. M. Albertson, "Possible Explanation of the Large Initial Production of Some Wells of the Haynesville Field," Bull. Amer. Assoc. Petrol. Geol., Vol. 17 (1932), pp. 295–296.

15. Karl ver Steeg, "Black Hand Sandstone and Conglomerate in Ohio," Bull. Geol. Soc. Amer., Vol. 58 (August 1947), p. 718.

16. Harold E. McNeil, "Wherry Pool, Rice County, Kansas," in *Stratigraphic Type Oil Fields*, Amer. Assoc. Petrol. Geol., Tulsa, Oklahoma (1941), pp. 126–131.

17. Lloyde H. Metzner, *Playa del Rey Oil Field*, Bull. 118, Geol. Branch, Calif. Div. Mines (1943), pp. 292–294.

18. James S. Hudnall, "East Texas Field," in *Occurrence of Oil and Gas in Northeast Texas*, Bull. 5116, Bur. Econ. Geol., Austin, Texas (1951), pp. 113–118.

19. J. E. Eaton, "The By-Passing and Discontinuous Deposition of Sedimentary Materials," Bull. Amer. Assoc. Petrol. Geol., Vol. 13 (July 1929), pp. 713–761.
R. D. Reed and J. S. Hollister, *Structural Evolution of Southern California*, Amer. Assoc. Petrol. Geol., Tulsa, Okla. (1936), 157 pages; also Bull. Amer. Assoc. Petrol. Geol., Vol. 20 (1936), pp. 1529–1692.

20. Robert M. Bailey, "Sedimentation," Oil Weekly, May 27, 1940, pp. 27–40.
S. W. Lowman, "Sedimentary Facies in Gulf Coast," Bull. Amer. Assoc. Petrol. Geol., Vol. 33 (December 1949), pp. 1939–1991. 37 references. An important contribution to near-shore and strand-area sedimentary conditions.
W. D. Keller, "The Energy Factor in Sedimentation," Jour. Sed. Petrol., Vol. 24 (March 1954), pp. 62–68.

21. James F. Swain, "Geology and Occurrence of Oil in Medina Sand of Blue Rock-Salt Creek Pool, Ohio," Bull. Amer. Assoc. Petrol. Geol., Vol. 34 (September 1950), pp. 1874–1886.

22. Hugh Stevens Bell, "Density Currents as Agents for Transporting Sediments," Jour. Geol., Vol. 50 (1942), pp. 512–547.
Rhodes W. Fairbridge, "Coarse Sediments on the Edge of the Continental Shelf," Amer. Jour. Sci., Vol. 245 (March 1947), pp. 146–152. 24 references.
Ph. H. Kuenen and C. I. Migliorini, "Turbidity Currents as a Cause of Graded Bedding," Jour. Geol., Vol. 58 (1950), pp. 91–127. 36 references.
Turbidity Currents, a symposium, Spec. Public. 2, Soc. Econ. Paleontologists and Mineralogists, Tulsa, Oklahoma (1951), 107 pages. 7 articles by 9 authors.
Ph. H. Kuenen, "Significant Features of Graded Bedding," Bull. Amer. Assoc. Petrol. Geol., Vol. 37 (May 1953), pp. 1044–1066. 19 references.

23. Bruce C. Heezen and Maurice Ewing, "Turbidity Currents and Submarine Slumps, and the 1929 Grand Banks Earthquake," Amer. Jour. Sci., Vol. 250 (December 1952), pp. 849–873. 45 references listed.

24. Ph. H. Kuenen, "Estimated Size of the Grand Banks Turbidity Current," Amer. Jour. Sci., Vol. 250 (December 1952), pp. 874–884.

25. John S. Barwick, "The Salina Basin of North Central Kansas," Bull. Amer. Assoc. Petrol. Geol., Vol. 12 (1928), pp. 177–189. Discussion to p. 199.
Harold E. McNeil, *op. cit.* (note 16), p. 126.
Stuart K. Clark, C. L. Arnett, and James S. Royds, "Geneseo Uplift, Rice, Ellsworth,

and McPherson Counties, Kansas," in *Structural Type Oil Fields,* Amer. Assoc. Petrol. Geol., Tulsa, Okla. (1948), Vol. 3, pp. 225-248.

26. Garvin L. Taylor and Duane H. Reno, "Magnetic Properties of 'Granite' Wash and Unweathered Granite," Geophysics, Vol. 13 (February 1948), pp. 163-181. See also Charles J. Deegan, O. &. G. Jour., September 2, 1948, pp. 84-88.

27. Henry Rogatz, "Geology of Texas Panhandle Oil and Gas Field," Bull. Amer. Assoc. Petrol. Geol., Vol. 23 (July 1939), pp. 983-1053. Bibliog. 7 items.

28. Ralph E. Grim, "Properties of Clay," in Parker D. Trask (ed.), *Recent Marine Sediments,* Amer. Assoc. Petrol. Geol., Tulsa, Okla. (1939), pp. 466-495. 161 references.

Ralph E. Grim, "Modern Concepts of Clay Materials," Jour. Geol., Vol. 50 (April-May 1942), pp. 225-275. 264 references cited.

C. S. Ross and S. B. Hendricks, *Minerals of the Montmorillonite Group, Their Origin and Relation to Soils and Clays,* Prof. Paper 205-B, U.S. Geol. Surv. (1945), pp. 23-79.

Ralph E. Grim, "Relations of Clay Mineralogy to Origin and Recovery of Petroleum," Bull. Amer. Assoc. Petrol. Geol., Vol. 31 (August 1947), pp. 1491-1499. Bibliog. 12 items.

Richard V. Hughes, "The Application of Modern Clay Concepts to Oilfield Development and Exploration," in *Drilling and Production Practice,* Amer. Petrol. Inst., Dallas, Texas (1950), pp. 151-163. Discussion to p. 167. 50 references.

Ben B. Cox, "Influence of Clay in Oil Production," World Oil, December 1950, pp. 174-182.

Ralph E. Grim, "Clay Mineralogy," Science, Vol. 135 (March 1962), pp. 890-898.

H. van Olphen, "An Introduction to Clay Colloid Chemistry," Interscience Publishers, New York-London (1963), 301 pages.

29. W. D. Keller, "Clay Colloids as a Cause of Bedding in Sedimentary Rocks," Jour. Geol., Vol. 44 (January-February 1936), pp. 52-59.

30. R. E. Grim, J. E. Lamar, and W. F. Bradley, "The Clay Minerals in Illinois Limestones and Dolomites," Jour. Geol., Vol. 45 (1937), pp. 829-843.

31. W. P. Kelley, "Base-Exchange in Relation to Sediments," in Parker D. Trask (ed.), *Recent Marine Sediments,* Amer. Assoc. Petrol. Geol., Tulsa, Okla. (1939), pp. 454-465. 19 references.

32. Ronald K. DeFord, "Surface Structure, Florence Oil Field, Fremont County, Colorado," in *Structure of Typical American Oil Fields,* Amer. Assoc. Petrol. Geol., Tulsa, Okla. (1929), Vol. 2, pp. 75-92.

33. Homer H. Charles and James H. Page, "Shale-gas Industry of Eastern Kansas," Bull. Amer. Assoc. Petrol. Geol., Vol. 13 (1929), pp. 367-381.

34. William L. Russell, "Notes on Origin of Oil in Kentucky," Bull. Amer. Assoc. Petrol. Geol., Vol. 18 (September 1934), pp. 1126-1131.

Coleman D. Hunter, "Economics Influences East Kentucky Gas Future," Oil and Gas Journal (July 9, 1962) pp. 170-174.

35. Louis J. Regan, Jr., and Aden W. Hughes, "Fractured Reservoirs of the Santa Maria District, California," Bull. Amer. Assoc. Petrol. Geol., Vol. 33 (January 1949), pp. 32-51.

36. John C. Wells, *Elk Hills Field* (Guidebook, Los Angeles Meeting), Amer. Assoc. Petrol. Geol. (March 1952), pp. 241-245.

37. G. Frederick Warn and Raymond Sidwell, "Petrology of the Spraberry Sands of West Texas," Jour. Sed. Petrol., Vol. 23 (June 1953), pp. 67-74.

38. *Classification of Carbonate Rocks,* a symposium, Amer. Assoc. Petrol. Geol., Tulsa, Okla., Memoir 1, William E. Ham, Editor (1962), 279 pages, 20 authors.

39. R. C. Murray, "Origin of Porosity in Carbonate Rocks," Jour. Sed. Petrol., Vol. 30 (March 1960), pp. 59-84.

40. John Rodgers, "Terminology of Limestone and Related Rocks: An Interim Study," Jour. Sed. Petrol., Vol. 24 (December 1954), pp. 225-234.

41. F. M. Van Tuyl, "The Origin of Dolomite," Ann. Report Iowa Geol. Surv., Vol. 25 (1916), pp. 249-421.

Edward Steidtmann, "Origin of Dolomite as Disclosed by Stains and Other Methods," Bull. Geol. Soc. Amer., Vol. 28 (1917), pp. 431–450.

F. J. Pettijohn, *op. cit.* (note 1), p. 312.

Virginia Edith Clee, *Bibliography on Dolomite,* Comm. on Sedimentation, Division of Geology and Geography, Nat. Research Council (1950), 41 pages.

42. J. Harlan Johnson, "Geologic Importance of Calcareous Algae with Annotated Bibliography," Quart. Colo. Sch. Mines, Vol. 38 (January 1943), 102 pages.

J. Harlan Johnson, "Limestones Formed by Plants," The Mines Magazine, Colo. Sch. Mines, Golden, Colorado (October 1943), pp. 527–533. 17 references.

J. Harlan Johnson, "Limestone Building Algae and Algal Limestones," Dept. Publications, Colo. Sch. Mines Foundation (1961), 297 pages.

F. J. Pettijohn, *op. cit.* (note 1), p. 163.

43. N. E. Baker and F. R. S. Henson, "Geological Conditions of Oil Occurrence in the Middle East Fields," Bull. Amer. Assoc. Petrol. Geol., Vol. 36 (October 1952), pp. 1885–1901.

G. M. Lees, *op. cit.* (note 10), pp. 67–72.

R. W. Powers, "Arabian Upper Jurassic Carbonate Reservoir Rocks," Amer. Assoc. Petrol. Geol., Tulsa, Okla., Memoir 1 (1962), pp. 122–192.

44. E. Mencher et al., *op. cit.* (note 9); the Maracaibo-Mara fields are described by H. J. Fichter and H. H. Renz, pp. 718–725.

45. E. R. Cummings and R. R. Shrock, *The Geology of the Silurian Rocks of Northern Indiana,* Pub. 75, Ind. Dept. Conserv. (1928), 226 pages.

Heinz A. Lowenstam, *Niagaran Reefs in Illinois and Their Relation to Oil Accumulation,* No. 145, Ill. Geol. Surv. (1949), 36 pages. 50 references cited.

46. John V. Morgan, "Correlation of Radioactive Logs of the Lansing and Kansas City Groups in Central Kansas," Tech. Paper 3311 Petrol. Technol., Trans. Amer. Inst. Min. Met. Engrs., Vol. 195 (1952), pp. 111–118.

47. Preston E. Cloud and Virginia E. Barnes, *The Ellenburger Group of Central Texas,* Pub. 4621, University of Texas (1946), 473 pages.

Samuel S. Goldich and E. Bruce Parmalee, "Physical and Chemical Properties of Ellenburger Rocks, Llano County, Texas," Bull. Amer. Assoc. Petrol. Geol., Vol. 31 (November 1947), pp. 1982–2020.

John G. Bartram, W. C. Imbt, and E. F. Shea, "Oil and Gas in Arbuckle and Ellenburger Formations, Mid-Continent Region," Bull. Amer. Assoc. Petrol. Geol., Vol. 34 (April 1950), pp. 682–700.

John P. Hobson, Jr., "How Lithologic Zoning of Dolomites Helps Find Oil," World Oil (July 1961), pp. 63–67.

48. World Oil (Feb. 15, 1964), p. 102.

49. W. A. Tarr, *Terminology of the Chemical Siliceous Sediments,* Report of Comm. on Sedimentation, Div. of Geol. and Geog., Nat. Research Council (1937–1938), pp. 8–27.

50. W. A. Tarr and W. H. Twenhofel, "Chert and Flint," in *Treatise on Sedimentation,* 2nd ed., Williams & Wilkins Co. (1932), pp. 519–546.

51. M. N. Bramlette, *The Monterey Formation of California and the Origins of Its Siliceous Rocks,* Prof. Paper 212, U.S. Geol. Surv. (1946), 57 pages.

52. Max David, "Devonian (?) Producing Zone, TXL Pool, Ector County, Texas," Bull. Amer. Assoc. Petrol. Geol., Vol. 30 (January 1946), pp. 118–119.

53. Sidney Powers (ed.), "Symposium on Occurrence of Petroleum in Igneous and Metamorphic Rocks," Bull. Amer. Assoc. Petrol. Geol., Vol. 16 (August 1932), pp. 717–858. Numerous occurrences of petroleum and bitumen associated with igneous and metamorphic rocks are described.

54. John C. May, *Conejo Oil Field,* Bull. 118, Calif. Div. Mines (1943), p. 424.

55. J. H. Beach, "Geology of Edison Oil Field, Kern County, California," in *Structure of Typical American Oil Fields,* Amer. Assoc. Petrol. Geol., Tulsa, Okla. (1948), Vol. 3, pp. 58–85.

56. L. E. Porter, "El Segundo Oil Field, California," Reprinted Trans. Amer. Inst. Min. Met. Engrs., Vols. 127 & 132 (1938–1939), p. 451.

57. Robert F. Walters, "Oil Production from Fractured Pre-Cambrian Basement Rocks in Central Kansas,'.' Bull. Amer. Assoc. Petrol. Geol., Vol. 37 (1953), pp. 300–313.

58. A. A. Hammer, "Rattlesnake Hills Gas Field, Benton County, Washington," Bull. Amer. Assoc. Petrol. Geol., Vol. 18 (July 1934), pp. 847–859.

59. E. H. Sellards, "Oil Accumulation in Igneous Rocks," in *The Science of Petroleum*, Oxford University Press, London and New York (1938), Vol. 1, pp. 261–265. 11 references.

60. Harry L. Baldwin, "Tupungato Oil Field, Mendoza, Argentina," Bull. Amer. Assoc. Petrol. Geol., Vol. 28 (1944), pp. 1455–1484.

61. Carl A. Moore (ed.), *A Symposium on Subsurface Logging Techniques*, University Book Exchange, Norman, Okla. (1949). 11 articles on logging and subsurface techniques.

Subsurface Geologic Methods, 2nd ed., Colo. Sch. Mines, Golden, Colo. (1951), Chap. 5, "Subsurface Logging Methods," and Chap. 6, pp. 344–503, "Miscellaneous Subsurface Methods." Contains many authoritative articles describing the common logging techniques.

M. P. Tixier, "Modern Log Analysis," Jour. Petrol. Technol. (December 1962), pp. 1327–1336.

M. R. J. Wyllie, "The Fundamentals of Well Log Interpretation," Academic Press, New York and London, 3rd ed. (1963), 238 pages. A standard reference book on well logging.

Sylvain J. Pirson, "Handbook of Well Log Analysis," Prentice-Hall, Inc., Englewood Cliffs (1963), 326 pages. A comprehensive statement of modern well logging and the interpretations of well logs in terms of the reservoir and its fluids.

See also, Reports and Transactions, Society of Petroleum Well Log Analysts, P.O. Box 4713, Tulsa, Oklahoma.

The different logging service companies publish privately many articles, brochures, documents, and reports, which describe their techniques and contain numerous drawings and technical explanations. These are generally available on request. Companies include the Schlumberger Well Surveying Corporation, Houston, Texas, the Halliburton Well Logging Service, Duncan, Okla., and the Lane-Wells Well Surveying Company, Los Angeles, Calif., and Houston, Texas.

62. *Recommended Practice for Standard Electrical Log Form*, American Petroleum Institute, Dallas, Texas (March 1948).

63. C. Schlumberger, M. Schlumberger, and E. G. Leonardon, "Electrical Coring: a Method of Determining Bottom-hole Data by Electrical Measurements," Trans. Amer. Inst. Min. Met. Engrs., Vol. 110 (1934), pp. 237–272.

L. A. Puzin, "New Well Logging Developments," World Oil: Part I, December 1953, pp. 134–142, 11 references; Part II, January 1954, pp. 124–135. A summary discussion of modern logging techniques and principles.

64. H. G. Doll, "Introduction to Induction Logging and Application to Logging of Wells Drilled with Oil Base Mud," Tech. Paper 2641, Trans. Amer. Inst. Min. Met. Engrs., Vol. 86 (June 1949), pp. 148–162.

65. H. G. Doll, "The MicroLog—a New Electrical Logging Method for Detailed Determinations of Permeable Beds," Tech. Paper 2880, Petrol. Technol., Trans. Amer. Inst. Min. Met. Engrs., Vol. 189 (1950), pp. 155–164. 6 references. "MicroLog" and "MicroLogging" are trade marks owned by the Schlumberger Well Surveying Corporation of Houston, Texas.

66. R. D. Ford, "The Laterolog," Tulsa Geological Society Digest, Vol. 20 (1952), pp. 95–99. "Laterolog" is a trade mark owned by the Schlumberger Well Surveying Corporation of Houston, Texas.

67. William L. Russell, "Well Logging by Radioactivity," Bull. Amer. Assoc. Petrol. Geol., Vol. 25 (September 1941), pp. 1768–1788.

68. William L. Russell, "Interpretation of Neutron Well Logs," Bull. Amer. Assoc. Petrol. Geol., Vol. 36 (February 1952), pp. 312–341. 11 references.

69. Robert E. Bush and E. S. Mardock, "Some Preliminary Investigations of Quantitative Interpretations of Radioactivity Logs," Tech. Paper 2780, Petrol. Technol., Trans. Amer. Inst. Min. Met. Engrs., Vol. 189 (1950), pp. 19–34. 8 references.

70. G. Frederick Shepherd, "Drilling-Time Logging," in *Subsurface Geologic Methods,* 2nd ed., Colo. Sch. Mines (1951), pp. 455–475.

71. R. W. Wilson, "Methods and Applications of Mud-Analysis and Cutting-Analysis Well Logging," in *Symposium on Subsurface Logging Techniques,* University of Oklahoma Book Exchange, Norman, Okla. (1949), pp. 81–87.

72. Wilfred Tapper, "Caliper and Temperature Logging," in *Subsurface Geologic Methods,* 2nd ed., Colo. Sch. Mines (1951), pp. 439–449.

73. *Ibid.,* pp. 444–449.

74. C. A. Doh and L. A. Puzin, "The Sonic Log," World Petroleum (March 1961), pp. 45–47, 82.

75. R. J. S. Brown, B. W. Gamson, "Nuclear Magnetism Logging," Jour. Petrol. Technol., Vol. 219 (1960), pp. 201–209.

76. A. J. Crowley, "Possible Criteria for Distinguishing Marine and Non-marine Sediments," Bull. Amer. Assoc. Petrol. Geol., Vol. 23 (November 1929), pp. 1716–1720.
Willis G. Meyer, "Stratigraphy and Historical Geology of Gulf Coastal Plain in Vicinity of Harris County, Texas," Bull. Amer. Assoc. Petrol. Geol., Vol. 23 (February 1939), pp. 145–211. See the section "Criteria for Distinguishing Marine from Non-marine Deposits," pp. 151–155.

77. Thomas L. Bailey, "Origin and Migration of Oil into Sespe Redbeds, California," Bull. Amer. Assoc. Petrol. Geol., Vol. 31 (November 1947), pp. 1913–1935.

78. W. T. Nightingale, "Petroleum and Natural Gas in Non-marine Sediments of Powder Wash Field in Northwest Colorado," Bull. Amer. Assoc. Petrol. Geol., Vol. 22 (August 1938), pp. 1020–1047.
W. T. Nightingale, "Geology of Vermilion Creek Gas Area in Southwest Wyoming and Northwest Colorado," Bull. Amer. Assoc. Petrol. Geol., Vol. 14 (August 1930), pp. 1013–1040.

79. W. H. Bradley, "Limnology and the Eocene Lakes of the Rocky Mountain Region," Bull. Geol. Soc. America, Vol. 59 (July 1948), pp. 635–648. 11 references.
A. M. Current, "Review of Geology and Activities in the Uinta Basin," Tulsa Geol. Soc. Digest, Vol. 21 (1953), pp. 52–61.

80. H. D. Borger, "Case History of Quirequire Field, Venezuela," Bull. Amer. Assoc. Petrol. Geol., Vol. 36 (December 1952), pp. 2304–2305.

81. Lyman C. Reed, "San Pedro Oil Field, Province of Salta, Northern Argentina," Bull. Amer. Assoc. Petrol. Geol., Vol. 30 (April 1946), pp. 591–605.

82. V. P. Baturin, "Genesis of the Productive Formation of the Apsheron Peninsula and of the Neighboring Regions," XVIIth Int. Geol. Cong., Moscow, USSR (1937), Vol. 4, pp. 279–293.

83. H. G. Kugler, "A Visit to Russian Oil Districts," Jour. Inst. Petrol. Technol., Vol. 25 (1939), p. 83.

84. K. D. Goguitidze, XVIIth Int. Geol. Cong., Moscow, USSR (1937), Fascicle 4, "The Petroleum Excursion—the Georgian Soviet Socialist Republic," table on pp. 22–25.

85. T. K. Huang, *Report on Geological Investigation of Some Oilfields in Sinkiang,* Series A, No. 21, Nat. Geol. Surv. of China (published in English in 1947).
C. H. Pan, "Non-marine Origin of Petroleum in North Shensi, and the Cretaceous of Szechuan, China," Bull. Amer. Assoc. Petrol. Geol., Vol. 25 (November 1941), pp. 2058–2068.

86. Arthur Wade, "Madagascar and Its Oil Lands," Jour. Inst. Petrol. Technol., Vol. 15 (1929), pp. 2–29. Discussion to p. 33.

V. Hourcq, "Madagascar," in *The Science of Petroleum,* Oxford University Press, London and New York, Vol. 6 (1953), Part I, pp. 166–168.

87. Robert Van Vleck Anderson, "Tertiary Stratigraphy and Orogeny of the Northern Punjab," Bull. Geol. Soc. Amer., Vol. 38 (December 1927), pp. 665–720.

E. S. Pinfold, "North-West India," in *The Science of Petroleum,* Oxford University Press, London and New York (1938), Vol. 1, p. 138.

88. G. W. Lepper, "Burma and Assam," in *The Science of Petroleum,* Oxford University Press, London and New York (1938), Vol. 1, pp. 133–136.

89. L. F. Ivanhoe, "Brackish, Glacial Strata Are Likely Oil Source Beds," World Oil (February 1, 1965), pp. 75–78.

CHAPTER 4

The Reservoir Pore Space

Porosity: measurement. Permeability: measurement – effective and relative permeability. Classification and origin of pore space: primary, or intergranular – secondary, or intermediate – relation between porosity and permeability.

THE FIRST essential element of a petroleum reservoir is a reservoir rock, and the essential feature of a reservoir rock is *porosity:* the rock must contain pores, or voids, of such size and character as to permit the storage of oil and gas in pools that are large enough to justify exploitation. Porosity, however, is not enough; the pores must be interconnected, to permit the passage of oil and gas through the rock. That is, the rock must be permeable (it is said to have *permeability*); otherwise there would be little if any accumulation into pools, nor could any petroleum that accumulated be produced by drilling wells, for it would not move into the wells fast enough. A pumice rock, for example, would not make a good reservoir even though the greater part of it might consist of pore space, for the pores are not interconnected and the porosity is not effective. The average shale cannot become a reservoir rock, for the pores are so minute that the capillary attraction of the fluids for the mineral grains effectively holds the fluids in the rock. To try to get oil out of a shale would be like trying to remove ink from a blotter.

There is a wide variation among reservoir rocks in the size of the individual pores and in the arrangement of the pores with respect to one another. These variations are called *primary* if they are controlled by (1) the depositional environment of the rock, (2) the degree of uniformity of particle size, and (3) the nature of the materials that make up the rock. The variations are called *secondary* if they depend on things that have happened to the rock since it was deposited; these may include (1) fracturing and shattering, (2) solution, (3) redeposition and cementation, and (4) compaction because of increased load.

Each pore of the reservoir rock may be thought of as a microspecimen of the reservoir and its petroleum pool, or as a microphysical and chemical laboratory where many physical relations and chemical reactions occur. The individual pore, with its fluid content and other associated phenomena, is the building unit, which, when multiplied countless trillions of times, becomes the pool and the reservoir. As such it becomes extremely important to both the exploration geologist and the petroleum engineer. The study of the pore space and its characteristics is termed *petrophysics*.[1]

The shape and size of some individual pores may be observed in well cuttings and cores by the unaided eye. Many pores, however, can be seen only with a binocular or petrographic microscope, and much of the reservoir pore space is submicroscopic in size. Pores filled with oil may also be observed under ultraviolet rays. The fluorescence of oil trapped in minute fractures and intercrystalline pores, not visible to the eye, stands out prominently, and pools have been discovered when this was the only way the oil was observed. Casts of the interconnected pores may be made by forcing wax or plastic material under pressure into a core or rock fragment and then dissolving away the surrounding rock material. Such a pore cast of the average reservoir sandstone looks much like a piece of bread, while the cast of a rock with angular grains and crystals has the appearance of rock candy. Photomicrographs of pore casts viewed stereoscopically offer a good means of observing the pore structure.

The *pore pattern* results from the complex interplay of the various factors that influence the porosity of the reservoir rock. The pattern is comprised of the pore size, the pore shape, the nature of the connections between pores, the character of the pore wall, and the distribution and number of larger pores and their relations to one another. The size of individual pores ranges from subcapillary and submicroscopic openings through capillary-sized openings to solution cavities of all shapes and sizes, including caverns formed in carbonate rocks. The individual pore may be tubular, like a capillary tube; or it may be nodular and may feather out into the bounding constrictions between nodules; or it may be a thin, intercrystalline, tabular opening that is 50–100 or more times as wide as it is thick. The wall of the pore may be clean quartz, chert, or calcite, or it may be coated with clay-mineral particles, platy accessory minerals, or rock fragments. The crookedness of the pore pattern, called the *tortuosity,* is the ratio of the distance between two points by way of the connected pores to the straight-line distance. *Pinpoint porosity,* as the name implies, consists of minute, isolated pores visible under the binocular microscope or, when filled with oil, by ultraviolet rays.

Porosity and permeability are mass properties of a rock. The pore pattern of a clastic reservoir rock is a function of several petrographic characteristics. These include: (1) grains—sizes, shapes, sorting, chemical composition,

RESERVOIR PORE SPACE [CHAPTER 4] 99

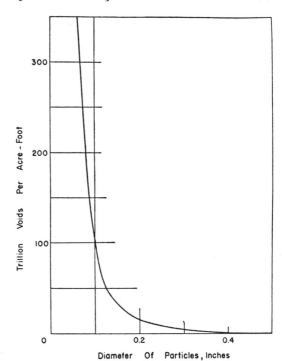

FIGURE 4-1

The probable number of pores per acre-foot of reservoir rock for varying particle sizes. [Redrawn from Jones, Petroleum Production, Reinhold Publishing Corp., New York, Vol. 1, p. 14, Fig. 2-2.]

mineral composition; (2) matrix—amounts of each mineral, how distributed, mineral and chemical composition; (3) cement—character, composition, amount, distribution with respect to grains and matrix.[2] The pore pattern of chemical reservoir rocks is dependent upon such factors as (1) fossil content, (2) fracturing and jointing, (3) solution and redeposition, (4) dolomite content, (5) recrystallization, (6) clay content, (7) bedding planes.

The number of separate pores in an acre-foot of average reservoir rock is enormous,[3] as may be seen in Figure 4-1. Since the average diameter of particles in most clastic reservoir rocks is between 0.002 and 0.01 inch (0.05–0.25 mm), the number of pores per acre-foot* of reservoir rock may be between 1 trillion and 1,000 trillion. Pores in most sandstone reservoirs have radii between 20 and 200 microns. It should be remembered that the calculations for the chart shown in Figure 4-1 are based on uniform, rhombohedrally packed particles and that the particles in the average clastic rock, being far from uniform in size, may give either larger or smaller figures. Carbonate rocks have a higher percentage of solution voids than sandstones of equal porosity, and probably contain a smaller number of pores per unit volume.

The surface area of the rock material in contact with the pore space increases greatly as the size of the particles diminishes. Jones[3] estimates that in

---

* An acre-foot is the volume of a one-acre area (43,560 square feet) one foot thick, or 43,560 cubic feet.

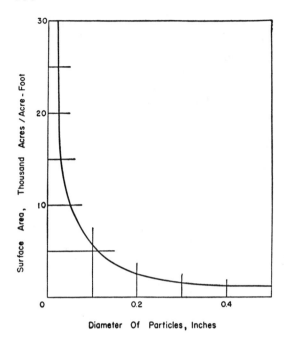

FIGURE 4-2

The surface area in acres per acre-foot of reservoir rock, depending upon the particle size. [Redrawn from Jones, Petroleum Production, *Reinhold Publishing Corp., New York, Vol. 1, p. 15, Fig. 2-3.*]

a sandstone consisting of rhombohedrally packed, medium-to-fine sand, with a particle diameter of 0.01 inch (0.25 mm), it is about 5,000 acres per acre-foot. The surface area may reach up to 30,000 acres per acre-foot in sandstone and siltstone. (See Fig. 4-2.) The large surface areas of the mineral material in the rocks with finer particle sizes become important in an understanding of such reservoir phenomena as wettability, adsorption, capillarity, solubility, and free surface energy. (See Chap. 10.)

POROSITY

Porosity is defined as the ratio of pore space to total volume of reservoir rock and is commonly expressed as a percentage. Two measurements, pore volume and bulk volume, are required to obtain the percentage porosity in accordance with the equation

$$\text{percentage porosity} = \frac{\text{pore volume}}{\text{bulk volume}} \times 100$$

Porosity varies greatly within most reservoirs, both laterally and vertically. If porosity is measured for each foot of core taken from the reservoir rock, as is the common practice, even some of the most uniform-appearing rocks show rapid and marked changes in porosity. The MicroLog (see pp. 80–81), especially, shows in detail the variable porosity that characterizes most reservoirs. The Springhill sand of the Manantiales field, Tierra del Fuego, Chile, is an example of such variable porosity. (See Fig. 4-3.) Another example of rapid variation in both porosity and permeability is in the Cedar Lake field of western Texas, where a dolomite in the San Andres limestone (Permian) is the reservoir rock; a section through a small part of the field is shown in Figure 4-4.

While porosity is generally stated as the percentage of pore space in the reservoir rock, it is frequently stated in reservoir estimates as acre-feet of pore space, or as volume in barrels per acre-foot of reservoir rock. Since there are 5.6146 cubic feet per U.S. barrel of 42 gallons, an acre-foot has a volume of 7,758 barrels. A rock with 10 percent porosity, then, contains 775.8 barrels of pore capacity per acre-foot.

The ratio of total volume of pore space to total volume of rock is called the *absolute* or *total porosity*. It includes *all* of the interstices or voids, whether interconnected or not. The porosity measurement ordinarily used in reservoir studies, however, is the ratio of the interconnected pore spaces to the total bulk volume of the rock, and is termed the *effective porosity*. It is commonly 5–10 percent less than the total porosity. The permeability of a rock depends on the effective porosity. The effective porosity may also be termed the *available pore space*, since oil and gas, to be recovered, must pass through interconnected voids. A pumice or scoria, for example, though it has a high total porosity, has a low effective porosity.

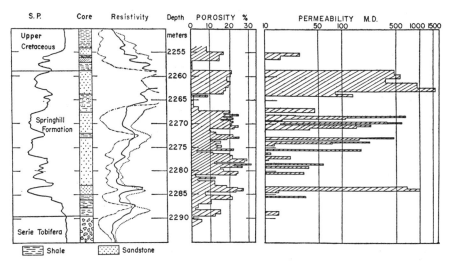

FIGURE 4-3 *Section through the Springhill sand (Cretaceous) producing in the Manantiales field, Magallanes Province, Tierra del Fuego, Chile. The sand is bounded by unconformities at the base and top, and the oil is of 42° Baumé gravity. This is an example of the variableness of the porosity, permeability, and character of a typical producing sand formation.* [Redrawn from Thomas, Bull. Amer. Assoc. Petrol. Geol., Vol. 33, p. 1582, Fig. 3.]

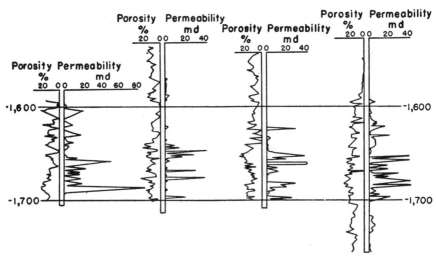

FIGURE 4-4 *Section showing the variable porosity and permeability in offsetting wells (1,320 feet apart) in the Cedar Lake field, western Texas. The producing formation is a crystalline dolomite within the San Andres formation (Permian).* [*Redrawn from Liebrook, Hiltz, and Huzarevich, Trans. Amer. Inst. Min. Met. Engrs., Vol. 192, p. 359, Fig. 5.*]

The porosity of most reservoirs ranges from 5 to 30 percent and is most commonly between 10 and 20 percent. Carbonate reservoirs generally have slightly less porosity than sandstone reservoirs, but the permeability of carbonate rocks may be higher. A reservoir having a porosity of less than 5 percent is generally considered noncommercial or marginal unless there are some compensating factors, such as fractures, fissures, vugs, and caverns, that are not revealed in the small sections of the rock cut by the core or the well bore. Representative porosities of some reservoir rocks are listed in Table 4-1, page 106. A rough field appraisal of porosities is (in percent):

Negligible	0–5
Poor	5–10
Fair	10–15
Good	15–20
Very good	20–25

Measurement

The measurements required for calculating porosity are made in the laboratory, either on small pieces of the rock cut from well cores or on well cuttings, and a number of methods for making them quickly and accurately have been devised and described.[4] Several qualitative methods of estimating porosity are also used, either to supplement the core analyses or to replace them if they are not available. These include:

The Electric Log. This is a measurement in millivolts of the natural electric potential of the rocks (the spontaneous potential, or SP). The low potentials are opposite the impervious beds, whereas the higher potentials are opposite the porous layers. (See pp. 77–81.)

Radioactivity Logs. Gamma-ray logs measure the natural emission of gamma rays from the formations logged, and neutron logs measure the emission of gamma rays induced from the formation by the action of neutrons. (See pp. 81–83.) The neutron log is primarily influenced by the presence of hydrogen and consequently by the presence in the formation of the reservoir fluids, gas, oil and water. The presence of fluids indicates that the rock has porosity. The gamma-ray log and the neutron log are widely used to indicate the porosity of limestone and dolomite reservoirs.

Other Logs. MicroLog and Sonic log measurements are very useful in determining porosity. Caliper logs often give a qualitative indication of porous zones and provide data for quantitative porosity determinations from other logs.

Microscopic Examination of Well Cuttings. If cores are not available, the examination of well cuttings through a binocular microscope is often the only way of directly observing porosity. Oil in minute openings may be seen to fluoresce when under ultraviolet light. An experienced microscopist can quickly determine the nature of the porosity and give a qualitative estimate of its relative amount, using such terms as "tight," "dense," "vugs," "pinpoint," "porous," "cavernous," "intercrystalline," and "intergranular." The absence of individual pores visible under the microscope generally means that the rock has too low a porosity to store appreciable amounts of oil.

Low pore space in a reservoir may be noted under the microscope as being due to various factors: the rock may be a dense, finely crystalline, lithographic limestone or dolomite; it may consist of fine and very fine sand particles; it may contain much clay as matrix and as wall coating on the sand grains; it may contain too much cementing material; or it may contain a high percentage of material that squeezes into the pore space under compression.[5]

Drilling-time Logs. A sudden increase of footage in a drilling-time log—or a sudden increase in the speed of drilling as the bit "falls away"—frequently means a porous formation; the more porous a rock is, the less dense it is, and the easier it drills. Such a change is often regarded as indicative of a porous pay formation, and a core is cut to determine the character of the rock.

Loss of Core. The cores recovered in the ordinary rotary core barrel may add up to less than the distance cored. Quite often this loss of core is due to the fractured, porous, and unconsolidated nature of the reservoir rock, which the core barrel is unable to retain, and which, consequently, is pumped up as well cuttings. The fact that zones of poor core recovery may represent rocks of abnormally high porosity explains the rule of thumb: "no core recovery, good well." There is no way of telling definitely whether poor core recovery actually indicates high porosity, but the drilling-time log may help, since such a formation will probably drill faster than denser rocks. The advent of diamond core heads has resulted in nearly 100 percent core recovery; where they are used, it is generally possible to obtain a continuous record of the porosity.

PERMEABILITY

Permeability is the property that permits the passage of a fluid through the interconnected pores of a rock—its effective porosity—without damage to or displacement of the rock particles. Permeability is, in other words, a measure of the fluid conductivity of a rock, and it is probably the most important single property of a reservoir rock. Permeability, in the geology of petroleum, is not absolute but relative; a rock is termed permeable if an appreciable quantity of fluid will pass through it in a short time (for example, an hour); it is termed impermeable if the rate of passage is negligible. It is recognized, however, that nearly all rocks have some permeability when considered in terms of long periods of geologic time and low-viscosity gases and liquids.

The unit of measurement of the permeability of a rock in the CGS system has been named the darcy after Henri Darcy,[6] who experimented with the passage of liquids through porous media in 1856. Darcy's law is expressed by the equation

$$q = \frac{kA}{\mu} \times \frac{dp}{dx}$$

in which q is the volume flux (volume per unit time) in centimeters per second for horizontal flow, k is the permeability constant in darcys, A is the cross-sectional area in square centimeters, μ is the fluid viscosity in centipoises, and dp/dx is the hydraulic gradient (the difference in pressure, p, in the direction of flow, x) in atmospheres per centimeter. This equation defines completely the viscous or laminar flow of homogeneous fluids through porous media of uniform packing and of uniform cross section.* For a given value of k the flow rate through any porous block of rock is thus proportional to the difference in pressure across the block and to the area of the block, and inversely proportional to the viscosity of the fluid and to the length of the block.

The darcy has been arbitrarily standardized by the American Petroleum Institute in terms of CGS units for use in the petroleum industry as follows: "A porous medium has a permeability of one darcy when a single-phase fluid of one centipoise viscosity that completely fills the voids of the medium will

---

* In reservoir rocks we are not concerned with turbulent flow. A rigorous statement of Darcy's law requires that the acceleration of gravity and the direction of flow be taken into account. For extended discussions of the derivation and limitations of Darcy's law the reader is referred to Hubbert, Jour. Geol., Vol. 48 (1940), pp. 787–826, and to Morris Muskat, *Physical Principles of Oil Production*, McGraw-Hill Book Co., New York (1949), pp. 123–131.

flow through it under 'conditions of viscous flow' at a rate of one centimeter per second per square centimeter of cross-sectional area under a pressure or equivalent hydraulic gradient of one atmosphere (76.0 cm of Hg) per centimeter." [7] Under "conditions of viscous flow" the rate of flow is so low as to be directly proportional to the hydraulic gradient. The darcy is the coefficient of proportionality between the quantities, and the particular numerical value of the permeability is a property or attribute of the medium alone and not of the fluid.

The permeabilities of average reservoir rocks generally range between 5 and 1,000 millidarcys.* A rough idea of one darcy is obtained if one considers a cube of sand one foot on a side. If the sand has a permeability of one darcy (1,000 md), this one-foot cube will pass approximately one barrel of oil per day with a one-pound pressure drop. Commercial production has been obtained from rocks whose permeabilities were as low as 0.1 md, but such rocks may have highly permeable fracture systems that are not revealed in the standard laboratory analysis. Permeability, along with porosity, varies greatly both laterally and vertically in the average reservoir rock. The examples from Chile and Texas, shown in Figures 4-3 and 4-4 on pages 101 and 102, are typical of most reservoir formations. A reservoir rock whose permeability is 5 md or less is called a *tight sand* or a *dense limestone,* according to its composition. A rough field appraisal of reservoir permeabilities is:

Fair	1.0–10 md
Good	10–100 md
Very good	100–1,000 md

A few representative porosities and permeabilities of oil pools are given in Table 4-1.

Measurement

The permeability of a reservoir rock is commonly determined in the laboratory by testing cores and pieces of core in a *permeameter*. Permeameters differ in design but generally consist of a core holder, a pump for forcing fluid through the core, manometers to measure the pressure drop across the core, and a flow meter for measuring the rate of flow of the fluid through the core. Laboratory methods are standardized so that measurements may be made rapidly and yet with sufficient accuracy for most reservoir problems. The test samples are generally cylindrical cores, 2 cm in diameter and 2–3 cm in length. Several methods have been devised and described.[8]

The fluid used in measuring the permeability of a reservoir rock is generally air or dry gas, and the pressure applied is the lowest that will cause a measurable rate of flow; serious errors result from pressures that cause turbulent flow. Air is most commonly used because it has little or no reaction with the rock

---

* A millidarcy (md) = 0.001 darcy.

TABLE 4-1 Representative Porosities and Permeabilities

Pool	Porosity		Permeability		Reference
Location, reservoir	Range %	Average %	Range md	Average md	
Bradford, Pennsylvanian Bradford sand (Devonian). (See also Fig. 4-17.)	2–26	15	0.1–500	50	1
12 pools producing from Smackover (Jurassic) limestone in southern Arkansas	12.5–21.3	16.9	50–2,000	737	2
Masjid-i-Sulaiman oil field, Iran, Asmari limestone		2		0.0005	
		5		0.007	3
		10		0.05	
		15		0.5	
Rangely oil field, Colorado, Weber sandstone (Pennsylvanian)		16		20	4
East Texas pool, Texas, Woodbine sand (Upper Cretaceous)		25	Up to 4,600	1,500	5
Ten Sections, Kern Co., California, Stevens sand (Upper Miocene)	15–30	20	10–3,000	140	6
Glenn pool, Oklahoma, Glenn sand (Pennsylvanian)		16		125	7
Oklahoma City field, Oklahoma, "Wilcox" sand (Ordovician), 9 analyses	8–22	16	79–2,497	688	8
Cumarebo oil field, Falcón, Venezuela, sands (Miocene)	3–39	21.7	1–3,397	200–300	9

1. C. R. Fettke, "The Bradford Oil Field," Pa. Geol. Surv., 4th series (1938), pp. 214–228.
2. Morris Muskat, *Physical Principles of Oil Production*, McGraw-Hill Book Co., New York (1949), p. 585.
3. H. S. Gibson, "Oil Production in Southwestern Iran," World Oil, May 1948, p. 273. Averages for four different Asmari limestone types.
4. W. Y. Pickering and C. L. Dorn, "Rangely Oil Field, Rio Blanco County, Colorado," in *Structure of Typical American Oil Fields*, Vol. 3 (1948), p. 143.
5. H. E. Minor and Marcus A. Hanna, "East Texas Field, Rusk, Cherokee, Smith, Gregg, and Upshur Counties, Texas," in *Stratigraphic Type Oil Fields*, Amer. Assoc. Petrol. Geol., Tulsa, Okla. (1941), pp. 625 and 626.
6. W. Tempelaar Lietz, "The Performance of the Ten Section Oil Field," Tech. Paper 2643 (September 1949), Trans. Amer. Inst. Min. Met. Engrs., Vol. 186 (1949), pp. 251–258.
7. K. B. Barnes and J. F. Sage, "Gas Repressuring at Glenn Pool," in *Production Practice*, Amer. Petrol. Inst. (1943), p. 57.
8. H. B. Hill, E. L. Rawlins, and C. R. Bopp, *Engineering Report on Oklahoma City Oil Field, Oklahoma*, RI 3330, U.S. Bur. Mines (January 1937), p. 199.
9. A. L. Payne, "Cumarebo Oil Field, Falcón, Venezuela," Bull. Amer. Assoc. Petrol. Geol., Vol. 35 (August 1951), p. 1869.

and does not cause any permanent change in the permeability. Furthermore, permeability measurements made with air are comparable with one another. It is recognized, however, that the permeability of a specimen of reservoir rock to air in the laboratory is not necessarily the same as the permeability of the rock to oil, gas, or salt water under reservoir conditions. Several factors tend to make permeability measurements using air higher than permeability measurements using reservoir fluids:

1. The sample is dried, and all gas, oil, and water are extracted from it before permeability measurements are made. Since most reservoir rocks are water-wet (that is, a thin film of reservoir water envelops each particle), the permeability of the dried specimen to air will be different from that of the normal water-wet specimen to gas or oil.

2. Reservoir rocks almost always contain some clay minerals, many chemically unstable. Some of these, especially the montmorillonites, take up water and swell to an extent depending on the character of the water. As the water is eliminated in the laboratory preparation of the sample, the clay minerals may either lose their water or break up into smaller particles, and either of these changes modifies the permeability measurement of the rock. Colloidal clay material in the reservoir rock may become loosened within the sample as it is being cleaned and dried. Thus fine pores may become clogged, or at least the pore pattern may be changed from what it was in the reservoir.

Where it is planned to flush the reservoir rock with water, as in water flooding for secondary recovery operations, it is desirable to make special permeability measurements with the same water that is to be used. What would then be measured is *permeability to water,* and that is generally lower than permeability to air. The efficiency and success of secondary recovery operations, in which water is pumped into the reservoir to drive or flush oil out of the pores, depend largely on using a water that does not swell the clay materials present in the rock.

3. Incomplete desaturation of the core may cause air trapping, or Jamin action. Thus the resistance to flow is markedly increased where gas and liquid globules alternate in a capillary-sized channel, as, for example, air globules in water or gas globules in oil.[9] When liquids are used for permeability tests, therefore, extreme care must be taken to eliminate all gas and air from the specimen tested; otherwise the permeability will be unduly low.

4. Permeability is independent of the fluid passing through the rock and also of the differential pressure. The permeability of a rock to gas, however, is greater than its permeability to liquid, probably largely due to the slippage of the gas along the rock wall, which does not occur with liquids. When air or other gas is used for the measurement, the higher the pressure, the smaller the volume, and consequently the mean free path of the gas molecules is greatly reduced until at high pressures gases and liquids are very similar. In order to correct for this difference between air and liquid, the *Klinkenberg*

scale has been devised.[10] It is based on the idea that "permeability to a gas is a function of the mean free path of the gas particles," which means that permeability depends on factors such as temperature, pressure, and composition of the gas. Of these factors, pressure is the most flexible. Low pressure results in the maximum mean free path, and it is at low pressure also that maximum slippage occurs. The Klinkenberg permeability factor (b) is obtained by measuring the permeability to air at several differing pressures, and extrapolating the curve to infinite pressure, at which the permeability will approximate the permeability to a liquid. The Klinkenberg permeability equivalent to air permeabilities in tight sands (below 1 md) may be as much as 100 percent higher, the correction approaching zero for high permeabilities. The Klinkenberg permeability factor, then, is a measure of the fractional error that comes from the slippage when low-pressure gas is used instead of a liquid. The relation is linear on log-log paper and is approximately[11]

$$\text{Klinkenberg permeability factor } (b) = 0.777 \, k_{ua.}^{-0.39}$$

Permeability is usually measured parallel to the bedding planes of the reservoir rock. Along this *horizontal*, or *lateral, permeability*, is found the main path of fluids flowing into the bore hole. Permeability across the bedding planes, or *vertical permeability*, is also frequently measured and is usually less than the horizontal permeability. High vertical permeability may permit bypassing and channeling of the water from below or the gas from above, thus changing the relative saturations at the bore hole and adversely affecting the productivity of the well.

The reason why horizontal permeability is generally higher than vertical permeability lies largely in the arrangement and packing of the rock particles during deposition. As flat grains tend to align and overlap parallel to the depositional surface, dissolving solutions move most easily in this direction, and in so far as the solutions have a solvent action on the minerals, they increase the horizontal permeability. Minor partings within a formation, and layering due to size-grading of particles, are more likely to extend parallel to the bedding than across it, so that they also tend to make permeability greater in the horizontal direction. It should be noted that what we generally measure in the laboratory is the permeability along the bedding planes. Where the reservoir rock has a steep or vertical dip, however, the direction of greater permeability may be more nearly parallel to the well bore than normal to it.

High vertical permeabilities are chiefly the result of fractures and of solution along fracture and joint planes that cut across the bedding. They are most commonly encountered in carbonate rocks and other brittle rocks and in clastic rocks with a high content of soluble material. They may also characterize loosely packed and uncemented sandstones.

If enough cores are taken and examined, sufficient permeability data are obtained by standard laboratory methods within the accuracy called for by the various geologic, engineering, and production problems. Field methods of obtaining permeability data, though not as accurate as the laboratory methods, are yet extremely useful and often furnish the only information one has on the permeability of a particular rock.

1. If there is so much free water in a formation that it enters the hole and dilutes the drilling mud in a rotary-drilled well, or partly fills the hole in a cable-tool well, it indicates that the formation is permeable. In cable-tool wells the rate at which the water enters the hole gives an even better idea of the gross permeability of the formation drilled than a laboratory examination of the core.

2. In a rotary operation the drilling mud is pumped down the well bore inside the drill pipe and out through the drill bit, and the cuttings are circulated up from the bottom of the hole with the drill mud, and brought to the surface in the annular space between the drill pipe and the wall of the well. When the drilling mud does not return to the surface, or only a part of it returns, the well is said to have *lost circulation*. When this happens, it means that the mud is draining off from the well bore and is entering a formation that has high permeability and a pressure less than the pressure of the drilling mud.

3. When there is a sudden decrease in the time it takes to drill a given thickness of formation, it indicates a softening of the formation; the bit has probably entered a bed of high porosity and presumably of high permeability.

4. One of the best over-all, or gross, measures of permeability is a production test, in which the decrease in bottom-hole pressure is measured against the production rate. If the formation is highly permeable, the rate of decrease in bottom-hole pressure with increasing production rate will be low, but if the formation is relatively impermeable, the decrease in bottom-hole pressure with increasing production rate will be high. The rate of recovery of reservoir pressure after a production test is also significant, for this gives an idea of the volume of the system and whether it is sealed or not. An open-flow test of the producing formation is a standard procedure when the total effect of the permeability, or fluid conductivity, of the reservoir plus the method of completing the well is needed for making accurate estimates of reserves and producing capacities.

5. A permeability profile of the reservoir rock may be made with an electric pilot,[12] which is an apparatus for locating, in a bore hole, the interface between two liquids of dissimilar electric conductivity. Salt water is run into the hole until the reservoir rock is completely covered. It is then forced into the formation by oil pumped into the well. The rate of fall of the oil–salt-water interface is measured by an electric pilot as it moves down the hole with the interface. The permeability of any part of the section of the reservoir is determined as a percentage of the permeability of the entire section.

6. Permeability may be determined qualitatively by pumping radioactive mud into a reservoir rock and then passing a Geiger counter down the well.

The counter registers the radioactivity opposite the reservoir rock. High radioactivity shows where the greatest amount of radioactive mud entered the formation and thus corresponds to zones of high permeability.

7. The permeability of drill cuttings or core fragments may also be calculated by the relationship that exists between the permeability of a rock and its capillary pressure curve.[13] (See also pp. 451–457.)

Effective and Relative Permeability

Darcy's law governing the flow of fluids through a porous material is based on the assumption that only one fluid is present and that it completely saturates the rock. In nature, however, the reservoir pore spaces contain gas, oil, and water in varying amounts, and each interferes with and impedes the flow of the others. Where a fluid does not completely saturate the rock, as is generally the case, the ability of the rock to conduct that fluid in the presence of other fluids is called its *effective permeability* to that fluid.[14] The effective permeabilities to air, gas, oil, and water, are designated, respectively, as k_a, k_g, k_o, and k_w.* It has been found that a given value of fluid saturation† bears a constant relation to the effective permeability; if the one changes, the other changes proportionately. This relation, however, differs for different rocks and must be determined experimentally. The factors that seem to influence the relation are clay swelling, adsorbed films, hydrophobic and hydrophilic surfaces, the presence of other, immiscible fluids, and the gas pressure.

The ratio between the effective permeability to a given fluid at a partial saturation and the permeability at 100 percent saturation (the absolute permeability) is known as the *relative permeability*.[15] It is expressed as k_g/k, k_o/k, or k_w/k (or as k_{rg}, k_{ro}, or k_{rw}), meaning the relative permeability to gas, oil, or water, respectively, and ranges from zero at a low saturation to 1.0 at a saturation of 100 percent. It is the ratio of the amount of a specific fluid that will flow at the given saturation, in the presence of other fluids, to the amount that would flow at a saturation of 100 percent, the pressure gradient and the other fluids being the same. Since the pore space of all reservoirs is full of gas, oil, and water, in varying proportions, the relative permeability of the rock to one fluid is dependent upon the amount (saturation) and nature of the other fluids present. It is always necessary, in fact, to use relative permeabilities rather than single-fluid permeabilities in reservoir studies. The relative permeability of a rock to any fluid increases as its saturation with that

---

* It is suggested (API RP No. 27, p. 4) that effective permeabilities always be written in the same order. Thus $k_{o(60,13)}$ means: "The effective permeability of the medium to oil when the percentage fluid saturation of the medium is 60 percent oil, 13 percent water, and 27 percent gas (millidarcys or darcys). The gas concentration is obviously derived by difference." And $k_{w(50,40)}$ means: "The effective permeability of the medium to water when the percentage saturation is 50 percent oil, 40 percent water, and 10 percent gas."

† The saturation is the ratio of a fluid to the total pore volume.

RESERVOIR PORE SPACE [CHAPTER 4] 111

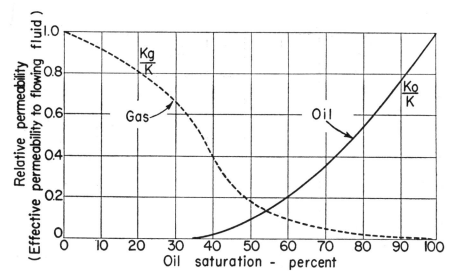

FIGURE 4-5 *Typical relative permeability relations with varying saturations of gas and oil.*

fluid increases, until finally, at 100 percent saturation, the full value of k is reached.

The relative permeability must be determined experimentally for each rock and each combination of fluid saturations. During production these ratios are continually changing. Charts of relative permeability are generally similar in pattern to the typical charts shown in Figures 4-5 and 4-6.[16] It may be observed in Figure 4-5 that there is no permeability to oil until the oil saturation has reached 30 percent or more. The reason is that the oil preferentially wets the rock surface and therefore clings to it and fills the smaller pore spaces. (See pp. 445–451 for a discussion of wettability.) During this period when the relative permeability ranges between 1.0 and 0.63, the gas has been moving freely. In other words, as long as the oil saturation ranges between zero and 30 percent, and the gas saturation between 100 and 70 percent, only gas will move through the rocks. At the point where the lines cross, the relative permeability is the same for gas and oil, and both should flow equally well. Above that point, the oil saturation increases to 100 percent, and the relative permeability of the rock to oil increases to 1.0 (which is k for the rock). During this period the gas saturation decreases from 0.15 to zero, the gas occurs as discontinuous bubbles, and the relative permeability of the rock to gas finally reaches zero.

The situation shown in Figure 4-6 differs in that the wetting fluid is not oil, but water. There is always some residual water in all pore spaces; but, as this chart shows, it does not begin to flow through the rock until the saturation is above 20 percent. Water at the low saturation is interstitial or "connate"

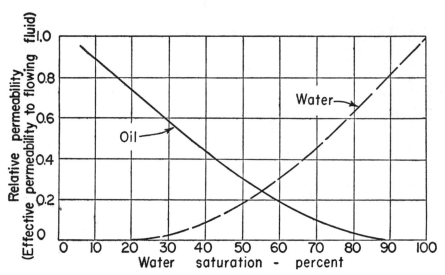

FIGURE 4-6 *Typical relative permeability relations with varying saturations of water and oil.*

water, which preferentially wets the rock and fills the finer pores. (For a discussion of connate and interstitial water see pp. 152–156.) As the water saturation increases from 5 to 20 percent, the oil saturation decreases from 95 to 80 percent; yet up to this point the rock permits the oil to flow, but not the water. Where the lines cross, at a saturation of 56 percent for water and 44 percent for oil, the permeability of the rock is the same for both fluids, and both flow equally well. As the water saturation rises above that level, the water flows more freely, and when the oil saturation gets down to about 10 percent, the oil stops moving; in other words, the rock then shows no permeability to the oil, and only water flows through it.

The relations shown in Figures 4-5 and 4-6 have a wide application to problems of fluid flow through permeable rocks. Probably the most important application they have to the geology of petroleum is the conclusion that there must be at least 5–10 percent saturation with the nonwetting fluid before that fluid begins to flow, and 20–40 percent saturation with the wetting fluid. This means that for oil or gas (as the nonwetting fluid) there must be a minimum of 5–10 percent saturation of the pore space before the fluid can begin to move through the water-saturated rock and accumulate into pools—provided, of course, that these relations, which prevail in the laboratory and through the life of an oil and gas pool, can be extrapolated to geologic time. Conversely, every oil pool has a residual oil saturation of 5–10 percent, which is not recoverable by ordinary production methods. These features will be considered in more detail later, under migration and accumulation of petroleum (Chap. 12).

CLASSIFICATION AND ORIGIN OF PORE SPACE

Two distinct general types of pore space in sedimentary rocks are recognized.[17] They are *primary, or intergranular, porosity* and *secondary, or intermediate, porosity*.

Primary, or Intergranular, Porosity

This is sometimes called *original porosity* because it is an inherent characteristic of the rock, established when the sediment was deposited. A sandstone is a permeable rock having primary porosity. Less permeable examples are shales, chalks, and crystalline rocks. The character of primary porosity is determined by the arrangement and form of the pores, the degree to which they are interconnected, and their distribution in the sedimentary rock unit.[18]

The term "packing" refers to the manner in which the particles of a clastic rock are fitted together.[19] The primary porosity of a rock is largely dependent upon its packing characteristics, which in turn depend largely on the uniformity or lack of uniformity of grain size. If all the grains in a sandstone were perfect spheres of uniform size, the porosity would range from 47.6 percent if the spheres were cubically packed to 25.9 percent if the spheres were rhombohedrally packed, with a mean of 36.7 percent. (See Fig. 4-7.) The porosity of aggregates of spheres, all packed in the same way, is theoretically independent of the size of the spheres, provided the spheres are all of the same size. Thus a sandstone composed of uniformly round, large grains will have the same porosity as a sandstone composed of uniformly round, small grains, if the packing is the same in both. Uniformity of grain size is never completely reached in clastic reservoir rocks, however, as may be seen from the porosity of an average sandstone, which is about 20 percent. The pores formed between the larger grains are filled with smaller, or matrix, particles. The resulting rock occupies a minimum space, for normal sedimentary processes, such as wave and current action, shake the particles down until the tightest packing possible for the particular combination of varying grain sizes and shapes being deposited has been reached.

FIGURE 4-7

Comparison of porosity in cubic packing of spheres (left) and rhombohedral packing (right). [Redrawn from Graton, Jour. Geol., Vol. 43 (1953), p. 800, Fig. 5.]

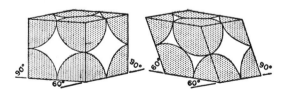

The nearest approach to uniform grain size is found in clastic rocks made up of well-sorted, well-rounded sand grains or oolites. The grains of very fine clastics are generally less well rounded than the grains of coarser materials because the natural agencies that produce rounding are not effective on very small particles.[20] In rock of uniform grain size, the smaller the grains are, the greater is the porosity; this effect is due to such factors as friction, adhesion, and bridging, which are greater with smaller grains because of the higher ratio of surface area to volume and mass.[21] The shape of clastic rock particles commonly varies from round through angular to flat and mica-like, and the size from coarse to fine or even colloidal; and there are widely varying amounts of cementing material between the individual grains. The porosity in the average clastic reservoir rock, therefore, is the combined effect of many variables, such as particle size, particle shape, sorting, packing, and character and amount of cementing material. Porosity declines rapidly with the addition of fine matrix particles that fill in the interstitial spaces. Many of the minor variations in porosity and permeability commonly encountered in the average clastic reservoir rock result from changes in depositional environment, to which mineral particles are extremely sensitive.

The porosity of most sandstone reservoir rocks is chiefly primary. Where porosity is extremely high, the sand grains are frequently loose and uncemented and may come up with the oil, sometimes in large quantities. For example, the loose sand grains that came up with the oil from the "Wilcox" formation (Ordovician), in the Oklahoma City field, had a porosity of 30 percent when tightly packed, whereas consolidated cores from the same field averaged about 16 percent in porosity.[22] Extremely low porosity is generally due to dirty sand, irregularity of grain size, and a high proportion of matrix material, and sometimes to a tight cementing of these constituents with silica, calcite, or dolomite.

The conditions that affect permeability differ considerably from those that affect porosity.[23] Some of the geologic conditions that have a bearing on the permeability of potential reservoir rocks are:

1. *Temperature.* Increase in temperature decreases the viscosity of a liquid, and the permeability varies inversely with the viscosity.

2. *Hydraulic gradient.* The rate of flow is directly proportional to the hydraulic gradient. Probably all rocks are permeable in some degree if the pressure difference is sufficiently high and the viscosity of the fluids sufficiently low. We are chiefly concerned, however, with low gradients, most of which will average less than 50 pounds per square inch pressure drop per mile, and few exceed 500 pounds per square inch pressure drop per mile.

3. *Grain shape and packing.* It has been found that with variable grain size, the permeability increases as the shapes of grains depart from that of true spheres. Thus the permeability of a sand composed of angular grains is greater than that of a sand composed chiefly of spherical grains of similar

size, largely because the angular grains are packed more loosely and also develop bridging. Rocks composed mainly of flat, mica-shaped particles and needle-like crystals pack loosely, have a high porosity, and in general, probably, have a high permeability. Decreasing grain size, on the other hand, increases porosity; but, because of the greater tortuosity and the higher capillary pressures, with a consequent higher saturation by the wetting liquid, the relative permeabilities are lower.

Compaction and cementation obviously reduce permeability based on primary porosity, whereas solution channels increase permeability. Fracturing, shattering, joint planes, and bedding planes, especially, increase permeability greatly by the large cross-sectional area of the tabular openings they produce. Permeability varies inversely with the length of flow and therefore inversely with tortuosity; so whatever shortens the path increases the permeability.

Carbonate reservoir rocks commonly have more secondary porosity than sandstones. There is no sharp boundary, and it is frequently difficult to distinguish between primary porosity and secondary porosity in carbonate rocks, but primary porosity is evident in some carbonate rocks, in such forms as (1) pores within and between fossil shells, fossil casts, coquina rock, fossil fragments, foraminifera, and algae; (2) pores between carbonate crystals and on cleavage planes within the crystals—called *intercrystalline porosity;* (3) pores associated with oolites and oolitic limestone; (4) pores along bedding planes, due to changes in depositional conditions at the bedding planes, to clastic material such as clay and silt, and possibly to crystal structures different from those in the body of the rock; (5) fractures due to desiccation or shrinkage occurring at the time of deposition.

Some of the diagenetic processes in carbonate rocks[24] continue after the rock has become lithified, and these contribute to the secondary porosity. Compaction, cementation, solution, recrystallization, and dolomitization are all common in diagenetic as well as post-diagenetic changes. The porosity that results from these processes, when it can be determined, may be said to be secondary, whereas that which merely modifies primary depositional characteristics may be said to be primary. Intercrystalline porosity may be either primary or secondary, and it is often difficult to determine which it is. The same is true of fractures, which may be formed either from secondary deformation or from primary shrinkage. Examples of some different types of limestone porosity are shown in the thin sections in Figure 4-8.

Examples of reservoirs that are chiefly in carbonate rocks of primary porosity are the oolitic and coquina lens of the Lisbon field, in Louisiana[25] (see also p. 311) and the Pennsylvanian coquina formation of the Todd field, in western Texas[26] (see pp. 308–310). The oil in the Southern fields of Mexico is produced from the El Abra limestone (Cretaceous). The pores in this reservoir are chiefly in shell fragments and in hollow casts of corals and of many types of mollusks, including rudistids, with intercommunication

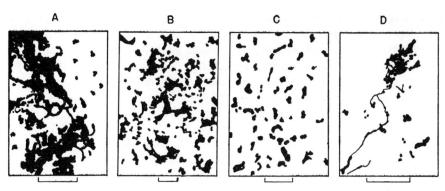

FIGURE 4-8 *Thin sections of plastic-impregnated limestone reservoir rocks showing the pores in black; scale, 1/100 inch. A, a reef type of porosity; B, primary porosity; C, pinpoint porosity; D, fracture porosity.* [Redrawn from Stewart, Craig, and Morse, Tech. Paper 3517, Trans. Amer. Inst. Min. Met. Engrs., Vol. 198 (1953), pp. 93–102.]

largely due to fracturing and shattering. The interiors of the casts are drusy; some of them contain bitumen (albertite) at the outcrop, whereas they contain oil, gas, or asphalt where penetrated by wells.[27] The El Abra formation is a reef limestone, and much of its porosity is primary—a characteristic of many organic reefs. (See also p. 361.)

Oolites are small round accretions, generally of calcite but also of silica. These accretions grow concentrically around a nucleus of foreign matter. Their diameters range from 0.25 to 2.0 mm and are commonly between 0.5 and 1.0 mm. Their name, which means "eggstone," is based on their similarity in appearance to fish roe. Larger accretions, of pea size, are called *pisolites*. The pore space associated with oolites is of two kinds. The greater part of it is between the oolites and is similar to that between the grains of any clastic rock. In some rocks, however, the oolites may have dissolved out, leaving voids, which are separated by undissolved matrix and cementing material. A rock consisting only of oolites or oolite casts frequently forms an excellent reservoir, highly porous and permeable.

There are many pools in oolitic limestone and dolomite reservoirs whose porosity is chiefly primary. In the Magnolia field, Columbia County, Arkansas,[28] more than 120 million barrels of oil has been produced from the dense, brown, oolitic Smackover limestone (Jurassic). The porosity of the clean oolite rock is about 20 percent and its permeability about 1,000 md, but the calcareous and chalky parts of the reservoir are less porous and permeable. The trap is a symmetrical, elongate dome fold, with a structural closure of 270 feet, covering an area of about 4,700 acres. The Reynolds oolite rock, in the upper part of the Smackover limestone, occurs extensively through the subsurface of southern Arkansas. It is the lowest producing formation in the Schuler field,[29] where the oolite rock consists of typical spherical oolites, ranging in size from very small to large, and a few pisolites. The oolites are loosely cemented, and the porosity is therefore high, being over 23 percent in places and averaging 16.7 percent. Both the horizontal and the vertical permeability locally exceed 15,000 md and have an overall average of 1,176 md. In the Carthage gas field of Panola County, Texas (one of the great gas fields of the United States), one of the important pay formations is the Upper Pettet limestone (Lower Cretaceous). This consists of a gray, fossiliferous, partly crystalline, porous, oolitic limestone varying up to 32 feet in thickness, which pinches out up the dip across a broad fold.[30] (See Fig. 8-4.) The McCloskey "sand" of the Illinois Basin, which is the equivalent of the Ste. Genevieve formation (Upper Mississippian) on the outcrop, is one of the most widespread and productive oolitic formations. The oil in many pools is in the pores between the oolites or in the casts left by dissolved oolites, or in both.[31]

Numerous examples of pools in which the reservoir rock is a limestone with both primary and secondary porosity could be cited. One of the larger pools of this kind is the Redwater pool, in Alberta, Canada,[32] which underlies an area of nearly sixty square miles. This is estimated to contain 1.5 billion barrels of oil in place, of which at least 600 million barrels is considered recoverable. The reservoir rock is an organic limestone, the Leduc (D-3) member of the Woodbend formation (Upper Devonian), and the oil has accumulated along its up-dip northeast edge, which wedges out within a shale formation as a stratigraphic trap. (See Fig. 7-49, p. 331.) The limestone, which is of the biostrome type, has an effective porosity of about 7 percent and an average permeability of 800 md. The pore space consists of a variable combination of spaces among fragments of limestone debris, fossil casts, fractures, and intercrystalline openings. A section through a dolomitic core from the nearby and similar Leduc field shows that it also has a combination of several types of porosity. (See Fig. 4-13, p. 127.)

Secondary, or Intermediate, Porosity

In secondary porosity the shape and size of the pores, their position in the rock, and their mode of interconnection bear no direct relation to the form of the sedimentary particles. It has also been called *induced porosity*.[33] Such porosity is found, for example, in a cavernous limestone and also in a fractured chert or a siliceous shale. Most of the reservoirs characterized by secondary porosity are in the carbonate rocks—limestones and dolomites; and this type of porosity is therefore often termed "limestone porosity" or "carbonate porosity." Secondary porosity may result from and be modified by (1) solution; (2) fractures and joints; (3) recrystallization and dolomitization; (4) cementation and compaction.

Solution. Percolating surface waters containing carbonic acid and organic acids penetrate the rock along various kinds of openings,[34] such as primary pores, fissures, fractures, joint planes, intercrystalline openings, and bedding planes. As these acid waters pass through the rock, they dissolve out the more soluble cations, including the carbonates of calcium and magnesium and salts of sodium and potassium, thereby still further opening the channels and increasing the porosity. Connected calcite crystals in a dolomite, preferentially dissolved out, may form new channels and expose additional mineral grains to the dissolving solutions. The process of solution goes on as long as the solvents are moving through the rock, and it continually changes the nature of the porosity and permeability.

The order of solubility of the commoner carbonates in acid solution is: (1) aragonite, (2) calcite, (3) dolomite, and (4) magnesite. Where dolomite and calcite are mixed in rocks of uniform grain size, twenty-four parts of calcite are dissolved for each part of dolomite until all of the calcite has gone into solution.[35] In less homogeneous rocks the ratio varies with the relative sizes of the crystals and with the way in which the carbonates are distributed. Dolomite is scarcely affected on a weathered surface, whereas calcite is strongly attacked.

Increased porosity develops in those parts of the rock where solution goes on more rapidly than redeposition. Some of the dissolved matter, however, is precipitated in other parts of the rock, thus forming a cement that reduces the porosity. Part of the dissolved matter, carried to the surface, may escape into the streams of the region. Solution phenomena are especially important in limestone and dolomite reservoirs, and, since most clastic rocks contain varying amounts of calcium and magnesium carbonates and other soluble minerals, they affect nearly all reservoir rocks to some extent.

Solutions of organic acids are formed mainly in the zone of weathering,[36] and largely through the decay of organic matter. Howard and David [37] added elm and maple leaves to water, soil bacteria, and limestone, and left the mixture exposed to open air. They obtained CO_2 in increasing amounts for the first two months and then in declining amounts for eleven months. Bacterial action also is strongest at the weathering surface. Countless numbers of organisms, both plant and animal, grow and later decompose in the soil and at the surface of the ground, and combine to assist in the chemical breakdown of many of the minerals found there.

At a weathering surface, therefore, both chemical and biochemical activity are high, and both help to supply organic and inorganic acids to the surface waters. The large springs, sink holes, and karst topography found in many limestone regions today give evidence of the importance of solution in the development of porosity. Because of the combination of solvent action with weathering, carbonate rocks immediately below unconformities are commonly very porous. An example of buried pre-Pennsylvanian karst topography, developed in Ordovician limestones in western Kansas, is shown in Figure 7-58, page 338.

Unconformities are common in most areas of sedimentary deposits. So far as they represent subaerial surfaces that have undergone weathering and erosion, they generally mark zones of solution porosity that may serve as reservoirs, and their position can be predicted in advance of drilling where the stratigraphy and geologic history have been worked out. The close relationship between porous, permeable limestone or dolomite reservoirs and overlying unconformity surfaces makes every unconformity an especially interesting target for a wildcat well. Many pools occurring in solution porosity formed below widespread unconformities might be cited. A few are described on pages 336–341.

Several examples of cavernous porosity have been described in the geologic literature.[38] A striking example is the large cavern found in the Dollarhide pool of Andrews County in western Texas.[39] Nine wells put down on forty-acre spacing within an area of one square mile encountered the cavern, as shown by the sudden dropping of the drilling bit. The cavern, which was full of oil, occurs in the Fusselman limestone (Silurian). Its height varied up to sixteen feet. There is approximately a thousand feet of limestone and dolomite of Silurian and Devonian age between the cavernous section and the pre-Permian unconformity. It is believed, however, that the cavern was formed during the pre-Permian erosion period.

In the highly productive Arab zone (Upper Jurassic) of the oil pools of Saudi Arabia, the porosity of the limestones is in their dolomitized and oolitic portions, which contain, besides many vugs and openings of finger size, cavernous openings up to three feet across, in which the drilling tools dropped. The result, in the Abqaiq field, at least,[40] is a permeability so high that the oil comes up almost as if it were being drawn from a tank.

Fractures and Joints. Fractures and joints in brittle rocks afford common and important types of secondary porosity.[41] The brittle reservoir rocks include limestones, dolomites, cherts, shales, siliceous sedimentary rocks, igneous rocks, and metamorphic rocks. Interbedded shales, sandstones, and limestones may show selective fracturing in certain beds. Since fractures afford channels for the movement of water, they are likely to be enlarged and modified by solution. They frequently combine with other types of both primary and secondary porosity to make a complex porous pattern; in fact, the presence of fractures often changes the permeability from millidarcys to darcys. The influence of fractures in determining the location of solution channels is shown diagrammatically in Figure 4-9.

Three causes are considered to account for most fractures. The first is diastrophism, such as folding and faulting. Some fractures may form at depth, where they accompany an increase in the volume of the rock resulting from the dilatant effect of the folding and bending of the strata.[42] The second cause is the removal of overburden by erosion in the zone of weathering.[36] As sediments are unloaded through erosion, the upper parts expand, and incipient weaknesses in the rocks become joints, fractures, and fissures. An increase of fracturing below an unconformity is therefore to be expected. Probably much

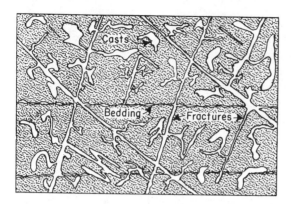

FIGURE 4-9

Idealized section showing how fractures aid in increasing permeability. Solutions passing through dissolve the wall material and widen the fractures, which connect otherwise isolated cavities and pores.

of the initial solution channeling through which surface waters percolate is the result of the gradual increase in jointing and fracturing that accompanies weathering. The third cause of fracturing is a reduction in volume of shales, while in place in the ground, due to diagenetic mineral changes coupled with a loss of water during compaction. Where intervening layers do not shrink, but act as struts or dividers, the loss of volume in the shales and siltstones is expressed in fractures, many of which have irregular and conchoidal forms, in contrast to the regular planes in fracture systems accompanying diastrophism.[43]

Nearly all reservoirs in limestone, dolomite, and siliceous rocks have at least some fracture porosity. The fracture planes combine with whatever porosity already existed to form an interconnecting system that greatly increases the permeability of the rock. Thus two systems of permeability are involved in many dense, fractured reservoirs: (1) the low-permeability blocks between fractures, where the oil moves slowly through short distances into (2) the high-permeability fractures that eventually lead to the well bore.[44] The percentage of water in fracture porosity is generally lower than in intergranular porosity, which means a higher percentage of the porosity in fractured reservoirs will consist of oil and gas. Even heavy, viscous oils that will not flow through rocks of low permeability may accumulate and move along fractures and be produced commercially. Thus any brittle rock, no matter how dense and compact it may appear on its outcrop or in well cuttings, may become a reservoir rock as a result of fracturing, fissuring, and shattering.

Some of the pools where fracture porosity in the reservoir rock is of great importance are mentioned below.

Pools in Basement Rocks. In California nearly 16,000 barrels of oil per day was produced from reservoirs in fractured igneous and metamorphic basement rocks. The most important producer is the Edison pool, near Bakersfield.[45] (See also p. 74.) Others, in fractured Franciscan (Jurassic?) schist, are the Torrance, Wilmington, Venice, El Segundo, and Playa del Rey pools, south and southwest of Los Angeles.[46]

Several wells produce from fresh basement granite in the Amarillo field, in the Texas Panhandle. One such well alone* has produced more than a million barrels of oil. Such production is undoubtedly the result of fractures in the granite, since nearby wells were either dry or nearly so. Presumably the oil has entered the granite from the sediments that were deposited along the flanks of the buried mountain range along which the field runs. (See also Fig. 3-8.)

Pools in Fractured Sediments. The Santa Maria district (see Fig. 7-62, p. 341), in California, has produced nearly 400 million barrels of oil, of which 75 percent came from fractured shale reservoir rocks and 25 percent from sands. Individual wells with initial productions of 2,500 barrels of oil per day were common, and some wells produced as much as 10,000 barrels per day.[47] Individual wells have produced more than a million barrels. The fractured reservoir rocks consist chiefly of cherts of the Monterey formation (Miocene), cherts interbedded with some calcareous shales, and sandstones, all of which are hard and brittle and generally fracture conchoidally. The oil occurs in fracture planes. It is difficult to obtain evidence of an oil pay because the core recoveries are poor in the brittle, fractured rock, and the oil is washed out with the drilling fluid. Loss of drilling mud is the best indication of a potential pay zone. The fractured chert reservoirs of the Santa Maria region are characterized by low porosities but high permeabilities, which make them satisfactory reservoirs for the heavy oils of the region (6–37° API gravity, but averaging less than 18°).

The Spraberry field of western Texas (see Fig. 13-11, p. 604) consists of a number of pools, some of which may later be found to connect with one another, extending over an area 150 miles long and up to 75 miles wide.[48] The pay formations extend through a vertical distance of nearly a thousand feet of Permian rocks, and the field has been variously estimated to contain a billion or more barrels of oil in place. The amount of oil recoverable under the present technology, however, may be small because of the low permeability of the reservoir. The trap is formed by a stratigraphic gradation up-dip into less permeable rocks. The reservoir rock consists chiefly of black, brittle shale, silty shale, varied sandy shales, calcareous and noncalcareous siltstone, and minor amounts of fine sand. Its particle diameters vary from $\frac{1}{30}$ to $\frac{1}{45}$ mm (the lower limit for sandstone is $\frac{1}{16}$ mm). Porosity is generally less than 10 percent, and the average permeability is $\frac{1}{2}$ md. Oil pools are unusual in reservoir rocks of such low porosity and low permeability. The effective permeability, therefore, is due almost entirely to fractures and conchoidal openings, which extend through the finer materials in all directions, but chiefly vertically. An isopotential map of the Tex-Harvey pool in the Spraberry field is shown in Figure 13-21, page 617, and indicates the trend of the major fractures by the size of the wells.[42] Fractured shale also forms the trap in the Florence field, Colorado (discussed on p. 282).

The Oriskany sand (Lower Devonian) of Pennsylvania and New York is a

---

* Richard B. Rutledge in a personal communication.

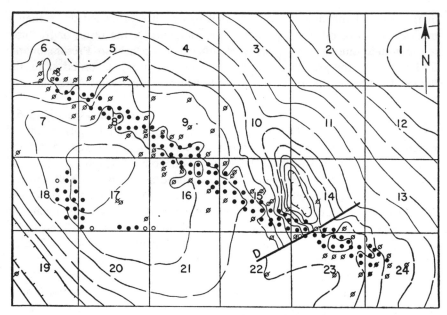

FIGURE 4-10 *Structural map of the Deep River field, Michigan. Contour interval 20 feet. Production is obtained from a dolomitic zone enclosed in the Rogers City limestone (Devonian). The dolomite is presumably related to a fracture permitting magnesium-bearing solutions to enter the limestone. The trap is formed by the dolomite zone, as well as the porosity. Note that the top of the fold is nonproductive. [Redrawn from Hunt, Independent Petroleum Association of America, Vol. 19 (April 1949), p. 42.]*

tight, fine-grained sand whose permeability averages less than 500 md and whose porosity ranges from 2 to 10 percent. It contains many gas pools, generally on anticlines,[49] and the gas from the more productive wells comes from small open fractures that appear to be joint planes.[50]

Permeability in carbonate reservoir rocks is often chiefly due to fractures. The reservoir of the West Edmond pool of Oklahoma (see Fig. 14-7, p. 644), for example, has been carefully studied and reported on.[51] The main producing formation of the field is the Bois d'Arc limestone formation at the top of the Hunton group (Devonian-Silurian). The trap is of the stratigraphic type, a permeable bed being truncated up the dip to the east and northeast by both the pre-Mississippian and the pre-Pennsylvanian unconformities. The porosity is partly primary, resulting from intercrystalline openings, fossil casts, and an oolitic member, the authors cited differing on the percentage. The permeability of the limestone and dolomite is low, but the entire pay section is cut by numerous fractures, which divide it into blocks of varying sizes. The oil has drained from the rocks into the highly permeable fractures, and moves along them to the producing wells.

RESERVOIR PORE SPACE [CHAPTER 4]

Fractures are indirectly responsible for the Deep River field in Michigan.[52] (See Fig. 4-10.) The oil is contained in a narrow, elongate body of porous dolomite within an impervious limestone at the top of the Rogers City formation (Devonian). The dolomite is believed to have been formed by magnesium-bearing ground waters moving along fractures through the normal limestones of the formation. The dry holes find neither dolomite nor oil. Vertical fractures have been found to be common in the Ellenburger dolomite (Cambro-Ordovician) of western Texas. One core thirty-six feet long contained fractures extending vertically throughout its entire length; the smallest were barely visible and the largest about one millimeter wide.[53]

The Scipio-Albion field in Michigan forms a remarkably straight and narrow belt for more than 25 miles and averages 3,500 feet in width. (See Fig. 4-11.) Oil and gas production is obtained from a dolomitized zone in the Trenton limestone (Ordovician). The Trenton is commonly contoured as a sag or slight syncline, but the probabilities are that it represents fracturing and minor faulting over a deep-seated fault in the basement. Presumably the narrow belt of dolomite is secondary and came in along the fault and fracture system. The field is estimated to contain more than 100 million barrels of oil

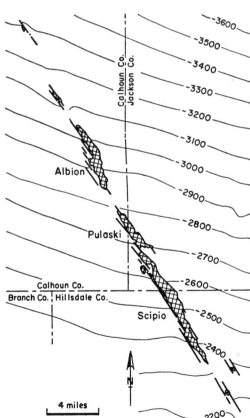

FIGURE 4-11

Map showing the Scipio–Pulaski–Albion trend field in southwestern Michigan. Producing area cross-hatched.

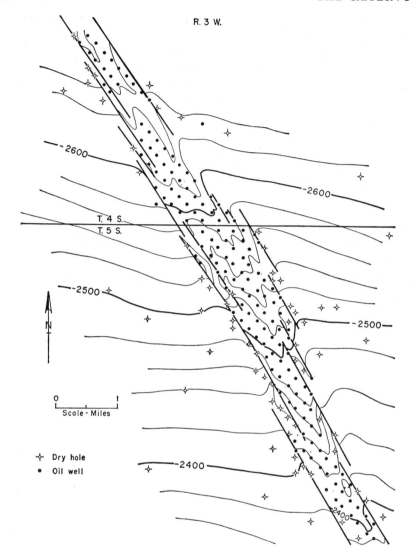

FIGURE 4-12 *Structure map of the Scipio pool in Michigan. Structure is on top of the Trenton formation (Ordovician) and between 5 and 10 of the holes show direct evidence of crossing faults. [Courtesy of Dan A. Busch.]*

and more than 200 billion feet of gas.[54] A detail of the Scipio area is shown in Figure 4-12.

In the Northern fields (Panuco area) of Mexico the reservoir rock consists of the Tamaulipas, Agua Nueva, and San Felipe limestones (Cretaceous).[54] These are all tight, close-grained, and compact, and have low porosity except for that afforded by induced openings, such as joints, fossil casts, and fractures, which give them high permeability. They are cut by faults, which, though of small displacement, have caused considerable shattering, as shown in cores and in fragments blown from the wells. Crevices large enough for drilling tools to drop through were found associated with the faults. The production of the Northern fields has reached nearly one billion barrels of oil of 12.5° API gravity. The heaviness of the oil prevents it from penetrating the low intercrystalline porosity of the limestone, and it occurs almost entirely in the fracture openings. Production is erratic, and wells with production ranging from mere showings up to 30,000 barrels per day or more are found within a few hundred feet of each other. The high permeability of portions of the reservoir rock is shown by a salt-water gusher that produced over 100,000 barrels of water per day.

Extensive fracturing and shattering characterize the porosity and permeability of the Cretaceous limestones and basement granite reservoirs of the Mara and La Paz oil fields in western Venezuela.[55] (See also Fig. 6-31, p. 263.) The oil column in the Mara field is more than 5,000 feet thick, and the producing limestone, with a low permeability, much of it below 0.1 md, and a total thickness of about 1,800 feet, is highly productive anywhere within the section where fracturing develops. The fracturing is frequently in closely spaced, vertical planes, probably related to the sharp folding and faulting that form the traps, and extends into the underlying granite, which also contains oil above the oil-water contact.*

Oil from the rich oil pools of southwestern Iran is found in the Asmari limestone (from Upper Oligocene to Lower Miocene). The Masjid-i-Sulaiman (see Fig. 6-12, p. 248) and Haft Kel (see Fig. 2-2, p. 18) pools have been studied carefully,[56] and it is found that, no matter how rich the well cuttings appeared to be, no appreciable production was obtained until a fissure was penetrated. Samples of cuttings showed porosity ranging from 2 to 15 percent, but the permeability was only between 0.00005 and 0.5 millidarcy, depending upon the amount of recrystallization (the more complete the recrystallization, the higher the porosity and permeability). The well productivity varies with the amount of the fissuring; if there were no fissuring, there would be no production. Two stages of fracturing have been observed.[57] The earlier, oil-free fractures are filled with calcite and other minerals, and many are vug-lined, with well-formed crystals. These apparently had nothing to do with the movement of oil, which is confined to a later set of fractures. The Agha Jari pool is also characterized by fracture permeability[58] (see Fig. 6-21), the reservoir being freely connected throughout even though the Asmari limestone is generally of low permeability. The highly fissured and fractured limestone is overlain by a salt formation 50–150 feet thick, which acts as a cover, or roof rock. Much of the fracturing of the Asmari limestone in the oil fields of Iran, and in its outcrop in the mountains, is along the pitch of the folds, and evidence of underground interconnection of points fifty miles apart has been observed.

* Walter S. Olson in a personal communication.

Recrystallization—Dolomitization Phenomena. Some carbonate reservoir rocks are nearly pure limestone, and some are nearly pure dolomite, but more are intimate and variable mixtures of the two. Where oil and gas occur in reservoirs consisting of both limestone and dolomite, the dolomite and dolomitic rocks are generally the more prolific producers of petroleum, chiefly because of their greater porosity. The origin of dolomite[59] and the reason for its greater porosity have long challenged investigators,[60] and because of the large amounts of petroleum associated with dolomites these problems are of interest to the petroleum geologist.

The basis for much of the discussion of the porosity of dolomite was the theory proposed by Elie de Beaumont in 1836. He pointed out that the molecular replacement of limestone by dolomite would result in a volume shrinkage of 12–13 percent. The chemical equation that explains the replacement is written as

$$2CaCO_3 + MgCl_2 \longrightarrow CaMg(CO_3)_2 + CaCl_2$$

Orton[61] used this idea in 1886 to explain the porosity in the Lima-Indiana field. Later workers[62] discounted the molecule-for-molecule mechanism of the replacement of limestone by dolomite, believing that the patches of solid dolomite in limestone indicated a volume-for-volume replacement.

Petrographic studies by Hohlt,[63] however, have revived the molecular replacement theory. He has shown that there is a pronounced tendency for calcite crystals in limestone to orient their c-axes in the plane of the bedding, presumably in response to pressure. In dolomites, however, the crystals are always oriented completely at random. Hohlt's explanation of this fact is that the shrinkage of the c-axis of the crystal during transformation from calcite to dolomite creates voids in the rock; the packing, in effect, is looser than in limestone. Hohlt also utilizes the random orientation of the crystals in dolomites to explain why solvent waters penetrate dolomites more readily than limestones. Dolomites offer much larger intercrystalline space for the passage of dissolving solutions and so present a much greater area of attack. Thus, in spite of their lower solubility, dolomites may be dissolved to as great an extent as limestone if attacked by more solution over a longer time.

Carbonate rocks also deform, in part, by distortion of the individual grains; when the material is recrystallized, the original textures may be obscured. Cloos,[64] by measuring the distortion that occurred in a certain folded oolitic rock, has shown the extent of the internal distortion that may occur in many folded carbonate rocks. Individual oolites, which had originally been spheres, were flattened and elongated until the entire rock mass was almost completely recast internally.

Many oil and gas pools in dolomite and dolomitic limestone could be cited. The oil in several of the organic-reef oil pools in western Canada, for example, occurs in dolomite. A section through a core from the producing D-3 reservoir (Upper Devonian) in the Leduc field southwest of Edmonton[65] is shown in

Figure 4-13. The rock contains vugs connected by minute cracks or crevices, and it has a variable amount of primary porosity. The original rock was chiefly an organic limestone, but it has now been dolomitized, and all of its primary biological structures have been obliterated. A part of the production is from detrital material, for, as in most organic-reef deposits, the reservoir material is a complex of many lithologic facies and rock types.

The oil in the Lima-Indiana field of Ohio and Indiana[66] occurs in porous dolomitic zones in the Trenton (Ordovician) limestone, which extends over an area 160 miles long and up to 40 miles wide. (See also Fig. 7-23, p. 307.) More than 500 million barrels of oil has been produced from this field since its discovery in 1884. It includes many separate pools, each associated with a separate, porous, dolomitized portion of the limestone, and the whole field lying across the Cincinnati and Findlay arches. The dolomitic producing rock grades into dense limestone up-dip toward the south, forming a stratigraphic trap. The porous dolomite, which is generally in the upper twenty or thirty feet of the Trenton formation, may be either a primary dolomite or one that was formed by replacement of limestone; it is crystalline, and in places it contains an abundance of solution cavities. The Deep River and the Scipio-Albion dolomitized trends in southwestern Michigan have been described. (See Figs. 4-11 and 4-12 and pages 123–124.) Sections through other pools producing from dolomite rocks are shown for the Belcher field of Ontario, Canada (Fig. 7-24, p. 308), the Ellenburger dolomite of the Apco field of western Texas (Figs. 7-60 and 7-61), and the Arbuckle dolomite of the Kraft-Prusa field of Kansas (Fig. 7-58, p. 338), the last two of Cambro-Ordovician age. Dolomitized Tamabra limestone (Cretaceous) forms the reservoir rock in the Poza Rica oil field of Mexico. (See Fig. 8-10, p. 356.)

Cementation and Compaction. After a pore space or pore pattern has been formed, whether it is primary, secondary, or both, it is generally modified by one or both of two common secondary changes, cementation and compaction. Both tend to reduce the percentage of pore space and the permea-

FIGURE 4-13

Section through a typical core from the D-3 dolomitized reef (Middle Devonian) of the Leduc field in Alberta, Canada, showing the intermingling of different kinds of porosity. The blank areas are nonporous and barren of oil. [Redrawn from Waring and Layer, Bull. Amer. Assoc. Petrol. Geol., Vol. 34, p. 307, Fig. 13.]

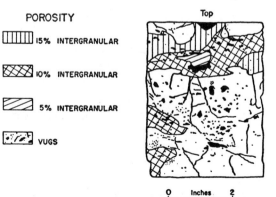

bility of the rock. These two processes may occur at any time during or after the deposition of the rock. In general the porosity of sedimentary rocks tends to decrease with an increase in depth of burial, increase in temperature, and increase in age.[67]

Cementation. Some cementation is primary; the cement may be precipitated, or it may be deposited along with the clastic material. Silica, carbonates, and other soluble materials may be precipitated contemporaneously with the deposition of detrital material. Primary cementing material is subject to recrystallization later, and it is then difficult to distinguish recrystallized cementing material from that introduced after the sediment became consolidated. Sandstones containing a siliceous cement deposited with the sand grains, or precipitated during the diagenesis, are called orthoquartzites to distinguish them from the metaquartzites, which result from metamorphic processes. Krynine has estimated that 90–95 percent of the quartzitic sandstones of the Appalachian region have a primary quartz cement.[68] If this is true, it improves the prospects for petroleum production in that region, which might otherwise be considered poor because of the opinion that quartzitic sandstones are always the result of regional diastrophism and metamorphism, and hence that all the potential reservoir rocks were impermeable, and all the oil has been squeezed out.

Materials that are insoluble and are therefore not precipitated may act like precipitated cements in that they fill voids, compact, and hold the grains together. Clay minerals, especially, are not soluble, but they are physically unstable, and they respond quickly to changes in pressure, temperature, and character of the water. They are deposited as interstitial detritus in varying form and amounts in nearly all sediments, and they form a common cementing material. Some clays alter to chlorites, sericites, and carbonates. As the water is squeezed out of the clays and muds, they are compressed tightly into the finer spaces between grains, where they act as a binding material holding the sand grains together. Graywackes are clastic rocks held together chiefly by primary detrital bonding material. Clay formed from weathering of feldspar grains, for example, clogs the pores in the Chanac formation (Tertiary) on the east side of the San Joaquin Basin of California, where it forms a bonding material as well as an up-dip obstruction that traps several oil pools. A different kind of detrital cement occurs in the sands of the McMurray formation (Cretaceous), near Athabaska Landing in northeastern Alberta. These sands are cemented by a viscous, heavy oil that may have been deposited along with the sand grains. When the oil is removed, the sand separates into loose grains.

The precipitation of cementing materials in the pores of clastic rocks, either during or after their diagenesis, is a secondary factor that modifies their porosity and permeability. The most common cementing materials in clastic reservoir rocks are, in the order of their abundance, quartz, calcite, dolomite, siderite, opal-chalcedony, anhydrite, and pyrite. Several minerals are frequently present in the same rock as cementing materials.[69] Most sandstones show evidence of cementation, together with more or less interlocking of grains, as a result of solution of grains in contact, solution of fine-grained siliceous matrix material, or introduction of silica from outside sources. (See Fig. 3-3.) A variety of minerals may act as cementing material. Forty core samples of feldspathic sandstones from wells in central and southern California showed the following secondary minerals deposited in the open pores or formed within the detrital clay matrix:[70] quartz, albite, orthoclase, microcline, dolomite, calcite, anatase, kaolinite, glauconite, barite, and pyrite.

Quartz, which is the most important precipitated cementing material of many clastic reservoir rocks, is also the first to be precipitated.[71] Silica is not found in the reservoir waters, and hence the source of the large amounts of silica cementing material, and the manner in which it was precipitated, have been the subject of much study, but the mechanisms are not fully understood.[72] The sources that have been suggested include: (1) silica precipitated from silica-bearing surface or meteoric waters; (2) silica carried by streams into the ocean, where it was precipitated along with the sand being deposited; (3) precipitated silica dissolved from small grains of silica-bearing minerals at the points of contact of sand grains as a result of pressure or rubbing during deposition, or of pressure during and after diagenesis (the Riecke principle);[73] (4) silica dissolved out of the clay minerals,[74] and carried in water squeezed out of the shales during loading and compaction. The nature of secondary silica overgrowth and the effects on a sandstone are shown graphically in Figure 4-14.

A spectacular sandstone reservoir rock showing secondary silica crystal growth occurs in the Petrólea field of eastern Colombia, where most of the oil obtained from the Barco formation (Lower Eocene) is found in the so-called "sparkling sandstone."[75] (See Fig. 6-37, p. 269.) These rocks derive their name from the fact that, where they crop out, the myriads of crystal faces on the secondary quartz sparkle in the sun. These sands average 12.5

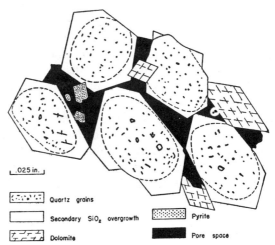

FIGURE 4-14

Thin section of an orthoquartzite showing secondary growth and recrystallization resulting in a marked change from the original pore pattern of the rock. [Redrawn from Krynine, Jour. Geol., Vol. 56 (1948), p. 152, Fig. 12.]

percent in porosity and 79 md in permeability, and the porosity is chiefly primary.

The source of carbonate cements is more readily explained than that of silica cements, for even sandstones commonly contain some carbonate, which may be dissolved and reprecipitated elsewhere. Carbonate cementing material in sandstones may be in the form of euhedral crystals of calcite or dolomite interspersed with the sand particles; or it may be plastered on the surfaces as a binder between sand grains; or it may be in the form of residues from former carbonate fossils, either preserved as recognizable fossils or concentrated in small residual patches.

As cementation often results from solution in place, the two processes work in opposite directions. Where solution is greater than the deposition of cement, porosity increases. Where deposition is greater than solution, porosity decreases. Both solution and cementation change the pore pattern unpredictably, especially the permeability.[76] Once a petroleum pool has formed, there is no further movement of the interstitial waters, and solution and cementation have reached a standstill. We may therefore conclude that solution and cementation in the reservoir occurred almost wholly before the petroleum accumulated.

Compaction. Three effects of load pressure are important in the geology of petroleum: (1) compaction of the reservoir sediments; (2) compaction of the nonreservoir sediments, especially the shales; (3) compression of the reservoir fluids. We are concerned here only with the compaction of the reservoir sediments.

Compaction of a reservoir rock is due chiefly to the increasing weight of the overburden. Its effect, like that of cementation, is to reduce porosity. The reduction of pore space by compaction in a sealed reservoir system causes an increase in reservoir fluid pressure. Compaction is especially significant in reservoir sediments containing shales or clays and colloidal material. Large amounts of adsorbed water are squeezed out of these by an increase in load pressure, and because the clays and colloids are highly plastic, they flow between the grains to form a cementing or bonding agent and thereby reduce the porosity. Clean sandstones found in some of the deepest wells, drilled below 15,000 feet, show no evidence of crushing,* which indicates that such rocks may well prove productive at great depths, whereas muddy or dirty sandstones would be made impermeable by pressure at far shallower depths. Even in clean sandstones, however, there is evidence that the number of grain contact points increases with depth, which means that pore space decreases downward.[77]

Compaction of a reservoir rock is of two kinds, plastic and elastic. Plastic compaction is the squeezing of the soft accessory minerals of the matrix, such as clays, weathered products, and colloids, into the open pores as the pressure increases and the water is driven out. The result is a loss of porosity, a reduction of permeability, and an overall lessening of the rock volume. (See Fig.

---

* R. B. Hutchinson, of the Superior Oil Company, Bakersfield, California, in a personal communication.

RESERVOIR PORE SPACE [CHAPTER 4] 131

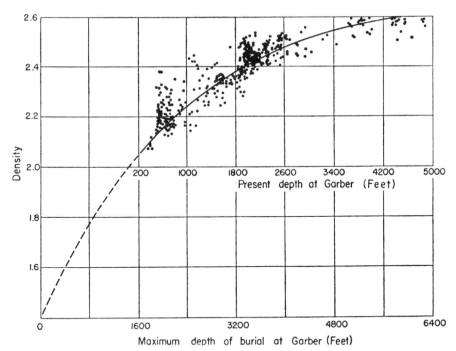

FIGURE 4-15 *Chart showing the progressive increase in density of the rocks with depth in the oil field of Garber, Oklahoma. [Redrawn from Athy, Problems of Petroleum Geology, Amer. Assoc. Petrol. Geol. (1934), p. 815, Fig. 1.]*

9-13.) Most plastic compaction occurs during the early diagenesis of the rock, when the high water content is being removed. Long-continued load pressures, however, undoubtedly maintain the process of plastic pore reduction long after diagenesis, though at a progressively slower rate. Figure 4-15 shows how the density increases with depth in the Garber field of Oklahoma. Cementation, as well as compaction, plays a part in this increase, and it becomes difficult if not impossible to separate the two processes. In sandstones, plastic compaction is evidenced by the squeezed, strained, and deformed soft minerals, by rearrangement of the grains, by closer packing, by breaking of the edges of the grains, and by closer adjustment of the same grains to the matrix material. A rock plastically deformed does not return, even in part, to the original volume. The volume of such a rock is therefore a function of the highest load pressure it has undergone during its geologic life.

A rock that has undergone elastic compaction, on the other hand, can, when the load pressure is reduced, return at least partially to the original volume. Such return is most likely to occur in a firm sandstone. Its occurrence means that energy was stored in the compressed sand grains while the load pressure was increasing and was released when the pressure diminished. It might be thought of as somewhat analogous to the energy stored in a coiled bedspring, which goes down when you go to bed and returns to its normal height when you get up. A layer of sandstone, however, because it contains some plastic minerals and because the load pressure may have caused closer packing, which is not elastic, can never fully recover its original thickness. How much elastic compression reservoir rocks can undergo, and how much energy can be stored in them, are questions on which opinion is divided and concrete evidence is scanty.

Elastic compression of an aquifer was regarded by Meinzer[78] as the source of the energy that caused artesian flow in certain wells. His argument was based on the fact that the weight of a column of water from the piezometric* surface to the aquifer is less than the weight of the equivalent thickness of rock overburden. The pressure of the water within the aquifer is exerted in all directions and helps support the weight of the overburden. Usually, as water is withdrawn from an aquifer, the piezometric surface is lowered, but Meinzer believed that in this case the loss of upward hydraulic pressure within the aquifer was compensated by a sinking of the overburden. In other words, the touching grains in the sandstone then supported a greater part of the load than when the water pressure was higher. As the load pressure on them increased, the grains were elastically compressed, and presumably would expand if the load pressure should again decline. The compression reduced the pore space, raised the pressure on the fluids, and drove the water to the surface in flowing artesian wells. The problem, however, of distinguishing the effect of compression of the reservoir solids from that of compression of the reservoir fluids, such as air, gas, and water, is difficult because the production of fluids would be similar under either. The concept of a continuous fluid phase extending from the groundwater table to great depths, with the resulting transmission through that fluid phase of pressures in accordance with the particular fluid-pressure gradients, offers a simpler and more realistic explanation of the origin of most subsurface fluid pressures. Moreover, in comparison with the compressibility of the various fluids, elastic compression of rocks is a negligible factor in subsurface fluid-flow calculations.

The same idea was used by Gilluly and Grant[79] in attempting to explain the subsidence at Long Beach, California. They suggest that the loss of reservoir pressure on the fluids, as a result of oil production, was sufficient to cause an equivalent increase in the effective weight of the overburden. The added weight, applied to the sand grains, causes an elastic compression of the sand, which reduces the volume of the sand and causes settling of the overburden all the way to the surface.

One of the problems connected with the compression of sandstones is that of distinguishing the plastic effect from the elastic effect. The two would be

---

* The piezometric surface is the surface connecting the highest points to which water from a single aquifer will rise in wells. The piezometric surface is equivalent to the potentiometric surface if the water everywhere has a constant density and that density is used to calculate the potentiometric surface (see pp. 395–397).

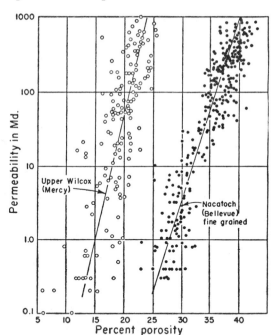

FIGURE 4-16

Relation between porosity and permeability in two reservoir rocks: at the left, the Upper Wilcox sandstone (Eocene) at Mercy, Texas; at the right, the Nacatoch sandstone (Upper Cretaceous) at Bellevue, Louisiana. There is a general increase in permeability with an increase in porosity. [Redrawn from Archie, Bull. Amer. Assoc. Petrol. Geol., Vol. 34, p. 945, Fig. 1.]

expected to go together in the average sandstone or graywacke containing more or less clay and other plastic material, but the relative importance of each in an underground reservoir rock is practically impossible to determine. The low compressibility of quartz and sandstone[80] leads to the conclusion that much if not most of the compaction undergone by clastic reservoir rocks is plastic rather than elastic. Likewise, the ease with which carbonate minerals and rocks recrystallize and fill the available pores suggests that most carbonate rock compaction is more plastic than elastic.

Relation Between Porosity and Permeability

The quantitative relation between porosity and permeability is obscure and variable.[81] Beyond the fact that a permeable rock must also be porous, there seems to be only a general connection. Two illustrations of the lack of a close relationship are shown in Figure 4-16. A great many measurements of porosity and permeability have been made on the Bradford sand (Lower Devonian), in Pennsylvania, which, as it has a fine, uniform-grained texture, might be expected to show a fairly constant ratio between the two properties if any exists. Yet Figure 4-17 shows that even in this rock there is no more than a rough relationship.[82] The effect of fracturing on the porosity and permeability of limestone and dolomite, as revealed by several hundred measurements, is shown graphically in Figure 4-18. The line $k_{90°}$ (permeability normal to the fracture system) represents the primary and vug porosity in the

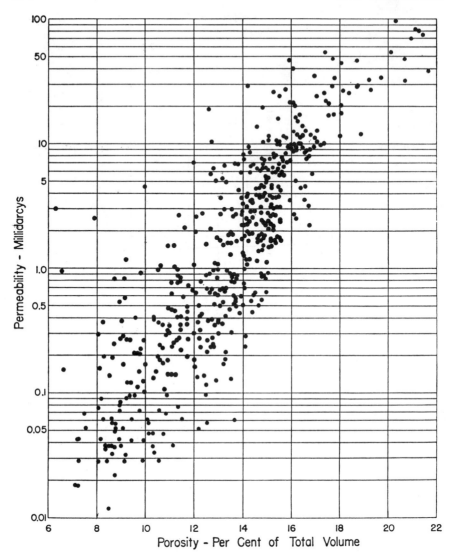

FIGURE 4-17 Porosity-permeability values for about 500 samples of the Bradford sandstone (Middle Devonian) from the Bradford field, northwestern Pennsylvania. The samples are all at one-foot intervals from cores taken from twenty-nine wells in a small area. The Bradford sand is considered of uniform character, and the chart, aside from showing the general increase of permeability with an increase in porosity, illustrates the wide spread, or lack of close relation, between porosity and permeability. For example, for any value of permeability there is variation in porosity of from six to ten percentage points. [Redrawn from Ryder, World Oil, May 1948, p. 174.]

RESERVOIR PORE SPACE [CHAPTER 4]

fractured limestone; the k_{max} line represents the maximum permeability found by measurements parallel to the fracture system. The permeability of unfractured dolomite increases in both directions uniformly with porosity.

While porosity is a dimensionless quantity, permeability reflects in measurable units the resistance that a rock sets up against the flow of a homogeneous fluid through it. Theoretically, permeability can be related to the geometry of the rock texture according to the equation[83]

$$k = 2 \times 10^7 \times \frac{\Sigma^3}{(1 - \Sigma)^2} \times \frac{1}{S^2}$$

in which k = permeability, Σ = fractional porosity, and S = the specific surface of the solid material. S is the specific surface of the pores and is equal to the surface of the solid material contained in one cubic centimeter of rock. The equation suggests that the higher the porosity and the smaller the specific surface, the higher the permeability. The smaller the individual pores, the greater becomes the specific surface, and consequently the less the permeability.

Artificial, or Man-made, Porosity and Permeability. Various methods of forming or increasing the pore space and the permeability of reservoir rocks have been devised. The early method was to *shoot* a well—that is, to explode a charge of nitroglycerin in the well bore opposite the reservoir. The fracturing of the reservoir increases the effective radius of the well bore and increases the porosity and permeability surrounding the hole and consequently permits more oil and gas to flow into the well. The response varies with different rocks, depending largely on whether the explosion packs the particles tighter or fractures the rock. Forcing acids into the reservoir rock under pressure is called *acidization*.[84] The acid enters the reservoir rock along connected porosity

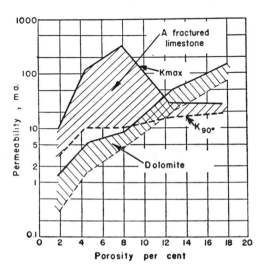

FIGURE 4-18

The relation between permeability and porosity in a group of Paleozoic limestone and dolomite reservoir rocks of western Texas. Kmax is parallel to the fracture system and $K_{90°}$ is at right angles. The difference (shaded) between Kmax and $K_{90°}$ shows the amount and distribution of the secondary porosity between fractures and vugs. [Redrawn from Kelton, O. & G. Jour., November 24, 1949, p. 119.]

openings and dissolves the soluble material as it penetrates the rock, thereby increasing both the permeability and the porosity. The permeability of limestone reservoirs, especially, is increased, and some sand reservoirs with a carbonate or other acid-soluble cement show a decided increase in production as a result of acidization. Various methods of hydraulic fracturing—driving a liquid containing sand grains into the pores of a reservoir rock by the use of extremely high injection pressure—have been developed.[85] Later, when the pressure is relieved, the liquid drains out, leaving the sand grains behind to hold the fractures open. These methods are known by various trade names: *Hydrafrac, Stratafrac, Sandfrac,* etc. Where the initial reservoir pressures are still available within the reservoir, many phenomenal increases in oil production have resulted.

CONCLUSION

The two essential mass properties of reservoir rock are effective porosity and permeability. Effective porosity provides storage space for oil and gas, and permeability permits them to move through the rock. Much progress has been made by the production engineer and in the core laboratories toward a better understanding of the factors that influence the porosity and permeability of reservoir rocks—particularly the role of clays, the effects of differing reservoir waters, and the petrophysics of different kinds of openings. A vast number of detailed quantitative data have been accumulated on the porosity and permeability of specific reservoirs, on the varying effect of each on oil and gas production, and on the interrelations of the pore space and the fluid content. There is yet much to learn, however, particularly on improving oil recoveries from rocks of extremely low porosity and low permeability, on the role of grain size in determining porosity and permeability, and on relating the different kinds of porosity and permeability to depositional environment—all of which will give a better understanding of underground conditions and aid in making more accurate predictions.

Selected General Readings

P. G. Nutting, "Some Physical and Chemical Properties of Reservoir Rocks Bearing on the Accumulation and Discharge of Oil," in *Problems of Petroleum Geology,* Amer. Assoc. Petrol. Geol., Tulsa, Okla. (1934), pp. 825–832. A summary of five years of laboratory work in the U.S. Geological Survey.

H. J. Fraser, "Experimental Study of the Porosity and Permeability of Clastic Sediments," Jour. Geol., Vol. 43 (November-December 1935), pp. 910–1010.

One of the most thorough discussions of the geological aspects of porosity and permeability.

Morris Muskat, *Flow of Homogeneous Fluids Through Porous Media,* McGraw-Hill Book Co., New York (1937), 763 pages. A standard work.

M. King Hubbert, "The Theory of Ground-water Motion," Jour. Geol., Vol. 48 (November-December 1940), pp. 785–944. A classic article largely concerned with theoretical fundamentals of porosity and permeability phenomena and fluid behavior under ground.

Richard B. Hohlt, "The Nature and Origin of Limestone Porosity," Quart. Colo. Sch. Mines, Vol. 43, No. 4 (October 1948), 51 pages. Bibliog. 67 items. A concise statement of the principles of the development of limestone and dolomite porosity.

J. M. Dallavalle, *Micromeritics, the Technology of Fine Particles,* 2nd ed., Pitman Publishing Corp., New York and London (1948), 543 pages. Extensive bibliography. Considers the behavior and characteristics of particles ranging from 10^{-1} to 10^5 microns in diameter.

Morris Muskat, *Physical Principles of Oil Production,* McGraw-Hill Book Co., New York (1949), 922 pages. Porosity and permeability phenomena and their underlying principles are considered in Chaps. 3 (pp. 114–176) and 7 (pp. 271–332).

Charles D. Russell and Parke A. Dickey, "Porosity, Permeability, and Capillary Properties of Petroleum Reservoirs," in Parker D. Trask (ed.), *Applied Sedimentation,* John Wiley & Sons, New York (1950), pp. 579–615. Bibliog. 52 items. An excellent summary of the phenomena associated with porosity and permeability in reservoirs.

G. E. Archie, "Introduction to Petrophysics of Reservoir Rocks," Bull. Amer. Assoc. Petrol. Geol., Vol. 34 (May 1950), pp. 943–961.

G. E. Archie, "Classification of Carbonate Reservoir Rocks and Petrophysical Considerations," Bull. Amer. Assoc. Petrol. Geol., Vol. 36 (February 1952), pp. 278–298. This and the next preceding article present concise classic discussions of the fundamentals of reservoir rock petrophysics.

American Petroleum Institute, New York and Dallas, Texas, *Recommended Practice for Determining Permeability of Porous Media,* API RP No. 27 (September 1952), 27 pages. Definitions, theory, measurements, and calculations concerning permeability, with bibliography of 51 references. An authoritative statement for the petroleum industry.

M. R. J. Wyllie, "The Fundamentals of Well Log Interpretation," 3rd ed. Academic Press, New York (1963), 238 pages. A standard reference book on modern well logging.

Sylvain J. Pirson, "Handbook of Well Log Analysis," Prentice-Hall, Inc., Englewood Cliffs (1963), 326 pages. A standard reference book.

Reference Notes

1. G. E. Archie, "Introduction to Petrophysics of Reservoir Rock," Bull. Amer. Assoc. Petrol. Geol., Vol. 34 (May 1950), pp. 943–961.

2. Paul D. Krynine, "Petrologic Aspects of Prospecting for Deep Oil Horizons in Pennsylvania," Prod. Monthly, Vol. 12 (January 1948), pp. 28–33.

3. Park J. Jones, *Petroleum Production: Vol. 1, Mechanics of Production*, Reinhold Publishing Corp., New York (1946), pp. 13–15.

4. A. F. Melcher, "Determination of Pore Space of Oil and Gas Sands," Trans. Amer. Inst. Min. Met. Engrs., Vol. 65 (1921), pp. 469–489. Discussion to p. 497. Melcher was the pioneer in the study of porosity, and this is his first published progress report.

Sylvain J. Pirson, *Elements of Oil Reservoir Engineering*, McGraw-Hill Book Co., New York (1950), pp. 20–34. 48 selected references on porosity.

T. A. Pollard and Paul P. Reichertz, "Core-Analysis Practices—Basic Methods and New Developments," Bull. Amer. Assoc. Petrol. Geol., Vol. 36 (February 1952), pp. 230–252.

5. Paul D. Krynine, "Reservoir Petrography of Sandstones," U.S. Geol. Surv., Oil and Gas Investig., May OM 126, "Geology of the Arctic Slope of Alaska" (3 sheets), from sheet 2.

6. Henri Darcy, *Les Fontaines publiques de la ville de Dijon*, Victor Dalmont, Paris (1856).

7. American Petroleum Institute, New York and Dallas, Texas, *Recommended Practice for Determining Permeability of Porous Media*, API RP No. 27 (September 1952), 27 pages.

8. Morris Muskat, *Physical Principles of Oil Production*, McGraw-Hill Book Co., New York (1949), pp. 123–149.

Sylvain J. Pirson, *op. cit.* (note 4), pp. 45–73. 71 selected references on permeability.
T. A. Pollard and Paul P. Reichertz, *op. cit.* (note 4).
American Petroleum Institute, *op. cit.*, pp. 12–16.

9. Ionel I. Gardescu, "Behavior of Gas Bubbles in Capillary Spaces," Trans. Amer. Inst. Min. Met. Engrs. (1930), pp. 351–368. Discussion to p. 370.

Stanley C. Herold, *Oil Well Drainage*, Stanford University Press, Stanford, Calif. (1941), 407 pages. Jamin action is discussed on pp. 25–33.

John C. Calhoun, "The Jamin Effect," O. & G. Jour., June 2, 1949, p. 79.

10. L. J. Klinkenberg, "The Permeability of Porous Media to Liquids and Gases," API Drill. and Prod. Pract., 1941, pp. 200–211. Discussion to p. 213.

11. Heid, McMahon, Nielsen, and Yuster, "Study of the Permeability of Rocks to Homogeneous Fluids," API Drill. and Prod. Pract., 1950, p. 238.

12. Dana G. Hefley and P. E. Fitzgerald, "Selective Acidizing and Permeability Determination by an Electrical Method," Trans. Amer. Inst. Min. Met. Engrs., Vol. 155 (1944), pp. 223–231.

P. N. Hardin, "The Electric Pilot in Selective Acidizing, Permeability Determinations, and Water Locating," in *Subsurface Geologic Methods*, Colo. Sch. Mines (1951), pp. 676–685.

13. W. R. Purcell, "Capillary Pressures—Their Measurement Using Mercury and the Calculation of Permeability Therefrom," Tech. Paper 2544, Trans. Amer. Inst. Min. Met. Engrs., Vol. 186 (February 1949), pp. 39–46. Discussion to p. 48. 8 references.

14. R. D. Wycoff and H. G. Botset, "The Flow of Gas-Liquid Mixtures Through Unconsolidated Sands," Physics, Vol. 7 (1938), p. 325.

Harry M. Ryder, "Permeability, Absolute, Effective, Measured," World Oil, May 1948, pp. 173–176.

Amer. Petrol. Inst., *op. cit.* (note 7), p. 4.

15. J. S. Osoba, "Relative Permeability, What It Is and How to Put It to Use in the Field," O. & G. Jour., July 27, 1953, pp. 326–333.

16. M. C. Leverett, "Flow of Oil-Water Mixtures Through Unconsolidated Sands," Tech. Pub. 1003 (November 1938), 21 pages, 8 references cited, and Trans. Amer. Inst. Min. Met. Engrs., Vol. 132 (1939), pp. 149–171.

M. C. Leverett and W. B. Lewis, "Steady Flow of Gas-Oil-Water Mixtures Through Unconsolidated Sands," Tech. Pub. 1206 (May 1940), 9 pages, and Trans. Amer. Inst. Min. Met. Engrs., Vol. 142 (1941), pp. 107–116.

17. H. J. Fraser, "Experimental Study of the Porosity and Permeability of Clastic Sediments," Jour. Geol., Vol. 43 (November-December 1935), pp. 910–1010. 81 references cited. This is a thorough discussion of the factors that influence both porosity and permeability.

18. W. C. Krumbein and L. L. Sloss, *Stratigraphy and Sedimentation,* 2nd ed., W. H. Freeman and Company, San Francisco (1963), pp. 118–123.

19. L. C. Graton and H. J. Fraser, "Systematic Packing of Spheres—with Particular Relation to Porosity and Permeability," Jour. Geol., Vol. 43 (1935), pp. 785–909. 69 references.

J. M. Dallavalle, *Micromeritics, the Technology of Fine Particles,* 2nd ed., Pitman Publishing Corp., New York and London (1948), Chap. 6, "Characteristics of Packings," pp. 123–148.

20. Oscar Edward Meinzer, "Compressibility and Elasticity of Artesian Aquifers," Econ. Geol., Vol. 23 (1928), p. 267.

21. H. J. Fraser, *op. cit.* (note 17), pp. 917–918.

22. H. B. Hill, E. L. Rawlins, and C. R. Bopp, *Engineering Report on Oklahoma City Field, Oklahoma,* RI 3330 (January 1937), Bur. Mines, pp. 199 and 201.

23. H. J. Fraser, *op. cit.* (note 17), pp. 959–1010.

24. Geo. A. Thiel, "Diagenetic Changes in Calcareous Sediments," Report of Comm. on Sedimentation, Nat. Research Council (March 1942), pp. 81–110. Bibliog. 174 items.

25. V. P. Grage and E. F. Warren, Jr., "Lisbon Oil Field, Claiborne and Lincoln Parishes, Louisiana," Bull. Amer. Assoc. Petrol. Geol., Vol. 23 (March 1939), pp. 281–324.

26. Robert F. Imbt and S. V. McCollum, "Todd Deep Field, Crockett County, Texas," Bull. Amer. Assoc. Petrol. Geol., Vol. 34 (February 1950), pp. 239–262.

27. John M. Muir, *Geology of the Tampico Region, Mexico,* Amer. Assoc. Petrol. Geol., Tulsa, Okla. (1936), p. 165.

28. George H. Fancher and Donald M. K. Mackay, "Magnolia," in *Secondary Recovery of Petroleum in Arkansas—A Survey,* Arkansas Oil and Gas Commission, Eldorado, Ark. (1946), Chap. 29, pp. 187–194.

29. Warren B. Weeks and Clyde W. Alexander, "Schuler Field, Union County, Arkansas," Bull. Amer. Assoc. Petrol. Geol., Vol. 26 (September 1942), pp. 1467–1516.

30. G. C. Clark and Jack M. DeLong, "Carthage Field, Panola County, Texas," in *Occurrence of Oil and Gas in Northeast Texas,* Bur. Econ. Geol., Austin, Texas (1951), pp. 55–63.

31. Lynn K. Lee, "Geology of Basin Fields in Southeastern Illinois," Bull. Amer. Assoc. Petrol. Geol., Vol. 23 (October 1939), pp. 1493–1506. McCloskey "sand" discussed on pp. 1500–1505.

32. Irene Haskett, "Reservoir Analysis of the Redwater Pool," Trans. Can. Inst. Min. & Met., April 1951; Canadian Oil and Gas Industries, July 1951, pp. 39–49.

33. A. W. Lauer, "The Petrology of Reservoir Rocks and Its Influence on the Accumulation of Petroleum," Econ. Geol., Vol. 12 (1917), pp. 435–472.

34. A. N. Murray and W. W. Love, "Action of Organic Acids upon Limestone," Bull. Amer. Assoc. Petrol. Geol., Vol. 13 (November 1929), pp. 1467–1475.

William C. Imbt and Samuel P. Ellison, Jr., "Porosity in Limestone and Dolomite Petroleum Reservoirs," API Drill. and Prod. Pract., 1946, pp. 364–372. 23 references. Contains photographs of plastic models of carbonate porosity.

35. W. V. Howard and Max W. David, "Development of Porosity in Limestone," Bull. Amer. Assoc. Petrol. Geol., Vol. 20 (November 1936), p. 1395.

36. Parry Reiche, "A Survey of Weathering Processes and Products," University of New Mexico Publications, Geology, No. 1 (1945), 87 pages. A concise discussion of weathering phenomena.

37. W. V. Howard and Max W. David, op. cit. (note 35), p. 1392.

38. John Emery Adams, "Origin, Migration, and Accumulation of Petroleum in Limestone Reservoirs in the Western United States and Canada," in *Problems of Petroleum Geology*, Amer. Assoc. Petrol. Geol. (1934), pp. 347–363. Cavernous porosity in the dolomitic limestone reservoir in Yates pool, Texas, is discussed on page 351.

C. D. Cordry, "Ordovician Development, Sand Hills Structure, Crane County, Texas" (Geologic Note), Bull. Amer. Assoc. Petrol. Geol., Vol. 21 (December 1937), pp. 1575–1591. Describes cavernous porosity in Ellenburger limestone (Ordovician).

39. D. H. Stormont, "Huge Caverns Encountered in Dollarhide Field," O. & G. Jour., April 7, 1949, pp. 66–68 and 94.

40. Philip C. McConnell, "Drilling and Producing Techniques that Yield Nearly 850,000 Barrels per Day in Saudi Arabia's Fabulous Abquaiq Field," O. & G. Jour., December 20, 1951, p. 197.

41. M. King Hubbert and David G. Willis, "Important Fractured Reservoirs in the United States," Fourth World Petrol. Congr. (June 1955), Section I/A/1 Proceed., pp. 57–84.

42. Warren J. Mead, "The Geologic Rôle of Dilatancy," Jour. Geol., Vol. 33 (1925), pp. 685–698.

Duncan A. MacNaughton, "Dilatancy in Migration and Accumulation of Oil in Metamorphic Rocks," Bull. Amer. Assoc. Petrol. Geol., Vol. 37 (February 1953), pp. 217–231.

43. George R. Gibson, "Relation of Fractures to the Accumulation of Oil," Spraberry Symposium, O. & G. Jour., November 29, 1951, pp. 107–117.

44. Sylvain J. Pirson, "Performance of Fractured Oil Reservoirs," Bull. Amer. Assoc. Petrol. Geol., Vol. 37 (February 1953), pp. 232–247.

45. J. H. Beach, "Geology of Edison Oil Field, Kern County, California," in *Structure of Typical American Oil Fields*, Amer. Assoc. Petrol. Geol., Tulsa, Okla. (1948), Vol. 3, pp. 58–85.

46. L. E. Porter, "El Segundo Oil Field, California," Trans. Amer. Inst. Min. Met. Engrs., reprinted Vol. 127 & 132, p. 451.

W. S. Eggleston, "Summary of Oil Production from Fractured Rock Reservoirs in California," Bull. Amer. Assoc. Petrol. Geol., Vol. 32 (July 1948), pp. 1352–1355.

Louis J. Regan, Jr., and Aden W. Hughes, "Fractured Reservoirs of Santa Maria District, California," Bull. Amer. Assoc. Petrol. Geol., Vol. 33 (January 1949), pp. 32–51.

Louis J. Regan, Jr., "Fractured Shale Reservoirs of California," Bull. Amer. Assoc. Petrol. Geol., Vol. 37 (February 1953), pp. 201–216.

Oil and Gas Journal, "Estimated Remaining Reserves of Crude Oil and Condensate in Larger U.S. Fields," O. & G. Jour., January 27, 1964, p. 160.

47. Louis J. Regan, Jr., and Aden W. Hughes, op. cit. (note 46), pp. 47–48.

48. Walter M. Wilkinson, "Fracturing in Spraberry Reservoir, West Texas," Bull. Amer. Assoc. Petrol. Geol., Vol. 37 (February 1953), pp. 250–265.

49. Charles R. Fettke, "Oriskany as a Source of Gas and Oil in Pennsylvania and Adjacent Areas," Bull. Amer. Assoc. Petrol. Geol., Vol. 22 (March 1938), pp. 241–266.

50. Fenton H. Finn, "Geology and Occurrence of Natural Gas in Oriskany Sandstone;" Bull. Amer. Assoc. Petrol. Geol., Vol. 33 (March 1949), pp. 303–335.

51. Max Littlefield, L. L. Gray, and A. C. Godbold, "A Reservoir Study of the West Edmond Hunton Pool, Oklahoma," Tech. Pub. 2203, Petrol. Technol., and Trans. Amer. Inst. Min. Met. Engrs., Vol. 174 (November 1947), pp. 131–155. Discussion on pp. 155–164.

Robert Malcolm Swesnik, "Geology of West Edmond Oil Field, Oklahoma," in *Structure of Typical American Oil Fields,* Amer. Assoc. Petrol. Geol., Tulsa, Okla. (1948), Vol. 3, pp. 359–398.

52. Kenneth K. Landes, "Porosity Through Dolomitization," Bull. Amer. Assoc. Petrol. Geol., Vol. 30 (March 1946), pp. 305–318. 41 references cited.

Kenneth K. Landes, "Deep River Oil Field, Arenac County, Michigan," in *Structure of Typical American Oil Fields,* Amer. Assoc. Petrol. Geol., Tulsa, Okla. (1948), Vol. 3, pp. 299–304.

53. Burton Atkinson and David Johnston, "Core Analysis of Fractured Dolomite in the Permian Basin," Petrol. Technol. & Develop., Trans. Amer. Inst. Min. Met. Engrs., Vol. 179 (1949), pp. 128–132.

54. R. J. Burgess, "Oil in Trenton Synclines," O. & G. Jour., Vol. 58, No. 33 (August 15, 1960), pp. 124–131.

55. John M. Muir, "Limestone Reservoir Rocks in the Mexican Oil Fields," in *Problems of Petroleum Geology,* Amer. Assoc. Petrol. Geol., Tulsa, Okla. (1934), pp. 377–398.

John M. Muir, *op. cit.* in note 27, pp. 165–171. Discusses porosity in the Northern fields.

Antonio García Rojas, "Mexican Oil Fields," Bull. Amer. Assoc. Petrol. Geol., Vol. 33 (August 1949), pp. 1336–1350.

56. J. E. Smith, "The Cretaceous Limestone Producing Area of the Mara and Maracaibo District, Venezuela," Proc. Third World Petrol. Congress, The Hague (1951), Sec. 1, pp. 56–72. 10 references.

57. H. S. Gibson, "Oil Production in Southwestern Iran," World Oil, May 1948. pp. 271–280, and June, 1948, pp. 217–226.

58. G. M. Lees, "Reservoir Rocks of Persian Oil Fields," Bull. Amer. Assoc. Petrol. Geol., Vol. 17 (March 1933), pp. 229–240.

59. D. C. Ion, S. Elder, and A. E. Pedder, "The Agha Jari Oil Field, Southwestern Persia," Proc. Third World Petrol. Congress, The Hague (1951), Sec. 1, pp. 162–186.

F. M. Van Tuyl, "The Origin of Dolomite," Ann. Rept. Iowa Geol. Surv. Vol. 25 (1916), pp. 251–421.

60. F. W. Clarke, *The Data of Geochemistry,* 5th ed., Bull. 770, U.S. Geol. Surv. (1924), pp. 565–580.

61. Edward Orton, "The Trenton Limestone as a Source of Petroleum and Inflammable Gas in Ohio and Indiana," 8th Ann. Rept. U.S. Geol. Surv. (1886–1887), Part II, pp. 642–645, 653–662.

62. Waldemar Lindgren, "The Nature of Replacement," Econ. Geol., Vol. 7 (1912), pp. 521–535.

Edward Steidtmann, "Origin of Dolomite as Disclosed by Stains and Other Methods," Bull. Geol. Soc. Amer., Vol. 28 (1917), pp. 431–450.

63. Richard B. Hohlt, "The Nature and Origin of Limestone Porosity," Quart. Colo. Sch. Mines, Vol. 43 (1948), No. 4, 51 pages.

64. Ernest Cloos, "Oolite Deformation in the South Mountain Fold, Maryland," Bull. Geol. Soc. Amer., Vol. 58 (September 1947), pp. 843–918.

65. W. W. Waring and D. B. Layer, "Devonian Dolomitized Reef, D-3 Reservoir, Leduc Field, Alberta, Canada," Bull. Amer. Assoc. Petrol. Geol., Vol 34 (February 1950), pp. 295–312. Contains many photographs of cores showing porosity.

66. Edward Orton, *op. cit.* (note 61), pp. 483–662.

J. Ernest Carman and Wilbur Stout, "Relationship of Accumulation of Oil to Structure and Porosity in the Lima-Indiana Field," in *Problems of Petroleum Geology,* Amer. Assoc. Petrol. Geol., Tulsa, Okla. (1934), pp. 521–529.

67. John C. Maxwell, "Influence of Depth, Temperature, and Geologic Age on Porosity of Quartzose Sandstone," Bull. Amer. Assoc. Petrol. Geol., Vol. 48 (May 1964), pp. 697–709.

68. Conference on Sedimentation, Research Committee, Amer. Assoc. Petrol. Geol., Tulsa, Okla. (1942), p. 66.

69. W. A. Waldschmidt, "Cementing Materials in Sandstones and Their Probable Influence on Migration and Accumulation of Oil and Gas," Bull. Amer. Assoc. Petrol. Geol., Vol. 25 (1941), pp. 1839–1879.

70. Charles M. Gilbert, "Cementation of Some California Tertiary Reservoir Sands," Jour. Geol., Vol. 57 (January 1949), pp. 1–17.

71. F. J. Pettijohn, *Sedimentary Rocks,* Harper & Brothers, New York (1949), pp. 480–485.

72. W. A. Tarr, *The Origin of Chert and Flint,* University of Missouri Studies, Vol. 1, No. 2 (1926), 46 pages.

C. B. Hood, "Theories for the Mechanism of the Settling of Silicic Acid Gels," Chem. Rev., Vol. 22 (1938), pp. 403–422.

73. John Johnston and Paul Niggli, "The General Principles Underlying Metamorphic Processes," Jour. Geol., Vol. 21 (1913), pp. 610–612.

74. Kenneth M. Trowe, "Clay Mineral Diagenesis as a Possible Source of Silica Cement in Sedimentary Rocks," Jour. Sed. Petrol., Vol. 23 (March 1962), pp. 26–28.

75. Frank B. Notestein, Carl W. Hubman, and James W. Bowler, "Geology of the Barco Concession, Republic of Colombia, South America," Bull. Geol. Soc. Amer., Vol. 55 (October 1944), pp. 1165–1216.

76. Melvin A. Rosenfeld, "Some Aspects of Porosity and Cementation," Prod. Monthly, Vol. 13 (May 1949), pp. 39–42. Bibliog. 18 items.

77. Jane M. Taylor, "Pore-Space Reduction in Sandstones," Bull. Amer. Assoc. Petrol. Geol., Vol. 34 (April 1950), pp. 701–716.

78. Oscar Edward Meinzer, *op. cit.* (note 20), pp. 268–291. See also the discussion on pp. 683–696.

79. James Gilluly and U. S. Grant, "Subsidence in the Long Beach Harbor Area, California," Bull. Geol. Soc. Amer., Vol. 60 (March 1949), pp. 461–529. 58 references listed.

80. L. H. Adams and E. D. Williamson, "The Compressibility of Minerals and Rocks at High Pressures," Paper 484, Geophysics Laboratory, Carnegie Inst., Washington, reprinted from Jour. Franklin Inst., Vol. 195 (April 1923), pp. 475–529. At a pressure of 8,000 psi the coefficient of compressibility of quartz is 0.000,002,681 per megabar (1,000 atmospheres).

81. H. G. Botset and D. W. Reed, "Experiment on Compressibility of Sand," Bull. Amer. Assoc. Petrol. Geol., Vol. 19 (July 1935), pp. 1053–1060. The experiment resulted in a decrease in porosity of 2 percent between no load and a full load of 3,000 psi, much of which was accounted for by rearrangement of grains and crushing of grains.

Charles B. Carpenter and George B. Spencer, "Measurements of Compressibility of Consolidated Oil-bearing Sandstones," RI 3540, U.S. Bur. Mines (1940), 20 pages. Contains a good historical summary of the work and offers ideas concerning compaction and compressibility.

82. A. C. Bulnes and R. U. Fitting, "An Introductory Discussion of the Reservoir Performance of Limestone Formations," Trans. Amer. Inst. Min. Met. Engrs., Vol. 160 (1945), pp. 179–201. 15 references. Comparison of sandstone and limestone porosity and permeability.

83. Harry M. Ryder, "Permeability, Absolute, Effective, Measured," World Oil, May 1948, pp. 173–177. 13 references.

84. W. V. Engelhardt and H. Pitter, *Relationships Between Porosity, Permeability and Grain Size of Sands and Sandstones,* Heidelberger Beiträge zur Mineralogie, Vol. 2 (1951), pp. 477–491. Contains derivation and applications of the equation.

85. W. W. Love and P. E. Fitzgerald, "Importance of Geological Data in Acidizing of Wells," Bull. Amer. Assoc. Petrol. Geol., Vol. 21 (May 1937), pp. 616–626.

P. E. Fitzgerald, J. R. James, and Ray L. Austin, "Laboratory and Field Observations of Effect of Acidizing Oil Reservoirs Composed of Sands," Bull. Amer. Assoc. Petrol. Geol., Vol. 25 (May 1941), pp. 850–870. 25 references.

P. E. Fitzgerald, "A Review of the Chemical Treatment of Wells," Jour. Petrol. Technol. (September 1953), Sec. 1, pp. 11–13. 7 references.

86. J. B. Clark, "A Hydraulic Process for Increasing the Productivity of Wells," TP 2510, Petrol. Technol. (January 1949), Trans. Amer. Inst. Min. Met. Engrs., Vol. 186 (1950), pp. 1–8.

CHAPTER 5

Reservoir Fluids – Water, Oil, Gas

Fluid content: source of data – fluid distribution – fluid contacts. Water: classification – character – oil-field brines. Oil: measurement – chemical properties – physical properties. Gas: measurement – composition – impurities.

THE RESERVOIR is that portion of the rock or rock layers that contains the pool of petroleum. So far we have discussed the reservoir rock and the spaces within the reservoir rock that contain the reservoir fluids; now we will consider the fluid content of the reservoir: the water, oil, and gas. Except for the relatively small volumes of permeable rock containing oil and gas pools —oil and gas reservoirs—nearly all of the pore space of the upper few miles of the earth's crust is filled with water; the oil and gas occur in a water environment. Not only did they pass along water-lined permeability highways to accumulate into pools, but the pools of oil and gas generally are separated from the rock walls of the reservoir by a film of water. Consequently, petroleum geologists are much interested in the water content of the rocks, even though their objective is to find oil and gas. In this chapter we will consider the fluids water, oil, and gas as they occur in oil and gas pools. The disseminated occurrences, migration, and accumulation of petroleum into pools—the pre-pool history—will be considered later, in Chapter 12.

FLUID CONTENT OF RESERVOIRS

The fluid content of a gas pool consists of gas and water, and that of an oil pool consists of gas, oil, and water. Each of these fluids occurs in varying proportions, or saturations, and each may vary widely in composition and physical properties from pool to pool. The physical properties of the fluids, moreover, are very different, at the higher temperatures and pressures that prevail in the deeper reservoirs, from those of the same fluids when recovered at the surface, or of chemically similar fluids in shallow reservoirs. The two dominant physical properties of oil and gas are (1) their relative immiscibility in water, and (2) their lower density than water, enabling the oil and gas to rise in water because of their buoyancy. The geology of petroleum is largely the geology of fluids.

Source of Data on Reservoir Fluids

While the fluid content of a petroleum reservoir consists of gas, oil, and water, there is an almost infinite variation in the compositions, relative amounts, and properties of these fluids in various reservoirs. Our knowledge of the reservoir content must be obtained by indirect methods, since it is impossible to see an oil and gas pool in place. We generally obtain our knowledge of the reservoir content by (1) examination of the reservoir fluids contained in (a) cores and well cuttings, (b) bottom-hole reservoir fluids, either as samples brought to the surface by testing devices or tested in place by electronic and similar equipment placed at the surface, (c) surface samples of reservoir fluids obtained from producing wells; (2) study of the producing history of the reservoir.

Examination of Reservoir Fluids. The fluids in a reservoir may be studied indirectly by means of the effects that they have on the various logs and on the drilling mud during drilling. Fluids may be observed directly when they are bailed up in cable-tool wells, or as they are obtained in drill-stem tests and other methods of obtaining samples of fluids from the bottom of the hole. Reservoir fluids may be observed in cores and well cuttings that are brought to the surface for examination; tasting a core for salt, for example, is a standard procedure at a well. Samples of the fluids may also be taken at the bottom of the well and kept under reservoir conditions until examined in the laboratory. Or the character of the fluids in the reservoir may be calculated, or restored, from study and analysis of samples taken at the top of the well.[1]

Modern practice calls for the careful sampling of all reservoir rocks in each well, and either continuous or closely spaced cores are therefore taken through each potential reservoir. After the core material is recovered, it is taken to the laboratory and examined for its lithology, porosity, and permeability, and the character and relative amounts of the gas, oil, and water that it contains are determined.[2] These core and sample determinations are compared with the electric log, which supplies a check on the fluid content, thickness, and variation of the pay formation and on the volume of its pore space.

When a core from a petroleum reservoir is cut and brought to the surface, its temperature and pressure decline to the levels that exist at the surface. The gas it contained comes out of solution and expands, and most of it escapes, along with a part of the oil. The oil and gas measurements made under surface conditions, therefore, tend to be minimum values. The interstitial water, however, being held in place by capillary forces, remains almost undisturbed in place in the core, though it may be more or less contaminated by water from the drilling mud. The water content of a core, therefore, tends to be a good deal closer to what it was originally than the oil content, and the gas content is much less than it was originally.

Several methods are used for reducing these discrepancies. One is the use of oil-base drilling mud instead of water-base mud when drilling through a reservoir. Oil-base muds prevent the flushing action and contamination of the water content of the core, thereby permitting a more accurate determination of the percentage of water. Since the use of oil-base mud contaminates the hydrocarbon content of the core, the difference between the water content and the pore space is considered to be hydrocarbons. Other methods for obtaining accurate core-fluid determinations attempt, as far as practical, to bring the reservoir conditions into the laboratory. One is to freeze the core quickly in its container as soon as it reaches the surface, and thereby keep in as much of its fluid content as possible until the core is examined in the laboratory. Another method is to remove the core from the reservoir under its natural pressure, by sealing it in the core barrel before lifting the barrel to the surface. Since this method is expensive, it is only occasionally used—chiefly in the early stages of the development of a pool. Truck-mounted laboratories, driven directly to the well being tested, eliminate many of the changes in core material that may occur between the well and the ordinary laboratory.

While the quantity of core and well-sample data available on the reservoir of an oil or gas pool may seem large, it must be remembered that the diameter of a hole is only from four to eight inches, so that the area of its cross section is infinitesimal compared with the area to be drained by a single well, which is normally from ten to eighty acres for an oil well and up to 160 acres or more for a gas well. For this reason measurements of the fluid content, permeability, and porosity of small core fragments, made in the laboratory, may fall far short of being correct when applied to the entire body of reservoir rock drained. Frequently the errors of measurement are compensating, however, so that it is often possible by the use of these data to make remarkably close approximations to the ultimate reserves of recoverable oil in a pool at an early date in its producing history.

Producing History of the Pool. Much knowledge of the reservoir's fluid content can be gained from study and interpretation of the changes that develop during the producing life of the pool, chiefly from the effects of fluid withdrawal and declining reservoir pressure. Some of the things that change are the relationships between fluids, especially the relative saturations with gas, oil, and water, which are affected by changing reservoir pressures; changing chemical composition and physical properties of the fluids; and the changing rates at which gas, oil, and water are produced. Daily, weekly, monthly, and yearly records of these changes supply the basic data regarding the fluid content of the reservoir.

The amount of fluid that may be contained in a rock depends upon its porosity and upon the pressure. The rate at which the fluids may be withdrawn depends on the rock's permeability, the reservoir pressure gradient, and the viscosity of the fluids. Three separate fluids exist in the reservoir; the gas and the oil are soluble in each other, but both are nearly immiscible with the water. The quantity of any one of these fluids present, and the rate at which it will flow through the rocks, is dependent not only on porosity, permeability, and reservoir pressure, but also on how much of each of the other fluids is present. Since water is present in all porous sediments, including petroleum reservoir rocks, it will be considered first.

Distribution of Gas, Oil, and Water in the Reservoir

The distribution of gas, oil, and water in the petroleum reservoir is based upon the interrelations of such factors as relative buoyancy, relative saturations of the pore space with each of the fluids, capillary and displacement pressures (see pp. 451–457), the hydrodynamic condition of the reservoir, and its porosity, permeability, and composition. In traps that contain gas, oil, and water, the fluids are roughly arranged in layers. Gas, being the lightest, fills the pores near the top of the trap. Below the gas is found a layer in which the pore-filling fluid is chiefly oil, and below that water alone occurs, the contact being the oil-water table. Where there is gas but no oil, the gas is immediately underlain by water, and the contact is the gas-water table. What has just been said applies to the characteristic fluid in each layer. Interstitial water, however, is present throughout the reservoir; it may occupy from a few percent up to 50 percent, but generally between 10 and 30 percent, of the pore space. It is not produced until the proportion of gas and oil to water has declined to the point where the reservoir rock has become more permeable to water than to the other fluids. The relations of gas, oil, and water in the average reservoir are shown diagrammatically in Figure 5-1, which presumes that the reservoir fluids consist of water, free gas, oil with dissolved gas, and oil. Some apparent exceptions to the ordinarily uniform layering of a petroleum reservoir occur. These exceptions are possibly explained by irregularities in porosity and permeability, local faulting, lensing, and other anomalous conditions, which usually cannot be detected by the available evidence.

Something about the trap, its geologic history and its relation to the oil and gas accumulation, may sometimes be learned from the nature of the oil-water contact in a pool. The Conroe field of Texas, for example (see Fig. 5-2), shows that the several producing sands of the Cookfield member of the Claiborne group (Eocene) are interbedded with shales, sandy shales, and sandy beds of low permeability. Despite this complexity, the oil-water contact is approximately level throughout the field, at the elevation of −4,990 feet (below sea level). The gas-oil contact ranges between −4,860 feet and −4,850 feet, but is occasionally as high as −4,835 feet. The uniformity of these fluid contacts indicates that there is a permeable interconnection between all of the sands and that hydrostatic conditions prevail. A section across the Ten Section field of California, on the other hand, suggests that each of the five sands within the Stevens zone (Miocene) is a separate reservoir, not

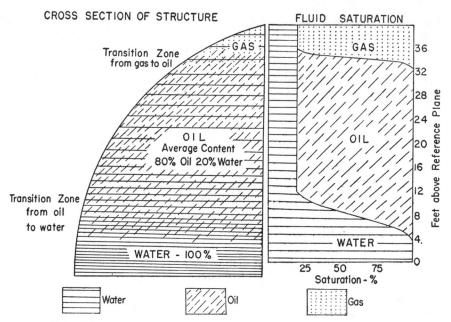

FIGURE 5-1 *Diagrammatic section to show the relative positions of gas, oil, and water in a typical petroleum reservoir.* [*Redrawn from Jersey-Humble report, Committee on Reservoir Development and Operations (1942), p. 12, Fig. 2.*]

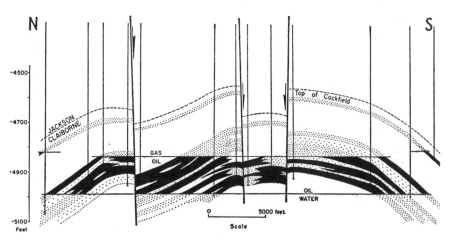

FIGURE 5-2 *Section through the producing formation of the Conroe field, Montgomery County, Texas. The common water-oil contact and gas-oil contact indicate a permeable connection between the various producing sands in the Cockfield member of the Claiborne (Eocene).* [*Redrawn from Michaux and Buck, in* Gulf Coast Oil Fields *(1936), p. 802, Fig. 5.*]

connected with the others, for each sand has its own oil-water table, and three of the sands originally had gas-oil contacts, each containing a cap of free gas at the top of the fold. The situation is shown in the structural section of Figure 5-3.

A section across the Van field, in eastern Texas, is especially revealing. (See Fig. 5-4.) The oil-water contact is at the same level in all of the fault blocks into which the dominant structure, a domal uplift overlying a deep-seated salt plug, has been divided. When the pool was discovered, however, there was a free-gas cap in the lower block marked A in the figure. This suggests that the accumulation occurred before the faulting; that the faulting lowered the previously higher part of the trap, block A, in relation to the remainder of the dome; that a permeable connection within the oil column between the various fault blocks permitted the oil-water table to adjust to a common plane throughout the pool while the gas-oil contact in the down-faulted block was protected by an impermeable roof-rock boundary, sealed by the underlying oil, and thereby left intact.[3] (See also Fig. 6-36, p. 268.) In the Salt Creek field of Wyoming each of the sands has a separate oil-water table, with no intercommunication, yet each oil-water table shows a tilt toward the north in accordance with the hydrodynamic gradient of the region. The relations are shown in Figure 5-5.

When the content of the entire pool or reservoir is considered in a broad

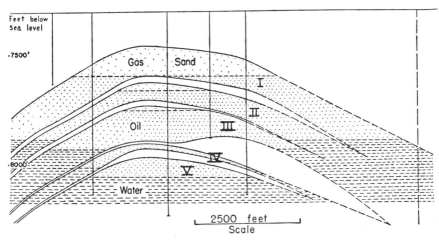

FIGURE 5-3 *Section across the Ten Section field, Kern County, California, showing within the Stevens zone (Miocene) five sands that produce oil and gas. The three upper sands produce both free gas and oil, while the two lower sands produce only undersaturated oil. The different water-oil contact levels and the different gas and oil content of each sand indicate that the sands are separate reservoirs and not interconnected.* [Redrawn from Wyatt and Baptie, Trans. Amer. Inst. Min. Met. Engrs., Vol. 132 (1939), p. 464, Fig. 3.]

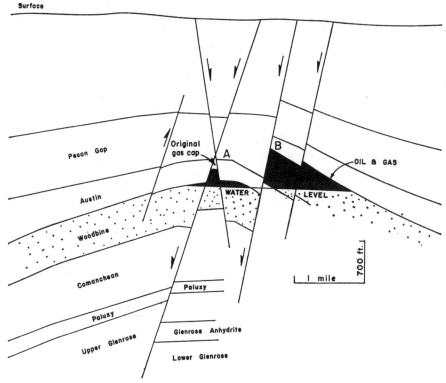

FIGURE 5-4 *Section through the Van field of Van Zandt County, Texas, showing the common water-oil contact that is found even though the field is broken up into a number of fault blocks, suggesting a deeply buried salt intrusion as the cause of the fold. (See Fig. 6-36, p. 268, for structural map.)* [Redrawn from Liddle, University of Texas Bull. 3601, Pl. 23 (1936).]

way, the contacts between the gas, oil, and water appear to be fairly well-defined plane surfaces, either level or inclined. These sharp boundaries are only apparent, however; when examined in detail, they are found not to be sharp, but to consist of transition zones extending through vertical distances of several feet—in some cases ten feet or more. (See Fig. 5-1 and pp. 563–567.)

WATER

The waters associated with oil and gas pools are called *oil-field waters*.[5] Wells that produce only water, or water with noncommercial oil and gas showings, when drilled into a potential reservoir rock (i.e., wells that miss finding an oil

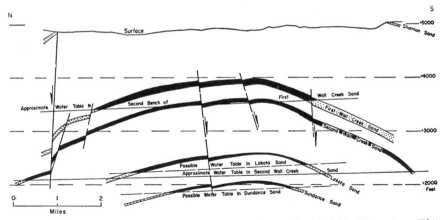

FIGURE 5-5 *Structural section through the Salt Creek oil field, Wyoming. This is a large dome fold with several producing sands in the Cretaceous and the Sundance sand (Jurassic). Each has a separate oil-water contact, indicating a lack of intercommunication. Also each oil-water contact is inclined toward the north in harmony with the hydraulic gradient (piezometric surface) of the region.* [Redrawn from Beck, in The Structure of Typical American Oil Fields (*1929*), *p. 600*.]

or gas pool), are called *dry holes, wet wells, dusters,* or *failures.* As noted previously, the lower edge or down-dip boundary of most oil and gas pools is marked by an oil-water or gas-water contact or boundary plane. The free water that bounds the pool and fills the pores below or around it is called *bottom water* or *edge water,* depending on its relation to the pool. The relationship is shown in Figure 5-6. As oil and gas production declines, most wells produce increasing amounts of free water. This water is interstitial water, bottom water, or edge water. In some pools water comes with the oil in the early stages of development, whereas in others appreciable water never comes up with the oil. Waters from overlying formations, above and separate from the oil or gas reservoir, are called *top waters.* Waters from water-bearing formations intervening between productive formations are called *intermediary waters.*

The Music Mountain pool in Pennsylvania, however (see pp. 298–300

FIGURE 5-6

Idealized section showing the position of bottom water and edge water in relation to the oil pool. A well is said to drill through the oil and into bottom water, or to drill outside the oil and into edge water.

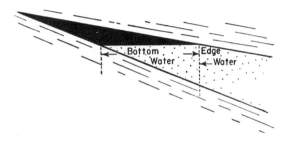

and Fig. 7-17, p. 300), has produced 5 million barrels of oil and is now nearly exhausted, yet has produced no free water. The reservoir is a sand body of the closed, shoestring type, and no information is available as to the interstitial water content. The Appalachian region contains other pools in which no edge water or free water has been encountered—in West Virginia, for example, the Cabin Creek field,[6] the Copely pool,[7] and the Griffithsville, Granny Creek, and Robinson Syncline pools.[8]

Classification of Oil-field Waters

Oil-field waters may be classified genetically into three groups, meteoric, connate, and mixed waters:

Meteoric water is the water that has fallen as rain and has filled up the porous and permeable shallow rocks, or percolated through them along bedding planes, fractures, and permeable layers. Waters of this type contain combined oxygen, chiefly as carbon dioxide. These waters are carried into the ground from the vadose zone above the groundwater table, where the oxygen reacts with the sulfides to produce sulfates, and the carbon dioxide reacts to produce carbonates and bicarbonates. The presence of carbonates, bicarbonates (hydrogen carbonates), and sulfates in an oil-field water suggests that at least some of the water had come from the surface. Such meteoric water may be related to the present ground surface and may reflect the mixing of ground water with oil-field waters, or it may be related to a buried unconformity surface that was once exposed to rain. The generally dilute nature of many Rocky Mountain oil-field waters, for example, is thought to be due to dilution by meteoric waters.[9]

Connate water was originally intended to mean the sea water in which marine sediments were deposited; presumably it originally filled all of the pores. It is doubtful, however, whether oil-field waters are actually the original water in place. The current usage is that "connate water is that interstitial water existing in the reservoir rock prior to the disturbance of the reservoir rock by drilling.[10] The reason for the change in usage is that most reservoir waters are quite different in chemical composition from sea water; they have undoubtedly circulated and moved and, in fact, have probably been completely replaced since the sediments were deposited. Most oil-field waters are brines, characterized by an abundance of chlorides, especially sodium chloride, and they often have concentrations of dissolved solids many times greater than that of modern sea water. This means, if the dissolved mineral content of the ancient seas was approximately the same as that of the present seas, that the original water has acquired some additional mineral matter since it entered the rock.

Mixed waters are characterized by both a chloride and a sulfate-carbonate-bicarbonate content. This suggests a multiple origin: presumably meteoric water mixed with or partially displaced the connate water of the rock. Mixed waters may occur near the present ground surface or may be found below unconformities.

Oil-field waters may also be classified, by their manner of occurrence, as *free water* and *interstitial water*.

Free Water. Most oil and gas pools occur in aquifers—that is, in water-saturated, permeable rocks. The water may be meteoric, connate, or mixed in origin. The water confined in the interconnected pores of a reservoir rock may be considered a continuous and interconnected body of water in which mineral particles have been deposited. It paves the highway along which petroleum has passed to concentrate into pools. Such water is confined, as in the water system of a city, and is ready to flow toward any point of pressure release. It is called free water to distinguish it from the interstitial, or attached, water.

Interstitial Water. It was originally thought that the entire pore space of a petroleum reservoir was filled only with oil or gas. As cores were examined, it was learned, however, that a variable amount of interstitial water coexisted with the oil and gas of every pool.[11] It is now believed that interstitial water has remained in the rock since the rock was deposited, and that it has clung so tightly to the rock surfaces that it was not displaced when the oil and gas accumulated. Interstitial water, in fact, is often referred to as "connate water," but the term "interstitial water" is preferred as it does not imply a knowledge of the water's origin. In a petroleum reservoir, most of the interstitial water is absorbed on the mineral surfaces or held in the finer capillary openings by capillary pressures. Interstitial water is present in all reservoirs and, as the water saturation increases toward the bottom of the pool, it grades into free water. The water displaced by the incoming oil and gas is free water.

The amount of interstitial water in an oil or gas reservoir is seldom less than 10 percent and ranges up to 50 percent or more of the pore space. The presence of interstitial water throughout practically all accumulations of oil and gas is reasonably well established.[12] It has been found at the highest point in a reservoir where the vertical oil column reaches as high as 2,000 feet above the oil-water contact. In cores from 600 feet above the oil-water contact in the Rangeley pool of Colorado almost 50 percent of the pore space is occupied by interstitial water.

The intimate association of interstitial water with oil and gas makes it of special importance in the accumulation and production of petroleum. The following are some of the effects of the interstitial water content of the reservoir.

1. There seems to be a general relationship between the nature of the porosity, permeability, and grain size of the reservoir rock and the amount of interstitial water present. Figure 5-7 shows how, in a number of reservoirs, the percentage of water generally increases with a decrease in permeability.

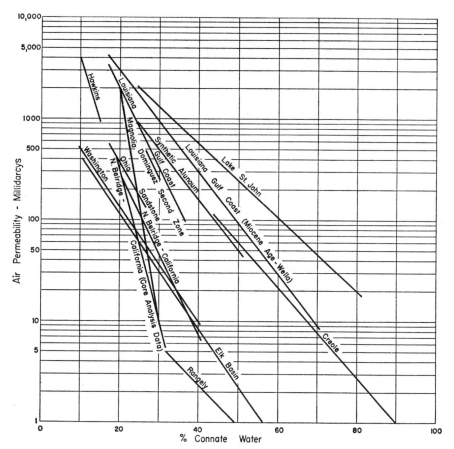

FIGURE 5-7 Relation between the percentage of interstitial water and the permeabilities of various oil reservoirs. [Redrawn from Bruce and Welge, in Drilling and Production Practice (*1947*), p. 170, Fig. 9.]

The amount of interstitial water generally increases also with a decrease in porosity. These phenomena may be accounted for in part by the fact that, since most sediments are preferentially water-wet (see pp. 445–451), more water is adsorbed on the finer sediments with their much greater surface area per unit volume. Furthermore, the capillary pressures holding the water in place are greater in the finer pores (see pp. 451–457). This relationship of high water content in the fine-grained sediments is also shown by Trask,[13] who finds that the initial amount of water when the sediments are being deposited approximates 45 percent of the pore volume in well-sorted, fine-grained sandstone, 50 percent in silts, 80 percent in clays, and more than 90 percent in colloids.

If a reservoir rock has variable porosity, ranging from fine to coarse, the greater percentage of interstitial water is commonly found in the smaller capillary openings and finer pores, characteristic of dirty or clayey sands, while the oil occupies the coarser pores. Two examples of Tertiary and Cretaceous reservoirs are shown in Figure 5-8 (see Fig. 10-11, p. 455, also). Adsorbed and interstitial water is not as important a factor in oil reservoirs where the pores are large or in reservoirs with fracture porosity. Crystalline limestones and dolomites are more likely than clean sandstones to have a high interstitial water content, and yet to produce only clean oil, because intercrystalline pore spaces are generally smaller.

2. The recoverable oil and gas reserve of a pool is reduced by the pore space that is occupied by water. A water-volume determination is therefore necessary before the pore space available for oil and gas is estimated.

3. Reservoir water, with its dissolved mineral content, is a chemical agent and may exert a profound chemical and physical influence on the min-

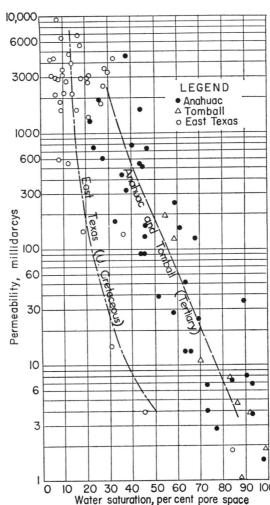

FIGURE 5-8

Chart showing the increase in water saturation with a decrease in permeability for the Woodbine sand (Upper Cretaceous) in the East Texas field and for the Tertiary sands in the Anahuac and Tomball fields of the Texas Gulf Coast. [Redrawn from Schilthuis, Trans. Amer. Inst. Min. Met. Engrs., reprinted Vols. 127 & 132, p. 269.]

eral content of the reservoir rock. Such an influence is especially strong on the clay and colloidal material of the rock, some of which may be quite unstable chemically. Over periods of geologic time, the relations between the water and the clays have become stabilized, and an equilibrium of the fluid content with its environment is reached. This equilibrium may be upset by any change in the character of the water content of the reservoir, as a result either of geologic phenomena or of water artificially injected by man during drilling or in secondary recovery operations. If reservoir conditions are modified, some of the clays may swell, permeability may be reduced or even blocked, production may decline, and the entire operation may suffer. Another result, almost wholly unknown but undoubtedly present, is the catalytic effect of the mineralized reservoir water on all the complex chemical and biochemical reactions that occur in the reservoir rocks.

4. Because of the adsorbed water film around the water-wet mineral grains, as shown in Figure 5-9, the oil in the pores of such reservoir rocks does not touch the rock walls, but is in contact only with the water; in such reservoirs, indeed, oil-rock interfacial relations do not exist, and only oil-water interfaces occur. The water, in effect, "lubricates" the oil.

5. Salt (NaCl), in the form of minute crystals, is common in most oil produced. It probably is to be explained, in part, as salt precipitated from the interstitial reservoir water as the fluid pressure in the reservoir drops during production. The remaining interstitial water would therefore have a somewhat different chemical content from the original interstitial water, and it might require some time to re-establish the equilibrium of the fluid content of the pore space. Decline in temperature and pressure between the reservoir and the surface would also permit the precipitation of salt from the water being produced. Salt precipitated from water after leaving the reservoir, however, would not affect the concentration of the remaining reservoir water.

6. Oil-field waters, with their dissolved salts, are electrolytes, and their electrical resistivity therefore decreases as their salinity increases. This phenomenon forms the main basis for the interpretation of the resistivity portion

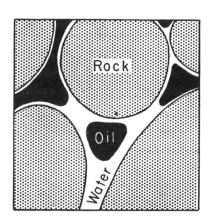

FIGURE 5-9

Idealized magnified section through sandstone showing the sand grains enclosed by a film of water and the oil with its dissolved gas occupying the inner spaces within the larger pores.

of the electric log. Rocks saturated with saline waters will generally have a low resistance because the high electrical conductivity of the interconnected pore water will overcome the low conductivity of the rock. Even in some reservoirs containing oil and gas with low conductivity, a high percentage of low-resistance interstitial water may overcome the high resistance of the oil and gas, and result in a misinterpretation of the reservoir content. Where the interstitial water saturation of an oil and gas reservoir is below 10 percent, which is rare, the electrical resistance of the formation approaches infinity. The measured resistivity is high because the amount of water is not sufficient to provide continuity, and therefore electrical conductivity, between the pores, for the water is in isolated droplets. As the water saturation increases, more water is available to form connected films around the mineral grains, and the water droplets make contact with one another; thus the interconnected water forms conducting paths for electricity extending out in all directions.

Character of Oil-field Waters

The character of oil-field waters is generally determined by three measurements: (1) The amount of interstitial water in a petroleum reservoir is commonly measured in percentage of effective pore space and is also known as the *water saturation*. (2) The total mineral solids dissolved in the water are commonly measured in parts per million or in density of the water. (3) The mineral constituents dissolved in the water are chemically analyzed, commonly both qualitatively and quantitatively.

Water Saturation. Two general methods of measuring the interstitial water saturation are common: (1) the laboratory analysis of cores, and (2) the calculation of the water saturation in place in the reservoir by use of the formation factor and the electric log.

1. The laboratory analysis is generally made by heating the samples and distilling off the water and oil, collecting and weighing the condensed liquids, and applying the amounts extracted to the porosity of the sample for a percentage of the porosity that is oil and water. The result is only approximate, for the core is contaminated by drilling water, and some of its own water was removed when gas came out of solution and left it as the pressure dropped from that of the reservoir to that of the surface. The reader is referred to the large amount of literature on the subject of core analysis, a few articles on which are listed.[14]

2. The *formation resistivity factor*, also called the *formation factor* (F), is a useful measurement in analyzing reservoir fluids, and is defined as the electrical resistance of a rock that is saturated with a conducting electrolyte, such as a brine, divided by the resistivity of the electrolyte.[15] When the resistivity factor of sandstone reservoirs is plotted against the porosity on a log-log grid, it is seen to increase as the porosity decreases. (See Fig. 5-10.) The

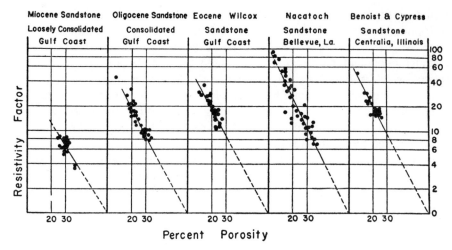

FIGURE 5-10 *Relation of porosity to the formation resistivity factor for several producing sands. [Redrawn from Archie, Bull. Amer. Assoc. Petrol. Geol., Vol. 34 (1950), p. 956, Fig. 8.]*

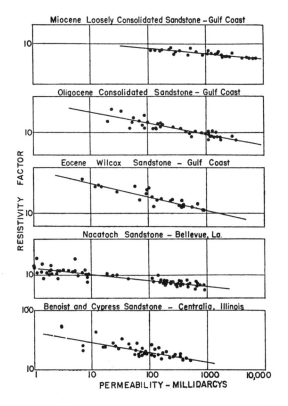

FIGURE 5-11

Relation of permeability to the formation resistivity factor for several producing sands. [Redrawn from Archie, Bull. Amer. Assoc. Petrol. Geol., Vol. 34 (1950), p. 957, Fig. 9.]

porosity in limestone reservoirs shows a similar relation to the formation resistivity factor.[16] There is likewise a linear relation between permeability and the formation resistivity factor; the formation resistivity factor increases as the permeability decreases. The relations are seen in Figure 5-11.

The relation of the formation factor to the resistivity of the rock and the brine may be expressed as

$$R_t = FR_w$$

where R_t is the true resistivity of the rock with its pores saturated with brine, F is the formation factor, and R_w is the resistivity of the brine. F also may be expressed as

$$F = P^{-m}$$

where P is the porosity and m is the cementation factor (m generally ranges between 1 and 3 for noncemented and highly cemented rocks respectively). (See Fig. 5-12.)

One important application of the electric log is a quantitative or semiquantitative estimate of the relative amounts of oil, gas, and water that are present in the logged formation from an examination of the self-potential and the resistivities obtained. The relations are expressed as

$$R_t = \frac{R_w}{P^m \times S^n} \quad \text{or} \quad S = \sqrt[n]{\frac{FR_w}{R_t}}$$

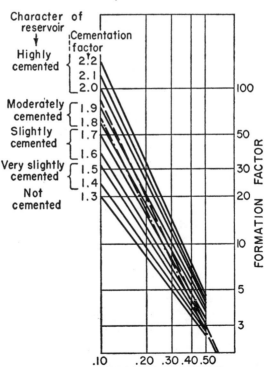

FIGURE 5-12

Relation of porosity to the formation factor for sands of differing cementation character. For example, a sand with a porosity of 15 percent may have a formation factor of 12 if unconsolidated; but if it is highly cemented, the formation factor would rise to 60. [Redrawn from Martin, O. & G. Jour., June 29, 1953, p. 111.]

The problem is to obtain the water saturation, S, as the rest of the pore space will be filled with oil or gas or both. The true resistivity, R_t, can be obtained from the log in ohm-meters by using tables; the porosity, P, can be obtained from core analysis, from a MicroLog or a Microlaterolog, from a Sonic log, or from the SP curve; the resistivity of the formation water, R_w, can be obtained from chemical analyses of the water from the same or similar sands, from the SP log, or from tables;[17] the cementation factor, m, and the saturation factor, n (ranges between 1.9 and 2.0), are generally constant for the same formation, and will be learned after some experience in the field. Under favorable circumstances, therefore, the value of S can be obtained; where it is low (below 40 percent), the interpretation is that the oil or gas saturation is high (enough to make an oil or gas producer). Re-examination of logs in the light of this equation has resulted in a number of pool discoveries.

Total Dissolved Salts, or Concentration. The total amount of mineral matter commonly found dissolved in oil-field waters ranges from a few hundred parts per million (ppm)* in practically fresh water up to 300,000 ppm in a heavy brine. Higher concentrations are recorded; the highest is that noted by Case,[18] a brine (Spec. Gr. 1.458) from the Salina dolomite (Silurian) of Michigan that contained 642,798 ppm, or 64.3 percent, of dissolved mineral matter. The dissolved mineral matter in sea water, as shown in Table 5-2, page 166 is about 35,000 ppm (3.5%). Some oil-field waters, particularly those of meteoric origin, and those in areas where the piezometric surface slopes considerably, contain less dissolved mineral matter than sea waters; other contain more. The relative concentration of different minerals also varies greatly from that of sea water.

The density† of water increases with the addition of salts in solution. An approximate relation between the specific gravity of an average brine and the total mineral matter in solution is shown in Table A-1 in the Appendix.

This table of the density of water provides a useful approximation of the amount of mineral matter when no other concentration determinations are known. Another approximate approach to the total dissolved mineral content is used in the Chester sand (Upper Mississippian) waters of Illinois, where the chloride values run approximately 60 percent of the total solids.[19] An example of the plotting of the approximate compositions of brines may be seen in Figure 5-13, a graph for rapidly estimating the composition of a Kansas brine from a determination of its density.

---

* Any unit volume of solution is taken to have a million parts of weight (million parts of weight = 100%). The number of parts of weight contributed to the solution (the % weight $\times$ 1,000,000) is the parts by weight per million (ppm). The concentration of dissolved mineral matter is usually recorded as ppm, as milligrams per liter (mg/liter), as grains per gallon (gr/gal), or as a percentage (%). One grain per U.S. gallon equals 17.12 ppm and per Imperial gallon equals 14.3 ppm. The value in parts per million multiplied by 0.0583 equals the number of grains per U.S. gallon.

† The density of pure water at standard conditions of 14.73 psi and 60°F is 62.34 pounds per cubic foot. This equals 0.433 pound per square inch per foot, or 43.33 psi/100 feet. Pure water weighs 350 pounds per barrel of 42 gallons. The specific gravity of pure water at 14.73 psi and 60°F is 1.00 (API 10°).

RESERVOIR FLUIDS—WATER, OIL, GAS [CHAPTER 5]   161

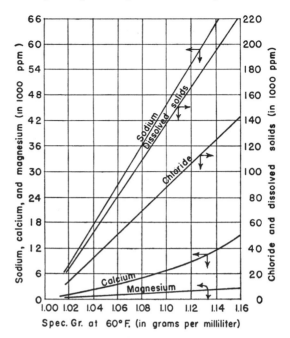

FIGURE 5-13

Graph for estimating the composition of a Kansas oil-field water from the specific gravity. [Redrawn from Jeffords, Bull. 76, Kansas Geol. Surv. (1948), Part I, p. 11, Fig. 6.]

Low-concentration waters generally have ionic analyses similar to those of high-concentration waters. This similarity suggests that the former are dilute waters and that the low concentrations may result from the infiltration of meteoric waters, either under the present surface conditions or under pre-unconformity geologic conditions. The presence of fresh or low-concentration waters in some places indicates a permeable connection with a surface intake or recharge area. Such a relation suggests, in turn, that hydrodynamic conditions might well have caused oil and gas to be flushed out of some structural traps. However, there are many oil fields throughout the world in which fresh or low-concentration water is produced along with the oil, and the presence of such water does not preclude the presence of oil or gas in suitable traps: for example, the Quirequire field of eastern Venezuela[20] (see p. 335 and Fig. 7-57), where the concentration ranges from a trace to 2,300 ppm and averages about 1,400 ppm; the Las Cruces field of western Venezuela, where the water is fresh, with only 323 ppm of salt in solution, and is suitable for domestic purposes;[21] and many pools in the Rocky Mountain region of the United States.[22]

Regional changes in the water of a formation are readily seen in *isoconcentration*, or *isocon*, maps, which, by contouring points of equal concentration, show the changing concentration of water in a continuous sand.[23] An example is shown in Figure 5-14, which is an isoconcentration map of the St. Peter–Simpson sands (Ordovician) of the Central States of the United States.*

---

* A number of isocon maps may be seen in the work cited in reference note 18, Figs. 4-13.

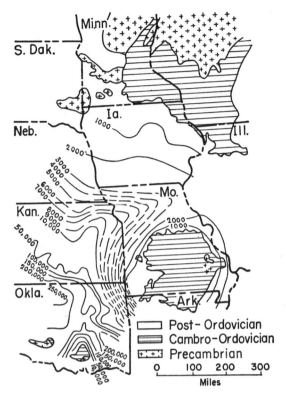

FIGURE 5-14

An isoconcentration map of the St. Peter sand waters of the central states of the United States. The values contoured are in ppm. [*Redrawn from Dott and Ginter, Bull. Amer. Assoc. Petrol. Geol., Vol. 14 (1930), p. 1216, Fig. 1.*]

Isocon maps also furnish rough checks of the information given by potentiometric surface maps on the general direction of water movement between the areas of intake and the areas of outlet of a reservoir rock.

The formation waters of most sedimentary areas increase in mineral concentration with depth. This increase may be due to the fact that salt water is heavier than fresh water and that, whenever salt water is elevated above fresh water by diastrophism, the heavier water eventually finds its equilibrium position as low as possible in the aquifer. Or, more likely, it may be due to the longer exposure of the deeper waters to phenomena that increase the concentration. An exception to the increase in concentration with depth is found in some of the fields in the Baku area and in the New Grozny oil field of the USSR. The upper waters of the New Grozny area are salty, with concentrations on the order of 60 grams per liter (60,000 ppm), whereas the waters in the deeper oil reservoirs have concentrations ranging from 0.80 to 5.0 grams per liter (800 to 5,000 ppm).[24] Another exception to the orderly increase in concentration with depth is in Kansas (see Table 5-4, p. 168), where the Pennsylvanian waters have concentrations of 10,000–125,000 ppm, whereas the underlying pre-Pennsylvanian Ordovician waters have concentrations of 5,000–35,000 ppm. The reason for the decrease there may be that the dilution occurred during the pre-Pennsylvanian erosion and that a permeability barrier to fluid migration exists across the unconformity. The sudden change in concentration marks the position of the unconformity. The existence of an unconformity may be the correct explanation, in other areas, of sudden decreases in concentration with depth. Another explanation for some regions, including Kansas, is that the lower formation waters have better access to the recharge areas and to dilution by meteoric waters than do the shallower formations.

Chemical Composition of Oil-field Waters. Most oil-field water analyses[25] are given in the ionic form, which usually is considered best, and the "Palmer System" [26] of interpretation is based on the ionic statement. Palmer groups together radicals that are either chemically similar or geologically associated. Thus the common metallic bases sodium (Na^+) and potassium (K^+), and the alkaline earths calcium (Ca^{++}) and magnesium (Mg^{++}), are grouped as alkalies, and all are positive ($+$) radicals. The acids or negative ($-$) radicals consist of two groups—the strong acids, sulfates ($SO_4^=$), and chlorides (Cl^-), and the weak acids, carbonates ($CO_3^=$), and bicarbonates (HCO^-).

The character of a brine may be described, in terms of its *reaction value*,* by the combination of four properties. These are:

(1) *Primary salinity.* Strong acids [sulfate ($SO_4^=$), nitrate (NO_3^-), and chloride (Cl^-)] combined with primary bases [sodium (Na^+) and potassium (K^+)].

(2) *Secondary salinity.* Strong acids combined with secondary bases [alkaline earths—calcium (Ca^{++}), barium (Ba^{++}), and magnesium (Mg^{++})]. Secondary salinity is also known as permanent hardness.

(3) *Primary alkalinity.* Weak acids [carbonate ($CO_3^=$), bicarbonate (HCO_3^-), and sulfide (S_2)] combined with the primary bases. Waters of primary alkalinity generally contain silica also. These waters are termed soft.

---

* The reaction values of the various ions in oil-field water are not in proportion to their various weights. The reaction value for each ion may be expressed in milligrams per liter or as a percentage of the sum of all the reaction values in the analysis. The reaction value is found by the following formula:

reaction value = amount by weight (mg/liter) × reaction coefficient[27]

$$= \text{amount by weight (mg/liter)} \times \frac{\text{valence}}{\text{atomic weight}}$$

= equivalents per million (epm)

= milliequivalents per liter (meq/l)

When the reaction value of a water is reduced to percentages, the character of the water is indicated without use of the concentration. The reaction coefficients for the ions usually determined in water analysis are:

Sodium (Na^+)	0.0434	Sulfate (SO_4^-)	0.0208
Potassium (K^+)	0.0256	Chloride (Cl^-)	0.0282
Calcium (Ca^{++})	0.0499	Nitrate (NO_3^-)	0.1061
Magnesium (Mg^{++})	0.821	Carbonate (CO_3^-)	0.0333
Hydrogen (H^+)	0.992	Bicarbonate (HCO_3^-)	0.0164

(4) *Secondary alkalinity.* Weak acids combined with secondary bases. Secondary alkalinity is characteristic of calcareous formations. It is also known as temporary hardness.

Of these properties, primary salinity and secondary alkalinity will always be present. If the strong acids exceed in amount the primary bases, the third property will be secondary salinity; if the strong acids do not exceed the primary bases in amount, the third property will be primary alkalinity.

The ionic statement and the derivation of the Palmer properties, as applied to oil-field water analyses, are illustrated in Table 5-1. It may be noted that the weights of positive and negative ions do not balance, but that the reaction

TABLE 5-1 Ionic Statement and Reaction Properties

	Milligrams per Liter	Equivalents per Million (epm) or Reaction Value	Reaction Value in Percentage (Palmer)
Positive Ions			
Na^+ and K^+ (by difference as Na^+)	17,000	765.0	38.67
Ca^{++}	2,960	148.2	7.48
Mg^{++}	927	76.2	3.85
Total	21,497	989.4	50.00
Negative Ions			
SO_4^-	2,620	54.5	2.75
Cl^-	34,000	932.0	47.10
HCO_3^-	177	2.9	.15
Total	36,797	989.4	50.00
Grand Total	58,294	1,978.8	100.00
REACTION PROPERTIES (PALMER)			
Primary salinity (Na^+ and K^+ chlorides and sulfates)			$38.67 \times 2 = 77.34$
Secondary salinity (Ca^{++} and Mg^{++} sulfates)			$(47.10 + 2.75 - 38.67) \times 2 = 22.36$
Primary alkalinity (Na^+ and K^+ carbonates)			00.00
Secondary alkalinity (Ca^{++} and Mg^{++} carbonates)			$(11.33 - 11.18) \times 2 = 0.30$

Source: Data from L. C. Case, "Application of Oil Field Water to Geology and Production," *Oil Weekly,* October 29, 1945, pp. 48–54.

values of positive and negative ions are exactly equal. Comingling of waters may cause precipitation or scale formation in casing and tubing.

In addition to the common elements sodium, potassium, calcium, and magnesium, minor amounts or traces of various elements have been found when complete analyses are performed. These minor amounts are erratic in quantity and are not ordinarily determined in the average water analysis. The minor elements include barium, strontium, iodine, bromine, boron, copper, manganese, silver, tin, vanadium, and iron. Barium, for example, seems to be found in many of the Paleozoic brines of the Appalachian region. It is thought that the barium appeared in the sediments originally as barite, which was precipitated in reactions induced by the meteoric waters entering the formations along their outcrop.[28] Radioactive salts have also been precipitated from oil-field brines.[29]

A few representative examples of oil-field water analyses, together with an analysis of an average sea water, are given in Table 5-2. Most descriptions of oil fields contain analyses of the reservoir waters, and numerous lists of oil-field water analyses have been published.[30]

When the water analyses are given in hypothetical combinations, it becomes necessary to reduce these to the ionic form. The factors used for calculating the amount of the positive radical in various salts are given in Table 5-3.

Water analyses and total solids show considerable variation from sand to sand even in the same well. Table 5-4 shows some of the differences in one well in Russell County, Kansas, in which Permian rocks at the surface are underlain by rocks of Pennsylvanian age, which rest unconformably on the Arbuckle limestone of Cambro-Ordovician age.

A variety of diagrams have been devised by which the chemical analysis of an oil-field water may be plotted so that its character can be readily visualized and the comparison with other waters easily seen. This is effective when plotted at the points of occurrence on stratigraphic and structural cross sections. Three such diagrams are shown in Figure 5-15. Part A shows the widely used Tickell method.[31] Its chief disadvantage is that the concentrations are not indicated. Part B shows the Parker method,[32] which has the advantage of concentrating more detail into a small space than most diagrams. Part C shows the Stiff method,[33] which has the advantage of showing the salt concentrations so that dilution effects are reduced, and of making a distinctive pattern by which the different water types can be readily distinguished. The unit milliequivalents per liter is used.

Two chemical differences between ordinary sea water and oil-field brines are (1) the absence of the sulfate radical ($SO_4^=$) from some oil-field brines and its presence in sea water, and (2) the absence of alkaline earths (Ca and Mg) from some oil-field waters and their presence in sea water.[34] The absence of sulfate from the waters of the San Joaquin Valley, California, has been explained by its reduction to sulfide, with a corresponding formation of carbonate,[35] and this is also believed to explain in part, at least, the sulfate-

TABLE 5-2 Representative Oil-field Water Analyses (ppm)

Pool	Reservoir Rock, Age	Cl⁻	SO_4^-	CO_3^-	HCO_3^-	$Na^+ + K^+$	Ca^{++}	Mg^{++}	Total ppm	References
Sea water, ppm		19,350	2,690	150		11,000	420	1,300	35,000	1
Sea water, percent		55.3	7.7	0.2		31.7	1.2	3.8		
Lagunillas, western Venezuela	2,000–3,000 ft Miocene	89	—	120	5,263	2,003	10	63	7,548	2
Conroe, Texas	Conroe sands Eocene	47,100	42	288		27,620	1,865	553	77,468	3
East Texas	Woodbine sand U. Cretaceous	40,598	259	387		24,653	1,432	335	68,964	4
Burgan, Kuwait	Sandstone Cretaceous	95,275	198	—	360	46,191	10,158	2,206	154,388	5
Rodessa, Texas-La.	Oolitic limestone L. Cretaceous	140,063	284	—	73	61,538	20,917	2,874	225,749	6
Davenport, Okla.	Prue sand Pennsylvanian	119,855	132	—	122	62,724	9,977	1,926	194,736	7
Bradford, Penn.	Bradford sand Devonian	77,340	730	—	—	32,600	13,260	1,940	125,870	8
Oklahoma City, Okla.	Simpson sand Ordovician	184,387	268	—	18	91,603	18,753	3,468	298,497	9
Garber, Okla.	Arbuckle limestone Ordovician	139,496	352	—	43	60,733	21,453	2,791	224,868	10

1. W. Dittmar, "Report on Researches into the Composition of Ocean Water, Collected by H. M. S. Challenger," *Challenger Reports*, Vol. 1, *Physics and Chemistry* (1884), pp. 1–251. Average of 77 water samples representative of all oceans.

2. Staff of Caribbean Petroleum Company, "Oil Fields of Royal Dutch-Shell Group in Western Venezuela," Bull. Amer. Assoc. Petrol. Geol., Vol. 32 (April 1948), p. 557.

3. Frank W. Michaux, Jr., and E. O. Buck, "Conroe Oil Field, Montgomery County, Texas," in *Gulf Coast Oil Fields*, Amer. Assoc. Petrol. Geol., Tulsa, Okla. (1936). p. 810. Sun Company Stewart No. 3.

4. H. E. Minor and Marcus A. Hanna, "East Texas Oil Field, Rusk, Cherokee, Smith, Gregg and Upshur Counties, Texas," in *Stratigraphic Type Oil Fields*, Amer. Assoc. Petrol. Geol., Tulsa, Okla. (1941), p. 639. Stanolind No. 1 Everetts.

5. By the courtesy of the Gulf Oil Corp. A representative analysis.

6. H. B. Hill and R. K. Guthrie, *Engineering Study of the Rodessa Oil Fields in Louisiana, Texas, and Arkansas*, RI 3715, U.S. Bur. Mines (August 1943), p. 90, No. H.

7. Stanley B. White, "Davenport Field, Lincoln County, Oklahoma," in *Stratigraphic Type Oil Fields*, Amer. Assoc. Petrol. Geol., Tulsa, Okla. (1941), p. 403. Texas Co., Patterson No. 1.

8. Jerry B. Newby et al., "Bradford Oil Field, McLean County, Pennsylvania, and Cattaraugus County, New York," in *Structure of Typical American Oil Fields*, Amer. Assoc. Petrol. Geol., Tulsa, Okla., Vol. 2 (1929), p. 435.

9. H. B. Hill, E. L. Rawlins and C. R. Bopp, "Engineering Report on Oklahoma City Field, Oklahoma," RI 3330, U.S. Bur. Mines (January 1937), p. 214, Analysis J. Carter Oil Co. Dunniven No. 1, at 6,454 feet.

10. Wesley G. Gish and Raymond M. Carr, "Garber Field, Garfield County, Oklahoma," in *Structure of Typical American Oil Fields*, Amer. Assoc. Petrol. Geol., Tulsa, Okla., Vol. 1 (1929), p. 191. Cosden-Marland No. 41, at 4,383 feet, flowing approximately 10,000 barrels of water per day.

TABLE 5-3 Factors Used to Convert Hypothetical Combinations to Ionic Form (Water Analyses)

Given	To Find	Factor	Given	To Find	Factor
KCl	K	0.524	$CaSO_4$	Ca	0.294
NaCl	Na	0.394	$MgSO_4$	Mg	0.202
$CaCl_2$	Ca	0.361	K_2CO_3	K	0.569
$MgCl_2$	Mg	0.255	Na_2CO_3	Na	0.434
K_2SO_4	K	0.449	$CaCO_3$	Ca	0.400
Na_2SO_4	Na	0.324	$MgCO_3$	Mg	0.288

Source: F. G. Tickell, "A Method for the Graphical Interpretation of Water Analyses," Summary of Operations, California Oil Fields, Vol. 6, No. 9 (1921), p. 7.

free waters of the Upper Cretaceous formations of the Rocky Mountains.[34]

The sulfate may be reduced either by living organisms, such as bacteria, or by inanimate organic matter.[36] Whether the reduction took place during the diagenesis of the sediments or is taking place now in the oil-field brines is still an unsettled question.

Where sulfate waters are present in the Rocky Mountains, as in the Carboniferous rocks, they are frequently associated with naphthenic base crude oils, the so-called black oils, and hydrogen sulfide (H_2S) is commonly present. This suggests either that active reduction of sulfate is now taking place in the more deeply buried formations, or that it may have occurred during

TABLE 5-4 Analysis of Water Sands from One Well, Russell County, Kansas

Approximate Depth in Feet	mg per liter, or ppm		
	Cl	SO_4	Total Solids
70	50	48	419
250	2,300	510	5,000
360	3,240	720	7,000
600	28,250	7,500	59,000
780	33,500	9,000	68,000
2,510	97,000	1,600	157,000
2,625	98,600	1,100	160,000
2,990	88,000	nil	141,000
3,300[1]	11,000	300	20,000

Source: Adapted from L. C. Case, Oil Weekly, October 29, 1945, p. 54.
[1] Arbuckle oil zone. This is the only water in the section containing H_2S.

FIGURE 5-15

Examples of graphs used to show the chemical analysis of oil-field waters. These are useful in making quick visual comparisons, as on logs or maps. [A, Redrawn from F. G. Tickell, Summ. of Oper., Calif. Div. Mines, Vol. 6, No. 9 (1921), p. 10, Fig. 3. B, Redrawn from J. S. Parker and C. A. Southwell, Jour. Inst. Petrol. Technol., Vol. 15 (1929), p. 158, Fig. 5. C, Redrawn from Henry A. Stiff, Tech. Note 84, Trans. Amer. Inst. Min. Met. Engrs., Vol. 192 (1951), p. 377, Fig. 3.]

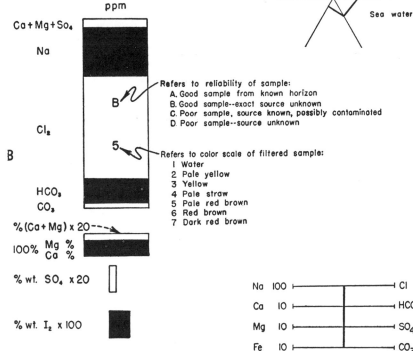

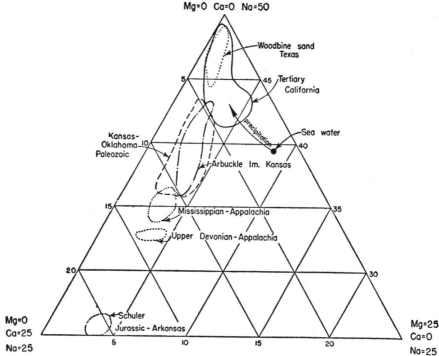

FIGURE 5-16 *The composition of various groups of oil-field waters in reaction-value percentages of the dissolved salt of* $Na^+ + K^+$, Ca^{++}, *and* Mg^{++}. *The high* Mg^{++} *content (8.5%) of sea water contrasts strongly with the low* Mg^{++} *content of the oil-field waters (generally 2–5%). The nearest approach to sea water is in the analyses of the youngest Pliocene waters, which apparently represent sea waters in an early state of evaporation.* [*Redrawn from de Sitter, Bull. Amer. Assoc. Petrol. Geol., Vol. 31 (1947), p. 2033, Fig. 1.*]

some earlier erosion period now marked by an unconformity. The removal of the calcium and magnesium from the Cretaceous waters in the Rocky Mountains is thought to be due to ion exchange with the bentonites.[37]

Some of the relations between oil-field waters are shown in Figure 5-16, which gives the composition of the dissolved salts in percentage reaction values of $Na^+ + K^+$, Ca^{++}, and Mg^{++}. The normal evaporation direction of sea water, shown by the solid line, trends directly into the youngest of the Tertiary water analyses, suggesting that these waters represent the early stages in the diagenesis of sea water.[38] This early phase is marked by the precipitation of magnesium and calcium sulfates and carbonates. The concentrations of oil-field brines compared with the Na^+ reaction values are shown in Figure 5-17. Here the theoretical change in the concentration of sea water, when all other salts than NaCl are extracted, is shown by the heavy line. The young

Tertiary water analyses of California follow this line and again suggest that these oil-field waters represent "fossil" sea waters in an early stage of their diagenesis. A later stage is an increasing concentration with a gradual increase in the calcium and magnesium ions. The low concentration of the Arbuckle waters of Kansas may be attributed to their dilution during the exposure to pre-Pennsylvanian erosion of the Arbuckle limestone or may be due to current hydrodynamic phenomena.

Uses of Water Analyses. Water analyses are used in a variety of problems connected with the exploration and development of oil and gas pools.[39] Some of these problems are:

1. Probably the most important geologic use of oil-field water analyses is their application to the interpretation of electric well logging. Electrical resistivities of oil-field waters in ohm-meters are generally determined at several

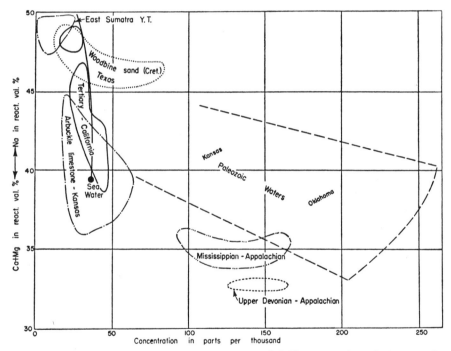

FIGURE 5-17 *The concentration of various oil-field waters in relation to the Na^+ values. The heavy line extending upwards from the position of sea water represents the change in concentration of sea water as all other salts than NaCl are gradually extracted, thus decreasing from the original 3.5 percent concentration to a theoretical 2.9 percent. The proximity of the younger Tertiary waters to this line and the increase in concentration with age are noticeable.* [Redrawn from de Sitter, Bull. Amer. Assoc. Petrol. Geol., Vol. 31 (1947), p. 2034, Fig. 2.]

temperatures when the chemical analysis is made.[40] Resistivities that have not been determined may be calculated from the mineral analyses of the waters.[41]

2. Reservoirs in multipay fields may frequently be distinguished and correlated by their water analyses. This is particularly useful in lenticular sand reservoirs. The detailed correlation of formations by water analysis is rarely dependable except locally. Regionally, however, certain formations may have certain general diagnostic characteristics that identify them and thereby aid in subsurface correlation. Thus the post-Chugwater (Triassic) waters of the Rocky Mountains are essentially solutions of sodium salts, whereas calcium and magnesium sulfates predominate in pre-Chugwater waters.[42]

3. Radical changes in the concentrations or other characteristics of a series of waters passing from a shallow to a deeper formation indicate a different geologic environment. In the absence of mechanical leaks, a sudden reduction in the concentration of the brine may suggest the crossing of an unconformity. Increase in salinity toward one side of a sedimentary basin and a decrease in salinity toward another side suggests the existence and location of a recharge or intake area and gives indirect evidence of the direction of fluid flow.

4. Water analyses may tell whether the water produced with the oil comes from the bottom of the well and is part of the formation water of the reservoir rock, or whether it is shallower formation water entering the well because of improper cementing of the casing or leaks due to breaks in the casing.

5. An increase in the salinity of shallow ground waters has been noted

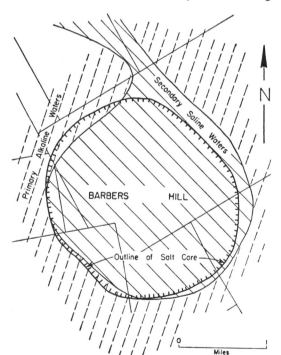

FIGURE 5-18

Map showing the reflection of a buried salt dome in the shallow ground water typical of a number of domes along the Gulf Coast. The example is the Barbers Hill dome in Chambers County, Texas. [Redrawn from Minor, in The Structure of Typical American Oil Fields, Vol. 3 (1934), p. 894, Fig. 2.]

within areas affected by salt domes. The effect on ground waters overlying the salt dome at Barbers Hill, Texas, where an inlier of secondary salinity is surrounded by normal waters showing primary alkalinity,[43] is shown in Figure 5-18. Anomalous water occurrences such as these have aided in prospecting for buried salt domes.

6. It is necessary to know the effect of injected water, both on the minerals in the reservoir rock and on the equipment used in water flooding and salt-water disposal. This calls for water analysis.

7. Corrosion of equipment, particularly by the presence of H_2S, is common. "Hard scale," composed of sulfates of barium, strontium, and calcium, or "soft scale," composed mainly of calcium carbonate, forms when unlike waters are mixed. The presence of scale within the oil-field equipment is often explained, therefore, by a leaky casing. "Soft scale" is mainly due to the loss of CO_2 from the water, which in turn is due to pressure reduction.

Origin of Oil-field Brines

We see that oil-field waters differ greatly from modern sea waters, both in the amounts of the dissolved salts and in the chemical composition of the salts. Yet reservoir rocks are presumed to have been deposited in sea water. A number of explanations have been offered to account for the changes, several of which are discussed below.

1. The most logical explanation of the high mineral concentration of many oil-field waters is in the phenomena associated with the adsorption of water to the clay mineral particles of both shales and reservoir rocks.[44] Calcium clays carried into the oceans by streams exchange their calcium ions for sodium ions through ion exchange. These sodium ions, along with other elements, become loosely attached, as unsaturated ions, to the corners and edges of the broken clay-building units. Saline water adsorbed on these unsaturated ions is called *broken-bond water* (most of the adsorbed water on illite is of this type). The capacity of a clay to adsorb broken-bond water is increased as particle size diminishes. Ions adsorbed to satisfy unsaturated clay ions could be either positive or negative, depending on the charges on the clays. Thus both Na^+ and Cl^- and other components of various salts could be adsorbed and concentrated on the clay particles.

These adsorbed ions may be freed in part by clay recrystallization, by cementation, and by diastrophism. Pressure increases the power of adsorption, whereas heat decreases it. Both pressure and heat increase the solubility of salts in water. The net result that might be expected from these various phenomena is to transfer the adsorbed ions to the water solution as it is squeezed out of the clay during burial and as it enters the more permeable formations.

2. The evaporation of water during deposition of sediments in enclosed basins would leave a heavier and more concentrated brine behind, thus possibly accounting for some of the observed high saline content. Wind-blown

salts from neighboring desert regions may have added salt to enclosed basins or semienclosed arms of the sea that later became buried and furnished a water of primary high salinity. While such factors might explain a primary source of wide salinity variation, they do not explain the high-salinity water in formations that exhibit no evidence of salt deposition, a relationship that would normally be expected.

3. It has been suggested that as pressures decrease, free gas expands and water evaporates into the gas, thus concentrating the remaining brines.[45] There is abundant evidence that the precipitation of salts and minerals occurs along with the reduction of reservoir pressure that accompanies production of gas and, to a lesser extent, of oil. However, the amounts of gas necessary to effect the high concentrations that frequently occur in reservoir rocks are so enormous as to discount this as a source of any substantial concentration of salts in oil-field brines.[46]

4. The numerous unconformities found in the geologic columns of most sedimentary areas might explain local variations in the saline content, since each represents a boundary change in geologic conditions and permits a mixing of waters. Some unconformities mark positions in the geologic sequence where meteoric waters entered the rocks and diluted the pre-existing waters. Other surfaces of unconformity may represent periods when the evaporation of water left salt to be reconcentrated by winds and currents into waters that later entered the underlying rocks or when the underlying rocks were otherwise exposed to waters of high salinity.

5. If the reservoir waters of high salinity are connate, the high salinity may be explained by a high original saline content of the sea water; but there is, in fact, little evidence of the composition of the geologically ancient sea waters. Volcanic dust or submarine volcanic gases and debris, for example, may have entered the ocean in large quantities at various times in the geologic past, thereby giving rise to a primary but local variation in the character and concentration of the water, and this may explain some of the observed occurrences. The variability of the waters within the same geologic formation casts doubt on this as a valid general explanation, but it may help explain some characteristics.

6. Most oil-field brines contain chiefly sodium chloride; hence the common name "salt water." A notable exception is in the South Mountain and Shiells Canyon pools of California, where the brine contains chiefly calcium chloride.[47] The presence of this salt has been attributed to modifications of the waters after their original burial within the sediments. Suggested causes are reactions with the oil; reactions with the sediments of the enclosing Sespe formation (Eocene-Oligocene), which is of continental origin; reactions with volcanic matter that may have been deposited with the Sespe rocks or with igneous bodies intruding the Sespe formation; and reactions with lake waters, the calcium chloride brines of which may have been concentrated by evaporation during the deposition of the sediments.

OIL

Only a minute fraction of the fluids of the reservoir rock is petroleum, but to petroleum geologists and to the petroleum industry the discovery and production of this small fraction are of supreme importance. Oil is commercially by far the most important form of petroleum throughout the world; next comes natural gas, and then the natural-gas liquids. The solid and semisolid petroleums are of minor importance, on the whole, although they may be important locally. Liquid petroleum, as it comes from the wells, or *crude oil,* varies widely in chemical composition, especially in its hydrocarbon compounds, and also in such physical properties as color, density, and viscosity. Most crude oils produced have an oily feel. They vary from opaque to translucent in thin layers, and in reflected light the color ranges from brown through reddish to slightly greenish yellow. Typical crude oils have a consistency between that of milk and that of cream; but, when all crude oils are considered, their consistency has a wide range of variation; at one extreme we have colorless liquids, as thin as gasoline, and at the other we have thick, viscous, black asphalts. Crude oil is immiscible with water, and—except for the oil from a few pools, which is denser than water or heavily contaminated with mineral matter—it floats on water. Numerous bubbles of gas arise in most oil when it first reaches the well head, the dissolved gas coming out of solution as the pressure is reduced. Petroleum oils are soluble in ether, acetone, carbon disulfide, benzol, benzene, chloroform, and boiling alcohol.

Measurement of Crude Oil

Crude oil is measured in barrels, tons, percentage of pore space, acre-feet of oil, or barrels per acre-foot of reservoir rock.[48] The most common unit is the barrel of 42 gallons (U.S.), the average weight of which is 310 pounds. Other units are listed in the conversion tables of the Appendix (see Table A-3).

When oil and gas are both being produced from a flowing well, the mixture goes first to a separator, where the gas is separated from the oil. The oil then goes to a tank in which it is measured. A pumping well yields only oil or oil with a small amount of gas, and this oil is pumped directly into a tank that has been calibrated so that the amount of oil, in gallons or barrels per vertical inch of fill-up, is known. The rate of production may thus be determined, in barrels per hour, per day, or per month. Where the flow of highly productive wells is to be measured, and sufficient tankage or pipe lines to handle the full flow are not available, the flow is measured for short periods, such as one or four or six hours, and the number is multiplied by the proper factor to give the capacity of the well per day. The number of barrels of oil a well first produces or is capable of producing in twenty-four hours is known as its *initial production.*

Oil in place in the ground is measured in several ways,[49] depending upon the purpose of the measurement. Where the purpose is to measure the volume of *oil in place,* the general method is to multiply the acre-feet of pore space, as calculated from core analyses and electric logs of the wells, by the percentage of oil saturation of the cores. Where the purpose is to measure the amount of recoverable oil or recoverable reserve in barrels of oil at the surface, the amount of oil in place is multiplied by the *shrinkage factor* (see p. 199), which is a measure of the decline in the volume of the oil as the gas comes out of solution when the oil is produced (see p. 200), and by the *recovery factor,* a measure of the percentage of oil in place that can be recovered, which depends upon porosity, permeability, type of reservoir energy,* and past experience under similar conditions. These methods are called *volumetric* or *saturation methods,* and they can be used early in the life of the property because they are not affected by proration or artificial curtailment of production.

The other common method of measuring oil in place is called the *decline-curve method;* this is used where records of free and uncurtailed production are available.[50] It plots the production of a well or group of wells during a considerable time and extrapolates the production curve into the future. The curves may be plotted on rectangular, semi-log, or log-log paper; the log papers have the advantage of permitting the curves and their extensions to be projected as straight lines. The estimated future production of all the wells for the anticipated economic life of the property may then be added together to give an estimate of the total recoverable oil left in the reservoir. This is the most reliable method of estimating recoverable reserves when a property has had a few years of experience upon which to base the curve. It is generally unsatisfactory, however, where the production has been prorated or artificially curtailed.

Chemical Properties of Crude Oil

Crude oil and natural gas, when underground and in their natural state, are at higher temperatures and under greater pressures than at the surface. All crude oils have some natural gas dissolved in them, and if there is more than enough gas to saturate the crude oil at the pressure and temperature that exist in the reservoir, the excess free gas accumulates as a free gas cap. (See p. 463.) The changes in pressure and temperature that occur during the production or chemical analysis of oil vaporize, release, or break down some of the hydrocarbons. It is therefore difficult or even impossible to obtain an accurate analysis of the thousands of compounds that are found in a crude oil as it exists under ground. The original composition of a crude oil can at

---

* Reservoir energy is the energy in the reservoir that causes the oil and gas to move into the well. See also pp. 458–459.

best be determined only approximately. The difficulty in separating the individual hydrocarbons in a crude oil may be seen in the fact that it has taken thirty-seven years of study to isolate and analyze 234 compounds.[51] The tremendous advances that are being made in hydrocarbon anlaysis—gas chromatography, mass spectrometry techniques, and hydrocarbon isotope geochemistry—make possible the rapid analysis of hydrocarbon molecules and a much more accurate and precise understanding of the compositions of many of the petroleum fractions.

The geologist is chiefly interested in the chemical and physical properties of petroleum substances as they occur underground: their chemical nature, and the changes in their composition that may result from the repeated changes in temperature and pressure which have been occuring throughout geologic time and which bear on their origin, migration, and accumulation. The refiner, on the other hand, is more interested in the numerous commercially valuable compounds that can be formed artificially in the refinery. Many, possibly most, of these artificial compounds have no counterpart in the naturally occurring oils and gases, but much can be learned about how the naturally occurring compounds may have formed by understanding some of the processes developed in the laboratories and refineries.

Although they may appear to be the same, no two petroleums occurring in different reservoirs are exactly alike, for each consists of mixtures of countless different hydrocarbons.[52] The typical chemical analyses, however, of crude oil, natural gas, and natural asphalt are rather uniform and fall within a general pattern, as given in Table 5-5.

TABLE 5-5 Chemical Composition of Typical Petroleum

Element	Crude Oil % Weight	Asphalt % Weight	Natural Gas % Weight
Carbon	82.2–87.1	80–85	65–80
Hydrogen	11.7–14.7	8.5–11	1–25
Sulfur	0.1– 5.5	2– 8	trace–0.2
Nitrogen	0.1– 1.5	0– 2	1–15
Oxygen	0.1– 4.5	—	—

The chemistry of petroleum is a part of organic chemistry, which is essentially the chemistry of carbon compounds. Organic chemistry is a broad and complex subject; nearly half a million different compounds of carbon have been found so far, and many more doubtless await discovery. The simplest organic compounds are those that contain only carbon and hydrogen and are known as hydrocarbons. And it is these that constitute the bulk of the componds found in most petroleums—natural gas, crude oil, and natural asphalts. Such minor elements as sulfur, nitrogen, and oxygen are present in most petroleums, but they are generally combined with organic carbon and hydrogen in complex molecules.

Before we discuss the chemistry of petroleum, it will be useful to give a brief review of some of the principles and terms used in the chemistry of hydrocarbons. For a more complete understanding of the subject, the reader is referred to any elementary book on organic chemistry.

A *saturated* hydrocarbon (sometimes called an alkane) is one in which the valence of all the carbon atoms is satisfied by single bonds. All the paraffins, for example, are saturated hydrocarbons, for each carbon atom is connected to each other carbon atom by a single covalent bond, and the remaining electrons in the carbon atom are each connected by a single covalent bond to a hydrogen atom. Saturated hydrocarbons are more stable, and are chemically less reactive, because the outer electron shells of both the carbon atoms and the hydrogen atoms have been filled through the sharing of electron pairs. Thus the compound achieves the electron structure of the stable and inert noble gases.

An *unsaturated* hydrocarbon is one in which the valence of some of the carbon atoms is not satisfied with single bonds, so that these atoms are connected to one another with two or more covalent bonds. An example is benzene, C_6H_6, in which there is not sufficient hydrogen to satisfy the electron requirement of the carbon atoms. Three of the carbon atoms, therefore, are each joined to another by two bonds. Unsaturated hydrocarbons are less stable than saturated hydrocarbons and have greater chemical activity. As a result, compounds with double and triple bonds combine readily with other compounds, and when heated they break up to become more nearly saturated compounds (single-bond compounds).

Isomers are substances of the same composition that have different molecular structure and therefore different properties. The first isomer in the paraffin series is the branched-chain isobutane, C_4H_{10} (*iso* meaning isomer), an isomer of normal butane (*n*-butane, C_4H_{10}). The structural formulas of these two compounds are:

n-butane: $CH_3(CH_2)_2CH_3$
B.P. $-0.5°C$

```
    H  H  H  H
    |  |  |  |
H — C — C — C — C — H
    |  |  |  |
    H  H  H  H
```

*iso*butane: $(CH_3)_2CHCH_3$
B.P. $-10.2°C$

```
    H  H  H
    |  |  |
H — C — C — C — H
    |  |  |
    H  |  H
       |
   H — C — H
       |
       H
```

There are three pentanes, each having the molecular formula C_5H_{12}, and containing 83.33 percent carbon and 16.67 percent hydrogen, and all having a molecular weight of 72.15 but each having a different boiling point:

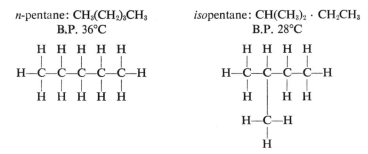

The number of isomers increases rapidly for higher members of the paraffin series. There are 5 possible isomers of hexane (C_6H_{14}), 18 of octane (C_8H_{18}), 75 of decane ($C_{10}H_{22}$), and 802 of tridecane ($C_{13}H_{28}$). In the olefin (C_nH_{2n}) series, structural isomerism begins with the third member, or butene (C_4H_8), and increases to 13 with hexene (C_6H_{12}), 27 with heptene (C_7H_{14}), and so on.

It has been shown mathematically that for paraffin hydrocarbons containing 18 atoms of carbon and 38 atoms of hydrogen per molecule there are 60,523 possible isomers, the number more than doubling for each added carbon atom in the molecule. Although it is probable that only a small fraction of these substances actually occur in measurable quantities in petroleum, one cannot fail to be impressed by the enormous complexity of the family of hydrocarbons.

Cracking is the process whereby the less volatile components of petroleum undergo complex changes when heated to high temperatures and put under high pressures, either in the presence or in the absence of catalysts. Carbon-to-carbon bonds are broken under these conditions, producing several compounds with lower boiling points. In this way, molecules that have boiling points too high to fall into the gasoline fraction are broken down and formed into new compounds that belong in the gasoline fraction. Thus complex molecules of high molecular weight are "cracked," or divided, into simpler compounds. Cracking has enabled the refiner to obtain a much larger amount of gasoline from each barrel of crude oil than it contains in its natural state and also to form many new compounds.

Polymerization is essentially the reverse of cracking, inasmuch as it causes a number of small molecules to unite and form a single larger molecule. It is the combining of simple molecules to form complex molecules.

Hydrogenation is the addition of hydrogen to double- and triple-bonded carbon atoms. Hydrogenation transforms unsaturated hydrocarbons into compounds containing more hydrogen atoms to the molecule until they finally become saturated (single covalent bonds). Hydrogen is always added one molecule (H_2) at a time. Examples are:

$$C_2H_2 + H_2 \longrightarrow C_2H_4 + H_2 \longrightarrow C_2H_6$$
$$\text{acetylene} \qquad \text{ethylene} \qquad \text{ethane}$$
$$\text{unsaturated} \qquad \text{unsaturated} \qquad \text{saturated}$$

$$H-C\equiv C-H + H_2 \longrightarrow \begin{array}{c} H \;\; H \\ | \;\; | \\ C=C \\ | \;\; | \\ H \;\; H \end{array} + H_2 \longrightarrow H- \begin{array}{c} H \;\; H \\ | \;\; | \\ C-C \\ | \;\; | \\ H \;\; H \end{array} -H$$

The source of the hydrogen necessary to the formation of the saturated hydrocarbons of petroleum is not known.[53] It may be derived from volcanic processes in the interior of the earth, from the decomposition of organic matter through bacterial action, and from the cracking and degradation of heavier oils caused by increasing heat and pressure under deep burial, perhaps aided by some catalyst or by some bacterial or chemical reaction whereby H_2S is broken down into free sulfur and H_2.

Hydrocarbon Series

The hydrocarbons have been divided into various *series*, differing in chemical properties and relationships. The four that comprise most of the naturally occurring petroleums are the normal paraffin (or alkane) series, the isoparaffin series (or branched-chain paraffins), the naphthene (or cycloparaffin) series, and the aromatic (or benzene) series. Crude oils are referred to according to their relative richness in hydrocarbons of these groups, as paraffinic-base, naphthenic-base, or mixed-base (naphthenic-paraffinic) oils. The aromatics are rarely the dominant group. The naphthenes include the complex residues of the high-boiling range (750°F and above) of all petroleums; if the residue consists largely of asphalts, the crude is said to be asphaltic. The approximate relations of the different major constituents of a number of crude oils are graphically shown in Figure 5-19.

The basic chemical classes of crude oil, as determined by the relative amounts of paraffins, naphthenes, asphalts, and aromatics, may be summarized for purposes of graphic comparison and then plotted on the simple four-way diagrams shown in Figure 5-20 or on the pattern diagrams shown in Figure 5-21. (See also the charts for graphically showing oil-field water compositions, pp. 168–169.) The approximate composition of the chief products resulting from the fractional distillation of crude oil is shown in Figure 5-22.

RESERVOIR FLUIDS—WATER, OIL, GAS [CHAPTER 5] 181

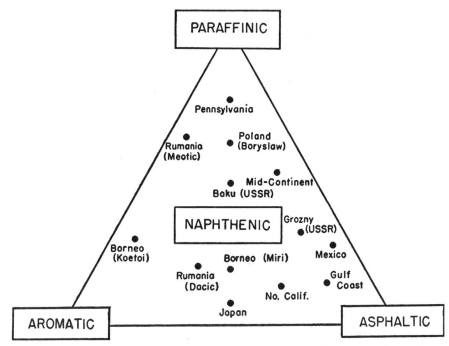

FIGURE 5-19 *The relations of constituents of various crude oils.* [*Redrawn from Gruse and Stevens,* Chemical Technology of Petroleum, *2nd ed (1942), pp. 6 and 7, Figs. 1 and 2.*]

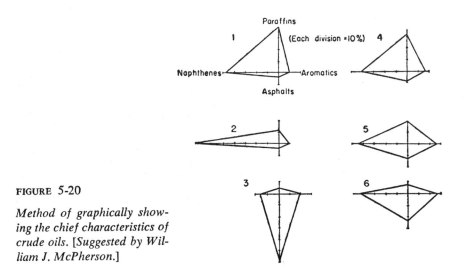

FIGURE 5-20

Method of graphically showing the chief characteristics of crude oils. [*Suggested by William J. McPherson.*]

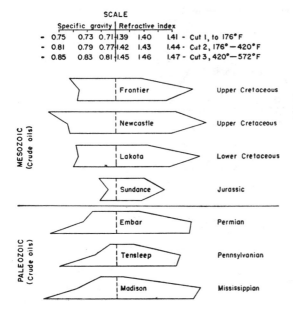

FIGURE 5-21

Pattern diagrams that may be used to correlate oil characteristics with reservoir rocks. Refractive indices and specific gravities for cuts 1, 2, and 3 (see p. 193) are plotted on opposite sides of a zero point and the data connected to give a closed pattern, the scale being at the top of the chart. Diagrams such as these are useful in making visual comparisons and are effective when plotted directly on long cross sections to show the changing characteristics of the oil in each of the reservoir rocks mapped. [Redrawn from Hunt, Bull. Amer. Assoc. Petrol. Geol., Vol. 37 (1953), p. 1849, Fig. 4.]

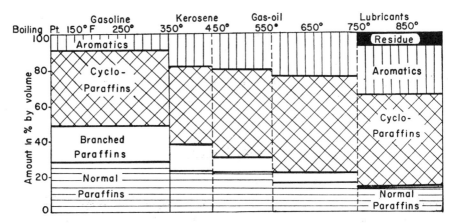

FIGURE 5-22 *The percentage composition by volume of the chief products obtained from United States crude oils. [Redrawn from Shaffer and Rossini Proc. Amer. Petrol. Inst., Vol. 32 (1952), p. 64.]*

Most petroleum oils consist of hundreds or even thousands of members of a few *homologous series** of hydrocarbons. They also contain large numbers of hydrocarbon compounds containing sulfur, nitrogen, or oxygen, and these compounds are found to occur in each fraction of the whole. Since the molecular weights of these hydrocarbon compounds are large, the presence of small volumes, 1 percent or less, may indicate that a large weight percentage of a crude oil is made up of compounds containing these elements.

Paraffin (Alkane) Series. The paraffin series of hydrocarbons is a saturated, straight-chain (aliphatic) series having the general composition C_nH_{2n+2}. It is a homologous series, progressing by a CH_2 increment from its simplest member, methane (CH_4), up to complex molecules with over 60 carbon atoms. Isomers with branched-chain structure occur in increasing numbers in addition to the normal, or n, members, beginning with butane (C_4H_{10}), until isomers with large numbers of carbon atoms are theoretically possible; however, only a few isomers of the paraffin members higher than the octanes have been identified. The commoner paraffin hydrocarbons are shown in Table 5-6.

The paraffin hydrocarbons—sometimes called the *methane series*—are chemically inactive; in fact, the name "paraffin" means "having little affinity." Methane (CH_4) is the simplest of all the hydrocarbons and is also the most stable. It forms in swamps from decaying vegetable matter, as "marsh gas," and is the chief constituent of natural gas.

Members of the paraffin series are generally the most abundant hydrocarbons present in both gaseous and liquid petroleums. All the members below pentane (C_5H_{12}) are gaseous at ordinary temperatures and are the chief constituents of natural gas. Paraffins between pentane and pentadecane ($C_{15}H_{32}$) are liquid and are the chief constituents of straight-run (uncracked) gasoline. The higher members of the paraffin series are waxy solids. Gasoline is generally composed of the hydrocarbons that boil within the temperature range from 40° to 200°C (from approx. 100° to 400°F), and its composition varies with the crude oil from which it is obtained. In addition to the large number of paraffin hydrocarbons found in crude oil, others are formed during the high-temperature cracking that also forms gasolines. The relative amounts of the different hydrocarbon series in different straight-run (uncracked) gasolines are shown in Table 5-7.

Some crude oils contain paraffin waxes, both amorphous and microcrystalline, which are obtained from the higher boiling-point fractions.[54] Many wells give trouble when paraffin wax is deposited on the walls of the tubing as the oil cools in rising to the surface. Paraffin wax may also precipitate and clog

---

* A homologous series is one in which each member differs from the next member by the same increment. Thus, in the paraffin series, each member differs from the next by the increment of CH_2. The members of a homologous series are said to be *homologs* of one another. Thus methane (CH_4) is a homolog of all compounds in the paraffin series from which it differs by a multiple of CH_2.

TABLE 5-6 Physical Properties of Some of the More Common Hydrocarbons of the Paraffin Series

Name	Chemical Formula	Physical State at 60°F and 14.65 psi	Molecular Weight	Boiling Point (°C) at Normal Conditions	Critical Temp. (°C)	Critical Pressure (Atm.)	Density Gas (Air = 1)	Density Liquid (Water = 1) Spec. Gr.
Methane	CH_4	Gas	16.04	−161.4	−82.4	45.8	0.554	$0.415^{-164°}$
Ethane	C_2H_6	Gas	30.07	−89.0	32.3	48.2	1.038	$0.546^{-88°}$
Propane[1]	C_3H_8	Gas	44.09	−42.1	96.8	42.0	1.522	$0.585^{-44.5°}$
n-Butane[1]	C_4H_{10}	Gas	58.12	0.55	153.1	36.0	2.006	$0.601^{0°}$
Isobutane[1]	C_4H_{10}	Gas	58.12	−11.72	134.0	36.9	2.006	0.557
n-Pentane[1]	C_5H_{12}	Liquid	72.15	36.0	197.2	33.0	2.491	0.626
Isopentane[1]	C_5H_{12}	Liquid	72.15	27.89	187.8	32.9	2.491	0.6197
n-Hexane	C_6H_{14}	Liquid	86.17	68.75	234.8	29.5	2.975	0.6594
Isohexane	C_6H_{14}	Liquid	86.17	60.30	228.0	—	2.975	0.6536
n-Heptane	C_7H_{16}	Liquid	100.20	98.42	267.0	27.0	3.459	0.6837
Isoheptane	C_7H_{16}	Liquid	100.20	90.10	257.9	27.2	3.459	0.6787
n-Octane	C_8H_{18}	Liquid	114.22	125.6	295.9	25.2	3.940	0.7028
Isooctane[2]	C_8H_{18}	Liquid	114.22	118.1	285.5	—	—	0.6976

[1] Compounds that comprise the liquefied petroleum gases (L.P.G.) and are extracted and stored under pressure in liquid form and sold commercially.
[2] The isooctane shown is the one commonly used as the standard for rating "octane numbers" of gasoline. Its "octane number" is 100.

TABLE 5-7 Percentages of Hydrocarbons in Various Straight-run Gasolines

Source	Paraffins	Naphthenes	Olefins[1]	Aromatics
Mid-Continent	72.9	22.0	1.9	3.2
Mixed California	58.9	31.6	2.2	7.3
Pennsylvania	82.5	15.3	2.1	trace
Mexico	82.3	10.9	1.5	5.3
Venezuela	71.0	20.4	0.	8.6
Michigan	85.2	7.4	2.9	4.5

Source: Adapted from Gruse and Stevens, op. cit. (reference note 53), p. 452.
[1] The olefins may be formed by minor cracking effects during distillation.

the pores at the face of the reservoir when the expanding gas cools as it enters the well. Some wells produce a Vaseline-like substance that can be shoveled except during the hot summer months when the temperature is above the melting point. A still more nearly solid phase of the paraffin series occurring naturally is ozokerite, a plastic, wax-like, paraffin vein material found in Utah and near Boryslaw, Poland.

Naphthene (Cycloparaffin) Series. The naphthene series of hydrocarbons, also known as the cycloparaffin series, is a saturated (single covalent bonds), homologous, closed-ring series, the members of which have the general formula C_nH_{2n}. It is isomerous with the olefin (alkane) series of the same composition, but members of the olefin series are structurally open-chain and unsaturated. For example, the cyclopropane member (C_3H_6) of the naphthene series has the structural formula

$CH_2CH_2CH_2$
Mol. wt. 42.08
B.P. $-34.4°C$

whereas its counterpart (isomer) among the olefins, propylene (also C_3H_6), has the structural formula

$CH_2 : CH \cdot CH_3$
Mol. wt. 42.08
B.P. $-47.0°C$

and is an open-chain, unsaturated compound. There is some doubt whether the olefins are present in crude petroleums underground, but they are present in petroleums under ordinary surface conditions and are common in the compounds produced in refinery operations. The naphthenes resemble the paraffins in physical and chemical characteristics, but are more stable than their isomers, the olefins.

Cyclopentane (C_5H_{10}) and cyclohexane (C_6H_{12})* are the chief members of the naphthene series found in petroleum,[55] although many naphthenes with from three to more than thirty carbon atoms in the rings are known. The naphthenes cyclopropane (C_3H_6) and methylcyclopropane (C_4H_8) are gases at ordinary temperatures and pressures, but all the other monocyclic naphthenes are liquids.

As seen in Table 5-7 (p. 185), Figure 5-22 (p. 182), and Figure 5-19 (p. 181), the naphthenes (cycloparaffins) constitute an important portion of petroleums as well as of most products, ranging between 7 and 31 percent in the straight-run gasolines from the fields listed. Crude oils with high percentages of naphthenic members are also called "asphalt-base crudes" because they include not only the simple naphthenic members but also many complex asphaltic members from the higher boiling-point ranges.

Aromatic (Benzene) Series. The aromatic (or benzene) series of hydrocarbons, so named because many of its members have a strong or aromatic odor, is an unsaturated, closed-ring (carbocyclic) series, having the general formula C_nH_{2n-6}. Benzene (C_6H_6), a colorless, volatile liquid, is the parent and most common member of the series found in petroleums. Other members commonly found in petroleum are toluene (methylbenzene, $C_6H_5CH_3$) and xylene (dimethylbenzene, $C_6H_4CH_3CH_3$). While aromatics are present in all petroleums, the percentage is generally small, as seen in Figure 5-19, page 181, ranging from about 10 percent in the Pennsylvania crude oils up to an exceptionally high 39 percent in the Borneo crudes. Benzene and its derivatives also occur extensively in the light oil fractions of tars obtained from the dry distillation of coals at temperatures above 1,000°C. Crude oils high in aromatics yield fractions of high octane rating.

---

* Cyclopentane, C_5H_{10}
 $CH_2CH_2CH_2CH_2CH_2$
 Mol. wt. 70.13
 B.P. 49.5°C

 Cyclohexane, C_6H_{12}
 $CH_2CH_2CH_2CH_2CH_2CH_2$
 Mol. wt. 84.16
 B.P. 81.4°C

Benzene, C_6H_6

Mol. wt. 78.11

B.P. 80°C

The structure of benzene, called a Kekule structure,* is a ring with alternating single and double bonds between the six carbon atoms, and with each carbon atom linked to one hydrogen atom. The benzene structure has the hexagonal pattern shown above. Derivatives can be obtained by replacing one or more H atoms with methyl (CH_3) or some similar group.

Other Constituents

Asphalt. Asphalt is a brown-to-black, solid-to-semisolid mixture of high-boiling point and high-molecular-weight hydrocarbon compounds that occurs either naturally or as a residue in the refining of some petroleums. It generally contains appreciable amounts of sulfur, oxygen, and nitrogen together with varying amounts of inert matter. The asphalts are closely associated with the naphthenes (cycloparaffins). See pages 185, 186.

Sulfur. Sulfur occurs to some extent (0.1–5.5% by weight) in practically all crude oils[56] and in each of the fractions that make up the oil. It may be in any one or more of the following forms: (1) free sulfur (S); (2) hydrogen sulfide (H_2S); (3) organic sulfur compounds, such as thiols, or mercaptans, which contain the SH group (an example is propanethiol, or propyl mercaptan, C_3H_8S), and the disulfides, which contain S_2 (an example is 2,3-dithiabutane, $C_2H_6S_2$). Many sulfur hydrocarbons are found in cracked distillates, but it is not known whether they are formed during the high-temperature distillation or whether they were originally present in the crude oil. No compounds carrying more than one atom of sulfur, except the disulfides, have been isolated from crude oil. Sulfur hydrocarbons form polar compounds, and they play an important role in boundary tensions, as discussed on pages 443–445.

The presence of sulfur and sulfur compounds in gasoline causes corrosion, bad odor, and poor explosion. Before the development of modern cracking processes by refineries, the presence of sulfur made petroleum less desirable and consequently worth less per barrel. Since sulfur can now be removed from oil, this price differential has been largely eliminated, and sulfur-bearing crude oils are nearly equal in value to nonsulfur crudes.

---

* Named after the German chemist Friedrich August Kekule, who first visualized the molecule as a group of little balls (atoms) joined by sticks. The benzene ring was discovered by Kekule after a dream of a monkey chasing its tail.

Crude oils of low API (American Petroleum Institute) gravity, or high specific gravity, generally contain more sulfur than others. Sulfur content has a wide range: at the low extreme are the high-gravity Pennsylvania crude oils carrying 0.07 or 0.08 percent sulfur, and at the high extreme are some heavy Mexican crude oils carrying from 3 to 5 percent sulfur. Many asphalt and bitumen seepages and oil shales have a high sulfur content. The Mexican heavy-oil seepages, locally called "chapopotes," contain from 6.15 to 10.75 percent sulfur. Crude oils carrying less than 0.5 percent sulfur are called "low-sulfur crudes," whereas those carrying more than 0.5 percent are called "high-sulfur crudes." Forty-two percent of the crude oil produced in the United States in 1946 was low-sulfur, and 58 percent was high-sulfur.[57]

The sulfur content of crude oils may vary greatly even within the same producing region. The sulfur content of each boiling range varies, moreover, for each kind of oil. The gasolines of western Texas, for example, are high in sulfur, whereas the gasolines from other high-sulfur crude oils, such as those of the Middle East, contain very little sulfur, the sulfur being concentrated in the residues. It has been found in Wyoming, for example, that the high-sulfur, low-gasoline, aromatic-naphthene-base crude oils are likely to be associated with the limestone and dolomite reservoir rocks, and that the low-sulfur, high-gasoline, paraffin-based oils are found in the sand reservoirs.[58] The approximate average sulfur content of crude oils of various gravities is shown in Figure 5-23. The chart shows the general increase in sulfur content with decrease in API gravity (increase in specific gravity).

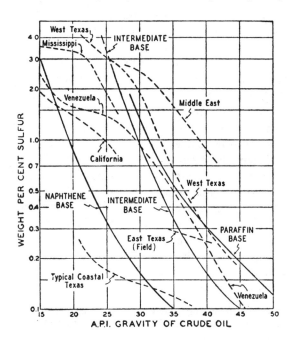

FIGURE 5-23

A Comparison of the sulfur content with the API gravity of various crude oils. [From Nelson, O. & G. Jour., November 23, 1953, p. 118.]

Nitrogen. Nearly all crude oils contain small quantities of nitrogen. Nothing is known of the nature of the nitrogen compounds in undistilled crude oil, but the nitrogen compounds in the distillates are frequently of the general type known as pyridines (C_5H_5N) and quinolines (C_9H_7N). Since nitrogen is a common, inert constituent of natural gas, it may be that the nitrogen content of the crude oil is contained in the dissolved gases. Nitrogen is an unwanted component of both crude oil and natural gas. About one-fifth of all American crude oil is classified as high-nitrogen, with more than 0.25 percent nitrogen, and the weighted average for all crude oils in the United States is 0.148 percent.[59] The highest known nitrogen content of crude oil in the United States is found in certain California oils, where a maximum of 0.82 percent has been observed.[60]

Oxygen. Oxygen is found in crude oil in various forms, generally averaging under 2 percent by weight, and ranging between 0.1 and 4.0 percent. It occurs in various forms[61] such as the following:
1. Free oxygen.
2. Phenols (C_6H_5OH).
3. Fatty acids and their derivatives [$C_6H_5O_6(R)$*].
4. Naphthenic acids having the general formula $C_nH_{2n-1} \cdot (COOH)$. The organic (naphthenic) acids add the carboxyl group to the hydrocarbons. The carboxyl formula is

$$COOH \text{ or } -C\begin{array}{c} \diagup O \\ \diagdown OH \end{array}$$

and the group has the properties of a weak acid.

5. Resinous and asphaltic substances. These are thought to be formed in part by the oxidation and polymerization of certain hydrocarbons in the crude. For example, the nonwaxy crude of Grozny, USSR,[62] contains 8.2 percent natural resins of specific gravity 1.04, empirical formula $C_{41}H_{57}O_2$, and molecular weight 589.

Asphaltenes differ from the resins in being colloidal solutions, although highly dispersed and stable. They are insoluble in petroleum naphthas but soluble in benzene and chloroform; and, instead of melting when heated, they swell and decompose into coke-like material. Their apparent molecular weights are on the order of several thousand, and their chemistry and molecular structure are indefinite. On analysis they show approximately the following composition: C, 85.2; H, 7.4; S, 0.7; and O, 6.7 percent. Asphaltenes are the chief constituents of the solid bitumens gilsonite and glance pitch.

---
* R stands for any alkyl group radical, such as methyl (CH_3-), ethyl (CH_3CH_2-), propyl ($CH_3CH_2CH_2-$), etc.

Miscellaneous Substances. Crude oil commonly contains minute amounts of a wide variety of miscellaneous substances, some organic and some inorganic. The material of organic origin, as seen under the microscope, includes such decay-resistant matter as siliceous skeletal tests, petrified wood fragments, spores, spines, cuticles, resins, coal and lignite fragments, algae, unicellular organisms, spore coats, insect scales, and barbules.[63]

The inorganic material may be observed in the ash. The ash from 113 pools in West Virginia[64] varied from 0.04 to 400 parts of ash per million, but most of the values ranged between 1 and 10 parts per million. The ash content of crude oils from Mexico, South America, and the Middle East ranges between 0.003 and 0.72 percent.[65]

The elements that have been identified in crude-oil ash include silicon, iron, aluminum, calcium, magnesium, copper, lead, tin, arsenic, antimony, zinc, silver, nickel, chromium, molybdenum, and vanadium.[66] Most of these elements are found in sea water and may have been derived from it, either as compounds in colloidal suspension or as materials secreted by algae and other marine organisms, which may also have provided the material from which the petroleum was formed. It is known that vanadium and nickel are concentrated in the porphyrins and replace the magnesium in chlorophyll, with the result that the vanadium and nickel content of crude oil is frequently several thousand times greater than the concentration in the earth's crust. Vanadium and nickel have been used to correlate crude oils.[66] Clay minerals sometimes come up along with crude oil. They settle out with the water that accompanies the oil, and this indicates that they are probably associated with the water rather than with the oil.

Uranium is found in many petroleums, and radioactive decay products are found in many natural gases and oil-field brines. Most uranium deposits, in fact, either are associated with or contain some carbonaceous material; probably the carbonaceous material, in some manner, aids in the precipitation of uranium. It is not known how the uranium enters petroleum. It may be carried along with migrating oil and gas, it may result from the radiation of radioactive sediments, or it may have been concentrated by either plant or vegetable matter, which in turn became the source material of petroleum.[67]

Most crude oils contain sodium chloride, which is measured in pounds per thousand barrels.[68] Desalting is required when the amount exceeds 15–25 lb per 1,000 bbls. An excess (more than 0.7 or 0.8%) is like an excess of sulfur; it corrodes equipment. Some of the salt occurs as crystals in the oil, and some is dissolved in the reservoir water that is generally produced along with the oil, possibly in part as an emulsion.

The composition in mole fractions of the typical reservoir hydrocarbon fluids is shown in Table 5-8.

Several partial analyses of the composition of various crude oils have been made. One of the oils from the Bradford pool of Pennsylvania is shown in Table 5-9. This is one of the more extensive published analyses, yet even here more than 58 percent of the compounds are lumped together in the groups having high molecular weight, and these contain most of the almost infinite number of separate compounds that are probably present in the average crude oil.

Analyses of crude oils are commonly made by the Hempel method of the U.S. Bureau of Mines. A sample analysis is shown in Table 5-10. The method consists in distilling a 300-ml charge of crude oil under definite and carefully controlled conditions. Distillation begins at atmospheric pressure (760 mm of mercury) and 25°C (77°F). The temperature is then gradually raised, without change of pressure, and ten *fractions,* or *cuts,** are taken off at intervals of 25°C,

TABLE 5-8 Composition (in Mole Fractions) of Typical Reservoir Hydrocarbons

Reservoir Type	Dry Gas	High-pressure Gas	High-pressure Oil	Low-pressure Oil
Methane	0.91	0.72	0.56	0.14
Ethane	0.05	0.08	0.06	0.08
Propane	0.03	0.05	0.06	0.08
Butanes	0.01	0.04	0.05	0.08
Pentanes	trace	0.02	0.04	0.05
Hexanes	trace	0.02	0.03	0.05
Heptanes plus	—	0.07	0.20	0.53

Source: Donald A. Katz and Brymer Williams, "Reservoir Fluids and Their Behavior," Bull. Amer. Assoc. Petrol. Geol., Vol. 36 (February 1952), p. 345, Table I.

or 45°F. The pressure of the distillation system is then reduced to 40 mm of mercury and the distillation continued until five more fractions of 25°C each have been obtained. The final temperature is 300°C. A large number of crude-oil analyses are available in the literature.[69] While the more elaborate techniques of chromatographic analysis and infrared analysis are necessary for determining the refinery value of a crude oil, the Hempel analysis, which is widely used, is less expensive and is simple, for it characterizes the entire crude.

The Correlation Index (CI).[70]   This index is a useful means of classifying oils on a qualitative basis. The correlation index is a number whose magnitude indicates certain characteristics of a crude-oil distillation fraction. The paraffins are given a CI value of 0, and benzene a CI value of 100. The lower the CI value of an analysis fraction, the greater the concentration of paraffin hydrocarbons in the fraction; the higher the CI value, the greater the concentration of naphthenic and aromatic hydrocarbons. A CI curve of a crude oil may be made by plotting the correlation index against the fifteen analysis fractions. (See Table 5-10.) Such a curve may be compared with other curves, and the relation between different oils, and between oils from different formations, may be readily visualized.[71] A chart showing correlation index curves of some crude oils in the United States is shown in Figure 5-24.

* A fraction, or cut, is the petroleum product that vaporizes between two temperatures during distillation.

TABLE 5-9 Composition of Crude Oil from the Bradford Field, Pennsylvania

Component	Weight, % of Crude Oil	Component	Weight, % of Crude Oil
Air	0.01	Dimethyl sulfide	0.006
Methane	0.0001	Methyl ethyl sulfide	0.003
Ethane	0.11	Diethyl sulfide	0.012
Propane	0.73	Ethyl n-propyl sulfide	0.012
n-Butane	1.71	Di-n-propyl sulfide	0.009
Isobutane	0.58	Di-n-butyl sulfide	0.009
n-Pentane	0.85	C_9 Paraffins and naphthenes, boiling	
Isopentane	2.18	ranges up to 225°C	11.5
Hexanes	3.40	C_8 Aromatics boiling to 225°C	1.84
Heptanes	3.37	Oxygen-nitrogen-sulfur compounds,	
Octanes	3.04	boiling range 40–225°C	0.788
Nonanes	2.69	Fraction, boiling range from	
Cyclopentane	0.049	225°C/740 mm to	
Methylcyclopentane	0.349	280°C/40 mm	29.9
Cyclohexane	0.518	High-molecular-weight components:	
Dimethylcyclopentanes	0.587	Av. mol. wt. 340	3.8
Methylcyclohexane	1.55	Av. mol. wt. 380	2.9
Ethylcyclohexane	0.36	Av. mol. wt. 410	3.3
C_8 naphthenes	2.07	Av. mol. wt. 460	3.6
C_9 naphthenes	1.68	Av. mol. wt. 550	3.6
Benzene	0.0389	Av. mol. wt. 890	9.0
Toluene	0.572	Losses and unaccounted for	2.3
Ethylbenzene	0.0398	Total	100.00
O-Xylene	0.1426		
m-Xylene	0.580		
p-Xylene	0.176		

Source: Julian Feldman, Lawrence Scarpino, Gus Pentazopolos, and Milton Orchin (Synthetic Fuels Research, U.S. Bureau of Mines, Bruceton, Pa.), "Composition of Crude Oil from the Bradford Field, Pennsylvania," Prod. Monthly, Vol. 16, No. 6 (April 1952), pp. 14–16.

Physical Properties of Petroleum Oils

The physical properties most commonly determined for petroleum are: (1) density, (2) volume, (3) viscosity, (4) refractive index, (5) fluorescence, (6) optical activity, (7) color, (8) odor, (9) pour and cloud points, (10) flash and burning points, (11) coefficient of expansion. Surface and interfacial

RESERVOIR FLUIDS—WATER, OIL, GAS [CHAPTER 5]

TABLE 5-10 Typical Hempel Analysis of Crude Oil by U.S. Bureau of Mines

Elk Basin Field Tensleep 5,620–5,697 feet	WELL 9	Wyoming Park County NE $\frac{1}{4}$ SE $\frac{1}{4}$ SE $\frac{1}{4}$ Sec. 36 T58N–R100W

GENERAL CHARACTERISTICS

Specific gravity, 0.892
Sulfur, 1.92 percent
Saybolt Universal viscosity
   at 77°F, 100 sec; at 100°F, 75 sec

API gravity, 27.1
Pour point, below 50°F
Color, black

DISTILLATION, BUREAU OF MINES HEMPEL METHOD

Distillation at atmospheric pressure, 588 mm First drop 25°C (77°F)

Fraction No.	Cut at °C	Cut at °F	Per-cent	Sum per-cent	Spec. Gr. 60/60°F	°API 60°F	CI	SU visc. 100°F	Cloud test °F
1	50	122	2.3	2.3	0.644	88.2			
2	75	167	2.1	4.4	0.674	78.4	6.1		
3	100	212	3.0	7.4	0.711	67.5	14		
4	125	257	3.5	10.9	0.739	60.0	18		
5	150	302	3.7	14.6	0.763	54.0	22		
6	175	347	3.7	18.3	0.781	50.0	24		
7	200	392	3.8	22.1	0.801	45.2	28		
8	225	437	4.5	26.6	0.821	40.9	32		
9	250	482	4.8	31.4	0.840	37.0	36		
10	275	527	6.0	37.4	0.860	33.0	40		

Distillation continued at 40 mm

Fraction No.	Cut at °C	Cut at °F	Per-cent	Sum per-cent	Spec. Gr. 60/60°F	°API 60°F	CI	SU visc. 100°F	Cloud test °F
11	200	392	1.4	38.8	0.873	30.6	43	43	Below 5°F
12	225	437	6.1	44.9	0.882	28.9	43	49	15
13	250	482	7.7	52.6	0.903	25.2	50	69	35
14	275	527	6.0	58.6	0.917	22.8	53	130	55
15	300	572	8.6	67.2	0.933	20.2	58	290	75
Residuum			33.0	100.2	1.005	9.2			

Carbon residue of residuum 15.6 percent; carbon residue of crude 5.1 percent

APPROXIMATE SUMMARY

	Percent	Spec. Gr.	°API	Viscosity
Total gasoline and naphtha	22.1	0.741	59.5	
Light gasoline	(7.4)	0.680	76.6	
Kerosene distillate	4.5	0.821	40.9	
Gas oil	15.6	0.859	33.2	
Non-viscous lubricating distillate	10.1	0.893–0.910	28.8–24.0	50–100
Medium lubricating distillate	6.5	0.910–0.924	24.0–21.6	100–200
Viscous lubricating distillate	8.4	0.924–0.942	21.6–18.7	Above 200
Residuum	33.0	1.005	9.2	
Distillation loss	0.0			

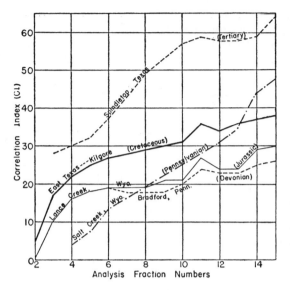

FIGURE 5-24

Correlation index curves of a few characteristic crude oils of the United States. [Redrawn from Wenger and Lanum, Petrol. Engr., September 1952, p. A-69, Figs. 4 and 5.]

tension, capillarity, adsorption, and wettability are considered more fully in Chapter 10, under reservoir mechanics.

Density (Gravity). The density of a substance is the weight of a given volume, such as pounds per cubic foot. A convenient method of expressing the same physical property is the specific gravity, in which no units of measurement need to be specified. Specific gravity is the ratio of the weights of equal volumes of the substance in question and pure water. Since volume is affected by temperature and pressure, these conditions must be specified. The practice in the United States is to compare the weight of unit volumes of oil and water at 60°F and one atmosphere pressure. Tables are available for converting measurements made at any other temperature.* Since the price of crude oil is commonly based on "gravity," these measurements are important.

The API gravity scale is an arbitrary one, which has the advantage of simplifying the construction of hydrometers, because it enables the stems to be calibrated linearly. API gravity does not have a straight-line relationship with specific gravity, nor with the other physical properties correlated with specific gravity, such as viscosity. High values of API gravity correspond to low specific gravity, and low values of API gravity to high specific gravity; so the scale cannot be used directly in engineering calculations.

A similar scale is the European Baumé gravity scale. These two arbitrary scales are related to specific gravity by the following formulas:

---

* U.S. Bureau of Standards, Washington, D.C., National Standard Petroleum Tables, Circular No. C-410. Since January 1, 1954, the petroleum industry uses American Society for Testing Materials, 1916 Race Street, Philadelphia 3, Pa., Petroleum Measurement Tables (ASTM D-1250), Table No. 5, "Reduction of Observed API Gravity to API Gravity at 60°F," and Table No. 7, "Reduction of Volume to 60°F. against API Gravity at 60°F."

RESERVOIR FLUIDS—WATER, OIL, GAS [CHAPTER 5]

$$\text{Degrees API} = \frac{141.5}{\text{Spec. Gr. at } 60°F} - 131.5$$

$$\text{Degrees Baumé} = \frac{140}{\text{Spec. Gr. at } 60°F} - 130$$

TABLE 5-11 Comparison of Gravity Scales (No Straight-line Relationship)

Spec. Gr. at 60°F	Baumé Gravity	API Degrees	Spec. Gr. at 60°F	Baumé Gravity	API Degrees
1.0000 (pure water)	10.0	10.0	0.8485	35.0	35.3
0.9655	15.0	15.1	0.8235	40.0	40.3
0.9333	20.0	20.1	0.8000	45.0	45.4
0.9032	25.0	25.2	0.7778	50.0	50.4
0.8750	30.0	30.2			

The conversion of specific gravity to Baumé gravity and API gravity is shown in Table 5-11. Gravities of some representative crude oils are given in Table 5-12. The effect of temperature on the specific gravity of crude oil is shown in Table 5-13. Table 5-14 expresses the effect of temperature differently, showing the change in specific gravity for each 1°F change in temperature of crude oils.

TABLE 5-12 Representative Crude-oil Gravities

Area	Spec. Gr. (Water = 1.0)	API Degrees
Canada: Alberta	0.9792–0.7507	13.0–57.0
Indonesia and New Guinea	0.9725–0.7507	14.0–57.0
Mexico: Tampico, Golden Lane, Panuco	0.9861–0.9218	12.0–22.0
Poza Rica	0.8762	30.0
Near East: Iran, Iraq, Kuwait, Saudi Arabia	0.8927–0.8109	27.0–43.0
Trinidad	0.9529–0.8203	17.0–41.0
United States, over all	1.0217–0.7351	7.0–61.0
Gulf Coast (Tertiary salt domes chiefly)	0.9402–0.7796	19.0–50.0
California (Tertiary)	1.0217–0.7796	7.0–50.0
Mid-Continent (Paleozoic chiefly)	0.934 –0.8017	20.0–45.0
USSR: Grozny and Baku districts	0.934 –0.835	20.0–38.0
Venezuela: eastern	0.9529–0.8203	17.0–41.0
western	1.000 –0.7507	10.0–57.0

Source: Data from O. & G. Jour., December 22, 1952, pp. 278–302, and *The Science of Petroleum*, Oxford University Press, London and New York, Vol. 2, pp. 840–930.

TABLE 5-13   Effect of Temperature on Specific Gravity

Specific Gravity 60°F	API Gravity 60°F	Gravity at Average Temperature					
		100°F		200°F		300°F	
		Spec. Gr.	°API	Spec. Gr.	°API	Spec. Gr.	°API
1.0	10.0	0.98	12.9	0.96	15.9	0.92	22.3
0.9	25.7	0.88	29.3	0.85	35.0	0.82	41.0
0.8	45.4	0.78	49.9	0.74	59.8	0.69	73.6
0.7	70.6	0.67	69.0				

Source: Adapted from H. S. Bell, *American Petroleum Refining*, 3rd ed., D. Van Nostrand Co., New York (1945), p. 66.

The gravity of two crude oils may differ considerably even though the oils seem to be closely related. There may be a difference in gravity between oils in adjacent reservoirs within the same field or geologic environment, between oils in the same reservoir rocks but in separate traps, and between oils within the same reservoir but of differing structural position. A few examples of local variations in the gravity of crude oils follow:

The oil in the Tensleep sand (Pennsylvanian) of the Elk Basin field, in Wyoming, ranged in specific gravity from 0.867 (API 31.8°) on the top of the fold to 0.892 (API 27.1°) at the base of the oil column at the edge of the pool.[72] The difference in gravity is largely explained by the fact that the oil near the top contains 460–490 cubic feet of gas per barrel of oil, whereas the oil at the lower edge of the pool contains only 134 cubic feet of gas. A similar relationship has been found in the Rangeley field, Colorado,[73] where the specific gravity of the crude

TABLE 5-14   Gravity of Crude Oil at Different Temperatures

Gravity at 60°F		Change in Gravity for Each 1°F Change in Temperature
Spec. Gr.	API	Spec. Gr.
0.90	25.7	0.00036
0.80	45.4	0.00039
0.70	70.6	0.00049

Source: Adapted from Anderson, *Ind. Chem.*, Vol. 12 (1920), p. 1011.

oil ranged from 0.849 (API 35.2°) on the top of the structure down to 0.869 (API 31.3°) at the base of the oil column at the edge of the field.* Oils in the Bartlesville and Red Fork sandstones (Pennsylvanian), in northeastern Oklahoma, generally become lighter with increasing depth. The relationship is roughly as follows:[75]

DEPTH IN FEET	GRAVITY API	GRAVITY SPEC. GR.
500–2,000	30–35	0.88–0.85
2,000–5,000	35–40	0.85–0.82
5,000–6,000	40–45	0.82–0.80

The averages of a large number of gravity determinations in the Tertiary oils of the Gulf Coast show the following changes with depth:[76]

DEPTH IN FEET	GRAVITY API	GRAVITY SPEC. GR.	DEPTH IN FEET	GRAVITY API	GRAVITY SPEC. GR.
0–1,000	19	0.94	6,000–7,000	46	0.80
1,000–2,000	21	0.93	7,000–8,000	47	0.79
2,000–3,000	29	0.88	8,000–9,000	51	0.77
3,000–4,000	33	0.86	9,000–10,000	49	0.78
4,000–5,000	34	0.85	10,000–11,000	50	0.78
5,000–6,000	44	0.81			

Production in the Burgan field of Kuwait, which probably contains more oil than any other field in the world, is from three sands of Middle Cretaceous age extending through a productive section of 1,100 feet. Each of the sands is separated from the others by shale and sandy shale intervals, and within the upper interval is a layer of Orbitolina limestone that forms a stratigraphic marker. The API gravity at the surface averages 31.8° but varies considerably in the reservoir, depending on the depth. The gravity of the oil in each of the sands is approximately the same for the same depth below sea level, but decreases approximately 1° for each 200 feet increase in depth. The oil-water contact is approximately the same for all sands, indicating an interconnection, probably through fracturing.[77] The lower gravity with depth is a reversal of the general rule that the heavier oils are in the higher and younger reservoirs. A similar situation exists in many of the oil fields of the Apsheron Peninsula, USSR.[78] For example, the shallow pay formations of the Surakhany field contain asphalt-free, light-yellow oils of a gravity of 0.720 (API 65°); at 700–900 meters (2,300–3,000 feet) the oils contain 7–8 percent resinous substances; at 1,450 meters (4,000 feet) the oils contain 12 percent resinous substances; and in reservoirs below 1,800 meters (6,000 feet) the oils contain 30 percent resinous substances and have a specific gravity of 0.90 (API 25.7°). There are seventeen productive sands in

---

* The Hawkins pool, in northeastern Texas, where oil is found in the Woodbine sand (Upper Cretaceous), is unusual in that the gradation is from an API gravity of 31° (Spec. Gr. 0.87) at the top of the oil column down to a heavy asphaltic oil of 16° gravity (Spec. Gr. 0.96) at the base of the oil column—in fact, the lower 3–10 feet is too viscous to flow.[74]

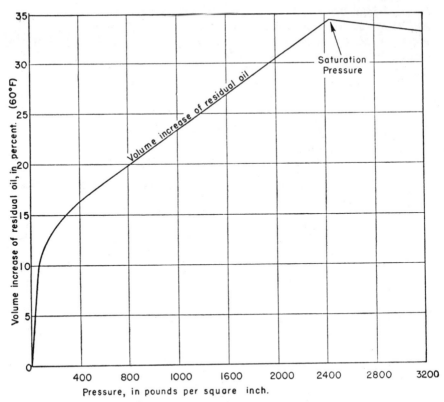

FIGURE 5-25 *A typical increase in volume of oil as gas is dissolved in it with increasing pressure (depth). The combined oil and gas show a marked increase in volume until no more free gas is in contact with the oil. From then on the volume of the oil-gas mixture decreases with a further increase in pressure, in accordance with Henry's law and the compressibility of fluids.* [Redrawn from H. B. Hill and R. K. Guthrie, U.S. Bur. Mines, RI 3715, p. 86 (1943).]

the Productive series (Pliocene) of the Bibi Eibat field, Apsheron, USSR, to a depth of 1,800 meters, and they too become heavier with depth, the shallower oils ranging from 0.840 to 0.860 Spec. Gr. (API 37–33°) and the oils in deeper sands ranging from 0.900 to 0.907 Spec. Gr. (API 25.7–24.5°). Water-bearing sands occur between some of the pay sands.

Crude oil density ranges from the oil being produced in the Oxnard field, Ventura County, California, which is heavier than water (API 5–7°) through the oil of 10° API gravity found in the Boscan field of western Venezuela, and the oil of 12° API gravity found in the Panuco district of Mexico, up to colorless distillates and condensates with an API gravity of 57° and above. Crude oils of from 27° to 35° API gravity are the most common and constitute the bulk of the world's production.

Volume. Oil in the reservoir contains dissolved gas, and the volume of the solution depends upon the *formation gas-oil ratio** and the reservoir pressure. The gas that may be dissolved in oil under increasing pressure increases the volume of the solution until the saturation pressure (bubble point) is reached, after which the volume decreases with increased pressure. (See Fig. 5-25.) Thus 0.5–0.8 of a barrel of gas-free oil at the surface of the ground, known as *stock-tank oil,* may represent one barrel of oil in the reservoir at the saturation pressure.† Data on a specific oil, that of Kettleman Hills field, California, are shown in Table 5-15. On the other hand, a thousand cubic feet of gas at

TABLE 5-15 Approximate Formation Volumes of One Barrel of Stock-tank 36° API Oil and Accompanying Gas at Reservoir Pressures and at a Temperature of 220°F, Kettleman Hills Field, California

Pressure (lb/sq in)	Approx. Depth (equivalent ft)	Approximate Formation Volume at Three Gas-Oil Ratios		
		1,000 cu ft/bbl	2,500 cu ft/bbl	5,000 cu ft/bbl
500	1,100	6.90	17.80	36.00
1,000	2,200	3.50	8.50	16.80
2,000	4,400	2.00	4.45	8.45
3,000	6,600	1.60[1]	3.20	5.70

Source: Adapted from McAllister, Trans. Amer. Inst. Min. Met. Engrs., Vol. 142 (1941), p. 53, Table III.
[1] *Example:* One barrel of oil and 1,000 cubic feet of gas measured at standard conditions (60°F and 14.7 psia) occupy 1.60 barrels of space under reservoir conditions of 3,000 psi, or approximately 6,600 feet below the surface.

the surface may represent only a few cubic feet of compressed gas in the reservoir, where the pressures are higher. The changes between one barrel of petroleum in the reservoir and the same petroleum at the surface are graphically shown in Figure 5-26.

The volume of liquid petroleums, at constant pressure, increases with increased temperature, but at a much lower rate than that of the gases. The coefficient of thermal expansion of liquid petroleums increases with an increase in API gravity (decrease in Spec. Gr.) and also increases with an increase in

---
* The formation gas-oil ratio, generally called the gas-oil ratio, is the number of cubic feet of gas per barrel of oil as found in the reservoir. The producing gas-oil ratio is the gas-oil ratio of oil as it is produced and is generally higher than the formation ratio.
† The factor that must be applied to a barrel of reservoir oil to convert it into an equivalent amount of stock-tank oil is known as the *shrinkage factor,* and the factor that is applied to a barrel of stock-tank oil to convert it into an equivalent amount of reservoir oil is known as the *formation-volume factor.* Shrinkage factors generally range between 0.63 and 0.88 and formation-volume factors between 1.14 and 1.60.

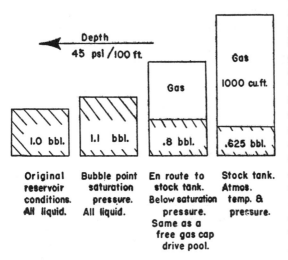

FIGURE 5-26

The changing volume of one barrel of reservoir oil in which 1,000 cubic feet of gas is dissolved as it passes from the reservoir, where all the gas is in solution, until it reaches the stock tank at the surface of the ground. There the original barrel of reservoir oil is seen to consist of 0.625 barrel of oil and 1,000 cubic feet of gas. The fractional volume of oil in the stock tank obtainable from one barrel of oil and its dissolved gas in the reservoir is called the shrinkage factor. The volume of reservoir oil necessary to yield one barrel of stock-tank oil is called the formation-volume factor.

temperature. The coefficient of expansion at 60°F of several oils is shown in Table 5-16. (See also pp. 210, 211.)

The volume of surface-equivalent gas that will dissolve in a unit volume of reservoir oil increases as the reservoir pressure increases, until the oil is finally saturated with gas and no more gas will dissolve in the oil. Generally this means a progressive increase with depth. In some deep-seated high-pressure reservoirs, the oil may contain as much as 150 times its volume of dissolved

TABLE 5-16 Coefficient of Expansion

Specific Gravity	API Gravity	Mean Coefficient of Expansion (vol/vol/°)
0.67	79	0.0008
0.67–0.72	78–65	0.0007
0.72–0.77	64–51	0.0006
0.78–0.85	50–35	0.0005
0.85–0.97	34–15	0.0004
0.97–1.076	14–0	0.00036

Source: H. S. Bell, *American Petroleum Refining*, 3rd ed., D. Van Nostrand Co., New York, p. 66.

FIGURE 5-27

The relation between the amount of gas dissolved in a barrel of oil, as measured by liberation at the surface, and the formation-volume factor of the oil in the reservoir. The graph is based on the data from a number of different pools. [Redrawn from Buckley (ed.), Petroleum Conservation *(1951)*, p. 101, Fig. 9.]

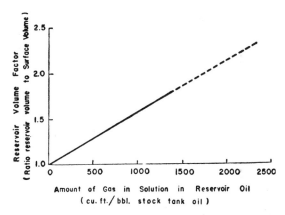

surface-equivalent gas.[79] The relation between pressure and the amount of gas in solution per barrel of crude oil is shown in Figure 5-27, and for a specific oil, from the Coldwater field, Michigan, in Figure 5-28. The solubility of gas in oil increases directly with pressure, in a straight-line relationship, in accordance with Henry's law.* However, the capacity to hold gas in solution is much lower in the heavy crudes (low API gravity) than in the lighter crudes. The relationships are shown in the chart in Figure 5-29.

At a given temperature, the volume of surface-equivalent gas that will go into solution in a given oil increases with an increase in pressure. An increase in the reservoir temperature, however, decreases the amount of gas that will go into solution, at an average rate of about 2 percent for each degree Fahrenheit. The relations are shown diagrammatically in Figure 5-30. The pressure-volume-temperature relations of a specific crude oil, from the Weber sandstone (Pennsylvanian) of the Rangely field, in Colorado, are given in Figure 5-31.

Viscosity. Viscosity is an inverse measure of the ability of a substance to flow; the greater the viscosity of a fluid, the less readily it flows. A viscous liquid will yield under the slightest stress if subjected to the stress for a sufficient length of time. Some substances, however, are capable of resisting small stresses for a practically infinite time, and will flow only if the stress is increased beyond a certain amount; these substances are called *plastic*, though for practical purposes they may be regarded as solid. Crude petroleums vary greatly in viscosity. Some, such as natural gas and light oils, are very mobile; others are highly viscous, and these grade into the semisolid petroleums, although the latter are, more strictly speaking, plastic.

The viscosity of a crude oil is generally dependent chiefly on the amount of gas dissolved in it and on the temperature (the more gas in solution and the

---
* Henry's law is that the mass of a slightly soluble gas that dissolves in a definite mass of a liquid at a given temperature is very nearly proportional to the partial pressure of that gas.

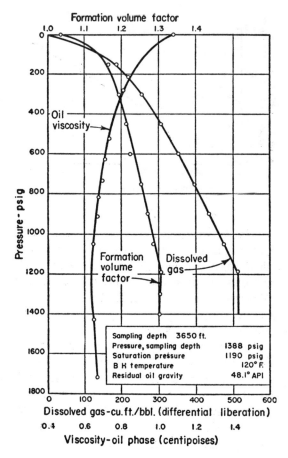

FIGURE 5-28

The characteristic change in viscosity, formation-volume factor, and amount of dissolved gas with increase in pressure in the oil from the Coldwater field, Isabella County, Michigan. [Redrawn from Criss, Jour. Petrol. Technol., T.P. 3748 (Feb. 1954), p. 26, Fig. 5.]

FIGURE 5-29

The general effect of increasing pressure (increasing depth) on the capacity of crude oils of different gravity to hold gas in solution. [Redrawn from Beal. Trans. Amer. Inst. Min. Met. Engrs., Vol. 165.]

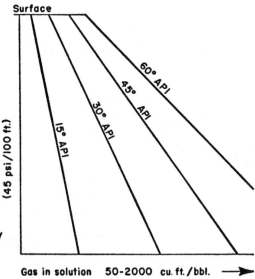

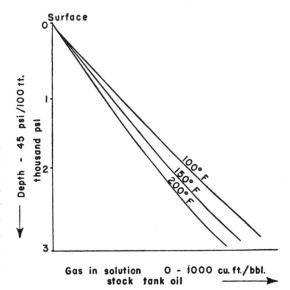

FIGURE 5-30

Effect of increasing temperature in reducing the solubility of natural gas in crude oil. The average decrease in solubility is about 2 percent for each degree Fahrenheit of increase in temperature. [Redrawn from Standing, O. & G. Jour., May 17, 1947, p. 95.]

higher the temperature, the lower the viscosity) and varies only slightly with changing pressure.[81] (See Fig. 5-32.) The reason the viscosity of a liquid decreases with an increase in temperature is that heating increases the molecular agitation (or velocity), which, in the absence of confining pressure maintaining a fixed volume, increases the intermolecular distance and the volume (expansion). An increase in intermolecular distance reduces both the intermolecular attraction and the friction caused by collisions of molecules. The effect of an increase in temperature on some Oklahoma and Kansas crude

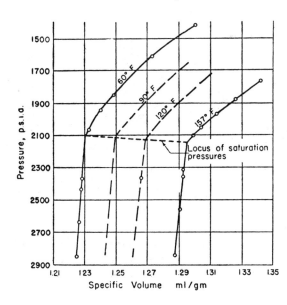

FIGURE 5-31

Pressure-volume-temperature relations of a subsurface sample of crude oil from the Rangely field, Colorado. The increase in volume with increase in temperature and the decrease in volume with increase in pressure are shown for a specific oil. [Redrawn from Cupps, Lipstate, and Fry, U.S. Bur. Mines, RI 4761, Fig. 28, opp. p. 46 (1951).]

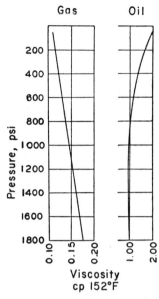

FIGURE 5-32

The increasing viscosity of natural gas and the decreasing viscosity of gas-free crude oil with increasing pressure (increasing depth) in the Buena Vista field, Kern County, California. The temperature is constant at 152° F. [Redrawn from Brubecker and Stutsman, O. & G. Jour., August 23, 1951, p. 118, Fig. 9.]

oils is shown in Figure 5-33. The viscosity of a gas, on the other hand, increases with an increase in temperature if, because of confining pressure, there is no increase in volume. The increase in molecular agitation due to heating, when

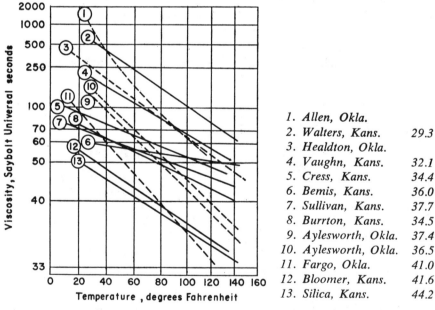

1. Allen, Okla.	
2. Walters, Kans.	29.3
3. Healdton, Okla.	
4. Vaughn, Kans.	32.1
5. Cress, Kans.	34.4
6. Bemis, Kans.	36.0
7. Sullivan, Kans.	37.7
8. Burrton, Kans.	34.5
9. Aylesworth, Okla.	37.4
10. Aylesworth, Okla.	36.5
11. Fargo, Okla.	41.0
12. Bloomer, Kans.	41.6
13. Silica, Kans.	44.2

FIGURE 5-33 *Relation of viscosity to temperature in different crude oils of Oklahoma and Kansas. The crude oils and their API gravities are listed beside the figure.* [Redrawn from Nelson, O. & G. Jour., January 5, 1946, p. 70, Fig. 2.]

there is no increase in the intermolecular distances, causes an increase in the frequency of the collisions between molecules and thus an increase in friction.

As more and more gas is dissolved in crude oil, the viscosity of the oil is progressively reduced. This is one of the most important effects of the presence of dissolved gas in oil. As more gas goes into solution in a crude oil, the API gravity of the oil increases and the specific gravity decreases. The effect of dissolved gas on both the viscosity and the gravity of a crude oil is shown in Figure 5-34. The viscosity of an oil is at a minimum at the saturation pressure —that is, the pressure at which the oil contains in solution all the gas it is able to hold and the excess gas is first released from solution: the "bubble-point pressure" (see pp. 437–438).[82] As the pressure is reduced, an additional amount of gas is released from solution, and the viscosity of the residual oil is increased. The increase in viscosity, due to release of gas, is greater than the normal decrease in viscosity that a decline in pressure causes in a gas-free oil or in an oil containing gas at pressures below the bubble point. The net result is an increase in the viscosity of the oil as the pressure is reduced below the saturation pressure and gas is released from solution. The importance of the viscosity of oil is evident when we remember that, if the viscosity should be reduced by half, either twice as much oil would flow through the same sand or it would require only half as much pressure to force an equal amount through. The change in viscosity of a specific oil, that of the West Edmond field, Oklahoma, as the pressure changes and in relation to the shrinkage factor and the solubility of gas in solution, is given in Figure 5-35.

Viscosities also vary directly with the densities of oils, and the densities vary with the composition. Thus, the greater the number of carbon atoms in a member of a hydrocarbon series, the greater will be its viscosity as well as its density. Some heavy crudes require heating to make them flow through the

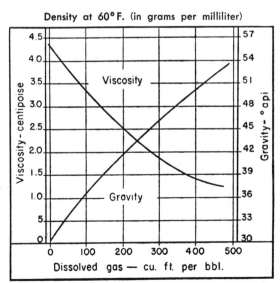

FIGURE 5-34

The effect of dissolved gas on the viscosity and gravity of a crude oil. [Redrawn from O. & G. Jour., January 13, 1944, p. 37.]

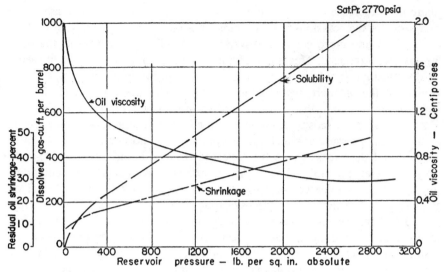

FIGURE 5-35 *The effect of gas in solution and of a drop in pressure on oil from the West Edmond oil field, Oklahoma. The graph is based on a bottom-hole sample analysis.* [Redrawn from Littlefield, Gray, and Godbold, *Trans. Amer. Inst. Min. Met. Engrs.*, Vol. 174 (1948), p. 147, Fig. 8.]

pipe lines. The combined relation of temperature and gravity to the viscosities of different crude oils is shown in Figure 5-36 for a group of Oklahoma oils.

Viscosities are measured by viscosimeters, of which a number have been developed. Each of the common commercial types—the Saybolt Universal, Saybolt Furol, Redwood No. 1, Redwood No. 2, and Engler—is calibrated with its own

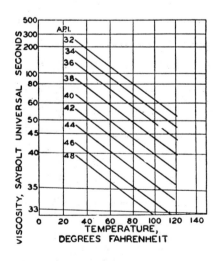

FIGURE 5-36

The progressive decrease in viscosity with an increase in API gravity and an increase in temperature for a group of typical Oklahoma crude oils. [Redrawn from Nelson, *O. & G. Jour.*, January 5, 1946, p. 70.]

scale, and the scales may be converted to poises or stokes* by the use of conversion tables (see the Appendix).[80] The Saybolt Universal Viscosimeter is commonly used in the United States, whereas the Redwood and Engler instruments, which are similar to it, are used in Europe. The measurement is purely arbitrary, the reading being the number of seconds (Saybolt Universal, or S.U., seconds) required for a definite quantity of petroleum under controlled temperature and pressure to flow through a special tube. It is desirable to obtain the viscosity of oil under reservoir conditions of temperature and pressure and with varying amounts of gas in solution. The rolling-ball type of viscosimeter, designed for this purpose, has a steel ball that rolls through an accurately bored barrel filled with oil and standing at an angle. The instrument is sealed at the temperature and pressure desired, and the time the ball takes to roll the length of the tube is electrically timed and calibrated to centistokes.

Refractive Index. The absolute refractive index (RI) of a substance is the inverse ratio of the speed of light in that substance to its speed in a vacuum. The absolute refractive index can be obtained by conversion from the RI measured in air. The refractive index is defined as the ratio of the sine of the angle of incidence to the sine of the angle of refraction, both angles being determined with respect to a normal to the surface. When a ray of light passes from a lighter to a denser substance, it is bent toward the normal, owing to decrease in velocity; when passing from a denser to a lighter medium, it is bent away from the normal. The range of refractive indices for petroleum is from 1.39 to 1.49. It is readily determined with an Abbe refractometer. The measurement offers a quick and fairly accurate method of determining the character of the oil from minute amounts that may be extracted from cores or drill cuttings.[83] The refractive index is also widely used in refinery operations to determine the character of petroleum fractions.

Since the refractive index is dependent on the density of the oil, the heavier (lower API gravity) oils have the higher indices. Table 5-17 shows some representative relations between density and refractive index. The indices of a group of Venezuelan crude oils of varying API gravity are shown in Figure 5-37.

Fluorescence. All oils exhibit more or less fluorescence, also called "bloom" (see p. 84), the aromatic oils being the most fluorescent. The fluorescent colors of crude oils range continuously from yellow through green to blue. This

---

* The CGS unit of viscosity is the *poise*, and the *centipoise* (cp) equals 1/100 poise. A fluid has a viscosity of one poise when a tangential force of one dyne causes a plane surface of one square centimeter area, spaced one centimeter from a stationary plane surface, to move with a constant velocity of one centimeter per second, the space between the planes being filled with the viscous fluid (API Bull. 228, 1941). Air has a viscosity of 1.8×10^{-4}, water of 1×10^{-2}, and gasoline of 0.6×10^{-2} poise. The absolute, or kinematic, viscosity, which is the ratio of the viscosity in poises to the density of the fluid, is expressed in *stokes* and *centistokes* and is used for precise engineering calculations.

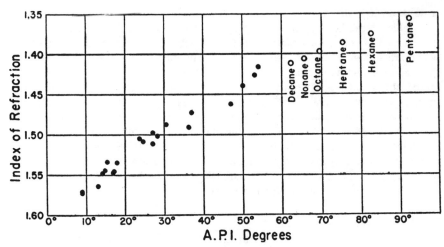

FIGURE 5-37 *The relation of the index of refraction to the API gravity for seventeen Venezuelan crude oils and for some paraffin constituents of light oils. [Redrawn from Hedberg, Bull. Amer. Assoc. Petrol. Geol., Vol. 21 (1937), p. 1473, Fig. 3.]*

property is used in the logging of wells to locate oil showings in the cores, cuttings, and drilling mud.[84] Fluorescence is rapidly reduced by aging, so that fresh oil is easily distinguished from oil previously caught in the drilling mud. Fluorescence is observed under ultraviolet radiation, that most generally used for petroleum having wavelengths of 2,537 and 3,650 angstrom units. Fluorescence permits the detection by the unaided eye of one part of oil in

TABLE 5-17 Refractive Indices of Representative Oils

API Degrees	Density	Refractive Index ($n20D^1$)
6	1.029	1.566
22	0.918	1.509
44	0.802	1.448
58	0.742	1.417
72	0.691	1.390

Source: A. L. Ward, S. S. Kurtz, Jr., and W. H. Fulweiler, "Determination of Density and Refractive Index of Hydrocarbons and Petroleum Products," in *The Science of Petroleum*, Oxford University Press, London and New York, Vol. 2 (1938), p. 1147, Table IV.

[1] The usual symbol for refractive index is n, or velocity in space/velocity in substance. The 20 means 20°C, and the D means that the refractive index is determined for the sodium D line.

100,000 parts of carbon tetrachloride, and with calibrated photographic methods one part in hundreds of millions can be detected.

Optical Activity. Most petroleum has optical activity: the power to rotate the plane of polarization of polarized light. This is measured with a polariscope in degrees per millimeter, and the average range is from 0 to 1.2 degrees. If the plane is rotated to the right, the substance is said to be dextrorotary; if to the left, levorotary. All crudes either are optically active or contain optically active distillation fractions, particularly in the intermediate range (250–300°C at 12 mm Hg). Fractions boiling below 200°C have not been observed to show active optical rotary power,[85] and apparently the property is lost at high temperatures.

It is thought that optical rotary power is confined to organic materials and is caused by the presence of a cholesterin-like substance. Cholesterin (cholesterol), which is an alcohol with the formula $C_{26}H_{45}OH$, is found in both vegetable and animal matter and is a constituent of new milk from fresh cattle. Optical activity is commonly given as an argument for the origin of petroleum from plant or animal remains, because, as far as we know, optically active oils cannot be synthesized inorganically.

Color. The color of petroleum by transmitted light varies from light yellow to red; some very dark or black oils are opaque. The higher the specific gravity (or lower the API gravity), the darker the oil. The cause of the color is not known, but it is thought to be related to the aromatic series of compounds. By reflected light, crude oil is usually green because of its fluorescence. Special refining will produce nearly colorless oils. Color is commonly determined with the Saybolt colorimeter.

Odor. The agreeable (to the producer!) gasoline-like odor of some oils, such as those of Pennsylvania, is due to their content of light hydrocarbons, paraffins, and naphthenes. Some oils of East India, California, and Russia contain a large percentage of aromatics, imparting a pleasant odor. Unsaturated hydrocarbons, sulfur, and certain nitrogen compounds usually impart a disagreeable odor. Among these are the mercaptans, which contain sulfur, but mercaptans are often added because their warning stench helps to detect leakage in pipe lines carrying commercial natural gas. Hydrogen sulfide (a gas) is a common offender, and oils with an odor of H_2S, probably in the gas produced with the oil, characterize some pools in southern Texas and Mexico.

Cloud and Pour Points. Tests are essential to determine the influence of low temperatures on crude oils and also to indicate the amount of solid paraffin waxes present. About 35 cc of the oil in a small glass bottle, with a thermometer inserted in the top, is immersed in a freezing bath or freezing mixture, and from time to time the bottle is removed and tilted. The *cloud point* is the temperature at which the first cloud appears in the oil. It is due to the settling out of the solid paraffin waxes; wax-free naphthenic oils show no cloud point. The temperature attained at the *pour point,* from 2 to 5 degrees lower than the cloud point, is the temperature at which the oil is last fluid and will not flow. Some typical pour points are shown in Table 5-19. If an oil's pour point is above the surface temperature, as it may be during the winter months, if not at other times, the oil will precipitate its paraffins as it approaches the surface and cease to flow until it is heated. Such oils are frequently expensive to handle, as much time and effort are required to keep the wells producing. The pour point of crude oil ranges from $-70°F$ to $+90°F$ or higher. An unusual crude oil of $34°$ API gravity and a pour point between $105°$ and $110°F$ occurs in the Lirik oilfield in central Sumatra. This pour point is $10°$ above the average atmospheric temperature and special pipe line facilities had to be designed to transport the crude to the sea terminal.[86]

Flash and Burning Points. The *flash point* is the temperature at which the vapors rising off the surface of the heated oil will ignite with a flash of very short duration when a flame is passed over the surface. When the oil is heated to a higher temperature, it will ignite and burn with a steady flame at the surface. The lowest temperature at which this will occur is known as the *burning point*. These measurements are a measure of the hazard involved in handling and storing petroleum and petroleum products, and their limits are generally fixed by state law.

Coefficient of Expansion. The coefficient of expansion for an increase in temperature of $1°F$ varies for crude oils from 0.00036 to 0.00096; for most crude oils it ranges between 0.00040 and 0.00065. The average coefficient of expansion for Pennsylvania crude oils is given as 0.000840 and for Baku

TABLE 5-18 Relation Between Gravity and Calorific Value of Crude Oil

Gravity		Calorific Value
Specific Gravity	API Gravity	(Calories per Gram)
0.7 –0.75	70.6–57.2	11,700–11,350
0.75–0.80	57.2–45.4	11,350–11,100
0.80–0.85	45.4–35.0	11,100–10,875
0.85–0.90	35.0–25.7	10,875–10,675
0.90–0.95	25.7–17.5	10,675–10,500

Source: Sherman and Kropff, Jour. Amer. Chem. Soc., Vol. 30, No. 2 (1908), p. 1630; H. S. Bell, *American Petroleum Refining*, 3rd ed., D. Van Nostrand Co., New York (1945), p. 45.

crude oils as 0.000817. In general, the heavier crude oils (low API gravity) have the lower coefficients of expansion, and the lighter crude oils (high API gravity) have the higher coefficients. (See also Table 5-16, p. 200.)

Calorific Value. A calorie is the quantity of heat that will change the temperature of one gram of water from 3.5°C to 4.5°C. This unit is sometimes called the small calorie, and the large calorie equals 1,000 small calories. The British thermal unit (B.T.U.) is the amount of heat necessary to raise the temperature of one pound of water 1°F and is equal to 252 small calories. The calorific value of crude oil decreases as the specific gravity increases (or as API gravity decreases). A rough summary of the relation between the gravities and calorific values of crude oils is given in Table 5-18. The B.T.U. value per pound of crude oil is about 18,300–19,500, compared with 10,200–14,600 per pound of bituminous coal.

The physical characteristics of some representative crude oils are summarized in Table 5-19.

NATURAL GAS

The natural gas of a petroleum reservoir consists of the low-boiling-point hydrocarbon gases and may range from minute quantities dissolved in the oil up to 100 percent of the petroleum content. In addition to the hydrocarbon gases there are gaseous impurities in varying amounts and consisting of hydrogen sulfide, nitrogen, and carbon dioxide. None of these gases is of much commercial value, but helium, which forms as much as 8 percent of some reservoir gases, may be commercially important.

Natural gas may be broadly classified as *associated* when it occurs with oil, and as *nonassociated* when it occurs alone. The natural gas in a petroleum reservoir may occur as *free gas*, as *gas dissolved in oil*, as *gas dissolved in water*, or as *liquefied gas*.

Free Gas. Free gas, when present, occupies the upper part of the reservoir and may be underlain either by oil (associated gas) or by water (nonassociated gas).*

Gas Dissolved in Oil. When oil and gas are in intimate contact, a certain amount of gas dissolves in the oil. The amount of gas in solution depends on the physical characteristics of both the gas and the oil, and on the pressure and temperature in the reservoir. With few exceptions, all oil occurring in pools contains some gas in solution, ranging from a few cubic feet up to thousands of cubic feet per barrel. Where the amount of gas is small, it is permitted to

---

* Natural gas is nearly everywhere considered to be the gaseous phase of petroleum. In Canada, however, the traditional usage has been legally changed, and nonassociated natural gas is considered not petroleum but a separate substance, whereas associated gas is considered as a petroleum substance. (See Borys v. Can. Pac. Ry and Imperial Oil Ltd., Jud. Comm. Privy Council Judgment given January 1953.)

TABLE 5-19 Physical Characteristics of Some Typical Crude Oils

Field, Location, Producing Formation	Gravity (Average)		Viscosity, Saybolt Universal, Seconds, 100°F	Pour Point	Reference
	Spec. Gr.	°API			
Powell, Northeastern Texas, Woodbine sand (Cretaceous)		37	42		1
Bradford, Pennsylvania, Bradford sand (Devonian)	0.801	45.2	38	below 5°F	2
Oklahoma City, Oklahoma, Simpson sands (Ordovician)	0.835	38	45	5°F	3
Rangely, Colorado, Weber sand (Pennsylvanian)	0.85–0.87	31.3–35.2	45–53	below 5°F	4
Rodessa, Texas-Louisiana, oolitic lime (L. Cretaceous)	0.812	42.8	39	below 5°F	5
Kirkuk, Iraq, Asmari limestone (Oligocene-Miocene)	0.844	36.1	350	below 0°F	6
Abqaiq, Saudi Arabia, "D" Member, Arab Zone (Jurassic)	0.84	37	40.2	−15°F	7
Lagunillas, Venezuela, La Rosa (L. Miocene)	0.948	17.8	992	−20°F	8
Spring Creek, Park County, Wyoming, Madison limestone (Mississippian)		12.6	6,000+	30°F	9

1. A. J. Kraemer and Gustav Wade, *Tabulated Analyses of Texas Crude Oils*, Tech. Paper 607, U.S. Bur. Mines (1939), p. 19, No. 48.
2. E. C. Lane and E. L. Graton, RI 3385, U.S. Bur. Mines (1938), p. 19.
3. H. B. Hill, E. L. Ramlios, and C. R. Bopp, *Engineering Report on Oklahoma City Oil Field, Oklahoma*, RI 3330, U.S. Bur. Mines (January 1937), p. 207, Table 65, No. 3.
4. Cecil Q. Cupps, Philip H. Lipstate, and Joseph Fry, *Variance in Characteristics of the Oil in the Weber Sandstone Reservoir, Rangely Field, Colorado*, RI 4761, U.S. Bur. Mines (April 1951), pp. 60–68.
5. H. B. Hill and R. K. Guthrie, *Engineering Study of the Rodessa Oil Field in Louisiana, Texas, and Arkansas*, RI 3715, U.S. Bur. Mines (August 1943), p. 8.
6. "Crude Oils," in *The Science of Petroleum*, Oxford University Press, London and New York, Vol. 5 (1951), Part I, p. 31.
7. *Ibid.*, p. 23.
8. *Ibid.*, p. 21.
9. W. J. Wenger and W. J. Lanum, "Characteristics of Crude Oils from Big Horn Basin Fields," *Petrol. Eng.*, February 1953, A-56, Table 6.

pass off into the air as the oil is produced. Where it is somewhat larger, the custom is to separate the oil and the gas and to burn the gas in flares, so as to prevent its accumulating in dangerous quantities in low places at the surface of the ground. Still larger quantities are saved and sold, used as a source of power on the property, or reinjected into the petroleum reservoir.

Pools in which all the gas is dissolved in the oil are called *undersaturated pools* to distingush them from *saturated pools,* in which an excess of gas forms a free-gas cap. The temperature and pressure at which the gas first begins to come out of solution are called the *bubble point*. Since the temperature remains more or less constant in the reservoir during production, whereas the pressure changes, the bubble-point pressure often is called the *saturation pressure* (same as vapor pressure). Where the saturation pressure is the same as the reservoir pressure, the oil in the pool contains all the gas it can hold in solution without any bubbling. Where the reservoir pressure is above the saturation pressure, the oil could hold more gas in solution than it actually does. Saturation pressures range from reservoir pressures down to atmospheric pressure. The relation in saturated and undersaturated pools is shown diagrammatically in Figure 5-38.

Gas Dissolved in Water. Gas is also dissolved in oil-field water. The solubilities of natural gas in oil-field water may be as much as 20 cu ft/bbl

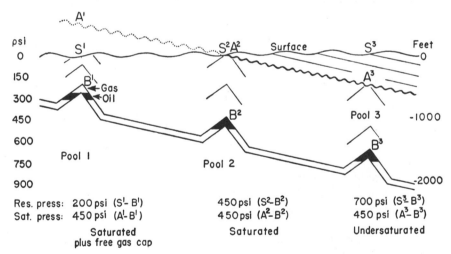

FIGURE 5-38 *Diagrammatic sketch showing the relations of the same amount of gas and oil under different conditions of pressure. In* Pool 1 *the reservoir pressure (based on depth* S^1B^1*) is less than the saturation pressure* A^1B^1 *and consequently there is a free-gas cap, which holds the gas in excess of that necessary to saturate the oil. The saturation pressure and the reservoir pressure are equal in* Pool 2, *and the pool is said to be saturated. The reservoir pressure* S^3B^3 *is greater than the saturation pressure* A^3B^3 *in* Pool 3, *and the pool is said to be undersaturated—it could hold more gas in solution.*

at 5,000 psi in the temperature range of many oil reservoirs; there is a reduction of about 5 percent in solubility per 1 percent increase (weight) of dissolved mineral matter when the water is a brine.[87] Field evidence suggests that the solubility of gas in reservoir water is about 6 percent of its solubility in oil.* When it is remembered that as much as 50 percent of the pore space of a reservoir is occupied by interstitial water, the amount of dissolved gas in the reservoir water may be important.

Liquefied Gas. Under reservoir conditions of high pressure, generally greater than 5,000 to 6,000 psia, and high temperature, natural gas and crude oil become physically indistinguishable. A new set of conditions controls their relations. These conditions are considered more fully in Chapter 9.

Measurement of Natural Gas

Natural gas is commonly measured [88] in cubic feet.† Since gas always fills its container, the amount of gas present depends upon the pressure and temperature. The measurement is therefore dependent upon a constant set of conditions. The standard conditions for reference are a temperature of 60°F and a pressure of 30 inches of mercury (approximately 14.73 psi), but a reference temperature of 20°C (68°F) is sometimes used. Gas volumes are written in multiples of 1,000, abbreviated as M; thus 3,540,000 cubic feet of gas would be 3,540 Mcf.

Various devices for measuring the quantity (volume) of gas flowing through a pipe have been developed.[89] Most measurements of gas being produced from wells are made by orifice meters,[90] which determine the difference in pressure between opposite sides of a constriction or orifice in the pipe line. The rate of flow may be calculated from the differential pressure, combined with the characteristics of the orifice, which is generally a round hole in a thin plate. Where gas is flowing at low rates and near atmospheric pressure, displacement meters are generally used. These meters measure the gas volumetrically as chambers are alternately filled and emptied, and the number of times the chambers are filled and emptied is recorded by a counter. The small amount of gas entrained in the drilling mud or contained in well cuttings is commonly identified by the use of gas detectors. (See p. 84.)

The volume of gas in place in a gas reservoir is measured in terms of the volume of gas at the surface by either of two general methods, somewhat similar to those used for the measurement of equivalent stock-tank oil in the oil reservoir.[91] The volumetric, or saturation, method consists in multiplying the acre-feet of volume of pore space containing gas by the ratio between the reservoir pressure and the surface pressure in atmospheres, and by a temperature correction

---

* I. I. Gardescu in a personal communication.

† In some countries, particularly the USSR, natural gas measurements are often converted into metric tons of oil of equivalent B.T.U. value; 1,000 cubic meters of natural gas = 0.824 metric ton of crude oil.

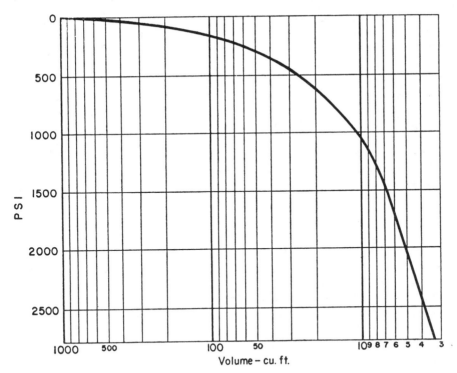

FIGURE 5-39 *Generalized chart of the change in volume of gas as the pressure is increased at constant temperature.*

that depends on how much the temperature differs from 60°F. The pressure ratio is expressed by the law of gases, according to which the volume of a perfect gas varies inversely with the pressure at constant temperature. (See Fig. 5-39.) With an assumed atmospheric pressure of 14.7 pounds per square inch, the correction necessary to bring to atmospheric pressure the gas in a reservoir under a pressure of 3,000 psi would be calculated by multiplying the volume by a pressure ratio factor of

$$\frac{3{,}000 + 14.7}{14.7} = 209.3 \text{ atmospheres.}$$

The volume also varies directly with the absolute temperature. The volume of a surface-equivalent gas at a reservoir temperature of 140°F would be reduced, when it reached a surface temperature of 60°F, according to the temperature correction factor

$$\frac{460 + 60}{460 + 140} = \frac{520}{600} = 0.867.$$

The pressure-volume method of estimating the amount of equivalent surface gas in a gas reservoir is based on the fact that, as the gas is produced from a reservoir, the reservoir pressure drops. The drop in pressure per unit volume of surface-equivalent gas removed from the reservoir is proportional to the volume of surface-equivalent gas remaining in the reservoir. For example, if the original pressure on a gas reservoir was 2,880 psi, and if after production of 400 million cubic feet of gas during a period of several years the pressure had declined to 2,720 psi, the pressure drop of 160 psi took place at the rate of $\frac{400,000,000}{160}$, or at the rate of 2,500,000 cubic feet per unit of pressure drop. The indicated remaining surface-equivalent gas in the reservoir is then 2.5 million cubic feet multiplied by 2,720, or 6.8 billion cubic feet. If the pressure of the reservoir at exhaustion of the reserve is taken to be 250 psi, for example, the recoverable reserve would be 2,500,000 cubic feet $\times$ (2,720 − 250), or 6,175,000,000 cubic feet of surface-equivalent gas. Measurements of this kind are valid only after some production history is available.

Composition of Natural Gas

Methane (CH_4), which is the most stable of all the petroleum hydrocarbons, makes up the greater part of the hydrocarbon content of natural gas. Variable but small amounts of other paraffin hydrocarbons, such as ethane (C_2H_6), propane (C_3H_8), butane (C_4H_{10}), pentane (C_5H_{12}), and hexane (C_6H_{14}), are often present, and in some cases even heptane (C_7H_{16}), octane (C_8H_{18}), and nonane (C_9H_{20}). Free hydrogen is only rarely found in natural gas, as in some volcanic regions and around some of the German salt mines. Carbon monoxide and unsaturated gases are present only in very minor amounts in some gases. Methane is not condensable at temperatures and pressures possible within the oil reservoirs, so that methane is always present as a gas; at elevated pressures it may be dissolved in the liquid hydrocarbons. The critical temperatures of the other hydrocarbons in natural gas are such that they may exist, under the conditions normally found in oil and gas reservoirs, either in the liquid or in the vapor phases. The total amount of carbon in a typical natural gas is about 35 pounds per thousand cubic feet. About a third of this may be recovered as *carbon black,* which consists of extremely fine particles of carbon from 10 to 150 microns in diameter, when the gas is burned in a limited supply of air. The location of many of the large gas pools may be identified from a long distance by the great clouds of black smoke rising from the carbon-black plants. The density of natural gas ranges from that of methane, which is 0.554 relative to air, up to densities higher than air for some wet gases. Most natural gases have a density between 0.65 and 0.90 (air = 1.0). Analyses of gas from some of the large gas pools are shown in Table 5-20.

Natural gas as it comes from the well is classified in the field as *dry gas, lean gas,* or *wet gas,* according to the amount of natural gas liquid vapors it contains.[92] A dry gas contains less than 0.1 gallon natural gas liquid vapors per 1,000 cubic feet, and a wet gas 0.3 or more. *Residue gas* is natural gas from which the vapors of natural gas liquids have been extracted. Gas recov-

TABLE 5-20 Average Composition of Various Commercial Natural Gases

Pool and Location	Spec. Gr. (Air = 1.0)	Methane	Ethane C_2H_6	Propane C_3H_8	Butane C_4H_{10}	Pentane and Heavier	CO_2	Reference
United States								
Panhandle-Amarillo		91.3	3.2	1.7	0.9	0.56	0.1	1
Hugoton, Kansas		74.3	5.8	3.5	1.5	0.6+		2
Carthage field, Texas	0.616	92.54	4.7	1.3	0.8	0.6		3
Velma, Oklahoma		82.41	6.34	4.91	2.16	1.18		4
Canada								
Turner Valley, Alta.		92.6	4.1	2.5	0.7	0.13		5
Trinidad								
Barrackapore (Oligocene)		95.65	2.25	1.55	0.80	0.25		6
Venezuela								
La Concepción	0.89	70.9	8.2	8.2	6.2	3.7	2.8	7
Cumarebo	0.965	63.89	9.49	14.41	8.80	5.4		8
USSR, Baku								
3 fields, av.	0.661	88.0	2.26	0.7	0.7	0.5	6.5	9
Grozny-Staror-Groznenskii	0.975	69.4	6.5	10.4	6.1	7.6		9
Kuban–Black Sea–Shirokaya Balka	0.67	93.3	1.4	0.7	2.1	2.5		9
Emba-Kuslary	0.757	82.7	2.3	4.2	5.6	2.2	1.5	9

1. Victor Cotner and H. E. Crum, "Natural Gas in Amarillo District, Texas," in *Geology of Natural Gas*, Amer. Assoc. Petrol. Geol., Tulsa, Okla. (1935), p. 409. Average of three samples from Wheeler and Carson counties.
2. C. H. Keplinger, J. R. Wanemacher, and K. R. Burns, "Hugoton, World's Largest Dry Gas Field in Amazing Development," O. & G. Jour., January 6, 1949, pp. 84–88.
3. G. C. Clark and Jack M. DeLong, "Carthage Field, Panola County, Texas," in *Occurrence of Oil and Gas*, Pub. 5116, University of Texas (August 1951), p. 59.
4. Richard B. Rutledge, "Velma Oil Field, Stephens County, Oklahoma," Petrol. Engr., December 1953, pp. 8–47.
5. S. E. Slipper, "Natural Gas in Alberta," in *Geology of Natural Gas*, Amer. Assoc. Petrol. Geol., Tulsa, Okla. (1935), p. 51.
6. H. H. Suter, "The General and Economic Geology of Trinidad, BWI," in *Colonial Geol. and Min. Res.*, Vol. 4 (1952), p. 28.
7. Staff of Caribbean Petroleum Company, "Oil Fields of Royal Dutch–Shell Group in Western Venezuela," Bull. Amer. Assoc. Petrol. Geol., Vol. 32 (April 1948), p. 595.
8. A. L. Payne, "Cumarebo Oil Field, Falcón, Venezuela," Bull. Amer. Assoc. Petrol. Geol., Vol. 35 (August 1951), p. 1870.
9. Adapted from Demitri B. Shimkin, "Is Petroleum a Soviet Weakness?" O. & G. Jour., December 21, 1950, pp. 214–226. 34 references. The average is from the Lenin, Bibi Eibat, and Surakhan fields.

ered at the surface from an oil well is called *casinghead gas*. The terms *sweet gas* and *sour gas* are also used in the field to designate gases that are low or high, respectively, in hydrogen sulfide. The natural gas of commerce, as delivered to the pipe lines, usually ranges between 900 and 1,200 B.T.U. per cubic foot and has the following general composition:

Methane (CH_4)	72.3 percent (active)
Ethane (C_2H_6)	14.4 percent (active)
Carbon dioxide (CO_2)	0.5 percent (inert)
Nitrogen (N_2)	12.8 percent (inert)

Impurities in Natural Gas

The principal gaseous impurities in natural gas are nitrogen, carbon dioxide, and hydrogen sulfide. They are called the *inerts*. Helium is an impurity that may be present in comparatively small amounts, yet it has considerable importance.

The presence of any large amount of carbon dioxide and nitrogen in natural gas decreases its inflammability and thus lowers its B.T.U. value. Nitrogen and carbon dioxide also raise the temperature necessary for combustion. The limit of inflammability of propane gas is reached when the weight ratio of carbon dioxide to propane is about 8:1, and for nitrogen when the ratio is about 15:1. For butane the limit of inflammability is reached when the weight ratio is about 9:1 for carbon dioxide and about 16:1 for nitrogen. Natural gases under high pressure containing so much carbon dioxide and nitrogen that they are noninflammable are sometimes used, however, as a source of power in oil-field operations, being fed into the boilers, where their expansive force acts like that of steam. Natural gases that are noninflammable are referred to as "wind" gases.

Helium. Helium (He) is a light, colorless, odorless, chemically inert element,[93] which exists as a gas at normal conditions of pressure and temperature. Helium is one of the noble gases (helium, neon, argon, krypton, xenon, and radon). Helium was first detected in 1868 by means of an unidentified yellow line in the sun's spectrum; it was first recognized as an element on earth in 1895. Helium is found in the atmosphere in the proportion of five parts per million by volume; in some uranium minerals; and in measurable volumes in the gases from some mines, fumaroles, and mineral spring waters; but by far the largest volumes of helium are found in natural gases, in some of which the proportion is as much as 8 percent by volume.

The origin of such large amounts of this chemically inert gas in natural gas deposits is still a problem. The fact that helium is given off in the disintegration of such radioactive elements as uranium, radium, and thorium has led to the belief that radioactivity is the original source of helium. Helium ions are discharged as positive, doubly charged particles, the nuclei of which are alpha rays. The rate of discharge of these alpha rays from a known amount of radioactive element can be calculated, and this permits a volumetric calculation of the amount of helium so formed in a given time. And, conversely, geologic age is measured by the "helium method." Table 5-21 shows the rate of production of helium by

TABLE 5-21 Rate of He Production by Various Radioactive Elements

Radioactive Substance	No. of Alpha Particles per Gram per Second	Helium Produced per Gram per Year (cu mm)
Uranium	2.37×10^{-4}	2.75×10^{-5}
Uranium in equilibrium with all its products	$9.7 \times 10_4$	11.0×10^{-5}
Thorium in equilibrium with all its products	$2.7 \times 10_4$	3.1×10^{-5}
Radium in equilibrium with emanation, radium A, and radium C	13.6×10^{10}	158

the different radioactive elements. From the estimated average content of radioactive substances in the rocks of the earth's crust, Rogers calculates that from 282 to 1,060 million cubic feet of helium is generated annually.[94] Wells has shown that the permeability of silica-rich igneous rocks at moderate temperatures— from 200° to 500°C—is much higher for helium than for other gases.[95] The conclusion follows that most helium probably originated from radioactive disintegration in the rocks not far beneath the formations in which helium-bearing gas occurs in the sediments. It is noteworthy that radioactive emanations are extremely soluble in oil. Boyle[96] found that radium emanations are about fifty times as soluble in refined petroleum oils as in water; helium could, therefore, have been liberated from the oils, as helium is very insoluble in oils, and it would have been expelled almost as fast as it was formed.

As there is no conclusive evidence that all helium originated in radioactive substances, large amounts of helium are considered by many to be of primordial origin. Nitrogen, moreover, is a common accessory of gases containing helium, and the origin of the high nitrogen content of natural gases rich in helium, as well as the origin of the helium, may be explained by some primordial origin. (See nitrogen, below.)

The only gas fields that have hitherto yielded natural gas rich enough in helium to warrant its extraction have been found in the United States. In these fields the helium content varies from less than 1 to 8 percent by volume.[97] It is possible, however, that helium-bearing gas pools have been discovered, and that helium is being extracted, in the USSR.

The helium-bearing gas found in the Ouray formation (Mississippian and Devonian) in the Rattlesnake oil and gas field, New Mexico, has an unusually high nitrogen content.[98] An analysis of it follows (amounts in percent):

Carbon dioxide	2.8
Oxygen	0.0
Methane	14.2
Ethane	2.8
Nitrogen	72.6
Helium	7.6
	100.0

Nitrogen. The element nitrogen (N_2) is a colorless, tasteless, odorless gas composing about 78 percent of dry air. It is found in natural gases in amounts up to 99 percent by volume, in mine gases up to almost 100 percent, in igneous rocks, in gases from mineral springs and geysers, in fumaroles, and dissolved in sea, rain, and mineral waters. There are at least two possible sources for the nitrogen in natural gas. The high percentage of nitrogen in air, coupled with its general chemical inactivity, suggests that the nitrogen of natural gases merely represents the nitrogen content of air that has been trapped in the sediments, perhaps with additions from igneous sources and from the decomposition of nitric organic compounds. The absence of oxygen is accredited to its removal through the oxidation of minerals included within the formations. A second possible source is suggested by the high nitrogen content in gases rich in helium. Perhaps the two gases in such cases had a common origin. Goslin found, for example, that there was a rapid liberation of nitrogen from animal and vegetable proteins when radium was added to flasks filled with water, fish, water plants, and soil.[99] High nitrogen-gas content in natural gases is not related to the nitrogen-compound content of associated petroleum deposits. Gases rich in nitrogen have been found in association with oils low in nitrogen compounds, and vice versa.

Large amounts of nitrogen are common in many gas pools in the Mid-Continent and Rocky Mountain regions, the nitrogen content varying from a few percent to almost 100 percent. The Westbrook field of Mitchell County, Texas, for example, produces gas that is 85–95 percent nitrogen. As the nitrogen content of air is high, there is very little use for the nitrogen from natural gas.

Carbon Dioxide. Carbon Dioxide (CO_2) is a colorless, odorless, noninflammable gas about one and a half times as heavy as air. It is easily soluble in water, one volume of water being capable of dissolving one volume of gas at surface temperature and pressure. At atmospheric temperatures and pressure it is inert; in concentrations over 8 percent it is toxic, producing unconsciousness by smothering. Carbon dioxide is generated in nature by the action of acids on carbonates and bicarbonates in igneous, sedimentary, and metamorphic rocks; by oxidation of hydrocarbons through contact with mineralized waters; by heating of carbonates and bicarbonates; and by the action of certain anaerobic bacteria in attacking hydrocarbons. Carbon dioxide is a normal constituent of atmospheric air to the extent of 3 volumes per 10,000 volumes of air. It is found in various amounts in all kinds of igneous, metamorphic, and sedimentary rocks. It is discharged in tremendous quantities in volcanic emanations, and is carried in solution in rain water, ocean water, and the waters of mineral springs. Natural gases rich in carbon dioxide occur principally in the western states of Montana, Colorado, Utah, and New Mexico, and oil is produced along with carbon dioxide gas in North Park, Colorado. Some California natural gases contain up to 49 percent carbon dioxide. The highest concentrations of carbon dioxide are found in pools in New Mexico, where large volumes of 99+ percent pure carbon dioxide occur, and where a number of wells capable of producing from 12 to 26 million cubic feet of carbon dioxide gas have been found.

The large carbon dioxide pools in New Mexico and Mexico are believed to have been given off from volcanic emanations and to have been formed in part by the heating action of igneous rocks coming in contact with limestones, the carbon dioxide being driven off as it is in a lime kiln. The great majority of the New Mexico pools rich in carbon dioxide gas lie within a few miles of regions of recent igneous activity.[100] The carbon dioxide content of certain California natural gases is thought to have resulted from the oxidation of the hydrocarbons through contact with mineralized waters.[101]

An interesting result of drilling wells rich in carbon dioxide is that the sudden expansion of the gas in the bore hole chills the pipe, drilling tools, and surface equipment to very low temperatures. The low temperatures consequent on the production of gas at the McCallum field, in Jackson County, Colorado, for example, which is 92 percent carbon dioxide, results in the covering of the pipes and tanks with a thick coating of ice or snow, even on the hottest days.[102]

Below is the analysis (in percent) of a composite sample of a gas high in carbon dioxide produced with the oil from the Tensleep sandstone (Pennsylvanian) in the Wertz Dome field, Wyoming:

Oxygen	0	Propane	3.35
Nitrogen	4.09	Isobutane	0.79
Carbon dioxide	42.00	normal Butane	2.05
Hydrogen sulfide	1.11	Isopentane	0.49
Methane	26.80	n-Pentane	0.54
Ethane	6.35	Hexane	+0.43

This gas has a B.T.U. value of 677, is produced with an oil of API gravity 35.3°, and is carrying 1.33 percent sulfur. The average natural gas of commerce has 1,075 B.T.U. per 1,000 cubic feet.

Hydrogen Sulfide. Hydrogen sulfide (H_2S) is a colorless gas with a characteristic foul odor. It is very soluble in water, one volume of water dissolving 4.3 volumes of the gas at 0°C and one atmosphere pressure, and is generally even more soluble in hydrocarbons than in water. Hydrogen sulfide has a decided corrosive effect upon metals, whether as a free gas or in solution in petroleum or reservoir waters. Small concentrations of the gas are toxic: concentrations as low as 0.005 percent cause sub-acute poisoning when breathed over a long period of time; concentrations of 0.06–0.08 percent in experiments with dogs caused immediate poisoning followed by suspension of breathing, heart failure, and death.[103] Natural gases with even smaller amounts of hydrogen sulfide are unsuitable for use as public fuels; some states have laws requiring that, if a natural gas contains more than 20 or 30 grains of H_2S per 100 cubic feet, the H_2S must be removed before the gas can be used commercially.[104] These problems arise, then, in the production of natural gases, petroleum, or reservoir waters rich in hydrogen sulfide: (1) a safety problem, (2) corrosion of equipment, and (3) necessary treatment to remove the hydrogen sulfide from the oil or gas.

Although hydrogen sulfide occurs in volcanic emanations and in the gases from certain mineral springs, and through the putrefaction of vegetable and animal matter, the hydrogen sulfide found in natural gas and petroleum is believed to have been formed by the reduction of sulfates to sulfides through organic and inorganic means. In brackish or stagnant water that contains no dissolved oxygen, bacteria begin to act upon salts containing chemically combined oxygen, such as sulfates in organic matter and those derived from weathered minerals and dissolved in the water. Various microorganisms have been isolated that have the power to form hydrogen sulfide by reducing sulfates to sulfides from the ooze found in brackish lakes. Bacteria capable of this action have been found in vegetable mold, in ditch mud, in the bottom water of inland seas, in lake sediments, in borings from shallow wells, and in some oil-well waters. The temperatures most favorable to the development of these bacteria are between 25° and 50°C. An inorganic mechanism for the production of hydrogen sulfide might consist in the natural reduction of sulfates according to the equation

$$2C + MeSO_4 + H_2O \longrightarrow MeCO_3 + CO_2 + H_2S$$

where C is an organic compound and Me a metal. There is considerable dispute over which of these two possible sources is the more important; the present trend, however, seems to favor the reduction of sulfates by organic means.

Natural gas high in hydrogen sulfide is found at many localities.[105] Some of the better-known areas are the Texas Panhandle, western Texas, southeastern New Mexico (where the gas occurs in Permian and Pennsylvanian sediments), the Tampico-Tuxpan area of Mexico (where gas highly charged with H_2S and CO_2 caused many fatalities among the workmen), the Texas and Louisiana Gulf Coast salt domes, and the oil fields of Iran. The presence of abundant gypsum in both the Gulf Coast and Iran suggests an origin due to the reducing action of bituminous matter on gypsum.

An exceptionally high hydrogen sulfide content of natural gases has been encountered in eastern Texas, Arkansas, and Wyoming. An analysis (in percent) of one such high hydrogen sulfide natural gas, found near Emory, northeastern Texas,[106] follows:

Carbon dioxide	4.50	n-Butane	1.28
Hydrogen sulfide	42.40	Isopentane	0.28
Methane	39.56	Pentane	0.30
Ethane	6.17	Hexanes (plus)	0.13
Propane	2.89		100.00
Isobutane	0.92		

The calculated heat value of this gas is 956 B.T.U. per 1,000 cubic feet. Its specific gravity is 0.973 (air = 1.0), and it is estimated that the gas will yield, through processing of the hydrogen sulfide, fifteen tons of sulfur per million cubic feet of gas.

The gas dissolved in the oil from the Eocene marl of the Majid-i-Sulaiman oil field, in Iran, contains 40 percent H_2S.[107]

CONCLUSION

We may summarize some of the more significant features of reservoir fluid content as follows:

1. Each of the fluids present—water, crude oil, and natural gas—is widely variable in the details of its chemical and physical characteristics.

2. The chemical and physical properties and the relative saturations of the fluids enter into all problems involving the migration and accumulation of oil and gas into a pool and the efficient production of the oil and gas content of the pool.

3. The more important of the measurements commonly made of the fluids include, for oil-field waters, the saturation, the rate of production, the concentration of dissolved salts, and the chemical composition; for crude oil, the saturation, the rate of production per day per pound of reservoir pressure decline, the chemical composition, the gravity, and the viscosity; for natural gas, the volume, the rate of flow per pound of reservoir pressure decline, the gasoline vapors present, the impurities present, and the sulfur content.

Selected General Readings

OIL-FIELD WATERS

Chase Palmer, *The Geochemical Interpretation of Water Analysis,* Bull. 479, U.S. Geol. Surv. (1911), 31 pages. Original discussion of the "Palmer system" of water analysis.

G. Sherburne Rogers, *Chemical Relations of the Oil Field Waters in San Joaquin Valley, California,* Bull. 653, U.S. Geol. Surv. (1917), 119 pages.

G. Sherburne Rogers, *Geochemical Relations of the Oil, Gas and Water, the Sunset-Midway Oil Field, California,* Prof. Paper 117, U.S. Geol. Surv. (1919),

103 pages. An extensive analysis of some of the relations of oil-field water to specific geologic situations.

James G. Crawford, "Oil-Field Waters of Wyoming and Their Relation to Geological Formations," Bull. Amer. Assoc. Petrol. Geol., Vol. 24 (July 1940), pp. 1214–1329. Contains a discussion of many characteristics and interpretations of oil-field water analyses.

L. C. Case, "Application of Oil Field Water to Geology and Production," Oil Weekly, October 29, 1945, pp. 48–54. An authoritative discussion of the applications of oil-field water analyses.

L. U. de Sitter, "Diagenesis of Oil-Field Brines," Bull. Amer. Assoc. Petrol. Geol., Vol. 31 (November 1947), pp. 2030–2040. Contains graphic presentation of many oil-field water analyses and their interpretation.

T. A. Pollard and Paul P. Reichertz, "Core-Analysis Practices—Basic Methods and New Developments," Bull. Amer. Assoc. Petrol. Geol., Vol. 36 (February 1952), pp. 230–252. 31 references. Describes various methods for determining liquid saturation of cores.

OIL

William A. Gruse and Donald R. Stevens, *The Chemical Technology of Petroleum,* 2nd ed., McGraw-Hill Book Co., New York (1942), 733 pages. A standard reference work.

H. S. Bell, *American Petroleum Refining,* 3rd ed., D. Van Nostrand Co., New York (1945), 619 pages. A standard reference work.

C. M. McKinney and O. C. Blade, *Analyses of Crude Oil from 283 Important Oil Fields in the United States,* RI 4289, U.S. Bur. Mines (May 1948).

O. C. Blade, E. L. Garton, and C. M. McKinney, *Analyses of Some Crude Oils from the Middle East, South America, and Canada,* RI 4657, U.S. Bur. Mines (1948), 45 pages.

Benjamin T. Brooks and A. E. Dunstan (editors), "Crude Oils, Chemical and Physical Properties," in *The Science of Petroleum,* Oxford University Press, London and New York (1950), Vol. 5, Part I, 200 pages. Many analyses and descriptive articles on crude oil.

Emil J. Burcik, *Properties of Petroleum Reservoir Fluids,* John Wiley & Sons, New York and London (1957), 190 pages.

Frederick D. Rossini, "Hydrocarbons in Petroleum," Jour. Chem. Education, Vol. 37, No. 11 (November 1960), pp. 554–561.

GAS

Henry A. Ley (editor), *Geology of Natural Gas,* Amer. Assoc. Petrol. Geol., Tulsa, Okla. (1935), 1,227 pages. 38 articles by 47 authors. A standard reference work.

C. E. Dobbin, "Geology of Natural Gases Rich in Helium, Nitrogen, Carbon

Dioxide, and Hydrogen Sulphide," in *Geology of Natural Gas,* Amer. Assoc. Petrol. Geol., Tulsa, Okla. (1935), pp. 1053–1072.

Norman J. Clark, "It Pays to Know Your Petroleum," World Oil: Part I, March 1953, pp. 165–172; Part II, April 1953, pp. 208–213. A clear explanation of the effects of changing reservoir conditions on oil and gas.

Reference Notes

1. Peter Grandone and Alton B. Cook, *Collecting and Examining Subsurface Samples of Petroleum,* Tech. Paper 629, U.S. Bur. Mines (1941), 67 pages.

2. Howard C. Pyle and John E. Sherborne, "Core Analysis," Tech. Pub. 1024, Petrol. Technol. (February 1939), and Trans., Amer. Inst. Min. Met. Engrs., Vol. 132 (1939), pp. 33–61. 15 references.

T. A. Pollard and Paul P. Reichertz, "Core-Analysis Practices—Basic Methods and New Developments," Bull. Amer. Assoc. Petrol. Geol., Vol. 36 (February 1952), pp. 230–252. 31 references.

3. Ralph Alexander Liddle, *The Van Oil Field, Van Zandt County, Texas,* Bull. 3601, University of Texas (January 1936), 82 pages, 26 maps and sections.

4. Cecil Q. Cupps, Philip H. Lipstate, Jr., and Joseph Fry, *Variance in Characteristics of the Oil in the Weber Sandstone Reservoir, Rangely Field, Colorado,* RI 4761, U.S. Bur. Mines (April 1951), 68 pages.

Ralph H. Espach and Joseph Fry, *Variable Characteristics of the Oil in the Tensleep Sandstone Reservoir,* Elk Basin Field, Wyoming and Montana, RI 4768, U.S. Bur. Mines (April 1951), 24 pages.

5. L. C. Case *et al.,* "Selected Annotated Bibliography on Oil-Field Waters," Bull. Amer. Assoc. Petrol. Geol., Vol. 26 (May 1942), pp. 865–881.

6. Theron Wasson and Isabel B. Wasson, "Cabin Creek Field, West Virginia," Bull. Amer. Assoc. Petrol. Geol., Vol. 11 (1927), pp. 705–719.

7. David B. Reger, "The Copley Oil Pool of West Virginia," Bull. Amer. Assoc. Petrol. Geol., Vol 11 (1927), pp. 581–599.

8. Ralph E. Davis and Eugene A. Stephenson, "Synclinal Oil Fields in Southern West Virginia," *Structure of Typical American Oil Fields,* Amer. Assoc. Petrol. Geol., Tulsa, Okla., Vol. 2 (1929), pp. 571–576.

9. James C. Crawford, "Waters of Producing Fields in the Rocky Mountain Region," Tech. Pub. 2383, Trans. Amer. Inst. Min. Met. Engrs., Vol. 179 (1949), pp. 264–285. Bibliog. 12 items.

10. L. C. Case, "The Contrast in Initial and Present Application of the Term 'Connate' Water," Jour. Petrol. Technol., Vol. 8, No. 4 (April 1956), p. 12.

11. Charles R. Fettke, "Core Studies of the Second Sand of the Venango Group from Oil City, Pa.," Trans. Amer. Inst. Min. Met. Engrs. (1927), pp. 219–230. Fettke was the first person to point out, in 1926, that an oil sand was not originally saturated with oil but contained appreciable quantities of water.

12. W. A. Bruce and H. J. Welge, "The Restored-state Method for Determination of Oil in Place and Connate Water," in *Production Practice and Technology,* Amer. Petrol. Inst. (1947), pp. 166–174.

13. Parker D. Trask, "Compaction of Sediments," Bull. Amer. Assoc. Petrol. Geol., Vol. 15 (1931), pp. 271–276.

14. Cleo Griffith Rall and D. B. Taliaferro, *A Method for Determining Simultaneously the Oil and Water Saturations of Oil Sands,* RI 4004, U.S. Bur. Mines (December 1946), 16 pages. Bibliog. 34 items.

John G. Caran, "Core Analysis—an Aid to Profitable Completions," Mines Magazine (February 1947), pp. 19–24. 15 references.

T. A. Pollard and Paul. P. Reichertz, *op. cit.* (note 2). Discussion of several methods of determining water saturation of reservoir rocks.

15. G. E. Archie, "The Electrical Resistivity Log as an Aid in Determining Some Reservoir Characteristics," Trans. Amer. Inst. Min. Met. Engrs., Vol. 146 (1942), pp. 54–62.

G. E. Archie, "Introduction to Petrophysics of Reservoir Rocks," Bull. Amer. Assoc. Petrol. Geol., Vol. 34 (May 1950), pp. 943–961.

W. O. Winsauer, H. M. Shearin, P. H. Masson, and M. Williams, "Resistivity of Brine-Saturated Sands in Relation to Pore Geometry," Bull. Amer. Assoc. Petrol. Geol., Vol. 36 (February 1952), pp. 253–277. 14 references.

R. G. Hamilton, "Exploitation of Geologic, Lithologic, Reservoir and Drilling Data to Improve Electric Log Interpretation," Tulsa Geol. Soc. Digest, Vol. 21 (1953), pp. 34–36.

16. G. E. Archie, "Classification of Carbonate Reservoir Rocks and Petrophysical Considerations," Bull. Amer. Assoc. Petrol. Geol., Vol. 36 (February 1952), p. 290.

17. M. P. Tixier, "Electric Log Analysis in the Rocky Mountains," O. & G. Jour., Vol. 48 (June 23, 1949), pp. 143 ff.

H. F. Dunlap and R. R. Hawthorne, "The Calculation of Water Resistivities from Chemical Analyses" (Tech. Note 67), Jour. Petrol. Technol., March 1951, Sec. 1, p. 71.

L. A. Puzin, "Connate Water Resistivity in Oklahoma," Petrol. Engr., August 1952, pp. B-67–78.

Marion L. Ayres, Rollie P. Dobyns, and Robert Q. Bussell, "Resistivity of Water from Subsurface Formations," Petrol. Engr., December 1952, pp. B-36–48.

Murphy E. Hawkins and J. L. Moore, "Electrical Resistivities of Oil Field Brines in South Arkansas and North Louisiana," Petrol. Engr., Vol. 28 (July–August 1956).

Murphy E. Hawkins, "Electrical Resistivities of Oil Field Brines in Northeast Texas," Petrol. Engr. (July 1957), pp. B-52, B-68.

18. L. C. Case, "Exceptional Silurian Brine near Bay City, Michigan" (Geol. Note), Bull. Amer. Assoc. Petrol. Geol., Vol. 29 (1945), pp. 567–570.

19. Wayne F. Meents *et al., Illinois Oil-Field Brines,* Ill. Petrol. No. 66, Ill. Geol. Surv. (1952), 38 pages, p. 7.

20. J. H. Reagan, "Notes on the Quirequire Oil Field, District of Pian, State of Monagas, Venezuela," Boletín de geología y minería, Tomo 11, Nos. 2, 3, 4 (1938).

21. Staff of Caribbean Petroleum Company, "Oil Fields of Royal Dutch Shell Group in Western Venezuela," Bull. Amer. Assoc. Petrol. Geol., Vol. 32 (April 1948), p. 622.

22. James G. Crawford, "Oil-Field Waters of Wyoming and Their Relation to Geological Formations," Bull. Amer. Assoc. Petrol. Geol., Vol. 24 (July 1940), pp. 1214–1329.

23. Wayne F. Meents *et al., op. cit.* (note 19). Numerous analyses and ten isocon maps of formation waters from various oil-producing formations.

24. Norbert T. Lindtrop, "Outline of Water Problems in New Grozny Oil Field, Russia," Bull. Amer. Assoc. Petrol. Geol., Vol. 11 (October 1927), p. 1037.

25. C. E. Reistle, *Identification of Oil Field Waters by Chemical Analysis,* Tech. Paper 404, U.S. Bur. Mines (1927), 25 pages.

C. E. Wood, "Methods of Analysis of Oil Field Waters," in *The Science of Petroleum,* Oxford University Press, London and New York, Vol. 1 (1938), pp. 646–652. Bibliog. 115 items. A general discussion of water analyses.

L. C. Case and D. M. Riggen, "Analysis of Mineral Waters," O. &. G. Jour., Part I, December 23, 1948, pp. 74–75, and Part II, January 6, 1949, pp. 72–88. Details of the procedure used in modern oil-field water analyses.

26. Chase Palmer, *The Geochemical Interpretation of Water Analyses,* Bull. 479, U.S. Geol. Surv. (1911), 31 pages.

27. E. M. Parks, "Water Analyses in Oil Production and Some Analyses from Poison

Spider, Wyoming," Bull. Amer. Assoc. Petrol. Geol., Vol. 9 (September 1925), pp. 927-946.

28. E. T. Heck, "Barium in Appalachian Salt Brines," Bull. Amer. Assoc. Petrol. Geol., Vol. 24 (1940), pp. 486-493.

29. Garland B. Gott and James W. Hill, *Radioactivity in Some Oil Fields of Southeastern Kansas,* Bull. 988-E, U.S. Geol. Surv. (1953), 122 pages.

30. Joseph Jensen, "California Oil-field Waters," in *Problems of Petroleum Geology,* Amer. Assoc. Petrol. Geol., Tulsa, Okla. (1934), pp. 953-985. 123 analyses.

L. C. Case, "Subsurface Water Characteristics in Oklahoma and Kansas." *op. cit.,* pp. 855-868. 64 analyses.

Walter R. Berger and Ralph H. Fash, "Relation of Water Analyses to Structure and Porosity in the West Texas Permian Basin," *op. cit.,* pp. 869-889. 42 analyses.

James G. Crawford, *op. cit.* in note 21. 512 analyses.

James G. Crawford, "Oil Field Waters of Montana Plains," Bull. Amer. Assoc. Petrol. Geol., Vol. 26 (August 1942), pp. 1317-1374. 279 analyses.

Wayne F. Meents *et al., op. cit.* (note 19). 750 analyses and discussion of method of analysis.

Cleo G. Rall and Jack Wright, *Analysis of Formation Brines in Kansas,* RI 4973, U.S. Bur. Mines (May 1953). 600 analyses and discussion of method of analysis.

31. F. G. Tickell, "A Method for the Graphical Interpretation of Water Analyses," Summ. of Operations, Calif. Oil Fields, Vol. 6, No. 9 (1921), pp. 5-11.

32. J. S. Parker and C. A. Southwell, "Chemical Investigation of Trinidad Well Waters and Its Geological and Economical Significance," Jour. Inst. Petrol. Technol., Vol. 15 (1929), pp. 138-173. Discussion to p. 182.

33. Henry A. Stiff, Jr., "The Interpretation of Chemical Water Analysis by Means of Patterns," Tech. Note 84, Petrol. Technol., Trans. Amer. Inst. Min. Met. Engrs., Vol. 192 (1951), pp. 376-379.

34. James C. Crawford, *op. cit.* (note 21), p. 1221.

35. G. Sherburne Rogers, *Chemical Relations of the Oil Field Waters in the San Joaquin Valley, California,* Bull. 653, U.S. Geol. Surv. (1917), p. 46.

36. Edson S. Bastin, "The Problem of the Natural Reduction of Sulphates," Bull. Amer. Assoc. Petrol. Geol., Vol. 10 (1926), pp. 1270-1299.

T. L. Ginsberg-Karagitschewa, "Origin of the Small Sulphate Content of Oil Field Waters," Petr. Z., Vol. 33, No. 1 (1937), pp. 7-12; reviewed by Jour. Inst. Petrol. Technol., Vol. 23 (1937), p. 227A.

Margaret D. Foster, "Base-Exchange and Sulphate Reduction in Salty Ground Waters along Atlantic and Gulf Coasts," Bull. Amer. Assoc. Petrol. Geol., Vol. 26 (May 1942), pp. 383-851.

37. James G. Crawford, *op. cit.* (note 22), p. 1222.

38. L. U. de Sitter, "Diagenesis of Oil-Field Brines," Bull. Amer. Assoc. Petrol. Geol., Vol. 31 (November 1947), pp. 2030-2040. 18 references cited.

39. John F. Sage, "Geological Interpretation of Water Analyses," Tulsa Geol. Soc. Digest, Vol. 21 (1953), pp. 98-104.

40. Cleo G. Rall and Jack Wright, *op. cit.* (note 30). The Bureau of Mines method of water resistivity measurements is described on pp. 5-6.

41. H. F. Dunlap and R. R. Hawthorne, *op. cit.* (note 17), Sec. 1, p. 17, and Sec. 2, p. 7.

42. James G. Crawford, *op. cit.* (note 22), p. 1318. Note especially the geologic relationships given in the Appendix, pp. 1319-1329.

43. H. E. Minor, "Oil Field Waters of the Gulf Coastal Plain," in *Problems of Petroleum Geology,* Amer. Assoc. Petrol. Geol., Tulsa, Okla. (1934), pp. 891-905.

44. William L. Russell, "Subsurface Concentration of Chloride Brines," Bull. Amer. Assoc. Petrol. Geol., Vol. 17 (October 1933), pp. 1213-1228.

W. P. Kelley and G. F. Liebig, Jr., "Base Exchange in Relation to Composition of

Clay with Special Reference to Effect of Sea Water," Bull. Amer. Assoc. Petrol. Geol., Vol. 18 (March 1934), pp. 358–367.

L. U. de Sitter, *op. cit.* (note 38), p. 2040.

45. R. Van A. Mills and Roger C. Wells, *The Evaporation and Concentration of Waters Associated with Petroleum and Natural Gas,* Bull. 693, U.S. Geol. Surv. (1919), 104 pages.

46. William L. Russell, *op. cit.* (note 44), pp. 1214–1221.

47. F. S. Hudson and N. L. Taliaferro, "Calcium Chloride Waters from Certain Oil Fields in Ventura County, California," Bull. Amer. Assoc. Petrol. Geol., Vol. 9 (October 1925), pp. 1071–1088.

48. American Petroleum Institute, *API Code for Measuring, Sampling and Testing Crude Oil,* Dallas, Texas (June 1948), 35 pages.

49. John F. Dodge, Howard C. Pyle, and Everett G. Trostel, "Estimation by Volumetric Method of Recoverable Oil and Gas from Sands," Bull. Amer. Assoc. Petrol. Geol., Vol. 25 (July 1941), pp. 1302–1326. Bibliog. 29 items.

Paul Paine, *Oil Property Valuation,* John Wiley & Sons, New York (1942), 204 pages.

Walter L. Whitehead, *Oil Property Valuation* (3rd ed. of *Examination and Valuation of Mineral Property*), Addison-Wesley Press, Cambridge, Mass. (1949), pp. 299–343.

Robin Willis, "How to Determine the Recoverable Reserves of an Oil Property," Petrol. World and Oil, Part I, November 1953, pp. 14–19, and Part II, December 1953, pp. 14–18.

50. W. W. Cutler, Jr., *Estimation of Underground Oil Reserves by Oil-Well Production Curves,* Bull. 228, U.S. Bur. Mines (1924), 114 pages.

51. Beveridge J. Mair, "Hydrocarbons Isolated from Petroleum," O. & G. Jour., Tulsa, Okla., September 14, 1964, pp. 130–134.

52. H. D. Wilde, "Why Crudes Differ in Value," Bull. Amer. Assoc. Petrol. Geol., Vol. 25 (June 1941), pp. 1167–1174.

J. M. Hunt and J. P. Forsman, "Relation of Crude Oil Composition to Stratigraphy in the Wind River Basin," Wyoming Geol. Assn. Guidebook, 12th Ann. Field Conf. (1957), pp. 105–112.

53. Wallace E. Pratt, "Hydrogenation and the Origin of Oil," in *Problems of Petroleum Geology,* Amer. Assoc. Petrol. Geol., Tulsa, Okla. (1934), pp. 235–245.

54. William A. Gruse and Donald R. Stevens, *The Chemical Technology of Petroleum,* 2nd ed., McGraw-Hill Book Co., New York (1942), pp. 566–580.

55. Harold M. Smith, "The Story of the Composition of Petroleum," O. & G. Jour., November 4, 1948, pp. 60–61.

56. E. Emmet Reid, "The Sulphur Compounds in Petroleum," in *The Science of Petroleum,* Oxford University Press, London and New York, Vol. 2 (1938), pp. 1033–1041. 132 references.

William E. Haines, Welton J. Wenger, R. Vernon Helm, and John S. Ball, *Sulfur in Petroleum,* RI 4060, U.S. Bur. Mines (December 1946), 42 pages. Discussion of occurrence and many analyses.

57. H. M. Smith and O. C. Blade, "Trends in Supply of High-Sulphur Crude Oils in the United States," O. & G. Jour., November 29, 1947, pp. 73–78.

58. John M. Hunt, "Composition of Crude Oil and Its Relation to Stratigraphy in Wyoming," Bull. Amer. Assoc. Petrol. Geol., Vol. 37 (August 1953), pp. 1837–1872. 24 references.

59. J. R. Bailey, "The Nitrogen Bases of Petroleum Distillates," in *The Science of Petroleum,* Oxford University Press, London and New York, Vol. 2 (1938), pp. 1047–1052. Bibliog. and references, 37 items.

William A. Gruse and Donald R. Stevens, *op. cit.* (note 54), pp. 115–118.

Oil and Gas Journal, "How to Lick the Nitrogen Problem (September 16, 1957), p. 114.

60. E. J. Poth et al., "The Estimation of Nitrogen in Petroleum and Bitumen," Ind. & Engr. Chem., Vol. 20 (1928), pp. 83–85.

61. Julius von Braun, "Napthenic Acids, Oxy-Compounds, etc.," in *The Science of Petroleum,* Oxford University Press, London and New York, Vol. 2 (1938), pp. 1007–1015. 82 references.

William A. Gruse and Donald R. Stevens, *op. cit.* (note 54), pp. 93–108.

H. S. Bell, *American Petroleum Refining,* 3rd ed., D. Van Nostrand Co., New York (1945), pp. 26–29.

62. William A. Gruse and Donald R. Stevens, *op. cit.* (note 54), p. 107. (Analysis of Sachanen and Vasil'ev.)

63. J. McConnell Sanders, "The Microscopical Examination of Crude Petroleum," Jour. Inst. Petrol. Technol., Vol. 23 (1937): Part I, "Object, Scope and Methods of Investigation," pp. 525–556; Part II, "Practical Results and Observations," pp. 557–573. References and bibliog. 51 items.

64. A. J. Headlee and Richard D. Hunter, "The Composition of Ash from West Virginia Petroleum," Prod. Monthly, Vol. 16 (December 1951), pp. 34–38.

65. W. H. Thomas, "Inorganic Constituents of Petroleum," in *The Science of Petroleum,* Oxford University Press, London and New York, Vol. 2 (1938), pp. 1053–1056. Bibliog. 24 items.

66. W. L. Nelson, "Vanadium in Crude Oils," O. &. G. Jour.: Part I, "Occurrence," November 30, 1950, p. 88; Part II, "Ash Analysis," December 7, 1950, p. 105.

W. L. Nelson, "Metal Contaminants in Petroleum," O. &. G. Jour., Vol. 56, No. 51, December 22, 1954, pp. 75–76.

Gordon W. Hodgson and Bruce L. Baker, "Geochemical Aspects of Petroleum Migration in Pembina, Redwater, Joffre, and Lloydminster Oil Fields of Alberta and Saskatchewan, Canada," Bull. Amer. Assoc. Petrol. Geol., Vol. 43 (February 1959), pp. 311–328. Discusses variations in vanadium and nickel content of four pools in Canada.

67. V. E. McKelvey, "Search for Uranium in Western United States," presented before Rocky Mountain Section Meeting, Amer. Assoc. Petrol. Geol., Casper, Wyoming, April 23, 1953.

68. W. L. Nelson, "Salt Content of Crude," O. &. G. Jour., January 20, 1958, p. 155.

69. Research Committee, Tulsa Geological Society, "Relationship of Crude Oils and Stratigraphy in Parts of Oklahoma and Kansas," Bull. Amer. Assoc. Petrol. Geol., Vol. 31 (January 1947), pp. 92–148. Hempel analysis characteristics of hundreds of crude oils, related to their stratigraphic occurrence.

C. M. McKinney and O. C. Blade, *Analyses of Crude Oils from 283 Important Oil Fields in the United States,* RI 4289, U.S. Bur. Mines (May 1948). 283 Hempel analyses.

Various authors, "Crude Oils," in *The Science of Petroleum,* Oxford University Press, London and New York, Vol. 5 (1950), 200 pages. Oils described are from the U.S.A., Venezuela, Saudi Arabia, Bahrein Island, and the Middle East.

70. Harold M. Smith, *Correlation Index to Aid in Interpreting Crude-Oil Analyses,* Tech. Paper 610, U.S. Bur. Mines (1940). 34 pages. See also Petroleum Engineer, September 1953, pp. E-31–35, for correlation index values.

71. L. Murray Neumann et al., "Relationship of Crude Oils and Stratigraphy in Parts of Oklahoma and Kansas," Bull. Amer. Assoc. Petrol. Geol., Vol. 25 (September 1941), pp. 1801–1809.

72. Ralph H. Espach and Joseph Fry, *op. cit.* (note 4).

73. Cecil Q. Cupps, Philip H. Lipstate, Jr., and Joseph Fry, *op. cit.* (note 4), p. 13.

74. E. A. Wendlandt, T. H. Shelby, Jr., and John S. Bell, "Hawkins Field, Wood County, Texas," Bull. Amer. Assoc. Petrol. Geol., Vol. 30 (November 1946), pp. 1846 and 1855.

75. Thomas Eugene Weirich, "Shelf Principle of Oil Origin, Migration and Accumulation," Bull. Amer. Assoc. Petrol. Geol., Vol. 37 (August 1953), p. 2031.

76. Fred R. Haeberle, "Relationship of Hydrocarbon Gravities to Facies in Gulf Coast," Bull. Amer. Assoc. Petrol. Geol., Vol. 35 (October 1951), p. 2241.

77. E. Boaden and E. C. Masterson, "Some Aspects of Field Operations in Kuwait," Jour. Inst. Petrol. Technol., Vol. 38 (June 1952), p. 406.

78. A. Velikovsky, "USSR," in *The Science of Petroleum*, Oxford University Press, London and New York, Vol. 2 (1938), p. 903.

79. Lester Charles Uren, *Petroleum Production Engineering, Oil Field Exploitation*, 2nd ed., McGraw-Hill Book Co., New York (1939), p. 17.

80. D. P. Barnard, "The Viscosity Characteristics of Petroleum Products and Their Determination," in *The Science of Petroleum*, Oxford University Press, London and New York, Vol. 2 (1938), pp. 1071–1079. Description of methods and conversion tables. 28 references.

81. Carlton Beal, Jr., "The Viscosity of Air, Water, Natural Gas, Crude Oil, and Its Associated Gases at Oil Field Temperatures and Pressures," Trans. Amer. Inst. Min. Met. Engrs., Vol. 165 (1946), pp. 94–115.

82. H. C. Miller, *Function of Natural Gas in the Production of Oil*, U.S. Bur. Mines and Amer. Petrol. Inst. (1929), 267 pages.

C. R. Hocott and S. C. Buckley, "Measurements of the Viscosities of Oils under Reservoir Conditions," Tech. Pub. 1220, Petrol. Technol., Trans. Amer. Inst. Min. Met. Engrs., Vol. 160 (1941), reprinted 1940–1941, pp. 320–321.

83. Hollis D. Hedberg, "Evaluation of Petroleum in Oil Sands by Its Index of Refraction," Bull. Amer. Assoc. Petrol. Geol., Vol. 21 (November 1937), pp. 1464–1476.

84. Jack De Ment, "Fluorescent Techniques in Petroleum Exploration," Geophysics, January 1947, pp. 72–98. See also *Subsurface Geologic Methods*, Colo. Sch. Mines (1951), pp. 320–329.

Arthur E. Messersmith, "The Fluoroscope as a Drilling Aide," World Oil, January 1950, pp. 95–96.

85. Benjamin T. Brooks, *The Non-benzenoid Hydrocarbons*, Chemical Catalog Co., New York (1922), p. 566.

86. J. M. Rupert and B. N. van Dieman del Jel, "Pumping Crude That Solidifies at Atmospheric Temperatures," International Oilman (November 1957), pp. 366–369.

87. C. R. Dodson and M. B. Standing, "Pressure-, Volume-, Temperature and Solubility Relations for Natural Gas-Water Mixtures," Calif. Oil World, Vol. 37 (December 15, 1944), pp. 21–27.

O. L. Culberson and J. J. McKetta, Jr., "The Solubility of Methane in Water at Pressures to 10,000 psia," Tech. Pub. 3082, Trans. Amer. Inst. Min. Met. Engrs., Vol. 192 (1951), pp. 223–226. 7 references.

88. American Petroleum Institute, *Recommended Practice for Measuring, Sampling, and Testing Natural Gas*, Dallas, Texas (April 1953), 18 pages. 27 references.

89. Howard S. Bean, "The Measurement of Gas with Particular Reference to Natural Petroleum Gases," in *The Science of Petroleum*, Oxford University Press, London and New York, Vol. 1 (1938), pp. 704–710. 13 references.

90. E. S. L. Beale, "The Principles of Practical Orifice Metering," *op. cit.* (note 18), pp. 688–703. 14 references.

91. Eugene A. Stephenson, "Valuation of Natural Gas Properties," in *Geology of Natural Gas*, Amer. Assoc. Petrol. Geol., Tulsa, Okla. (1935), pp. 1011–1033.

Paul Paine, *Oil Property Valuation*, John Wiley & Sons, New York (1942), 204 pages.

Henry J. Gruy and Jack Crichton, "A Critical Review of the Methods Used in Estimation of Natural Gas Reserves," O. & G. Jour., October 25, 1947, pp. 88–89 and 116.

92. Henry A. Ley, "Natural Gas," in *Geology of Natural Gas*, Amer. Assoc. Petrol. Geol., Tulsa, Okla. (1935), pp. 1073–1149.

93. G. Sherburne Rogers, *Helium-Bearing Natural Gas*, U.S. Geol. Surv. (1921), 113 pages.

94. *Ibid.*, p. 60.

95. R. C. Wells, "Origin of Helium Rich Natural Gas," Jour. Washington Acad. Sci., Vol. 19, No. 15 (1929), pp. 321–327.

96. R. W. Boyle, "Note on the Solubility of Radium Emanations in Liquids," Trans. Roy. Soc. Canada, 3rd ser., Vol. 3, No. 3 (1909), p. 75.

97. C. E. Dobbin, "Geology of Natural Gases Rich in Helium, Nitrogen, Carbon Dioxide and Hydrogen Sulphide," in *Geology of Natural Gas,* Amer. Assoc. Petrol. Geol., Tulsa, Okla. (1935), pp. 1053–1064.

98. H. H. Hinson, "Reservoir Characteristics of Rattlesnake Oil and Gas Field, San Juan County, N. Mex.," Bull. Amer. Assoc. Petrol. Geol., Vol. 31 (April 1947), pp. 731–771.

99. H. Goslin, "The Argon-Nitrogen Ratio of Natural Gas," Comptes-rendus de l'Académie des Sciences, Paris, No. 200 (1935), pp. 1137–1139.

100. Robert Latimer Bates (compiler) *et al.,* "The Oil and Gas Resources of New Mexico," Bull. 18, New Mexico School of Mines (1942), p. 302.

101. G. Sherburne Rogers, *op. cit.* (note 93), p. 37.

102. Dean E. Winchester, "Natural Gas in Colorado, Northern New Mexico, and Utah," in *Geology of Natural Gas,* Amer. Assoc. Petrol. Geol., Tulsa, Okla. (1935), p. 377.

103. R. R. Sayers *et al., Investigation of Toxic Gases from Mexican and Other High-sulphur Petroleums and Products,* Bull. 231, U.S. Bur. Mines (1925), 108 pages.

104. John M. Devine and C. J. Wilhelm, *Hydrogen-Sulphide Content of the Gas in Some Producing Oil Fields,* RI 3128, U.S. Bur. Mines (1931), 15 pages.

105. C. E. Dobbin, *op. cit.* (note 97), pp. 1069–1072.

106. O. & G. Jour., May 24, 1951, p. 82.

107. G. M. Lees, "Persia," in *The Science of Petroleum,* Oxford University Press, London and New York, Vol. 6 (1953), Part I, p. 80.

CHAPTER 6

Reservoir Traps – General and Structural

The anticlinal theory. Classification of traps. Structural traps: caused by folding—by faulting—by fracturing.

THE FIRST essential element of a petroleum reservoir, you will recall, is the reservoir rock, and the second is the existence of connected pore spaces that are collectively capable of holding and storing petroleum. The third element is the oil, water, and gas—which are either in motion or are capable of moving—that occupy the connected pore spaces. The fourth element is the trap—the place where oil and gas are barred from further movement.

Since oil and gas are lighter than water, and since the reservoir rocks generally have a regional slope, though often slight, the petroleum moves through the water both vertically and laterally until it is barred by an impervious or less pervious rock. The impervious stratum that overlies the reservoir rock is called the *roof rock*.* A roof rock that is concave† as viewed from below prevents the oil and gas from escaping either vertically or laterally, thus localizing the pool of oil and gas. Such an external barrier is a *structural trap*. A lateral lessening of permeability due to facies changes, truncations, and other stratigraphic changes will, together with the roof rock, form an interior barrier, or *stratigraphic trap*.

Structural traps are the result of changes in the form of the reservoir rock; stratigraphic traps are the result of changes in the continuity of the rock. *Fluid barriers* occur where a difference in fluid potentials causes a down-dip

---

* The term "cap rock" is also used, but it is better reserved for the cap rock of a salt plug.

† The term "concave" is used to describe the shapes of such structures as folds, domes, arches, and the intersecting planes that are peaked, or roof-shaped.

flow of water to oppose the up-dip migration of petroleum. Increased fluid potential gradients usually exist where the flow space is constricted; for example, where a formation thins or where permeability is reduced.

Reservoirs are infinitely variable in detail. Practically every sedimentary area has been deformed to some degree, and most areas have been deformed several times. Lateral change in the rock properties is the rule rather than the exception, and the direction of fluid flow has undoubtedly varied throughout geologic time owing to the continual change in the potentiometric surface. The relative importance of each of these elements in trapping a specific pool of petroleum may not be known until the field is completely developed and years of production history are available.

But the job of the petroleum geologist is to try to locate a favorable combination of these elements *before* a pool is found. His data are generally fragmentary: geophysical measurements are often inconclusive, well records for subsurface controls are lacking or widely scattered, fluid pressures are unknown, and outcrops are poor or distant. Small amounts of information must be combined and projected for long distances both vertically and horizontally. Of the essential elements, the most readily determined in advance of drilling is the presence of traps that result from the structural features of the reservoir rock. The structural geology may be mapped in a variety of ways, such as surface mapping, core drilling, subsurface mapping, and geophysical surveying. Since most reservoirs show at least some deformation, structural mapping adds information of value to most predictions; it becomes the basis of prediction where the trap is controlled by the deformation of the reservoir rock. Determination of the nature of the reservoir rock, on the other hand, including such characteristics as extent, porosity, and permeability, is more difficult; in fact, these characteristics cannot be fully determined without adequate subsurface data, which are obtained only from well records. Such data are based on studies of well logs, well cuttings, and cores, and on cross sections and subsurface maps that show the distribution of rocks, their correlations, and their unconformities. Several kinds of maps are needed to show the data upon which to base predictions of the position of favorable reservoir characteristics; these include areal-geology maps, structure maps, sand-distribution maps, formation-thickness (isopach) maps, subcrop and paleogeologic maps, productivity maps, isopotential maps, lithofacies maps, and other maps. They are discussed in more detail in the section on subsurface mapping, pages 591–618.

The simplest and commonest way for a permeable underground formation to become a trap is to be folded into an anticline. An anticline is the most readily mapped of the common traps and can frequently be mapped at the surface. The close association of oil and gas pools with anticlinal folds was noted early and led to what has long been known as the *anticlinal theory* of oil and gas occurrence. Geologists everywhere searched for anticlines and domes on which to drill—almost to the exclusion of any other kind of trap. Consequently the anticlinal theory has had such a predominant place in exploration that it warrants a brief review of its development and of its gradual change into the more modern *trap theory*.

THE ANTICLINAL THEORY

Although petroleum in various forms has been known since man made his earliest records, it was not until petroleum attained economic importance that much serious thought was given to why, how, and where it accumulated into pools. The discovery of oil by E. L. Drake in western Pennsylvania, in 1859, marked the beginning of the modern oil industry even though oil had previously been mined and obtained from wells at a number of places in various parts of the world. Drake demonstrated that drilling for oil could be successful, and the amount of oil he produced, though small compared with modern production, showed the world that drilling was the most effective way to obtain oil. From that time forth, the main question in the minds of every prospector for oil was where to put down the drill. Thought on this subject was greatly stimulated by the increasing demand for petroleum, and down to the present time it has constituted the core of petroleum geology.

A variety of theories have attempted to explain the conditions that result in an oil and gas pool in order to predict where such conditions may be found and where discoveries may be made. Of these the anticlinal theory[1] has received the most serious and continued attention. The development of the anticlinal theory in geologic thinking is itself an interesting subject, but it can only be touched on here. The theory was conceived even before Drake's discovery, and, having been modified and expanded through the years, it is fundamentally as true today as when first proposed.

One of the earliest references to a relation between oil and anticlinal folding was made before Drake's time, when oil seepages near Gaspé, at the mouth of the St. Lawrence River in eastern Canada, were visited by Sir William Logan in 1842 and were described in 1844 as occurring on anticlines.[2] Interest in the subject was greatly stimulated by Drake's discovery, and within a short period thereafter several geologists working independently arrived at a similar conclusion: anticlines are the place to look for petroleum.

I. C. White, one of many students of oil and gas accumulation, who had been an assistant on the Second Geological Survey of Pennsylvania (the Lesley Survey) since 1875, resigned in 1883 to engage in commercial work. An early assignment of his was to determine whether or not it was possible to predict the presence or absence of gas by a knowledge of the geologic structure. He had learned that gas wells in western Pennsylvania were located close to anticlines, and he inferred that there must be some connection between the position of the gas pools and the structure. His statement of the anticlinal theory, published in 1885, is as good even now as it was then.

"After visiting all the great gas-wells that had been struck in western Pennsylvania and West Virginia, and carefully examining the geological surroundings of each, I found that every one of them was situated either directly on, or near, the crown of an anticlinal axis, while wells that had been bored in the synclines on either side furnished little or no gas, but in many cases large quantities of salt water. . . . During the last two years, I have submitted it to all manner of tests, both in locating and condemning gas territory, and the general result has been to confirm the anticlinal theory beyond a reasonable doubt.

"But while we can state with confidence that all great gas wells are found on the anticlinal axes, the converse of this is not true; viz., that great gas-wells may be found on all anticlinals.

". . . The reason why natural gas should collect under the arches of the rocks is sufficiently plain, from a consideration of its volatile nature. Then too, the extensive fissuring of the rock, which appears necessary to form a capacious reservoir for a large gas-well, would take place most readily along the anticlinals, where the tension in bending would be greatest." [3]

Because of the wide attention this article on the occurrence of gas received from the oil operators of the time, I. C. White has often erroneously been called the "father of the anticlinal theory." He made no such claim for himself, but he did take credit for "the work which the writer has especially accomplished, and in the doing of it, so enforced the lessons of geology upon the minds of the men engaged in the practical work of drilling for oil, that the acceptance of the structural theory is now almost universal among them as well as among geologists. In this the writer has been ably assisted by Dr. Edward Orton." [4]

Many arguments for and against the anticlinal theory were published in the various scientific journals of the 1890's. But the practical success that White achieved in advising the drilling of wells for gas on anticlinal structures finally led him to state in 1892: "As is well known, it was formerly a popular saying among practical oil men that 'Geology has never filled an oil tank'; and to such a low estate had geology fallen that a prominent producer of oil and gas, disgusted with geology and geologists, was once heard to remark that if he wanted to make sure of a dry hole he would employ a geologist to select the location. It has been my pleasant task during the last eight years to assist in removing this stigma from our profession, so that with the valuable assistance of Ohio's distinguished geologist, Professor Orton, Dr. Phinney of Indiana, and others, the battle against popular as well as scientific prejudice has been fought and won and this long standing reproach to geology in great part removed." [5]

McCoy and Keyte recognized the importance of nonstructural accumulations and of traps that were not anticlines. In 1934 they used the term "structural theory" instead of "anticlinal theory," remarking that the term was "broad enough to include the occurrence of oil on the flanks of anticlines, in lenses, and in synclines, as well as the occurrence of oil on anticlines, domes, and noses." They summarized the structural theory as follows: ". . . commercial deposits of oil and gas are associated with structural irregularities in porous sedimentary rocks, the most important irregularity being the anticline or dome." [6]

Even though the term "anticlinal theory" may now have fallen into disuse, the fundamental principle upon which it was based—that oil and gas accumulate as high within the reservoir as possible—is as good today as it ever was. The only difference is that we now recognize a large number of conditions other than anticlines that will form a bounding rock surface that is concave as viewed from below.

The term "trap" was first introduced by McCollough in 1934 and applied to containers as diverse in character as those due to asphalt seals, lenses, local porosity variations, truncation, and overlap, especially in homoclinal dip areas, as well as to folding and faulting.[7] Continued recognition of the fact that structural deformation is only one of several ways in which a trap may be formed, and that a large number of pools have accumulated in traps formed in other ways, increases the desirability of using a more inclusive term than "anticlinal theory." The term "trap theory" is now more commonly used, and the geometrical configuration that holds in the oil is now commonly called a "trap," whatever its shape or cause may be. Its essential characteristic is that the oil and gas are capable of accumulating and being held in it.

The importance of the oil-water contact in defining the trap has led to some modification of former ideas that assumed the water table to be either level or nearly level. It is now recognized that in some regions, where there is a fluid potential gradient, the oil-water contact is distinctly tilted. This tilt affects the position of the pool with respect to the rock boundaries, displacing it varying distances in the direction of water movement, even to the point of flushing the pool completely out of the trap. The fundamental concepts of the effect of moving water on substances of less density are contained in the classic article by Hubbert[8] describing the physics of water movement. The effect of a tilted water table on the position of the pool was shown by Russell,[9] and his demonstration was expanded by Hubbert.[10] The whole idea was explored independently by Hill,* who arrived at the same conclusion as Hubbert —namely, that the trap for oil and gas in a reservoir rock is at the position of least potential energy, and that this position is determined by the difference in hydraulic head or fluid potential across the trap.

CLASSIFICATION OF TRAPS

Numerous classifications of reservoir traps have been proposed. Clapp's final classification contained the following main headings: (1) anticlinal structures, (2) synclinal structures, (3) homoclinal structures, (4) quaquaversal structures, or "domes," (5) unconformities, (6) lenticular sands, (7) crevices or cavities irrespective of other structure, (8) structures due to faulting.[11] Heroy classifies traps as (1) depositional traps, (2) diagenetic traps, (3) deformational traps.[12] Wilson classifies traps as (1) closed reservoirs: (a) reservoirs closed by local deformation of strata, (b) reservoirs closed because of varying porosity of the rock (no deformation of strata necessary other than re-

---

* Gilman A. Hill in an unpublished manuscript, Stanford University, 1951.

gional tilting), (c) reservoirs closed by combination of folding and varying porosity, (d) reservoirs closed by combination of faulting and varying porosity; (2) open reservoirs (none of commercial importance).[13] Heald sets up two groups of reservoirs: (1) closed by local deformation of strata, (2) closed because of varying permeability of the rock.[14] Wilhelm's classification is quite detailed and attempts to take into account all trap-forming factors participating in the boundaries of petroleum reservoirs. His chief classification headings are: (1) convex trap reservoirs, (2) permeability trap reservoirs, (3) pinch-out trap reservoirs, (4) salt trap reservoirs, (5) piercement trap reservoirs.[15]

No classification is entirely satisfactory, for many traps are unique and would not fit readily into any but an extremely detailed classification. The following scheme, however, is believed to be as useful as any. Although simple, it includes a place for most kinds of traps known to contain oil and gas in commercial quantities. It is not all-inclusive, but the exceptions will be pointed out as they occur. The classification divides the traps broadly into three basic types: (1) structural traps, (2) stratigraphic traps, (3) combinations of these two.

1. A *structural trap* is one whose upper boundary has been made concave, as viewed from below, by some local deformation, such as folding or faulting or both, of the reservoir rock. The edges of a pool occurring in a structural trap are determined wholly or in part by the intersection of the underlying water table with the roof rock overlying the deformed reservoir rock.

2. A *stratigraphic trap* is one in which the chief trap-making element is some variation in the stratigraphy or lithology, or both, of the reservoir rock, such as a facies change, variable local porosity and permeability, or an up-structure termination of the reservoir rock, irrespective of the cause. The areal extent of a pool occurring in a stratigraphic trap is determined wholly, or in large part, by some stratigraphic variation associated with the reservoir rock. The pool may rest on an underlying water table, which may be either level or tilted, or it may completely fill the voids in the reservoir rock with no produceable underlying reservoir water. The flow of water down-dip through the restricted permeability that forms the stratigraphic barrier to up-dip movement of the petroleum probably is an important element in the trapping of petroleum in many stratigraphic traps. (See also Chap. 9.)

3. *Combination traps*. Between these extremes—there is an almost complete gradation—traps are found that illustrate almost every imaginable combination of structure and stratigraphy. Traps in which structure or stratigraphy is clearly the most influential factor can readily be classified as structural or stratigraphic types. As the middle ground between a structural and stratigraphic trap is approached, however, it becomes increasingly difficult to decide the relative importance of each. Traps in this middle group—traps

formed by both structural and stratigraphic causes in roughly equal proportions—are best called combination traps.

In practice, when we speak of a "trap," we commonly mean its rock boundaries, and we use such a term as "structural trap," "stratigraphic trap," "anticlinal trap," or "combination trap" when we wish to indicate its cause. The position of the pool within the trap may depend in part on the movement of the formation water. Where there is no movement, the pool is as high as possible in the trap; if the water is in motion, the pool may be displaced for varying distances down the side of the trap. The movement of the water is determined by the fluid potential gradient in the reservoir rock unit that contains the trap. (See Chap. 9.) A trap, therefore, though present, may be ineffective because of particular fluid, temperature, and pressure conditions of the present or of the geologic past.

STRUCTURAL TRAPS

Traps that are formed chiefly as a result of folding and faulting are the ones that are most apparent from surface mapping and most readily located underground; they are also the ones that give the most help to the discovery of oil and gas. Consequently, structural traps have received the most attention by geologists, as indicated by the persistent use of the term "anticlinal theory" and the general custom of designating any trap resulting from deformed rocks as a "structure." Nearly all of the surface mapping, shallow core drilling, and geophysical mapping, and most of the subsurface mapping, has as its objective the location of traps that result wholly or in part from deformation of the reservoir rock.

One important aspect of structural features such as anticlines is that the structure generally extends vertically through a considerable thickness of sedimentary formations, thereby causing traps to form in all of the potential reservoir rocks affected by them. For this reason the drilling of structural traps involving a good thickness of sediments is considered good prospecting, even though specific reservoir rocks or other features of subsurface stratigraphy may not be known in advance. It is reasoned that, if the geologic section to be explored contains any reservoir rocks, they are most likely to produce where they are deformed into traps. One of many possible examples in which folding affects a great thickness of strata is the Santa Fe Springs field, of Los Angeles County, California, where a smooth, nearly circular dome fold extends down to form traps in twenty-five or more reservoir rocks, each of which contains an oil pool.[16] A section through the field is shown in Figure 6-1.

The deformation of the reservoir rock may be determined in several ways. Some folds and faults extend from the reservoir rock to the surface of the ground, where they may be mapped by ordinary surface mapping methods. A modification of surface mapping is the use of shallow core drilling, to

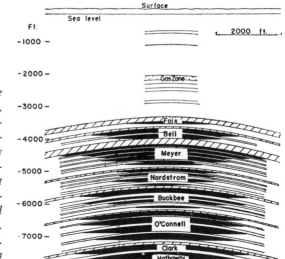

FIGURE 6-1

Section through the Santa Fe Springs oil field, California, an example of how many separate traps, holding many separate pools, are formed by one fold. The lack of connection between the different reservoirs is shown in the different water-oil contact level for each productive sand. [Redrawn from Winter, Bull. 118, Calif. Div. Mines (1943), p. 345, Fig. 142.]

depths of one or two thousand feet, for example, by means of which structural mapping is extended below the surface. Core drilling is especially useful in regions where the stratified rocks are covered with alluvium and other unconsolidated materials, which not only conceal the structure but also interfere with geophysical measurements. Still deeper mapping—called subsurface mapping—is done with the data available from various kinds of well logs. (See pp. 75–85.) How detailed the resulting map will be depends on how close together the wells are drilled and how much information there is on the logs. Subsurface mapping of geologic structure is also done without logs, by interpretation of geophysical data. These data include measurements of various physical properties of the rocks, such as their elasticity, magnetic susceptibility, and density; all of these physical properties are measured at the surface, but the surface measurement determines them down to depths of several miles, often with great accuracy. In this way it is possible to map the structure of formations at or near the anticipated reservoir formation. Since the trap *must be in the reservoir rock,* mapping on or close to the reservoir formation is most desirable, for it eliminates many of the errors that come from projecting shallow deformation down to the producing formation. If the folding of the reservoir rock does not extend far above the reservoir horizon, as where it is terminated by an unconformity surface, the first evidence of deformation may be found thousands of feet below the surface. Where this occurs, both well data and geophysical surveys may be required for a mapping of the reservoir rock. The location of such hidden traps is often very difficult to predict.

Some deformation of the reservoir rock has been involved in the formation of nearly all petroleum traps (there are some exceptions among traps of the lens and reef types). Since deformation is commonly associated in some manner or other with nearly all kinds of traps, structural mapping is of prime importance as a method of searching for oil and gas pools. The structure alone may be enough to form a trap, or it may be combined with favorable stratigraphic and fluid conditions. The standard exploration procedure is to aim the first test well at what is believed, on the basis of structural mapping, to be the highest part of the potential reservoir rock.

The great hazard in all structural mapping of deep-seated rocks is inadequacy of the data; the data recorded on the maps, whether geological, geophysical, or geochemical, are frequently indefinite, hazy, or inconclusive. Geologic reasoning and predictions are no better than the validity of the evidence upon which they are based. Another hazard is that the structure of the reservoir rock may be quite different from that of the overlying formations. It may shift laterally, or it may lose closure with depth and thereby become ineffective as a trap; or the pool may be displaced down the lee flank of the fold by moving water in the reservoir rock. Prospective traps that develop these unforeseen shifts in position are sometimes facetiously described as "rubber structures" or "structures that have slipped out from under." For reasons such as these, a second or third test well is often justified in the attempt to discover a pool in a trap, the new location being determined by the information obtained from the first hole.

Underground traps formed chiefly by structural deformation result from nearly all of those types of folding, faulting, and other deformation that can be observed at the surface of the ground. These structural features can be combined in a great variety of ways; when structural and stratigraphic phenomena are combined, the number of possible variations is almost infinite. Through the growing accumulation of geophysical surveys, through close drilling, accurate sampling, improved logging, and careful study of stratigraphy and paleontology, geologists have become increasingly aware of the innumerable ways in which traps can differ. Reservoir structure is as complex in places as any surface structure. The problem is to predict, in advance of drilling and from whatever data are available, the nature and character of the deformation. The geologist's predictions are based on the coordination and piecing together of many kinds of surface, subsurface, and geophysical information, which differ in value and whose relative value is one of the things that have to be judged.

Traps that are predominantly structural in origin may be classified, according to the chief kinds of deformation involved, as caused by (1) folding; (2) faulting, both normal and reverse; (3) fracturing; (4) intrusion of a plug, usually of salt; (5) combinations of the preceding.

Traps Caused by Folding

Structural traps that result chiefly or altogether from folding are of widely varied shapes; they include everything from low domes essentially circular

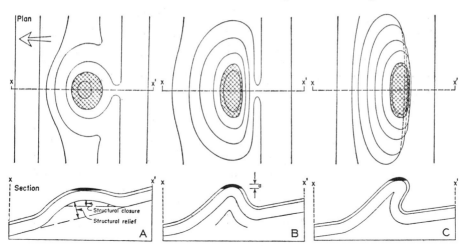

FIGURE 6-2 *Idealized structural maps and sections of typical anticlinal dome folds; such folds are characteristic of many traps containing oil and gas pools (oil, shaded and black). (Arrow shows direction of dip.)*

in plan (Fig. 6-2, A) to long, narrow anticlines (B), which may be symmetrical or asymmetrical or even overturned (C). The amount of *structural closure*—the vertical distance from the highest point down to the lowest closed contour—may range from a few feet up to thousands of feet. The *structural relief* of a fold, which is generally greater than the structural closure, is the height to which a folded bed rises above the regional slope, and is measured from the highest point to the projection of the regional slope below. (See Fig. 6-2, A.) The structural closure of a trap is referred to sea level as a datum. It does not necessarily represent the structural relief, as may be seen in the profiles in Figure 6-3, where the same fold may have different amounts of structural closure, depending on the regional dip of the formation. The capacity of a fold trap to hold oil and gas depends chiefly on the structural closure, the thickness of the reservoir rock, the rock's effective porosity, the reservoir pressure, and the conditions of fluid flow through the

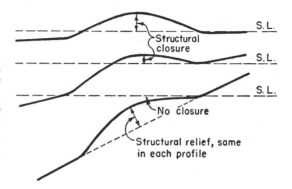

FIGURE 6-3

Diagrammatic profiles showing how the same amount of structural relief will have differing amounts of structural closure according to the rate of the regional dip (i.e., the angle with a horizontal reference plane).

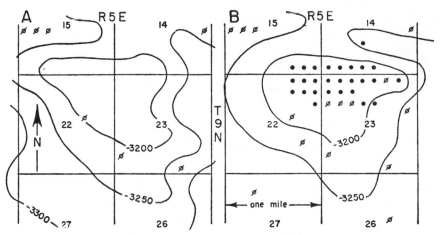

FIGURE 6-4 A, *A seismic map and the dry holes of 1929;* B, *the South Earlsboro pool, Seminole County, Oklahoma, following discovery in the Wilcox sand, and the structure on the Viola limestone (Ordovician) that overlies the pay formation.* [Redrawn from Weatherby, Geophysical Case Histories, Soc. Explor. Geophysicists (1949), p. 289, Fig. 4, and p. 290, Fig. 5.]

rock. The actual volume of the reservoir is that of the effective pore space between the underlying water table and the overlying roof rock. It is useful, in stating the size of the pool, to give the height of the *oil column* or the *gas column* or the *oil and gas column*. This is the vertical distance from the underlying oil-water or gas-water contact to the highest point to which either the oil or the gas extends in the trap. It is, in other words, the maximum vertical thickness of the oil, of the gas, or of the combined oil and gas in the pool. (See *a*, Fig. 6-2, B.)

Folding, as evidenced by some bending or tilting of the reservoir rock, is present to a greater or lesser extent in practically every trap, whether stratigraphic or structural. A trap in which folding is the predominant trap-forming factor is classed as a folded trap even though faulting and stratigraphic factors may have helped in some degree to form the closure.

The causes of folding are, of course, the same for reservoir rocks as for rocks at the surface. They include such varied mechanisms as horizontal compression, tangential or couple pressures, drag folding (possibly on a large scale), initial dips around buried hill, settling around buried hills, diapiric folding, and domes caused by deeply buried intrusive salt plugs. The folding may all take place at one time, or it may be an aggregate result of a series of episodes, each causing the fold to become more acute with increasing depth. A folded trap is seldom completely free from faulting. A reservoir may be cut by a fault that cannot be observed at the surface of the ground, and it is often extremely difficult to detect such a fault by subsurface methods unless a test well actually crosses the fault plane.

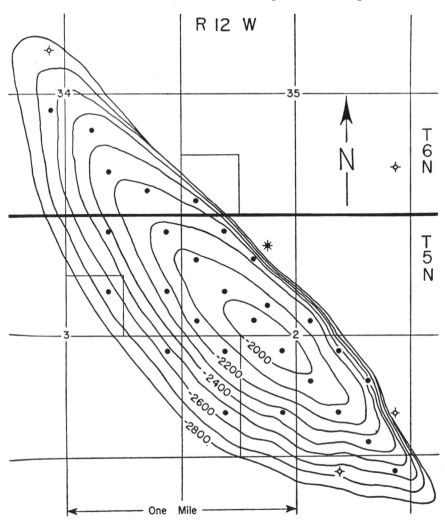

FIGURE 6-5 *Structure on top of the Bromide sand (Ordovician) in the Apache pool, Caddo County, Oklahoma. The contour interval is 100 feet. This fold formed at some time after the Caney (Mississippian) was deposited and before the Permian was laid down. The evidence is shown in Figures 6-19 and 13-12. [Redrawn from V. C. Scott, Bull. Amer. Assoc. Petrol. Geol., Vol. 29, p. 103, Fig. 3.]*

Some traps that are chiefly the result of folding are shown in the accompanying figures. Figure 6-4 illustrates the structure of the South Earlsboro pool, Oklahoma. This is one of the early discoveries made by seismic mapping of the characteristic low-relief, nearly circular dome folds found throughout the Mid-Continent region. The producing "Wilcox" sand (Ordovician) is a blanket sand through this area, and the fold determines the trap. The struc-

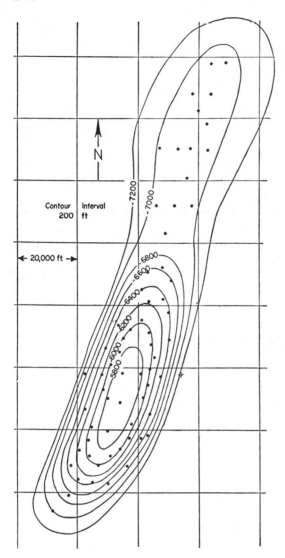

FIGURE 6-6

Structure of the Jurassic oolitic and dolomitic producing formation, the Arab zone, in the Abqaiq pool of Saudi Arabia. This great pool is 30 miles long and 6 miles wide and has a maximum oil column of over 1,500 feet; the average initial production of each of the sixty-six wells was 17,000 barrels per day. [Redrawn from McConnell, O. & G. Jour., December 20, 1951, p. 199.]

ture of the Apache field, in Caddo County, Oklahoma, is shown in Figure 6-5. (See also Figs. 6-19 and 13-12 for structural section and paleogeologic map of the Apache field.) Figure 6-6 is a section of the elongated dome fold containing the Abqaiq pool, in Saudi Arabia, which is typical of many traps in the Near East, although some are overlain by complexly folded incompetent formations. (See Figs. 6-12, p. 248, and 6-21, p. 254.) The structure of the Vallezza field, in northern Italy, is a faulted, overturned, and recumbent anticline. Such a fold is unusual as a trap, but it does show the extent to which overturning can go and still form a trap. A section through the field is shown in Figure 6-7.

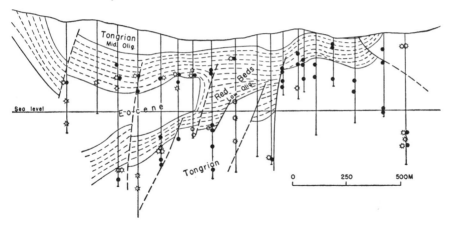

FIGURE 6-7 *Section through the Vallezza oil field, northern Italy, showing the surface occupied by a syncline, which overlies a recumbent fold of Eocene and Oligocene formations. Oil production occurs at the solid circles and gas at the hatchured circles. This is an extreme example of an overturned fold forming a trap. [Redrawn from Reeves, Bull. Amer. Assoc. Petrol. Geol., Vol. 37, p. 612, Fig. 4 (1953), from section by Grieg.]*

Changes with Depth. Many folds and other structures change in shape, size, or amplitude, or shift their position laterally, in passing from the surface or shallow depths down to the reservoir rock. Folding at the surface or at shallow depth is therefore not always a reliable guide in searching for petroleum pools that are trapped in reservoir rocks at great depths, for it frequently does not parallel the deeper folding. Seismic mapping in advance of drilling, applied to the prospective reservoir rock or to some formation close to it, may show the deep structure to be completely out of harmony with the known shallow structure. Or a test well located on the crest of a surface fold may

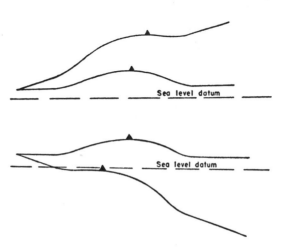

FIGURE 6-8

Diagrammatic sketches showing the gain or loss of structural closure and the shift in position of the crest of a fold when projected through converging strata. The fold has the same amplitude in all profiles. Black triangles mark the crest of the structure in different positions.

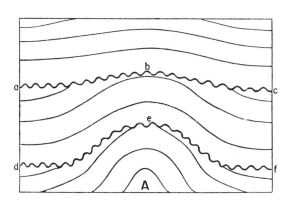

FIGURE 6-9

Diagrammatic sketch showing the increase in structural relief with depth as a result of superimposed, repeated folding. The evidence is in the thinning over the crest of the deep fold. In the example shown, three periods of folding add up to make the final fold, A: one period of folding before unconformity surface def, a second period after def and before unconformity abc, and a third period after the deposition of the youngest formations shown.

find no evidence of folding at the level of the reservoir rock. In either case doubt is thrown on the accuracy of the data, and many such differences are not necessarily due to the poor quality of the geophysical measurements or the well records; they are sometimes due to actual downward changes of structure. Many discrepancies between shallow and deep structure cannot be foretold, but some can be anticipated, or at least suggested as probable, if the geologic history of the region is known.

The discrepancies between the shallow and the deep folding, and therefore in the position of the pool, may result from various causes, such as (1) convergence of the intervening strata; (2) repeated folding; (3) parallel folding; (4) discordant and diapiric folding; (5) buried hills, anticlines, or fault blocks; (6) asymmetric folding; (7) shallow and surface weathering phenomena—deposition, solution, slumping; (8) pre-unconformity deformation; (9) overriding thrust faults; (10) displaced pool.

Convergence of the Intervening Strata. Where the strata between a shallow or surface formation and the reservoir rock are converging regionally,[17] the deep folding on the reservoir rock may be expected to shift its crest with depth in the direction of the convergence, and either to increase or to decrease the amount of structural closure with respect to sea level, depending on the direction of regional dip compared with the direction of the convergence. For example, a fold that is mapped as a terrace at the surface may become a closed dome at the reservoir, or a fold mapped as a dome at the surface may lose its closure at depth and therefore not be a trap. The diagrams in Figure 6-8 show how the structure may change and shift its crest with depth, even where the amount of deformation is actually the same in both the shallow and the deep strata.

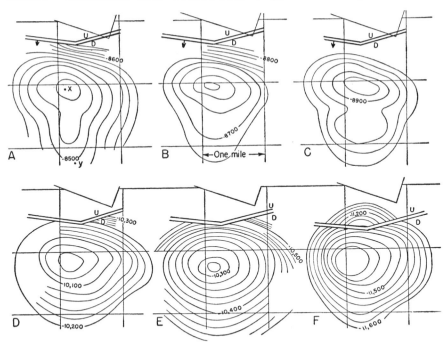

FIGURE 6-10 *Structural maps of various producing formations in the Erath oil field, Louisiana, showing the growth of the deep fold by recurrent periods of folding. The approximate amounts of closure between comparable points, as* x *and* y, *are:*

A	8,500-foot sand	45 feet of closure
B	8,700-foot sand	71 feet of closure
C	8,900-foot sand	83 feet of closure
D	10,100-foot sand	153 feet of closure
E	10,400-foot sand	155 feet of closure
F	11,600-foot sand	172 feet of closure

Isopach maps of these sands would show 17 feet of closure between F *and* E, *70 feet of closure between* D *and* C, *12 feet of closure between* C *and* B, *26 feet of closure between* B *and* A, *and 45 feet of closure after* A, *or a total of 170 feet (actually 172 feet) of closure on the 11,600-foot sand. [Redrawn from Steig et al., Bull. Amer. Assoc. Petrol. Geol., Vol. 35: Figs. 8, 11, 13, 15, 17, and 27.]*

Repeated Folding. Some folds have grown in intensity during the geologic life of the reservoir rock.* Such folds become increasingly more acute downward; that is, they show more and more structural relief with increasing depth. Under these conditions, formations commonly become thinner on the crest of the fold than on the flanks; the thinning may occur intermittently at several

---

* The geologic life of the reservoir rock is the time between its origin, such as its diagenesis or lithification, and the present.

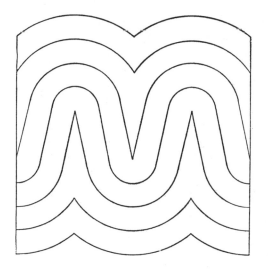

FIGURE 6-11

Parallel folding may be important as a cause of increased structural closure over anticlines with depth, especially in areas where the reservoir rock is several miles below the surface. With sufficient depth, anticlines that result from parallel folding will eventually disappear.

periods, or it may be continuous through a part of the geologic column. Sudden increases in amount of folding commonly appear just below unconformities. The time at which the increased folding occurred is readily shown by cross sections and by isopach maps (maps that show variation in stratigraphic thickness by means of thickness contours), the evidence of the folding being the thinning of the rocks over the fold. (See also pp. 596–598.) Figure 6-9 shows some of the evidence of repeated folding that is found at depth. The fact that the unconformity surface *def* shows more of a fold than surface *abc* shows that folding occurred during the time interval represented by the intervening rocks. Furthermore, the folding of *def* took place after the

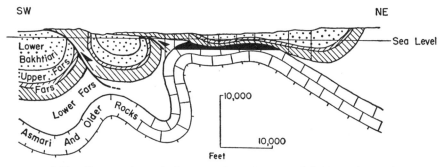

FIGURE 6-12 *Section through the Masjid-i-Sulaiman oil field, Iran. Production is from the Asmari limestone (Lower Miocene and Upper Oligocene). The section shows the great discordance in structure between the incompetent, shallow sands, shales, and evaporites of the Upper Fars (Miocene-Pliocene) and the underlying, competent Asmari limestone. More than one billion barrels of oil has been produced from this reservoir. [Redrawn from Lees, in* The Science of Petroleum, *Vol. 1, p. 147, Fig. 4.]*

FIGURE 6-13

Diagrammatic section through a diapir fold showing how the incompetent formations below are squeezed up and out in the folding of the overlying competent formations. Contrast this type of folding with the folding in Iran shown in Figures 6-12 and 6-21.

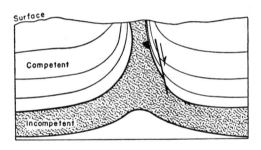

youngest of the rocks below it were folded, later truncated by erosion, and finally overlapped by the succeeding formation. The intermittent downward increase of folding shown here is common to a great many folded traps. The Oklahoma City fold, in Oklahoma[18] (Fig. 14-7, p. 644), and the Hawkins fold, in Texas,[19] are examples of folds that have grown repeatedly since the time the reservoir rock was formed. Several repeated periods of folding may be seen in the dome structure of the Erath field, Louisiana, by comparing structure maps on progressively deeper producing formations. (See Fig. 6-10.)

Parallel Folding. Thick sections of sedimentary rocks, especially of shales that are folded into low-relief anticlines and domes, may be expected to fold normally—that is, in such a way that the fold in the reservoir rock is virtually parallel to that at the surface and to those in the intervening formations. Parallel folding means that the thickness of the beds did not change materially during folding and that the folding becomes more acute with depth.[20] The effect of parallel folding is shown in Figure 6-11. In areas where the reservoir rock is two or three miles below the surface, the folding on it from this cause is likely to be considerably sharper than on the shallow rocks.

Discordant and Diapiric Folding. The nature of folding depends, in part, on whether an incompetent formation underlies or overlies a competent for-

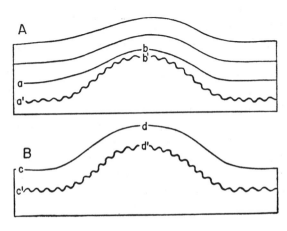

FIGURE 6-14

The difference between a buried hill, A, and no buried hill, B. The difference between aa' and bb' in A is a measure of the topographic relief of the unconformable surface, or the height of a buried hill, which localized the folding of the younger formations. In B the thicknesses cc' and dd' are the same, showing no topographic relief of the unconformable surface.

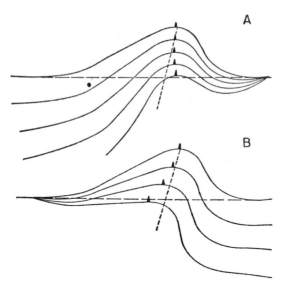

FIGURE 6-15

Diagrammatic sections showing the effect of converging formation thicknesses when added to an asymmetrical folding. The dashed line is the axis of the asymmetrical fold if the formations are of equal thickness. In A the formations are thinning in the direction of the steep-dip side of the fold, and the crest of the fold shifts in the direction of thinning. In B the beds are converging toward the low-dip side of the asymmetrical fold, and the crest of the fold shifts in the direction of thinning.

mation. Where a competent reservoir rock is *overlain* by a great thickness of soft, incompetent rocks, there may be a great discrepancy between the folding at the surface or at shallow depths and the folding on the reservoir rock. Conspicuous examples are some of the anticlinal folds containing the great oil pools of the Middle East. (See the cross section in Fig. 6-12.) The competent Asmari limestone reservoir rock (Miocene and Oligocene) is overlain by a series of interbedded evaporites and shales, which are extremely incompetent and whose deformation is entirely discordant with that of the Asmari limestone. Similar relations are found in the foothill belt of the Carpathians, where the oil pools occur in flysch sediments of Cretaceous-to-Oligocene age, and in the Polish oil fields. Here the oil is trapped in folds that are completely out of harmony with those in the shallower formations,[21] and the shallow structure gives little help to one who is trying to predict the deep structure.

Where the incompetent formations *underlie* the competent formations, a different kind of fold develops—one characterized by a central core of older, incompetent material, which has been squeezed or injected up through the crest of an anticline. Such folds, which are called *diapir folds* or *piercement folds* (see Fig. 6-13), are common in Europe in Aquitaine and the Carpathian region and in the USSR in the Caucasus.[22]

The fold through which the underlying material is extruded is generally tightly compressed and has vertical or nearly vertical dips near its axis. Where the extruded material is salt, as it commonly is in Romania, these structures are sometimes called "salt anticlines." Where the extruded material is mud, they sometimes form mud volcanoes and mud flows. (See Fig. 2-5, p. 23.) Other injected material includes sands and clays, anhydrite, mud breccias, and bitumen or asphalt. The oil associated with diapir folding is generally trapped

TRAPS—GENERAL AND STRUCTURAL [CHAPTER 6]   251

in the gently dipping formations alongside the steeply dipping axial rocks. It was formerly believed that the oil was squeezed out of the central core and deposited in the flanking reservoirs. It is now thought, however, that diapir folding is only a manifestation of the deformation of a combination of competent and incompetent rocks, and has no genetic relation to petroleum. Diapir folding may form traps, and that is its important relation to petroleum.

Buried Hills. Some folds arch over buried topographic highs, called "buried hills." [23] These buried hills have been investigated by many geologists, and the part they play in the formation of the overlying dome fold or anticline

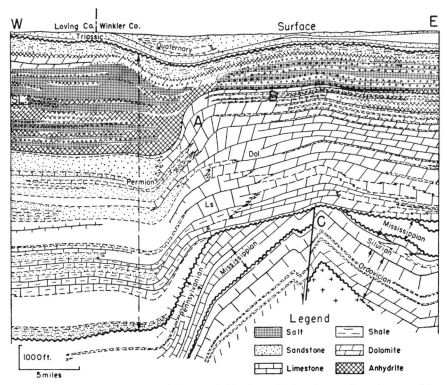

FIGURE 6-16 *Section XX' of Figure 3-10 through the slumped section resulting from solution of the underlying salt formations. Dips at the surface are meaningless in interpreting the deep structure in areas such as this. A is the location of the Hendrick pool producing from the Permian limestone reef complex shown as BB' in Figure 3-10, B is the location of the line of sand pools shown as AA' in Figure 3-10; and C is typical of the pre-Permian and pre-Pennsylvanian anticlines that contain pools in Devonian, Silurian, and Ordovician formations. This section is typical of the geologic conditions that occur in the province of western Texas and southeastern New Mexico shown in Figure 3-10, p. 73. [Redrawn from West Texas Geol. Soc. (1949).]*

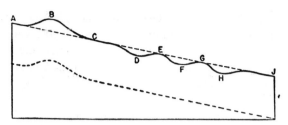

FIGURE 6-17 *Profile showing how the fold that is present at depth in the stratigraphic section, ABC, rises above the regional dip, whereas the superficial folds, DEF and FGH, do not. These superficial folds are probably the result of shallow solution and slumping.*

has been variously explained. It has commonly been supposed that the overlying sediments have become differentially compacted or drape down over their sides, thus giving rise to a shallow fold overlying a sharper fold in the deeper potential reservoir rock.[24]

Many of the "buried hills" were actually not hills or topographic highs when the overlapping sediments were being deposited. They may appear from casual examination to have been at one time elevated areas around which the sediments were compacted into domes, but they are frequently, in fact, merely folded surfaces of unconformity. When the beds are flattened out and placed on a stratigraphic section (see Fig. 6-14), the hills vanish. Many are buried anticlines, which were eroded and left little or no topographic relief to be covered by the succeeding formation, and were later refolded along with overlying formations.

A true buried hill may consist of a topographic high, a bioherm, or organic reef, or a resistant lens composed of some such material as a sand or gravel, surrounded by clays and shales. Where a fold overlies a buried hill or an uncompactable lens of sediments in the stratigraphic section, the folding may be explained in one of two ways:

1. It may be due to more compaction around the edges of the hill, where the shales are thicker, than over the top of the hill, where the shales are

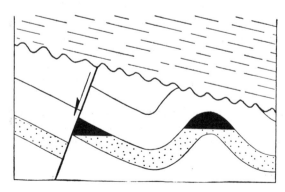

FIGURE 6-18

Idealized section showing characteristic pre-unconformity traps that are obscured by post-unconformity formations.

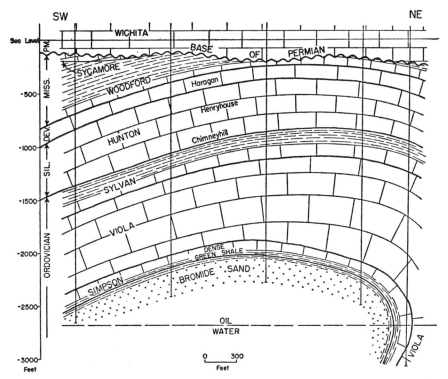

FIGURE 6-19 *Section across the Apache pool, Oklahoma, showing (1) production from a trap that is an overturned anticline, and (2) an example of a wide discordance between the structure of the shallow and that of the deep formations. The area distribution of the pre-Permian rocks below the unconformity is shown in Figure 13-12, page 605, and the structure of the Bromide sand is shown in Figure 6-5, page 243.* [Redrawn from V. C. Scott, Bull. Amer. Assoc. Petrol. Geol., Vol. 29, p. 101, ig. 1.]

thinner. This would cause the overlying sediments to drape over the edges of the hill and thereby form a dome or anticline. It is difficult to see how this folding could extend vertically much beyond the immediate overlying and surrounding shale formations, for diagenesis would be expected to lithify and solidify the clays into shales, and erosion would probably bevel off any uplifted areas, so that later formations would be of uniform thickness and would consequently compact uniformly.

2. The hill may have localized later folding in the region. This localization of folding may be the result of initial dips surrounding the hill or the bioherm, which could cause the overlying sediments to dip as much as 30° away from the center or core of the hill.[25] Where the buried hill is caused by a buried anticline or fault, it may be expected to localize later folding and cause a fold to extend vertically upward into much younger formations.

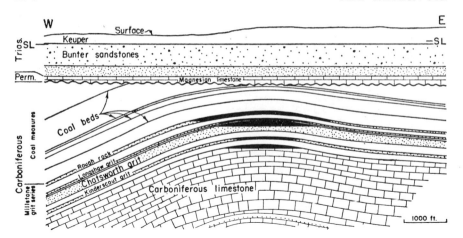

FIGURE 6-20 Section through the Eakring oil field, Nottinghamshire, England. This is a multiple-pay field with the traps formed by an anticline completely masked by flat-lying Permian and Triassic formations. [Redrawn from Kent, in The Science of Petroleum, Oxford University Press (1953), Vol. 6, Part I, p. 56, Fig. 2.]

The presence of a buried hill can be revealed by an isopach map showing the interval between the surface of unconformity and the first marker bed above it. In some places, as shown in Figure 6-14, B (where dd' is as thick as cc'), little or no evidence is found of any appreciable topographic relief on surfaces of unconformity. The topographic relief is shown, as in Figure 6-14, A, as the difference between aa' and bb'. Buried hills, as the cause of folding by compaction and draping over the sides, have probably been overemphasized. It seems more reasonable to regard many of the folds associated with such buried topographic features as due chiefly to normal folding processes that were localized by the irregularities in the rocks, such as folds, faults, initial dips, and igneous intrusions.

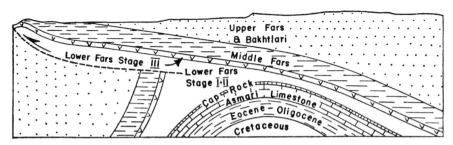

FIGURE 6-21 Section across the Agha Jari structure in Iran: an example of how thrust faulting at times obscures the underlying structure, in this case containing one of the world's great oil fields. [Redrawn from Lane, O. & G. Jour., August 4, 1949, p. 57, Fig. 2.]

TRAPS—GENERAL AND STRUCTURAL [CHAPTER 6]   255

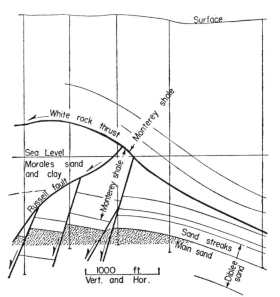

FIGURE 6-22

Section across the Russell Ranch field, in the Cuyuma Valley, California. The trap is formed by a faulted fold, which was later covered by an overthrust sheet. Oil pool shaded. [Redrawn from Eckis, AAPG Guidebook (1952), Los Angeles, p. 91.]

Asymmetric Folding. Asymmetric folding causes a shift of the crest of an anticline with depth, in the direction of the flank with the lower dip. The shift may be considerable where the depth to the potential reservoir formation is two or three miles and there is a large difference in dip between the opposing flanks. The position of the crest at depth may be calculated [26] from the shallow or surface evidence, or it may be determined by seismic surveys. Where asymmetric folding occurs in an area of rapid convergence of the formations involved in the folding, the two factors may combine to cause a much greater

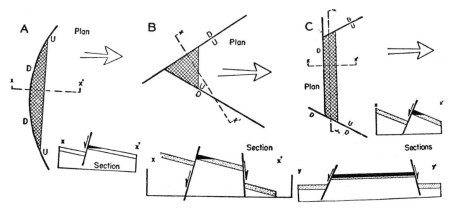

FIGURE 6-23 *Idealized diagrams showing characteristic traps formed chiefly by normal faulting, coupled with regional homoclinal dip: A, a trap formed by a single curved fault; B, a trap formed by two intersecting faults; C, a trap formed by several intersecting faults. Arrows show dip.*

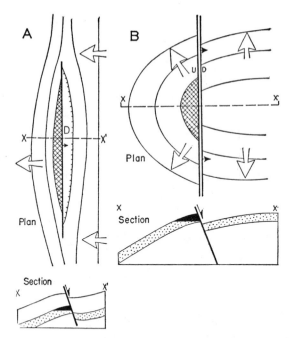

FIGURE 6-24

A, *trap formed by intersection of a low fold with a normal fault;* B, *trap formed by intersection of a normal fault with a more acute fold (Arrows show direction of dip.)*

shift than would normally be calculated from either one, or they may nullify each other. These two effects are shown diagrammatically in Figure 6-15.

Shallow and Surface Weathering Phenomena. Wherever highly soluble formations are exposed by erosion or come within the zone of circulating ground water, the resultant surface structure is often completely at variance with the deeper structure.[27] Solution of salt and other evaporites may throw the rocks at the surface into a jumbled mass of high dips, irregular folds, and erratic structures that have no meaning in relation to the deeper structure. Several large areas of salt solution, slumping, and collapse, for example, occur in western Texas, and the shallow structure is entirely unrelated to that of the underlying rocks. A section through the collapse over the Hendrick pool is shown in Figure 6-16. Because of the wide distribution of evaporites,[28] fossil slumps may occur in some areas, and some of the thinning, now ascribed to original causes, of the salt beds found at the edges of depositional basins may well be due to solution by surface waters circulating during the time interval represented. The swelling of bentonitic and montmorillonite clays also may give rise to misleading surface folds. Surface dips in caliche have been mistaken for true formation dips and have caused structures to be mapped—and drilled—that do not affect the deeper formations.[29] At times the fold that carries down may be distinguished from the superficial fold by plotting a profile. The true fold, *ABC* in Figure 6-17, rises above the regional dip; the superficial folds, *DEF* and *FGH,* do not—they are, in fact, more like residual folds between two synclines. Folding such as *CEGJ* is sometimes spoken of as "pan-of-biscuit" folding.

TRAPS—GENERAL AND STRUCTURAL [CHAPTER 6]   257

Pre-unconformity Deformation. Folding and faulting that occur below buried unconformities are frequently not indicated at the surface, as the idealized section in Figure 6-18 shows. That is so over the Apache pool in Caddo County, Oklahoma, for example, and the reason is apparent from a structural section through the pool. (See Fig. 6-19.) A structural map of the overturned fold in the producing formation is shown in Figure 6-5, page 243, and a paleogeologic map of the pre-Pontotoc surface of unconformity in Figure 13-12, page 605. In the Eakring field of England, also, several traps containing oil and gas pools are concealed below an unconformity surface. (See Fig. 6-20.)

Overriding by Thrust Faults. Thrust faults may obscure the underlying structure, and a number of pools have been trapped in structures concealed by overriding sediments. An example is the Agha Jari anticline in Iran, shown in Figure 6-21, where the almost homoclinal dip above the overthrust fault gives no evidence of the underlying anticline. The Russell Ranch field, in the Cuyama Valley of California, is another example. (See Fig. 6-22.) The pre-thrust, normal faulting and folding, which localize both the trap and the pool, are completely hidden by the overriding sediments.

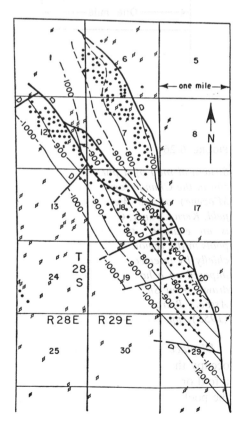

FIGURE 6-25

Structural map of the Vedder sand (Lower Miocene) of the Round Mountain field, Kern County, California. This is an example of a trap formed by a curved fault intersecting a homoclinal dip. [Redrawn from Brooks, AAPG Guidebook (1952), *Los Angeles, p. 148.*]

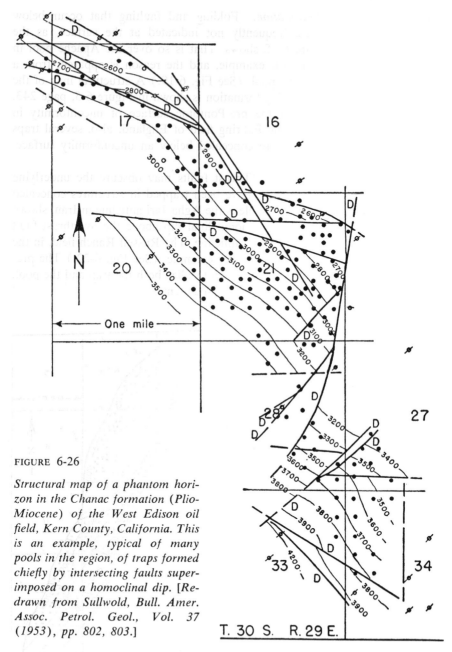

FIGURE 6-26

Structural map of a phantom horizon in the Chanac formation (Plio-Miocene) of the West Edison oil field, Kern County, California. This is an example, typical of many pools in the region, of traps formed chiefly by intersecting faults superimposed on a homoclinal dip. [Redrawn from Sullwold, *Bull. Amer. Assoc. Petrol. Geol.*, Vol. 37 (1953), pp. 802, 803.]

Displaced Pool. In most traps, if a pool of oil or gas is present, it will occupy the structurally highest position in the reservoir rock, as the crest of a fold or the peak of a fault trap. There are some exceptions, however, where the pool is displaced for varying distances down one side of the trap. In most

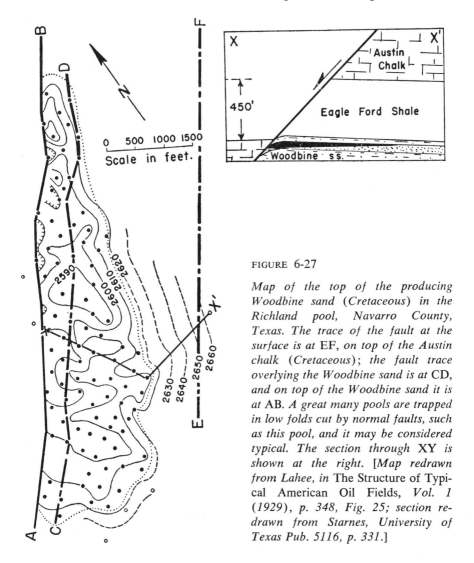

FIGURE 6-27

Map of the top of the producing Woodbine sand (Cretaceous) in the Richland pool, Navarro County, Texas. The trace of the fault at the surface is at EF, on top of the Austin chalk (Cretaceous); the fault trace overlying the Woodbine sand is at CD, and on top of the Woodbine sand it is at AB. A great many pools are trapped in low folds cut by normal faults, such as this pool, and it may be considered typical. The section through XY is shown at the right. [Map redrawn from Lahee, in The Structure of Typical American Oil Fields, *Vol. 1 (1929), p. 348, Fig. 25; section redrawn from Starnes, University of Texas Pub. 5116, p. 331.]*

cases the crest of the structure will still be productive, and drilling into the highest point of the trap will make a discovery, but occasionally the displacement is enough to leave the crest of the structure barren. Displacement of pools is generally due to fluid potential gradients that result in the movement of water through the reservoir rock; if this condition is suspected, the fluid potential gradients of the area and the densities of the water and the expected oil should be studied, and the test well should be drilled where the shape of the trap indicates the pool will be trapped under these conditions. The subject is considered in more detail in Chapter 12.

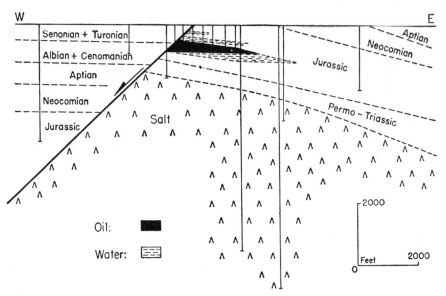

FIGURE 6-28 *Section through the Dossor oil pool in the Emba province of the USSR. Petroleum is trapped in a faulted Jurassic sand above a thick, folded salt mass. There is no evidence of intrusion of the salt. One well, at a depth of 750 feet, produced over 75,000 barrels of oil in 30 hours in 1911.* [*Redrawn from T. Jeremenko,* Neftyonoye Khozyaystro (Petroleum Economy), *Moscow (1939), and C. W. Sanders, Bull. Amer. Assoc. Petrol. Geol., Vol. 23 (1939), p. 505.*]

Traps Caused by Faulting

Normal, or gravity, faults and reverse and thrust faults[30] in the reservoir rock have wholly or partly formed the trap for many oil and gas pools; most pools found in structural traps, in fact, are modified by faults. Faulting may be the sole cause of the formation of a trap, but more commonly faults form traps in combination with other structural features, such as folding, tilting, and arching of the strata, or with variations in the stratigraphy or permeability. Faulting has been a minor trap-forming element in many pools, where it modifies the trap and causes local variations in the production characteristics.

Seepages of oil and gas are often associated with fault outcrops. Thus faults are commonly thought of as vertical channels permitting migration between reservoirs and to the surface. The presence of seepages at the surface suggests that the potentiometric surface of the aquifer is above the level of the ground. Lack of seepage, on the other hand, may indicate that the potentiometric surface is below the level of the ground. Many faults form the boundary plane of a pool of oil and gas, and this may be due to the fact that the fault is tightly sealed and holds the petroleum from further migration; or it probably more commonly is due to higher fluid potentials within the fault

TRAPS—GENERAL AND STRUCTURAL [CHAPTER 6] 261

channels and up-dip across the fault which act as an added barrier to the up-dip movement of petroleum. The combination of the fault and the hydrodynamic conditions forms a trap that holds a pool.

Normal Faulting. Normal, or gravity, faulting, combined with a regional homoclinal dip, may form traps. There may be a single curved fault, as in Figure 6-23 (A), the intersection of two faults (B), or a combination of several faults (C). Normal faulting, combined with low folding, forms many pools, as in Figure 6-24 (A). As the folding becomes more acute, the trap becomes more definite (B); traps such as these are common on many elongated anticlines and domes.

The Round Mountain field, Kern County, California, contains pools trapped by the intersection of curved faults with a homoclinal regional dip. (See Fig. 6-25.) The West Edison field, nearby, contains traps formed by the intersection of normal faults with a homoclinal regional dip. (See Fig. 6-26.)

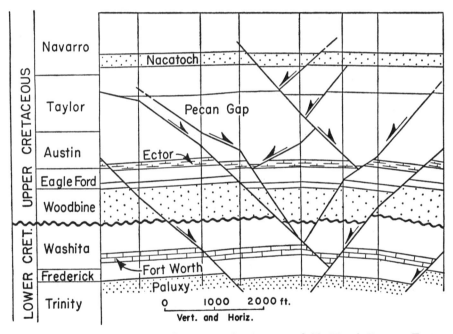

FIGURE 6-29 *Section across the top of the Quitman field, Wood County, Texas. An example of a complex fault pattern developed along the crest of an elongate anticline. The Eagle Ford and Paluxy sands are productive in the high portion of nearly every fault block. The Woodbine sand is productive in one fault block. The oil-water contact is 104 feet higher in one Paluxy producing fault block than in the others, suggesting some post-accumulation faulting. [Redrawn from Scott, in* The Structure of Typical American Oil Fields, *Vol. 3, pp. 426–427, and Smith, University of Texas Pub. 5116, p. 318.]*

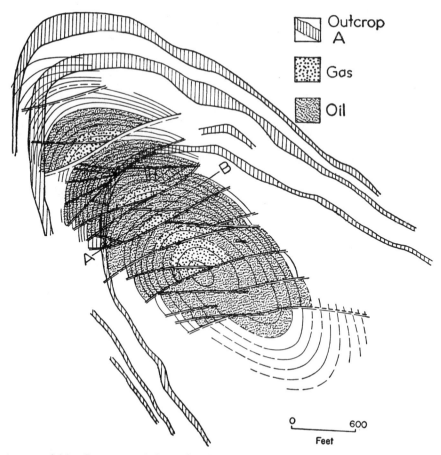

FIGURE 6-30 Structure of the Kala oil and gas field, Azizbekovo Oil Trust, Baku region, Apsheron Peninsula, USSR. (See Fig. 2-3, p. 21, for location.) The contour interval is 10 meters (33 feet). The distribution of the oil and gas in the Sabunchi series (Productive formation, Pliocene) is shown. The cross faults form separate pools along the axis of the fold. The Kala fold is one of the simpler folds of the great Baku producing province. [Redrawn from "Oil Deposits of Azerbaidjan," XVIIth Int. Geol. Cong., Moscow (1937), Vol. 4, p. 116.]

Many pools in this part of California are trapped, in large part, by normal faults.

The Richland pool, in Navarro County, Texas, is one of a great many pools trapped by a normal fault cutting across an arch or low fold in the reservoir rock.[31] A structural map and section of the reservoir are seen in Figure 6-27, which shows not only the trace of the fault at the surface but its intersection with the producing sand. In the Dossor oil field, in the Emba province of the USSR,[32] shown in Figure 6-28, production is obtained from Jurassic sands that are arched and faulted; here, as in the Richland pool, the fault plane forms one side of the reservoir.

Faulting often breaks up a field into separate pools; where that happens, the fault planes may become the boundary of a pool and tightly seal it off. The Quitman field, in Wood County, Texas, is one among many in which an anticlinal dome fold is broken into a number of fault blocks, each containing a trap in which oil and gas have accumulated. A section across the field is shown in Figure 6-29. Another is the Kala field, in the Baku district of the USSR, a structural map of which forms Figure 6-30. This combination of folding and faulting is common to many anticlinal traps. In some the transverse faults are not large enough to separate the pools; but in others, as in the Kala fold, many of the faults form boundaries of separate pools, distinguished by their differing oil-water contacts. Thrust faulting combines with anticlinal folding to form the traps in some of the prolific Cretaceous limestone reservoirs of western Venezuela. A section across the Mara field is shown in Figure 6-31 and shows the thrust horst along the axial part of the fold, a characteristic of several of these fields.

The Inglewood field, in California, is an example of a trap composed of a dome that has been modified by normal faulting and forming several pools. A structural map of the reservoir and two sections across it are shown in Figure 6-32.

The Creole field, in the Gulf of Mexico off the coast of Louisiana, consists of four separate pools trapped by normal faults associated with a dome fold caused by a deep-seated salt plug. The field is also of interest for another reason: it was all drilled from a single off-shore platform by means of directed drilling.[33] The traces of the drill holes and their relations to the faults are

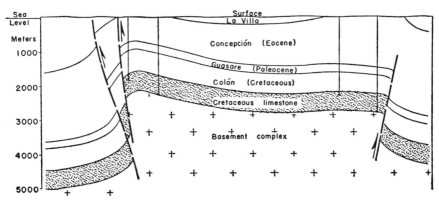

FIGURE 6-31 *Section across the Mara field, western Venezuela. Large production occurs from the Cretaceous limestone and from the fractured granites of the basement complex. Length of section about six miles.* [Redrawn from Mencher et al., Bull. Amer. Assoc. Petrol. Geol., Vol. 37, p. 725, Fig. 3.]

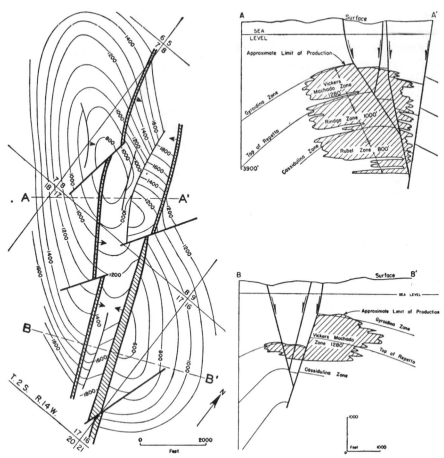

FIGURE 6-32 *Structure of the Inglewood field, Los Angeles basin, California, contoured on top of the Gyroidina zone, and sections along AA' and BB'. Movement along the main fault is chiefly horizontal, the movement being south along the east side (right lateral). The distribution of the oil is shown as hatching in the sections.* [Redrawn from Driver, Bull. 118, Calif. Div. Mines (1943), p. 307.]

shown in Figure 6-33. The distribution of the gas, oil, and water in the deeper pools is shown in Figure 6-34, and a section through the pool in Figure 6-35. The structure of the producing Woodbine formation (Upper Cretaceous), in the Van field, northeastern Texas, is shown in Figure 6-36. Two separate pools, *A* and *B,* are trapped by a fault superimposed on a dome fold. A section from northeast to southwest across the field is shown in Figure 5-4, page 150, where an analysis is made of the time the accumulation occurred.

Minor faulting may in some cases follow incipient fracturing and the shallow effect of subsurface stresses related to the folding; as erosion and removal

TRAPS—GENERAL AND STRUCTURAL [CHAPTER 6] 265

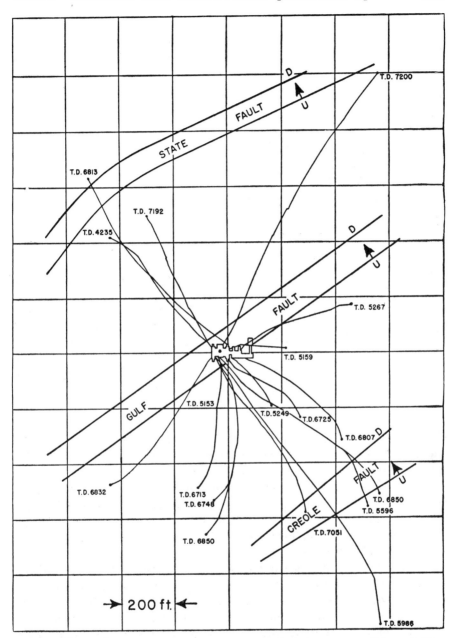

FIGURE 6-33 *Horizontal traces of nineteen holes drilled from one offshore drilling platform in the Creole oil field, Louisiana, Gulf of Mexico. The structural conditions encountered are shown in Figures 6-34 and 6-35.* [Redrawn from Wasson, in The Structure of Typical American Oil Fields, *Vol. 3, Amer. Assoc. Petrol. Geol. (1948), p. 288, Fig. 5.*]

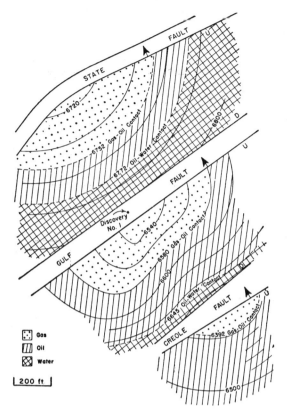

FIGURE 6-34

Structure and distribution of gas, oil, and water in the chief producing sand, the Gulf sand (Miocene), of the Creole oil field, found at about 6,700 feet. [Redrawn from Wasson, in The Structure of Typical American Oil Fields, *Vol. 3 (1948), p. 292, Fig. 7.]*

of the overburden bring these stressed rocks closer to the surface, the stress caused by the folding may be relieved by minor faulting. Thus many fold traps are accompanied by minor faults, which may or may not reach down to the reservoir, and faults that do reach the reservoir merely change the outline of the pool without affecting the water-oil contact, which remains planar throughout the pool. Such a pool is in the Petrólea field, in the Barco region of Colombia,[34] shown in Figure 6-37. The minor shallow faulting is both of a normal and a thrust type.

Pools trapped by normal faulting are almost always on the upper side of the fault. One might expect, on looking at a cross section, that the lower side of a fault would form a trap, but it seldom does so, apparently because the oil and gas escape up-dip around the ends of the fault. Pools in which petroleum is trapped on the lower side of a fault are exceptional and are generally to be explained by a combination of minor faulting, permeability variations, hydrodynamic forces directed down-dip folding along the lower side, and truncation of the reservoir by the lower side of the fault. Many of the pools along the Gulf Coast of southern Texas and Louisiana, however, lie on the lower side of normal faults. These pools are found immediately to the south of a series of

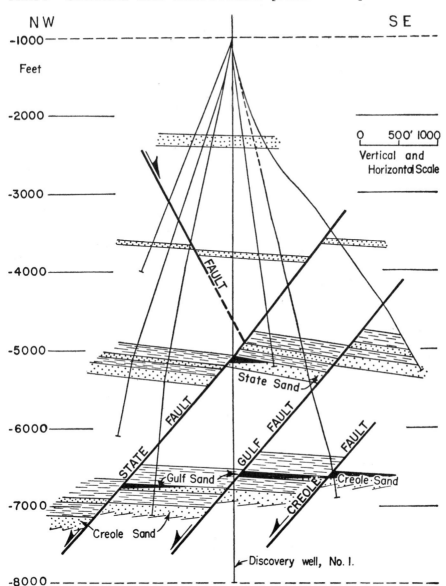

FIGURE 6-35 *Section through the Creole field showing the manner in which the field was developed by directed drilling from one platform standing in the Gulf of Mexico. While the overall structure is a dome fold over a deeply buried salt intrusion, the separate pools are defined by normal faulting associated with the fold.* [Redrawn from Wasson, in The Structure of Typical American Oil Fields, *Vol. 3 (1948), p. 286, Fig. 4.*]

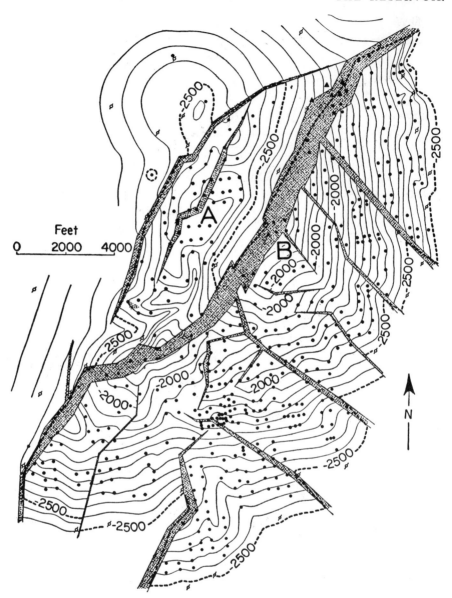

FIGURE 6-36 *Structure of the top of the productive Woodbine sand (Cretaceous) in the Van field, Van Zandt County, Texas. The shaded areas show where the top of the Woodbine sand is absent because of faulting. The dashed line marks the boundaries of the pool. A section across the field is shown in Figure 5-4, page 150. The Van field, it is estimated, will ultimately produce 400 million barrels of oil. It is typical of faulted folds caused by the deep intrusion of salt masses. (See Chap. 8.)* [Redrawn from Liddle, University of Texas Bull. 3601, Fig. 19, and Betts, University of Texas Bull. 5116, p. 401.]

TRAPS—GENERAL AND STRUCTURAL [CHAPTER 6]   269

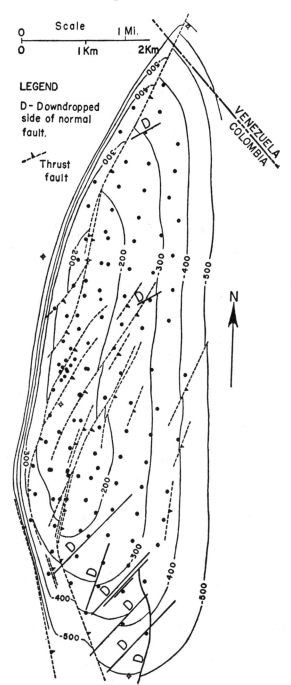

FIGURE 6-37

Structure of the Petrólea North Dome oil field, Barco Concession, northern Colombia, contoured on the productive Zone 3 (Cretaceous). Contour interval is 100 meters (328 feet). This trap is a dome fold with steep or overturned beds on the west flank, numerous thrust faults, and some normal faults. Oil seepages are common at the surface, and prolific oil production has been found at the shallow depths of 26–513 meters (about 85–1700 feet). (See Fig. 14-12 for location.) [Redrawn from Notestein, Hubman, and Bowler, Bull. Geol. Soc. Amer., Vol. 55 (1944), p. 1208, Pl. 6.]

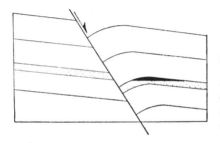

FIGURE 6-38

Diagrammatic section through down-dropped fault and fold, characteristic of many traps along the Gulf Coast of Texas and Louisiana.

south-dipping normal faults parallel to the Gulf, and dipping, as the rocks of the region do, toward the Gulf. When these pools are examined more closely, however, it is found that they are nearly all trapped by closed anticlines, which occur along and parallel to the lower or down-dropped side of these faults. The situation is diagrammatically shown in Figure 6-38. The north closure of the trap is in the north dip into the fault plane, which in itself is very unusual, and the fault seldom has anything to do with the actual bounding of the pool; the trap is formed by the anticline. The faulting, however, must be genetically related to the folding, and a number of theories have been advanced to explain the unusual relationship. Some of these explanations are:

1. Subsidence of the basement along hinge lines parallel to the coast, causing an abruptly steeper dip, along which the overlying incompetent formations were pulled apart.

2. Slipping of the formations toward the coast along bedding planes, causing breaks to form, into which the lower-side beds dip.

3. Deeply buried salt ridges paralleling the coast, which plastically intruded the overlying formations, causing normal faults.[35] The mechanics of the intrusion and the formation of the faults are similar to those believed to cause the graben faulting over salt plugs, which is discussed on pages 361–363. These fault breaks may be, in fact, the incipient up-slope edge of future great landslides into the Gulf of Mexico.

The trap in the Amelia field, in Jefferson County, Texas, is a fold on the lower side of a normal fault. A map of the structure on the Langham producing sand (Frio formation, Oligocene) is shown in Figure 6-39, and a section through the field is shown in Figure 6-40.

Reverse and Thrust Faulting. Traps associated with thrust and reverse faulting* may form either above or below the fault plane. The trap may be bounded on one side by the fault, but more often it is formed by folding asso-

---

* When the hanging wall has apparently moved up with respect to the footwall, a fault is termed a *reverse fault* if the fault plane makes a high angle (above 45°) with the horizontal, a *thrust fault* if the fault plane makes a low angle (less than 45°) with the horizontal.

TRAPS—GENERAL AND STRUCTURAL [CHAPTER 6]   271

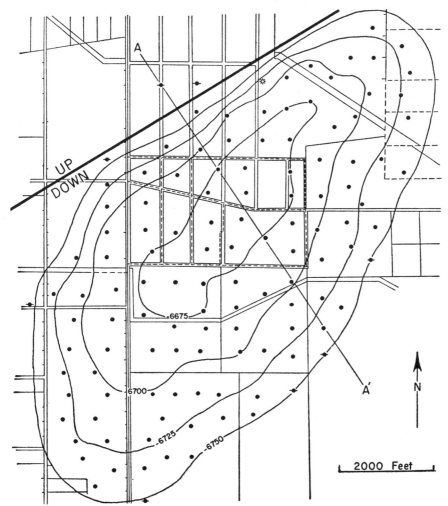

FIGURE 6-39 *Structure of the Amelia field, Jefferson County, Texas, contoured on top of the producing Langham sand (Frio formation, Oligocene). This dome fold on the lower side of a normal fault is typical of many pools in southern Texas and Louisiana. Section AA' is shown in Figure 6-40. Folds such as these are probably not due to an intrusive salt plug, although regional deep-seated salt movement may account for their unusual character, the rocks dipping into the down-faulted side of a normal fault in a way that is not found in most faulting.* [Redrawn from Hamner, Bull. Amer. Assoc. Petrol. Geol., Vol. 23, p. 1646, Fig. 3.]

ciated with the thrust faulting. Some traps characteristically associated with thrust faults are diagrammatically shown in Figure 6-41.

The Talang Akar pool, in Sumatra, is trapped in an overthrust elongated fold. Maps and sections are shown in Figure 6-42. The South Mountain field,

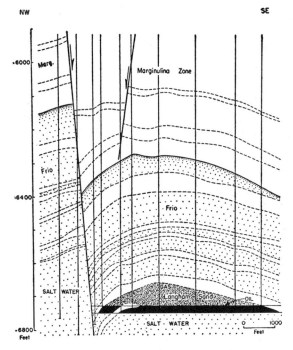

FIGURE 6-40

Section through the Amelia field, Texas, showing the relations between the dome fold and the fault. The dip into the down side of the fault is contrary to most faulting, except in the system of faults parallel to the Gulf Coast in Louisiana and Texas, of which this is an example. The discordance in dip between the two upper members in the section and the Frio formation suggests an intervening period of deformation. [Redrawn from Hamner, Bull. Amer. Assoc. Petrol. Geol., Vol. 23, p. 1647, Fig. 4.]

in Ventura County, California, is also found in folded sandstone reservoir rocks, which have been thrust up over the underlying formations along the South Mountain thrust fault. A short distance away, several pools occur below the thrust plane of the Timber Canyon overthrust fault, which acts as a trap boundary. A section through these two areas may be seen in Figure 6-43.

Several pools producing gas from the Oriskany sand (Lower Devonian), in northern Pennsylvania and southern New York, occur in folds associated with reverse faulting. One such field consists of the Woodhall and Tuscarora[36] pools, in southern New York, of which a structural map is shown in Figure 6-44 and a section in Figure 6-45.

An overthrust field is found in the Turner Valley gas and oil field, in Alberta, Canada, where oil and gas accumulated within the overthrust mass and were later carried forward by the overriding rocks.[37] During the thrust movement the folding was intensified, and the anticline was overturned in the

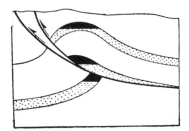

FIGURE 6-41

Diagrammatic section showing characteristic positions of traps associated with thrust faulting.

TRAPS—GENERAL AND STRUCTURAL [CHAPTER 6] 273

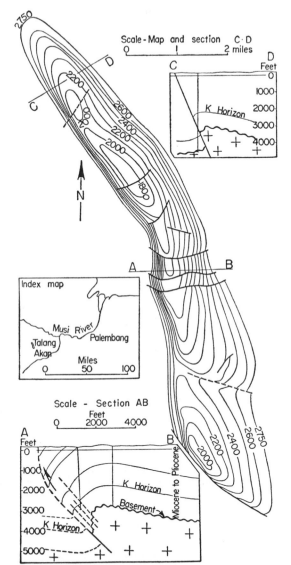

FIGURE 6-42

Structural map (contour interval 100 feet) and sections of the Talang Akar overthrust anticline, Sumatra. Production is from the Talang Akar sandstone (Miocene) at depths between 2,100 and 2,750 feet, and nearly 200 million barrels of oil of 38 API gravity has been produced from this field. [Redrawn from George Martin Lees, Quart. Jour. Geol. Soc. London, Vol. 108 (December 1952), p. 15, Fig. 11, a map after George Barnwell.]

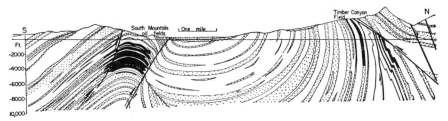

FIGURE 6-43 *Structural section across South Mountain and Timber Canyon oil fields, Ventura County, California. The trap in the South Mountain field is an overthrust fold and reverse fault, and the trap in the Timber Canyon field is formed by the faulted edges of the reservoir rocks where they are overridden and sealed by an overthrust fault. This section shows examples of accumulations both above and below thrust-fault planes.* [Redrawn from Bailey, AAPG Guidebook (1952), Los Angeles, p. 68.]

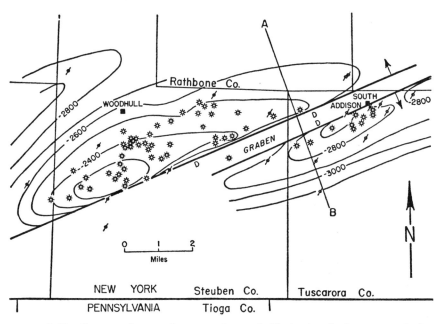

FIGURE 6-44 *Structural map (contour interval 50 feet) of the gas-producing Oriskany sandstone (Lower Devonian) in the Woodhull and Tuscarora gas pools of southern New York State. The traps are associated with reverse faulting, which is common to the traps of gas pools in this region. Fault blocks, such as shown in Figure 6-45, seem to be necessary for commercial production, possibly because of the porosity and permeability developed by the intense shattering and fracturing of the brittle Oriskany sandstone. (See Finn, Bull. Amer. Assoc. Petrol. Geol., Vol. 35, p. 306.)* [Redrawn from Finn, Bull. Amer. Assoc. Petrol. Geol., Vol. 33, p. 332, Fig. 14.]

TRAPS—GENERAL AND STRUCTURAL [CHAPTER 6]   275

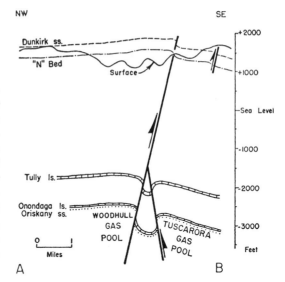

FIGURE 6-45

Section through the Woodhull-Tuscarora gas pools (AB in Fig. 6-44). Reverse faults such as these are common in the Oriskany gas region of Pennsylvania and New York, where they either modify or form the boundaries of a number of gas pools. [Redrawn from Finn, Bull. Amer. Assoc. Petrol. Geol., Vol. 33, p. 333, Fig. 15.]

direction of movement. A section across this feature is shown in Figure 6-46. (See also Fig. 14-14, p. 651, for an account of the development of the fold and the pool.) Traps defined by reverse faults occur in the Circle Ridge field, Fremont County, Wyoming, where the Permian has been thrust over the Triassic to form traps, in which several pools occur,[38] both above and below the overthrust fault plane. The relations are shown in the structural map and section of Figure 6-47. In the Aliso Canyon field, in California, traps associated with overthrust faults contain oil pools. A map of the surface relations is shown in Figure 6-48 and a section across the area in Figure 6-49. The oil in the Achi-Su field, in the Caucasus region of the USSR, which is typical of several fields in the Grozny area, is found in an elongated fold associated with an overthrust fault. A structural map and sections are shown in Figure 6-50.

Complex thrust faulting may be associated with incompetent formations, such as soft clays and shales, and with anhydrites and other salts of evaporite rocks. The faulting may be directly responsible for the trap, but more often the faults merely obscure the trap, as shown in a section across the Agha Jari structure in Iran, which is typical of several of the great oil pools of the Near East. (See Fig. 2-2, p. 18.) The oil fields in the Ploesti region of Romania are characterized by traps associated with reverse faulting and salt injection.[21] The section across the Moreni-Bana field, shown in Figure 6-51, is typical of many traps in this region. See Figure 8-11, page 357, for a section through the Mene Grande field. The Bitkow field, in the North Carpathian province of Poland (now in the USSR), is another in which complex folding and thrust faulting combine to form traps.[39] A section across the field is shown in Figure 6-52. Traps such as these are common throughout the entire North Carpathian province.

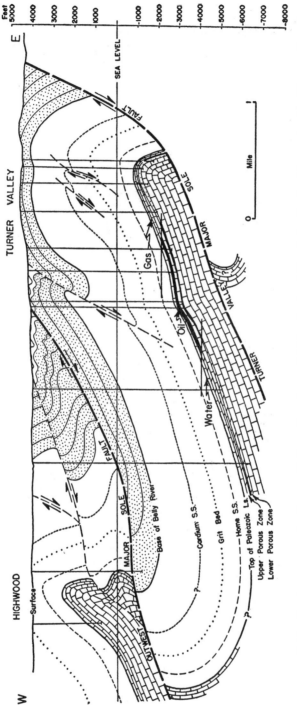

FIGURE 6-46 Section across the Turner Valley oil and gas field in the foothills of the Rocky Mountains in Alberta, Canada. The trap is a fold thrust out over a major sole fault plane. Numerous minor thrust and reverse faults are not shown. The next similar fold to the west, the Highwood, showed only water when drilled. A possible explanation for the difference in productivity is suggested in Figure 14-14, showing the prefault history of the region. [Redrawn from Link, Bull. Amer. Assoc. Petrol. Geol., Vol. 33, p. 1479, Fig. 5.]

TRAPS—GENERAL AND STRUCTURAL [CHAPTER 6] 277

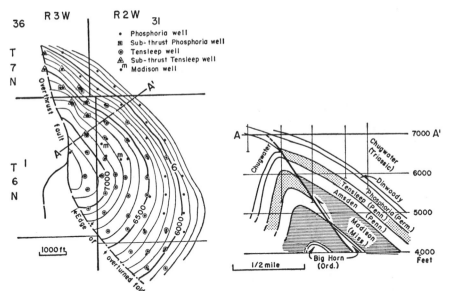

FIGURE 6-47 *Structural map of and section across the Circle Ridge field, Fremont County, Wyoming. Contours on top of the Phosphoria formation (Permian) at an interval of 100 feet. Oil is obtained from several formations both above and below the thrust-fault plane, which merges with an overturned fold to form the traps.* [Redrawn from Beebe, O. & G. Jour., Sept. 14, 1953, pp. 110 and 112.]

FIGURE 6-48

Map showing the Aliso Canyon oil pool and the thrust faults that crop out in the vicinity. The ruled areas show the formation below the Santa Susana fault plane, one area of which is at the east end of the field. See Figure 6-49 for a N-S section through the field. [Redrawn from Leach, in The Structure of Typical American Oil Fields, *Vol. 3 (1948)*, p. 26, Fig. 2.]

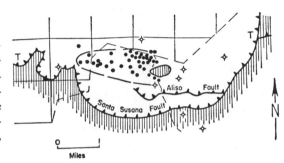

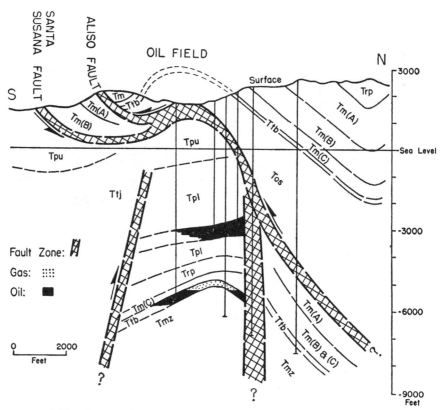

FIGURE 6-49 *Structural section through the Aliso Canyon oil field, in the Ventura basin, California. This field is an example of pools producing from complexly folded and thrust-faulted traps. [Redrawn from Leach, in* The Structure of Typical American Oil Fields, *Vol. 3 (1948), p. 32, Fig. 5.]*

Complex thrust faulting occurs in the Ventura field of California.[40] (See Fig. 9-12, p. 410.) The trap is localized mainly by the Ventura anticline, but at depth it is cut by many large reverse and thrust faults. These separate the field into a number of pools, each contained within a separate fault block of sediments completely or partly bounded by the fault planes. Each pool, when produced, has its own peculiar reservoir pressure and oil and gas content, showing it to be completely separate from the pools in adjacent blocks.

Traps Caused by Fracturing

Fracturing of reservoir rocks is a common cause of porosity and permeability and undoubtedly is an accessory cause in many more pools.[41] (See pp. 119–125.) It may also be regarded as the chief cause of traps in a few instances where special conditions prevail. In the Florence field, in Colorado,[42]

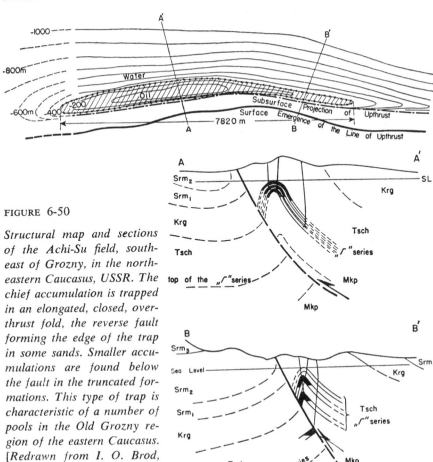

FIGURE 6-50

Structural map and sections of the Achi-Su field, southeast of Grozny, in the northeastern Caucasus, USSR. The chief accumulation is trapped in an elongated, closed, overthrust fold, the reverse fault forming the edge of the trap in some sands. Smaller accumulations are found below the fault in the truncated formations. This type of trap is characteristic of a number of pools in the Old Grozny region of the eastern Caucasus. [Redrawn from I. O. Brod, XVIIth Int. Geol. Cong. (Moscow, 1937), Vol. 4, p. 28, Figs. 3 and 4.]

which was one of the first oil pools discovered in the United States, the oil is trapped within the fractured portion of the Pierre shale (Cretaceous), which is nearly flat-lying and of uniform texture over a wide area beyond the field. Apparently fracturing alone is responsible for localizing the trap, for there is no folding or stratigraphic change associated with the pool, and where the fracturing plays out there is no oil accumulation. A section through the pool is shown in Figure 6-53. Production is erratic, and one well is reported to have produced nearly a million and a half barrels of oil.[43]

Similar traps contain the gas found in the Cherokee shales (Pennsylvanian) of eastern Kansas.[44] The gas occurs in a widespread, uniform, black, carbonaceous shale, and salt water is generally produced along with it. The long life and slow decline of these pools are best accounted for by the slow feed-

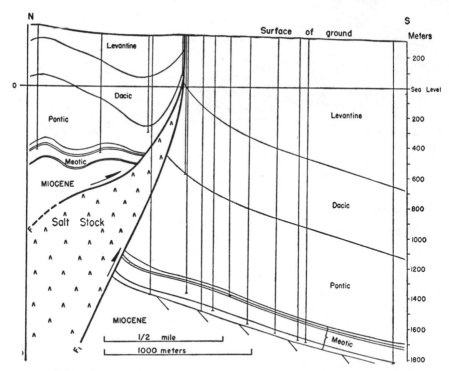

FIGURE 6-51 *Section through the Moreni-Bana oil field in the Ploesti district of southern Rumania. Reverse faulting, associated with incompetent salt formations, makes the deformation in many fields of this area extremely complicated. Presumably the salt in the Moreni-Bana field did not form the trap by intruding, as did the Gulf Coast salt domes, but was squeezed into its present position along a reverse fault plane as a result of the deformation. [Redrawn from Walters, Bull. Amer. Assoc. Petrol. Geol., Vol. 30 (1946), p. 334, Fig. 12.]*

ing of the gas into the wells from fractures, bed partings, laminae, and microscopic pores in the shales. Most of the gas in eastern Kentucky comes from the Black, or Ohio, shale (Devonian and Mississippian?), a fissile, finely laminated, bituminous, black-to-brown shale.[45] While fracturing of the shales is not mentioned by those who have described the pools, it requires shooting with nitroglycerin to obtain production.

In the Mount Calm pool, Hill County, northeastern Texas, the oil is produced from the Austin chalk (upper Cretaceous), from a trap formed by fractures associated with faulting. A section through this pool is shown in Figure 6-54.

If we compare the petroleum found in different pools within a single province, we generally find that the oil in structural traps differs from that in other traps in several respects: it is commonly of higher API gravity, it is

TRAPS—GENERAL AND STRUCTURAL [CHAPTER 6]   281

more likely to be covered with a free gas cap and to be defined at the base by a sharp water-oil contact, its pressure is more nearly normal for the depth than that in other kinds of pools, and the reservoir energy is more commonly an active water drive. These characteristics combine to give pools in structural traps higher average yields per acre, but probably fewer productive acres, than lenticular and reef-type reservoirs. Another important characteristic of such a pool is that its limits are more safely predicted after discovery, if the structure has been mapped, than those of a lenticular reservoir, which gives no clue to its size. Sharp differences in lease prices within short distances may therefore occur, before the limits of a pool are determined, solely because of the position of the leased land in relation to the geologic structure.

CONCLUSION

Most oil and gas has been found in traps that might be classed as either wholly or partly structural, and many examples have been shown. The two

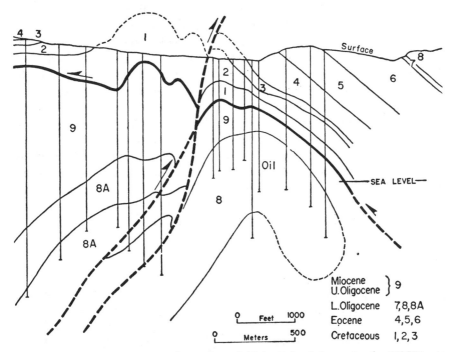

FIGURE 6-52 *Section across the Bitkow field in Poland (now in the USSR), in the northern foothills of the Carpathian Mountains. This type of complicated trap, made up of close folding and thrust faulting, is characteristic of many of the fields in this region. [Redrawn from Walters, Bull. Amer. Assoc. Petrol. Geol., Vol. 30 (1946), p. 326, Fig. 5.]*

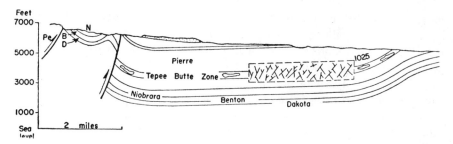

FIGURE 6-53 Section through the Florence field, Colorado, showing an oil pool trapped in a system of intersecting fractures and joints (outlined by dashed lines) within the Pierre shale (Cretaceous). No free water is found with the oil, and the pool has been producing since 1876. [Redrawn from McCoy, Sielaff, Downs, Bass, and Maxson, Bull. Amer. Assoc. Petrol. Geol., Vol. 35 (1951), p. 1008, Fig. 7.]

most important features of structural traps are the wide variety of structural conditions that may form traps, and the fact that a structural trap may extend vertically through thick sections of potentially productive rocks. Structural mapping of all kinds has been the most consistently successful method of locating traps. There are several ways of mapping structure—surface, subsurface, core-drill, and geophysical; each has as its objective the finding of locally high structural conditions in underground reservoir rocks that might prove to be traps in which a pool of oil or gas has accumulated. Where clean, widespread, or blanket sands occur, the regional dips are high, and where sloping potentiometric surfaces are known, the structural traps generally require a large structural closure to be effective; where the reservoir rocks are lenticular and variable, minor local deformation may be sufficient. Both conditions may occur in the same area where different kinds of reservoir rocks are found.

(For the Selected General Readings, see Chap. 8.)

FIGURE 6-54

Section through the Mount Calm pool, Hill County, Texas, showing a small pool trapped in a faulted and brecciated area in the Austin chalk (Upper Cretaceous). None of the wells produced water. [Redrawn from Hood, University of Texas Bull. 5116, p. 241 (1951).]

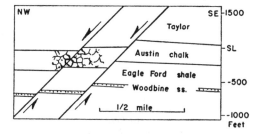

Reference Notes

1. J. V. Howell, "Historical Development of the Structural Theory of Accumulation of Oil and Gas," in *Problems of Petroleum Geology*, Amer. Assoc. Petrol. Geol., Tulsa, Okla. (1934), pp. 1–23. Bibliog. 85 items.
Paul H. Price, "Anticlinal Theory and Later Developments in West Virginia," Bull. Amer. Assoc. Petrol. Geol., Vol. 22 (August 1938), pp. 1097–1100.

2. W. E. Logan, *Report of Progress (1844)*, Canada Geol. Surv. (1846), p. 41.

3. I. C. White, "The Geology of Natural Gas," Science, Vol. 5 (June 26, 1885), pp. 521–522; excerpted in J. V. Howell, *op. cit.* (note 1), pp. 15–16.

4. I. C. White, *West Virginia Geol. Survey*, Vol. 1 (1899), pp. 176–177. Also contains a good summary of his early use of the anticlinal theory. See also *op. cit.* in note 5, appendix to p. 216.

5. I. C. White, "The Mannington Oil Field and the History of Its Development," Bull. Geol. Soc. Amer., Vol. 3 (1892), p. 193.

6. Alex W. McCoy and W. Ross Keyte, "Present Interpretations of the Structural Theory for Oil and Gas Migration and Accumulation," in *Problems of Petroleum Geology*, Amer. Assoc. Petrol. Geol., Tulsa, Okla. (1934), pp. 253–307.

7. E. H. McCollough, "Structural Influence on the Accumulation of Petroleum in California," *Problems of Petroleum Geology*, Amer. Assoc. Petrol. Geol., Tulsa, Okla. (1934), pp. 735–760.

8. M. King Hubbert, "The Theory of Ground-water Motion," Jour. Geol., Vol. 48 (November–December 1940), pp. 785–944.

9. William L. Russell, *Principles of Petroleum Geology*, McGraw-Hill Book Co., New York (1951), pp. 207–209.

10. M. King Hubbert, "Entrapment of Petroleum under Hydrodynamic Conditions," Bull. Amer. Assoc. Petrol. Geol., Vol. 37 (August 1953), pp. 1954–2026.

11. Frederick G. Clapp, "Role of Geologic Structure in the Accumulation of Petroleum," *Structure of Typical American Oil Fields*, Amer. Assoc. Petrol. Geol., Tulsa, Okla., Vol. 2 (1929), pp. 667–716. 97 references cited.

12. William B. Heroy, "Petroleum Geology," in *Geology 1888–1938*, 50th Anniv. Vol. Geol. Soc. Amer. (1941), pp. 535–536.

13. W. B. Wilson, "Proposed Classification of Oil and Gas Reservoirs," in *Problems of Petroleum Geology*, Amer. Assoc. Petrol. Geol., Tulsa, Okla. (1934), pp. 433–445.

14. K. C. Heald, "Essentials for Oil Pools," in *Elements of the Petroleum Industry*, Amer. Inst. Min. Met. Engrs., New York (1940), pp. 47–55.

15. O. Wilhelm, "Classification of Petroleum Reservoirs," Bull. Amer. Assoc. Petrol. Geol., Vol. 29 (November 1945), pp. 1537–1579.

16. H. E. Winter, "Santa Fe Springs Oil Field," Bull. 118, Calif. Div. Mines (April 1943), pp. 343–346.

17. Clifton S. Corbett, "Method for Projecting Structure Through an Angular Unconformity," Econ. Geol., Vol. 14 (1919), pp. 610–618.
A. I. Levorsen, "Convergence Studies in the Mid-Continent Region," Bull. Amer. Assoc. Petrol. Geol., Vol. 11 (July 1927), pp. 657–682.
Frederic H. Lahee, *Field Geology*, 5th ed., McGraw-Hill Book Co., New York (1952), pp. 667–671.

18. D. A. McGee and W. W. Clawson, Jr., "Geology and Development of Oklahoma City Field, Oklahoma County, Oklahoma," Bull. Amer. Assoc. Petrol. Geol., Vol. 16 (October 1932), pp. 957–1020.

19. E. A. Wendlandt, T. H. Shelby, Jr., and John S. Bell, "Hawkins Field, Wood County, Texas," Bull. Amer. Assoc. Petrol. Geol., Vol. 30 (November 1946), pp. 1830–1856.

20. D. F. Hewett, "Measurements of Folded Beds," Econ. Geol., Vol. 15 (1920), pp. 367–385.

John B. Mertie, Jr., "Delineation of Parallel Folds and Measurement of Stratigraphic Dimensions," Bull. Geol. Soc. Amer., Vol. 58 (August 1947), pp. 779–802.

21. Ray P. Walters, "Oil Fields of Carpathian Region," Bull. Amer. Assoc. Petrol. Geol., Vol. 30 (March 1946), pp. 319–336.

22. L. Mrazec, "Les plis diapirs et le diapirisme en général," Comptes-rendus des Séances de l'Institut Géologique de Roumanie, Vol. 6 (1927), pp. 226–272.

Karol Bohdanowicz, "Geology and Mining of Petroleum in Poland," Bull. Amer. Assoc. Petrol. Geol., Vol. 16 (November 1932), pp. 1061–1091.

23. Sidney Powers, "Reflected Buried Hills in the Oil Fields of Persia, Egypt, and Mexico," Bull. Amer. Assoc. Petrol. Geol., Vol. 10 (1926), pp. 422–442.

24. Eliot Blackwelder, "The Origin of Central Kansas Oil Domes," Bull. Amer. Assoc. Petrol. Geol., Vol. 4 (1920), pp. 89–94.

Ivan F. Wilson, "Buried Topography, Initial Structures and Sedimentation in Santa Rosalía Area, Baja California, Mexico," Bull. Amer. Assoc. Petrol. Geol., Vol. 32 (September 1948), pp. 1762–1807. Includes bibliography of 63 items on differential compaction over initial structures.

25. C. L. Dake and Josiah Bridge, "Buried and Resurrected Hills of Central Ozarks," Bull. Amer. Assoc. Petrol. Geol., Vol. 16 (July 1932), pp. 629–652.

26. H. G. Busk, *Earth Flexures, Their Geometry,* Cambridge University Press, London (1929), 106 pages.

William Daniel Gill, "Construction of Geological Sections with Steep-limb Attenuation," Bull. Amer. Assoc. Petrol. Geol., Vol. 37 (October 1953), pp. 2289–2406.

27. W. G. Woolnough, "Pseudo-tectonic Structures," Bull. Amer. Assoc. Petrol. Geol., Vol. 17 (September 1933), pp. 1098–1106.

28. W. C. Krumbein, "Occurrence and Lithologic Associations of Evaporites in the United States," Jour. Sed. Petrol., Vol. 21 (June 1951), pp. 63–81. 36 references listed.

29. W. Armstrong Price, "Caliche and Pseudo-anticlines," Bull. Amer. Assoc. Petrol. Geol., Vol. 9 (1925), pp. 1009–1017.

30. Mason L. Hill, "Classification of Faults," Bull. Amer. Assoc. Petrol. Geol., Vol. 31 (September 1947), pp. 1669–1673. 16 references listed.

31. Frederic H. Lahee, "Oil and Gas Fields of the Mexia and Tehuacana Fault Zones, Texas," in *Structure of Typical American Oil Fields,* Amer. Assoc. Petrol. Geol., Tulsa, Okla., Vol. 1 (1929), pp. 304–388. Describes a number of oil pools similar to the Richland.

32. C. W. Sanders, "Emba Salt Dome Region, USSR, and Some Comparisons with Other Salt-Dome Regions," Bull. Amer. Assoc. Petrol. Geol., Vol. 23 (April 1939), pp. 492–516.

33. Theron Wasson, "Creole Field, Gulf of Mexico, Coast of Louisiana," in *Structure of Typical American Oil Fields,* Amer. Assoc. Petrol. Geol., Tulsa, Okla., Vol. 3 (1948), pp. 281–298.

34. Frank B. Notestein, Carl W. Hubman, and James W. Bowler, "Geology of the Barco Concession, Republic of Colombia, South America," Bull. Geol. Soc. Amer., Vol. 55 (October 1944), pp. 1165–1216.

35. Miller Quarles, Jr., "Salt Ridge Hypothesis on Origin of Texas Gulf Coast Type Faulting," Bull. Amer. Assoc. Petrol. Geol., Vol. 37 (March 1953), pp. 489–508.

36. Fenton H. Finn, "Geology and Occurrence of Natural Gas in Oriskany Sandstone in Pennsylvania and New York," Bull. Amer. Assoc. Petrol. Geol., Vol. 33 (March 1949), pp. 303–335. 13 references listed. Numerous pool and regional maps of the Oriskany gas province.

37. Theodore A. Link and P. D. Moore, "Structure of Turner Valley Gas and Oil Field, Alberta," Bull. Amer. Assoc. Petrol. Geol., Vol. 18 (November 1934), pp. 1417–1453.

Theodore A. Link, "Interpretations of Foothills Structures, Alberta, Canada," Bull. Amer. Assoc. Petrol. Geol., Vol. 33 (September 1949), pp. 1475–1501.

38. Alex W. McCoy III *et al.,* "Types of Oil and Gas Traps in Rocky Mountain Region," Bull. Amer. Assoc. Petrol. Geol., Vol. 35 (May 1951), p. 1005. The Circle Ridge field described by W. G. Olson.

Lewis F. Beebe, "Wyoming's Circle Ridge Field," O. & G. Jour., September 14, 1953, pp. 109–114.

39. Ray P. Walters, *op. cit.* (note 21), p. 326.

40. E. V. Watts, "Some Aspects of High Pressures in the D-7 Zone of the Ventura Avenue Field," Trans. Amer. Inst. Min. Met. Engrs., Vol. 174, pp. 191–200. Discussion to p. 205.

41. M. King Hubbert and David G. Willis, "Important Fractured Reservoirs in the United States," Fourth World Petrol. Congr. (June 1955), Section I/A/1 Proceed., pp. 57–84.

42. Ronald K. deFord, "Surface Structure, Florence Oil Field, Fremont County, Colorado," in *Structure of Typical American Oil Fields,* Amer. Assoc. Petrol. Geol., Tulsa, Okla., Vol. 2 (1929), pp. 75–92. Bibliog. 22 items.

43. Harry W. *Oborne,* "Symposium on Fractured Reservoirs" (Discussion), Bull. Amer. Assoc. Petrol. Geol., Vol. 37 (February 1953), p. 318.

44. Homer H. Charles and James H. Page, "Shale-Gas Industry of Eastern Kansas," Bull. Amer. Assoc. Petrol. Geol., Vol. 13 (1929), pp. 367–381.

45. Coleman D. Hunter, "Natural Gas in Eastern Kentucky," in *Geology of Natural Gas,* Amer. Assoc. Petrol. Geol., Tulsa, Okla. (1935), pp. 915–947.

CHAPTER 7

Reservoir Traps (continued) – Stratigraphic and Fluid

Primary stratigraphic traps: in clastic rocks – in chemical rocks. Secondary stratigraphic traps. Fluid traps.

STRATIGRAPHIC TRAP [1] is a general term for traps that are chiefly the result of a lateral variation in the lithology of the reservoir rock, or a break in its continuity. A permeable reservoir rock changes to a less permeable or to an impermeable rock; it is truncated by an unconformity, and overlapped; or it changes along its bedding; and the boundary between the two kinds of rocks chiefly determines the extent of the reservoir. This boundary may be sharp, or it may be gradational; the condition may be either local or regional in extent; and the change in permeability may be wholly responsible for the trap or only partly so.

Nearly all stratigraphic traps, indeed, have some structural elements, the only exceptions being some of those in isolated lenses and organic reefs, which generally are traps without regard to the regional dip or to any arching or deformation. There is no sharp demarcation between structural traps and stratigraphic traps, and some traps are determined in about equal measure by stratigraphic and structural causes; they might be classed as stratigraphic by some geologists and as structural by others. For that reason, it is useful to set up an intermediate class of combination traps.

Stratigraphic traps may be conveniently divided into two general classes. *Primary stratigraphic traps,* formed during the deposition or diagenesis of the rock, include those formed by lenses, facies changes, shoestring sands, and reefs. *Secondary stratigraphic* traps have resulted from later causes, such as solution and cementation, but chiefly from unconformities.

PRIMARY STRATIGRAPHIC TRAPS

Primary stratigraphic traps are a direct product of the depositional environment—that is, of the character of the material in the reservoir rock and the conditions under which it was being deposited. The impervious, concave upper boundary surface of these traps, as well as the effective pore space, is essentially the result of primary sedimentary processes. Such traps have also been called "depositional" traps and "diagenetic" traps.[2]

The effectiveness of a primary stratigraphic trap is chiefly determined by the shape and attitude of the reservoir formation. The trap may be completely localized by the lenticular shape of the porous and permeable body of rock that becomes the reservoir, as shown in Figure 7-1 (A), or it may be partly localized by the configuration of the impermeable up-dip edge of a portion of the reservoir rock, superimposed on a homoclinal dip (B). All gradations and combinations of these general conditions are found.

Primary stratigraphic traps may be divided into two general groups: (1) lenses and facies of clastic and igneous rocks; (2) lenses and facies of chemical rocks, including biostromes, organic reefs, and bioherms.

Lenses and Facies in Clastic Rocks

Some reservoirs are in thin lenticular bodies of porous and permeable clastic rock enclosed in impermeable sediments. Their areal extent generally does

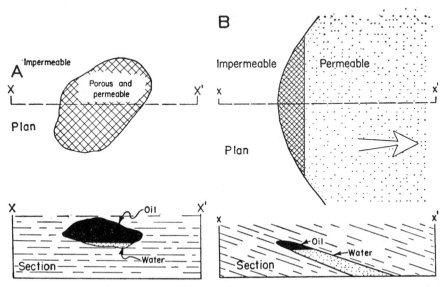

FIGURE 7-1 *Sketches showing* (A) *typical lens-type traps completely surrounded by impermeable rocks, and* (B) *an irregular up-dip edge of permeability on a homoclinal regional dip. Arrow shows direction of dip.*

not exceed a few square miles, although there are a number of exceptionally large ones. Most commonly the lens consists of clastic material—sandstones, arkoses, coquinas, and the special class of weathered, brecciated, and redeposited igneous and metamorphic rocks known as "basalt" and "serpentine." Turbidites, while often difficult to recognize, probably account for many lenticular sandy deposits, especially on the seaward side of rapidly filling depositional basins.

The boundary between a lens and the enclosing rock may be either sharp or gradational. Lenses may be contemporaneous with the enclosing sediments, or they may have formed slightly earlier. By an increase in area, lenses grade into what might be termed restricted facies deposits, which, in turn, grade into deposits of normal facies.

A facies change is a lateral gradation (or less commonly an abrupt change) within a formation or group of rocks, resulting from the contemporaneous deposition of rocks of differing character.[3] If the difference is lithologic, we have a *lithofacies change;* if the difference is in the fossil content, we have a *biofacies change*. Lithofacies gradations from permeable to impermeable rocks are the cause of many traps that contain oil and gas pools. Since lithofacies gradations are more widespread than lenses, often being regional in extent, they are more important in regional analyses. "Facies" may be used as a group term to include several formations, individual sand lenses, or organic reef deposits, such as the sandy facies of the Cherokee formation (Pennsylvanian) of Oklahoma and Kansas, and the reef facies of the Upper Devonian of Alberta. When a sandstone grades into a shale up a homoclinal dip, or a permeable dolomite gives way, up the dip, to an impervious limestone, the up-dip edge of permeability may mark the critical edge of a single trap or of a group of traps.

The oil and gas pool may completely fill the porous part of a sand lens; it may occupy only the high portion of the lens; or, if the regional structure is monoclinal, it may accumulate in irregularities along the regional up-dip edge of permeability. One thing to remember in exploration is that, where one such primary stratigraphic trap is found to contain oil, there may be others like it nearby, for the conditions that determine the presence of facies and lens traps are commonly of regional extent, and the local phenomena are likely to be repeated over wide areas.

In many regions where petroleum is trapped in these sands, the pattern of pools reflects a completely random distribution of sand patches, lenses, sandy zones, bars, and channels. An example is the Third Stray sand of the Venango Group (Devonian), which is in northwestern Pennsylvania near the site of the historically famous Drake well.[4] Many prolific pools were found in this region in an irregularly distributed Third Stray sand, as shown in Figure 7-2. The region contains similar sand zones, some shallower and some deeper. (See also Fig. 7-17, p. 300.)

The lateral variation of sands in the Upper Mississippian of southern Illi-

TRAPS—STRATIGRAPHIC AND FLUID [CHAPTER 7] 289

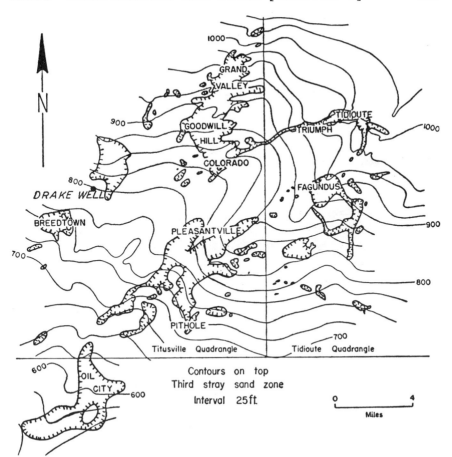

FIGURE 7-2 *Map showing the distribution of the producing and abandoned pools in the Third Stray sand of the Venango Group (Devonian), one of many irregular and patchy producing sands of northwestern Pennsylvania. The pools generally coincide with the distribution of porosity and permeability in the sands, which are lenticular in shape, and range from conglomerate to fine-grained in texture. The Triumph Streak was discovered in 1860, or only one year after the Drake well, and wells were found that produced as much as 1,000 barrels per day. Contours on top of the Third Stray sand horizon at 25-foot interval above sea level.* [Redrawn from Sherrill, Dickey, and Matteson, in Stratigraphic Type Oil Fields *(1941), p. 526.*]

nois is shown in a stratigraphic section across that region. (See Fig. 7-3.) These Mississippian rocks are characterized by sand patches, lenses, bars, channels, and facies changes, and in addition they are truncated toward the north by overlapping Pennsylvanian formations, which also contain lenticular sands. A great many oil pools are found in these Mississippian and Pennsylvanian sands;

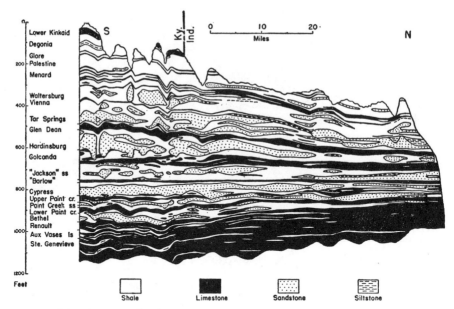

FIGURE 7-3 *Stratigraphic section, using the Barlow–Beech Creek limestone as datum horizon, showing the relations of the Chester (Upper Mississippian) strata across the southern end of the Illinois basin. The section extends for seventy miles from the Poole oil field in northwestern Kentucky to the outcrop of the formations in southern Indiana. Note the great vertical exaggeration in scale. The extreme lateral variableness in the section helps explain many of the oil pools found in these rocks in the Illinois basin. At some place or other nearly every sand and limestone in the column has been found productive.* [*From Swann and Atherton, Jour. Geol., Vol. 56, opposite p. 272, Fig. 6.*]

most of them are associated with folding, but many are limited on one or more sides by the edges of permeability.[5]

A great many pools in sand lenses and patches are found throughout the Pennsylvanian rocks extending from Pennsylvania to Texas. One example is illustrated in Figure 7-4, an isometric diagram of the Dora pool, in Seminole County, Oklahoma. The sand lens, which is almost completely surrounded by impervious shales, is expected to yield an ultimate total of about 10 million barrels of oil.[6] A somewhat different type of patchy sand reservoir is shown in Figure 7-5, a section through the Hull-Silk field in Archer County, Texas. The interbedded Pennsylvanian sandstones and limestones found there are characteristic of many pools in the Mid-Continent region.

Regional facies changes from permeable to impermeable rocks determine the location of the edges of a great many oil and gas pools, which sometimes are called *strandline pools* when they are associated with shore phenomena. Some patchy and lenticular sand formations may trap series or groups of pools

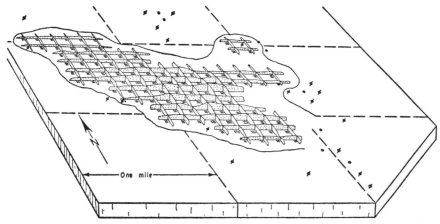

FIGURE 7-4 *Isometric block diagram of the Dora sand (Pennsylvanian) in the Dora pool of Seminole County, Oklahoma. The shape of the sand body suggests an origin associated with offshore phenomena. The sand is composed of well-sorted quartz grains and varies up to 100 feet in thickness. The trap is an example of a sand lens or sand patch almost completely enclosed by shales.* [Redrawn from Ingham, in Stratigraphic Type Oil Fields (1941), p. 418.]

located in long linear "trends." Two outstanding examples are the Yegua-Jackson (Upper Eocene) trend (AA' in Fig. 7-6) and the Frio-Vicksburg trend (Oligocene) (BB'). The pools along these trends are trapped in various combinations of lenticular sands, up-dip pinch-outs, and local folding and

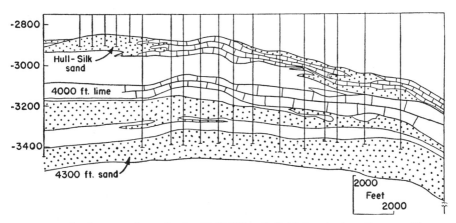

FIGURE 7-5 *Section through the Hull-Silk oil field in Archer County, Texas, showing the complex interbedding of sandstones and limestones that forms many traps in the Pennsylvanian rocks. The Hull-Silk pool is expected to have an ultimate production of 40 million barrels.* [Redrawn from Thompson, in Stratigraphic Type Oil Fields (1941), p. 668.]

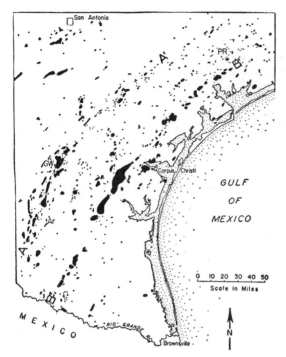

FIGURE 7-6

The South Texas subprovince of the Gulf Coast province of Texas, showing the trend pattern (AA') *of oil pools trapped at and near the up-dip wedge edge of the Yegua-Jackson sands* (Eocene), *and* (BB') *of oil pools trapped at and near the up-dip wedge edge of the Frio-Vicksburg sands* (Oligocene). (*See Fig. 14-11, p. 648, also.*)

faulting, all associated with shore features. The trap may be the irregular edge of a sand on a homoclinal slope, as in the Armstrong pool of southwestern Texas,[7] shown in Figure 7-7 (near the southern end of AA' in Fig. 7-6). In the Government Wells field of the same region (see Fig. 7-8) there are two sands, one above the other, and both contain pools where they wedge out along the up-dip edge of permeability. Folding aids in forming the trap, but the pools are really trapped in stratigraphic changes along the former strand lines of the two sands. The Pickett Ridge field of Wharton County, Texas, in the Frio trend, is shown in Figure 7-9 and a section of it in Figure 7-10. The Maikop region, along the northwestern Caucasus Mountains, in the USSR, contains many pools located along the up-dip edges of sands on a homoclinal slope. A section through the Khadyzhinskaya field is shown in Figure 7-11. Many pools are trapped in the irregularities in the wedged-out, up-dip termination of the Bartlesville sand in northeastern Oklahoma and in the adjacent sand patches. The Pennsylvanian rocks all through this region dip rather uniformly to the west-southwest at 40–50 feet per mile. (See Fig. 7-12.) Another long, linear sand-patch trend comprises the Sand Belt pools of western Texas, shown as AA' in Figure 3-10. There the sands of the Whitehorse formation (Permian) extend along the eastern, or lagoonal, side of the great Permian reef trend (BB' in Fig. 3-10) and form the traps for many pools. The sands grade out to the west into the pure dolomite of the reef, and to the east into anhydrite and dolomite.[8]

The Bolivar Coastal fields along the eastern side of Lake Maracaibo, Venezuela, make up another well-known producing region where a number of pools are found in the strandline sands overlying the unconformity below which the chief production is obtained. (See pp. 333–336 and see Fig. 14-12 for location.) A typical section across this area is seen in Figure 7-13. Once a producing trend such as those that have been mentioned has been established, exploration is largely directed within the projected trend lines, where either random drilling or drilling on every real or suspected minor structural anomaly is justified.

Shoestring-sand Traps. As the name implies, these consist of long, narrow sand deposits, and may be considered as sand lenses of a special type. They range up to a half or three-quarters of a mile in width and up to many miles in length. Except at their ends they are completely surrounded by impervious clays and shales. Geologists believe that some sand traps of this character are channel fillings, and that others are offshore sand bars, and we can sometimes determine in which class a given trap belongs from certain distinctive characteristics. Since the rather straight areal pattern of offshore bar pools differs from that of the sinuous channel-filling pools, an early determination

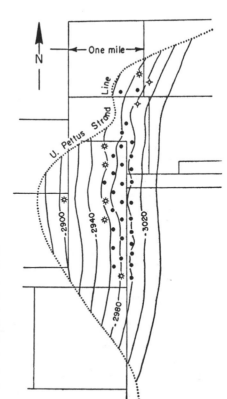

FIGURE 7-7

Map of the Armstrong pool, in Jim Hogg County, Texas (located near the south end of trend AA', at Ar, of Fig. 7-6). The structure is a rather uniform east dip, and closure is determined by the irregular configuration of the up-dip edge of permeability, in this case the edge of the Upper Pettus sand (Eocene). [Redrawn from Freeman, Bull. Amer. Assoc. Petrol. Geol., Vol. 33, p. 1268, Fig. 6.]

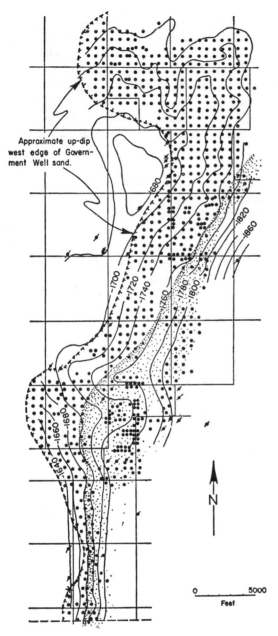

FIGURE 7-8

Structure and sand distribution of the lower producing sand, the Government Wells sand, of Jackson age (Eocene), of the Government Wells field in Duval County, Texas (shown as GW in Fig. 7-6, along the Jackson-Yegua trend AA' of Fig. 7-6). Two sands pinch out up the dip, the more prolific lower Government Wells sand and the upper Government Wells sand (shown by fine dots in map). This field is characteristic of many producing areas along the trend AA'. Contours are on the upper Government Wells sand and, where it is absent, on its projected position in the geologic section. [Redrawn from Trenchard and Whisenant, in Gulf Coast Oil Fields (1936), p. 637, Fig. 3.]

of the origin of the trap becomes of considerable importance in exploration. For that reason much thought has been given to the criteria for each type of origin, and comparisons have been made with known bars and channels, both those that are being formed at the present time and those that have been formed in earlier periods and exposed by erosion.

TRAPS—STRATIGRAPHIC AND FLUID [CHAPTER 7]   295

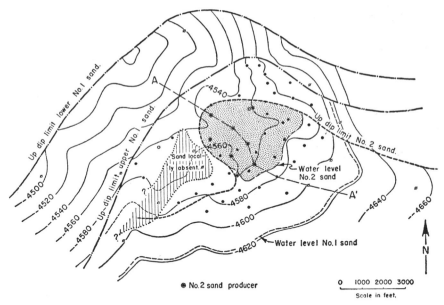

FIGURE 7-9 Map showing distribution of producing sands in Pickett Ridge oil field, Wharton County, Texas (located at PR in Fig. 7-6). Section AA' across the field is shown in Figure 7-10. A low-relief arching and a shaling out of the sands form the traps in this multiple-pool field (contour interval 20 feet.) [Redrawn from Gulf Coast Oil Fields, Houston Geol. Soc. (1941), Introduction.]

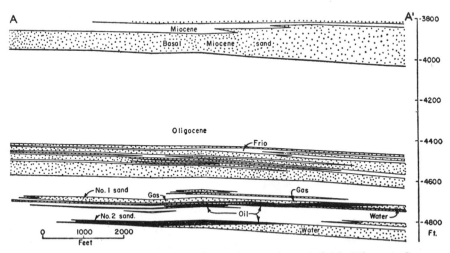

FIGURE 7-10 Section AA' across the Pickett Ridge oil field, Wharton County, Texas, showing the nature of the pinch-out of the two productive sandstones. [Redrawn from Gulf Coast Oil Fields, Houston Geol. Soc. (1941), Introduction.]

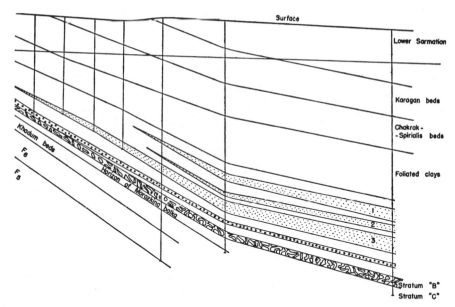

FIGURE 7-11 *Section through the Khadyzhinskaya oil field in the Maikop region, USSR, typical of many pools in the northwestern Caucasus region. The producing sands are in the Maikop series (Oligocene). The structure is a homoclinal dip of 10–15° to the north, and the traps are formed by irregularities along the up-dip wedge-out of the various sands.* [Redrawn from Prokopov, XVIIth Int. Geol. Cong. (1937), Vol. 4, p. 68.]

Offshore Bars. Bass,[9] who has largely supplied the criteria used for distinguishing sand-bar deposits from channel-filling deposits, has pointed out the close similarity between the shoestring-sand deposits of the Pennsylvanian in eastern Kansas and the modern shore deposits along the eastern coast of

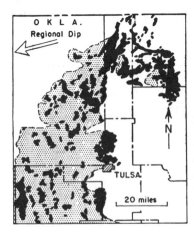

FIGURE 7-12

Map of northeastern Oklahoma, showing the distribution of the oil pools (black) in the Bartlesville sandstone (stippled). Note the accumulations along the irregular up-dip eastern edge of the sand. [Redrawn from Weirich, Bull. Amer. Assoc. Petrol. Geol., Vol. 37, p. 2041.]

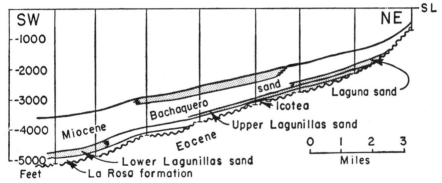

FIGURE 7-13 *Section across the Lagunillas field, in the Bolivar Coastal region, eastern Lake Maracaibo, Venezuela, one of several large oil fields in which the oil is trapped at the up-dip terminations of post-Eocene sands along a strand-line trend. Folding and faulting are locally important in forming the traps.* [Redrawn from Sutton, Bull. Amer. Assoc. Petrol. Geol., Vol. 30 (1946), p. 1728.]

the United States. The identifying characteristics of off-shore bar deposits are: (1) They have a rather flat base and an upper surface that is concave as viewed from below. (2) The sides of individual deposits are straight. (3) Individual lenses are often slightly in echelon. (4) On one side—the seaward side—there is usually a sharp contact between sand and shale; on the opposite side—probably lagoonal—the sand grades into shales and clays, and is "muddy" and of low permeability. (5) Isolithologic contours—loci of similar sedimentary characteristics, such as sorting, composition, and texture—usually run parallel to the edges of the deposits. (6) Sorting is more uniform than in channel-filling deposits. (7) Production is more uniform than in channel-filling deposits.

The shoestring-sand pools of Greenwood and Butler Counties, Kansas, have been intensively studied [10] and are generally considered to represent offshore sand bars deposited along the western coast of a shallow Cherokee (Pennsylvanian) sea. (See Fig. 7-14.) Much high-grade oil has been produced from these sands throughout an area 100 miles long and 50 miles wide. The sands form elongated lenses, from 50 to more than 100 feet in thickness, from 2 to 6 miles long, and up to 1½ miles in width, systematically arranged end to end into trends from 25 to 45 miles in length. The Sallyards shoestring-sand trend of eastern Kansas is shown in Figure 7-15, where the sand trend is superimposed on a hypothetical map of the Cherokee sea and compared with the offshore sand bars of the modern New Jersey coast. The sands are composed chiefly of quartz, but contain a little feldspar and mica, and are loosely cemented with siderite.

In the central part of the southern peninsula of Michigan, important gas reserves are found in a group of pools in elongated sand lenses that were once

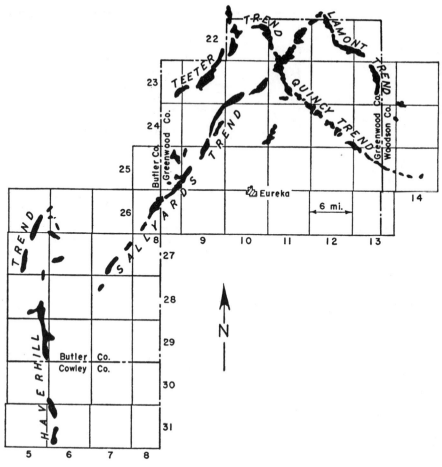

FIGURE 7-14

Map of the shoestring-sand pools in Kansas.

offshore bars in a shallow Mississippian sea.[11] The reservoir rock is the Michigan "Stray" sand, which is at the base of the Michigan formation and lies unconformably on the Upper Marshall (Napoleon) formation. The cleanest sands were deposited along the thickest part of the bars, and the muddier sands along the lee, or lagoon, side. The sand bars are thought to be localized by shoals or higher underlying topography, caused partly by erosion and partly by low anticlinal trends. A section through the Six Lakes pool (Fig. 7-16) shows that these deposits have rather flat bases and (from below) concave upper surfaces.

The Music Mountain pool,[12] in northwestern Pennsylvania, completely fills a buried Upper Devonian off-shore bar. It is shown in Figure 7-17. The producing formation, called the Silverville sand, consists of a gray or brownish-

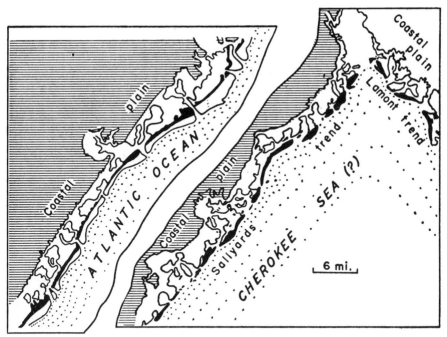

FIGURE 7-15 *Map of the Sallyards shoestring-sand trend in eastern Kansas (right), superimposed on a hypothetical Cherokee sea (Pennsylvanian) and compared with a map of a portion of the New Jersey coast (left). The offshore bars in both maps are in black. (See also Fig. 7-14.)* [Redrawn from Bass, *Bull. Amer. Assoc. Petrol. Geol.*, Vol. 18 (1934), p. 1330, Fig. 10.]

gray conglomeratic sand, of medium-to-coarse grain, up to 70 feet thick and from 800 to 2,000 feet wide, extending over a distance of four miles. Pieces of reservoir rock from wells contain fragments ranging from sandstone, completely cemented with carbonates, to friable sand containing thin clay seams and clay pellets. The shape of the reservoir in cross section suggests that it

FIGURE 7-16

Section across the Six Lakes gas pool in Mecosta and Montcalm counties, Michigan, showing the flat base and mound-like upper boundary of the Stray sand (Mississippian), characteristic of offshore sand-bar deposits. [Redrawn from Ball, Weaver, Crider, and Ball, in Stratigraphic Type Oil Fields (1941), p. 263, Fig. 13, Sec. BB'.]

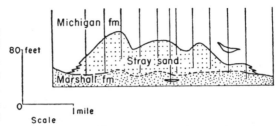

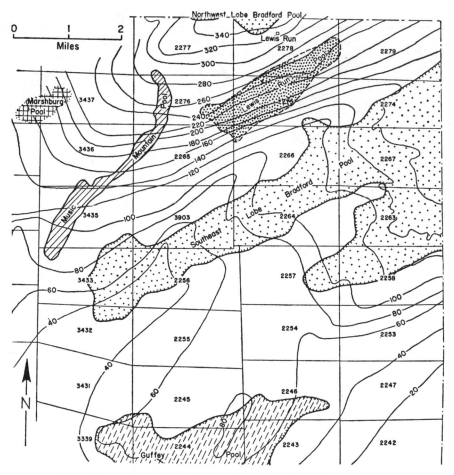

FIGURE 7-17 *Map showing pools in Lafayette Township, McKean County, Pennsylvania. Each pool fills a different sand in the Devonian, and each trap is independent of the structure of the rock, as shown in the contours on the Bradford Third sand horizon. The Music Mountain pool, which produces from the Silverville sand, is probably an ancient offshore bar, and wells producing as much as 500 barrels per hour were drilled into it.* [*Redrawn from Fettke, in* Stratigraphic Type Oil Fields *(1941), p. 493.*]

was formed as an offshore bar, for the sand thickens upward from a flat base to a surface that is concave as viewed from below.

The Ceres shoestring-sand pool, in Noble County, Oklahoma, has an average width of approximately 1,000 feet, a length of $10\frac{1}{2}$ miles, and an average sand thickness of 25 feet, and is of Pennsylvania age.[13] The outline of the pool and a section through the sand are shown in Figure 7-18. The porosity of the sand averages 20 percent and the permeability 100 md. The sand body is almost completely filled with oil, and no effective water drive is present,

TRAPS—STRATIGRAPHIC AND FLUID [CHAPTER 7] 301

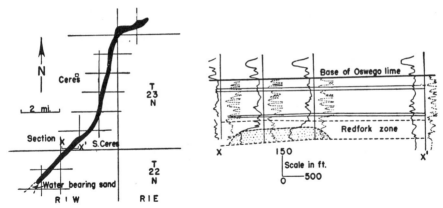

FIGURE 7-18 *Map of the South Ceres pool, Noble County, Oklahoma, and a typical section, XX', across the pool. The producing Red Fork sand, because of the shape of its cross section, seems to represent an offshore sand bar, and it has been found productive for a length of 10.5 miles. The bar has an average width of 1,000 feet, and it occurs in the Cherokee shales of Pennsylvanian age.* [Redrawn from Neal, *World Oil,* December 1951, p. 92.]

although the sand is water-bearing at its southwest end. From its form and character the sand body is believed to represent an ancient buried offshore bar.

Channel Fillings. Channels being filled with sand, gravel, and clastic debris

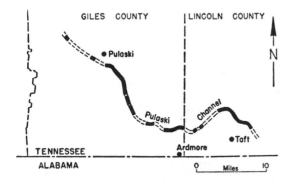

FIGURE 7-19

Map and typical section of the Pulaski Channel, south-central Tennessee. The exposed areas are shown in black. The channel was cut into rocks of Trenton age (late Ordovician). [Redrawn from Wilson, *Bull. Geol. Soc. Amer.,* Vol. 59, pp. 735, Fig. 1, and 755, Fig. 5.]

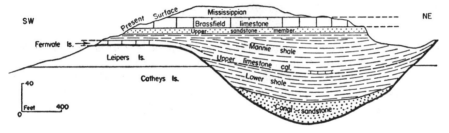

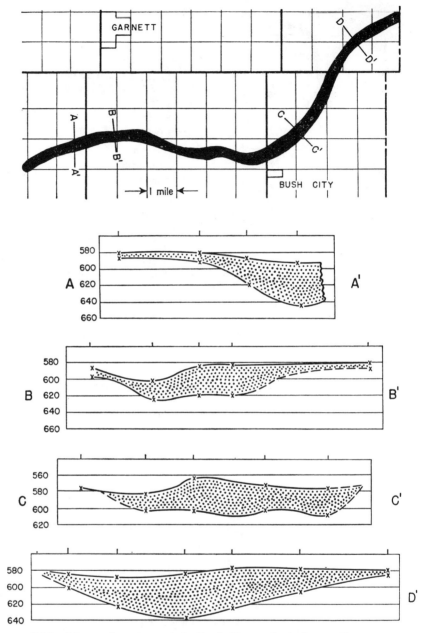

FIGURE 7-20 *Map and sections of the Bush City pool, Anderson County, eastern Kansas, characteristic of a number of pools in this region. The sinuous pattern and the central thickening of the sand suggest that this is a sand-filled channel deposit. Oil fills the channel.* [Redrawn from Charles, in Stratigraphic Type Oil Fields (1941), p. 47, Fig. 3.]

may be observed at many places. Streams meandering back and forth within their flood plain often leave the old channel, when it becomes clogged with sand and gravel, and start a new one. Streams entering the ocean along low shores may need several channels to distribute their water and its load of sand, gravel, and mud. Distributary systems may develop, such as the birdfoot pattern of the Mississippi River delta; when these become clogged with debris, new channels form. Tidal flats, deltas, and deltaic deposits may contain a succession of sand- and gravel-filled channels, for the later stream load must bypass the earlier deposits in order to enter the sea. Channel fillings may coalesce to form widespread deposits, some of which may leave traces of their origin.

Characteristics of channel fillings that may help to identify them underground, where the only evidence is from well records, are: (1) the base of the deposit is convex, as viewed from below, because the sand fills a pre-existing valley, and the top is more widespread than the base; (2) the deposit consists of a wide variety of materials, and the texture, composition, and grain size may vary abruptly; (3) the plan of the deposits shows a sinuous, meandering pattern, sometimes with curved outlines similar to those of ox-bows along rivers.

"Fossil" stream channels, cut into the underlying rocks and later filled with sands, conglomerates, and other previous clastic material, may be observed and studied at a number of places where they crop out. Two of the best-known and most interesting "fossil" channels to study are the Warrensburg and Moberly channels in Missouri, which were cut into Pennsylvanian rocks and are now exposed by erosion.[14] They have both been traced for more than forty miles, and their original lengths were probably much greater. The channels are long, narrow, and straight, and in places are more than 200 feet deep. The streams are thought to have flowed toward the north. Before the deposition of the Kansas City formation (Pennsylvanian), the channels were filled with poorly sorted sands and muds, a little peat, and fragments of conglomerate. Other channel sands crop out in the Leavenworth region, in northwestern Missouri and northeastern Kansas.[15]

A "fossil" channel in south-central Tennessee, known as the Pulaski channel,[16] is intermittently exposed through a distance of twenty-seven miles. This channel, which is as much as 100 feet deep and from 1,000 to 2,000 feet wide, is cut into rocks of Trenton age (late Ordovician) and is filled with material of Richmond age (late Ordovician). (See Fig. 7-19.) It follows a meandering course and is believed to have been cut by a subaerial stream. Along the bottom there are cross-bedded sands and conglomerates, which contain pebbles and cobbles of the Trenton limestone through which the channel cuts. One can readily imagine channels such as the Warrensburg and Pulaski becoming the sites of oil pools when they are completely buried.

Asphalt-saturated channel sands and conglomerates of Lower Pennsylvanian age crop out in central Kentucky, where they rest on truncated Upper and

Middle Mississippian formations on the southwestern side of the Cincinnati arch.[17] These deposits presumably represent inspissated oil pools that accumulated in channel-sand deposits and were later exposed by erosion. Farther west these channels contain many oil pools at depths of around 400 feet, and some wells produce as much as 900 barrels per day.[18]

In a number of localities oil and gas are obtained from pools trapped in what appear to have been stream channels similar to those found now exposed at the surface. One of these is the Bush City channel,[19] one of many such channels in southeastern Kansas, which has been carefully mapped and described. (See Fig. 7-20.) It consists of a sand body thirteen miles long and a quarter of a mile wide, near the top of the Cherokee shale (Pennsylvanian). The most productive wells of both oil and gas occurred where the channel crossed the tops of low anticlines, but the production was continuous throughout the length of the channel. The sand contained no free water, and production was ended toward the southwestern end of the pool by a sharp decrease in the API gravity of the oil from 35° Baumé (Spec. Gr. 0.84) to 14° Baumé (Spec. Gr. 0.9) and an increase in viscosity. Beyond that point the sand occurs in normal thickness but contains neither oil nor free water. The pool terminated to the northeast, or up the regional dip, by a shaling up of the sand and a decrease in oil to mere showings. Other pools in this region have been described [20] as occupying gently winding channels cut sharply into the underlying shale to depths of about fifty feet and filled with sands and silts.

Channel-sand pools also occur in the Maikop field, USSR, where production comes largely from Tertiary sands and conglomerates deposited in a meandering channel that extends for a distance of eight kilometers. (See Fig. 7-21.) The most productive sands are the uppermost formation, C_2, and are generally

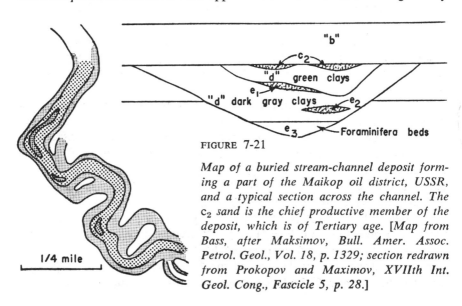

FIGURE 7-21

Map of a buried stream-channel deposit forming a part of the Maikop oil district, USSR, and a typical section across the channel. The c_2 sand is the chief productive member of the deposit, which is of Tertiary age. [Map from Bass, after Maksimov, Bull. Amer. Assoc. Petrol. Geol., Vol. 18, p. 1329; section redrawn from Prokopov and Maximov, XVIIth Int. Geol. Cong., Fascicle 5, p. 28.]

in narrow, sinuous channels, originally called "sleeves." These are superimposed on the larger channel, but in places they spread out beyond it. The area is remarkable for its large production; the discovery well, near Shirvansky, yielded 375,000 barrels of oil within a short time from a depth of only 281 feet.[21] Wells producing as much as 7,000 barrels of oil per day were found in the C_2 sand, and one well produced over $2\frac{1}{2}$ million barrels of oil during the first six years of its life.[22]

Exploration for shoestring pools is extremely difficult because their location is rarely indicated by any folding or deformation in the overlying formations. Successful exploration requires the drilling of many test wells, for until the relative position and trend of the pools have been established, there are few clues to their location. The chief guides are careful mapping of sands and close attention to oil and gas showings in thin sands that may develop into bars or channels. Generally, where one such pool is found, there will be others. Many shoestring-sand pools have been discovered accidentally during drilling on local folds in the hope of finding structural traps; most of them, however, seem to be independent of local folding, except that the gas-oil ratio may be somewhat higher where the structure is high.

Lenses of Volcanic Rock. A group of pools has been found in the Interior Coastal Plain region of Texas, in which the producing traps consist of lens-shaped masses of igneous rock enclosed within late Austin or early Taylor sediments, or both (Upper Cretaceous). Basaltic igneous rocks were erupted and intruded intermittently into several Upper Cretaceous formations over an extensive area.

Seventeen of these igneous masses form traps that have yielded commercial pools of oil and gas. The igneous masses consist partly of unaltered basic rocks of several varieties, such as olivine basalt, melilite basalt, limburgite, gabbro, and phonolite. "Serpentine" is the term commonly used for the altered products of these rocks, and drillers call *any* of the igneous material "serpentine." The rock has been identified in the Hilbig pool [23] as palagonite, a hydrated basic glass. In some pools this material has remained in place, but in others it has been more or less reworked and redeposited. The oil is found in the altered rocks, both that which is in place and that which has been reworked, in and around the original volcanic cones. Production from these altered rocks varies widely in abundance, the initial production* ranging from a few barrels to more than 5,000 barrels a day. The production is also spotty, highly productive wells often being close to wells of very small production. The oil is mostly light, with an API gravity of 36–39° (Spec. Gr. 0.84–0.83), but much oil of lower API gravity is also found. The oil is high in paraffin (ozokerite), and free water is rare in most fields. A section through a typical "serpentine pool" is shown in Figure 7-22. So far, "serpentine" pools are found in only

---

* Initial production is the actual or possible production of oil or gas during the first twenty-four hours after completion.

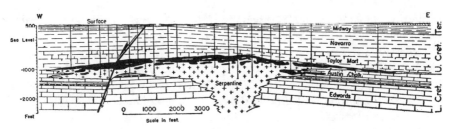

FIGURE 7-22 *Section through the Lytton Springs oil pool, Caldwell County, Texas. The reservoir is an altered igneous rock, called "serpentine," and the section is typical of several similar pools of this region. The pool is nearly circular in plan, no free water is encountered in the reservoir, and the oil is of 38° API gravity. [Redrawn from Collingwood and Rettger, Bull. Amer. Assoc. Petrol. Geol., Vol. 10, p. 959, Fig. 3.]*

this one area of Texas, where they have been discovered by random drilling or by drilling on magnetic anomalies. While the reservoir rock and the trap are unusual, the oil is trapped here, as in other pools, within permeable rocks covered with a concave impervious surface.

A reservoir rock that is true serpentine is found in the Jatibonico pool, about thirty kilometers southeast of Jarehueca, in the province of Santa Clara, Cuba.[24] Over 1,200 wells have been drilled in this pool, but the production from them is small.

Lenses and Facies in Chemical Rocks

Two general classes of primary stratigraphic traps occur in rocks of chemical origin, almost all of them carbonate rocks. Both are important as oil and gas producers. They are: (1) Porous facies, either lithofacies or biofacies, enclosed in or terminated by normal impervious shales, limestones, or dolomites; nearly tabular lenses composed of carbonate residues of organisms are called *biostromes*. (2) Porous mound- or lens-shaped carbonate masses consisting chiefly of debris from sedentary organisms and surrounded by impervious rocks; these are called *organic reefs* or *bioherms.*

Porous Carbonate Facies. Traps of this group may be either local or regional in extent. They resemble sand facies, except that the rock in them is of chemical origin and usually consists of a carbonate. The most common type is formed by the dolomitization of limestone, in which the magnesium carbonate deposited has less volume than the calcium carbonate removed by solution, so that the resulting rock is porous and permeable. (See also pp. 126–127.) Some traps of this class are in a permeable sandy or cherty facies enclosed within a carbonate rock. Others are in recrystallized clastic lenses consisting chiefly of shells, coquina, oolites, or carbonate fragments. Bedded layers of organic remains such as these are biostromes.[25] Some of them were probably formed in place by sessile (attached) animals; others may consist of remains of organisms washed into their present place by waves or currents.

The up-dip termination of the permeable dolomitic facies of the Trenton limestone (Middle Ordovician), which extends for 170 miles across northwestern Ohio and northern Indiana, determines the location of many oil and gas pools.[26] (See Fig. 7-23 and also p. 127.) The pools are nearly continuous throughout this distance but are especially concentrated in the more porous dolomitized parts of the formation. Since the origin and date of the dolomitization are unknown, the individual trap may be either primary or secondary. It is considered primary because the local dolomitization was probably due in part to some primary contemporaneous cause even though the process may not have been completed until some later date. Traps formed by local dolomitization within limestones are common. One of them is in the Belcher field, Ontario (Fig. 7-24). The dolomitized

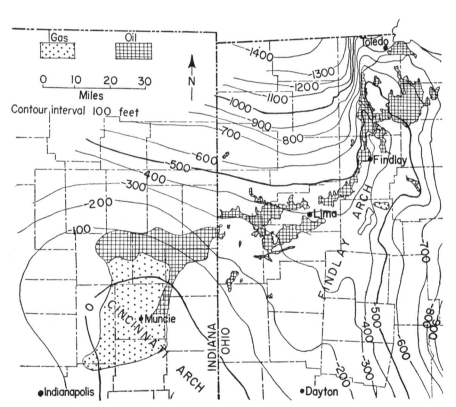

FIGURE 7-23 *Map showing the structure of the Lima-Indiana field in Indiana and Ohio and the distribution of its oil and gas pools. The pools are within the dolomitized portion of the Trenton limestone (Ordovician), and the up-dip edge is marked by the edge of the dolomite; water fills the pores down the dip.* [Redrawn from Carman and Stout, in Problems of Petroleum Geology, *p. 522, Fig. 1.*]

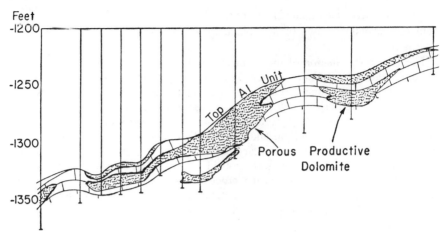

FIGURE 7-24 Section through the Belcher oil field, Ontario, Canada, showing traps formed by porous dolomite patches enclosed within an impervious limestone. The A1 producing formation is the Salina formation (Silurian). [Redrawn from Stuft, Proc. Geol. Assoc. Canada, Vol. 5 (1952), p. 81, Fig. 6.]

limestone along a fracture system of the Deep River pool in Michigan is described on page 123 (see Fig. 4-10).

A local porous-facies, or biostrome, trap is seen in the highly productive "crinoidal limestone" lens (Pennsylvanian) on the west flank of the Todd dome, in Crockett County, Texas.[27] (See Figs. 7-25 and 7-26.) The producing formation consists of a porous and locally cavernous limestone containing crinoid stems, bryozoans, and other fossil fragments and forming a biostromal lens that lies immediately above the unconformity separating the Pennsylvanian from the Ordovician. The crinoidal deposit apparently has some genetic relation to the dome and may be debris that formed as an offshore bar or as a reef of sessile organisms that grew on some elevated topographic feature of the sea floor. There is a similar productive reservoir in a coquina lens or biostrome on the southeast flank of the Lisbon dome, Louisiana,[28] whose relation to the structure is shown in Figure 7-27. The folding that formed the dome occurred after the deposition of the coquina rock; presumably the folding had little if any bearing on the location of the coquina reef, for there is no evidence of thinning of the pre-reef formation.

Organic Reefs. Many great oil pools, especially in North America, are trapped in organic reefs of various shapes and sizes.[29] Organic reefs are common, being known to occur in rocks of every geologic age from the Precambrian to Recent. Some of them consist wholly of organic material that grew in place, but most of them are complex mixtures of interbedded original and detrital organic material and debris.[30] Because they are difficult to discover (many, perhaps most, reef pools have been discovered by chance or in the search for other kinds of traps), it will be a long time before their ultimate possibilities are known. But because of their richness a substantial and increas-

TRAPS—STRATIGRAPHIC AND FLUID [CHAPTER 7]   309

ing part of our future exploration throughout the world will probably be in search of reef traps.

Reefs of this kind were called "coral reefs" until it was learned that in many of them corals either were absent or occurred only as minor constituents. *Organic reef* seems a better term[31] and is coming into common use among geologists. A broad, general term, it is applied to reefs of all shapes and sizes

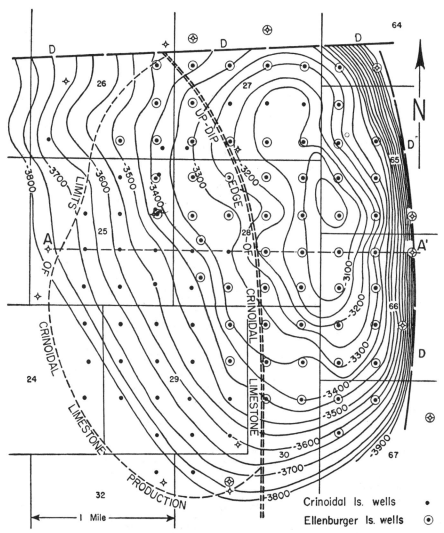

FIGURE 7-25 *Structure of the Ellenburger limestone (Ordovician) in the Todd field, western Texas. The distribution of the overlying crinoidal limestone reef (Pennsylvanian) and of the wells producing from it is also shown. A section is shown in Figure 7-26. [Redrawn from Imbt, Bull. Amer. Assoc. Petrol. Geol., Vol. 34 (1950), p. 247, Fig. 5.]*

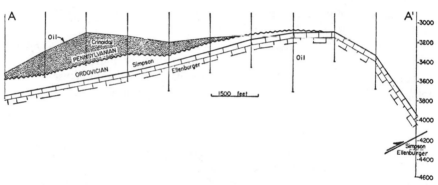

FIGURE 7-26 *Section AA' across the Todd field, Crockett County, Texas, showing the crinoidal reef of Pennsylvanian age that lies on the west flank of the Ellenburger fold. Both the crinoidal reef and the Ellenburger dome have trapped pools. The thinning of the pre-unconformity Simpson formation toward the dome suggests that there was a limestone island to the east, which may have caused the reef organisms to grow and accumulate at this location. Salt water characterizes the Pennsylvanian pool, whereas sulfur water accompanies the Ellenburger production.* [Redrawn from Imbt and McCollum, Bull. Amer. Assoc. Petrol. Geol., Vol. 34 (1950), p. 253, Fig. 9.]

that consist largely of organic debris embedded in sediments of differing character. Round, dome-like organic masses are called *bioherms*.[32] Cummings defines a bioherm as "consisting of any dome-like, mound-like or otherwise circumscribed mass, built exclusively or mainly by sedentary organisms such as corals, stromatoporoids, algae, brachiopods, mollusca, crinoids, and enclosed in normal rock of different lithologic character."[33] The term "bioherm" is useful in distinguishing the generally circular reefs from the great elongate reefs, tens or even hundreds of miles in length and only a few miles wide. Anomalous masses of limestone and dolomite that cannot be demonstrated to be chiefly of organic origin are called *limestone massifs, lime banks, lime reefs, reef masses, reef-like deposits,* or merely *reefs.*

After a productive organic reef has been completely drilled and developed, we can be quite sure of its geologic characteristics—its size, shape, and internal variations. But in searching for buried reefs, or even after the first few wells have encountered reef material underground, we can better predict what kind of exploration will be most effective and in what direction it should be pursued if we understand the possible reef patterns and habits of growth. It is therefore worthwhile, as a guide to organic reefs that are completely hidden and whose surrounding geology is known only through well records, to study both the reefs growing in the present oceans and the fossil reefs that can be seen and examined at the surface of the ground in many parts of the world. A logical sequence for study is: (1) modern organic reefs; (2) fossil organic reefs exposed by erosion; (3) buried and productive organic reefs.

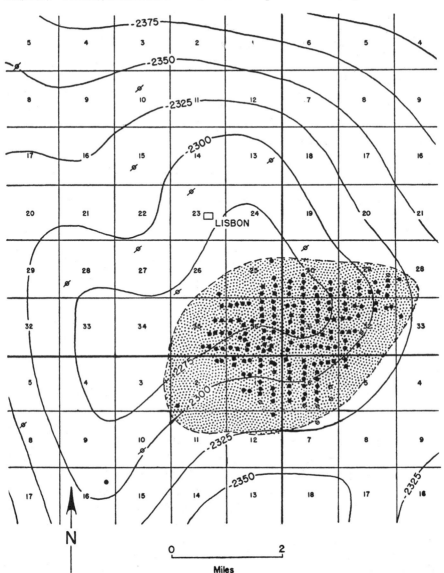

FIGURE 7-27 *Structure of the Annona chalk (Cretaceous) in the Lisbon pool area, Claiborne and Lincoln parishes, northern Louisiana. The outline of the porous, oolitic, and coquina-like Patton biostrome lens (Lower Cretaceous) is shown by the area of the producing wells.* [*Redrawn from Grage and Warren, Bull. Amer. Assoc. Petrol. Geol., Vol. 23 (1939), p. 301, Fig. 2.*]

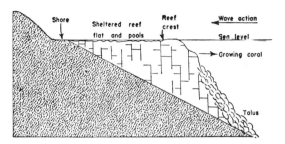

FIGURE 7-28

Section through a typical fringing reef, showing the different parts that make up the reef complex. [Redrawn from Yonge, Endeavour, July 1951, p. 144.]

Modern Reefs. Four main types of organic reefs are forming in the warm seas of the East Indies, the Indian Ocean, and the Florida-Bahama-West Indies region at the present time.[34] They are classified as fringing reefs, barrier reefs, atolls, and table reefs.

A *fringing reef* is closely attached to the shore and appears at low tide as an exposed shelf or flat from a few feet to a quarter of a mile wide. (See Fig. 7-28.) The longest fringing reef known is the one that extends practically without interruption for more than 2,700 miles along the coast of the Red Sea.[35] The fringing reef shelf consists mainly of dead reef rock, but living corals and other organisms are abundant on the seaward exterior slope and generally grow outward to a depth of about twenty fathoms. The seaward slope drops off rapidly from the narrow reef shelf or platform. Here the structures that are being built by various organisms are constantly being subjected to the destructive action of breakers. Fractured rock masses, called "coral heads," many of which are very large, are broken off from this edge and washed up on the shore behind it. Most of the broken and fractured bioclastic material, however, falls downward along the seaward slope, where, aided by undertow, it forms a talus sloping seaward. On the inner side of the reef there is commonly a boulder zone of accumulated reef fragments, and between this and the island or continental mass to which the reef is attached there may be an inner flat—a small flat-bottomed channel in which growth and decay are balanced.

As the reef grows seaward, this inner channel widens and deepens and a fringing reef may merge into a *barrier reef,* which forms a breakwater, or barrier, protecting an inner lagoon. The lagoon is generally not more than thirty fathoms in depth; it usually has a flat bottom that makes a sharp angle with the inside of the live reef but slopes gradually up the landward side. Near the edges of the lagoon the bottom is covered with a layer of lime "sand," and there is a luxuriant growth of large, branching corals in its quiet waters. In the central part of the deeper lagoons the coral sand may grade into an amorphous mud. Since deposition is not common in lagoons,[36] solution of calcareous materials and removal of mud particles in suspension are probably taking place. The lagoonal side of fossil reefs frequently contains evaporites, oolites, and red and green shales. Fringing reefs are likely to form on shores marked by cliffs or steep slopes, but barrier reefs are more likely to form several miles off gradually sloping shores. An asymmetric island may have a fringing reef on one side and a barrier on the other.

The largest and most impressive known modern reef is the Great Barrier Reef off the northeast coast of Australia, which extends outward from the shore for 30–90 miles and roughly parallels the coast for 1,200 miles, disappearing at the New Guinea reefs. It is wedge-shaped in cross section, thickening away from the shore, and at the outer edge it forms a great wall that rises in some places 1,800 feet above the ocean floor.[37] Between the barrier and the shore is an immense lagoon, in which the water, connected with the open sea by narrow passageways through the barrier, is generally 30–40 fathoms deep. The floor of the lagoon is covered with unconsolidated lime fragments and with clastic sediments composed largely of foraminifera.

An *atoll* is essentially a nearly circular barrier reef, at times surrounding a small island, from which it is separated by a lagoon. The exposed part of the atoll is generally not more than a mile in diameter and is almost always broken by cross channels that divide it into islets. An atoll may be very irregular in plan. Atolls of this kind dot the western Pacific and the Indian Ocean. They have a common feature in that each begins to grow from the ocean floor. Many probably developed on mountain tops, possibly on extinct volcanoes. They generally grow from hard bottoms, but some of those in the

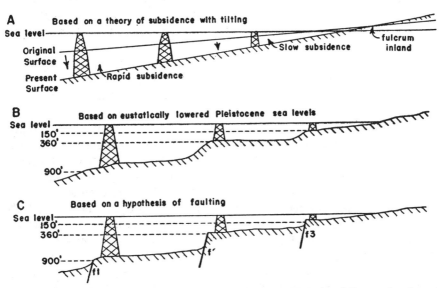

FIGURE 7-29 *Sections showing three possible causes of the different heights to which reefs grow within a limited area: A, subsidence with tilting; B, eustatic lowering of Pleistocene sea levels; C, successive fault escarpments. [Redrawn from Teichert, Geog. Review, Vol. 38 (1949), p. 248, Fig. 17.]*

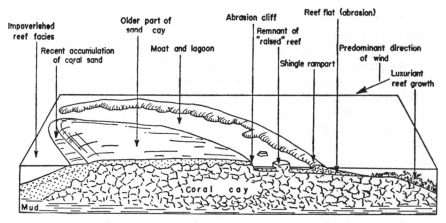

FIGURE 7-30   *Idealized section through a typical reef and cay in the Bay of Batavia, East Indies, showing the component parts.* [*Redrawn from Umbgrove, Bull. Geol. Soc. Amer., Vol. 58 (1947), p. 742, Fig. 9.*]

Java sea and the Bay of Batavia are rooted in muddy bottoms.[38] Three possible ways of accounting for the different heights to which they may grow within a limited area are indicated in Figure 7-29. The exposed parts of a typical atoll are shown diagrammatically in Figure 7-30. Of particular interest are the *ramparts,* or *shingle ramparts,* consisting of reworked reef debris washed up by the waves into linear deposits, and with high initial porosity and permeability.

Small, isolated reefs, either submerged or partly above water, that do not enclose lagoons have been called *table reefs*.[39] They are common in the coral seas, and a number of submerged table reefs have been found in the Pacific. They are also called *sea mounts,* and flat-topped, deep-lying sea mounts are called *guyots.* Most table reefs are thought to be remnants of former atolls.[40] Several borings have been made in order to find the foundation upon which modern reefs grow and to determine the depth of the coral-rock formation.[41] The boring on Bikini Island went through unconsolidated coral rock from the surface to a depth of 2,556 feet;[42] the lower half of this material was found to be of Miocene age. Miocene coral rock was found, also, in a boring on Kita-Daita-Zima (North Borodoni Island), which reached a depth of 1,416 feet.[43] The boring put down on Funafuti Island in 1899 showed the rock to consist of coralline limestone and dolomite for the total depth of the hole, 1,114 feet.[44] The rapid dolomitization of the reef rocks almost completely obscured the organic remains. Basement rock was reached in two deep holes on Eniwetok Island; one encountered olivine basalt beneath shallow-water limestone of Eocene age at 4,154 feet, and the other found hard basement rock at 4,610 feet, but no samples were recovered.[45] The top several hundred feet drilled were Quaternary reef limestones, and most of the rest of the sediments consisted of soft Tertiary reef limestones with minor amounts of dolomite. The atoll appears to be a limestone cap resting on the summit of a volcano that rises two miles above the ocean floor.

The organisms that build or add to the modern reef are of two kinds. Those that build the framework of the reef are chiefly colonial and gregarious organisms, including algae, sponges, and a few others, all of which pile themselves upon one another to form a strong structure that is rigid but porous. Those that live within the pores of the framework are noncolonial algae, corals, sponges, mollusks, brachiopods, bryozoans, and foraminifera. Two of the most important modern reef-builders in the Pacific[46] are the coral *Acropora* and the calcareous alga *Lithothamnium*.[47] Lithothamnium deposits form ridges, frequently on the windward side of the atoll, thereby protecting the *pinnacles* (erosional remnants of former reefs) on the lee side. Plans of two atolls showing *Lithothamnium* ridges are shown in Figure 7-31.

Modern organic reefs vary in shape and in their position with respect to land, but they all have several characteristics in common. One of these is that the organisms seem to require a temperature that is rarely below 68°F, and modern reefs are therefore confined to the warm waters of the Pacific and Indian Oceans. Algae, however, do not require warm water; they will grow even in Arctic waters. They form a substantial part of many reefs both fossil and modern, calcareous red and green algae being the most common. Algae, being plants, cannot grow without sunlight. The maximum depth to which light penetrates in clear water is about 50 fathoms, which thus becomes a limiting factor for most algal growth, the most vigorous growth taking place within

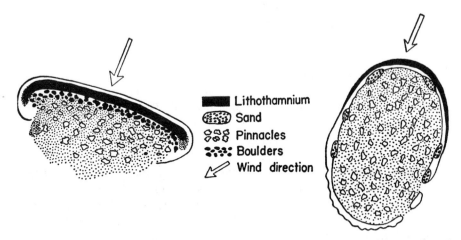

FIGURE 7-31 *Idealized plan showing the* Lithothamnium *ridges on the windward side of atoll-like reefs, where the seas are powerful. The pinnacles form in the shallow water on the lee of the ridges. Atoll reefs such as these generally vary from less than one to several square miles in area. The molding effect of the prevailing winds is evident from the drawings.* [Redrawn from Yonge, Endeavour, July 1951, p. 144, Fig. 14.]

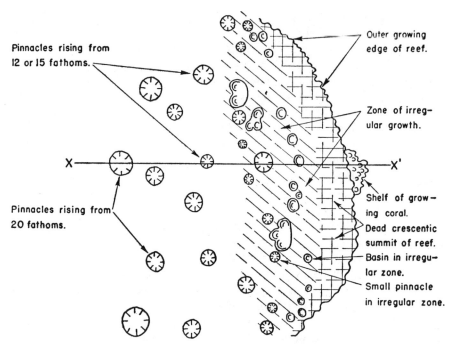

FIGURE 7-32 *Idealized plan of typical modern growing reef as found along the Great Barrier Reef off the northeast coast of Australia. This plan and section XX' in Figure 7-33 illustrate how some of the irregularities of the reef complex may develop.* [Redrawn from Report of Great Barrier Reef Committee, *Trans. Royal Geog. Soc. Australasia* (1925), p. 53.]

100 feet of the surface. Modern reef organisms consume great quantities of oxygen, and the reef growth is therefore most active where the waves and water currents bring in a continuous and abundant supply of oxygen together with the nutrients upon which the organisms feed. This means that the active growth is generally on the windward side. It also means that the organism must be able to erect a wave-resistant structure and bind it together so that it is able to withstand the waves and storms.

Another common characteristic of modern reefs is that the porosity in each of them varies widely from place to place. Some porosity is caused by boring organisms that so completely break up the rock as to leave only small fragments, which are redeposited to form highly permeable clastic deposits on the flanks of the reef. Other porosity is due to the voids within the framework of many of the organisms that build the reef—the spaces between the tabulae of corals, the coils of gastropods, and the diaphragms of protozoans, and the deserted chambers of cephalopods. Much of this empty space is sealed off until later tapped by boring organisms. Other porosity is developed along the active growing front, where the waves break off both large and small fragments,

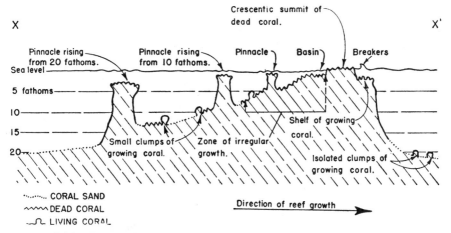

FIGURE 7-33 *Section XX' of Figure 7-32, showing details of various component parts of a growing reef such as the Great Barrier Reef off the northeast coast of Australia. [Redrawn from Report of Great Barrier Reef Committee, Trans. Royal Geog. Soc. Australasia (1925), p. 53.]*

which then contribute to the coarse, porous clastic deposits at the foot of the reef. Rampart ridges and bars of coarse reef debris, formed by wave action, make up other great masses of porous material. Lime sands, transported and sorted by the waves, form porous bars and patches. Some of these features are shown in Figures 7-32 and 7-33.

Cavernous openings are common in reefs. Some are openings under canopies formed by large branching corals, and others are hollowed out by wave action. Two rather common types of cavernous openings are found in the Bikini region of the Pacific.[48] One of these includes the "blowholes" through which water spouts 5–20 feet in the air after each wave. A blowhole begins as a surge channel—a large groove on the seaward side of the reef through which the water washes in and then drains away. A surge channel may be narrow but 50–75 feet long; when algae grow across its top, it becomes a tunnel, with only a small opening left at the shoreward end; water is driven through this by the pressure of each high incoming wave. The second type of cavern forms a "room-and-pillar" structure. Pillar-like algal masses 15–20 feet in diameter grow vertically from the sea floor until they reach the surf level, a foot or two below the low-tide level, where they begin to grow more vigorously. They then spread out laterally until they coalesce over the spaces intervening between the pillars; they finally enclose the pillars completely, and then they die. The structure they leave behind them resembles the "room-and-pillar" stopes in certain mines, and they have been named accordingly. The sequence of their development is shown in Figure 7-34.

Although most modern reef organisms provide initial porosity in some form, other organisms and chemical processes reduce the porosity. These pore-fillers

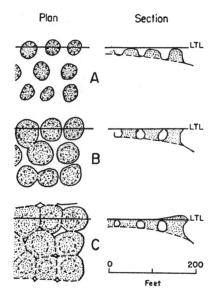

FIGURE 7-34

Diagrammatic sketch showing the development of "room-and-pillar" cavernous openings in modern reefs, such as the Bikini reef of the Pacific. The shaded part represents calcareous algal material; A, algal bosses or knobs; B, shelving and lateral growth in the surf zone; C, development of the reef floor, underlain by room-and-pillar structure; LTL, low-tide level. Cavernous openings such as these may account for cavernous openings in some producing reefs. [Redrawn from Tracey, Ladd, and Hoffmeister, Bull. Geol. Soc. Amer., Vol. 59 (1948), p. 875, Fig. 6.]

are the algae, the other small organisms that live and die in the larger openings, and the lichenous nullipores, which attack both living and dead corals and fill every interstitial opening they can reach, cementing the whole mass into a solid rock. Chemical precipitation of carbonate further reduces the porosity and further cements the shells together.

Scarcity of organic remains in reef rock may be due to (1) recrystallization during diagenesis, often involving dolomitization or being followed by it; (2) destruction by boring algae, which are common in nearly all reefs; (3) deposition of material that has been ground and pulverized by the waves and macerated by boring, predatory, mud-eating organisms.

Fossil Reefs. Organic reefs embedded in ancient sediments, especially those of Paleozoic and Mesozoic age, have been found in many parts of the world.[49] They range in size from small concentrations of organic debris, a few feet thick and a few square feet in area, up to the great organic reefs hundreds of feet thick and hundreds of miles long. Fossil reefs were first observed in outcrops but have since been recognized in the subsurface, where they have been found to be highly productive of oil in many localities.

The Silurian rocks of Wisconsin, Indiana, Ontario, and Illinois especially are noted for the large number of bioherms they contain; the known occurrences are shown in Figure 7-35. They are most common in the Niagaran series (Middle Silurian). These bioherms are generally structureless masses of uneven-textured dolomite that interrupt the bedding of the normal stratigraphic sequence. Some are almost wholly unfossiliferous, but others contain abundant fossils, which indicate that their chief builders were gastropods, brachiopods, and corals. Many have central cores of dense, unbedded dolomite

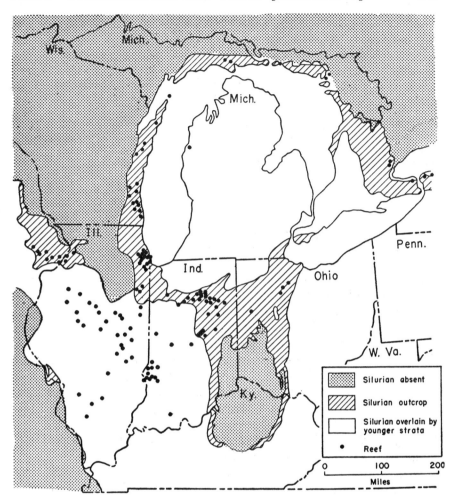

FIGURE 7-35 *Distribution of known organic reefs, in both outcrop and subsurface, in Silurian formations (chiefly Niagaran) in the east-central United States. [Redrawn from Lowenstam, RI 145, Illinois Geol. Surv. (1950).]*

that is porous and cavernous. The cavities vary from a few inches to a foot or more in diameter and are thought to result from original voids that were modified by solution and redeposition. The outer parts of bioherms generally grade into thin-bedded, dense, unfossiliferous dolomites containing scattered chert nodules, which dip away from the central mass at angles that are commonly as high as 40° and may even be as high as 70°. The high dips are thought to be the result of slumping and compaction of the lime mud called "drewite." [50] These dolomite beds surrounding the bioherm commonly overlap, wedge, and thicken toward the middle of the bioherm. All the evidence appears to indi-

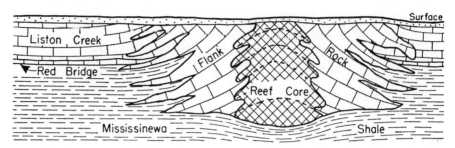

FIGURE 7-36 *Section across a Niagaran (Middle Silurian) organic reef (bioherm) found in the well-known reef locality of the Wabash Valley in northern Indiana. The diameter of the reef shown is approximately 2,000 feet, and the height of the section approximately 145 feet. This section illustrates a reef that began to grow on a shale floor, into which it later settled as the overlying weight increased. [Redrawn from Cumings and Shrock, Bull. Geol. Soc. Amer., Vol. 39 (1928), p. 598, Fig. 7.]*

cate that the Niagaran bioherms were built in fairly shallow water and stood as low mounds from a few feet to some tens of feet above a surrounding sea bottom on which fine calcareous muds were being deposited.

The organic reefs of the Niagaran epoch—in Illinois and Indiana, for example—consist of nearly pure carbonate, either limestone or dolomite, with higher electrical resistivity than the normal inter-reef rock; unlike many limestones, they contain no chert. They range from a few feet up to several miles in diameter, from a few feet up to a thousand feet in height, and from round through elliptical to ridge-like in shape. They may consist only of the massive core, in contact with the surrounding formations, or the core may be flanked by bedded reef-derived detritus that dips away in all directions. Details of one of these reefs are shown in Figure 7-36.

An exposed ancient reef that is analogous to present barrier reefs is the spectacular Capitan reef in the Permian Basin of western Texas and New Mexico,[51] shown in Figure 7-37. This reef, which is partly above and partly below the surface, dwarfs all other known fossil reefs, having a thickness of 1,200 feet or more and extending for a distance of at least 400 miles. (See Fig. 3-10, p. 73.) Many oil and gas pools are found along much of its buried portion (*BB′*). The recognition of this massive dolomitic limestone body, and of the equally thick underlying Goatseep reefy mass, was long delayed—chiefly, it would seem, because of misinterpretation of the organic remains in the rocks, which presented a difficult problem. The Capitan and Goatseep reefs occur in the Guadalupe series of the Permian, and mark a transition from the deep-water clastic sandstones and interbedded limestones of the Permian Basin to the limestones in the shallow-water evaporite deposits of the back-reef or lagoonal areas. Another great organic reef development is the Abo-Wichita Albany (Permian) reef in West Texas and New Mexico. This was a narrow reef-trend that rimmed the Midland basin to the west, north and east. More

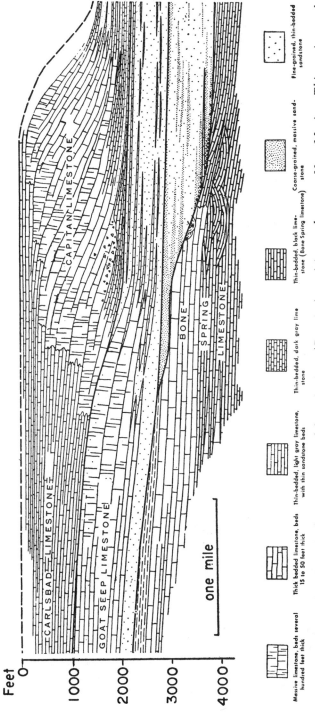

FIGURE 7-37 Section through the famous Guadalupe-Capitan (Permian) reef in southeastern New Mexico. This reef extends for hundreds of miles to the east and southeast, where many oil pools are trapped in the porous Capitan reef limestone. [Redrawn from King, U.S. Geol. Surv., Prof. Paper 215 (1948), Fig. 7.]

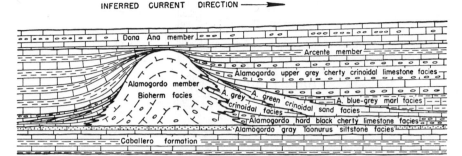

FIGURE 7-38 *Section through a bioherm, typical of many in the Alamogordo member of the Lake Valley formation (Mississippian), in the Sacramento Mountain region of southeastern New Mexico. It is believed that bioherms similar to these produce oil where buried in north-central Texas. [Redrawn from Laudon and Bowsher, Bull. Amer. Assoc. Petrol. Geol., Vol. 25 (1941), p. 2128, Fig. 10.]*

than sixty oil pools have been discovered in locally developed reef mounds, and more discoveries are anticipated.[52]

In the Sacramento Mountains of New Mexico[53] typical bioherms containing crinoidal remains are exposed in the Mississippian Lake Valley rocks. These bioherms are rounded, knob-like, flat-bottomed masses of hard featureless limestone, a typical example being shown in Figure 7-38. They are distinguished by their absence of bedding, particularly toward their centers. They vary in size, but some are as much as 200 feet thick and a mile across at the base. They have initial dips away from the center of as much as 40°. The only fossil remains are single, isolated, calcareous crinoid columns on the inferred lagoonal, or lee, side of the bihermal areas, where a loosely cemented, fragmental, and disintegrated crinoidal facies occurs, evidently a detrital deposit washed down from the growing reef. This is interbedded with the outer margins of the bioherms, from which it is separated by a definite lithologic break. Typical also of the lagoonal areas is a sand facies, and farther inward there is a calcareous fossiliferous marl; both of these are poorly developed, if at all, on the seaward side. Because of the thickening toward the bioherm of the adjacent talus-like beds and the very abrupt and pronounced thinning of the first overlying bed as it passes over the bioherms, Laudon and Bowsher believe that the individual bioherms stood as mounds above the sea floor.

In the Borden, or Knobstone, group (Lower Mississippian) of Indiana bioherms occur as disconnected limestone masses composed chiefly of crinoids and containing much chert; they are completely enclosed by clastic siltstones, fine-grained sandstones, and thin beds of shale.[54] The Borden bioherms were built up contemporaneously with the enclosing clastic sediments.

A reef of organisms in Permian red beds, in South Park, Colorado, has been described by Johnson.[55] The reef seems to have been built by one species of lime-secreting alga, a large colony of which must have been growing here for a considerable time. It is approximately 1,100 feet long, 80 feet thick, and 300 feet wide. Its outlines are not symmetrical but show irregular tongues of algal limestone interfingering with the red beds. When material from this bioherm was examined microscopically, it was found that the organic structure had been partly destroyed by the crystallization of dolomite; algae are nevertheless represented by numerous coupled and broken threads, tubes, and masses of cellular tissue.

Many reefs are found in the Permian of the Perm Basin, USSR. They are productive in some places.[56] One that crops out as a hill is shown in the section through Tra-Tau in Figure 7-39.

Organic reefs in two Gulf Coast salt domes, Damon Mound and West Columbia, have been described by Ellisor.[57] She found that the foraminiferal assemblage within the limestone lens of Oligocene age differs from that in the surrounding shales of equivalent age and stratigraphic position, and that many of the limestone genera are characteristic of warm, shallow seas around coral reefs. The diagnostic species *Heterostegina antillea,* which occurs throughout the limestone, is characteristic of the modern coral reefs of Antigua and the West Indies.

Stratigraphic relations similar to those described may be identified around many fossil reefs, and they form a useful means of identifying a facies or its position with respect to other parts of the reef complex.[58] The nomenclature is given on the idealized diagram through a reef in Figure 7-40. Many of these features may be correlated with the modern reef, as shown in Figure 7-30, page 314. The back-reef, or lagoonal, facies (A) consists of interbedded limestones, dolomites, sandstones, red shales, and evaporites such as anhydrites, which merge into the main reef (B). The fore-reef, or seaward, side consists of gray-to-black bedded limestones and shales (D), which grade into the reef

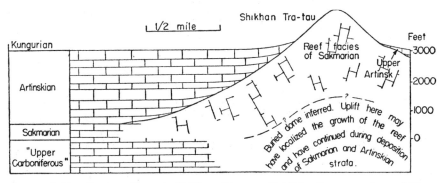

FIGURE 7-39 *Section through Shikhan Tra-tau ("shikhan" means sharp hill), a bioherm that is part of a chain of partially buried reefs, a few miles north of the Ishimbaevo oil field. (See Fig. 7-51.)* [Redrawn from Dunbar, Bull. Amer. Assoc. Petrol. Geol., Vol. 24, p. 256, Fig. 7.]

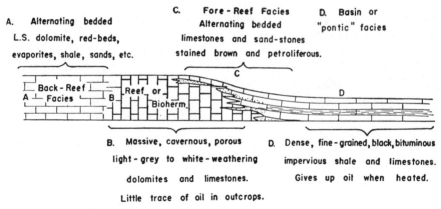

FIGURE 7-40 *Characteristic rock facies occurring in different parts of a reef.* [Redrawn from Link, Bull. Amer. Assoc. Petrol. Geol., Vol. 34 (1950), p. 287, Fig. 18.]

rock through a mixture of sand and reef debris (C). Coarse conglomerates with boulders up to 15 feet in diameter, and one transported block 900 feet long and 40 feet thick, have been observed as submarine landslide deposits that slid off the steep talus slopes of organic reefs in Leonard-Capitan formations (Permian) of western Texas.[58]

Fossil reefs occur in some regions at different elevations and at different stratigraphic positions within a thick sequence of rocks, thus forming a *reef complex,* possibly better described as a *reefy* or *reef-like complex.* Such conditions presumably result from alternating advances and retreats of the sea across favorable reef-building territory. The advancing sea permits bioherms to grow progressively nearer the shore, as at A_1, A_2, and A_3 in Figure 7-41. Each successive reef is overlain by foredeep and basin facies, which indicate the transgressive nature of the deposits. Later, as the sea retreats, successive reef growths form along the retreating shore line, possibly as shown diagrammatically in Figure 7-42, where the reefs are represented by R_1, R_2, and R_3.

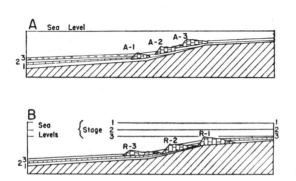

FIGURE 7-41

Idealized transgressive (A) *and regressive* (B) *shore lines and the reef deposits formed under these conditions. Reefs A-1 and B-2 are contemporaneous with the shore lines 1 and 2, etc.* [Redrawn from Link, Bull. Amer. Assoc. Petrol. Geol., Vol. 34 (1950), p. 278, Fig. 13.]

FIGURE 7-42

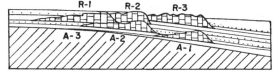

Idealized combination of reefs formed under advancing shore line, A-1, A-2, and A-3, and reefs formed as the shore line retreated, R-1, R-2, and R-3. Reefs A-3 and R-1 correspond to the farthest shoreward advance of the sea. [Redrawn from Link, Bull. Amer. Assoc. Petrol. Geol., Vol. 34 (1950), p. 279, Fig. 15.]

The identifying characteristic of the regressive type of bioherm is the burial of the reef by lagoonal or back-reef material. In Figure 7-41 the bioherms A_3 and R_1 are the same and constitute the reef nearest the shore.[59] A transgressive and regressive shore may also cause the reef core to slope as shown in Figure 7-43.

Productive Reefs. Organic reefs containing oil and gas are found at many levels within the geologic column; they are of widespread occurrence throughout the world, but appear to be especially common in North America—possibly because of more drilling there. They range through all gradations and combinations of atolls, table reefs, fringing reefs, barrier reefs, biostromes,

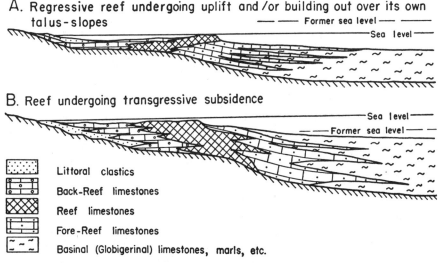

FIGURE 7-43 *Diagrammatic sketch illustrating (A) how organic barrier reefs may move seaward and overhang the oceanward talus slopes as the land rises and the waters shoal, and (B) how the reverse action may occur as the waters deepen. Here the reef transgresses over the lagoonal facies and approaches the land. The diagram, based on observations of organic reefs in the Middle East, shows how reef complexes may develop. [Redrawn from Henson, Bull. Amer. Assoc. Petrol. Geol., Vol. 34 (1950), p. 229, Fig. 13.]*

and bioherms; and although they are primarily formed from organisms growing in place, many of them consist of a bewildering mixture of detrital, clastic, chemical, and biochemical deposits. They may be associated with folding, but more commonly are not. This accounts for much of the difficulty in locating them by any of the ordinary exploration methods, which are chiefly designed to locate areas of rock deformation. The productive areas range in extent from a few acres up to many square miles, and may contain anything from minor noncommercial deposits up to single pools of half a billion or more barrels. A brief description of some of the more important or interesting organic-reef pools will give an idea of their characteristics and variations.

Western and central Texas and southeastern New Mexico contain many pools trapped in organic reefs of Mississippian, Pennsylvanian, and Permian age. The most spectacular of these is the Capitan barrier reef of western Texas and southeastern New Mexico, the outcrop of which has been described on page 321. The reef dips eastward and goes underground, where it becomes productive in a long trend of separate pools along the western edge of the Central Platform along which it grew. It is shown as *BB'* in Figure 3-10, page 73. The traps in many of these pools are of characteristic reef materials, and the porosity and permeability are not necessarily confined to the structurally high parts of the reefs. In several of the pools, therefore, notably the Hendrick pool,[60] the production comes from the crest and the western, or seaward, flank of the reef, where the porosity occurs, the impervious eastern flank being barren. The traps along the buried Capitan reef are actually combinations of reefs, folding, and a regional stratigraphic permeability barrier.

In north-central Texas there are productive Mississippian reefs that are thought to be similar to the Mississippian reefs that crop out in southeastern New Mexico. (See Fig. 7-38, p. 322.) They are small, massive knobs of limestone, generally of small area, covering from a few acres up to half a square mile. They are highly productive, and the oil is of good quality.

Between these two Permian and Mississippian reef areas in Texas, a large number of productive limestone reefs have been found in the Pennsylvanian rocks. They range in size from a few acres up to many square miles, and they occur as limestone masses as much as 1,500 feet thick, completely enclosed by shales containing thin beds of limestone and sandstone. Their upper surfaces are irregular, with high knobs and peaks projecting above the average reef level. The largest of these is the Scurry-Snyder reef, probably an atoll, contours on the top of which are shown in Figure 7-44. The reef is built on top of Mississippian limestone that overlies Ordovician limestone. The permeable connection between the reef mass and the widespread, permeable, underlying Ordovician formations is given as a reason why this reef is productive, whereas other reefs in the region are unproductive because they are separated from the Ordovician by impermeable shale formations.[61] Flanking bituminous shales may also have been the source of the oil. In the productive reefs the oil-water

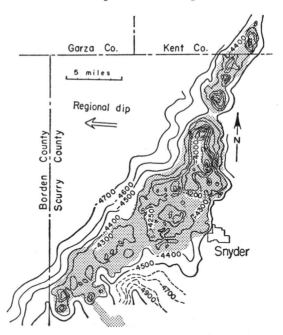

FIGURE 7-44

Map showing the Scurry-Snyder field (stippled) of Texas. The trap is formed by a mass of organic reef limestone projecting into homoclinal dipping shales and thin limestones, the contours (100-foot interval) showing the configuration of the top of the reef. (See Fig. 7-45 for section.) Recoverable reserves of a billion barrels are estimated. [Redrawn from World Oil, September 1951, p. 101.]

contacts are level planes that may be close to the highest point in the reef or much lower down, so that the oil column may be either thin or thick. The floor upon which the reefs grew is thought to have been a broad structural arch that extended through the area during early Pennsylvanian time and is now obscured by the regional westward tilting. A stratigraphic and a structural section across the area of reef development are shown in Figure 7-45. (See also Fig. 7-44.) The largest of the reefs occurs in Scurry County and is 23 miles long and 4–8 miles wide. The Scurry reef was formed from the hard parts of organisms that lived, died, and were buried in one locality. The reef rock is a mixture of organic shell debris, lime muds, and lime sands, welded together by calcite cement. Most of the porosity consists of solution cavities.[62] A section through a typical producing pool in the Scurry County reef complex is shown in Figure 7-46.

Under the plains of western Canada there is a great series of productive organic reefs extending from the Arctic regions almost to the northern boundary of the United States. These reefs are in Upper Devonian rocks. Some individual pools in them contain as much as half a billion barrels of recoverable oil. The producing reefs are knobs and masses of limestone and dolomite built up on a widespread limestone floor. The pores are of various kinds—vugs, fossil casts, fractures, and intergranular voids. Some of the reservoir rock is detrital debris on the flanks of the reefs. The regional setting is shown in Figure 7-47, an idealized section from west to east across the Alberta syn-

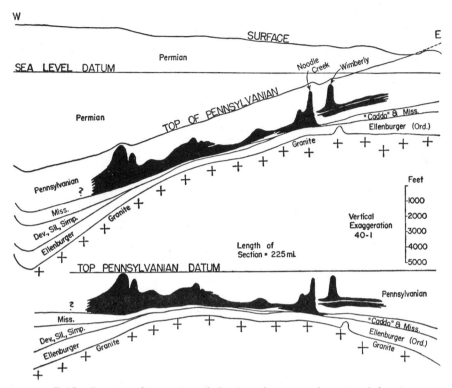

FIGURE 7-45 *Stratigraphic section (below) and structural section (above) across the Pennsylvanian reef area of western Texas. The lower section shows the regional structural arch that was present when the top of the Pennsylvanian was being deposited as an approximately level surface. Later, but before the deposition of the Cretaceous rocks, the area was tilted down toward the west, as shown in the upper section, thereby obscuring the earlier arching, which had presumably localized the development of the large and rich organic reef block of the Scurry County region. (See also Fig. 7-46.) [Redrawn from Richardson, Tulsa Geol. Soc. Digest (1950), p. 75, Fig. 1.]*

cline. Figure 7-48 shows the linear trends of many reefs. Figure 7-49 is a detailed section across the Leduc reef pool, which is typical of many of these pools.[63]

The bioherm at Norman Wells, on the Mackenzie River in northwestern Canada, near the Arctic Circle, contains a productive pool that is one of the farthest north in the world.[64] The limestone reef consists of a lenticular mass thinning in all directions from a maximum thickness of more than 400 feet. It is enclosed in the Fort Creek shales (Upper Devonian), which are on a monoclinal southwest-dipping structure. The bioherm has a lower, unbedded coralline limestone member, which is generally impermeable and nowhere productive. Overlying this, and sometimes separated from it by a thin shale member, is a productive, unbedded coralline limestone containing corals, bryozoans, stromatoporoids, and other fossil remains, embedded in sand. The similarity of the geologic conditions in the Fort Norman pool to those in the Alberta reef pools, nearly a thousand miles to the south, indicates that the Devonian reefs may be of vast extent and arouses a hope that the intervening country may contain tremendous petroleum deposits.

In the great Southern or Golden Lane, district of Mexico, more than a billion barrels of oil have been produced from an elongate limestone barrier reef of Middle Cretaceous age. Production is obtained along the crest of the reef, which extends for fifty miles but is less than a mile wide. (See Fig. 7-50.) The upper surface has an asymmetric cross section, dipping steeply (at 30°) to the west and gently to the east. The western slope, which may be in places

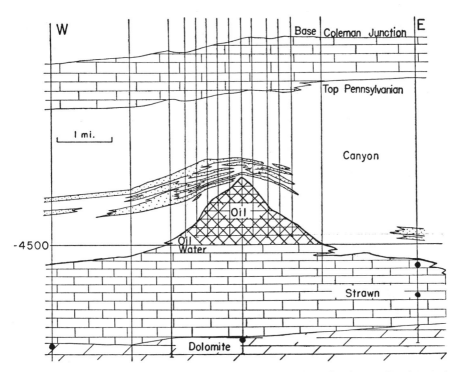

FIGURE 7-46 *Section through the North Snyder pool in the Scurry-Snyder reef area of western Texas. The trap is in an organic reef of Canyon age built up from a floor of Strawn limestone (Pennsylvanian). The slight folding of the sands above the reefs may be the result either of initial dips or of compaction of the shales intervening between the sands and the underlying reef. The lack of shallow structure is noticeable, a characteristic that makes exploration for reefs extremely difficult.* [*Redrawn from Keplinger and Wanenmacher, World Oil, September 1950, p. 182.*]

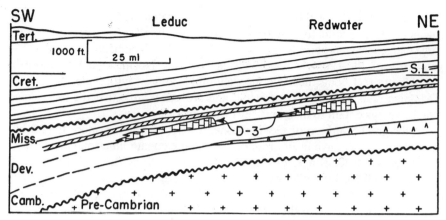

FIGURE 7-47 Diagrammatic section through the Leduc and Redwater fields in Alberta, Canada, showing the regional setting of the organic reefs that contain the pools. [Redrawn from Dr. A. W. Nauss, as reproduced by Aaring, Oil in Canada, October 20, 1952.]

a fault surface, is the seaward slope, and the eastern side, with its coral knolls, the lagoonal side. It is believed that the reef grew along the crest of a ridge or tectonic element such as a long fold or a tilted fault block.[65] Farther south in Mexico is the Poza Rica field, which has a potential capacity of nearly a billion barrels, also contained in reef-facies limestone.[66] (See Fig. 8-10.)

The Perm Basin, which is west of the Ural Mountains in the USSR, is strikingly similar in stratigraphy and general geologic conditions to the Permian basin of western Texas and New Mexico.[67] A number of pools have

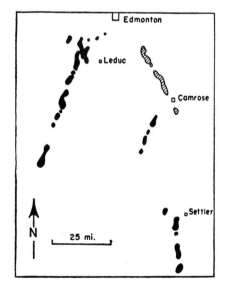

FIGURE 7-48

Map showing the linear trends of Devonian reef reservoirs (black) south of Edmonton, Alberta, Canada, and a linear trend of pools (Joarcam) in the Cretaceous Viking sand (ruled). Once trends such as these have become established, they justify intensive exploration.

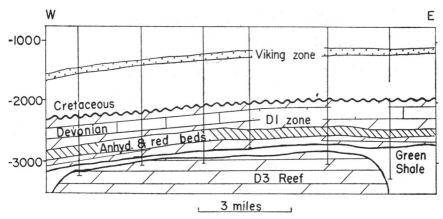

FIGURE 7-49 Section across the Leduc field, Alberta, Canada, showing the development of the dolomitic organic reef known as the D3 reef within the green shales of the Upper Devonian. Some of the Alberta producing reefs are limestone, and others are dolomite. A section through a core showing the nature of the porosity is shown in Figure 4-13, page 127. The unconformity in the section separates the overlying Cretaceous formations from the Devonian. [Redrawn from Waring and Layer, Bull. Amer. Assoc. Petrol. Geol., Vol. 34, p. 298, Fig. 4.]

been found in Permian limestone reefs in this basin, which, together with the Paleozoic platform area located eastward toward the Volga, gives promise of some day rivaling the Texas–New Mexico Permian basin in productivity. Typical of the Perm Basin reef occurrences is the Ishimbaevo field,[68] a section of which is shown in Figure 7-51. The oil occurs in limestone massifs, or reef structures (in the Artinskian stage of the Permian) that are similar to the isolated reef mountains ("shikhans") that crop out in the basin. (See Fig. 7-39, p. 323.) The porosity is in fossil casts, brecciated limestone, and detrital limestone and fossil fragments located near the higher parts of the reef. Mineralized water underlies the oil, which is high in aromatics and has a specific gravity of 0.87–0.88 (approximately 30° API). As in many limestone pools, the oil is high in sulfur (about 2.5 percent). Wells ranged from 100 to 600 tons per day each (700–4,200 barrels).

FIGURE 7-50

Section across the Golden Lane and Poza Rica fields, Mexico. (See also Fig. 8-10.) [Redrawn from Rockwell and Rojas, Bull. Amer. Assoc. Petrol. Geol., Vol. 37 (1953), p. 2555, Fig. 3.]

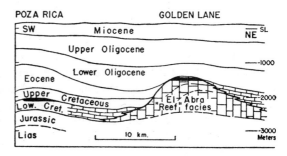

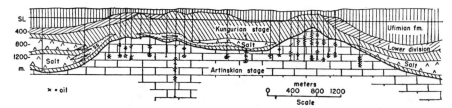

FIGURE 7-51 *Section through the Ishimbaevo oil field in the Bashkir ASSR, USSR. Oil seepages have been known in the vicinity since 1768, and the producing formations are Artinskian (Permian and Upper Carboniferous) limestones and dolomites. (See Fig. 7-39.) The producing formations are rich in organic remains and are thought to be organic reefs.* [Redrawn from A. Blokhin, *XVIIth Int. Geol. Cong.* (1937), Fascicle 1, Fig. 6.]

As can be seen from the examples shown, oil-producing organic reefs vary greatly; in fact, nearly every organic reef is different from the others. Some are long and narrow, others are slender and circular pinnacles; some are composed of limestone, and others of dolomites; some contain oil saturated with gas, and others contain undersaturated oil; nearly all contain water under the oil; but in any productive region not all reefs are productive. In some reefs the detritus is well developed and is often highly productive; in others little or no detrital material is recognizable.

The first discovery of an organic reef in a region is likely to be made by accident, for the location of a reef, unlike that of a structural trap, is not likely to be indicated in any obvious way at the surface. A reef may indeed give rise to structural anomalies, some of which can be mapped by geophysical and seismic methods, but these are generally so slight that they are not considered indicative of traps, and they would not normally justify drilling in a province in which no reefs had previously been found. But a region in which one reef has been found is likely to contain others, for reefs are rarely if ever isolated; they occur in groups, or trends, in many places related to ancient shore lines. They are similar in this respect, and in their random patterns, to strand-zone patches and lenses of sand. When one reef has been found, therefore, an aggressive search ought to be made in the hope of finding others nearby; and, in the present state of our knowledge of reefs and the like, we may sometimes be justified in looking for them in areas in which none have been found hitherto. Search for these features is likely to involve the drilling of many wildcat wells, put down on the basis of slight indications or even at random, and may thus be very expensive. This risk, however, would often be justified, especially in areas in which one or more reefs have already been discovered. For reef pools, in the aggregate, have yielded a substantial part of production in the past and may be expected to continue to do so in the future. One fact that makes reefs especially attractive is that many individual pools found in them have been of major size and have yielded large profits.

SECONDARY STRATIGRAPHIC TRAPS

Secondary stratigraphic traps are those that result from some stratigraphic anomaly or variation that developed *after* the deposition and diagenesis of the reservoir rock. They are almost always associated with unconformities, and might therefore be called *unconformity traps*.

An unconformity is a break in the geologic sequence and is marked by a surface of erosion, or at least of nondeposition, separating two groups of strata. Unconformities vary greatly in character and are accordingly designated by various terms.[69] The smaller breaks between strata have been called *diastems*. Where the beds above and below the unconformity surface are parallel or nearly parallel, the term *disconformity* has been used. Where the beds above and below the unconformity are not parallel but meet at an angle, the contact is called an *angular unconformity* or a *nonconformity*. The hiatus in the rock sequence, or the time interval, represented by one unconformity may be much greater than that represented by another, and it may even vary greatly along a single surface of unconformity.[70] In the center of a sedimentary basin, for example, beds above and below the unconformity may have been deposited in a nearly continuous sequence, whereas toward the edges of the basin the upper formation may overlap onto progressively older and older beds and finally onto the basement rocks. In the center there would be only a slight unconformity, or a mere disconformity, between rocks that were all Cretaceous, for example, while at the edges there would be a major unconformity, as between Cretaceous and Precambrian rocks.

Unconformities of varying areal extent and stratigraphic hiatus are common in the geologic sections of nearly all producing provinces. Reservoir rocks may occur immediately above the plane of unconformity or immediately below it, or within the weathered material that marks the unconformity itself. The effects of erosion along the plane of unconformity are of particular importance in provinces in which the reservoir rocks are more or less soluble, especially where there are limestones and dolomites. Here percolating meteoric waters descend through the weathering zone and dissolve out the more soluble parts of the underlying formations, so that high porosity and permeability are developed. Some of the places where oil and gas pools commonly occur in traps associated with surfaces of unconformity are shown in Figure 7-52.

A great many oil and gas pools, possibly most of them, are intimately associated with unconformities in some way or other. Unconformities are therefore important phenomena in the geology of petroleum. In some pools the trap itself is a direct result of some phenomenon connected with the development of the surface of unconformity. Traps bounded by an unconformity are broadly classed as stratigraphic, and they are also classed as secondary stratigraphic traps because they are formed after the lithification and diagenesis of the reservoir rock.

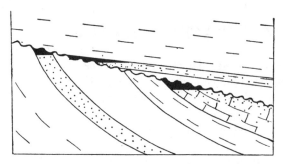

FIGURE 7-52

Diagrammatic section showing traps commonly associated with surfaces of unconformity.

Where each bed in the overlying formation extends successively beyond the edge of the next older or underlying bed, the position is called *overlap*.[71] The overlap may be transgressive, as in an advancing sea, the evidence being in finer sediments overlying coarser sediments, and it is then called *onlap* or *transgressive overlap;* or it may be regressive, as in a retreating sea, the evidence being an upward gradation into coarser sediments, and it is then called *offlap* or *regressive overlap*. (See Fig. 7-53.) The overlapping formation may or may not rest on a surface of unconformity; if it does, the unconformity may be either disconformable or nonconformable. Where the overlapping formation transgressively overlaps the edges of an erosionally truncated series of formations, the relation is called an *overstep*.[72]

It is difficult to identify unconformities from well data, but Krumbein[73] lists approximately forty criteria that can be of use in this regard. These are classified as sedimentary, paleontologic, and structural. Krumbein emphasizes that the association of several criteria greatly increases the chance that an unconformity is present; so a criterion that would not be of much weight by itself may become important when associated with others. Probably the most common criteria are these: a hiatus in the paleontologic sequence; a poorly sorted formation or a conglomerate in a sequence of generally well-sorted and fine-grained rocks, especially when associated with a thin red shale; evidence of a discordance in dip along a plane surface; evidence of weathering in the underlying formation; and evidence of regional truncation and overstepping, which must be gathered from several wells.

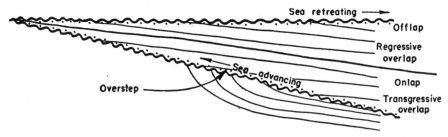

FIGURE 7-53 *The general relations of transgressive and regressive, overlap, offlap, and overstep depositional geometries.*

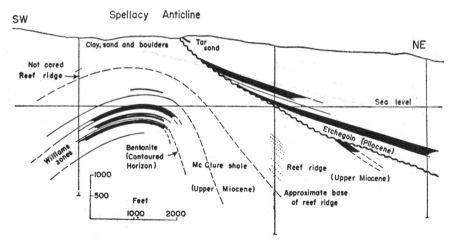

FIGURE 7-54 *Structural section across the Williams and Twenty-Five Hill area, in the Midway oil field, California, showing pools trapped above and below an unconformity plane. It also shows the multiple sand possibilities of an anticlinal trap.* [Redrawn from Hillis, Bull. 118, Calif. Div. Mines, p. 526, Fig. 225.]

A surface of unconformity may mark the boundary between a permeable and an impermeable formation, and thus form the upper or lower limit of a reservoir. More than 200 million barrels of oil was trapped in the Delhi pool in northern Louisiana, where the trap is formed by the unconformable overlap of the Eocene biostromal "gas rock" limestone onto the truncated Lower Cretaceous and Jurassic sands.[74] An example of pools trapped both above and below the unconformity plane is shown in Fig. 7-54. Sands that intersect an underlying plane of unconformity, as in this example, are called *buttress sands*. Additional examples of this are illustrated in Figure 7-55, a section across the Cutbank pool in Montana. In the Antelope Hills field of California the overlying buttress sand rests unconformably on an underlying truncated sand, and both contain oil. (See Fig. 7-56.) The Plio-Pleistocene sands of the Quirequire field in the Maturin basin of eastern Venezuela buttress out against the underlying unconformity and form the principal trap. This is one of the large oil fields of South America, having produced more than 600 million barrels of oil of 12–26° API gravity; and it is noteworthy that this oil has come from continental beds. The waters are erratic in character, and range in concentration from fresh to that of sea water. A section across the field is shown in Figure 7-57.

Weathering and groundwater circulation are generally attended by solution, cementation, and recrystallization,[75] and evidence of all these may be expected close below an unconformity. They often account for the porosity and permeability of a reservoir rock, and the irregular distribution of such permeability within impervious rocks may form the trap. (See Fig. 7-58.)

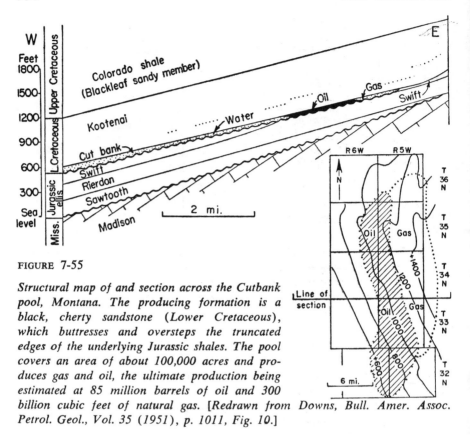

FIGURE 7-55

Structural map of and section across the Cutbank pool, Montana. The producing formation is a black, cherty sandstone (Lower Cretaceous), which buttresses and oversteps the truncated edges of the underlying Jurassic shales. The pool covers an area of about 100,000 acres and produces gas and oil, the ultimate production being estimated at 85 million barrels of oil and 300 billion cubic feet of natural gas. [Redrawn from Downs, Bull. Amer. Assoc. Petrol. Geol., Vol. 35 (1951), p. 1011, Fig. 10.]

The clogging of the porosity by oil residues at a place where the reservoir rock crops out and the lighter fractions are permitted to escape, leaving the heavier residues behind, is another way an unconformity or erosion surface may determine the position of a pool. The early development on the Coalinga anticline in California[76] was in a pool trapped by the clogging of the surface outcrops with asphalt and tar, which formed a stratigraphic obstruction to the escape of the oil and gas. Down the dip the oil became liquid and producible. Similar situations exist in the Reddin district, in the Ouachita Mountains of Oklahoma, where progressive decreases in viscosity from solid gilsonite to high-gravity oil may be observed. In the Oklahoma City field, basal Pennsylvanian (or weathered Ordovician) asphaltic material in the unconformity zone between the Pennsylvanian rocks and the underlying Ordovician reservoir clogged the porosity of the reservoir by surface seepage during the pre-Pennsylvanian erosion period.[77] (See also p. 643.) A somewhat similar type of clogging helped form the trap in the Pleasant Valley pool, California,[78] where the pores of the Gatchell sand (Middle Eocene) were filled with colloidal clayey material, probably in association with weathering on a surface of erosion. The

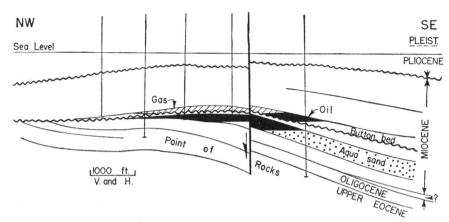

FIGURE 7-56 *Section through the Antelope Hills oil field, California. Two separate pools are shown, one above and one below the plane of unconformity.* [*Redrawn from Armbruster,* AAPG Guidebook (1952), *Los Angeles, pp. 206–207.*]

clayey material is white kaolin formed by the alteration of feldspars, and the loss of permeability up-dip in part causes the trap. (See also pp. 352–353.) Some of the pools in the Chanac formation (Upper Miocene?), along the east side of the San Joaquin Valley, in California, are also in traps caused in part by the up-dip clogging of the pores and a reduction of permeability due to the deposition of weathered feldspars, presumably on a surface of erosion.[79]

An up-dip intersection of two planes of unconformity determines, or helps to determine, the trap in a number of pools. The West Edmond pool, north of Oklahoma City, for example, is formed by an up-dip wedge-out of the permeable cherts and limestones of the Bois d'Arc member of the Hunton

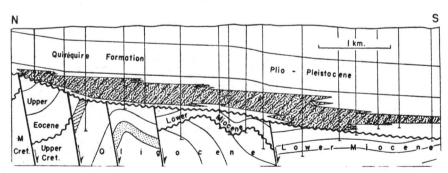

FIGURE 7-57 *Section through the Quirequire field, eastern Venezuela. The producing sand (shaded) buttresses out against the underlying unconformity and consists of continental, unsorted clastics, ranging from silty sands to boulders.* [*Redrawn from Borger,* Bull. Amer. Assoc. Petrol. Geol., *Vol. 36 (1952), p. 2322, Fig. 18.*]

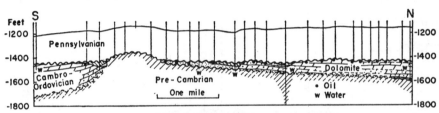

FIGURE 7-58 *Structural section through the Kraft-Prusa oil field, Barton County, Kansas, showing the irregular distribution of porosity in the Cambro-Ordovician dolomites resulting from pre-Pennsylvanian weathering. [Redrawn from Walters and Price, in* The Structure of Typical American Oil Fields, *Vol. 3 (1948), p. 258, Fig. 4.]*

formation (Devonian-Silurian) at the intersection of the pre-Pennsylvanian and the pre-Mississippian unconformities. (See Fig. 14-7.) The Tatums pool, in Carter County, Oklahoma, is trapped in a complex truncation of Pennsylvanian conglomerates and sandstones, which also buttress against an underlying pre-Pennsylvanian basement.[80] A section is shown in Figure 7-59.

The Apco pool, in Pecos County, Texas, is trapped at the intersection of two unconformities. A paleogeologic map of the pre-Permian geology forms Figure 7-60, and a section across the pool from northwest to southeast is shown in Figure 7-61. The porosity in the producing Ellenburger limestone (Ordovician) is due to fracturing and honeycombing, probably developed during the pre-Permian erosion period when the Ellenburger limestone formed an east-facing escarpment extending across the area. Traps formed by the intersection of unconformities also occur in the Santa Maria region of California, a section across which is shown in Figure 7-62. The porosity is the result of fracturing of the brittle Monterey shales (Miocene) and siltstones, and may have developed during the period when the unconformities were developing.

The intersection of two unconformities helps form the trap of the East Texas pool. In this case, however, the lower unconformity is merely incidental to the trap; the upper unconformity is the important one, as it truncates the

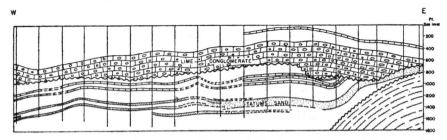

FIGURE 7-59 *Section through the Tatums field, Carter County, Oklahoma. The trap is formed by a complex of Pennsylvanian sand, limestone conglomerate, and intersecting unconformities buttressing against a pre-Pennsylvanian basement. [Redrawn from Grimes, Bull. Amer. Assoc. Petrol. Geol., Vol. 19 (1935), p. 403, Fig. 2.]*

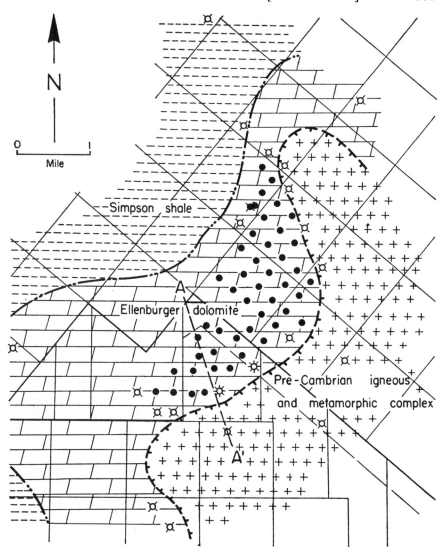

FIGURE 7-60 *Paleogeologic map of the pre-Permian geology in the area of the Apco field, Pecos County, Texas. The oil is found in the Ellenburger limestone (Ordovician), where it is truncated and overstepped by Permian shales and limestones. Section AA' is shown in Figure 7-61.* [*Redrawn from Ellison, in* The Structure of Typical American Oil Fields, *Vol. 3 (1948), p. 409, Fig. 6.*]

Woodbine sand, and the trap would have been formed whether the pre-Woodbine unconformity was present or not. Because the wedge-out combines with arching to form the trap, it is better classed as a combination trap. (See pp. 350–351.)

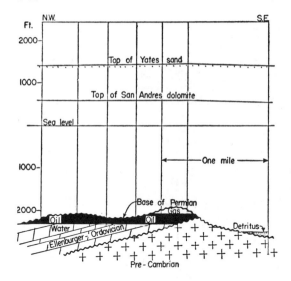

FIGURE 7-61

Section AA' through the Apco pool in Pecos County, western Texas. [Redrawn from Ellison, in The Structure of Typical American Oil Fields, *Vol. 3 (1948)*, *p. 416, Fig. 10.*]

FLUID TRAPS

During recent years another trapping agency has been discovered.[81] When a fluid potential gradient exists in an aquifer such that the flow of water is directed down the dip, the hydrodynamic force may bar the up-dip movement of any petroleum in the aquifer, in which case the buoyancy of the petroleum will cause it to accumulate into a pool. The size of the pool will depend on several variables, such as the density of the water, the density of the petroleum, the lithology of the aquifer, and the magnitude of the fluid potential gradient in the aquifer.

An analogy may help explain the principle. Let us imagine an inclined tube (Fig. 7-63,A) that is filled with water flowing downward through the tube. Within the tube there are corks, each with enough buoyancy to rise upward in the tube in opposition to the downward force caused by the downward flowing water. If the tube were to be slightly constricted, as in B, the fluid potential gradient would increase, the volume-flux of water through the constriction would be increased, and the corks would not have enough buoyancy to allow them to move through the constricted area even if there were plenty of room to pass through. The corks would accumulate below the constriction until enough total buoyancy developed to force the corks at the up-slope edge of the accumulation to pass through the constriction. The number of corks left below the constriction after the breakthrough occurred would depend on the magnitude of the hydrodynamic pressure gradient in the constricted zone and the total buoyancy developed by the accumulated corks. The corks might be likened to oil and gas; and the constriction, to the lessening of permeability along the up-dip edge of an aquifer. If the water were flowing upward through the tube, then the corks, with their natural buoyancy in this water system, would move up-slope with the water and would pass through the constriction at an even faster rate than they would under hydrostatic conditions, and there would be no accumulation.

In this analogy the water passes through the constricted area, and this is probably the situation in most aquifers. The up-dip loss of permeability in most aquifers is, therefore, not a total loss of permeability except in special situations. Surfaces of unconformity, for example, offer many avenues for the movement of fluid beyond the edge of the truncated aquifer even though they are small. Similarly, most sand formations may be expected to extend laterally beyond the main sand body in thin stringers, thin sandy patches, and small sandy remnants, all of which permit some permeability to water. The fact the fluid potential gradients extend up to the edge of aquifers would also indicate the movement of water through the edge. About the only really tight cover or barrier to fluid movement would be precipitates such as salt, anhydrite, and very dense limestones.

As more and more fluid pressure data become available, it is learned that fluid potential gradients exist laterally in many aquifers and also vertically across bedding and between aquifers separated by shales and limestones.[82] It is obvious that fluid potential gradients must be continually changing with

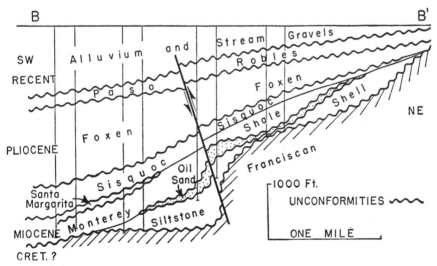

FIGURE 7-62 *Section across the Santa Maria oil field, Santa Barbara County, California. Many wells with a capacity of 10,000 barrels per day of oil of 16.5° API gravity were drilled into the Monterey (Miocene) fractured shales, cherts, and sands, which overlap onto the underlying "Franciscan" [Jurassic (?) or basement] formation and are in turn truncated out up-dip by the overlying Sisquoc sandy shales. Six disconformities have been identified within the Monterey producing zone.* [Redrawn from Canfield, Bull. Amer. Assoc. Petrol. Geol., Vol. 23, p. 71, Fig. 6.]

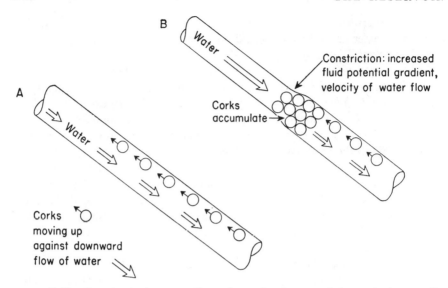

FIGURE 7-63 *In tube A the water flows down the slope, and the corks have sufficient buoyancy to rise against the downward flow. In tube B a constriction increases the fluid potential gradient and the velocity of flow, thus creating a local downward force greater than the buoyancy and causing the corks to accumulate below the constriction.*

the changing patterns of uplift, mountain building, erosion, and deposition. Faulting may seal off flow paths or may open up new channels. The study of fluid pressures of numerous aquifers in many sedimentary basins indicates that formation water is commonly in a dynamic state; this area of investigation is frequently termed *hydrodynamics*. The applications of hydrodynamics to petroleum exploration are many, and the more common and obvious applications will be considered in more detail where they apply.

CONCLUSION

Stratigraphic traps present different problems of exploration than structural traps. In a structural feature, for example, thick sections of rocks, possibly containing several potential reservoir rocks, are deformed and thereby have a number of opportunities to trap pools, one above another. Stratigraphic traps, on the other hand, seldom have any relation with either overlying or underlying reservoirs, but may be repeated laterally. A test well on a structural feature frequently has several potential objectives, whereas a test well on a stratigraphic feature generally has only one. Stratigraphic phenomena generally extend over wide areas or along elongated trends, and the discovery of one offshore sand bar, channel filling, sand patch, or organic reef therefore strongly suggests that others will be found in the region.

Structural features may be determined in many ways, either from the surface by surface mapping or geophysics, or at depth by subsurface mapping. Potential stratigraphic traps, however, generally have to wait to be visualized until enough wells have been drilled to supply the needed stratigraphic information. This means that more test wells have to be drilled in search of stratigraphic pools than in search of structural pools. In fact, careful attention to details of the sedimentation, oil and gas showings, and stratigraphy of an area offers about the best consistent approach to prospecting for pools in stratigraphic traps. This is not to say that the structure of such an area should not be mapped, for it may combine with stratigraphic variations to form traps; but most structural effects of local stratigraphic phenomena are small and difficult or impossible to interpret correctly.

Even after discovery, there are few clues to the size of the pool, or the direction in which it will extend; stratigraphic pools, in fact, are frequently a succession of surprises. An early understanding of the stratigraphic environment, sedimentary history, and the fluid potential environment of a newly productive area is of the most help; it may point to truncated reservoirs, shoestring sands, sand patches, facies changes, fluid barriers, or organic reefs as the cause of the trap, and from this information the pattern of the reservoir and the size of the pool may be anticipated and exploration guided with fewer dry holes.

(For the Selected General Readings, see Chap. 8.)

Reference Notes

1. W. E. Pugh and B. G. Preston (ed.), *Bibliography of Stratigraphic Traps*, Seismograph Service Corporation, Tulsa, Okla. (1951), 190 pages. Annotated.

2. William B. Heroy, "Petroleum Geology," in *Geology 1888–1938*, 50th Anniv. Vol. Geol. Soc. Amer. (1941), pp. 534–539.

3. Raymond C. Moore, *Meaning of Facies*, Memoir 39, Geol. Soc. Amer. (June 1949), pp. 1–34.
W. C. Krumbein and L. L. Sloss, *Stratigraphy and Sedimentation*, W. H. Freeman and Company, San Francisco (1963), pp. 316–331.

4. R. E. Sherrill, P. A. Dickey, and L. S. Matteson, "Types of Stratigraphic Oil Pools in Venango Sands of Northwestern Pennsylvania," in *Stratigraphic Type Oil Fields*, Amer. Assoc. Petrol. Geol., Tulsa, Okla. (1941), pp. 507–538. Bibliog. 12 references.

5. David H. Swann and Elwood Atherton, "Subsurface Correlations of Lower Chester Strata of the Eastern Interior Basin," Jour. Geol., Vol. 56 (July 1948), pp. 269–287. 25 references.

6. W. I. Ingham, "Dora Oil Pool, Seminole County, Oklahoma," in *Stratigraphic Type Oil Fields,* Amer. Assoc. Petrol. Geol., Tulsa, Okla. (1941), pp. 408–435.

7. James C. Freeman, "Strand-line Accumulation of Petroleum, Jim Hogg County, Texas," Bull. Amer. Assoc. Petrol. Geol., Vol. 33 (July 1949), pp. 1260–1270.

8. R. L. Denham and W. E. Dougherty, " 'Sand Belt' Area of Ward and Winkler

Counties, Texas, and Lea County, New Mexico," in *Stratigraphic Type Oil Fields*, Amer. Assoc. Petrol. Geol., Tulsa, Okla. (1941), pp. 750–759.

9. N. W. Bass, "Origin of Bartlesville Shoestring Sands, Greenwood and Butler Counties, Kansas," Bull. Amer. Assoc. Petrol. Geol., Vol. 18 (October 1934), pp. 1313–1345. 32 references.

N. W. Bass, Constance Leatherock, W. Reese Dillard, and Luther E. Kennedy, "Origin and Distribution of Bartlesville and Burbank Shoestring Oil Sands in Parts of Oklahoma and Kansas," Bull. Amer. Assoc. Petrol. Geol., Vol. 21 (January 1937), pp. 30–66. 50 references cited.

10. W. K. Cadman, "The Golden Lanes of Greenwood County, Kansas," Bull. Amer. Assoc. Petrol. Geol., Vol. 11 (November 1927), pp. 1151–1172.

A. E. Cheyney, "Madison Shoestring Pool, Greenwood County, Kansas," in *Structure of Typical American Oil Fields*, Amer. Assoc. Petrol. Geol., Vol. 2 (1929), pp. 150–159.

N. W. Bass, *op. cit.* (note 9).

11. Max W. Ball, T. J. Weaver, H. D. Crider, and Douglas S. Ball, "Shoestring Gas Fields of Michigan," in *Stratigraphic Type Oil Fields*, Amer. Assoc. Petrol. Geol., Tulsa, Okla. (1941), pp. 237–266.

12. Charles R. Fettke, "Music Mountain Oil Pool, McKean County, Pennsylvania," in *Stratigraphic Type Oil Fields*, Amer. Assoc. Petrol. Geol., Tulsa, Okla. (1941), pp. 492–506.

13. E. P. Neal, "South Ceres, Oklahoma's Oddest Shoestring Field?" World Oil, December 1951, pp. 92–98. 14 references listed.

14. C. F. Marbut, "Geological Descriptions of the Calhourn, Lexington, Richmond, and Huntsville Sheets," Mo. Geol. Surv., Vol. 12, Part II (1898), pp. 123–210, 270, 331. Describes the Warrensburg and Moberly channels.

Henry Hinds and F. C. Greene, "Stratigraphy of the Pennsylvanian of Missouri," Mo. Geol. Surv., Vol. 13 (1915), pp. 90–106.

Henry Hinds, "Unconformities in the Pennsylvanian," Geologic Note, Bull. Amer. Assoc. Petrol. Geol., Vol. 10 (1926), pp. 1303–1304.

15. Henry Hinds and F. C. Greene, "Leavenworth-Smithville, Missouri-Kansas," Atlas Folio 206, U.S. Geol. Surv. (1917), pp. 6 and 10.

16. Charles W. Wilson, Jr., "Channels and Channel-Filling Sediments of Richmond Age in South-Central Tennessee," Bull. Geol. Soc. Amer., Vol. 59 (August 1948), pp. 733–766.

17. James Marvin Weller, "The Geology of Edmonson County," Ky. Geol. Surv., Series VI, Vol. 24 (1927), 246 pages, pp. 199–208.

18. C. G. Strachan, "Pre-Pennsylvanian Channelling in Western Kentucky and Its Connection with Oil Accumulation," Tulsa Geol. Soc. Digest (1935), pp. 36–40.

19. Homer H. Charles, "Bush City Oil Field, Anderson County, Kansas," in *Stratigraphic Type Oil Fields*, Amer. Assoc. Petrol. Geol., Tulsa, Okla. (1941), pp. 43–56.

20. John L. Rich, "Further Observations on Shoe String Oil Pools of Eastern Kansas," Bull. Amer. Assoc. Petrol. Geol., Vol. 10 (June 1926), pp. 568–580.

21. A. Beeby Thompson, *Oil Field Exploration and Development* (Vol. 1 of *Oil Field Principles*), 2nd ed., D. Van Nostrand Co. (1950), pp. 67 and 429.

22. C. A. Prokopov and M. I. Maximov, "Oil Field of the Kuban–Black Sea Region," XVIIth Int. Geol. Cong., Fascicle 5 (1937), pp. 29–30.

23. Jerome S. Smiser and David Wintermann, "Character and Possible Origin of Producing Rock in Hilbig Oil Field, Bastrop County, Texas," Bull. Amer. Assoc. Petrol. Geol., Vol. 19 (1935), pp. 206–220.

24. O. & G. Jour., Vol. 44: June 9, 1945, p. 76; July 14, 1945, p. 92.

25. E. R. Cumings, Reefs or Bioherms?" Bull. Geol. Soc. Amer., Vol. 43 (March 1932), pp. 345–347.

26. J. Ernest Carman and Wilbur Stout, "Relationship of Accumulation of Oil to Structure and Porosity in the Lima-Indiana Field," in *Problems of Petroleum Geology*, Amer. Assoc. Petrol. Geol., Tulsa, Okla. (1934), pp. 521–529.

Henry A. Ley, "Lima-Indiana District, Indiana and Ohio," in *Geology of Natural Gas*, Amer. Assoc. Petrol. Geol., Tulsa, Okla. (1935), pp. 843–852. 12 references.

27. Robert F. Imbt and S. V. McCollum, "Todd Deep Field, Crockett County, Texas," Bull. Amer. Assoc. Petrol. Geol., Vol. 43 (February 1950), pp. 239–262.

28. V. P Grage and E. F. Warren, Jr., "Lisbon Oil Field, Claiborne and Lincoln Parishes, Louisiana," Bull. Amer. Assoc. Petrol. Geol., Vol. 23 (March 1939), pp. 281–324.

29. William Morris Davis, *The Coral Reef Problem*, Spec. Pub. No. 9, Amer. Geog. Soc., New York (1928), 596 pages. Extensive bibliography.

W. E. Pugh, *Bibliography of Organic Reefs, Bioherms, and Biostromes*, Seismograph Service Corporation, Tulsa, Okla. (1950), 139 pages.

30. Preston E. Cloud, Jr., "Facies Relationships of Organic Reefs," Bull. Amer. Assoc. Petrol. Geol., Vol. 36 (November 1952), pp. 2125–2149. 70 references.

31. W. B. Wilson, "Reef Definition," Bull. Amer. Assoc. Petrol. Geol., Vol. 34 (February 1950), p. 181.

W. H. Twenhofel, "Coral and Other Organic Reefs in Geologic Column," Bull. Amer. Assoc. Petrol. Geol., Vol. 34 (February 1950), pp. 182–202. Bibliog. 79 items.

32. Edgar R. Cumings and Robert R. Shrock, "Niagaran Coral Reefs of Indiana and Adjacent States and Their Stratigraphic Relations," Bull. Geol. Soc. Amer., Vol. 39 (June 1928), p. 599. Bibliog. 66 items.

33. E. R. Cumings, Proc. Indiana Acad. Sci., Vol. 39 (1930), p. 207.

34. Thomas Wayland Vaughn, "Corals and the Formation of Coral Reefs," Ann. Rept. Smithsonian Inst. for 1917 (1919), pp. 189–276.

a. Harry S. Ladd, "Recent Reefs," Bull. Amer. Assoc. Petrol. Geol., Vol. 34 (February 1950), pp. 203–214. 21 references.

b. Harry S. Ladd, "Reef Building," Science, Vol. 134, No. 3481 (September 15, 1961), pp. 703–716.

35. Harry S. Ladd, *op. cit.* (note 34a), p. 204.

36. J. Stanley Gardiner, *Coral Reefs and Atolls*, Macmillan Co., New York (1931), pp. 130–137. See also review by Lloyd, Bull. Amer. Assoc. Petrol. Geol., Vol. 17 (January 1933), pp. 85–87.

37. Great Barrier Reef Committee, "The Great Barrier Reef," Trans. Royal Geog. Soc. Australasia (1925).

38. J. H. F. Umbgrove, "Coral Reefs of the East Indies," Bull. Geol. Soc. Amer., Vol. 58 (August 1947), pp. 729–778. Bibliog. 116 items.

39. Risaburo Tayama, "Table Reefs, a Particular Type of Coral Reefs," Proc. Imperial Acad. Tokyo, Vol. 11 (July 1935), pp. 268–270.

40. Preston E. Cloud, Jr., *op. cit.* (note 30), p. 2141.

41. Harry S. Ladd, *op. cit.* (note 34a), pp. 211–214.

42. H. S. Ladd, J. I. Tracey, Jr., and G. G. Lill, "Drilling on Bikini Atoll, Marshall Islands," Science, Vol. 107 (January 16, 1948), pp. 51–55.

43. Shoshiro Hanzawa, "Micropaleontological Studies of Drill Cores from a Deep Well in Kita-Daito-Zima (North Borodoni Island)," in *Jubilee Publication in Commemoration of Prof. H. Yabe's 60th Birthday* (1940), Vol. 2, pp. 755–802.

44. Royal Society of London, *The Atoll of Funafuti*, Report of Coral Reef Committee (1904), 428 pages.

45. H. S. Ladd et al., "Drilling on Eniwetok Atoll, Marshall Islands," Bull. Amer. Assoc. Petrol. Geol., Vol. 37 (October 1953), pp. 2251–2280. 21 references listed.

46. Harry S. Ladd, *op. cit.* (note 34a), pp. 204–205.

47. J. Harlan Johnson, "The Algal Genus *Lithothamnium* and Its Fossil Representatives," Quart. Colo. Sch. Mines, Vol. 57, No. 1 (January 1962), 111 pages.

48. J. I. Tracey, Jr., H. S. Ladd, and J. E. Hoffmeister, "Reefs of Bikini, Marshall Islands," Bull. Geol. Soc. Amer., Vol. 59 (September 1948), pp. 861–878.

49. Amadeus W. Grabau, *Principles of Stratigraphy*, A. G. Seiler, New York (1932), Chap. X, "Coral and Other Reefs," pp. 384–449. Bibliog. 93 items.

W. H. Twenhofel, *op. cit.* (note 31), pp. 187–189.

Edgar R. Cumings and Robert R. Shrock, *op. cit.* (note 32), pp. 579–620.

H. A. Lowenstam, "Niagaran Reefs in Illinois and Their Relation to Oil Accumulation," RI 145, Ill. Geol. Surv. (1949), 36 pages; reprinted in O. & G. Jour. March 2, 1950, pp. 48 ff.

50. Ruth D. Terzaghi, "Compaction of Lime Mud as the Cause of Secondary Structure," Jour. Sed. Petrol., Vol. 10 (August 1940), pp. 78–90.

51. Philip B. King, *Geology of the Southern Guadalupe Mountains, Texas*, Prof. Paper 215, U.S. Geol. Surv. (1948), 183 pages. Extensive bibliography.

52. Floyd Wright, "Abo Reef: Prime West Texas Target," O. & G. Jour., Pt. I, July 30, 1962; Pt. II, August 6, 1962.

53. L. R. Laudon and A. L. Bowsher, "Mississippian Formations of Sacramento Mountains, New Mexico," Bull. Amer. Assoc. Petrol. Geol., Vol. 25 (December 1941), pp. 2107–2160. Bibliog. 43 items.

54. Paris B. Stockdale, "Bioherms in the Borden Group of Indiana," Bull. Geol. Soc. Amer., Vol. 42 (September 1931), pp. 707–718.

55. J. Harlan Johnson, "Permian Algal Reef in South Park, Colorado," Bull. Amer. Assoc. Petrol. Geol., Vol. 17 (July 1933), pp. 863–865.

56. Carl O. Dunbar, "The Type Permian: Its Classification and Correlation," Bull. Amer. Assoc. Petrol. Geol., Vol. 24 (February 1940), pp. 237–281. 71 references.

57. Alva Christine Ellisor, "Coral Reefs in the Oligocene of Texas," Bull. Amer. Assoc. Petrol. Geol., Vol. 10 (1926), pp. 976–985.

58. E. Russell Lloyd, "Capitan Limestone and Associated Formations of New Mexico and Texas," Bull. Amer. Assoc. Petrol. Geol., Vol. 13 (June 1929), pp. 645–658.

Philip B. King, *Geology of the Marathon Region, Texas*, Prof. Paper 187, U.S. Geol. Surv. (1938), 148 pages. Extensive references.

Philip B. King, *op. cit.* (note 51).

J. K. Rigby, "Subaqueous Landslides and Turbidity Currents in the Permian of West Texas," Amer. Assoc. Petrol. Geol., Houston meeting, March 24, 1953.

59. Theodore A. Link, "Theory of Transgressive and Regressive Reef (Bioherm) Development and Origin of Oil," Bull. Amer. Assoc. Petrol. Geol., Vol. 34 (February 1950), pp. 263–294.

60. A. L. Ackers and R. DeChicchis, "Hendrick Field, Winkler County, Texas," Bull. Amer. Assoc. Petrol. Geol., Vol. 14 (July 1930), pp. 923–944. Many sections, maps, water analyses.

61. Carl B. Richardson, "Regional Discussion of Pennsylvanian Reefs of Texas," Tulsa Geol. Soc. Digest, Vol. 18 (1950), pp. 74–75.

62. Richard E. Bergenback and Robert T. Terriere, "Petrography and Petrology of Scurry Reef, Scurry County, Texas," Bull. Amer. Assoc. Petrol. Geol., Vol. 37 (May 1953), pp. 1017–1029.

63. W. W. Waring and D. B. Layer, "Devonian Dolomitized Reef, D-3 Reservoir, Leduc Field, Alberta, Canada," Bull. Amer. Assoc. Petrol. Geol., Vol. 34 (February 1950), pp. 295–312.

R. P. Lockwood and O. A. Erdman, "Stettler Oil Field, Alberta, Canada," Bull. Amer. Assoc. Petrol. Geol., Vol. 35 (April 1951), pp. 865–884.

Irene Haskett, "Reservoir Analysis of the Redwater Pool," Can. Oil and Gas Industries (July 1951), pp. 39–49; also Can. Inst. Min. Met., April 1951; also O. & G. Jour., October 18, 1951, pp. 90–92 and 146.

64. J. S. Stewart, "Norman Wells Oil Field, Northwest Territories, Canada," in *Structure of Typical American Oil Fields*, Amer. Assoc. Petrol. Geol., Tulsa, Okla., Vol. 3 (1948), pp. 86–109.

65. John M. Muir, *Geology of the Tampico Region, Mexico*, Amer. Assoc. Petrol. Geol., Tulsa, Okla. (1936), 280 pages. Bibliog. 212 items.

John M. Muir, "Geology of the Tampico-Tuxpan Oilfield Region," in *The Science of Petroleum*, Oxford University Press, London and New York, Vol. 1 (1938), pp. 100–105. 24 references cited.

Antonio García Rojas, "Mexican Oil Fields," Bull. Amer. Assoc. Petrol. Geol., Vol. 33 (August 1949), pp. 1336–1350.

Manuel Alvarez, Jr., "Geological Significance of the Distribution of the Mexican Oil Fields," Proc. Third World Petrol. Congress, The Hague (1951), Sec. 1, pp. 73–85. Also contains many structural maps of Mexican oil fields.

66. Guillermo P. Sales, "Geology and Development of Poza Rica Oil Field, Vera Cruz, Mexico," Bull. Amer. Assoc. Petrol. Geol., Vol. 33 (August 1949), pp. 1385–1409.

67. Carl O. Dunbar, *op. cit.* (note 55). A different interpretation is presented by G. Marshall Kay, "Classification of the Artinskian Series in Russia," Bull. Amer. Assoc. Petrol. Geol., Vol. 25 (July 1941), pp. 1396–1404.

68. N. Guerassimov, "The Oil Fields of the Permian Prikamye"; A. Blokhin, "The Oil Fields of the Bashkirian ASSR"; and V. Boutrov, "The Oil Fields of the Samarskaya Luka (Samara Bend)"; XVIIth Int. Geol. Cong., USSR (1937), Fascicle 1, *The Petroleum Excursion*, 63 pages. Bibliog. 17 items.

H. G. Kugler, "A Visit to Russian Oil Districts," Jour. Inst. Petrol. Technol., Vol. 25 (1939), pp. 68–88.

69. Joseph Barrell, "Rhythms and the Measurements of Geologic Time," Bull. Geol. Soc. Amer., Vol. 28 (December 1917), pp. 745–904. A classic article on causes of bedding, diastems, and unconformities.

70. Eliot Blackwelder, "The Valuation of Unconformities," Jour. Geol., Vol. 17 (1909), pp. 289–299.

71. Amadeus W. Grabau, "Types of Sedimentary Overlap," Bull. Geol. Soc. Amer., Vol. 17 (December 1906), pp. 567–636. A classic article.

Frank A. Melton, "Onlap and Strike-overlap," Bull. Amer. Assoc. Petrol. Geol., Vol. 31 (October 1947), pp. 1868–1878.

Frederic H. Lahee, "Overlap and Non-Conformity" (Geol. Note), Bull. Amer. Assoc. Petrol. Geol., Vol. 33 (November 1949), p. 1901.

72. H. R. Lovely, "Onlap and Strike-overlap," Bull. Amer. Assoc. Petrol. Geol., Vol. 32 (December 1948), pp. 2295–2297.

Frederick M. Swain, "Onlap, Offlap, Overstep and Overlap," Bull. Amer. Assoc. Petrol. Geol., Vol. 33 (April 1949), pp. 634–636.

73. W. C. Krumbein, "Criteria for Subsurface Recognition of Unconformities," Bull. Amer. Assoc. Petrol. Geol., Vol. 26 (January 1942), pp. 36–62.

74. T. H. Philpott, "Paleofacies, the Geologist's Tool," Bull. Amer. Assoc. Petrol. Geol., Vol. 36 (July 1952), p. 1315.

75. Parry Reiche, *A Survey of Weathering Processes and Products*, Pub. in Geol. No. 1, University of New Mexico (1945), 87 pages.

76. Ralph Arnold and R. Anderson, *Geology and Oil Resources of the Coalinga District, California*, Bull. 398, U.S. Geol. Surv. (1910), 354 pages.

77. D. A. McGee and W. W. Clawson, Jr., "Geology and Development of Oklahoma City Field, Oklahoma County, Oklahoma," Bull. Amer. Assoc. Petrol. Geol., Vol. 16 (October 1932), p. 1020.

78. H. W. Weddle, "Pleasant Valley Oil Field, Fresno County, California," Bull. Amer. Assoc. Petrol. Geol., Vol. 35 (March 1951), pp. 619–623.

79. Everett C. Edwards, *Kern Front Area of the Kern River Oil Field*, Bull. 118, Calif. Div. Mines (1943), pp. 571–574.

80. Glenn Grimes, "Tatums Pool, Carter County, Oklahoma," Bull. Amer. Assoc. Petrol. Geol., Vol. 19 (March 1935), pp. 401–411.

81. Gilman A. Hill, William A. Colburn, and Jack W. Knight, "Reducing Oil-Finding Costs by Use of Hydrodynamic Evaluation," in *Economics of Petroleum, Exploration,*

Development, and Property Evaluation, Southwestern Legal Foundation, Dallas, Texas, Prentice-Hall, Inc., Englewood Cliffs (1961).

82. B. A. Tkhostov, *Initial Rock Pressures in Oil and Gas Deposits,* translated by R. A. Ledward, The Macmillan Company, Pergamon Press, (1963), 118 pages. Illustrates many vertical pressure gradients in oil fields of the USSR.

CHAPTER 8

Reservoir Traps (continued) – Combination and Salt Domes

Combination traps. Salt domes: occurrence – Gulf Coast salt plugs – cap rock – associated traps – origin.

SOME TRAPS combine structural, stratigraphic, and fluid barriers in varying proportions. The reader might, in fact, classify as combination traps some of the traps that have been described earlier as either structural or stratigraphic traps, for the contribution of each element to the trap is sometimes a matter of personal judgment.

COMBINATION TRAPS

A combination trap generally has a two- or three-stage history: (1) a stratigraphic element caused the edge of permeability of the reservoir rock, (2) a structural element caused the deformation that combines with the stratigraphic element to complete the rock portion of the trap, and (3) a down-dip flow of formation water increased the trapping effect. No one element alone, therefore, forms the trap, but each may contribute a share of the trap and all may be essential to the size and position of an existing pool of petroleum. The rock elements of a combination trap are more readily determined; consequently, they are the two elements most commonly used to identify a combination trap. The fluid-potential-gradient–fluid-flow element, it should be remembered, is probably present as an important trap-making phenomenon in many pools, even though the data necessary to prove its presence may be lacking. Because two or three episodes were essential to the formation of a trap,

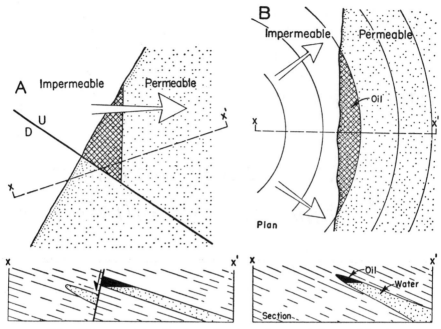

FIGURE 8-1 *Diagrammatic sketches showing characteristic combination traps: A, intersection of a fault with the up-dip edge of permeability; B, arching across an up-dip edge of permeability. Arrows show direction of fluid flow.*

there was no trap, and therefore no pool could have accumulated, until after the last episode had occurred. We are chiefly concerned in this chapter with the rock elements of combination traps, and two common types are shown in Figure 8-1.

A combination trap formed by the intersection of the up-dip edge of per-

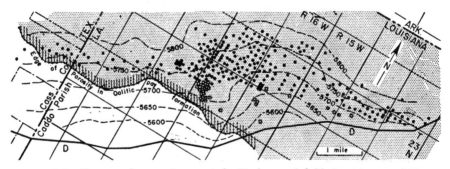

FIGURE 8-2 *Structural map of part of the Rodessa oil field, Louisiana and Texas, showing the pool in the oolitic limestone member of the Glen Rose formation (Lower Cretaceous). The trap is a combination of an up-dip pinch-out of permeability intersected by a normal fault. [Redrawn from Hill and Guthrie, U.S. Bur. Mines, RI 3715, p. 20, Fig. 25.]*

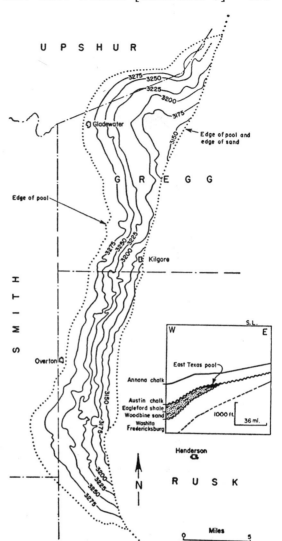

FIGURE 8-3

Structure on top of the Woodbine (Cretaceous) producing sand in the East Texas pool. The intersection of two unconformity planes along the east boundary marks the edge of the trap, as shown in the section. The pool is the largest in the Western Hemisphere, with a capacity of approximately 5 billion barrels of recoverable oil. [Redrawn from Minor and Hanna, in Stratigraphic Type Oil Fields (1941), p. 617, Fig. 6.]

meability with a fault may be seen in the Rodessa field of northeastern Texas, where the oolitic limestone member of the Glen Rose formation (Lower Cretaceous) is cut by the Rodessa fault and forms one of the pools in the field. A structural map of the formation is shown in Figure 8-2. The East Texas pool nearby,[1] the structure of which, with a section across it, is shown in Figure 8-3, is an outstanding combination trap. The intersection of two unconformity planes in the producing Woodbine sand (Upper Cretaceous) forms the wedge-edge of permeability, and subsequent folding of the large Sabine uplift, in northwestern Louisiana and northeastern Texas, arched the up-dip edge to form the trap in which the largest pool in the Western Hemi-

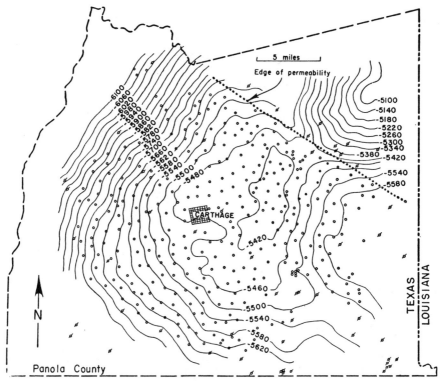

FIGURE 8-4 *Structure of the Pettet formation, an oolitic, fossiliferous, permeable limestone of the Glen Rose group (Lower Cretaceous) in the great Carthage gas field, Panola County, eastern Texas. The two Pettet pay zones extend over the field, but permeability grades out up the dip to the northeast, thereby forming a combination structural and stratigraphic trap.* [*Redrawn from Clark and Delong, University of Texas Pub. 5116 (1951), p. 56.*]

sphere accumulated. Another example near by is the great Carthage gas field, which occupies nearly all of Panola County, Texas. The producing Pettet oolitic and fossiliferous limestones (Lower Cretaceous) wedge out up the dip to the northeast by a loss of permeability; and across this wedge-out a low, broad arch developed, which combined with the wedge-out to form the trap. The field covers about 240,000 acres and is estimated to contain 5 trillion cubic feet of recoverable gas. A structural map of the pool is shown in Figure 8-4.

A structural map and a section across the Pleasant Valley pool, Fresno County, California, are shown in Figure 8-5. The up-dip clogging of permeability by weathered feldspars combines with an arching to form the trap. Farther west, along the same fold, the Gatchell sand again terminates, and the

East Coalinga pool is trapped in the combination of a fold with the up-dip edge of the sand. The structure of the Gatchell sand in the East Coalinga pool is shown in Figure 8-6 and a section across the pool in Figure 8-7. The Montebello field, in Los Angeles County, California, is a multiple sand field, in which a number of pools are found in combination traps composed of lenticular, wedged-out, faulted, and folded Tertiary sands. (See Fig. 8-8.) A field with a number of combination traps in folded buttress sands above an unconformity is the East Cat Canyon field, Santa Barbara County, California. (See Fig. 8-9.)

The Poza Rica pool in Mexico is another large combination trap. A plunging anticline crosses the up-dip wedge-out of permeability in the Tamabra limestone (Upper Cretaceous), which is due to a change from permeable dolomite to dense limestone, and the combination of the wedge-out of permeability and the folding forms the trap. A structural map of the pool, showing the position of the impermeable limestone barrier on the fold, is given in Figure 8-10. (See also Fig. 7-50.) The Mene Grande field, in western Venezuela (see Fig. 14-10 for location), consists of several pools trapped by complex combinations of folding, faulting, and up-dip edges of permeability, and especially by truncation and unconformable overlap. Some of the pools are in combi-

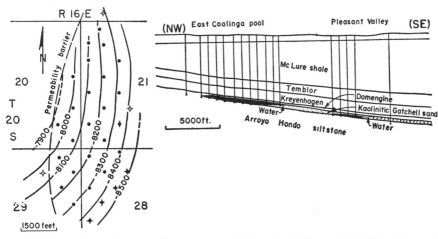

FIGURE 8-5 *Map showing the structure of the Gatchell sand (Middle Eocene) in the Pleasant Valley pool, Fresno County, California. The trap is formed by a combination of an arch and an up-dip loss of permeability due to clogging of the pores in the sand by kaolin. The section extends along the arch and shows the relations between the East Coalinga pool and the Pleasant Valley pool. Hydrodynamic conditions are also an important factor in the trapping of both of these oil pools; water flows down-dip. (See also Figs. 8-6 and 8-7.)* [Redrawn from Weddle, Bull. Amer. Assoc. Petrol. Geol., Vol. 35, pp. 620 and 622, Figs. 2 and 4.]

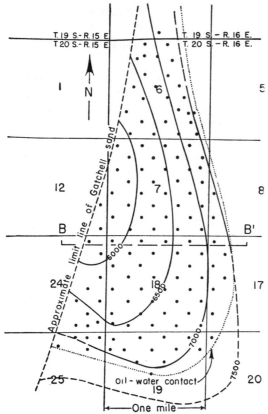

FIGURE 8-6

Structure of Gatchell sand (Miocene) in the East Coalinga pool, Fresno County, California. The up-dip wedge-out of permeability combined with the east-plunging arch forms the trap. Section BB' is in Figure 8-6. Note the tilted oil-water contact. Hydrodynamic conditions are an important factor in this trap; the direction of water flow is ENE. [Redrawn from Chambers, Bull. 118, Calif. Div. Mines (1943), p. 489, Fig. 207.]

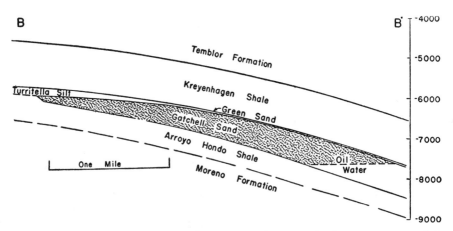

FIGURE 8-7 Section BB' across the East Coalinga pool, Fresno County, California. [Redrawn from Chambers, Bull. 118, Calif. Div. Mines (1943), p. 489, Fig. 207.]

TRAPS—COMBINATION AND SALT DOMES [CHAPTER 8]   355

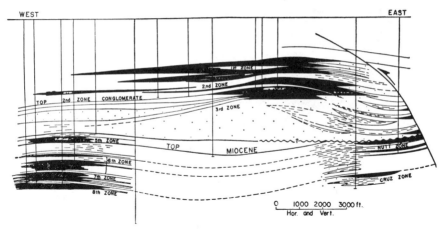

FIGURE 8-8 *Section through the Montebello field, Los Angeles County, California. This is one of many multiple-reservoir fields in California. The field is on a large dome fold, and thrust faulting along the east flank determines some of the traps. Several periods of earlier folding are suggested, which, when coupled with lenticular sands (Tertiary), formed a number of separate pools within the area of the field. [Redrawn from Atwill, Bull. Amer. Assoc. Petrol. Geol., Vol. 24 (1940), p. 1124, Fig. 8.]*

nation traps, notably the middle Pauji pool. A section through the field is shown in Figure 8-11. The oil ranges in gravity from 10° API (1.0) in the shallow Tar sands to 31° API in the deeper sands.

Formation pressure data are not available for many of the pools discovered in the past, and for that reason we are often unable to determine the local hydrodynamic conditions that existed in the reservoir rock prior to discovery of a pool, as described on pages 458–473. The presumption is, however, that

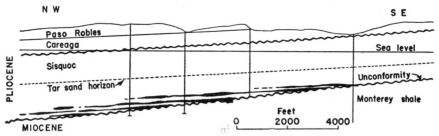

FIGURE 8-9 *Section across the East Cat Canyon oil field, Santa Barbara County, California, showing a whole series of pools trapped in the onlapping sands of the Sisquoc formation (Pliocene), where they intersect or buttress against the underlying unconformity. Such sands are also called "buttress sands." The section is along the axis of an anticlinal fold plunging toward the northwest. [Redrawn from Cross, Bull. 118, Calif. Div. Mines, p. 437, Fig. 180.]*

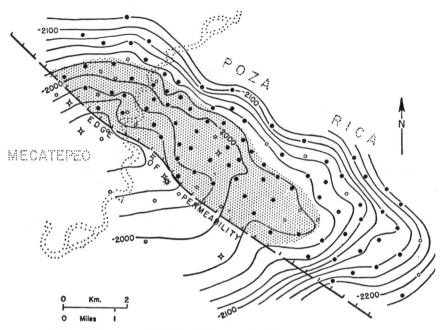

FIGURE 8-10 *Structure of the Poza Rica pool, Mexico, on top of the producing Tamabra limestone (Middle Cretaceous), showing a combination trap where a fold crosses the up-dip grading out of permeability. The area of the gas cap is shaded. The pool has produced over half a billion barrels of oil of 30° API gravity since its discovery in 1930.* [Redrawn from Salas, Bull. Amer. Assoc. Petrol Geol., Vol. 33 (1949), p. 1399, Fig. 9, and Bernetche, O. & G. Jour., April 21, 1949, p. 161.]

where there is a large pool, the hydrodynamic environment is favorable to petroleum entrapment and assists the geological trapping mechanisms of a barrier zone in barring the further migration of petroleum, thus causing a large accumulation. The rock barrier may result from a lessened permeability due to stratigraphic variation or it may be a structural feature, such as a fault, both of which would permit the flow of water in a hydrodynamic environment.

SALT DOMES

The intrusion of deep-seated rocks into the overlying sediments may form a great variety of traps, structural, stratigraphic, and combination. Some of these traps are associated with igneous rocks. The great majority, however, of the commercially important traps of this class are in sediments associated with the rock-salt intrusions of the Gulf Coastal region of the United States, northern Germany, the North Sea, and the Emba region of the USSR, so that the pools of this group are commonly called *salt-plug pools* or *salt-dome pools;* in fact, "salt-dome field" applies more often, as several pools around one plug are common.[2] Not all traps formed by intrusive rocks are productive; a great many unproductive traps have been formed in the sediments associated with the intrusion of salt as well as of igneous rocks. The general lack of productiveness of traps associated with igneous intrusions may be due to the high temperatures to which the rocks have been subjected, to the fact that intrusion generally occurred long after the sediments were deposited and any petroleum ever present had moved into other traps, to the absence of petroleum where the intrusion occurred, or only to a lack of drilling. Most of our knowledge of the principles and phenomena of traps caused by salt intrusion has been acquired through the intensive study that has been given by many geologists to the productive salt-dome structures on the Gulf Coast. For that reason special attention is given here to the geology of the Gulf Coast, where hundreds of salt domes occur both on the land side of the coast and far out beneath the waters of the Gulf. Offshore, or underwater, drilling was developed in the Gulf of Mexico off the shores of Louisiana and Texas and makes possible the drilling of domes under water up to 500–600 feet in depth. Wells may be drilled and completed in still greater water depths in the near future.

Salt may be said to have both an active and a passive role in its association

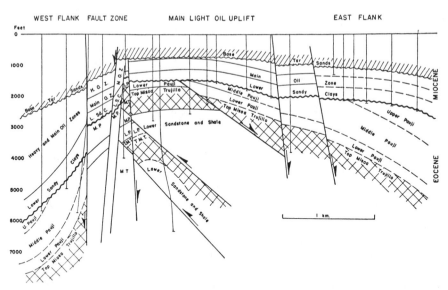

FIGURE 8-11 *Section across the Mene Grande oil field, western Venezuela. A number of separate structural and sedimentary episodes add up to the complex structure shown. Most of the formations are productive, and the field has produced about 400 million barrels of oil of 18° API gravity. [Redrawn from Staff of Caribbean Petroleum Co., Bull. Amer. Assoc. Petrol. Geol., Vol. 32 (1948), p. 574, Fig. 16.]*

with petroleum traps. The deformation of the enclosing sediments of the true salt plug, as found along the Gulf Coast of the United States, is caused by active intrusion of the rising salt mass and its uplifting of the adjacent strata. This active rising of salt should not be confused with the passive role played by salt in some of the "salt domes" of other areas, Romania,[3] Poland, and the Aquitaine plains of France,[4] for example, where salt and other evaporites are squeezed out and even extruded on the crests of anticlines or along large faults. In these cases the salt mass acts as any incompetent formation, rising as a result of the deformation of the enclosing rocks. The structures thus formed may best be called *salt anticlines*. A diapir fold with a central salt mass is sometimes hard to distinguish from a true salt dome. The source of the salt in the intrusive salt dome is deep-seated, whereas in the salt anticline the salt is interbedded with the folded sediments and is infolded with them. Its incompetence gives rise to the anomalous folding, although it is sometimes extruded from some salt plugs as erosion bares the fold. Where the climate is dry, the salt and evaporites may flow out on the surface and form a mass resembling a glacier,[5] but where the climate is wet, the salt dissolves away as fast as it reaches the surface.

Occurrence

Salt domes and stocks occur near the center of the Zechstein (Permian) area in the Northwest, or Hannover, basin of Germany, and oil has been found in traps associated with a number of these intrusions. The producing formations range from Rhaetic (Triassic) through Lower and Upper Cretaceous, and the traps, like those in the Gulf Coast region, consist of complex combinations of faulting, truncation, wedge-out, overlap, and salt overhangs.[6] In passing from the center of the Zechstein salt and gypsum basin to the edges, the salt becomes thinner and is more and more interbedded with shales, limestones, and marls. The salt and gypsum there act as incompetent formations and are folded along with the competent rocks. Another salt-dome region is the Emba district, north of the Caspian Sea, in the USSR. Approximately 300 salt domes and salt structures have been defined, and there may be a total of 1,200 when all have been mapped.[7] The salt occurs over an area of 500 square kilometers and is thought to be Permian. Oil production is chiefly from Jurassic sands flanking the domes, in the cap rock, and in faulted salt structures. (See Fig. 8-22.) Six small pools have been found, and the area has been known to contain oil since 1857. The sedimentary section is thin here, in contrast to the thick section in the Gulf Coast, which possibly explains why only a few pools have been found. The Paradox basin, in eastern Utah and western Colorado, contains a number of salt-dome structures, associated with the gypsum, anhydrite, and salt deposits of the Paradox formation (Pennsylvanian).[8] So far these salt domes have proved unproductive, though deeper petroleum accumulations have been found in

Mississippian and Devonian beds on folds that are discordant to the Paradox salt anticlines. The interior salt-dome structures of northern Texas, Louisiana, and Mississippi (not to be confused with the Gulf Coast region) are spectacular salt intrusions into otherwise low-dipping sediments, which include Cretaceous rocks overlain by Tertiary rocks at the surface.[9] Only a few of these salt domes have been found to develop traps that contain petroleum. In southern Iran[10] more than 100 salt plugs have been mapped, all of them unproductive. The material brought up by the salt includes fragments of Precambrian igneous rocks as well as fragments of sediments of all ages from Cambrian to late Tertiary. Unproductive salt plugs have been observed in other regions, such as the Red Sea[11] and the Dead Sea, Arabia,[12] North Africa,[13] and in the Aquitaine plain, which lies north of the Pyrenees, in southeastern France.[14] Some of the unproductive salt-plug intrusions may become productive when more wells are drilled and more is learned about the geologic phenomena that accompany their development.

Gulf Coast Salt Plugs

The massive salt plugs developed along the Gulf Coast, in Texas, Louisiana, and Mississippi, have moved up through the overlying sediments for distances of thousands of feet. They are, in fact, still moving, as is shown by their surface expression as either mounds or depressions in the topography of the Recent and late Tertiary surface of the coastal plain. The source of the salt is believed to be in southward extensions of the Jurassic and Permian salt formations that are found by drilling farther up the regional dip, in southern Arkansas and northeastern Texas. A large number of traps have been formed in the sediments alongside these salt plugs and also in the overlying faulted and domed sediments, and a great many of these traps have been found to contain petroleum. A section in southern Louisiana across the Mississippi delta region, shown in Figure 8-12, indicates the general nature of the formations encountered.

The tops of the salt plugs near the Gulf Coast lie at widely varying depths. Some of the plugs have completely pierced the overlying sediments, so that the salt has come to the surface and washed away; others are deeply buried, and the sediments overlying them are merely domed and faulted. Plugs may therefore be roughly classified according to the depth from the surface to the top of the salt as follows:

1. From the surface to 2,000 feet: piercement domes.
2. From 2,000 feet to 6,000 feet: intermediate domes.
3. Below 6,000 feet: deep domes.

They may also be classified according to age:

1. Young: Characterized by anticlines and domes underlain by salt cores. Relatively little deformation.

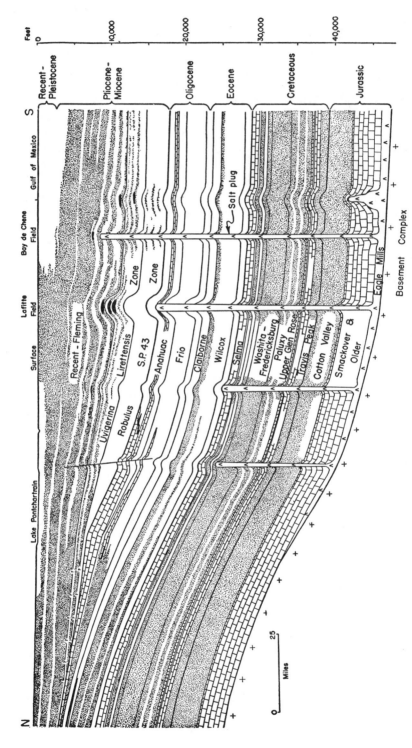

FIGURE 8-12 Section across the delta of the Mississippi River, showing the nature of the formations associated with some of the salt domes of the area. [Redrawn from Carsey, *Bull. Amer. Assoc. Petrol. Geol.*, Vol. 34 (1950), p. 362, Fig. 2.]

2. Mature: Salt cores become vertical-sided stocks, on which cap rock has begun to accumulate. (See also pp. 367–374.)

3. Old: A thick cap rock has formed, commonly with an overhang, and there is much solution breccia along the sides. The adjacent sediments are much faulted and fractured, and a well-defined rim syncline encircles the dome. The common shapes of the Gulf Coast salt domes are shown in the sections in Figure 8-13.

The surface expressions of piercement and intermediate salt domes on the Gulf Coast are (1) a topographic mound; (2) a surface depression, lake, or marsh; (3) a surface mound with a central depression; (4) sulfur or sour water springs; (5) surface evidences of petroleum, such as gas seepages, oil seepages, and "paraffin" dirt; (6) gas, oil, and increased salt content in water wells.

The Gulf Coast salt plugs consist of the mineral halite (NaCl) and variable amounts of insoluble matter ranging from 5 to 10 percent by weight.[15] Anhydrite ($CaSO_4$) is the chief insoluble mineral; those present in minor quantities are dolomite, calcite, pyrite, quartz, limonite, sulfur, and hauerite; and several other minerals occur sporadically. Brine and inclusions of oil and gas are also found in the salt plugs, and potash salts have been found in about half of the plugs examined. Fragments of sandstone are found occasionally, one of the larger pieces being a thin vertical slab in the salt mine on Avery Island, Louisiana. This is only 10 inches or less in thickness, but it extends over 80 feet vertically, and along the strike for 75 feet horizontally.[16] Petrographic studies of sandstone inclusions from the occurrences on Avery Island and Jefferson Island, Louisiana, show them to be different from the known Tertiary sands of the Gulf Coast;[17] so they are probably from Cretaceous or older formations. The salt core is sheathed in one or more layers of shale, gouge, or anhydrite, depending on the number of faults parallel to its surface. The relations of the sheaths and of the component parts of a typical Gulf Coast salt dome are shown in Figure 8-14. A detailed section across the side of the Broistedt salt dome, eleven miles southwest of Braunschweig, Germany, shows, in Figure 8-15, some of the sheathing phenomena that frequently occur along the sides of the salt plugs.

The internal structure of salt plugs has been studied and mapped by Balk[18] in a number of salt mines. He found the salt to have a layered structure, the layers consisting of gray and white salt. The white salt was almost pure halite, while the gray salt contained varying amounts of anhydrite and other insoluble materials. The white layers ranged from fractions of an inch up to several feet in thickness and greatly predominated in amount. The layering of the mass gave an opportunity to study the structure, which, exposed in all planes on the walls and roofs of the mine rooms, showed the extremely plastic nature of salt when deformed. The beds stand vertical around the edges of the

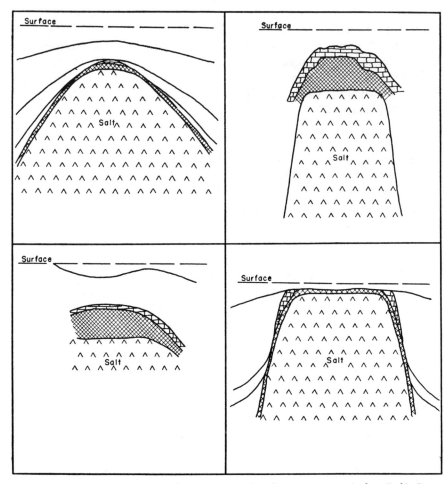

FIGURE 8-13 *Eight sections showing typical salt structures of the Gulf Coast province.* [*Redrawn from Hanna, in* Problems of Petroleum Geology *(1934), pp. 640–641, Figs. 5 and 6.*]

salt mass, but in the interior of the dome they are thrown into a remarkable array of folds, both large and small, both open and isoclinal.

Formations above the salt plugs, with the exception of the piercement plugs, are cut by numerous normal faults, often radiating from the center of the dome, which cause vertical displacements of as much as 1,000 feet. (Examples may be seen in Figs. 8-17, 8-18, 8-19, and 8-26.) Most of the faults dip from 45° to 65° and separate the reservoirs into many complex fault blocks. One or more central grabens, or down-faulted blocks, are especially characteristic of the formations overlying salt plugs.[19] A hypothesis for the sequence of their growth and development by tensional forms created by the intrusion of a salt plug is shown in Figure 8-16. A number of domes along the Gulf

TRAPS—COMBINATION AND SALT DOMES [CHAPTER 8] 363

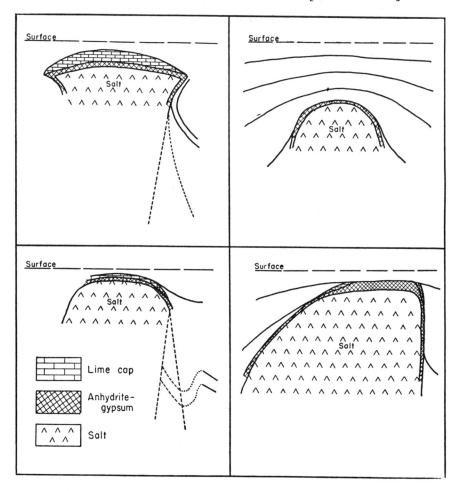

Coast of Texas and Louisiana are thought to be underlain by deep-seated salt plugs, although no salt has yet been found in them. Radiating and graben faulting in these domes is generally considered to indicate that they are underlain by deep-seated salt plugs that have lifted the overlying sediments but for some reason have not yet pierced them.

The Van field, in northeastern Texas,[20] the structure of which is shown in Figure 6-36, page 268, and the Hastings field, near Houston, Texas, shown in Figure 8-17, are over domes that are thought, because of their circular shape and graben-fault pattern, to have been caused by deep-seated salt intrusions, though the salt has not yet been reached. The Eola field, in Louisiana,[21] is presumed to be over another deep-seated salt plug overlain by graben-faulted sediments, in which several pools have accumulated. The structure of the producing sand (Cockfield, Eocene) is shown in Figure 8-18. The complexity of the faulting that may occur in the beds overlying a salt intrusive

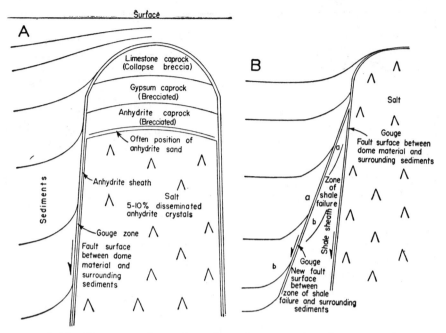

FIGURE 8-14 *Diagrammatic sections through a typical Gulf Coast salt dome, showing* (A) *the component parts and* (B) *the zone of shale and gouge sheathing.* [*Redrawn from Hanna, Bull. Amer. Assoc. Petrol. Geol., Vol. 37 (1953), p. 280, Fig. 13.*]

is shown in the section through the Reitbrook salt dome of Germany, Figure 8-19.

At many Gulf Coast salt domes the surrounding sediments thin toward the salt plug. This may be due to a decrease in the amount of material deposited as the dome neared the surface, to erosion during the growth of the dome, or to squeezing of the softer formations by the intruding salt mass. As some beds or formations are thinned more than others, it appears likely that the salt plug moved upward intermittently, and that the rate and the timing of upward movement were determined in part by the relations between the movement and the sedimentation. If, for example, the top of the salt plug reached the surface, the water at the surface would be expected to dissolve the salt away as fast as it moved upward. As the salt moved upward and maintained its position at the surface, the beds next to the salt would probably be eroded as fast as the salt dragged them up to the surface. If, on the other hand, the salt started its upward movement when overlain by a great thickness of sediments, it might pierce a part of the sediments and dome the remainder. Erosion might then occur and initiate a new cycle of uplift. At the present time we see the tops of the salt domes ranging in position from the surface to depths below any yet reached by the drill, and that situation has probably existed since the beginning of the Tertiary. The stratigraphic thinning toward the Port Neches

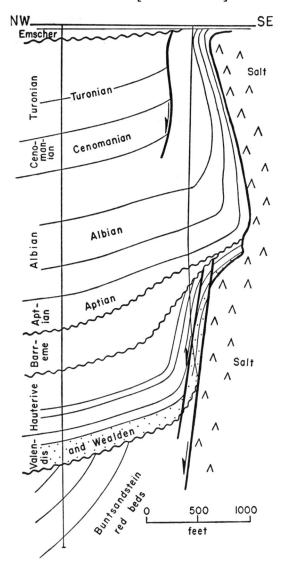

FIGURE 8-15

Section across the edge of the Broistedt salt plug, eleven miles southwest of Braunschweig, northern Germany. The section shows some of the complicated combinations of squeezing, sheathing, faulting, truncation, and overlap that form traps around salt plugs. [Redrawn from British Intelligence Objectives Subcommittee, Report 1014, opp. p. 11.]

dome, in Orange County, Texas, is shown in the section of Figure 8-20. The nature of the thinning over a group of productive salt domes in southeastern Texas is shown in Figure 8-21, where the thickness of the Frio formation (Oligocene) is shown by contours (isopach map; see pp. 596–600). The thinning of the Frio formation over the salt plugs is interpreted to mean that the domes were growing during or shortly after Frio time. An example of the complete loss of formation over a salt dome in another salt-dome province is the Iskine salt dome, in the Emba salt-dome region of the USSR. It is shown in the geologic map and section of Figure 8-22.

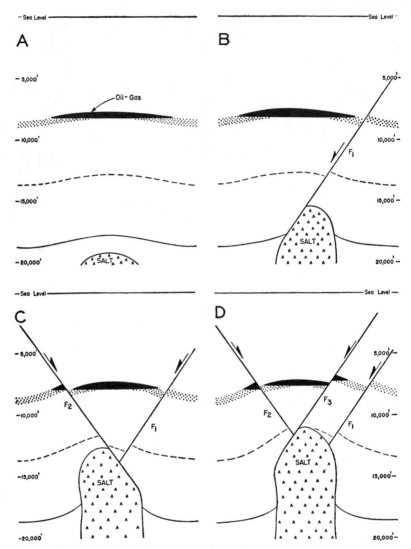

FIGURE 8-16 Sections illustrating a hypothesis for the progressive development of the graben-faulting pattern over deep-seated salt plugs. The horizontal and vertical scales are the same in each figure. A shows the first gentle doming of the overlying sediments as the salt plug begins to rise. Oil and gas (black) accumulate in the trap being formed. Normal fault F_1 forms in B as the intrusion of the plug into the overlying sediments continues. Later, in C, normal fault F_2 forms, and the salt mass continues to rise. Still later, normal fault F_3 forms, leaving a typical but simple fault-graben pattern over the salt plug. Figures 8-17 and 8-18 show two of the more complex graben systems that develop over deeply buried salt plugs. [Redrawn from Wallace, Bull. Amer. Assoc. Petrol. Geol., Vol. 28 (1944), p. 1306, Fig. 37.]

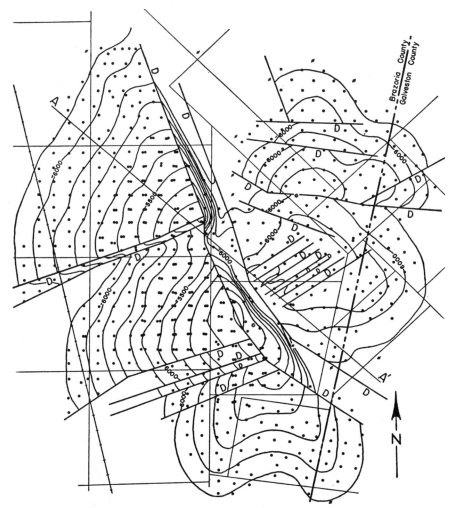

FIGURE 8-17 *Structure of the Frio producing sand (Oligocene) of the Hastings field, south of Houston, Texas. The intricate normal fault and graben pattern of the structure suggests a deeply buried salt core as the cause of the deformation. Contour interval is 100 feet. An ultimate production of one-half billion barrels is estimated for this field. See Figure 12-24 for section AA′. [Redrawn from Halbouty, Houston Geol. Soc., Guidebook (1953).]*

Cap Rock

Nearly all the salt domes of the Gulf Coast states and some in Germany have a disk-like plate, called the *cap rock,* composed of anhydrite, gypsum, limestone, dolomitic limestone, and occasionally sulfur, over all or part of their top surface.[22] Anhydrite forms the major part of all cap rock, and it

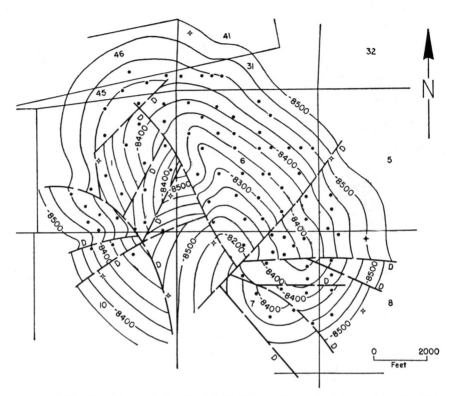

FIGURE 8-18 *Structure of the Cockfield (Eocene) producing sand in the Eola field, Avoyelles Parish, Louisiana. The principal production is from the lower Eocene Wilcox formation, about 2,000 feet deeper. The complex fault and graben system that coincides with the roughly circular fold is common to domes overlying deeply buried salt plugs.* [Redrawn from Bates, Bull. Amer. Assoc. Petrol. Geol., Vol. 25 (1941), p. 1385, Fig. 10.]

lies directly on the massive salt plug. Next above the anhydrite comes a zone of mixed gypsum and anhydrite. Where calcite is present, it lies on top. A transition zone of calcite, gypsum, and anhydrite generally occurs between the zone of gypsum and anhydrite and the carbonate zone. The average cap rock is between three and four hundred feet in thickness, but cap rocks up to one thousand feet thick are known. A "false cap" occurs above the true cap rock in some domes in which the gumbo (driller's term for sticky mud) and sands have been more or less calcified.

It was first thought that a cap rock was a limestone and anhydrite formation that overlaid the salt formation, and that, when the salt plug rose, it pushed a disk of the overlying formation ahead of it. The absence of sedimentary deposition in the cap rock, and the sharp truncation of the salt folds at the contact with the cap rock, do not fit such an explanation. It is now believed that the cap rock consists of the insoluble residual anhydrite and carbonate constituents of the salt that became concentrated at the top of the rising salt plug when the salt dissolved. A wide, toadstool-like cap rock, as in the Wienhausen-Eicklingen salt dome of Germany (see Fig. 8-23), obviously could not have been pushed up through the narrow base without being completely broken up. This example combines many salt-dome phenomena: cap rock, overhang, sheathing, faulting, and squeezing out of pierced formations.

Deep-seated rocks, however, are found in some salt plugs. For example, a mass of chalk twenty-five feet thick, containing an abundance of Upper Cretaceous foraminifera, was found at a depth of 5,300 feet in the limestone cap rock of the South Liberty salt dome, in Jefferson County, Texas.[24] It apparently became detached from its original position by the upward movement of the salt mass, and indicates that the deep-seated body of salt from which the salt plug rose is at least as old as Upper Cretaceous. In some domes the limestone from the cap rock has apparently been eroded and redeposited around the periphery of the salt during the time represented by some of the uncon-

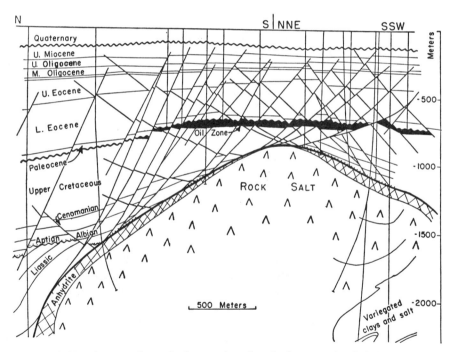

FIGURE 8-19 *Section through the Reitbrook salt dome and oil field, Germany, showing the characteristic faulting that frequently accompanies salt intrusion. The chief oil reservoir is in fractures and fissures in the Upper Cretaceous limestones. The chief gas reservoirs are in the Lower Eocene and Oligocene sands.* [*Redrawn from Behrmann, in British Intelligence Objectives Subcommittee,* Oil Fields Investigation, *Part III, Fig. 95.*]

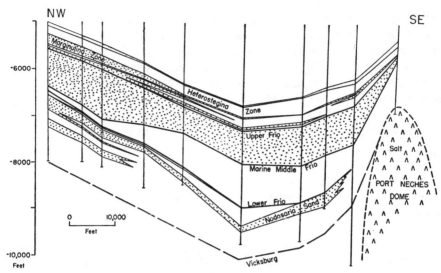

FIGURE 8-20 *Structural section across the Port Neches salt dome, Orange County, Texas, showing the thinning and wedging out of Oligocene sediments close to the dome, a feature common to many domes. See Figure 8-21 also. [Redrawn from Reedy, Bull. Amer. Assoc. Petrol. Geol., Vol. 33 (1949), p. 1851, Fig. 11.]*

formities,[25] for detrital deposits of limestone fragments have been found around the flanks of several of the domes on which the limestone cap rock is missing.

The cap rock of some salt domes, and sometimes the salt also, is found to *overhang,* or drape down, the side of the main mass of salt.[26] It is commonly found to overhang only one or two sides, but in a few domes, such as the Barbers Hill dome,[27] it overhangs all sides. The significance of the overhanging salt or cap rock is that many pools are found in traps formed by sands buttressed against the salt plugs below the overhang and are reached only by drilling either through the overhang or around it. The cause of this overhanging has been variously ascribed to: (1) a tipping of the axis of the salt dome during its growth; (2) the static weight of the cap; (3) a shearing off of slabs of cap rock from the top of the plug, which slid over the side as the plug rose; (4) solution of salt by circulating waters, which would tend to produce an overhang wherever the upper part of the salt is protected by a cap rock and the salt below is exposed to circulating waters. The last-named cause is generally thought to be the one most commonly operative. Details of cap-rock overhang in several domes are shown in Figure 8-24.

Free sulfur is present in the cap rock of nearly every dome, but it occurs commercially only in domes having a thick calcite cap. Considerable sulfur extends into the gypsum-calcite transition zone, but sulfur occurs only in minute quantities in the anhydrite zone below. Free sulfur is rarely found in massive salt. The free sulfur is secondary and later than the limestone, probably derived from the destruction of the anhydrite ($CaSO_4$).[23] A section through the Hoskins Mound, in Figure 8-25, shows a typical sulfur-bearing cap rock. The carbonate minerals strontianite ($SrCO_3$) and aragonite ($CaCO_3$) are found in the calcite zone, and dolomite is found chiefly in the anhydrite zone. The sulfates barite ($BaSO_4$) and celestite ($SrSO_4$) are found in the calcite zone. Pyrite is found in all zones. Galena (PbS), sphalerite (ZnS), and hauerite (MnS_2) are found in the calcite zone. Other minerals that have been observed in minute quantities in the cap rocks are siderite ($FeCO_3$), realgar (AsS_2), hematite (Fe_2O_3), smithsonite ($ZnCO_3$), and oolitic barite ($BaSO_4$).

Traps Associated with Salt Domes

When salt plugs have risen through soft and incompetent Tertiary shales and sands, as they have along the Gulf Coast, they have affected the stratig-

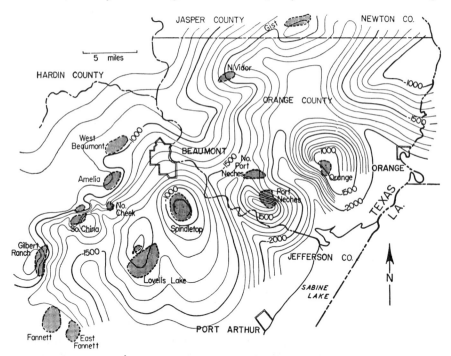

FIGURE 8-21 *Map showing the thickness of the upper and middle Frio formation (Oligocene) in an area of southeastern Texas in which a number of productive salt domes (shaded) occur. Such a map is called an isopach map. (See Chap. 13.) The thinning of the formation as many of the salt domes are approached is noticeable, and suggests their growth during or immediately after the Frio period, and later truncation of the Frio over the rising salt plugs. [Redrawn from Reedy, Bull. Amer. Assoc. Petrol. Geol., Vol. 33 (1940), p. 1846, Fig. 8.]*

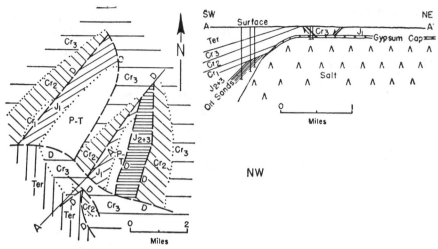

FIGURE 8-22 *Map of formations overlying the Iskine salt mass in the Emba salt-dome region of the USSR, north of the Caspian Sea. The Iskine oil field occurs in Jurassic sands on the southwest flank of the salt mass. Section AA', across the southwest flank of the dome, shows the production to be in steeply dipping, truncated Jurassic sands overlain unconformably by Lower Cretaceous beds.* Ter, *Tertiary;* Cr_3, *Upper Cretaceous;* Cr_2, *Middle Cretaceous;* Cr_1, *Lower Cretaceous;* J_{2+3}, *Upper and Middle Jurassic;* J_1, *Lower Jurassic;* P-T, *Permo-Triassic.* [*Redrawn from Krivolai, Petroleum Industry (1936, Moscow), and from Sanders, Bull. Amer. Assoc. Petrol. Geol., Vol. 23 (1939), p. 504, Fig. 4.*]

raphy and structure of the adjacent sediments in many ways that have enabled them to trap petroleum in pools. A rising plug punctures the reservoir formations, as a poker might pass through a stack of newspapers; and the ragged, upturned edges of the beds, pressed tightly against the salt, form numerous traps around its sides. Much fracturing, together with radial and rim faulting, accompanies the upward movement of the salt plug and cuts the flanking reservoir rocks, which are chiefly sands, into many triangular blocks, some of which form traps and contain petroleum. In some domes the rise of the salt plug was intermittent, and the upper surface of the plug was at or close to the surface of the ground at several times during the Tertiary period. Where this happened, arching of the shallow formations, followed by erosion and truncation and later by overlap of the succeeding formations before the next upward movement of the salt, left many flanking truncated wedges of sand, many of which are potential traps at some distance from the salt core. Figure 8-26, an idealized section through a salt plug, illustrates the common types of traps associated with salt plugs. Pools in one or more or even all of these positions may occur around a single salt plug; in fact, salt-dome fields with several pools are commoner than single pools associated with salt plugs.

The combination of faulting, squeezing out of sands by the salt, and trunca-

TRAPS—COMBINATION AND SALT DOMES [CHAPTER 8]   373

tion of sands makes the area around the flanks of the average Gulf Coast salt dome extremely complex structurally. A salt-dome field may contain five, ten, or more separate pools. Our present understanding of the complexity of the geology surrounding the salt domes, which is in striking contrast to the simple geology of the smooth Gulf Coast plain, is largely the result of the detailed evidence furnished by electric logs, coupled with careful well sampling, detailed paleontologic and mineral analysis, and some of the most precise and advanced geophysical work, including seismic, gravity, and magnetic interpretations, that has been done anywhere in the world. A peripheral section showing complex faulting around the southern flank of the Jennings salt dome of Louisiana, and typical of many salt domes, is presented in Figure 8-27.

A troublesome problem in exploring the traps around many Gulf Coast salt domes is that of *heaving shale*[28]—shale that is under such high pressure that it squeezes the sides of the hole together, or sloughs off into the hole until it is lost. Much of the trouble has been eliminated by the use of modern drilling muds and by drilling under high pressures. The heaving is thought to be due in part to the primary characteristics of the formations drilled, such as water adsorption and base-exchange phenomena in the clay minerals, for it is confined to five separate

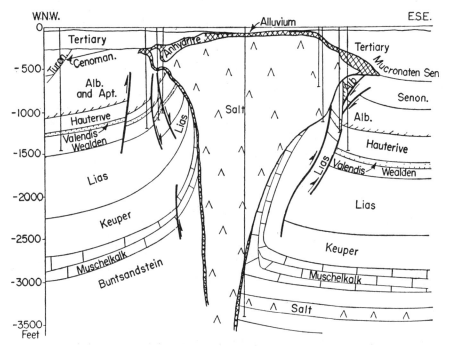

FIGURE 8-23 *Section through the Wienhausen-Eicklingen salt stock, seven miles southeast of Celle, northern Germany. Note the wide overhang on both sides, the sheathing, the faulting, and the squeezing out of formations.* [*Redrawn from A. Bentz (1949).*]

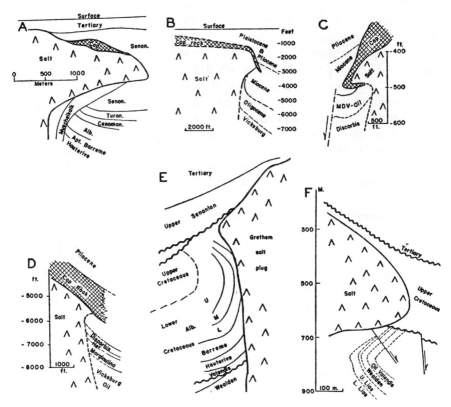

FIGURE 8-24 *Sections through salt domes showing the nature of some overhang phenomena and complex conditions that frequently characterize the edge of an intrusive salt plug:* A, *Wienhausen-Eicklingen salt stock, Germany;* B, *Barbers Hill, Texas (AAPG, 16:476);* C, *High Island, Louisiana (GC, AAPG, p. 933);* D, *Jennings, Louisiana (GC, AAPG, p. 973);* E, *Gretham-Hademstorf, Germany (BIOS, 1014, Fig. 49);* F, *Hanigsen, Germany (BIOS, 1014, Fig. 54).*

stratigraphic zones that extend parallel to the shore line and along the ancient strand lines.[29] It is partly caused by the excessively high pressures that develop where fault blocks containing these formations are squeezed and sealed tight around the intruding salt mass. The lower pressure within the drill hole permits the surrounding shale, with its content of high-pressure gas, to squeeze into the hole. (See also pp. 405–409.)

Origin

The origin of the Gulf Coast salt domes,[30] and probably of most other salt domes as well, has been ascribed to a number of causes, some of which have been discredited. Probably because the rock formations surrounding the salt domes were turned up, one of the early theories, called the "volcanic theory,"

TRAPS—COMBINATION AND SALT DOMES [CHAPTER 8]   375

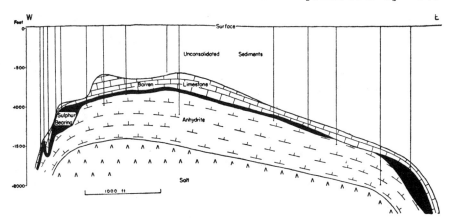

FIGURE 8-25 *Section through the Hoskins Mound salt dome, Brazoria County, Texas, showing the sulfur-bearing limestone (black) in the cap rock of the dome.* [Redrawn from Marx, in Gulf Coast Oil Fields, *p. 840, Fig. 3.*]

postulated an igneous intrusion from below. The deep-seated intrusive body was supposed to have given off volcanic gases, which caused salt to precipitate from solutions; the salt then pushed the strata upward and formed the dome. Since there is no evidence of volcanic activity along the Gulf Coast, this theory has been abandoned.

Later, several investigators thought the intrusion of the salt resulted from precipitation of salt dissolved in circulating waters.[31] It was supposed that waters entered along the outcrops far to the north of the Gulf Coast, and that in percolating southward down the dip they became heated, because of an increase in temperature with depth (see pp. 414–415), and picked up large

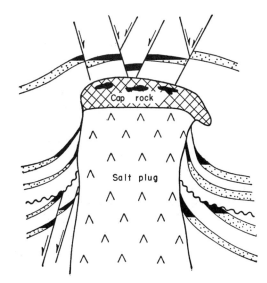

FIGURE 8-26

Idealized section through a Gulf Coast salt-dome field, showing some of the common types of pools (black) found in traps associated with the salt intrusion. Many of the flanking traps are wedge-shaped in plan.

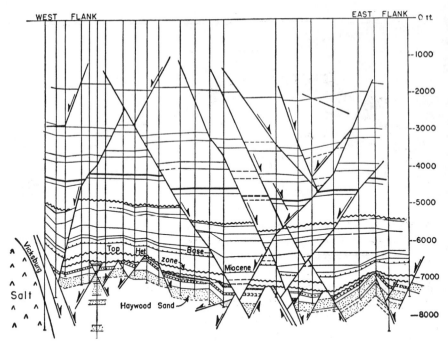

FIGURE 8-27 *Peripheral section around the south flank of the salt dome at Jennings, Louisiana. This shows the complex system of faults, the erosional truncations, the wedging out of reservoir rocks, and the many kinds of traps characteristic of the flanks of Gulf Coast salt domes. The basis for these intricate structural sections is the electric well log, coupled with careful micropaleontologic and lithologic analysis of well cuttings.* [*Redrawn from Roach, Bull. Amer. Assoc. Petrol. Geol., Vol. 27, p. 1106, Fig. 3.*]

quantities of salt in solution. Whenever these salt-saturated solutions crossed lines of weakness, such as intersecting faults, they would tend to rise; as they rose, they would cool at the average rate of approximately 1°C for each 100 feet of rise, and therefore precipitate part of the dissolved salt. The salt crystallized as it was precipitated, and the crystals continued to grow. The force exerted by the growing crystals was supposed to have supplied the power that lifted the dome, and it was calculated to be sufficient to lift more than three thousand feet of overburden. But this theory is descredited by the abundant evidence of flow structure within the salt plugs, which indicates that the salt moved *en masse*.

It is now commonly believed that the rise of the salt plugs along the Gulf Coast of the United States, and probably elsewhere as well, is best explained by the plastic-flow theory. The classic experiments of Nettleton[32] did much to win adherents to this view among geologists, for he made comparable intrusions at will in the laboratory.* The plastic-flow theory is based on the idea that both sediments and salt behave as highly viscous liquids or as plastic substances capable of flowing. Salt, under standard conditions, has a specific gravity of 2.2, which is not materially increased at higher load pressures caused by deep burial. Normal, shallow, Gulf Coast sediments, on the other hand, are lighter than salt at surface load pressures, but their bulk specific gravity increases with load pressure, owing to the reduction of pore volume; their bulk density equals that of salt at a depth of 2,000 feet, and equals or exceeds a specific gravity of 2.4 at a depth of about 10,000 feet or more. The difference in specific gravity of salt and sediments with depth is shown in Figure 8-28, and this difference is also the basis for prospecting for salt plugs by the use of gravity surveys. Certain combinations of critical conditions, if they affected the salt for a long time, would tend to make it more plastic and eventually cause it to pass its critical stage and flow into areas of lower pressure. They depend upon a complex set of highly variable factors: (1) the composition, character, thickness, and stratigraphic relation of the parent salt formation; (2) the temperature of the salt formation, which increases on an average of 1°C per 100 feet of depth; (3) the pressure on the salt formation, which increases about 1 lb per sq. in. per foot of depth; (4) the water content of the salt formation and the adjacent rocks, which would greatly influence the critical condition required for plastic flow, salt being highly soluble in water, and dry salt being more likely to be plastic than wet salt. Van Tuyl [33] calculated

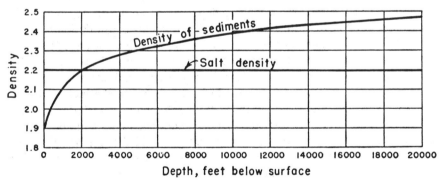

FIGURE 8-28 *The difference in specific gravity with depth between salt and typical sediments of the Gulf Coast.* [Redrawn from Nettleton, in Gulf Coast Salt Domes *(1936), p. 83, Fig. 1.*]

---

* Nettleton's experiments consisted in placing a layer of heavy asphaltic oil, with a Spec. Gr. of about 1.0, over a layer of heavy syrup, with a Spec. Gr. of about 1.4, in a glass jar and covering it with a sheet of rubber. To simulate salt, the jar is turned upside down and placed on a surface that has irregular projections which distort the rubber and "trigger" the upward movement of the lighter oil through the syrup. By photographs taken at regular time intervals, the progress of the vertically moving oil through the syrup could be shown, and the resulting forms look very much like salt plugs.

that, with an average depth of 12,000 feet, the resistance of the salt to plastic flow is finally overcome. The upper surface of the salt formation is probably irregular because of depositional and structural factors, and the first evidence of couple forces would be folds in the relatively incompetent salt. Once a portion of the salt is elevated above its surroundings, a differential lateral pressure develops because the sediments over the slightly thicker area surrounding the upward protruding salt are heavier than those over the area of the protuberance. The combination of the lateral pressure and the increased temperature with the vertical component of the irregularity in the upper salt surface eventually "triggers" the salt into moving up as a plastic mass. As the salt plug moves upward, the lateral pressure increases, and salt moves in from the surrounding region to replace the salt that moved out. The plug rises until stopped by a lack of salt at the source or by some more rigid formation.

Many of the Gulf Coast domes are separated from the surrounding regional structure by rings of depressed sediments,[34] which have been called *rim synclines*. It is believed that these synclines are due to the thinning of the underlying parent salt formation as the salt drains away from it and into the salt intrusion. Under this theory, where domes are closely spaced, the midpoint between any two domes would thus be underlain by the thickest salt, and the overlying structure would also be anticlinal. Anticlines formed in this way have been called *residual anticlines* or *residual domes*.

In summary, the history of the Gulf Coast salt domes appears to have been as follows: (1) the salt was originally deposited as an evaporite, in Jurassic or Permian time, or in both; (2) horizontal movements began during the time of deposition of Tertiary formations; (3) the upper surface of the salt formation was irregular because the upper contact was uneven and because of folding; (4) slight initial projections of low-density salt upward into higher-density sediments caused a differential force, which, together with the increased temperature, overcame the strength of the salt; the salt then began to rise vertically as a plastic mass; (5) once started, movement continued until equilibrium was established, after which either no more salt was added from below or the strength of the overburden formations became great enough to stop further upward movement; (6) the salt moved upward again as additional sediments added to the weight of the overburden and again increased the load pressure on the parent salt mass; many salt plugs moved repeatedly during Tertiary time.

Salt domes only account for the traps; they have nothing to do with the origin of the oil. The characteristics of petroleum accumulation and production associated with salt domes are as follows:

1. Accumulation has occurred in a wide variety of traps, such as:
   a. porous cap rocks, which often yield large gushers;
   b. folded overlying sands;

TRAPS—COMBINATION AND SALT DOMES [CHAPTER 8] 379

    c. flank sands: (1) truncated by the salt and in that case often of steep dip, or (2) pinched out by uplift, or (3) confined to fault blocks.
2. Production is often extremely rich over small areas owing to the steep dips and the great height of the oil column.
3. The small areas of many of the pools require accurate aiming of the wells to hit the target, at times through or around a salt or cap-rock overhang.
4. Pressures are often abnormally high in proportion to the depth.
5. The heaving of shales, due to swelling of bentonitic clays, gas in the shales, abnormally high pressures, and steep dips, often makes drilling expensive.
6. The oils are generally heavy (low API gravity), especially in the shallower pay formations.

CONCLUSION

The most important thing to remember, probably, from the three chapters on traps is the almost infinite variety of geologic conditions that can go together to form a trap. The serious student of petroleum exploration should examine every trap he can for the critical elements that are essential if a trap is to be present. He should ask himself, "What geologic conditions had to develop for this trap to form?" Hundreds of traps of all kinds have been described in the geologic literature and they may be studied in structural maps and cross sections. Each arose from a unique combination of conditions. Experience in analyzing the traps that have been found and described gives direction and confidence to one's analysis of the prospective trap, upon which money, time, and sweat must be expended if it is found to be worth testing.

    The question, "How does one look for oil and gas?" may be answered simply yet accurately: "Look for a trap and then drill into it." A person able to find traps *is,* in fact, a successful petroleum geologist; he really needs nothing more; for if he drills enough wildcat wells into traps in a petroliferous region, he is assured of discovering oil and gas. The difficulty is that it is not easy to find traps, and the difficulty increases as exploration gets deeper and the complexity of the geology increases. The obvious traps are drilled early, and only the obscure traps remain: traps for which the data are indefinite, the drilling is deep, the points of control far apart, the geologic history uncertain, or the fluid conditions unknown.

    The search for petroleum in a producing region generally follows a rough chronological order: (1) drilling structural traps, (2) drilling combination traps, and (3) drilling stratigraphic traps. Structural traps, being the more readily located, are drilled first; the information gained from drilling them is used to locate stratigraphic and fluid-flow combinations with structures where pools may be trapped; finally, when the remaining structures are small and obscure, the emphasis is on finding sand patches, reefs, shoestring sands, and other typically stratigraphic phenomena. The stages grade into one another.

Much of the Middle East is still in stage 1, Venezuela may be said to be largely in stage 2, and large areas in the Mid-Continent region of the United States are in stage 3.

Up to now this book has been concerned chiefly with the petroleum reservoir and especially with the traps, of which a variety have been illustrated. We have seen there are many kinds of traps, but, even so, many trap-forming combinations of geologic conditions are still unexplored; practically every new pool, in fact, is characterized by a new combination of conditions. Much of what follows also deals with the reservoir, but with emphasis on the background factors that one must understand if one is to look for traps effectively. These factors include the fluids in the reservoir, the pressure and temperature and their changing relations to each other as a pool is produced, the influence on exploration of one's ideas on the origin and migration of petroleum, and some of the methods of assembling the geologic data on which to do the speculative and imaginative thinking that must be done if one is to visualize in advance the conditions that prevail in the undiscovered trap buried deep under the ground.

Selected General Readings

STRUCTURAL, STRATIGRAPHIC, AND COMBINATION TRAPS

American Association of Petroleum Geologists, Tulsa, Oklahoma, monthly bulletin and many special volumes, containing great numbers of maps and sections of traps of all kinds throughout the world.

James F. Pepper, Wallace DeWitt, Jr., and others, a series of maps and reports on the Berea sands of Ohio and West Virginia, published by the U.S. Geological Survey as Oil and Gas Investigations, Preliminary Maps, Nos. 9, 39, and 79. Contain excellent maps and sections showing many productive patchy sands and sand trends in this region.

W. N. Bass, "Origin of Bartlesville Shoestring Sands, Greenwood and Butler Counties, Kansas," Bull. Amer. Assoc. Petrol. Geol., Vol. 18 (October 1934), pp. 1313–1345. 32 references. A thorough analysis of the origin of this type of sand deposit.

The Science of Petroleum, Oxford University Press, London and New York, Vol. 1 (1938), pp. 57–265, and Vol. 6 (1953), Part I, 174 pages. Contain maps and descriptions of traps and oil pools throughout the world.

Amer. Assoc. Petrol. Geol., *Stratigraphic Type Oil Fields,* Tulsa, Okla. (1941), 902 pages. Contains descriptions of 38 stratigraphic pools and a selected bibliography of articles describing other pools.

The State of California, Department of Conservation, Division of Oil and Gas, San Francisco, has published an excellent series of descriptions of oil and gas fields, maps, data, and geology. Pertinent examples are: *Geologic Formations and Economic Development of the Oil and Gas Fields of California,* Bull. 118 (1943, reprinted 1948); *California Oil and Gas Fields, Part I, San Joaquin, Sacramento, and Northern Coastal Regions* (1960), 493 pages; *California Oil and Gas Fields, Part II, Los Angeles-Ventura Basins and Central Coastal Regions* (1961), 409 pages; *Northern California,* Bull. 181 (1962), 412 pages.

L. S. Matteson and D. A. Busch, *Oil Bearing Sands in Southwestern Pennsylvania,* Special Bull. 1, Topographical and Geological Survey of Pennsylvania, 4th Series (1944). Numerous maps and sections of the many productive sands of this region.

L. I. Nettleton (ed.), *Geophysical Case Histories,* Society of Exploration Geophysicists, Tulsa, Oklahoma (1948). 60 articles, describing traps of many kinds, and their geophysical interpretations.

E. Mencher and others, "Geology of Venezuela and Its Oil Fields," Bull. Amer. Assoc. Petrol. Geol., Vol. 37 (April 1953), pp. 690–707. Contains many structural maps and sections of many traps containing oil fields in Venezuela. Bibliog. 24 items.

Frank A. Herald (editor), *Occurrence of Oil and Gas in Northeast Texas,* Bureau of Economic Geology, Austin, Texas, Publ. No. 5116, in cooperation with East Texas Geological Society (April 1951), 449 pages; and *Occurrence of Oil and Gas in West Texas,* Bureau of Economic Geology, Austin, Texas, Publ. No. 5716, in cooperation with West Texas Geological Society (August 1957), 442 pages. These two publications contain many authoritative descriptions, maps, and data concerning oil and gas pools of northeast and west Texas.

William Morris Davis, *The Coral Reef Problem,* Spec. Pub. No. 9, American Geological Society (1928), 596 pages. Extensive bibliography. The most complete account of the subject of coral reefs.

Heinz A. Lowenstam, "Marine Pool, Madison County, Illinois, Silurian Reef Producer," in *Structure of Typical American Oil Fields,* Amer. Assoc. Petrol. Geol., Tulsa, Okla. (1948), Vol. 3, pp. 153–188. A detailed description of a producing reef that is typical of many reefs in the central United States.

F. G. Walton Smith, *Atlantic Reef Corals,* University of Miami Press (1948), 111 pages. Bibliog. 20 items. Handbook including numerous enlarged photographs of actual reef coral structures.

Reef Issue, Jour. Geol., Vol. 58, No. 4 (July 1950), pp. 289–487. A group of seven papers, by eleven authors, on both modern and ancient organic reefs, containing numerous diagrams, photographs, and maps of reefs and extensive lists of references.

Norman D. Newell and others, *The Permian Reef Complex of the Guadalupe Mountains Region, Texas and New Mexico,* W. H. Freeman and Company, San Francisco (1953), 236 pages. A description of the environmental conditions under which these celebrated Permian reefs were formed. Extensive bibliography and numerous sections, maps, and photographs.

Harry S. Ladd, "Reef Building," Science, Vol. 134, No. 3481 (September 15, 1961), pp. 703–716. Extensive bibliography.

SALT DOMES

Various authors, "Symposium on Salt Domes," Jour. Inst. Petrol. Technol., Vol. 17 (1931), pp. 252–371. Many articles on salt-dome phenomena throughout the world.

Gulf Coast Oil Fields: A Symposium on the Gulf Coast Cenozoic, Amer. Assoc. Petrol. Geol., Tulsa, Okla. (1936), 1,070 pages. Authoritative articles by 52 authors on the geology of the Gulf Coast region of the United States.

Donald C. Barton, "Mechanics of Formation of Salt Domes with Special Reference to Gulf Coast Salt Domes of Texas and Louisiana," in *Gulf Coast Salt Domes,* Amer. Assoc. Petrol. Geol., Tulsa, Okla. (1936), pp. 20–78. A discussion of many of the problems and phenomena connected with the origin of salt plugs.

Reference Notes

1. H. E. Minor and Marcus A. Hanna, "East Texas Oil Field," in *Stratigraphic Type Oil Fields,* Amer. Assoc. Petrol. Geol., Tulsa, Okla. (1941), pp. 600–640.

2. *Geology of Salt Dome Oil Fields: A Symposium,* Amer. Assoc. Petrol. Geol., Tulsa, Okla. (1926), various authors, 797 pages.
Symposium on Salt Domes, Jour. Inst. Petrol. Technol., Vol. 17 (1931), various authors, pp. 252–371.
Gulf Coast Oil Fields: A Symposium, Amer. Assoc. Petrol. Geol., Tulsa, Okla. (1936), 52 authors, 1,070 pages.

3. I. P. Voitesti, "Geology of the Salt Domes in the Carpathian Region of Rumania," Bull. Amer. Assoc. Petrol. Geol., Vol. 9 (1925), pp. 1165–1200. Discussion to p. 1206. 53 references.

4. Jacques Dupouy-Camet, "Triassic Diapiric Salt Structures, Southwestern Aquitaine Basin, France," Bull. Amer. Assoc. Petrol. Geol., Vol. 37 (October 1953), pp. 2348–2388. 28 references.

5. J. V. Harrison, "Salt Domes in Persia," Jour. Inst. Petrol. Technol., Vol. 17 (1931), pp. 303–305.

6. Hans Stille, "The Upthrust of the Salt Masses of Germany," in *Geology of Salt Dome Oil Fields,* Amer. Assoc. Petrol. Geol., Tulsa, Okla. (1926), pp. 142–166. Bibliog. 35 items.
S. E. Coomber, "Germany," in *The Science of Petroleum,* Oxford University Press, London and New York, Vol. 1 (1938), pp. 184–188. References and bibliog. 12 items.
British Intelligence Objectives Sub-Committee (BIOS), *Oil Field Investigations, German Oil Industry, 1939–45,* Part III, Sec. 2 (W.I. Rept. No. 1014, 37 Bryanstone Square, London). Numerous maps and descriptions of oil fields.
Frank Reeves, "Status of German Oil Fields," Bull. Amer. Assoc. Petrol. Geol., Vol. 30 (September 1946), pp. 1546–1584. Bibliog. 48 items.

7. C. W. Sanders, "Emba Salt-Dome Region, USSR, and Some Comparisons with Other Salt-Dome Regions," Bull. Amer. Assoc. Petrol. Geol., Vol. 23 (April 1939), pp. 492–516.

N. I. Bouialor, "The Emba Oil-bearing Region," XVIIth Int. Geol. Cong., Moscow, Vol. 4 (1937), pp. 169–180.

8. Thomas S. Harrison, "Colorado-Utah Salt Domes," Bull. Amer. Assoc. Petrol. Geol., Vol. 11 (February 1927), pp. 111–133.

H. W. C. Prommel and H. E. Crum, "Salt Domes of Permian and Pennsylvanian Age in Southeastern Utah and Their Influence on Oil Accumulation," Bull. Amer. Assoc. Petrol. Geol., Vol. 11 (April 1927), pp. 373–393.

A. A. Baker, C. H. Dane, and John B. Reeside, Jr., "Paradox Formation of Eastern Utah and Western Colorado," Bull. Amer. Assoc. Petrol. Geol., Vol. 17 (August 1933), pp. 963–980. 32 references cited.

9. Sidney Powers, "Interior Salt Domes of Texas," in *Geology of Salt Dome Oil Fields,* Amer. Assoc. Petrol. Geol., Tulsa, Okla. (1926), pp. 209–268.

W. C. Spooner, "Interior Salt Domes of Louisiana," in *Geology of Salt Dome Oil Fields,* Amer. Assoc. Petrol. Geol. (1926), pp. 269–344. Bibliog. 15 items.

10. J. V. Harrison, *op. cit.* (note 5), pp. 300–320.

11. G. M. Lees—"Salt, Some Depositional and Deformational Problems," Jour. Inst. Petrol. Technol., Vol. 17 (1931), pp. 259–280. 24 references.

12. Arthur Wade, "Intrusive Salt Bodies in Coastal Asir, South Western Arabia," Jour. Inst. Petrol. Technol., Vol. 17 (1931), pp. 321–330. 14 references.

13. G. D. Hobson, "Salt Structures: Their Form, Origin, and Relationship to Oil Accumulation," in *The Science of Petroleum,* Oxford University Press, London and New York, Vol. 1 (1938), pp. 255–260. 20 references.

14. A. J. Eardley, "Petroleum Geology of Aquitaine Basin, France," Bull. Amer. Assoc. Petrol. Geol., Vol. 30 (September 1946), pp. 1517–1545.

Jacques Dupouy-Camet, *op. cit.* (note 4).

15. Ralph E. Taylor, "Water-insoluble Residues in Rock Salt of Louisiana Salt Plugs," Bull. Amer. Assoc. Petrol. Geol., Vol. 21 (October 1937), pp. 1268–1310. Bibliog. 73 items.

16. K. C. Heald, "Sandstone Inclusion in Salt in Mine on Avery's Island," Bull. Amer. Assoc. Petrol. Geol., Vol. 8 (1924), pp. 674–676.

17. Ralph E. Taylor, *op. cit.* (note 15), pp. 1268–1310, 1496, and 1594.

18. Robert Balk, "Salt Dome Structure," Bull. Amer. Assoc. Petrol. Geol., Vol. 31 (July 1947), pp. 1295–1299.

Robert Balk, "Structure of Grand Saline Salt Dome, Van Zandt County, Texas," Bull. Amer. Assoc. Petrol. Geol., Vol. 33 (November 1949), pp. 1791–1829. Bibliog. 89 items.

Robert Balk, "Salt Structure of Jefferson Island Salt Dome, Iberia and Vermilion Parishes, Louisiana," Bull. Amer. Assoc. Petrol. Geol., Vol. 37 (November 1953), pp. 2455–2474. 26 references.

19. Travis J. Parker and A. N. McDowell, "Scale Models as Guides to Interpretation of Salt-Dome Faulting," Bull. Amer. Assoc. Petrol. Geol., Vol. 35 (September 1951), p. 2076–2094.

W. E. Wallace, Jr., "Structure of South Louisiana Deep-seated Domes," Bull. Amer. Assoc. Petrol. Geol., Vol. 28 (September 1944), pp. 1249–1312. Bibliog. 17 items.

20. Ralph Alexander Liddle, *The Van Oil Field, Van Zandt County, Texas,* Bull. 3601, Texas University (January 1936), 82 pages.

21. Fred W. Bates, "Geology of Eola Oil Field, Avoyelles Parish, Louisiana," Bull. Amer. Assoc. Petrol. Geol., Vol. 25 (July 1941), pp. 1363–1395.

22. Marcus I. Goldman, "Petrography of Salt Dome Cap Rock," Bull. Amer. Assoc. Petrol. Geol., Vol. 9 (January 1925), pp. 42–78.

Levi S. Brown, "Cap-rock Petrography," Bull. Amer. Assoc. Petrol. Geol., Vol. 15 (May 1931), pp. 509–522. Discussion to p. 529.

Ralph Emerson Taylor, *Origin of the Cap Rock,* Louisiana Salt Domes, Geol. Bull. 11, La. Dept. of Conservation (August 1938), 191 pages.

23. Archer H. Marx, "Hoskins Mound Salt Dome, Brazoria County, Texas," in *Gulf Coast Oil Fields,* Amer. Assoc. Petrol. Geol., Tulsa, Okla. (1936), p. 851.

24. E. P. Tatum, "Upper Cretaceous Chalk in Caprock of McFaddin Beach Salt Dome, Jefferson County, Texas" (Geol. Note), Bull. Amer. Assoc. Petrol. Geol., Vol. 23 (March 1939), pp. 339–342.

25. Marcus A. Hanna, "Evidence of Erosion of Salt Stock in Gulf Coast Salt Plug in Late Oligocene," Bull. Amer. Assoc. Petrol. Geol., Vol. 23 (April 1939), pp. 604–607.

26. Sidney A. Judson and R. A. Stamey, "Overhanging Salt on Domes of Texas and Louisiana," Bull. Amer. Assoc. Petrol. Geol., Vol. 17 (December 1933), pp. 1492–1520.

27. Sidney A. Judson, P. C. Murphy, and R. A. Stamey, "Overhanging Cap Rock and Salt at Barbers Hill, Chambers County, Texas," Bull. Amer. Assoc. Petrol. Geol., Vol. 16 (May 1932), pp. 469–482.

28. Gustav Wade, *Review of the Heaving-shale Problem in the Gulf Coast Region*, RI 3618, U.S. Bur. Mines (1942), 64 pages.

29. J. M. Frost III, "Geologic Aspects of Heaving Shale in Texas Coastal Plain," Bull. Amer. Assoc. Petrol. Geol., Vol. 23 (February 1939), pp. 212–219. Bibliog. 52 items.

30. E. DeGolyer, "Origin of North American Salt Domes," Bull. Amer. Assoc. Petrol. Geol., Vol. 9 (August 1925), pp. 831–874.

Marcus A. Hanna, "Geology of the Gulf Coast Salt Domes," in *Problems of Petroleum Geology*, Amer. Assoc. Petrol. Geol., Tulsa, Okla. (1934), pp. 629–678.

Donald C. Barton, "Mechanics of Formation of Salt Domes with Special Reference to Gulf Coast Salt Domes of Texas and Louisiana," in *Gulf Coast Oil Fields*, Amer. Assoc. Petrol. Geol., Tulsa, Okla. (1936), pp. 20–78.

31. Robert T. Hill, "The Beaumont Oil Field with Notes on the Other Oil Fields of the Texas Region," Jour. Franklin Inst., Vol. 154 (1902), pp. 273–274.

G. D. Harris, "The Geological Occurrences of Rock Salt in Louisiana and East Texas," Econ. Geol., Vol. 4 (1907), pp. 12–34.

Bailey Willis, "Artesian Salt Formations," Bull. Amer. Assoc. Petrol. Geol., Vol. 32 (July 1948), pp. 1227–1264. 47 references listed.

32. L. L. Nettleton, "Fluid Mechanics of Salt Domes," in *Gulf Coast Oil Fields*, Amer. Assoc. Petrol. Geol., Tulsa, Okla. (1936), pp. 79–108. Bibliog. 23 items.

L. L. Nettleton, "Recent Experimental and Geophysical Evidence of Mechanics of Salt-Dome Formation," Bull. Amer. Assoc. Petrol. Geol., Vol. 27 (January 1943), pp. 51–63.

33. F. M. Van Tuyl, "Contribution to Salt-dome Problem," Bull. Amer. Assoc. Petrol. Geol., Vol. 14 (August 1930), pp. 1041–1047.

34. C. H. Ritz, "Geomorphology of Gulf Coast Salt Structures and Its Economic Application," Bull. Amer. Assoc. Petrol. Geol., Vol. 20 (November 1936), pp. 1413–1438.

PART THREE

Reservoir Dynamics

9. *Reservoir Conditions—Pressure and Temperature*
10. *Reservoir Mechanics*

INTRODUCTION TO PART THREE

THE HISTORY of every petroleum pool may be divided into a static and a dynamic period. So far we have been considering the static elements, such as the reservoir rock, the reservoir fluids, and the traps that hold the pool in place. These elements have become stabilized during a long period in which changes were slowly going on within the reservoir and its fluids in response to crustal movements, igneous activity, loading and unloading of the overburden, and changes in the rate and direction of the hydraulic circulation. Once a pool is discovered and withdrawal of the fluids begins, equilibria are upset and the pool may be said to enter a dynamic period, in which changes take place rapidly. The differences between the static and the dynamic state are thus differences of degree, chiefly the degree of stabilization of the fluids. The changes during the static period are so slow as to be imperceptible and can only be inferred from study of the conditions within the reservoir and of the changes that take place in the dynamic period.

Most of our knowledge of what happens within a reservoir when its fluids are withdrawn and its pressure and its temperature change come from the work of the petroleum engineers, who have written much on this subject.* What these engineers are chiefly concerned with is producing the greatest possible amount of oil and gas from a pool at the lowest possible cost. To do so they must study oil and gas movement in the tapped reservoir, under what the petroleum geologist might regard as laboratory conditions; the pool might be considered as the pilot plant for the larger basin or province. An understanding of what happens within a reservoir when the oil and gas are being removed will help us to understand what happened when the oil and gas were being concentrated into pools and thus help us to discover other pools.

---

* See especially the publications of the American Petroleum Institute, the American Institute of Mining and Metallurgical Engineers, Petroleum Division, and *The Petroleum Engineer*.

The next two chapters are concerned with the dynamic phenomena that accompany the withdrawal of fluids from the reservoir: the changes in pressure and temperature, the fluid potential gradients that develop, and the interrelated effects, or mechanics.

CHAPTER 9

Reservoir Conditions – Pressure and Temperature

Reservoir pressure: measurement – gradients – sources – variations. Temperature: measurement – geothermal gradient – uses of thermal measurements – sources of heat energy – effects of heat.

THE TWO DOMINANT variable conditions that affect every petroleum reservoir are pressure and temperature, and each of them is a form of stored and available energy. As either or both of them vary, the volumes of the rocks affected by them vary, but of most importance to us is the fact that the volumes of the fluids contained in the rocks vary. Differences in fluid pressure and temperature between one area and another determine *pressure* and *temperature gradients,* which are fundamental to most problems involving the movement of oil and gas through the rocks. This movement may be a migration that permits oil and gas to accumulate to form a pool, or it may be a movement of oil and gas from a pool into a well bore. Much of petroleum engineering has to do with relations between pressure, volume, and temperature —or, as they are commonly called, collectively, *PVT.* The volume changes are most pronounced in the gas content of the reservoir, but they also occur in the other fluids, the water and the oil, and to a lesser extent in the materials that constitute the reservoir rock. Abnormal fluid pressure gradients and any temperature gradients indicate that energy has changed from potential to kinetic, and that work is being done.

Viscosity and buoyancy are properties of oil and gas that have great effect on the ability of these fluids to move through the water-filled, connected pores of the rocks. Both vary with pressure and temperature, which generally increase with depth. Of the two reservoir conditions, pressure probably has a wider effect than temperature. Both influence the volumes, and therefore the relative buoyancies, of the fluids, but pressure has the greater influence, especially on natural gas. The chief effect of increasing temperature is a decrease in the viscosity of the liquids; this causes them to flow more readily.

RESERVOIR PRESSURE

The fluids confined in the pores of the reservoir rock occur under a certain degree of pressure, generally called *reservoir pressure,*[1] *fluid pressure,* or *formation pressure.* We can determine this pressure by measuring the force per unit area exerted by the fluids against the face of the reservoir rock where it has been penetrated by a well. We state the amount of pressure in pounds per square inch (psi) or in pounds per square inch absolute (*psia*)* or in atmospheres (multiples of 14.7 psi). Since all the fluids in the system are in contact with one another, they transmit pressures freely, and pressures measured on one fluid are actually the pressures on all fluids. Other terms used for reservoir pressure are *bottom-hole pressure, water-pressure, closed-in pressure, well pressure,* and *"rock" pressure.* The terms may or may not be intended as the exact equivalent of "reservoir pressure," but generally they may be considered as such. The term "rock" pressure was used in some earlier papers as an equivalent of fluid pressure; that term did not imply a measurement or calculation of the load pressure exerted by the overlying column of rocks.

The hydrostatic pressure gradient for a fresh-water (Spec. Gr. 1.0) system is 0.433 psi/ft. This relationship may be shown graphically by a pressure-depth diagram like the one in Figure 9-1,A. If the water contains dissolved salts its specific gravity will be higher than that of fresh water (see Table A-1, Appendix), and consequently the hydrostatic pressure gradient of such a system will be greater in terms of psi/foot than that of a fresh-water system and will be displayed on a pressure-depth plot by a pressure gradient line

---

* The abbreviation "psi" commonly means the same as "psig," which means pounds per square inch, gauge measurement. Psi or psig is converted to psia by adding the atmospheric pressure at the surface of the ground. The average atmospheric pressures for different altitudes are given in the following table:

Altitude, feet	Atmospheric pressure, psi	Altitude, feet	Atmospheric pressure, psi
0	14.7	6,000	12.0
1,000	14.2	7,000	11.7
2,000	13.6	8,000	11.3
3,000	13.1	9,000	10.9
4,000	12.6	10,000	10.1
5,000	12.1		

The pressure against which the reservoir pressure is measured is the surface pressure at the elevation at which the pressure bomb is sealed. Psia pressures are used in accurate engineering calculations, specially where gas volumes are involved.

RESERVOIR CONDITIONS [CHAPTER 9]   391

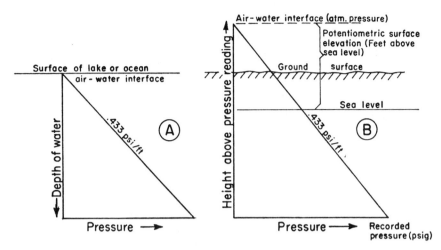

FIGURE 9-1 *Fluid pressure in an open system; A is the force per unit area ex- erted by a column of water extending up from the point of pressure meas- urement to the free air-water interface—for example, the surface of a lake or ocean. In a confined system, such as B, the pressure at any depth is a measure of the force per unit area exerted by a column of water extending up from the point of pressure measurement to the phantom air-water interface, the potentiometric surface elevation, which may be either above or below the surface of the ground. Pure water (Spec. Gr. 1.0) exerts a pressure of 0.433 psi per foot of depth in the water column.*

whose slope is less than that of the fresh-water hydrostatic pressure gradient. At the surface of an open system, such as a lake, the water pressure is zero psig; that is, the fluid pressure at the surface of an open system is the atmos- pheric pressure. For a closed hydrostatic system (Fig. 9-1, B) in which the water is homogeneous and has the same density as that in an open system, the water pressure gradient will be the same, but the elevation at which the fluid pressure is zero psig (atmospheric pressure) may be above or below the surface of the ground. The elevation, referred to a common datum such as sea level, at which a fluid pressure is measured is called the datum elevation. Fluid pressures, plotted on a pressure-depth diagram, at their respective datum elevations, provide a convenient method for investigating the fluid potential environment of reservoir units in a basin, or in a local prospect area.

Reservoir pressure, unless otherwise specified, is generally thought of as the *original,* or *virgin, pressure*—the pressure that existed before the *natural pres- sure* equilibrium of the formation had been disturbed by any production. The original pressure can be measured directly only by the first producing well drilled into the reservoir, for the pressure begins to decline as soon as oil and gas are withdrawn. When a producing well is shut in, the reservoir pressure begins to rise. The rise is rapid at first, and then gradually slows until finally the maximum pressure is reached. This maximum pressure is called the *static bottom-hole pressure,* the *shut-in pressure,* or the *static formation pressure.* If the well is not shut in long enough for the pressure to reach its maximum, the maximum may be extrapolated from pressure buildup curves. The static bottom-hole pressure in a producing well is generally lower than the original reservoir pressure of the pool. The difference between the shut-in pressure and the original reservoir pressure is a measure of the decline in reservoir pressure. *Flowing pressure,* also called *bottom-hole flowing pressure,* is measured while the well is producing; and the difference between flowing pressure and static pressure is called the *differential pressure. Casing pressure,* or *surface pressure,* is the static pressure exerted within the well casing at the top of the hole when the well has been shut in and the pressure allowed to build up as much as it can. From it, reservoir pressure may be calculated by adding the total weight of the column of whatever air, gas, oil, and water there may be in the hole. *Tubing pressure* is the pressure at the top of the hole within the tubing through which the well is producing. It may be static (i.e., measured when the well is shut in), in which case it equals the casing pressure, or it may be measured while the well is flowing. When gas bypasses the oil and breaks through, the tubing pressure increases because the density of the gas is less than the density of the oil. *Back pressure* is the pressure against which the well is producing at the reservoir—that is, the resistance to flowing pressure. It is the psig at the surface plus the pressure exerted by the fluid column within the well bore.

The changes in reservoir pressure that accompany production are of great importance to the production engineer. In general, pressure declines with fluid withdrawal, and the rate of decline per unit of gas or oil produced furnishes some of the best data upon which to base estimates of the character and amount of the reserves, the highest efficient rate of production, and the efficiency of the operation. Where the pressure decline per unit of petroleum produced from a pool is rapid, the volume of the reservoir is probably small; if the decline is slow, the reservoir may be extensive. An early knowledge of the pressure decline characteristics of a reservoir is thus of great importance for estimating its available energy and consequently its productive possibilities. Pressure measurements are an indispensable engineering tool. We are concerned here chiefly with the original pressures, their causes, and some of their effects. The changes in pressure that occur during production will be considered more fully in Chapter 10, on reservoir mechanics.

The original pressures are intimately associated with the history of the water content of the rocks. When looked at regionally, the amount of oil and gas in the rocks is infinitesimal compared with the amount of water present—in fact, most oil and gas pools may be said to occur in aquifers. Water acts not only as the medium through which the oil and gas must pass to accumulate into pools, but also as the chief transmission agency of the fluid pressures from one area to another. The water may be thought of as an interconnecting web that is in continuous phase throughout the permeable rocks. In the shales and finer-grained rocks of extremely low permeability, the water is present as a film a few molecules thick. In the more permeable rocks—those called aquifers—it occupies 10 to 40 percent of the rock volume. Since oil and gas pools are almost always intimately associated with reservoir formation water, many of the pressure phenomena that occur in the geology of ground water also occur in the geology of oil and gas.

Pressure Measurement

Reservoir pressure makes itself evident in several ways. In holes drilled with cable tools, any water pressure that exists is quite evident, for there is generally enough pressure in all water-bearing formations to support a column of water to some height in the hole, or even enough to make it flow out over the top of the hole. In the drilling of these holes, the formations are exposed only to atmospheric pressure plus the weight of a small amount of drilling water, so that the formation water has free access to the hole. The rate at which water enters the hole may be measured by bailing, in which case the rate is recorded on the well log as the number of bailerfuls per hour (No/B/Hr) required to remove the water as fast as it enters the hole. The height to which the water rises in the hole when bailing stops is recorded and is obviously a function of the pressure exerted by the water. It is known as the *static water level*. If the water fills the hole, it is recorded as a hole full of water (HFW). Many formations that produce water slowly, because of low permeability of the rock, would be capable of filling more of the hole, or even all of it, if given a longer time to fill up than the drillers generally allow during drilling operations.

In rotary-drilled holes filled with drilling mud the problem of obtaining pressure measurements is different. The mud is made considerably heavier than water to ensure that the pressure within the hole at the face of the reservoir will be higher than the reservoir pressure. If this were not done, the reservoir fluids would enter the hole and force the mud back out of the hole. Several devices have been developed for placing pressure-measuring instruments opposite the face of the reservoir, and recording the reservoir pressure, even when the hole is full of mud.[2] These are generally self-recording and self-contained sensitive pressure gauges, sometimes called *pressure bombs,* which are lowered into the hole on the testing tool. A packer is set above the tool to keep the drilling mud off the fluid-bearing formation, and the measurement of reservoir pressure is thereby taken with reference to atmospheric conditions at the surface of the ground. After the drilling mud has been removed and the well is producing, reservoir pressure may be measured by placing a pressure bomb in the tubing opposite the reservoir.

Shut-in pressures may also be calculated by measuring the casing pressure at the top of the hole and adding the weight of the fluid column that extends down from the surface to the reservoir. Where the top of the liquid is at some distance down in the hole, the weight of the column of fluid may be calculated by adding the weight of the liquid to the weight of the air between the liquid level and the surface. Where the entire hole is full of gas, the bottom-hole, or reservoir, pressure[3] is obtained by adding the weight of the gas in pounds per square inch, as determined for that particular gas and for the temperature of the hole, to the gauge pressure measured at the surface.

Pressure-measuring equipment has kept pace with the steadily increasing demands of measuring higher pressures at the greater depths of drilling. A gas-condensate pressure of 11,690 psi was measured in an 18-foot sand at 16,112 feet in southern Louisiana[4] and the bottom-hole pressure was estimated to be around 15,000 psi.

Pressure Gradients

We are concerned with two kinds of fluid pressure gradients. The first is the hydrostatic pressure gradient, the increase of fluid pressure with depth due to the water in the aquifer overlying the site at which the fluid pressure is measured. The second is the hydrodynamic pressure gradient, or fluid potential gradient, which exists in an aquifer in which the water is flowing. If the potentiometric surface of a given aquifer is horizontal, then the system is said to be in hydrostatic equilibrium; the formation water is at rest. If the potentiometric surface has a slope, then the system is said to be in hydrodynamic equilibrium; the water is in motion.

Static Gradients. In most reservoirs the static gradient averages about 45 psi per 100 feet of depth; this figure is for water containing 55,000 parts per million of dissolved salts. (See Table A-1, Appendix.) Gradients as high as 100 psi/100 feet are encountered in some reservoirs; such gradients are believed to be due not to the weight of the overlying water column but to the weight of the overlying rocks, for which the pressure gradient is about 100 psi/100 feet.

Some examples of reservoir fluid pressures are shown in Figure 9-2, which gives the fluid pressures measured in ten oil pools producing from the Smackover limestone in southern Arkansas. The gradients averaged about 52 psi/100 feet. The density of the brine in the Smackover is 1.22, which corresponds to 52 psi/100 feet and places the potentiometric elevation approximately at the surface of the ground. The average surface static gradient along the Gulf Coast of the United States is generally taken as 46.5 psi/100 feet, although many higher gradients have been measured. The reservoir pressures from a group of pools in the Greater Oficina district of eastern Venezuela are shown in Figure 9-15.

When the reservoir pressure causes the water confined in an aquifer to rise above the top of the aquifer when penetrated by a well, the water is said to be *artesian;* and if it flows over the top of the well, the well is called a *flowing artesian well*. The water in a well rises until its weight exerts a pressure equal

RESERVOIR CONDITIONS [CHAPTER 9]

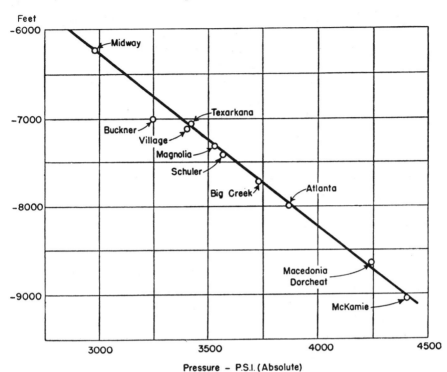

FIGURE 9-2 *Chart showing the relation of original reservoir pressure to depth below the surface in ten pools in southern Arkansas, producing from the Smackover limestone (Jurassic). The brine from the reservoirs has a specific gravity of 1.22, which corresponds to a pressure gradient of 0.52 psi/ft. The reservoir pressure is hydrostatic with reference to a column of brine whose density is the same as that of the reservoir brine.*

to the reservoir pressure; thus the well may be thought of as a manometer. The height to which confined water rises in wells drilled at different points into the same aquifer may be plotted as a phantom equilbrium surface (air-water surface of one atmosphere pressure), which is called the *potentiometric* ("potential-measuring") or *piezometric* ("pressure-measuring") *surface*.* Where

---

* The term "piezometric surface" has had a long usage in the geology of groundwater. Where the energy of the water is being considered, Hubbert (Bull. Amer. Assoc. Petrol. Geol., Vol. 37, pp. 1973-1974) has used the term "potentiometric surface." The *potentiometric surface* of an aquifer is the surface connecting all points of head (the elevation to which the water rises above a datum plane) and is a measure of the potential energy of the water at all points in the upper surface of the aquifer, regardless of the elevation of the aquifer. In usage, the piezometric surface corresponds to the calculated potentiometric surface only when the density of the water encountered at the particular pressure measurement site is employed to convert that fluid pressure to a potentiometric surface elevation. The piezometric surface corresponds to the surface that would be defined by a series of wells, viewed as manometers, whereas the potentiometric surface is calculated, and a deliberate assumption must be made as to the fluid density employed to convert the fluid pressure values to fluid potential values.

the height to which water will rise in a well cannot be observed directly, the elevation of the potentiometric surface at any location may be calculated from the density of the water and the reservoir pressure if both are known. A flowing artesian well occurs where the potentiometric surface is higher than the ground surface at the location of the well.

A sample situation is illustrated in Figure 9-3. The intake area of the aquifer is higher than the discharge area, and the potentiometric surface connects the two areas. An area where the potentiometric surface is above the surface of the ground is frequently termed an "area of excess pressure." In such an area water would flow out of a well and above the ground surface. In an area where the potentiometric surface is below the ground surface—an "area of subnormal pressure"—water would stand in the well.

Dynamic Pressure Gradients. In a discussion of hydrodynamic conditions the terminology involved can lead to confusion and misunderstanding. The difficulty occurs with respect to any concept involving pressures and gradients. The term "pressure" is often incorrectly used to mean fluid potential, as is the term "pressure gradient" to mean potential gradient. A rigorous treatment of hydrodynamic concepts and terminology is presented by M. King Hubbert in his classic paper.[43]

The relationship, as presented by Hubbert, between fluid pressure and fluid potential is:

$$\Phi = gz + \frac{p}{\rho}$$

where Φ = fluid potential, g = the gravitational constant, z = the datum ele-

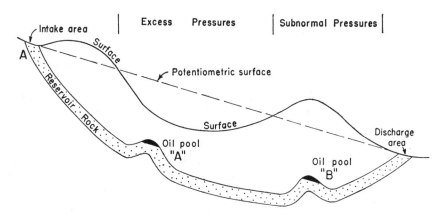

FIGURE 9-3 *Diagrammatic section showing the relation of the potentiometric surface to the surface of the ground as a cause of either apparent "excess" or deficient pressure for the depth of the reservoir below the surface of the ground. Length of section up to hundreds of miles. Hydrodynamic conditions exist in this simplified model; water flows from A to C, in the direction of the sloping potentiometric surface.*

vation at the site of pressure measurement, $p =$ the static fluid pressure, and $\rho =$ the density of the reference fluid—usually water. The fluid potential is related to the "head," or potentiometric surface (h), by the following relationship:

$$\Phi = gh$$

Upon dividing both relationships by g, the following equation is obtained:

$$\frac{\Phi}{g} = h = z + \frac{p}{\rho g}$$

It is this relationship that commonly is employed to convert fluid pressures to fluid potential values as represented by h, the potentiometric surface of head. As $\rho g = \text{grad } p$, the static pressure gradient of the reference fluid is employed for the conversion. The one critical assumption concerns the density of the reference fluid within real geologic situations, as the density of the water in any hydrogeologic system is never a constant. In practical applications, however, less error will be introduced if all fluid pressures in any given hydrogeologic system are converted to potentiometric surface values on the assumption that all of the water of that system is of a constant density. Appropriate corrections can then be made in local situations as regards the given density contrast between any two points of fluid pressure measurement.

Pressure is a measurement of force per unit area. Different fluid pressures will exist in a connected fluid system at each separate reference elevation even though there is no fluid movement—that is, when hydrostatic conditions exist. If differences in fluid pressure exist at the same reference elevation, then a difference in fluid potential exists and hydrodynamic conditions prevail. Fluid pressures measured at different reference elevations can be corrected to the fluid pressure that would exist at the same reference elevation by the use of the appropriate static pressure gradient (ρg) for the given fluid. If fluid pressure differences still exist when such a correction is made, then, again, hydrodynamic conditions prevail in this system.

In practice, regional and areal hydrodynamic studies of aquifers are normally made in terms of the potentiometric surface calculated from fluid pressure measurements wherever available in the area of interest. The interpretation is displayed as a potentiometric surface map, with contour lines connecting points of equal fluid potential or potentiometric surface elevations. The flow of water in this system is thereby depicted as movement in the direction normal to the potentiometric surface contours from the areas where the potentiometric surface is high to areas where the potentiometric surface is low. In other words, water in the aquifer flows down the slope of the potentiometric surface.

In the practical application of pressure data to petroleum exploration, if two or more fluid pressure measurements are made, the interpretation of the data depends on whether the measurements were made in the same reservoir, in the same well, or in the same aquifer, and whether at the same depth and at the same time or at different times. For example, static reservoir pressure measurements are often made in the same well at different times to provide fluid pressure data used in determining the reservoir's productivity characteristics, in terms of barrels of oil produced per psi drop in reservoir pressure. If fluid pressure measurements are made in different aquifers in the same well at essentially the same time, the aquifers are said to be in hydrostatic equilibrium locally if the calculated potentiometric surface elevations are the same for all of the measured formation pressures. But if the potentiometric surface elevations differ for the different aquifers in which the fluid pressures were measured, then a hydrodynamic gradient exists between these acquifers, and if there are permeable paths between the aquifers, such as may be provided by faults, fractures, or unconformity surfaces, the formation fluids will move along these paths from the aquifers with relatively high fluid potentials to those with relatively low fluid potentials. Similarly, if a regional pressure study of a single aquifer reveals that the potentiometric surface is horizontal, then the system is in hydrodynamic equilibrium; if the potentiometric surface slopes, then a hydrodynamic gradient exists, and the fluids will move along permeable paths from the areas of high fluid potential to areas of low fluid potential. (See Figs. 9-4 and 9-5.) An analogy can be found in the water system of a city; the potentiometric surface is horizontal when all faucets are closed, but when one is opened the pressure there is lowered, a fluid potential gradient is established, and water flows toward the open faucet.

Hydrodynamic conditions, when present in a sedimentary basin, may be envisaged as existing in two basic forms: (1) the occurrence of different fluid potentials within an aquifer, which causes flow of water in the aquifer along the bedding planes; and (2) the existence of differences in fluid potentials between aquifers in the geologic section, which will cause fluid movement, upward or downward, via permeable paths across bedding planes, from the aquifers with relatively high fluid potential to those of lower fluid potential. The reservoir system in which an oil or gas pool is trapped is subject to the hydrodynamic forces of one or both of these environments, depending upon the existing geologic conditions. The evidence for a fluid potential gradient in an aquifer is a sloping potentiometric surface. The evidence for a vertical fluid-potential gradient is a difference between the elevation of the potentiometric surface of one aquifer relative to that of another aquifer at a different depth within the same vertical section of rocks. (See Fig. 9-4.) There may also be recognized two general sources for fluid potential gradients within a reservoir rock: (1) the man-made fluid potential gradients in and surrounding a producing well or pool that form as a result of the withdrawal of fluids and consequent lowering of fluid pressure, and (2) the natural fluid potential gradients of the region. (See Fig. 9-5.)

The fluid potential gradient of a particular aquifer or reservoir system is normally described as the change of potentiometric surface in feet for a given horizontal distance, such as the vertical distance, bb' (measured in feet), for

RESERVOIR CONDITIONS [CHAPTER 9]   399

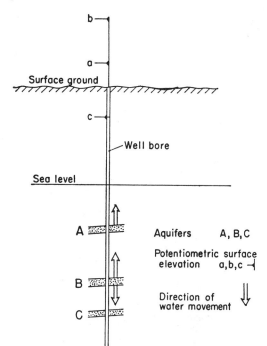

FIGURE 9-4

Three aquifers, A, B, and C within a single bore hole, with differing potentiometric surfaces, a, b, and c. Arrows show direction of fluid flow if permeable paths are available.

a horizontal distance XY (Fig. 9-6). Or it may be given in drop in head or potentiometric surface per mile (e.g., 25 ft/mi between X and Y). The hydrodynamic relationship between aquifers in the geologic section is shown by the difference in potentiometric surface elevations of the systems at the same location. These are shown in Figure 9-4 and 9-6 as a, b, and c, the potentiometric surface elevations for aquifers A, B, and C in the same well.

When fluids are withdrawn from a well, there is a decline in the reservoir

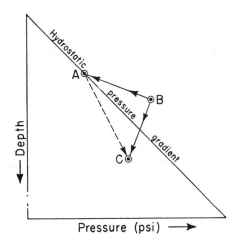

FIGURE 9-5

The fluid-pressure readings in aquifers A, B, and C of Figure 9-4, plotted at their respective datum elevations and with reference to a hydrostatic pressure gradient to show the differences in fluid potential between the aquifers. Water flow would occur from aquifer B to A and C, and from A to C if permeable paths connect the aquifers.

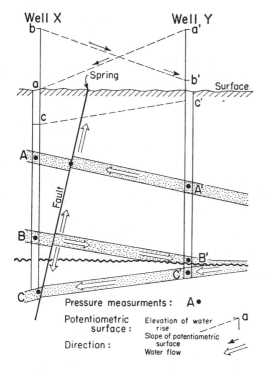

FIGURE 9-6

Idealized section showing two wells, X and Y, each with fluid pressure measurements at A, B, and C, and at A′, B′, and C′, showing the elevation of the potentiometric surface of each aquifer at a, b, c, a′, b′, and c′. Solid arrows show the direction of the slope of the potentiometric surface between wells in each aquifer, and open arrows show the direction of fluid flow where permeable paths permit intermixing of reservoir fluids.

pressure around that well. The area of lower pressure spreads out in all directions and develops a local fluid potential gradient toward the well. It is, in effect, an artificial, local potentiometric surface, sloping in toward the well. The reduced fluid potential around a single well joins with the low-pressure areas surrounding other wells, until the pressure of the entire pool is lower than the original reservoir pressure. The decline in pressure then spreads outward from the pool for varying distances and at varying rates, depending on the nature of the reservoir fluids and the permeability of the reservoir rock.

The measurable pressure gradient caused by producing a pool trapped in a sealed reservoir* (e.g., an isolated lens or porous rock surrounded by relatively impervious material) does not extend beyond the permeability barrier. More commonly, however, the pool is in a confined system and connected to a widespread aquifer; the pressure gradient developed by production from such a pool may extend outward for a considerable distance. The zone of reduced reservoir pressure extends for several miles in the Reynolds oolite pools (Smackover formation) in southern Arkansas and for many miles in the Asmari limestone pools of Iran. An outstanding example is found in the East Texas, or Typer, Basin, where the effects of the pressure decline in the East

---

* A *sealed reservoir* is sealed in on all sides by relatively impervious rocks, whereas an *unsealed reservoir* is open on one or more sides. A *confined system* is an artesian system, as in reservoir rock *AC* in Figure 9-5, where the reservoir fluids are confined in an unsealed reservoir rock connected to a source of fluid potential that is higher than that in the reservoir at the location under consideration.

Texas pool are evident throughout the entire basin, to a distance of seventy miles or more from the pool.[7] The decline in reservoir pressure in the East Texas Basin is shown in Figure 10-17, page 467.

An idealized illustration of some of the pressure components in two wells that penetrated several aquifers is shown in Figure 9-6. Three or more well-control points would be necessary to show the true rate and direction of water flow. Hydrodynamic conditions such as these have undoubtedly been present during geologic time and have been continually changing in terms of the magnitude of the fluid potential gradients and direction of water flow, along with changes caused by erosion, deformation, and deposition. The differences in fluid potential that exist between aquifers and within particular aquifers form a complex of hydrodynamic conditions that is normal for most sedimentary regions.

Sources of Reservoir Pressure

A variety of sources may cause the fluid pressure found underground in the reservoir rocks. The effect of some of the causes is to contribute to the present pressure system, whereas the effect of others is only temporary. The relative importance of the effects may also depend on whether the reservoir rock is sealed or confined by boundaries of low fluid transmissibility. The three major sources of the fluid pressures in the reservoir rocks are (1) the pressure exerted by the water above the point of pressure measurement, (2) the pressure exerted by the rock overburden, and (3) osmotic phenomena. There are lesser sources also, such as (4) temperature changes, (5) secondary precipitation or cementation phenomena, (6) earthquakes, (7) atmospheric and oceanic disturbances, and (8) chemical and biochemical reactions. All of these help to determine the original reservoir pressure, and it is often difficult or impossible to determine how much has been contributed by each one.

Pressure Due to a Column of Water. The interconnected pores of nearly every potential reservoir rock are filled with water, which exerts pressure. This is a major source of the permanent pressure system. When the water is at rest, it exerts hydrostatic pressure, which acts at right angles to the boundary surface, is the same in all directions at any point within the fluid, and is the same at all points of equal elevation. The pressure at any point in the water is, then, equal to the pressure gradient in psi per foot of depth multiplied by the height in feet of water above the confined surface at which the pressure is determined—that is, to the potentiometric surface. The energy stored and evidenced by reservoir pressure is potential energy, since it exists by virtue of its position.

Most oil-field waters are brines with varying concentrations of mineral salts.

The density of water varies with its dissolved mineral content; consequently, the hydrostatic pressure gradients range from 0.433 psi per foot of depth for pure water to 0.50 psi or more per foot for the most highly concentrated brines. The relation between brine concentration and static pressure gradient is shown in Table A-1, page 685. The average static pressure gradient of oil-field water is about 45 psi per 100 feet of depth, and most of the thousands of pools discovered have had original pressures consistent with this figure. Some of the deviations from it are probably due to varying salt concentrations, but we know that they may be due in part to other causes. It is frequently extremely difficult to distinguish and evaluate all the causes of variation that are at work in any particular case.

Compaction Phenomena. The fluid pressure measured in most petroleum pools is dominantly that resulting from the potential energy of a column of water extending from the reservoir to the elevation of the potentiometric surface of the reservoir at the location of the pool. There are many pools in which the fluid pressure cannot be fully explained in this way. In some instances pressure exerted by the enclosing rocks against the fluids, either by load pressure or by forces transmitted through the rocks as a result of diastrophism, appears to be a major contributing source of the fluid pressure system.

The load pressure exerted by a column of average rock is about one pound per square inch per foot of height,* or 100 psi per 100 feet of depth, which is slightly more than twice the average hydrostatic pressure gradient of water. This load pressure also is called the *lithostatic pressure, geostatic pressure, earth pressure, overburden pressure,* or *rock pressure* (though "rock pressure" is occasionally used to mean reservoir fluid pressure). The pressures that result from diastrophism and rock deformation may be called *geodynamic pressures,* and they are not ordinarily measurable.

The chief difference between fluid pressure and lithostatic pressure is that fluid pressure is transmitted through the fluids that fill the pores in the rock, whereas lithostatic pressure is transmitted through mineral particles in close contact—in short, through the rock itself. The touching mineral particles act as struts. Where the struts fail, as where the lithostatic pressure becomes great enough to squeeze the rock volume into a smaller space and reduce the pore space, then the lithostatic pressure is transmitted to the fluid pressure. For example, in the early stages of a rapidly filling basin, the increasing weight of the overburden permits the grains, especially of clays and shales, to touch and compact, thereby transmitting a force to the contained fluids, most of which are squeezed out of the sediments.

Osmosis. Clays act as semipermeable membranes and thereby permit osmotic and electro-osmotic pressures to develop across the bedding whenever there is a marked contrast in the concentrations of the dissolved salts on either side of

---

* The average natural specific gravity (all pores filled with water) of the typical sedimentary basin sediment is about 2.1 for sandstones, 2.3 for shales, and 2.4 for limestones. Using the specific gravity of 2.3 as an average and multiplying it by 0.43 psi/foot (the pressure gradient of pure water), we get 0.989 psi/foot as the pressure gradient due to the weight of the sediments. In round numbers an average figure of 1 psi/foot is generally used.

the clay. This is one of the three major sources of reservoir fluid pressure, and probably is responsible for many of the anomalous fluid potential gradients that are observed in underground aquifers. It contributes to the regional fluid pressure pattern, and its combination with the other fluid pressure sources explains in part the fluid pressures found in oil and gas pools. This will be discussed more fully on page 471.

The potentiometric surface of the Point Lookout sandstone of the Mesaverde group (Cretaceous) in the San Juan Basin of northwestern New Mexico and southwestern Colorado, as mapped by Berry, is especially revealing.[8] (See Fig. 9-7.) The lowest elevation of the Mesaverde outcrop around the edges of the basin is about 5,000 feet, but the potentiometric surface is in the form of a "sink" with a minimum elevation of 3,500 feet, or 1,500 feet below the lowest possible surface outlet. The potentiometric surface slopes inward from the flanks of the basin toward the apparent outlet area in the central part of the basin, which is below the lowest outcrop elevation. No other aquifer systems exist with lower fluid potentials that have regional continuity beyond the margins of this basin. In order for fluid flow to occur in this manner within this system it must, therefore, be discharged into some other acquifer, across bedding planes and by means of some force other than gravitational fluid flow. Osmotic phenomena provide a mechanism whereby water can flow through the shales that serve as semipermeable membranes from an area of low fluid potential to an area of high fluid potential. The large natural gas fields within the Mesaverde group of the San Juan Basin appear to have been trapped principally by this large fluid potential gradient within reservoir rocks of low transmissibility. This potential gradient is presumed by Berry to have originated via osmotic phenomena.

Temperature Changes. Changes in temperature change the fluid pressure. An increase in temperature causes the oil, gas, and water to expand more than the rocks, and if in a relatively sealed reservoir rock with no additional pore space available, an increase in the fluid pressure will result. Conversely, a decline in temperature causes a decline in fluid pressure. Causes of either local or regional significance, such as an approaching igneous mass, may temporarily warm up either a small or a large area. A decrease in the fluid potential away from the warmer region would be expected to develop, and a movement of fluids toward the colder region would occur until equilibrium was again established. Cooling of a region in a confined system, on the other hand, would reduce the fluid pressure in that area and cause the fluids to move toward it. Temperature changes are probably much more important in their effects on the viscosity of the subsurface fluids than on their volume, because of the relatively low coefficients of expansion of subsurface fluids and their surrounding rock framework.

Secondary Precipitation or Cementation. In many reservoir rocks the original minerals have been partly reconstituted and recrystallized, and secondary minerals have been formed. Where there has been more deposition than solution, a decrease in porosity has resulted, which in a sealed or in a confined space means an increase in fluid pressure. Where solution is greater than deposition, on the other hand, an increase in pore space results, and there is a decrease in fluid pressure. These differences are probably only temporary and soon become equalized.

Earthquakes. The advance compressive waves of earthquakes have been observed to cause elastic compression in shallow aquifers,[9] and earthquakes must in this way have caused many sudden changes in fluid pressure throughout past geologic time. The effects are a sudden rise and fall in the ground-water level, sometimes causing springs to start flowing again. If such a force operates in aquifers at shallow depths, it undoubtedly is present at greater depths and in the petroleum reservoirs. The Tehachapi-Bakersfield (California) earthquake of 1953, for example, caused the production in the near-by Mountain View pool to double over a period of several weeks following the disturbance. The enormous number of earthquakes, now averaging over one million per year, and their resultant compressions and expansions during geologic time could perhaps become an important "micro" pressure source; they may be likened to a "sonic" vibration of the earth.

Tides, Tsunamis, Atmospheric Pressures. Tidal and other disturbances of oceans undoubtedly cause minor and temporary elastic pressure effects in the underlying rocks. The shifting of water into the ice caps and back into the ocean at various times during geologic history shifted the load over large areas of the earth, a matter of hundreds of pounds per square inch. The great tsunamis,[10] or seismic sea waves, caused by submarine earthquakes, such as have been observed within recent years, must have occurred throughout past geologic time and certainly caused innumerable temporary changes in pressure. Even the usual changes in atmospheric pressures from night to day, and from high to low extremes in connection with storm movements, are sufficient to cause fluctuations in the rise of water from shallow aquifers in wells and springs.* Reservoir pressures in a pool in Illinois, for example, at a depth of 1,500 feet, are known to fluctuate whenever a train passes over the surface.† The subsurface effects of varying atmospheric pressure were noted as far back as 1753 when Hanway,[11] in describing an area of oil seepages in the Caspian Sea, stated: "When the weather is thick and heavy [low pressure] the springs boil up higher and the naphtha often takes fire at the surface . . . in clear weather [high pressure] the springs do not boil up over 2–3 feet."

Though many of these causes may seem minute, their cumulative importance over geologic time may be great. The significance of these countless minor pressure changes is not in adding to the permanent reservoir pressure

---

* Don E. White in a personal communication.
† John A. Murphy in a personal communication.

but in the fact that they may, when added together, "trigger" the release of oil and gas held in the rocks by capillary forces and friction and thus aid their migration and accumulation. During the early stages of burial, especially, these minor pressure changes may significantly influence the fluid pressure relationships within the sediments.

Chemical and Biochemical Causes. A decrease in fluid volumes means a decrease in the fluid pressure. The volume of a solution is far less than the combined volumes of the solute and the solvent. In fact, the addition of common salt to distilled water decreases the volume of the solvent because the molecules are forced closer together.[12] Solution and precipitation are common in all aqueous underground formations. The breakdown of the larger hydrocarbon molecules into simpler compounds tends to increase the volume; if the volume of the system is relatively fixed, the fluid pressure will increase. The change may be due to catalytic reactions, bacterial reactions, radioactive decay, or temperature changes. The decomposition of organic matter through bacterial actions, for example, frequently forms pockets of methane gas under pressure, as may be observed in swamps or along beaches. Although specific examples of such decay in deeply buried sediments are lacking, phenomena such as these may be a source of local fluid pressure increases. Where the void chamber has low permeability boundaries, the increase in fluid pressure would be maintained for a relatively long interval of time; in an unsealed or unconfined rock, it would merely furnish a temporary, local increase in fluid pressure, which would be quickly dissipated.

Variations from Expected Pressures—Anomalous Pressures

The potentiometric surface is generally close to the surface elevation of the ground. There are many exceptions, however, and in some areas the potentiometric surfaces are hundreds or even thousands of feet either above or below the surface of the ground. (See Figs. 9-6, 9-8, and 9-9.) Such areas are called areas of "excess" or of "subnormal" fluid pressure. One mode of origin of such anomalous fluid pressures is shown in Figure 9-3. Since many areas lack the topographic relief capable of generating such fluid pressure anomalies, some other explanation for anomalous fluid pressures must be attempted.

Excess reservoir pressures are difficult to measure because they are frequently encountered unexpectedly. If the pressure exerted by the column of drilling mud is not enough to hold the fluids in the rocks, they may blow the mud out of the hole, and, if gas is present, it may catch fire. Some idea of the fluid pressure may be had by comparing the pressure exerted by the mud column at the time of the blow-out with the pressure exerted by the mud column when the well is brought under control. When sufficiently heavy drilling mud is used to protect against excess pressure and blowouts, there

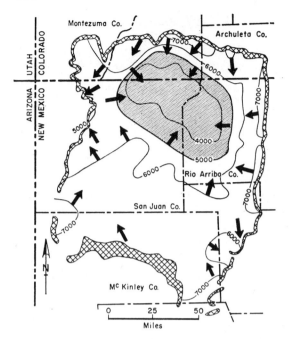

FIGURE 9-7

Potentiometric surface of the Point Lookout sand (Cretaceous) in the San Juan Basin, New Mexico and Colorado. Arrows show the direction of water flow, and contours show the potentiometric surface in feet above sea level. The pronounced, closed fluid potential depression (stippled)—a fluid pressure "sink"—indicates a striking fluid pressure anomaly, wherein non-Darcy Law fluid flow occurs across the bedding. [After Frederick A. F. Berry, Ph.D. thesis, Stanford University, 1959.]

is a danger of lost circulation when the high-pressure mud enters the low-pressure formations that may occur above those of excess fluid pressure. Lost circulation of drilling mud, in fact, is a possible indication of a low-pressure fluid-bearing formation. When such a formation is located within

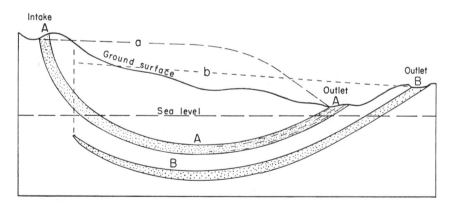

FIGURE 9-8 *Idealized section through a structural sedimentary basin, showing a profile of the potentiometric surface a of reservoir rock AA and the potentiometric surface b of reservoir rock BB. The height of the potentiometric surface above the reservoir rock at any point indicates the fluid pressure at that point. The change in slope of the potentiometric surface a occurs where the reservoir rock AA grades from sand to shaly sand.*

FIGURE 9-9

Diagrammatic sketch showing how the reservoir pressure may be atmospheric when the potentiometric surface intersects the trap below its highest point. In this case the oil floats on the water.

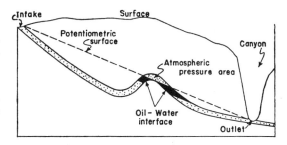

a sequence of higher-pressure formations, it is often of great significance in the location of a potential petroleum-producing reservoir.

The following list gives some of the more common suggestions offered to explain pressure anomalies.

1. What appears to be a local fluid pressure anomaly may merely be an extension into that area of a regional potentiometric surface that is unusually high or low. If a larger area is considered it will be found that the local fluid pressure is not anomalous but is in conformity with the fluid pressures of the region.

2. Local anomalously high potentiometric surfaces may be due to local diastrophic pressure associated with deformed strata, especially in soft shales and unconsolidated sands such as those that characterize many of the Tertiary productive regions. Salt intrusion, faulting, minor folding, lateral sliding and slipping, and squeezing that results from down-dropping of faulted blocks are some of the phenomena that might cause anomalous fluid pressures in recently deposited sediments.

3. The mechanical compression of water-filled sediments, such as clays, has been suggested by Hubbert and Rubey[13] as a mechanism that will produce abnormally high fluid pressures. Thus the progressive loading of a sedimentary section creates a high fluid pressure in clays undergoing compaction, and also in the porous sand lenses contained within them.

4. Anomalously high or low fluid pressures within a reservoir may be due to the presence of aquifers that are either above or below the reservoir and have higher or lower potentiometric surfaces than the reservoir. (See Fig. 9-6.) The connection may be via Darcy or channel flow through truncation and unconformable overlap, faults, fractures, or via diffusive flow through osmotic and electro-osmotic phenomena. Many large uplifts with anomalous fluid pressure systems are associated with faults and fracture systems in the basement rocks through which fluid pressures may be transmitted for long distances.

Abnormally high or low reservoir fluid pressures are not as important, per se, in the geology of petroleum as are the fluid potential gradients that exist in the reservoirs. Some examples of fluid pressures that are excessive for the depths of the reservoirs are shown in Figures 9-10,[14] 9-11, and 9-12.[15]

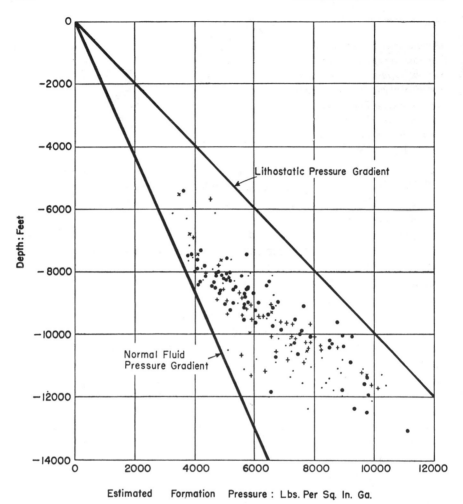

FIGURE 9-10 *The estimated reservoir pressure versus depth for over 100 wells in a group of Gulf Coast oil fields. Large circles, Frio formation; small dots, Cockfield, Miocene; crosses, Anahuac formation.* [*Redrawn from Cannon and Sullins, O. & G. Jour., May 25, 1946, p. 120, and Oil Weekly, May 27, 1946, p. 34.*]

The reservoirs associated with salt domes along the Gulf Coast of Texas and Louisiana, especially, are frequently characterized by excess fluid pressures, which are probably caused by the faulting and diastrophism accompanying the intrusion of the salt domes into the soft and incompetent Tertiary sediments of the region. Some examples are shown in Figure 9-10. Wedges or blocks of tightly compressed sediments with fluid pressures on the order of 50–80 psi per 100 feet of depth are common, and it requires great care in drilling to prevent the wells from "blowing out" and getting out of control. Isolated sand lenses, also in the soft Tertiary sediments, are often characterized by high fluid pressures and low fluid volumes, perhaps because they were sealed in by the load of the overlying sediments.

A notable example of excess fluid pressures is found in the isolated thrust-fault blocks of the Ventura oil field of California.[15] (See Fig. 9-12.) The best explanation seems to be that these fluid pressures are caused by diastrophic compression and then sealed in by the thrust faults bounding the zones or pools that occur in the field. The rocks are soft Tertiary formations. Another possible cause of excess fluid pressures in sealed reservoirs is the increase in volume of hydrocarbons of high molecular weight when they are converted to the lighter hydrocarbons. Another example of excess fluid pressures occurs in the Bayou St. Denis area of Jefferson Parish, Louisiana, on the northern edge of Barataria Bay. There the bottom-hole fluid pressure was 12,635 psi at a depth of 13,000 feet.[16] Gas wells in the Yates sand (Permian) of the East Wasson field in Yoakum County, western Texas, encountered fluid pressures up to 2,800 psi at a depth of 3,120 feet. The gas was non-inflammable and was 97 percent nitrogen.[17] Reservoir pressures in

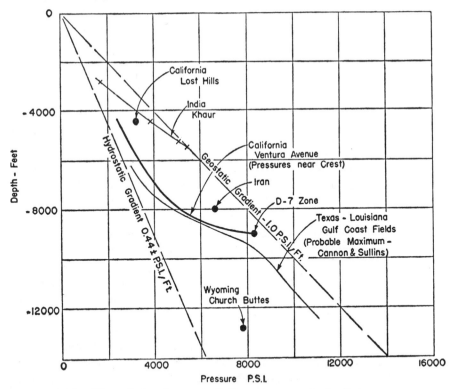

FIGURE 9-11 *The relation between fluid pressure and depth in some excess-pressure pools. [Redrawn from Watts, Trans. Amer. Inst. Min. Met. Engrs., Vol. 174 (1948), p. 194, Fig. 2.]*

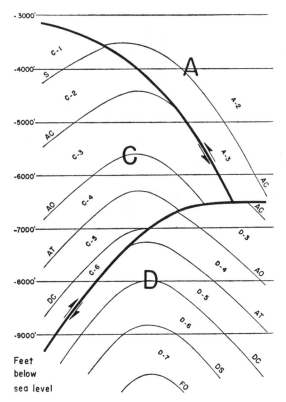

FIGURE 9-12

Diagrammatic section across the Ventura Avenue oil field, California, at a depth of from 3,000 to 9,000 feet below sea level. The letters and numbers represent local correlation and producing formations, and the large letters, A, C, *and* D, *represent separate fault blocks formed by a complex system of intersecting thrust faults superimposed on a large dome fold. An example of excess reservoir pressure is found in the* D-7 *zone in the* D *block, which has a reservoir pressure approaching the geostatic gradient of 1.0 psi per foot of depth.* [Redrawn from Watts, Tech. Paper 2204 (1947), Trans. Amer. Inst. Min. Met. Engrs., Vol. 174 (1948), p. 192, Fig. 1.]

the Khuar field, Punjab, India (now Pakistan), were found to be almost equal to the geostatic pressure; a fluid pressure of 5,060 psi was encountered at a depth of 5,212 feet, and a pressure of 5,420 psi at a depth of 5,748 feet.[18] These pressures are thought to be due to orogenic movements, which also sealed in the pool. The Forest Reserve oil field of Trinidad is reported to have a reservoir pressure that results in a pressure gradient of 0.89 psi per foot of depth from the surface to the reservoir.[19] The producing formations in both Pakistan and Trinidad are sands and shales of Tertiary age characterized by close folding and much faulting.

Anomalously high fluid pressures are also characteristic of the diapirically folded rocks in the Carpathian and Caucasus regions of eastern Europe. There, however, the amount of original reservoir pressure was not generally measured. Pressure-measuring devices may not have been available, and even if they had been they could hardly have been used, for the drilling methods were unable to cope with the reservoir pressures encountered, and many of the early wells blew wild for weeks and even months. Large quantities of loose sand, moreover, were produced along with the oil and gas from many of these wells and would have cut out and destroyed any equipment used underground.

Where high artificial fluid pressures are applied to rocks underground, as in water repressuring, hydraulic fracturing, acidizing, and cementing, it has been observed that, as the pressures approach the geostatic pressure for the given depth, the formations often "break down"; there is a sudden increase in the amount of fluid that enters the formation, apparently as a result of a sudden increase in space. Such a condition may result from the formation of new fractures, from the opening up of incipient fractures, or from parting along bedding planes. It has also been explained as being due to lifting of the overburden. Breakdown pressures of from 0.5 to 1.7 times the geostatic pressure normal for the given depth have been reported.[20]

In some areas, reservoir pressures that seem abnormally low for the depth below the surface may be due to the naturally low position of the piezometric surface. One well-known example of abnormally low reservoir pressure is found in the Amarillo–Texas Panhandle field of northwestern Texas. Here, at a surface elevation of 3,400 feet, a gas-oil contact at 200–400 feet above sea level, and an oil-water contact at 0–100 feet below sea level, the reservoir pressure was originally 430 psi, or about half of what might be normally expected for the depth of 2,000–2,500 feet to the top of the gas (1,000 feet above sea level) production. Two explanations have been offered.[21] One is that the pressure was sealed in at some previous geologic period when the producing formations were under hydrostatic pressure conditions. The other is more probable: that the granite-wash producing formation crops out to the east in the Wichita Mountains of Oklahoma at an elevation of about 1,000 feet above sea level, which would be approximately the potentiometric surface of the producing formation of the Amarillo field. An extreme example might be in plateau country cut by deep canyons wherein there is considerable structural relief of the various reservoir rocks. At many locations the reservoir rocks in the subsurface are situated above the groundwater table of the area, in which case the rock pore spaces are exposed to the atmosphere even though they might be encountered at considerable depths below the ground surface.

Although it is true that fluid pressures within the same interconnected reservoir rock show a more or less uniform increase with depth, an increase of fluid pressure with depth is not always found when fluid pressures are measured in different formations or reservoirs.[22] Fluid pressures are sometimes less in deeper formations than in shallower formations in the same hole.

Pools Trapped in Isolated Reservoirs. Many pools that are trapped in apparently sealed reservoirs—isolated lenses, bioherms, and biostromes—are found to have fluid pressures that are approximately hydrostatic, or equivalent to the pressure exerted by a column of water extending to the surface of the ground. If these reservoirs were truly sealed immediately after burial, the higher-than-atmospheric pressure would have to be due to some postdepositional phenomena.

Probably the best explanation for the increase in fluid pressure can be found by tracing the normal changes in volume and pressure that result from lithification and diagenesis. Since the sediments with which we are dealing were deposited in water, presumably their pores were originally filled with water. As the overburden increased, clays were compacted into shales and sands into sandstones, with a consequent loss of volume. Even though the permeability of shales is low, there is some connection between the pore spaces, permitting the water content to be in phase continuity. Thus the pressure due to loss in volume would have been transmitted to the water content and the water squeezed out of the shales into lenticular aquifers and potential reservoir rocks where it became a part of the measurable hydrodynamic environment of the sedimentary basin of which the aquifer is a part. A reduction in volume is expressed chiefly by reduction in porosity, which is the equivalent of a reduction in water content: water must be squeezed out of the sediments and out of the pores in which it was originally held. Where does it go?

Most of the reduction in volume comes within the clay and shale sediments. Freshly deposited clay may have a porosity of more than 50 percent. By the time the clays have become indurated into shales, the average porosity will have decreased to about 13 percent, largely as a result of the overburden pressure. As the overburden pressure increases and persists over geologic time, the average shale porosity will continue to decrease, although at a slower rate,[23] and at depths of 5,000–7,000 feet it may be expected to range between 5 and 10 percent.[24] The compaction is greater in clays and shales than in sands because of the plastic nature of the clays and the fact that many of them have been swollen by the water adsorbed onto their particles and by planar water contained within the molecular structure of their crystal plates. The removal of this water means an equivalent reduction in volume of the rock particles. The relative importance of the various processes that result in the compaction of clays and shales is shown graphically in Figure 9-13.

But sandstones also may lose pore space with an increase in the weight of the overburden, since they commonly contain some clay minerals. (See pp. 65–66.) These clays are squeezed into the pores held open by the touching sand grains, and a closer packing results. Thus a shaly or "dirty" sand may be expected to have suffered more reduction in pore volume for the same pressure than a clean sand. A clean sand would be expected to have preserved more of its porosity and permeability against the increase in overburden pressure during geologic time than a shaly sand. A deeply buried sand that is clean is therefore more attractive as a potential reservoir rock than a shaly or muddy sand.

The fluid, chiefly water, that is squeezed out of the clays and shales may be expected to move in the direction of lower fluid potential via the permeable paths that afford the least resistance to the flow of water. Wherever the fluid goes it must displace water that is already there, including water in the apparently sealed reservoirs, and this water in turn displaces other water, the

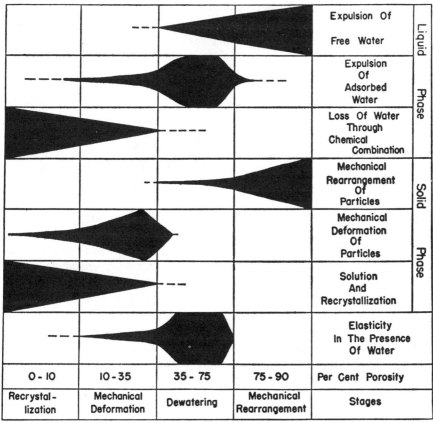

FIGURE 9-13 *The changing importance of different processes in the compaction of sediments from watery muds into dense shales and slates. For example, the expulsion of free water and the mechanical rearrangement of the rock particles dominate the compaction during the early stage, that of mechanical rearrangement, when the porosities range between 75 and 95 percent, the consistency of a watery mud. Later on, during the stage of mechanical deformation, when the porosity is between 10 and 35 percent, the expulsion of adsorbed water and the elasticity in the presence of water have declined, and the deformation of the particles is at a maximum. The final compaction process is the recrystallization of the mineral component grains and particles.* [Redrawn from Hedberg, Amer. Jour. Sci., 5th series, Vol. 31 (1936), p. 281, Fig. 7.]

end point being an escape to the surface of the ground or to another aquifer in which the fluid potential is lower. In other words, the fluid pressure developed during the compaction of the clays and shales is transmitted to the water and sets up a fluid potential gradient that causes fluid flow toward an environment of lower potential energy. Equilibrium is established when the compressive forces squeezing the fluids out of the shales and lenticular reservoir rocks equal the forces resisting further movement of water. These forces include (1) the relative difference between the fluid potential energy of the aquifers and reservoirs and that of the compacting shales, (2) the frictional resistance caused by the low permeabilities of the intervening rocks, (3) the jell strength, or molecular strength, of clay minerals holding sheets of water, and (4) the capillary forces holding the fluids in the pores. The establishment of a state of hydrostatic pressure equilibrium may be a long, slow process during which the opposing forces are nearly equal, and the isolated reservoir may have excess fluid pressures over long periods of time. In some cases, then, the fluid pressure may be sealed in, but with an imperfect seal, of extremely low permeability, that permits, over a period of time, a fluid pressure readjustment to the more normal fluid pressures found in the aquifers above and below the seale-in rock.

Faulting may not only cause excess fluid pressure but also provide for a release of pressure. Faulting offers a possible explanation of the hydrostatic pressures found in the Tertiary sands of the Gulf Coast plains region of the United States, where, although the reservoir sands fail by several thousands of feet in reaching a surface outcrop, the fluid pressures are generally normal for their depth. The region is characterized by a large number of normal faults, some associated with the salt plugs and others more or less parallel to the coast and to the thickness contours (isopachs) of the sands. These faults, possibly in conjunction with a low-permeability connection to the surface, may have provided a release of the excess fluid pressures that were transmitted to the reservoir fluids by the squeezing of the soft clays and shales as the overburden pressure increased.

It was probably in some such ways as these that normal pressures developed in the fluids of porous reservoirs, even where there is no known permeable fluid connection between the reservoir and the surface of the ground, and hence no active water-pressure head. Pressures in many sealed reservoirs often approach what might be called a hydrostatic depth equilibrium; this must mean that there is or has been some connection, that there is some permeability, even in rocks that appear impermeable to oil, gas, and water. The connection may be long and tortuous, along devious and variable paths, in rocks of highly irregular permeability. Attainment of a state of equilibrium may require long geologic time, or it may come suddenly, perhaps as a result of faulting. But the dominant, ultimate fluid pressure to be overcome by the squeezed-out water is approximately that necessary to push a column of water to the level of the potentiometric surface.

TEMPERATURE

The temperature underground normally increases with depth below the surface, and the rate of increase with depth is called the *geothermal gradient*. The gradient is approximately constant below an upper surface (50–400 feet), where the temperature is affected by atmospheric temperature changes and by the circulation of ground water. While the temperature gradient is generally constant within any one hole, it may vary greatly from area to area, even in the same stratigraphic sequence of rocks.[25] Unlike reservoir pressures, which generally decline with oil and gas production, the reservoir temperatures remain fairly constant.

Temperature Measurement

Continuous underground temperature measurements are made by lowering a recording thermometer into a well. The recording device may be within an instrument called a *temperature bomb,* which is lowered into the hole, or it may remain at the surface while only a thermometer is lowered.[26] If a measurement of temperature at the thermal equilibrium of the well is desired, the well may have to be left idle for days or even weeks in order to allow the effects of the various local conditions affecting the temperature to dissipate, such as setting cement, gas entering the hole, casing leaks, or water sands.

Geothermal Gradient

The geothermal gradient is obtained by dividing the difference between the temperature of the formation and the mean annual temperature by the depth of the formation:

$$\text{geothermal gradient} = \frac{\text{formation temperature} - \text{mean annual surface temperature}}{\text{depth in feet}}$$

It may be expressed in several ways. One way is to give the number of degrees Fahrenheit per 100 feet of depth (°F/100 ft), and the average is found to be two degrees Fahrenheit per 100 feet (2°F/100 ft, or 1.11°C/100 ft). Another and possibly more common method is to give the reciprocal gradient, or the number of feet per degree Fahrenheit (ft/°F). This is found to average 50 feet of increase in depth for each degree Fahrenheit (50 ft/°F) and to range from about 20 feet to 180 feet per degree Fahrenheit. A 5,000-foot hole may thus be expected to have, on the average, a bottom-hole temperature of 160°F (a surface temperature of 60° plus 100°, the average gradient for 5,000 feet). Typical geothermal gradients are given in Figures 9-14, 9-15, and 9-16. Abnormally high gradients are on the order of 20–40 feet per degree F, and abnormally low gradients are on the order of 120–180 feet per degree F.

An *isothermal surface* is the surface in the ground at which the temperature is everywhere the same. A profile section showing several isothermal surfaces between Oklahoma City and Tulsa, Oklahoma, is shown in Figure 9-17. The isothermal surface, for example, of 100°F is found at a depth of about 4,200 feet near Oklahoma City, but at only 1,800 feet near Tulsa, about 100

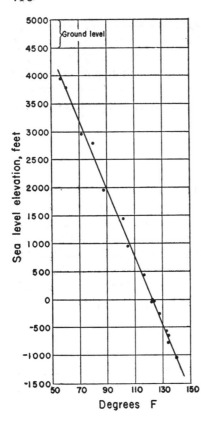

FIGURE 9-14

Geothermal gradient in the Elk Basin oil field, Wyoming, with points of control shown. The temperature increases 80°F to a depth of 5,000 feet, or at a geothermal gradient of 1°F/62.5 ft. [Redrawn from Espach and Fry, U.S. Bur. Mines, RI 4768, Fig. 12, opp. p. 10.]

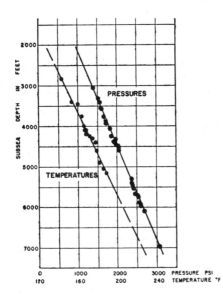

FIGURE 9-15

Temperature and fluid pressure gradients in a group of pools in the Greater Oficina district, eastern Venezuela. The control points are shown. The geothermal gradient is approximately 1°F/50 feet of depth. [Redrawn from Hedberg, Sass, and Funkhouser, Bull. Amer. Assoc. Petrol. Geol., Vol. 31 (1947), p. 2124, Fig. 11.]

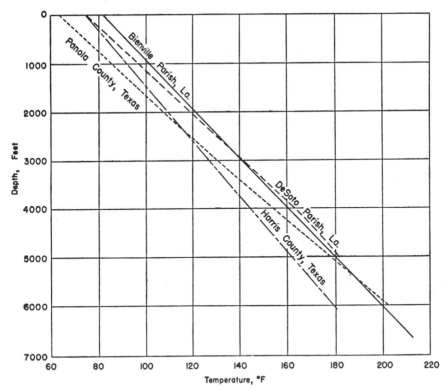

FIGURE 9-16 *Geothermal gradients of a group of oil pools in northeastern Texas and northwestern Louisiana. The gradients range from 1°F/44 feet to 1°F/50 feet. [Redrawn from Nichols, Tech. Paper 2114, Trans. Amer. Inst. Min. Met. Engrs., Vol. 170 (1947), p. 46, Fig. 2.]*

miles away.[27] This means that the geothermal gradient near Oklahoma City is around 1°F per 100 feet of depth, whereas it is 1°F per 36 feet of depth near Tulsa. All the rocks near Tulsa are older and nearer the Precambrian granites and metamorphic rocks of the basement than those at Oklahoma City, and, being older and different in character, they may possess a higher rate of heat conductivity.

Contours may be used to show the geothermal gradient of an area. Contours that show the position above or below sea level of planes of equal temperatures are called *isogeothermal contours,* and a single such contour is called an *isogeotherm.* Contours that show the gradient per 100 feet of depth are called *isogradient contours.* An isogradient contour map of parts of Texas, New Mexico, Oklahoma, Arkansas, and Louisiana, for example, shows variations of 0.4–2.2°F per 100 feet of depth.[28] (See Fig. 9-18.) Of interest is the abnormally low gradient in the form of a "basin" coinciding roughly with the central platform of the Midland Basin of New Mexico and western Texas.

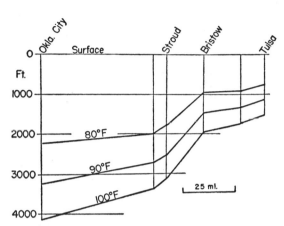

FIGURE 9-17

Section showing isothermal surfaces between Oklahoma City and Tulsa, Oklahoma, a distance of approximately 100 miles. The 100°F surface slopes at a rate varying from 1°/107 feet at the west to 1°/36.5 feet at the east, near Tulsa. The 100°F isotherm approximately parallels the dip of the formations; the older Paleozoic sediments and Precambrian granites are closer to the surface at the eastern end of the section. [Redrawn from McCutchin, Bull. Amer. Assoc. Petrol. Geol., Vol. 14, p. 542, Fig. 2.]

The growing interest in oil and gas production at depths below 15,000 feet opens up many problems involving the effects of the high pressures and temperatures that may be encountered at these depths. An extrapolation of temperature gradients in a large number of different areas of the United States showed that a temperature of 212°F was found at a depth of less than 7,000 feet in a third of the cases, and deeper than 10,000 feet in another third.[29] The critical temperature of water is 374°C (705°F), and it was calculated to occur at depths consistently greater than 30,000 feet. Pressure and temperature data on a few deep wells are shown in Table 9-1. It should be noted that the gradients in some of these deep holes are not as great as in many local high-gradient areas that are productive.

When the data are accurate and the gradient is plotted on a depth-temperature chart, it is found that the gradient rarely follows a straight line and is most often slightly convex toward the depth axis; that is, it usually becomes slightly steeper with depth, which indicates that the temperature gradient gradually increases with depth.[29]

The gradient in a hole or in an area may vary with the character of the rocks being measured. A well in southeastern New Mexico,[30] for example, shows the following gradients for different types of sediments:

	GRADIENT (ft/°F)	DEPTH (feet)
Saline and limestone section	188–210	0–1,500
Limestone	144	1,500–4,000
Sandstone	99.4	4,000–6,000

An average gradient of 131.8 feet per °F is present in this well, which has a total depth of 6,683 feet in Permian rock. Stated differently, the gradient changes from 2°F per 100 feet of depth between the surface and a depth of about 3,500 feet to 1.2°F per 100 feet between 3,500 feet and 6,300 feet. Similarly, a chart of the average temperature gradients for eight measured wells from the Rangely field, Colorado[31] (Fig. 9-19), indicates that a change in gradient occurs at 1,800 feet above sea level, where the overlying shales of the Mancos formation (Upper Cretaceous) overlie the sandstones of the Dakota formation (Upper Cretaceous). An example of an increase in the temperature gradient in a pool exists in the La Paz field of western Venezuela, as shown in Figure 9-20. At depths of about 4,000 feet the temperature gradient changes from 1.66°F/100 feet above this depth to 1.0°F/100 feet below this depth. This depth corresponds with a change from Tertiary sands and shales to Cretaceous limestones. In each of these examples the change in geothermal gradient is best explained by a change in the thermal conductivity (see p. 423) of the rocks drilled.

Temperature gradients may also vary with local structure. The gradient on

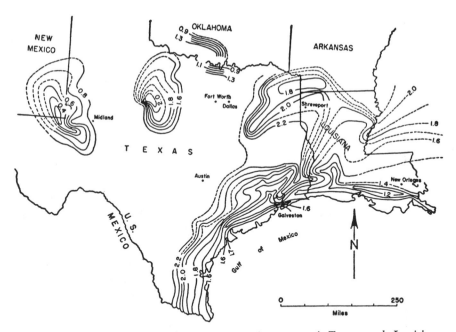

FIGURE 9-18 *Isothermal gradient contours in parts of Texas and Louisiana, showing the thermal gradient in degrees Fahrenheit per 100 feet of depth. The estimated subsurface temperature at any point may be determined by multiplying the thermal gradient shown on the map by the depth in hundreds of feet and adding the average surface temperature of 74°F. [Redrawn from Nichols, Trans. Amer. Inst. Min. Met. Engrs., Vol. 170 (1947), p. 46, Fig. 2.]*

TABLE 9-1 Fluid Pressures and Temperatures in Some Deep Wells

Name and Location of Well Deepest Formation Elevation above Sea Level	Total Depth (Feet)	Formation Fluid Pressure			Formation Temperature			Reference
		Depth	Psi	Gradient	Depth (Feet)	°F	Gradient (less 60° at Surface)	
Superior Oil Co., Well 34, #51-11, Sec. 11, T8N-R12W, Caddo Co., Okla., Springer Formation (Pennsylvanian), Elev. 1,440 ft	17,823	17,823	13,153	74 psi/100 ft	17,738	253	1.1°F/100 ft	1
Superior Oil Co., Pacific Creek, Sec. 27, T27N-R103W, Sublette Co., Wyo., Frontier sand (Upper Cretaceous), Elev. 7,090 ft	20,521	20,521	14,365	70 psi/100 ft	20,521	310	1.2°F/100 ft	1, 2
Superior Oil Co., Limoneira #1, Sec. 9, T2N-R22W, Ventura Co., Calif., Upper Miocene, Elev. 261 ft	18,734	18,734	10,678	57 psi/100 ft	18,734	250	1.0°F/100 ft	1
Bahamas Oil Ltd., Andros #1, Stafford Creek, Andros Is., BWI, Lower Cretaceous, Elev. 19 ft	14,585	14,585	6,315	43 psi/100 ft	10,670	98	0.3°F/100 ft	1

1. R. B. Hutcheson in a personal communication.
2. Charles J. Deegan, "Deepest Well Confirms Theories of Regional Conditions," O. & G. Jour., August 11, 1949, pp. 56–57.

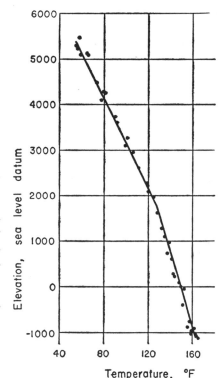

FIGURE 9-19

An average of eight temperature gradients in the Rangely oil field, Colorado. The change in rate of increase in temperature from 1°F/51 ft to 1°F/68 ft occurring at 1,800 feet above sea level (3,500 feet below the surface) coincides with the contact between the Mancos shale above and the Dakota sand below (both Cretaceous). The difference in thermal conductivity of the shales and sands probably best explains the change in gradient. [Redrawn from Cupps, Lipstate, and Fry, U.S. Bur. Mines, RI 4761 (1951), Fig. 5.]

the crests of fifty-seven anticlinal structures,[32] for example, was found to range between 47.4 and 50.6 feet per degree F, whereas the gradient on the flanks ranged from 51.3 to 62.1 feet per degree. A careful study of the reservoir temperatures in the Weber sandstone (Pennsylvanian) of the Rangely field of Colorado has shown that the reservoir temperatures are influenced by structure. The relations are shown in Figure 9-21, which indicates that the 160°F isogeotherm rises over the crest and steep flank of the anticline. Producing anticlines in Oklahoma have also been found to have higher temperatures than the surrounding areas.[26] The variations are small, but measurable and fairly consistent.

Studies in California[33] and in the Rhine valley[34] indicate that the presence of petroleum does not influence temperatures. Temperature varies, rather, with the structural position of the measurement, the structurally higher beds carrying the higher temperature, regardless of the presence of oil. High-temperature anomalies are also observed over salt domes. The explanation of higher temperatures over structural uplifts is best explained, generally, by the fact that rocks with higher temperatures are brought closer to the surface. Theoretically, the mapping of shallow temperatures would offer a means of exploring for high structural features, but the cost of such data is probably high in comparison with that of other mapping methods.

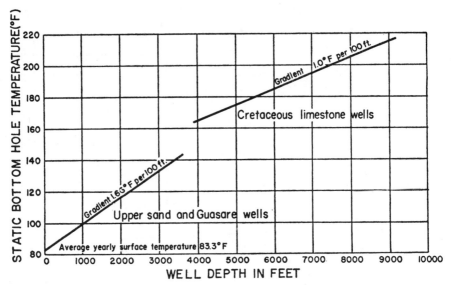

FIGURE 9-20 *The change in geothermal gradient from the surface to the bottom of the Guasare sands, at about 4,000 feet, and in the Cretaceous limestone from about 4,000 feet to 9,000 feet. [Redrawn from Rojas, Smith, and Austin, report of Ministry of Mines and Hydrocarbons, Venezuela, Nat. Petrol. Convention (1951), p. 214, Fig. 5.]*

Uses of Thermal Measurements

Most thermal measurements are made in order to locate different local heat effects within a well, and they show up as anomalous irregularities and abnormal gradients.[35] Such measurements are useful tools to the production engineer. The commonest determinations are the position of the cooling that accompanies the expansion of gas, thus locating the point where the gas enters the well, and the position of the heating that accompanies the setting of cement, thereby locating the top of the cement used in setting the casing. Temperature surveys aid in correlating strata between wells, since different strata have different specific heats and thermal conductivities. Temperature surveys are also used to detect gas leaks in wells from gas-storage reservoirs, and losses as low as 300 cubic feet of gas per day may be measured.

Sources of Heat Energy

Temperature measurements furnish evidence of (1) the presence of heat energy stored in the sediments and their fluids, and (2) the direction of the flow of heat, as shown by the thermal gradients. The geothermal gradient, coupled with the specific heat of the average sedimentary rock, shows that

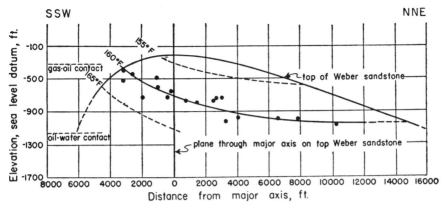

FIGURE 9-21 Isogeothermal section through the Rangely field, Colorado, showing the position of twenty thermal measurements across the structure, a dome fold, of the producing Weber sandstone (Pennsylvanian). The 160°F isogeotherm rises about 600 feet as it approaches the steep side of the fold. [Redrawn from Cupps, Lipstate, and Fry, U.S. Bur. Mines, RI 4761 (1951), Fig. 6, opp. p. 6.]

there is a vast amount of heat energy flowing both toward the surface and parallel to it through the upper few miles of the earth's crust.

The source of the heat of the upper few miles of the earth's crust may be in the outward flow of heat from the central core of the earth,[36] in the presence of igneous magmas that are cooling, in the disintegration of radioactive elements,[37] or in the heat of subcrustal thermal convection currents.[38] Lesser sources of heat include the frictional heat formed during diastrophism, when individual grains rub against one another, and the exothermal chemical reactions that take place within the permeable reservoir rocks, both of which sources, if present, are temporary and local in their effects.

The ability of a substance to transmit heat is known as its *thermal,* or *heat, conductivity.** The thermal conductivity of some common earth materials is shown in Table 9-2. Variations in geothermal gradient from one area to another may be explained by combinations of different heat sources with different thermal conductivities; changes in geothermal gradient in one well or in several wells in one area may best be explained by varying thermal conductivities of the different sediments penetrated.

It should be remembered that the observed underground temperature at any point may be the net effect of several causes. Moreover, any temperature reading, like any reservoir-pressure measurement, covers but a moment in geologic time; it marks only one point on a curve that would probably be

---

* Thermal conductivity is defined as the amount of heat in calories transmitted per second through a plate of the material one centimeter thick across an area of one square centimeter when the temperature difference is one degree centigrade.

TABLE 9-2 Thermal Conductivities of Various Substances

Substance	Conductivity* (Calories)— Approximate Values at Ordinary Temperatures	Substance	Conductivity* (Calories)— Approximate Values at Ordinary Temperatures
Basalt	0.0052	Sandstone	0.0055
Chalk	0.0020	Slate	0.0047
Earth's crust, average	0.004	Petroleum	0.000355
Granite	0.004–0.005	Water (0°C)	0.00120
Lime	0.00029	Water (20°C)	0.00143
Magnesium carbonate	0.00023–0.00025		

Source: Compiled from Bingham and Jackson, Bull. 14, U.S. Bur. Std. (1918), p. 75.

found, if it were possible to observe its direction over recent geologic time, to be either rising or falling rather than level. Since each of the ordinary geologic processes going on in the petroleum-bearing regions, such as deposition, erosion, diastrophism, and igneous intrusion, has its own heat effect, we may well conclude that the temperature of the sediments in the upper few miles of the earth's crust has been fluctuating up and down throughout geologic time, at least within the limits of present observed temperatures—that is, between 60° and 325°F. Whatever effects of increasing or decreasing heat we may observe in the present oil fields occurred also, we may therefore believe, at various times in the geologic past.

Effects of Heat

Temperature enters into many of the problems of the geology of petroleum, especially those having to do with production and appraisal and with the origin, migration, and accumulation of petroleum. In general, the problems involving temperature arise from the effects of heat on the physical properties of petroleum, of the associated fluids, of the dissolved salts, and of the reservoir rock. Some of the effects of an increase in temperature are summarized in the following section. A reverse effect may be expected to accompany a decline in temperature (loss of heat).

1. The most important effects of an increase in temperature, in the geology of petroleum, are an increase in the viscosity of gas (at low pressures) and a decrease in the viscosity of oil under conditions of constant pressure. (See pp. 201–207.) The effect of a temperature increase on the viscosity of pure water is given in Table 9-3. The combined effect of pressure and temperature changes on the viscosity of water is shown in Table 9-4.

2. Increase of temperature causes an increase in the volume of gas, oil, and

TABLE 9-3 Viscosity of Pure Water

Temperature		Viscosity Centipoises	Temperature		Viscosity Centipoises
°C	°F		°C	°F	
0	32	1.792	90.3	200	0.315
20.20	71.6	1.00	100	212	0.284
30.8	100	0.784	120.1	250	0.232
50	122	0.549	140.9	300	0.194
60.5	150	0.464	150	302	0.184

Source: Adapted from *Handbook of Chemistry and Physics*, 13th ed. (1948), pp. 1729–1730.

rock. The volume of a gas is directly proportional to its absolute temperature; hence the volume of the gas increases $\frac{1}{273}$ of its volume at 0°C for each degree centigrade of increase in temperature at constant pressure, or $\frac{1}{460}$ of its volume at 32°F for each degree Fahrenheit. The coefficient of expansion* of all gases at moderate pressures is approximately the same: 0.003665 per degree centigrade, or 0.002174 per degree Fahrenheit. These figures do not hold, however, for high pressures. The increase in the volume of a specific oil with an increase in temperature is shown graphically in Figure 5-31, page 203. The effect of temperature on the density of water is shown in Table 9-5. The coefficient of linear thermal expansion† of several sedimentary rocks and minerals is given in Table 9-6. Heating carbonate rocks in the laboratory to

TABLE 9-4 Viscosity of Water at High Pressures and Temperatures (at Constant Volume)

Pressure (psia)	Absolute Viscosity (Centipoises)			
	32°F	50.5°F	86°F	166.6°F
14.2	1.792	1.40	0.871	0.396
7,100	1.680	1.35	0.895	0.411
14,200	1.65	1.33	0.921	0.428
21,300	1.67	1.33	0.950	0.443
28,400	1.71	1.35	0.986	0.461

Source: Adapted from Percy W. Bridgman, *The Physics of High Pressures*, Macmillan Co., New York (1931), p. 346.

* The coefficient of expansion of gases at constant pressure is the change in volume per unit volume per degree centigrade.
† The cubical expansion of solids is approximately three times the linear coefficient and is the increase in volume per unit volume per degree centigrade.

TABLE 9-5 Temperature and the Density of Water

Degrees C	Degrees F	Grams per Cubic Centimeter
4	32	1.000000
10	50	0.99973
50	122	0.998807
100	212	0.95838

temperatures of 100–700°C, at confining pressures of one atmosphere, 5,000 psi, and 10,000 psi, not only increased their volume permanently but increased their permeabilities also.[39]

3. An increase in temperature causes an increase in fluid pressure where the fluids are confined. In natural petroleum reservoirs, since the fluids are either confined or sealed, there is little or no room for expansion, and a temporary increase of reservoir pressure in confined reservoirs, and a permanent increase in sealed reservoirs, would result. Increase of temperature increases the volume of the liquid and solid elements of the reservoir system as well as that of the gases, but only to a degree that is infinitesimal by comparison. Where the materials are confined, the effect is to increase the fluid pressure, since any expansion of the rock presumably takes place chiefly at the expense of its porosity, a decrease of which puts pressure on the fluids. An increase in the temperature of the reservoir fluids, therefore, may result in an increase in their volume or an increase in their pressure or in both. The decrease in specific gravity accompanying an increase in temperature accounts for the low boiling points of many of the lighter compounds of crude oils. At temperatures below 300°F many of them are in gaseous form at atmospheric pressures.

TABLE 9-6 Coefficients of Linear Expansion

Substance	Coefficient $\times 10^{-6}$	Substance	Coefficient $\times 10^{-6}$
Calcite, parallel to axis	25.14	Quartz, crystal	5.21
Granite	8.3	Rock salt	40.40
Marble		Sandstone	7–12
15–100°C	15	Slate	6–10
100–200°C	23	Limestone, oolitic	9
200–300°C	28	Flint, SiO_2	17.4

Source: From Handbook of Chemistry and Physics, 30th ed., Chem. Rubber Pub. Co., Cleveland (1948) pp. 1745–1750.

4. An increase in temperature causes a decrease in the solubility of gas in oil. For example, the amount of gas held in solution at a pressure of 2,000 psi is 670 cubic feet of gas per barrel of oil at a temperature of 140°F, but only 600 cubic feet per barrel at a temperature of 200°F. (See Fig. 5-30, p. 203.)

5. An increase in temperature causes an increase in the solubility of salts in water. Most salts show an increase in their solubility in water with an increase in temperature. Sodium chloride, however, which is one of the commonest salts of oil-field waters, changes its solubility only slightly with an increase in temperature.

Effects of Heat from Intrusive and Extrusive Rocks. The presence, in sediments that are being explored for petroleum, of dikes, sills, and other intrusive igneous rocks, or of extrusive volcanic rocks, is generally a cause for concern. (See also pp. 74 and 423.) Just what do they signify? Obviously there once was considerable heat; possibly it distilled the oil out of the shales. Frequently the intruded shales show only a fraction of an inch of metamorphism and baking, with consequent loss of permeability, at the contact with the intrusive. Even if oil and gas had accumulated before the intrusion, there is no reason why the heat should have driven it all away if the trap was sealed tight, and the heat should have had no effect on the petroleum if the accumulation took place after the intrusion or extrusion. One effect of the dike or sill, however, may have been to reduce the permeability of the reservoir rock locally, and thereby to change the flow pattern of the migrating oil and gas; or it may even have formed a trap where there was none before. Or the opening of fissures associated with the igneous activity may have permitted oil and gas to escape upward or into nearby lower-pressure areas; in effect the intrusion may have been similar to hydraulic fracturing. Tuffs and agglomerates, generally deposited parallel to the sedimentary bedding, have become reservoir rocks (see p. 74), but their feeder vents have probably become dikes across the bedding and might change the flow pattern of the reservoir fluids.

The association of igneous activity with petroliferous deposits has been noted at a number of places. The Tampico-Tuxpan area of Mexico, for example, contains a large number of seepages associated with igneous intrusions.[40] The producing Palestine and Tar Springs sands (Upper Mississippian) in the Omaha pool, Gallatin County, Illinois, are cut by mica-peridotite dikes and sills ranging from one foot to fifty feet in thickness.[41] The large seepages and oil sands in the Permian formations and in the Botacatu sandstone (Triassic) are closely associated with numerous dikes and sills, as well as with the great Serra Geral lava flows (Triassic), in the northern Paraná region (State of São Paulo), Brazil. They are apparently the remains of what once were large accumulations of oil. Triassic (Karroo) sands and grits near the village of Bemolanga in northwestern Malagasy (Madagascar) are bituminous over a large area. This coincides with an area of great vulcanicity of Tertiary age. The concentration of petroleum is greatest where the dikes and fracturing are the thickest; this strongly suggests that the petroleum may have been distilled from the surrounding shales by the heat of the intrusives.[42]

We may conclude, then, that the presence of igneous intrusives as dikes and sills, or of extrusives as tuffs and agglomerates, does not improve the chances of finding commercial oil and gas pools. There are enough examples of oil associated with such features, however, such that their presence should not condemn a region. Any rock with sufficient porosity and permeability to permit the flow of water may be expected to become a potential reservoir rock, and many intrusives and extrusives and sediments associated with them have this characteristic. Probably the chief effect of intrusives is to modify the existing traps in the sediments penetrated, or to form new traps, the intrusive dike or sill or the metamorphosed zone surrounding the intrusion becoming an impermeable boundary. These new traps would only accumulate such oil as either formed or migrated after the igneous action.

CONCLUSION

We are accustomed to use and observe many pressure and temperature phenomena in our daily living, our laboratories, and our factories, and now we find they are active elements in oil and gas reservoirs deep in the ground. Their influence is evident in almost every phase of the relations of the fluids to one another and to their environment in the reservoir. Like all other elements of an oil and gas pool, they are highly variable and subject to continual change. The effects of pressure and temperature, in general, are twofold; one effect is on the sediments, where compaction, loss of porosity and permeability, and the squeezing out of fluids are most important; and the other effect is on the fluids, where energy is stored and where changes in volume and viscosity are the dominant effects. The petroleum geologist must be acquainted with the applications of pressure and temperature to the sediments and to the fluids if he is to understand what goes on within the reservoir.

Pressure and temperature are probably as important to the geology of petroleum as they are to the geology of ore deposits, and many of the effects on the fluids in the earth are common to both. In much of the geology of petroleum we are dealing with generally lower temperatures and pressures than in the geology of ore deposits, but even so, at the increasing depths that drilling is capable of attaining—now below 25,000 feet—we are reaching into the realms of high pressure. Some of the producing formations now found at shallow depths were once buried below much greater thicknesses of sediments; but, as far as can be observed, the effects of the higher pressures and temperatures on these rocks and fluids are indeterminate. An important application of a knowledge of high-pressure and high-temperature conditions, then, is to learn their influence on the fluids, the porosity, and the permeability of rocks that become, or have been, deeply buried. Most important, however, is the effect of pressure and temperature gradients on the movement of the fluids, both into and out of the reservoir. The basis for this understanding is known as reservoir mechanics, and this is the subject of the next chapter.

Selected General Readings

PRESSURE

William B. Heroy, "Rock Pressure," Bull. Amer. Assoc. Petrol. Geol., Vol. 12 (April 1928), pp. 355–384. Discussion of principles involved in reservoir pressure determinations.

D. G. Hawthorn, "Subsurface Pressures in Oil Wells and Their Field of Application," Trans. Amer. Inst. Min. Met. Engrs., Vol. 103 (1933), pp. 147–165. Discussion to p. 169. Describes some of the uses and applications of subsurface pressures.

V. C. Illing, "The Origin of Pressure in Oil Pools," in *The Science of Petroleum*, Oxford University Press, London and New York, Vol. 1 (1934), pp. 224–229. A general discussion of the sources of reservoir pressures.

M. King Hubbert, "Entrapment of Petroleum under Hydrodynamic Conditions," Bull. Amer. Assoc. Petrol. Geol., Vol. 37 (August 1953), pp. 1954–2026. A classic treatment of hydrodynamic conditions.

Hollis D. Hedberg, "Gravitational Compaction of Clays and Shales," Amer. Jour. Sci., Fifth Series, No. 184 (April 1936), pp. 241–287. 88 references listed. A classic discussion of the effects of pressure on clays and shales.

M. King Hubbert and William W. Rubey, "Role of Fluid Pressure in Mechanics of Overthrust Faulting," Bull. Geol. Soc. Amer., Vol. 70 (February 1959), pp. 115–166.

William W. Rubey and M. King Hubbert, "Role of Fluid Pressure in Mechanics of Overthrust Faulting," Bull. Geol. Soc. Amer., Vol. 70 (February 1959), pp. 167–206. These two classic articles contain many examples of abnormal fluid pressures in oil fields, a discussion of the causes of abnormal pressures, and an extensive bibliography.

B. A. Tkhostov, *Initial Rock Pressures in Oil and Gas Deposits*, Translated by R. A. Ledward, the Macmillan Company, Pergamon Press (1963), 118 pages. A discussion of rock pressures in the USSR (1960).

Gilman A. Hill, William A. Colburn, and Jack W. Knight, "Reducing Oil-Finding Costs by Use of Hydrodynamic Evaluation," in *Economics of Petroleum, Exploration, Development, and Property Evaluation*, Southwestern Legal Foundation, Dallas, Texas, Prentice-Hall, Inc., Englewood Cliffs (1961). One of few published discussions of applications of hydrodynamic analysis to petroleum exploration.

Frederick A. F. Berry, "Hydrodynamics and Geochemistry of the Jurassic and Cretaceous Systems in the San Juan Basin, Northwestern New Mexico and Southwestern Colorado," Ph.D. thesis, Stanford University (1959), 192 pp. One of the few available studies of the hydrodynamics and related water chemistry of an entire geologic basin.

TEMPERATURE

K. C. Heald, et al., *Earth Temperatures in Oil Fields,* Bull. 205, Amer. Petrol. Inst. (1930), pp. 1–129. Contains several articles on underground temperature effects.

F. B. Plummer and E. C. Sargent, *Underground Waters and Subsurface Temperatures of the Woodbine Sand in Northeast Texas,* Bull. 3138, University of Texas (October 1931), 178 pages. Extensive bibliographies on underground waters and temperatures. Contains numerous maps, charts, thermal analyses, and an integrated thermal study of a geological unit, the Tyler Basin.

C. E. Van Orstrand, "Temperature Gradients," in *Problems of Petroleum Geology,* Amer. Assoc. Petrol. Geol., Tulsa, Okla. (1934), pp. 989–1021. Discusses the relation of temperature to geologic features and contains a long list of temperature measurements in the United States.

C. V. Millikan, "Temperature Surveys in Oil Wells," Tech. Paper 1258, Petrol. Technol., Trans. Amer. Inst. Min. Met. Engrs., Vol. 142 (1941), pp. 15–23; reprinted Vol. 136 & 142, pp. 77–85. Describes the interpretation of various geothermal anomalies.

Carlton Beal, "The Viscosity of Air, Water, Natural Gas, Crude Oil and Its Associated Gases at Oil Field Temperatures and Pressures," Tech. Paper 2018, Petrol. Technol., Trans. Amer. Inst. Min. Met. Engrs., Vol. 165 (1946), pp. 94–115. 40 references. Tables and charts of viscosity effects due to changing temperature and pressure.

Reference Notes

1. William B. Heroy, "Rock Pressure," Bull. Amer. Assoc. Petrol. Geol., Vol. 12 (April 1928), pp. 355–384.

2. L. A. Pym, "Bottom-Hole Pressure Measurement," in *The Science of Petroleum,* Oxford University Press, London and New York, Vol. 1 (1938), pp. 508–515. 8 references.

3. P. McDonald Biddison, "Estimation of Natural Gas Reserves," in *Geology of Natural Gas,* Amer. Assoc. Petrol. Geol., Tulsa, Okla. (1935), pp. 1035–1052. Contains formulas and tables for gas-pressure calculations. Bibliog. 12 items.

4. O. & G. Jour., Vol. 57, No. 17, April 20, 1959.

5. W. A. Bruce, "A Study of the Smackover Limestone Formation and the Reservoir Behavior of Its Oil and Condensate Pools," Trans. Amer. Inst. Min. Met. Engrs., Vol. 155 (1944), pp. 88–132; also in Petrol. Technol., Tech. Pub. 1728 (May 1944).

Lawrence A. Goebel, "Cairo Field, Union County, Arkansas," Bull. Amer. Assoc. Petrol. Geol., Vol. 34 (October 1950), pp. 1954–1980.

6. "Review of Middle East Oil" and "Reservoir Studies and Characteristics," Petroleum Times, London, June 1948, pp. 15–17.

7. R. C. Rumble, H. H. Spain, and H. E. Stamm, "A Reservoir Analyzer Study of the Woodbine Basin," Trans. Amer. Inst. Min. Met. Engrs., Vol. 192 (1951), pp. 331–340; or Tech. Pub. 3219.

8. Gilman A. Hill, William A. Colburn, and Jack W. Knight, "Reducing Oil-Finding Costs by Use of Hydrodynamic Evaluation," in *Economics of Petroleum, Exploration, Development, and Property Evaluation,* Southwestern Legal Foundation, Dallas, Texas, Prentice-Hall, Inc., Englewood Cliffs (1961).

Frederick A. F. Berry, "Hydrodynamics and Geochemistry of the Jurassic and Cretaceous Systems in the San Juan Basin, Northwestern New Mexico and Southwestern Colorado," Ph.D. thesis, Stanford University (1959), 192 pp.

9. Stephen Taber, "Effect of Earthquakes on Artesian Waters," Econ. Geol., Vol. 23 (1928), pp. 696–697.

10. Francis P. Shepard, *Submarine Geology,* Harper & Brothers, New York (1948), 348 pages, pp. 47–53.

11. Jonas Hanway, *An Historical Account of the British Trade over the Caspian Sea . . . ,* London (1753), Vol. 1.

12. R. E. Gibson, "The Nature of Solutions and Their Behavior under High Pressure," Scientific Monthly, Vol. 46 (February 1938), pp. 103–119.

13. M. King Hubbert and William B. Rubey, "Role of Fluid Pressure in Mechanics of Overthrust Faulting," Bull. Geol. Soc. Amer., Vol. 70, pp. 149–158.

14. G. E. Cannon and R. C. Craze, "Excessive Pressures and Pressure Variations with Depth of Petroleum Reservoirs in the Gulf Coast Region of Texas and Louisiana," Trans. Amer. Inst. Min. Met. Engrs., Vol. 127 (1938), pp. 31–38.

George Dickinson, "Geological Aspects of Abnormal Reservoir Pressures in Gulf Coast Louisiana," Bull. Amer. Assoc. Petrol. Geol., Vol. 37 (February 1953), pp. 410–432. Bibliog. 36 items.

15. E. V. Watts, "Some Aspects of High Pressure in the D7 Zone of the Ventura Avenue Field," Trans. Amer. Inst. Min. Met. Engrs., Vol. 174 (1948), pp. 191–200. Discussion to p. 205. 16 references cited. Also Tech. Paper 2204, Petrol. Technol. (May 1947).

16. Leigh S. McCaslin, Jr., O. & G. Jour., September 8, 1949, p. 56.

17. Alden S. Donnelly, "High-pressure Yates Sand Gas Problem, East Wasson Field, Yoakum County, West Texas," Bull. Amer. Assoc. Petrol. Geol., Vol. 25 (1941), pp. 1880–1897.

18. C. E. Keep and H. L. Ward, "Drilling Against High Rock Pressures with Particular Reference to Operations Conducted in the Khaur Field, Punjab," Jour. Inst. Petrol. Technol., Vol. 20 (1934), pp. 990–1013. Discussion to p. 1026.

19. H. H. Suter, "The General and Economic Geology of Trinidad, B.W.I.," Colonial Geol. and Min. Res., Vol. 3 (1952), p. 22.

20. John C. Calhoun, "Pressure Parting of Formations," O. & G. Jour., January 12, 1950, p. 85. Bibliog. 5 items.

21. Victor Cotner and H. E. Crum, "Geology and Occurrence of Natural Gas in Amarillo District, Texas," in *Geology of Natural Gas,* Amer. Assoc. Petrol. Geol., Tulsa, Okla. (1935), pp. 401–403.

22. C. V. Millikan, "The Geological Application of Bottom Hole Pressures," Bull. Amer. Assoc. Petrol. Geol., Vol. 16 (September 1932), pp. 891–906.

23. L. F. Athy, "Density, Porosity, and Compaction of Sedimentary Rocks," Bull. Amer. Assoc. Petrol. Geol., Vol. 14 (January 1930), pp. 1–35.

Parker D. Trask, "Compaction of Sediments," Bull. Amer. Assoc. Petrol. Geol., Vol. 15 (1931), pp. 271–276.

24. Hollis D. Hedberg, "The Effect of Gravitational Compaction on the Structure of Sedimentary Rocks," Bull. Amer. Assoc. Petrol. Geol., Vol. 10 (1926), pp. 1035–1072.

Hollis D. Hedberg, "Gravitational Compaction of Clays and Shales," Amer. Jour. Sci., 5th series, Vol. 31 (April 1936), pp. 240–287. 88 references cited. A classical article on the subject of compaction.

25. P. L. Moses, "Geothermal Gradients Now Known in Greater Detail, World Oil (May, 1961), pp. 79–82. Also, American Petroleum Institute, Shreveport meeting, March, 1961. Many American reservoir temperatures are listed.

26. Hubert Guyod, "Temperature Well Logging," Oil Weekly, October 21, 28, November 4, 11, December 2, 9, 16, 1946.

Henry G. Abadie, "Thermal Surveys Applied to Oil Field Problems," Petrol. Eng., June 1947, pp. 47–48.

27. John A. McCutchin, "Determination of Geothermal Gradients in Oklahoma," Bull. Amer. Assoc. Petrol. Geol., Vol. 14 (1930), pp. 535–557.

28. Earl A. Nichols, "Geothermal Gradients in Mid-Continent and Gulf Coast Oil Fields," Trans. Amer. Inst. Min. Met. Engrs., Vol. 170 (1947), pp. 44–50.

29. H. Cecil Spicer, "Rock Temperatures and Depths to Normal Boiling Point of Water in the United States," Bull. Amer. Assoc. Petrol. Geol., Vol. 20 (1936), pp. 270–279.

C. E. Van Orstrand, "Temperature Gradients," in *Problems of Petroleum Geology*, Amer. Assoc. Petrol. Geol., Tulsa, Okla. (1934), pp. 989–1021.

30. Walter B. Lang, "Note on Temperature Gradients in the Permian Basin," Jour. Wash. Acad. Sci., Vol. 20 (1930), pp. 121–123.

Walter B. Lang, "Geologic Significance of a Geothermal Gradient Curve," Bull. Amer. Assoc. Petrol. Geol., Vol. 21 (September 1937), pp. 1193–1205.

31. Cecil Q. Cupps, Philip H. Lipstate, and Joseph Fry, *Variance in Characteristics of the Oil in the Weber Sandstone Reservoir, Rangely Field, Colo.*, RI 4761, U.S. Bur. Mines (April 1951), pp. 6–8.

32. C. E. Van Orstrand, "Normal Geothermal Gradient in United States," Bull. Amer. Assoc. Petrol. Geol., Vol. 19 (January 1935), pp. 78–115. Bibliog. 12 items.

33. Anders J. Carlson, "Geothermal Variations in Coalinga Area, Fresno County, California," Bull. Amer. Assoc. Petrol. Geol., Vol. 15 (1931), pp. 829–836.

34. I. O. Haas and C. R. Hoffman, "Temperature Gradient in Pechelbronn Oil-bearing Region, Lower Alsace; Its Determination and Relation to Oil Reserves," Bull. Amer. Assoc. Petrol. Geol., Vol. 13 (1929), pp. 1257–1273. Bibliog. 31 items.

35. Alexander Deussen and Hubert Guyod, "Use of Temperature Measurements for Cementation Control and Correlations in Drill Holes," Bull. Amer. Assoc. Petrol. Geol., Vol. 21 (June 1937), pp. 789–805.

C. V. Millikan, *Temperature Surveys in Oil Wells*, Tech. Paper 1258. Petrol. Technol., Amer. Inst. Min. Met. Engrs. (1940); Trans., reprinted Vol. 136 & 142, pp. 77–86.

36. Louis B. Slichter, "Cooling of the Earth," Bull. Geol. Soc. Amer., Vol. 52 (April 1941), pp. 561–600. 38 references cited.

37. Robley D. Evans and Clark Goodman, "Radioactivity of Rocks," Bull. Geol. Soc. Amer., Vol. 52 (April 1941), pp. 459–490. 63 references cited.

38. Reginald Aldworth Daly, *Strength and Structure of the Earth*, Prentice-Hall, New York (1940), pp. 396–399.

39. John C. Maxwell and Peter Varrall, "Expansion and Increase in Permeability of Carbonate Rocks on Heating," Trans. Amer. Geophys. Union, Vol. 34 (February 1953), pp. 101–106.

40. E. DeGolyer, "The Effect of Igneous Intrusions on the Accumulation of Oil in the Tampico-Tuxpam Region, Mexico," Econ. Geol., Vol. 10 (November–December 1915), pp. 651–662.

41. R. M. English and R. M. Grogan, "Omaha Field and Mica-Peridotite Intrusives, Gallatin County, Illinois," in *Structure of Typical American Oil Fields*, Amer. Assoc. Petrol. Geol., Tulsa, Okla. (1948), pp. 189–212.

42. Arthur Wade, "Madagascar and Its Oil Lands," Jour. Inst. Petrol. Technol., Vol. 15, No. 72 (1929), pp. 2–33.

43. M. King Hubbert, "Entrapment of Petroleum under Hydrodynamic Conditions," Bull. Amer. Assoc. Petrol. Geol., Vol. 37 (August 1953), pp. 1954–2026.

CHAPTER 10

Reservoir Mechanics

Phase relationships. Interface phenomena. Capillary pressure. Reservoir energy. Movement of oil and gas in pool. Production phenomena. Secondary recovery. Gas production. Economic factors and regulations.

THE FIELD of reservoir mechanics[1] includes a study of those physical and chemical changes in the reservoir and the reservoir fluids that take place during the dynamic period when a petroleum reservoir is being produced. An understanding of these phenomena is essential to efficient and profitable production of oil and gas. In the field of reservoir mechanics, the interest of the petroleum geologist merges with the interest of the petroleum engineer, for it is here that many geologic observations are best explained by engineering concepts. Strictly speaking, reservoir mechanics may be said to come into operation when the pool is penetrated by its discovery well and the first oil and gas are removed, although the same physical laws must have been operating ever since the petroleum-bearing rocks were laid down. Most of what evidence we have as to what happens when fluids are moving under ground comes from reservoir studies. The movement of oil and gas from the pool to the well may be thought of as an example in reverse and in miniature—a hand specimen—of the larger geologic process, over a much greater time, of the regional migration of oil and gas from a source area and its accumulation into a pool. The petroleum geologist, moreover, is called on more and more to interpret reservoir conditions in terms of geologic concepts; that is another reason for his need to understand the phenomena surrounding the movement of oil and gas out of the pool and into the well.

A great mass of new data and many new ideas continue to come from engineering studies of reservoir conditions and reservoir mechanics. Some of these have a direct application to many problems of petroleum exploration. Reservoir mechanics deals with a wide variety of subjects; it reaches into organic chemistry, physics, physical chemistry, fluid mechanics, and engineering. The particular concepts and subjects in which we as exploration geologists are interested are briefly reviewed in the following order: (1) phase relationships of hydrocarbons; (2) interface phenomena; (3) capillary phenomena; (4) reservoir energy; (5) movement of oil and gas in a pool; (6) production phenomena; (7) secondary recovery.

PHASE RELATIONSHIPS

An understanding of many of the different equilibria that exist between the fluids within an oil and gas reservoir under changing conditions of pressure and temperature is best obtained through a study of phase behavior.[2] Pressures and temperatures change, at times rapidly, during the production of oil and gas from a pool. Much of modern production practice is based on a knowledge of how oil and gas act in the reservoir—their phase behavior—when the fluids are being produced. This knowledge has been obtained in the laboratories of the chemist, the physicist, and the refiner and by study of actual field conditions. The phenomena studied in this way are of interest to the petroleum geologist, for many of them have occurred underground during geologic time.

The different substances in a petroleum reservoir that are in a state of equilibrium when the pool is discovered, and with which we deal during its producing life, are reservoir rock, water, natural gas, and crude oil. There are many systems* to study in a reservoir, such as the rock-water, the rock-gas, the rock-oil, the water-oil, the gas-oil, the rock-water-oil, the water, the oil, and the gas system. A *phase* comprises all parts of a system that have the same properties and composition. A phase may be described as a homogeneous, physically distinct, and mechanically separable portion of a system.† A phase is separated from all other phases by sharp physical boundaries called *interfaces* or *boundary surfaces*. The number of *components* of a system is the lowest number of independently variable constituents that expresses the composition of each of the possible phases that may occur. For example, if water is the system, it may occur in three phases, vapor, liquid, and solid, depending upon the prevailing conditions of temperature and pressure. It is composed of one constituent, H_2O, in all phases and is therefore a *one-component system*. The crude-oil–water system may be thought of as a two-component sys-

---

* A *system* is the substance or mixture of substances that is being studied, isolated from all others.

† Strictly speaking, we would speak of the methane-ethane system and the phases in which they occur as gas and liquid. For practical purposes, however, many reservoir concepts are clearer if crude oil is considered as the system, with two phases, natural gas and crude oil. Or we may think of the crude-oil–water system commonly occurring in the petroleum reservoir,

RESERVOIR MECHANICS [CHAPTER 10] 435

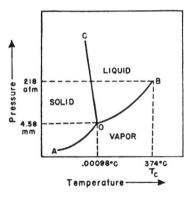

FIGURE 10-1

Phase diagram (schematic) of the system water. The triple point, O, corresponds to a pressure of 4.58 mm Hg and a temperature of 0.0098°C. Curve AO is the sublimation curve of ice and vapor, curve OC the equilibrium curve of ice and liquid, and curve OB the vapor-pressure curve of liquid and vapor. This diagram holds only for low pressures.

tem, and under reservoir and surface conditions of temperature and pressure three phases may occur: natural gas, oil, and water.

Water is a common example of a one-component system. A diagram of the system water under two variables, temperature and low pressure, is shown in Figure 10-1. The diagram shows the three possible single phases (ice, liquid, and vapor), the three two-phase equilibria (ice-vapor along line *AO*, ice-liquid along line *OC*, and vapor-liquid along line *OB*, which ends at the critical temperature, T_c), and the point *O*, where all three phases are in equilibrium. Point *O* is known as the triple point.

The phase in which a substance (or mixture of substances, such as crude oil) exists at any one time is determined by independent variables, commonly temperature and pressure, and also by such variables as concentration, density, and volume. The smallest number of variables that must be specified in order to completely define the remaining variables of the system and thereby determine the phase is known as the *degree of freedom*. For example, if the temperature, pressure, concentration, volume, etc. are such that a minute amount of gas (vapor) phase is beginning to form from a liquid phase of oil, the gas and the oil are in equilibrium with each other so long as the temperature, pressure, concentration, volume, etc. remain constant. If one is changed, all the others are changed, provided the gas and oil are to maintain their equilibrium relation. The system oil, therefore, is said to have one degree of freedom.

The generalized relation between the number of degrees of freedom, the number of components, and the number of phases in a system is expressed in the *phase rule*. The phase rule, as first stated by Willard Gibbs in 1876, can be expressed in symbols as

$$F = C - P + 2$$

where *F* is the number of degrees of freedom (pressure, temperature, density, etc.), *C* is the number of components in the system, and *P* is the number of phases present.

Applying this rule to a one-component system, such as water, we see that where all three phases, water, vapor, and ice, are present simultaneously, as at O in Figure 10-1,

$$F = C \text{ (one component)} - P \text{ (three phases)} + 2$$
$$F = 1 - 3 + 2$$
$$F = 0$$

There are no degrees of freedom. In other words, no change in temperature or pressure is possible if the three phases are to coexist.

Where two phases are present, such as liquid and vapor (see Fig. 10-1),

$$F = C \text{ (one component)} - P \text{ (two phases)} + 2$$
$$F = 1 - 2 + 2$$
$$F = 1$$

Either the temperature or the pressure (but not both) must be fixed if the two phases are to be maintained in equilibrium with each other at some point along line OB.

And if only one phase, such as the liquid phase, COB in Figure 10-1, is present in the water system, then

$$F = C \text{ (one component)} - P \text{ (one phase)} + 2$$
$$F = 1 - 1 + 2$$
$$F = 2$$

The water would still remain in the liquid phase even if both the temperature and the pressure were varied. The liquid phase of water, in other words, has two degrees of freedom, meaning that two conditions, as temperature and pressure, must be specified to define the system in this case.

If the temperature is high enough (above the critical temperature of 374°C in Fig. 10-1), only one phase, the vapor phase, is possible, and no increase in pressure will liquefy any part of the component. At these very high temperatures and pressures, the two phases cannot coexist, and the properties of a liquid and its vapor or gas become indistinguishable. The degrees of freedom of a phase do not apply above the critical temperatures and pressures. At intermediate temperatures and pressures, the liquid phase (water) appears and can coexist with the gaseous phase (water vapor).

The hydrocarbons of a reservoir belong to an extremely complex multicomponent system. When it is remembered that any system with three or more components becomes very difficult to understand, it is seen how impossible it will ever be to fully understand the petroleums with their thousands of hydrocarbons. Most studies of phase relations of hydrocarbons have been confined to one- or two-component systems, consisting of the simpler members such as methane, propane, and butane.

Phase diagrams for systems made up of two components, such as water and oil, can be constructed by using the mole compositions as abscissae and the temperature and pressure as ordinates, respectively. Three-component systems can be constructed by using an isosceles triangle for the composition, the

FIGURE 10-2

Isobaric temperature-composition diagram of a mixture of component A and component B. As the temperature is increased at a constant pressure (isobaric), a mixture such as X, containing 60 mole percent A and 40 mole percent B, may be traced along line X. Up to N it is in a liquid phase. At N, with a temperature Tb, an infinitesimal amount of gas with composition Q comes into equilibrium with the liquid. This is known as the bubble point, *and Tb is the bubble-point temperature. As the tempera-*

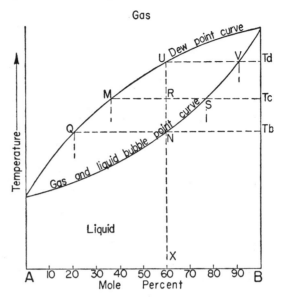

ture increases, more gas is evolved, and the composition of both the gas and the liquid changes, although the overall composition remains the same. Thus, at R, with a temperature Tc, the liquid phase has composition S and the gas composition M. At point U only an infinitesimal amount of the liquid of composition V is left. This is known as the dew point, *and Td is the dew-point temperature for mixture X. At temperatures above the dew point, only the gas phase is present, the gas being of the original composition of the mixture.*

phase boundaries being depicted by isothermal or isobaric contours. Such diagrams are often used by petrologists and metallurgists. For systems of more than three components, there is no simple method of constructing phase diagrams. Where only two phases coexist, the boundary is a surface; where three or more coexist, they have only a line in common. This means that isometric diagrams are necessary if we show the relations between three or more phases.

The isobaric temperature-composition chart in Figure 10-2 shows the effect of temperature changes, at a constant pressure, upon a mixture containing 60 mole percent of component A and 40 mole percent of component B.

At the lowest temperature on the diagram only the liquid phase is present. As the temperature of the mixture is raised at constant pressure (P_1), the volume will be increased, but no phase changes occur until the point N is reached. Here an infinitesimal amount of gas, having the composition Q, is in equilibrium with the liquid. This is known as the *bubble point*, and the corresponding temperature, T_b, is the bubble-point temperature for this mixture at the pressure P_1. As the temperature continues to rise, more and more gas is evolved, and the compositions of both the liquid (lower curve) and the gas (upper curve) change. For instance, at point R, corresponding to the temperature T_c, the liquid phase has the composition defined by point S, while the gas phase has the composition defined by point M. The relative amount of each phase present can be determined from the linear ratios of the lines MR and RS. At the point U only an infinitesimal amount of liquid of composition V remains in equilibrium with the gas. This is the *dew point,* and T_d is the dew-point temperature for the mixture X at constant pressure. At any temperature above the dew point, only the gaseous phase is present, the gas being of the original composition of the mixture. Both the bubble point and the dew point are changeable and depend on variables. For example, there is only one bubble point (saturation pressure) and one dew point for a given volume of liquid and a given volume of gas at a given pressure and temperature. If any variable is changed, both the bubble point and the dew point are changed. The *bubble-point curve* is the line along which gas first begins to come out of solution, and the *dew-point curve* is the line along which liquid first begins to condense out of gas.

In an analogous manner the effects on composition of changing pressure at a constant temperature may be shown for a similar mixture of liquid and gas. An isothermal pressure-composition chart is shown in Figure 10-3. The two diagrams might be combined by using an isometric diagram,[3] but this is beyond our needs.

A combined pressure-temperature diagram for a hypothetical mixture of crude oil and natural gas is shown in Figure 10-4. Temperature is used as the abscissa and pressure as the ordinate. This type of diagram does not show

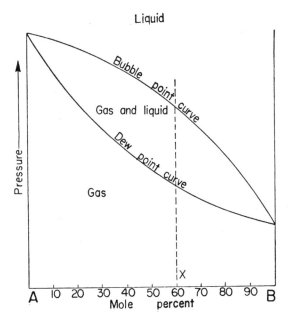

FIGURE 10-3

Isothermal pressure-composition diagram of a mixture of component A *and component* B. *As the pressure is increased at a constant temperature* (*isothermal*), *a mixture such as* X, *containing 60 mole percent* A *and 40 mole percent* B, *may be traced along vertical line* X *in a manner similar to that given beside Figure 10-2.*

FIGURE 10-4

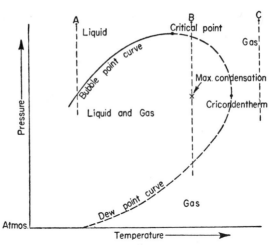

A pressure-temperature diagram for a hypothetical two-component system of crude oil and natural gas of constant composition. Points A, B, and C might be considered the contents of three different reservoirs. In A the mixture is in liquid form in the reservoir. As the liquid is produced, the pressure drops until the bubble point is crossed, and there gas begins to form. This continues until surface pressures are reached, and the two phases, gas and liquid, are separated in a separator at the well head. Reservoir B contains only the gas phase because the temperature is above the critical point. As it is produced and the pressure drops, a liquid begins to form as the dew-point curve (dashed) is crossed. Both gas and liquid continue to be present until the pressure reaches the dew point again, and then only the gas phase remains. The phenomenon of appearance and disappearance of the liquid by pressure reduction at constant temperature is called retrograde condensation. The fluid in reservoir C is in the gas phase and remains so until produced because no decrease in pressure causes any change in phase.

the composition changes of the phases under varying conditions as do the diagrams previously considered. At sufficiently high pressures there will be only one phase present. Under the conditions of temperature and pressure represented by point A, this single phase is liquid, but at the point B the single phase will be gaseous. This is because B is above the critical temperature.* These two points, A and B, might represent conditions in two reservoirs. As the reservoir fluids are produced, the pressure to which they are subjected drops, and the changes in their phase relations may be anticipated. The accompanying temperature drop that might be expected from expansion is probably insignificant, except in the immediate vicinity of the producing well, and can be neglected because the effects are considerably less than the pressure effects. Where the original reservoir conditions are represented by point A, the loss of pressure will result in the formation of a gas phase when the bubble-point curve on the diagram is crossed. Liquid and gas coexist and are produced together. The lower the final pressure, the more gas comes out of the reservoir fluid, and at surface temperatures the operator obtains both

---

* The critical temperature (T_c) is the temperature above which a gas cannot be liquefied by pressure alone. The critical pressure (P_c) is the pressure at which a gas exists in equilibrium with its liquid at the critical temperature.

natural gas and crude oil in the separator tanks at the well-head. The ratio of liquid to gas produced is the same at the surface as in the reservoir.

With the reservoir conditions represented by point B, a sufficient decrease in pressure to cross the dew-point curve, as when gas is being produced, results in the condensation of a liquid from the gaseous phase, and two phases are present. It will be noted here that the temperature exceeds the critical temperature, but the pressure is less than the critical pressure. The point of maximum temperature at which the two phases coexist is called the *cricondentherm*. As the pressure continues to drop, the volume of condensed liquid increases to a maximum value, and then the liquid begins to evaporate to gas as the dew point is reached through continued pressure decrease. As the dew-point curve is again crossed, the last drop of liquid disappears, and only the gaseous phase remains.

A pool under conditions represented by point B is known as a *retrograde condensate* pool, for the condensate is produced by pressure reduction at constant temperature. If the dew point is reached within the reservoir, the liquid oil is left behind and becomes nonrecoverable because, as it condenses out of the gas, it is bound as a film to the pore walls by molecular adsorption; it lags behind and becomes nonproducible. The gas, which would have redissolved in the oil at higher pressure, passes on to the well and is produced. There are producing pools of this type, and they require special methods of production to avoid condensing the liquid phase within the reservoir. Pressure maintenance, by recycling of gas under pressure through the reservoir, is the commonest production technique used to overcome the loss of recoverable liquid hydrocarbons in condensate pools. At sufficiently high temperatures, such as represented by the point C in Figure 10-4, a pressure drop does not result in any phase change. (See also p. 485.)

Geologic time, it may be supposed, has enabled the conditions across the various phase interfaces within the reservoir to approach a static equilibrium and become essentially stabilized. In other words, relations between the reservoir fluids are adjusted to such factors as the temperature, pressure, saturation, and character of the water, oil, and gas present within the reservoir. But, as soon as any one of these factors is changed, the equilibrium is upset, and the system again begins a readjustment toward equilibrium and stability. The conditions may be changed naturally within the reservoir, in the course of geologic time, by such factors as changes in pressure, addition or loss of dissolved mineral matter in the water content, changes in temperature, and changes in the character of the oil and gas, either by additions or by internal molecular changes that result from changes in temperature or pressure. Or phase equilibria of the reservoir may be altered artificially as liquids are withdrawn through wells. Thus the phase-interface equilibrium relations of the fluids in the reservoir are upset and continuously changing when one or more liquids are produced at disproportionate rates, when the pressures are lowered during production, or when water, air, or gas is injected into the reservoir in order to replenish the energy and thereby increase production.

It can be seen, therefore, that changes in reservoir conditions and in the relations between reservoir fluids force us to deal with a large and indefinite number of variable factors. Because of the innumerable unknown complexities existing within the reservoir, we can have but a fragment of understanding of what goes on there. Such as we have is chiefly the result of a growing comprehension of such concepts as the chemistry and physics of boundary surfaces, phase relationships, and production engineering, all of which may be applied to our understanding of the geology of the reservoir. There is, however, much yet to learn about reservoir conditions.

INTERFACE PHENOMENA

We are concerned with fluids contained in the pores of the rocks. The interfaces between fluids, and between fluids and solids, are the loci of many physical and chemical relations[4] that explain what has happened and what happens in the reservoir during both the formation and the production of a pool. Surface free energy, surface tension, interfacial tension, adsorption, adhesive forces, and wettability are examples of phenomena that occur at the interfaces. Because the pores in reservoir rocks are chiefly of capillary size (arbitrarily taken to be less than 0.5 mm in inside diameter), capillary phenomena help explain many of the relations between fluids and rocks that are found in the petroleum reservoir. Moreover, the interface phenomena are all present within openings of capillary size.

The fundamental basis for explaining many boundary relationships between and within gas, liquid, and solid, including capillary phenomena, is the fact that all molecules exert attractive forces on one another. These attractive forces, generally known as *van der Waals forces,* are opposed to the molecular agitation (kinetic energy of the molecules). Molecular agitation is increased by an increase in temperature. One tendency of such thermal agitation, for example, is to cause the molecules within a liquid to separate from one another and form a gas. Because van der Waals forces are considered to vary inversely as the sixth or seventh power of the distance between the centers of the molecules, they have a minor effect on gases. As the intermolecular distances become smaller, however, as in a liquid and even more so within a solid, the forces become much more powerful. The boiling point of a liquid, for example, is a measure of the molecular agitation required to overcome the van der Waals attractive forces. Van der Waals forces exist both within a single physical state (liquid, solid, or gas) and across phase boundaries (liquid-liquid, liquid-gas, solid-liquid, etc.).

Surface Energy, Surface Tension, Interfacial Tension

All the molecules in a liquid except those at the surface are acted upon by molecular attraction or pull from all sides (van der Waals forces), but those at the surface, being in contact with air, gas, or the liquid's vapor, are only partially surrounded by other molecules of the liquid, and are attracted only toward the body of the liquid. This attraction tends to draw the surface molecules inward as if the liquid were surrounded by an invisible contractile membrane.* In effect this tends to reduce the area of the surface to a minimum, and for a given volume the minimum is a sphere, as a rain drop in air. This spontaneous contraction of the surface of a liquid indicates that free energy is involved—that work must be done to extend such a surface. This energy is called the *surface free energy*. The amount of work necessary to form a square centimeter of surface area (ergs/cm^2)† is called the *surface energy* of the substance. The tension at a liquid surface in contact with either air or its own vapor is commonly called *surface tension* (σ) and is expressed as the force in dynes necessary to extend the contracted surface a distance of one centimeter (dynes/cm). The surface tension is numerically equal to the surface energy. The surface tensions of waters and crude oils are given in Table 10-1.

The surface tension of a liquid becomes lower when its temperature is raised because the increase in temperature agitates the molecules of the liquid and tends to push them through the surface of the liquid and into the vapor phase. This tendency is opposed, to some extent, by the surface free energy, which tends to pull the surface inward because of van der Waals forces. When the critical temperature is reached, the surface tension is finally reduced to zero, the liquid and its vapor become identical with each other, and the liquid-vapor meniscus disappears.

If the interface is between two liquids or between a liquid and a solid, the term *interfacial tension* (γ) is used for the force tending to reduce the area of contact, which is equivalent to the surface free energy of the boundary per unit area. The interfacial tension between reservoir water and crude oil has been measured for a group of Texas pools[5] and found to range between 15 and 35 dynes/cm for a temperature of 70°F, 8 and 25 dynes/cm for a temperature of 100°F, and 8 and 19 dynes/cm for a temperature of 130°F. The average value of the interfacial tension for the temperature coefficient was found to be 0.195 dyne/cm/°F at 70–100°F and 0.093 dyne/cm/°F at 100–130°F. Many reservoir phenomena are dependent on the interfacial tensions between the reservoir liquids and between the reservoir solids and the liquids.

---

* The thickness of a molecule is on the order of 10^{-8} cm.

† An *erg* is a unit of work in the CGS system of measurement and is equal to the energy of a 1 gram mass moving at the rate of 1 cm/sec. A *joule* is 10 million (10^7) ergs. A *dyne* is the unit of force in the CGS system of measurement and equals the force that will give an acceleration of 1 cm/sec/sec to a mass of 1 gram.

TABLE 10-1 Surface Tensions of Waters and Crude Oils

Substance	Surface Tension (σ)
Pure water	From 72.5 dynes/cm at 70°F to 60.1 dynes/cm at 200°F, or at an average gradient of 0.095 dyne/cm/°F. [c]
Mineralized waters	The range observed is from 59 to 76 dynes/cm. Dissolved inorganic salts tend to increase the surface tension; surface-active agents tend to decrease it. The effect of dissolved gas in subsurface waters is to reduce the surface tension, but not as much as in crude oils.
Crude oil	At 70°F the surface tensions in crude oils are found to lie between 24 and 38 dynes/cm, the lower surface tensions accompanying the lower-molecular-weight crudes (high API gravity). Higher temperatures and dissolved natural gas both tend to reduce the surface tension of crude oils. Values on the order of 1 dyne/cm may be expected at temperatures and pressures exceeding 150°F and 3,000 psi. At the critical state, surface tension vanishes completely.

Source: Adapted from Morris Muskat, *Physical Principles of Oil Production*, McGraw-Hill Book Co., New York (1949), p. 99–101.

Boundary tension is a general term used to designate all surface and interfacial tensions at boundary surfaces, such as liquid-gas, liquid-liquid, and solid-liquid. While the boundary tensions of liquid-gas and liquid-liquid may be readily measured, it is difficult or impossible to obtain accurate measurements between solids and either liquids or gases. Another difficulty is that the reservoir contains no air, and it is therefore necessary to measure interfacial tensions between crude oil and oil-field brines in the absence of air.[6] The reason is that a tough, tenacious film forms on crude oil where exposed to the air. The film is presumably the result of oxidation and of salts in the brine emulsified with oil, and is asphaltic in general character. The effect of the film is to reduce the interfacial tension, up to 50 percent in some cases, from what it is under oxygen-free reservoir conditions. This film is adsorbed * on the oil at the oil-water interface.

Where two immiscible liquids or a liquid and a solid are in phase contact, the surface molecules of each substance are attracted across the interface by van der Waals forces. This attraction across the interface, which is called the *energy of adhesion,* differs from interfacial tension, which is a property of the interface and is the differential resultant pull away from it. The lower

---

* *Adsorption* is the concentration of molecules of gas, liquid, or dissolved substance on the surface of a liquid or solid substance. They are held in place by van der Waals forces. Adsorption should not be confused with *absorption,* which is the process whereby one substance draws another substance into itself, the assimilation of molecules into a substance to form a solution or new compound. The term *sorption* is sometimes used for either of these phenomena.

the interfacial tension, the greater the energy of adhesion. The energy of adhesion is measured by the amount of work necessary to separate the two substances and is equal numerically to the sum of the surface tensions of the substances singly, minus the interfacial tension of the interface.*

The surface tension (σ) of a pure substance, and the interfacial tension (γ) between two pure substances, are characteristic properties. The magnitude of the attractive forces between a solid and a liquid is somewhat less than the sum of their surface tensions, and that between two liquids is lower than the surface tension of the liquid with the higher surface tension. A low interfacial tension means that phase interfaces are relatively easy to form. Soapy water and air, for example, with a low interfacial tension, readily form a foam. Along with a low interfacial tension goes a high adhesive energy, which makes the two phases difficult to separate. A simple example is seen in the interfacial tension between water and glass. Their interfacial tension is low, and the result is that water spreads out as a thin film, which adheres tightly to the glass and can only be removed by wiping the glass or evaporating the water. Consequently, the energy of adhesion is high. On the other hand, when mercury is placed on glass, the high interfacial tension means low adhesion force; the mercury collects in a droplet and rolls easily off the glass with little resistance.

Surface-active agents are substances that reduce the interfacial tension. They include a variety of substances, such as polar hydrocarbon compounds, salts dissolved in water, and complex hydrocarbon compounds containing metals such as zinc, copper, nickel, titanium, vanadium, calcium, and magnesium,[7] which are found both in petroleum ash and in reservoir waters. The effect of surface-active substances is to cause a more intimate mixture of two otherwise immiscible liquids, such as oil and water, or to increase the energy of adhesion of a liquid, such as oil, to a solid surface, such as a rock particle. Generally, one portion of each molecule of a surface-active agent is polarized and attaches itself to the ionic substance, as mineral or water, and another portion is nonionic and attaches itself to the non-ionic substance, as oil, causing the ionic and nonionic substances to adhere more closely than before. Surface-active agents are used extensively in industry, the most familiar example being the detergent soaps. Ordinary reservoir rock, because of its complex chemical and mineral composition, and reservoir waters, because of their variable composition, undoubtedly contain many surface-active agents, which

---

* The energy, or work, of adhesion (W) between the liquid water (w), for example, and the solid rock particle (s) may be expressed as

$$W_{sw} = \sigma_s + \sigma_w - \gamma_{sw}$$

and that between the liquid oil (o) and the solid rock as

$$W_{so} = \sigma_s + \sigma_o - \gamma_{so}$$

where W_{sw} and W_{so} represent the adhesion energy between the solid and water and the solid and oil respectively, σ_s, σ_o, and σ_w the surface tension in dynes/cm of the solid, oil, and water respectively, and γ_{sw} and γ_{so} the interfacial tension in dynes/cm between the solid and water and the solid and oil respectively.

have had an effect on the boundary tensions and therefore on the movement and distribution of oil and water in the rock. Not only are surface-active agents present in reservoir waters and reservoir rock, but the polar compounds within crude oil are also surface-active. They are adsorbed at the oil-water interface, and also at the oil-rock interface, where they tend to lower the interfacial tensions[8] and thereby increase the energy of adhesion.

An *emulsion* is the stable dispersion of one liquid within another, such as oil within water (O/W) or water within oil (W/O). Low interfacial tensions promote the formation of an emulsion. Agitation of the liquids is generally necessary; but, if the interfacial tensions are quite low, below 5–10 dynes/cm, emulsions may form with only slight shaking. The chief cause of low interfacial tension is the presence of an emulsifying agent—that is, a surface-active agent that is slightly soluble in one of the liquids and becomes absorbed on its surface as a film. The effect is to lower the interfacial tension, and thereby increase the amount of adhesion between the liquids, with the result that a large interfacial area is created between them. This surface takes the form of minute spheres, from 0.5 micron to 10 microns in diameter, as one liquid becomes dispersed in the other. If the two liquids are completely miscible in each other, the interfacial tension is reduced to zero, and the energy of adhesion would be extremely high. In fact, they would no longer exist as two separate liquids, but as a single phase. However, in two such immiscible liquids as oil and water, the interfacial tension never becomes zero, and therefore the tendency is to form emulsions of one in the other, rather than to become one completely miscible liquid phase. Whether an emulsion would be water in oil (W/O) or oil in water (O/W) depends primarily on which side of the adsorbed film is concave, and therefore has the higher tension. The O/W emulsifying agents are generally more soluble in water than in oil. Roughly, 35 percent of the oil produced comes out of the well carrying water in the form of a water-in-oil emulsion.[9] This emulsion may form after the oil leaves the reservoir and becomes agitated by pumping and other production methods. The possibility that emulsions form within the reservoir rock, however, agitated by diastrophism, earthquakes, and other causes, is worth considering in connection with some of the geologic problems of the reservoir.

Wettability. One effect of the energy of adhesion is known as wettability.[10] If water is poured on a glass plate, the water spreads out in a thin sheet and when it finally comes to rest its upper surface makes a very small angle with the glass. The adhesive attraction of the water for the glass is greater than the cohesive attraction* of the water molecules for one another. The glass is said to be *water-wet,* or hydrophilic (water-loving).† If, instead of water,

---

* The work of cohesion is the amount of work necessary to separate a column of a liquid one square centimeter in cross section into two, which results in producing two squares of liquid surface where there was no interface before. The work of cohesion is twice the surface tension, or 2σ. The term "cohesion" is applied to the attractive forces between like molecules, whereas "adhesion" is applied to forces between unlike molecules.

† Other terms in use include *hydrophobic* (water-hating), *oleophilic* (oil-loving), and *oleophobic* (oil-hating).

mercury, with a surface tension about seven times that of water, is used, the mercury contracts into a spherical droplet making an angle greater than 90° with the glass, and rolls freely around on the glass. Here the cohesive attraction of the molecules of mercury to one another is greater than the adhesive attraction of the mercury molecules to the glass. The mercury does not wet the glass and is said to be a *nonwetting* fluid to glass. Thus, in terms of interfacial tensions, the liquid that has the lower interfacial tension (higher adhesion) with the solid is the wetting liquid.

Wettability is, in part at least, a surface effect. If, in the example cited, the water is removed from the clean glass surface, the surface wiped with a towel, and the water returned to the surface, it forms droplets and runs off. Why? The answer is that the towel has left a trace of grease on the glass. The effect is clearly confined to the surface, as the grease cannot enter into the glass.[11]

In considering a solid and two immiscible liquids, such as rock, oil, and water, the contact angle that the interface between the liquids makes with the solid surface offers the means for a better understanding of some of the wettability phenomena. For an oil-water-rock system such as we encounter in a petroleum reservoir, the contact angle of the oil-water interface with the solid rock surface is measured through the water phase, as shown in Figure 10-5. Since water is generally the wetting liquid, the angle (θ) is generally less than 90°, because the energy of adhesion is greater through the acute angle made by the water-rock interface than through the obtuse angle of the oil-rock interface. In the diagram, the forces at X are neutralized and are at zero when the system is in equilibrium. In general, the solid is said to be nonwet by the water when the contact angle θ is greater than 90°, and water-wet when the angle is less than 90°. A question that has a bearing on the interfacial tension between a liquid and a solid, and therefore on wettability, is whether the liquid is advancing over a dry surface or receding from a surface previously wetted. The difference between the contact angle of an advancing liquid and the angle formed by the same liquid and the same solid as the liquid retreats is termed the *hysteresis* of the contact angle, and accounts for the spread found in contact-angle measurements.

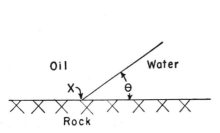

FIGURE 10-5

The contact angle θ is generally measured through the water. Since the angle of the water-oil interface with the rock surface is less than 90°, the diagram indicates that the water preferentially wets the rock surface. The angle θ approaches zero for most reservoir rocks. The system rock-oil-water shown is in equilibrium when the pressure at X is zero.

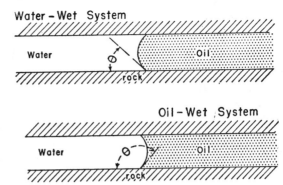

FIGURE 10-6

Examples of water-wet and oil-wet systems in a capillary opening. If the water is forced to move, it displaces the oil in the water-wet system but advances over the oil in the oil-wet system.

When two immiscible liquids, such as oil and water, fill the reservoir pores, they compete for a place on the solid surface, and then the question of which liquid preferentially wets the rock surfaces becomes important in many production problems. The relative ability of water and oil to wet is theoretically the same as the relative degree of adhesion between the two liquids and the rock minerals. A simplified example is shown in Figure 10-6, in which two immiscible liquids, as water and oil, are placed in a capillary opening. If the water is forced to move, in the water-wet system, it will displace the nonwetting liquid, in this case oil; the lower the angle (the greater the adhesion), the more efficiently it will displace the nonwetting oil. The reverse situation is shown in the oil-wet system, where the oil is the wetting phase. If water should now be forced to move, the obtuse angle would cause it to advance over the top of the oil, and leave the wetting liquid undisturbed. It is important for the efficiency of water flooding, for example, that water displace the oil and thereby enable it to be recovered. The liquid that preferentially wets the rock surface occupies the space next to the rock in the pores and in the fine interstices, whereas the nonwetting liquid occupies the interior spaces. The situation is ideally shown in Figure 10-7, where oil (black) occupies various positions within the water-wet pores, depending on the saturation. Natural gas is always nonwetting; oil may be considered a wetting phase relative to gas, and it is commonly the nonwetting phase relative to brine.

Any liquid will wet different solids to different degrees, and any solid will exhibit different degrees of wettability for different liquids. For that reason the minerals of most reservoir rocks, being widely varied, will have a wide variety of wettability relations to each of the liquids present;[13] the wettability characteristic of a reservoir rock and its fluids is unique. A microscopic section through a graywacke, shown in Figure 10-8, and other microscopic sections through sands, seen in Figure 3-4, show many different kinds of minerals with different wettability characteristics found in common reservoir rocks. Part of the difference in reservoirs is due to the fact that the gross effect of a solid-liquid interface increases with an increase in the area of the boundary interface: in the fine-grained rocks, with their finer pores and their

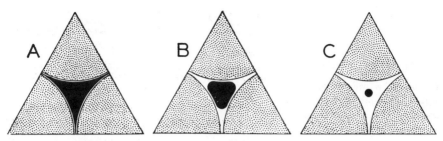

FIGURE 10-7 *The distribution of nonwetting oil (black) in a single water-wet pore (water blank).*
  A. *The oil saturation is in the range 80–90 percent, and the oil, in phase continuity, is a continuous interconnecting web through the adjoining pores (known as* funicular *saturation) and would move readily with a hydrodynamic pressure gradient. The water saturation is 10–20 percent; the water is in* pendular *position and adsorbed on the sand grains.*
  B. *The oil saturation is about 50 percent and the water saturation about 50 percent; the oil is an isolated,* insular *globule that might easily be changed to continuous phase with a slight drop in fluid pressure. Saturations such as these are at the borderline; they may or may not produce clean oil, probably oil and water. The water is in a pendular position, surrounding the grain contacts.*
  C. *The oil saturation is at its residual level, about 10–20 percent, and the oil occurs as a small, isolated insular globule. Any hydrodynamic pressure gradient would move only the water, leaving the oil behind as nonrecoverable.*

greatly increased surface area per cubic foot, there will be much more work of adhesion than in coarser rocks, with their lesser surface area per cubic foot.

A practical way of determining the gross wettability of a shale is to grind it to a fine powder in a nonmetal mortar and pour the powder into a glass with a mixture of oil and water that has been thoroughly mixed together. Then shake the whole together and let it settle. Observe if the water or the oil is attached to the shale particles. This crude method is highly useful when sitting on a well far from a laboratory.

Crude oils consist of many different combinations and mixtures of hydrocarbons, each of which possesses a different and definite energy of adhesion for sand particles. For example, the degrees of adhesion to a specific sand of twelve different types of crude oils were measured, and each crude gave a different value to the work of adhesion, ranging from 58.8 to 72.6 dynes.[12] From this it is inferred that different oils will behave quite differently when displaced by the same water and from the same reservoir rock.

The oil that is left behind in the movement of oil through a reservoir rock may occur as pendular rings,* or as a thin film adsorbed on the particle surfaces in the oil-wet portions of the rock, or as discontinuous insular globules, surrounded by water, in the water-wet portions. Presumably the adhesive films are on oil-wet particles that retained some of the oil after most of it had passed by in its migration toward a place where it accumulated. If, therefore, a permeable rock contains oil-stained mineral grains or scattered, discontinuous globules of oil, migrating oil has probably passed through it, and it is a potential reservoir rock.

Many of the factors that affect the wettability of reservoir rocks are not fully understood. In general, it is believed that most rock particles are preferentially wet by the first liquid present. A part of the first liquid present would be adsorbed as a thin, even monomolecular film on the mineral surfaces and would thereby, in most cases, determine the wettability. At least both the silica and the carbonate reservoir rocks appear to be generally water-wet, as would be expected if the first liquid present has a tendency to wet the rock preferentially, since they were deposited in water. A water-wet reservoir is also generally high in interstitial water saturation, as would be expected if

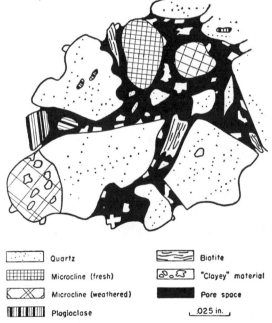

FIGURE 10-8

Section through an arkosic graywacke, showing the variety of minerals present. Each presents a different degree of wettability to each of the liquids present in the pores. Such a rock would not be either altogether water-wet or altogether oil-wet, but probably a combination of both. Through geologic time such a complex assemblage of mineral components has undoubtedly contributed to the surface-active agents of the fluids, thus affecting the relative wettability of the rock to these fluids. [Redrawn from Krynine, Jour. Geol., Vol. 56 (1948), p. 154, Fig. 14.]

* The relative amounts and location of the wetting and nonwetting fluids within a pore are identified as follows: the wetting fluid that clings to the rock surface at the contact points of the rock particles occurs as *pendular rings;* the interconnected nonwetting fluid filling the center of the pore is said to be *funicular;* and the separate globules existing when saturation by the nonwetting fluid has decreased so that the globules are not connected are said to be *insular*.[22] (See Fig. 10-7.)

water was the first liquid present.[13] A few reservoirs are thought to be oil-wet; one is the Bradford sand of the Bradford pool, Pennsylvania. A monomolecular film of oil appears to envelop each sand grain. One explanation[14] of how this happened is that some of the polar hydrocarbon compounds, such as those containing oxygen, nitrogen, sulfur, or some of the metals, are either mildly acidic or mildly basic in their reactions. The most effective surface-active agent for a sand, which is acidic, would be a basic compound. A basic polar hydrocarbon compound, being a surface-active agent, might be supplied from within the oil by one or more of its constituent hydrocarbon compounds and adsorbed on the mineral grains. The resulting decrease in the interfacial tension between rock and oil would cause the sand grains to become preferentially oil-wet rather than water-wet. The Ordovician sands of the Oklahoma City field are also thought to be oil-wet. They may have become oil-wet during the lowering of the ground-water table, when the rocks were exposed to weathering and erosion that is recorded by the unconformity separating the truncated Ordovician oil sands from the overlapping Pennsylvanian shales. Another explanation is that polar hydrocarbon compounds within the oil were able to displace the water from the sand grains and deposit a hydrocarbon film on the sand surfaces, thereby giving the sand its oil-wet property. This explanation also fits some of the Tensleep sandstone (Pennsylvanian) pools of Wyoming. There the quartz sand grains have adsorbed some of the heavier hydrocarbons (in layers about 0.7 micron or 1,000 molecules thick) so firmly that they cannot be washed off with gasoline or powerful solvents. Obviously, such grains are oil-wet to the reservoir fluids.[15]

In reservoir rocks that are water-wet the water clings to and envelops the rock particles as a thin film possibly only a few molecules thick. The oil does not touch the rock particles directly, but is separated from them by these films of water; most oil interfaces, therefore, are oil-water rather than oil-rock. Exceptions occur where scattered oil-wet mineral grains are present, in which case the oil clings directly to them.

Some large pools consist only of gas or of gas with a very minor amount of oil. The reason for such occurrences has puzzled geologists for a long time. Probably the most reasonable suggestion was made by Hill,[16] who called attention to the fact that some rocks, notably nonmarine sediments, are preferentially oil-wet. He suggests that when a mixture of oil and gas passes through such an environment, the oil wets the rock and is left behind, and the gas passes on to accumulate as a pool in some trap. Where the rocks are water-wet the oil and gas pass through together and form the more normal association of oil and gas in the same trap. A fine-grained nonmarine clay sediment would probably provide the most effective "blotter" to hold the oil.

We may summarize as follows some of the factors that may affect the interfacial tension between crude oil and reservoir waters in petroleum reservoirs:

1. Temperature: an increase in temperature decreases the interfacial tension.
2. Pressure: an increase in pressure decreases the interfacial tension.
3. Gas in solution in oil and water: the more gas in solution above the bubble-point pressure, the lower the interfacial tension; the more gas in solution below the bubble point, the higher the interfacial tension.
4. Viscosity: a decrease in viscosity difference between oil and water results in a decrease in interfacial tension.
5. Specific gravity: a decrease in the difference in specific gravity between oil and water generally results in a decrease in interfacial tension. A decrease in specific gravity generally means a decrease in viscosity, and the relation between specific gravity and interfacial tension may be connected with the relation between viscosity and interfacial tension.
6. Surface-active agents: a concentration of surface-active agents, either in the water or in the oil, lowers the interfacial tension, in some cases markedly.

CAPILLARY PRESSURE

The tendency for any curved surface of the interface between two immiscible fluid phases, as between gas and liquid or between liquid and liquid, is to contract into the smallest possible area per unit of volume. A difference in pressure exists between the two phases, measured at adjacent points on opposite sides of the curved interface, the fluid on the concave side of the curved surface being under a greater pressure than the fluid on the other side. In capillary openings this pressure difference becomes extremely significant;[17] it is known as the capillary pressure (P_c dynes/cm^2).

The amount of the capillary pressure depends on the interfacial tension and the degree of curvature according to the equation

$$P_c = p_1 - p_2 = \gamma \frac{1}{r_1} + \frac{1}{r_2}$$

where $p_1 - p_2$ is the difference in pressure between the concave and convex sides of the curved interface, γ is the interfacial tension in dynes/cm, and r_1 and r_2 are the principal radii of curvature, taken at right angles to each other. This is the fundamental equation of capillarity.

If the curvature is a section of a sphere, r_1 and r_2 become equal, and

$$P_c = \frac{2\gamma}{r}$$

It can be seen from the equation that, the smaller the radius of curvature of the interface, or the higher the interfacial tension, the greater the capillary pressure. The degree of curvature depends on the size of the rock pores and upon the relative proportions (saturations) of the fluids present. Conse-

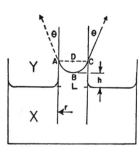

FIGURE 10-9

The rise of a wetting liquid in a dry capillary tube, which is the basis of capillary phenomena. The liquid X, which does not perfectly wet the tube, rises in the capillary tube against air, Y, along a curved surface, ABC. The tendency to contract the surface to a minimum, which in the diagram is shown as surface ADC, causes the liquid to rise in the tube until the weight of the liquid in the tube, BL, balances the contractile force at ABC. The smaller the tube, the greater the curvature and the higher the rise.

quently, the specific capillary pressure is the difference in pressure across the fluid interface at a particular saturation of the wetting phase—usually water in petroleum reservoirs.

The rise of a liquid in a capillary tube is a familiar example of the effect of capillary pressure. It bears on the problem of fluid movement in a reservoir because there also we are dealing with pores, openings, and tubes of capillary size. The rise may be explained in terms of either (1) the capillary pressure or (2) the work of adhesion.

1. The diagram in Figure 10-9 shows a capillary tube inserted in water. A meniscus, or curved interface, forms between the water, X, and the air, Y, at ABC, because surface free energy causes the interface to contract to the smallest surface area. In the diagram the smallest surface area would be ADC, or a plane surface normal to the walls of the tube. A pressure (P_c) is therefore directed from B toward D, causing the liquid to rise in the tube until the weight of the liquid within the tube is sufficient to balance the upward pressure exerted against the interface.

2. The water-air interface makes an acute angle, θ, with the walls of the tube; the lower the angle, the greater the work of adhesion. The tube is said to be "wet" by water. As the capillary pressure drives the meniscus upward, the angle θ becomes greater, and the adhesion of the water to the glass tube less. This causes the water to advance up the glass surface until equilibrium is again reached. The process is repeated until the capillary pressure acting upward is balanced by the weight of the water in the tube.

The forces involved in the rise of a liquid in a capillary tube are shown in the equations

$$\gamma \cos \theta = \frac{rhg(d_1 - d_2)}{2}, \qquad h = \frac{2\gamma \cos \theta}{rg(d_1 - d_2)}$$

where γ is the interfacial tension, θ the contact angle of the interface with the walls of the tube ($\cos \theta$ is approximately 1 in the average water-wet reservoir), r the radius of curvature of the tube, h the height to which the interface rises along the walls of the tube, g the acceleration of gravity, and $d_1 - d_2$ the difference in density between the liquid in the tube and the liquid outside.

When θ is greater than 90°, cos θ is negative, and the liquid falls in the tube, as mercury does in a glass tube. When the angle is less than 90°, cos θ is positive, and the liquid rises in the tube, as water does in a glass tube.

Two of the important effects of capillary pressure in oil and gas pools are that (1) it controls the original, static distribution of the fluids within an untapped reservoir, and (2) it provides the mechanism whereby oil and gas move through reservoir pore spaces until they are barred from moving farther. The static capillary pressure within the pores of a reservoir, which may for comparison be thought of as a bundle of capillary tubes, is in part a function of the relative fluid saturations; and the capillary pressure and the fluid saturations combine to determine the distribution of the fluids within the pore pattern of the rock. If two water-wet sand grains—as, for example, in Figure 10-10—are in contact with each other, and the water saturation of the rock is higher than the oil saturation, the water-oil interface will occupy the position a in the diagram as it encloses the grain contact in a pendular ring.[17] The capillary pressure of such a water-oil system may be plotted as a function of the saturation, as shown in Figure 10-11.

The minimum capillary pressure that will force the entry of a nonwetting fluid into capillary openings saturated with a wetting fluid is known as the *displacement pressure, entry pressure,* or *forefront pressure.*[18] In the common case of the water-wet sand, it is the resistance that must be overcome by the capillary pressure if it is to force the nonwetting oil or gas, or both, from one pore to the next through the intervening constriction. An analogy might be the force needed to change the shape of a rubber balloon (capillary pressure) so that it will pass through a knothole (entry pressure). As the pressure on the oil or gas globule increases, the globule is deformed, and the radius of curvature of the advancing oil-water or gas-water interface decreases until at some point the deformed and protruding forefront of the oil or gas will push through the constriction and make an advance forward into the next pore.[19]

FIGURE 10-10

Two water-wet sand grains in contact with each other in a rock of high water saturation compared with oil saturation. The interfacial curvature is low (a), and the capillary pressure is low. If the water is drained away until the oil-water interface is at b, the curvature is high and the capillary pressure is high. The

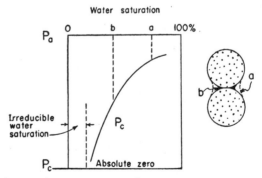

saturation at which no further drainage of water occurs is known as the irreducible water saturation. The relations may be plotted as in the chart. [Redrawn from Pirson, Elements of Petroleum Engineering, McGraw-Hill Book Co., p. 256, Fig. 4-12.]

(See also pp. 551–554.) The displacement pressure (P_d) is seen to vary inversely with the pore diameter.* This means that, for fluids of similar interfacial tension, the finer-grained low-porosity and low-permeability rocks require the greater capillary pressure to permit entry of the nonwetting phase into the pores.

A factor that affects the amount of petroleum that can be trapped in a pool (the oil or gas column) is the capillary pressure created by the difference in density between petroleum and water in an accumulation of oil and/or gas in a permeable and water-wet reservoir rock. Each increment of oil or gas that migrates into the accumulation increases the buoyancy of the total nonwetting phase accumulation, thereby increasing the capillary pressure at all elevations above the oil-water contact. In a hydrostatic environment the capillary pressure at the highest structural position of an accumulation is a maximum for a given petroleum accumulation. If the roof rock is of a heterogeneous type with regard to its permeability and displacement pressure properties, then as a petroleum accumulation increases in vertical extent the consequent increase in capillary pressure at any site within the oil column will force the oil and/or gas to enter smaller and smaller constrictions, until the buoyancy of the hydrocarbon phase is incapable of forcing further advance into the bounding barrier rock—or, stated differently, until the capillary pressure is not high enough to exceed the higher displacement pressure of the barrier rock. If the forces of buoyancy are sufficiently large, petroleum will penetrate the barrier rock and migration will occur. The resistance to migration of petroleum in a hydrostatic environment is due to decreased pore size and lack of interconnected pores, whereas in a hydrodynamic environment the direction of water flow across the barrier rock and its effect on capillary pressure determine the size of pores into which oil or gas will enter and consequently determine the amount of oil or gas that can accumulate in a given trap.

If hydrodynamic conditions at a particular trap cause water to move downward through the barrier rock and down the dip of the reservoir rock, the effect of this down-dip flow is to reduce the capillary pressure at the upper limit of an oil or gas accumulation, as well as throughout the accumulation,

---

* For a cylindrical capillary tube of radius R_c the displacement pressure is given in the equation

$$P_d = \frac{2\gamma \cos \theta}{R_c}$$

At the instant the interface enters the tube, the capillary pressure and the displacement pressure are equal, as expressed in the equation

$$\frac{2\gamma}{r} = P_c = P_d = \frac{2\gamma \cos \theta}{R_c}$$

where r is the radius of curvature of the interface, R_c is the radius of a cylindrical capillary tube, θ is the angle of contact, and γ is the interfacial tension.

FIGURE 10-11

Examples of capillary pressure curves. If oil, forced into a water-wet core or a rock fragment saturated with water, drives out and replaces the water, the graph of the relations between saturation and capillary pressure looks like this. In curve 1 the pressure needed to force the oil to enter the rock, the displacement, or entry, pressure, must rise from A to B. Curve 1 shows the rock to be a clean, uniform-grained, and well-sorted rock with a permeability of 100–200 md. The curve from B to E, being nearly flat, shows that it takes very little additional pressure to increase the oil saturation (decrease the water saturation)

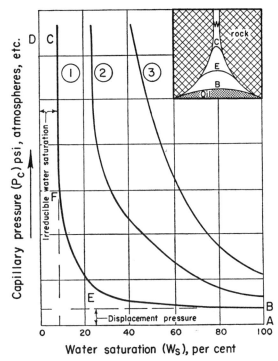

greatly. An increase in pressure from E to F, however, adds only about 10 percent oil, and, from there on to C, additional pressure does not reduce the water content appreciably. The water content of DC is called the residual water *(the irreducible water saturation). Most of it is confined to the pendular areas. The changing position of the oil-water interface within a single capillary pore as the saturation changes is shown in the inset, where the letters correspond to the same letters on the graph (W = water). Curve 2 is characteristic of a limestone or dolomite rock with 15–25 md permeability. Curve 3 is indicative of a gradation in size of grains and pores and of a variable permeability due to the presence of clay and matrix material; a steady increase in pressure is needed to force the oil into the rock, and the residual water content is high because of the larger amounts of water held in the finer pores.*

thus allowing a large accumulation to be trapped below a barrier zone that has fairly high permeability and relatively low displacement pressure properties. Conversely, if hydrodynamic conditions cause an up-dip flow of water through the reservoir and the barrier zone, the effect is to cause the capillary pressure within an accumulation to be increased, which will force oil or gas into smaller and smaller openings and will seriously limit the trapping ability of the barrier, or, depending upon the minimum displacement pressure of the barrier zone, may prevent the accumulation of an oil or gas pool.

If the pressure needed to force oil into a saturated water-wet fragment or core of sand, for example, is charted for different water saturations, curves similar to those shown in Figure 10-11 are obtained. Several laboratory techniques are used to obtain these measurements.[20] The shape of the curve varies with different rocks and also with such factors as the interfacial tensions, the porosity and permeability of the rock, the textural and pore pattern of the rock, and the angle of contact that the interface between the fluids makes with the rock surface.[21]

The *rock barrier* that keeps the oil and gas from moving farther, both vertically and laterally, is called an "impervious" barrier, but actually it does have some permeability. Water, being in continuous phase, will move through it, but the small size of the openings requires a greater capillary pressure to force the oil and gas into the pores than is available. (The capillary displacement pressure of water against oil in sediments of various grain sizes is given in Table 10-2.) Hence the oil is retained below an "impervious" cover. Even

TABLE 10-2 Capillary Displacement Pressure of Water Against Oil

Sediment	Grain Diameter (Millimeters)	Capillary Pressure (P_c) (Atmospheres)
Clay, very fine	10^{-4}	Approximately 40
Clay	Less than 1/256	Greater than 1
Silt	From 1/256 to 1/16	Between 1 and 1/16
Sand	From 1/16 to 2	Between 1/16 and 1/500
Granules	From 2 to 4	Between 1/500 and 1/1,000

Source: M. King Hubbert, "Entrapment of Petroleum under Hydrodynamic Conditions," Bull. Amer. Assoc. Petrol. Geol., Vol. 37 (August 1953), p. 1977, Table I.

within the reservoir, variations in porosity and permeability result in water remaining in the smaller pores for lack of a sufficient capillary pressure to force the free water out and permit the oil to replace it. Similar problems occur during the production of an oil and gas pool, and some of them will be considered more fully later.

The displacement problem is reversed when a pool is being produced, or when it is being flooded with water in a secondary recovery operation. (See pp. 481–484.) Then the problem is to displace the oil by water and free the oil so that it may be produced. The normal withdrawal of oil and gas during production develops a pressure gradient by a drop in reservoir pressure near the well and thereby causes the oil to move to the well. Secondary operations, on the other hand, provide for fluid movement toward the producing wells by forcing air, gas, or water into the reservoir under pressure through input wells located at some distance from the producing wells.

Interface phenomena enter into nearly all problems concerned with the migration and accumulation of petroleum into pools, and also into nearly all problems concerned with the production and recovery of oil from the reservoir—in other words, with all underground fluid movement. The movement of petroleum into a pool and out of it may be thought of as a summation of what happens within a single pore. The single pore then becomes the unit of study of the forces involved. Phenomena occurring within a pore of capillary size are individually slight, but they become significant when multiplied by the trillions upon trillions of pores in an acre-foot or a cubic mile of sedimentary rock, and they help to explain many of the problems of the petroleum geologist and petroleum engineer.

The first application of displacement pressure occurs at the time the pool is being formed. Since most reservoir rocks are water-wet, at that time the problem is the displacement of the original water by the migrating oil. (See p. 447.) Obviously, not all of the water was displaced, for all oil and gas reservoirs contain varying amounts of interstitial water, even though only oil or gas is being produced. We call the displaced water free water to distinguish it from the thin layer of adsorbed water held to the mineral grains and the discontinuous pendular water around the grain contacts and at the ends of the larger pores, held by capillary pressure. These make up the *residual water,* or the *irreducible water saturation.* (See Fig. 10-10.)

Many attempts have been made to explain reservoir phenomena by physical laws. Most of these laws are based on pure substances and ideal conditions, and most of them deal with only two or three of the possible variables, such as temperature, pressure, boundary tensions, and fluid phases. They are sound, theoretically, and many can be proved to be mathematically and physically correct. The trouble, however, as geologists especially know, is that the variables in the field are many times the number visualized in the laboratory, and accurate measurements, even of variables known to be involved, are difficult or even impossible. These quantitative explanations should generally be considered only as concepts that give direction to our thinking; they may or may not apply to the specific problem. If all of the data could be determined accurately, if all of the chemical and physical properties of the sediments were known, if the geology could be interpreted correctly, and if the location and size of all buried structural and sedimentary features were known (most of which are impossible in practically every case), then we could expect to make correct predictions based on sound physical laws. The theoretical approach is invaluable, but we should keep in mind its limitations, for the application of any equation or quantitative explanation to subsurface conditions is no better than the accuracy and validity of the data used. Multiplying and integrating a poor observation, or assigning a numerical quantity to a guess or an assumption, does not make it more accurate, even though it might have the appearance of mathematical infallibility as it appears on the printed page.

RESERVOIR ENERGY

Oil has no inherent energy with which to produce itself. The natural energy that is present in the reservoir and available to move the oil into the wells is the potential energy of reservoir pressure. This energy is stored mainly in the compressed fluids, and its amount depends largely on the fluid potential, or head, of the reservoir fluids. To a lesser extent it may be stored in the compressed rocks that form the reservoir rock. The reservoir performance—that is, the movement of the fluids from the reservoir into the well—depends on the amount and kind of energy present and on its efficient use.

To move the recoverable oil and gas into the well bore where it may be produced, the reservoir energy must be sufficient to overcome (1) the boundary forces holding the oil or gas within the pore system, and (2) the viscous resistance of the gas and oil to movement. A capillary pressure must be achieved that will force the free, or recoverable, oil into the next pore, and so on, until the well bore is reached. This is done by forming and maintaining a zone of decreased reservoir pressure at and surrounding the well. The oil adsorbed on mineral grains is nonrecoverable, as is some of the oil held in the smaller pores and as pendular rings at grain contacts. The energy needed to move gas, with its much lower viscosity, is obviously much less than that required to move oil.

A flowing well has enough additional pressure, when the oil and gas enter the well bore, to overcome the pressure of a column of oil, gas, and possibly some water, extending from the reservoir to the surface, and to force the oil out at the surface. When the pressure is only sufficient to move the oil into the well bore, the oil is pumped up. When the oil does not move into the hole, the pressure may be artificially restored by pumping air, gas, or water into the reservoir until the pressure gradient is again sufficient to move the oil into the well bore.

The moment a well taps the petroleum reservoir, energy begins to be released, and work is done. An area of lower pressure begins to form around the well; a fluid potential gradient is established and fluid begins to move toward the well. If the amount of available reservoir energy is small, the reservoir pressure declines perceptibly for each barrel of oil or cubic foot of gas produced. If the reservoir energy supply is large, great volumes of oil and gas will be produced before there is an appreciable loss in the reservoir pressure, and of course all gradations between large and small reservoir energies are found in oil and gas pools.

Most pools contain several kinds of energy sources. No one kind will predominate for the complete life of the pool, but each functions in proportion to its ability to maintain a pressure gradient toward the well. Production efficiencies—obtaining the most oil from the reservoir—depend largely on the kinds of reservoir energy present. The most efficient production is obtained when the drop in reservoir pressure is the lowest per unit of oil and/or gas produced at the surface. The dominant kind of reservoir energy is commonly used to classify the pool. The various forms of reservoir energy are difficult to separate, in many cases, especially during the early life of the pool, when there has not been sufficient time to determine the relation of each particular form of energy to production.

The early recognition of the kind of energy that is dominant is one of the essential problems of the petroleum engineer charged with the most efficient exploitation of the pool. The reservoir energy sources that may move the oil toward the well are: (1) gas dissolved in oil; (2) free gas under pressure: (a) reservoir that contains chiefly gas, and (b) oil reservoir with free-gas cap; (3) fluid pressure: (a) usually hydrostatic, at times in part hydrodynamic, and (b) compressed water, and oil and gas or both in gaseous or liquid phase; (4) elastically compressed reservoir rock; (5) gravity; (6) combinations of the above.

Gas Dissolved in Oil

In all oil pools at least a little natural gas is dissolved in the oil under pressure,[22] so that potential energy is present in varying amounts in all pools in the form of gas that requires pressure to remain in solution. Compressed and dissolved gas is commonly the dominant form of available energy in pools trapped in isolated and sealed reservoirs, such as are formed by lenses, fault blocks, or cemented sands. Dissolved-gas energy is freed by the expansion of the gas released from solution as the fluid pressure drops in the pool and in the fluid column in the well; and, as the gas expands, it moves in the direction of the lower fluid potential gradient, dragging or carrying or driving the oil along with it. Pools that produce solely as a result of the expansion of natural gas liberated from solution in the oil are variously called *dissolved-gas, solution-gas, internal-gas, solution-gas expansion,* and *depletion-drive pools.*

The pressure in a dissolved-gas oil pool is at its maximum when the pool is first tapped, and declines as the oil is produced. Since the energy is stored chiefly in the gas originally compressed and dissolved in the oil in the reservoir, the pressure decline is roughly proportional to the amount of gas withdrawn. As the reservoir pressure declines, the production rate also declines because there is no available energy to restore the reservoir pressure to its original value. Shutting in the well *does not* replenish the energy. When all of the gas has been exhausted, the reservoir pressure is reduced to atmospheric pressure, and the only energy left to move the remaining oil to the well is that of gravity, the use of which is frequently a slow and uneconomical process. (See pp. 471–472.) It is therefore important to conserve the natural energy of the dissolved-gas type of reservoir, for wasted energy will be reflected in higher costs and in lower ultimate recoveries of oil.

During the producing life of a dissolved-gas pool, there comes a time when the reservoir pressure is reduced to the saturation pressure (the bubble point) all through the reservoir, and free gas comes out of solution as minute bubbles dispersed through the oil. The gas may continue to be produced along with the oil as fast as the gas is formed, or it may collect at the top of the reservoir and become a *secondary free-gas cap*. This secondary gas cap adds little to the available energy or efficiency of production, however, and should not be confused with an *original free-gas cap*. Once a secondary free-gas cap is formed in a gas-expansion pool, it merely continues, because of the reduced pressure, to grow and expand into the space previously occupied by the oil that has been produced. Wells producing from that part of an oil pool where a secondary gas cap is forming develop a high gas-oil ratio and may eventually produce only gas.

As the gas content of a dissolved-gas oil pool declines or becomes exhausted, the reservoir energy may be renewed by injecting gas under pressure into the reservoir. Frequently the injected gas is gas that has been produced along with the oil, separated from it at the surface, and returned under pressure to the reservoir through neighboring intake wells. The reservoir pressure is thereby either restored or maintained, and, as the gas passes through the reservoir from the higher-pressure intake wells to the lower-pressure producing wells, it again moves in the direction of the lower fluid potential gradient, expands, and carries with it a load of oil.* The process of maintaining the reservoir pressure near its original value through the injection into the reservoir of gas under pressure is called *pressure maintenance* or *repressuring*.†
The pressure is also maintained through the injection of water under pressure along the down-dip edge of the reservoir; this is called *water flooding*. If the energy from the gas dissolved under pressure has become exhausted before new gas or water is injected under pressure, the process is called *secondary recovery*. (See also pp. 481–484.) The production mechanics that govern oil production from a natural dissolved-gas drive and from an artificial or repressured gas drive are similar.

The efficiency of gas expansion alone as an oil-recovery mechanism is lower than that of any other common energy source, ranging from 10 to 30

---

* Carll expressed the idea as early as 1880 when he stated: "Drawing oil from the rock may be compared to drawing beer from the barrel. The barrel is placed in the cellar and a bar pump installed—at first the liquor flows freely through the tube without using the pump, but presently the gas weakens and the pump is called into requisition; and finally the gas pressure in the barrel becomes so weak that a vent hole must be added to admit atmospheric pressure before the barrel can be completely emptied by the pump."[28]

† The process of removing wet gas from a reservoir, stripping it of its condensate content, and returning the dry gas to the reservoir, in order to maintain the original reservoir pressure and at the same time replace the wet gas with dry gas, is called *recycling*.

percent, and being, in the majority of cases, less than 20 percent of the oil in place in the reservoir.* This low percentage of recovery is due chiefly to the limited amount of gas originally available for the expulsion of the oil from the reservoir and to the ease with which gas channels through the oil and bypasses it. The existence of a range in recovery is due largely to the fact that the most efficient rate of production, called the MER (see p. 477), which is the rate that permits all or most of the energy stored in the compressed gas to be utilized in lifting its full load of oil, commonly calls for too slow a rate of production to be profitable. Therefore, in order to increase the rate of production until it becomes profitable, the pressure is permitted to decline rapidly, a high gas-oil ratio develops early, and the limited supply of energy in the form of compressed gas is soon depleted. The reason why the gas-oil ratio increases rapidly in some pools after only a small production of oil is that the relative permeability to oil shows a rapid decline at first; the production of one-quarter of the oil may, for example, reduce the permeability to oil to about one-tenth what it was originally. (See Fig. 4-6.) As the oil is removed and the permeability to the remaining oil decreases, the sand rapidly becomes more permeable to the low-viscosity gas, and consequently the gas-oil ratio increases. Figure 10-12 shows the characteristic decline in production that attends a decline in reservoir pressure, together with an increase in the gas-oil ratio, during the producing life of an oil pool that gets its drive from dissolved gas. A generalized decline curve of a pool with dissolved-gas energy is shown in Figure 10-13. A chart showing a comparison of yields of a group of pools in the United States, as related to the solution-gas–oil ratio, is shown in Figure 10-14.

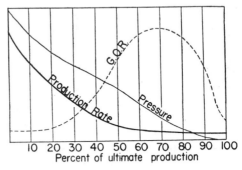

FIGURE 10-12

The production history characteristic of pools with a solution-gas drive as the source of energy. [Redrawn from Murphy, *Petrol. Engr.*, August 1952, p. B-92.]

---

* The *oil in place* in a reservoir consists of the fraction of the pore space filled with oil; the *recoverable oil* is the *stock-tank oil* (oil at the surface) that may be produced by all known methods, primary and secondary, under current economic conditions; the *physically recoverable oil* is the stock-tank oil that may be produced by all known methods irrespective of economics. The *residual oil* is the oil left behind in the reservoir after production. It is adsorbed on the rock surfaces and held by capillary pressure, as pendular oil, in the finest pores, and is *nonrecoverable*. *Primary recoverable oil* is oil produced by the use of the natural energy of the reservoir, and *secondary recoverable oil* is oil produced when the energy is artificially restored, as by water flooding or repressuring.

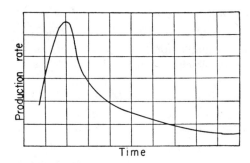

FIGURE 10-13

Generalized curve of declining production from a pool with solution-gas energy. [*Redrawn from Murphy, Petrol. Engr., August 1952, p. B-94.*]

Still another cause of the comparatively low efficiency of dissolved-gas-expansion is the increase in viscosity of the oil as the gas is removed. As the gas begins to come out of solution in the oil, the viscosity of the gas-oil mixture is lowered at first, when the gas is dispersed through the oil in finely divided bubbles. The result is that the mixture moves more freely. This effect is only temporary, however; as the gas bubbles consolidate into larger bubbles, phase continuity develops, and the pressure gradient permits the gas to flow more freely toward the well, finally leaving gas-free oil behind. Consequently,

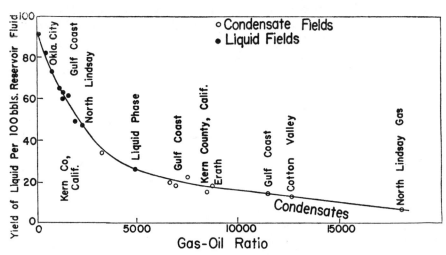

FIGURE 10-14 *A comparison of yields with the ratio of dissolved gas to oil for some pools in the United States. In the group of Gulf Coast pools, for example, the gas-oil ratio through the separator at the surface of the ground varies from 1,250 to 1,500 cubic feet of gas per barrel of oil. For each 100 barrels of reservoir fluid, there will be produced approximately 60–62 barrels of liquid along with the gas. The highest ratio of a well producing from a liquid phase in the reservoir is 4,900 cubic feet per barrel, while the lowest ratio for a condensate well is 3,300 cubic feet per barrel, showing some overlapping.* [*Redrawn from Katz and Williams, Bull. Amer. Assoc. Petrol. Geol., Vol. 36 (1952), p. 354, Fig. 12.*]

the remaining oil, being more viscous, is more difficult to move, and much of it becomes nonrecoverable. Moreover, little water is produced from pools with dissolved-gas drive, and the benefits of the flushing action of water as it sweeps oil ahead of it are lost.

Free-gas Cap

The presence of an original free-gas cap overlying the oil pool shows there is an excess of gas beyond that necessary to saturate the oil at reservoir temperature and pressure. Pools in which this condition exists are known as *saturated pools*. The free gas is compressed under reservoir pressure and forms a "cap" over the oil. In pools with a free-gas cap, as in those with a dissolved-gas drive, energy is supplied by gas dissolved in the oil, but additional energy is also stored in the compressed gas cap. The free gas also has a sweeping or cleaning effect as the cap expands with a decline in pressure, which attends the decrease in the volume of the underlying oil as it is withdrawn. As the free gas moves toward the wells, it separates more oil from the pores and carries or drives it to the wells. Pools that have an original gas cap, and that produce solely as a result of the expansion of both the dissolved gas and the free-gas cap, are said to have a *free-gas drive* or a *gas-cap drive*. Such pools range through all ratios of volume of gas cap to oil; some pools have only a small gas cap, and others only a small volume of oil compared with the gas.

The difference between a free-gas drive and a dissolved-gas drive frequently begins to fade out as the pressure declines to the bubble point in a pool with a dissolved-gas drive. In a dissolved-gas pool the pressure may decline until a free-gas cap forms, so that in the later stages, pools of both types act in a similar manner. The energy supply is at its maximum in either kind of pool when the pool is first penetrated, then declines along with the decline in pressure until it is finally exhausted and no more oil is recoverable.

Before the reservoir conditions of gas-drive pools were understood, it was common practice to permit the well to "blow" gas until the excess gas was exhausted and the oil finally came. Wells drilled into the free-gas cap began producing gas immediately, and many of those drilled into the saturated oil zone soon produced only gas if allowed to flow too rapidly. As a result, some of the energy in the compressed gas was wasted, which caused an ultimate loss of oil production and necessitated pumping at a much earlier date than necessary if this energy had been conserved. Modern production practice attempts to conserve the energy of the pool by making each unit of gas move the maximum amount of oil into the hole and, if possible, up to the surface of the ground.

The efficiency of a free-gas drive is substantially greater than that of a dissolved-gas drive alone, and it expels from 30 to 80 percent of the oil in place in the reservoir; most recoveries, however, are less than 60 percent.

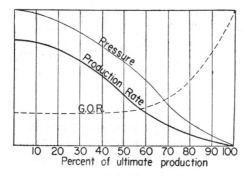

FIGURE 10-15

The production history characteristic of pools with a free-gas cap as source of energy. [Redrawn from Murphy, *Petrol. Engr., August 1952, p. B-92.*]

The chief limitation of the free-gas drive is that the low viscosity of the gas permits it to flow readily through the more permeable sections of the reservoir; if the pool is produced at too rapid a rate, the gas by-passes much of the oil and fails to displace it. The characteristics of the production decline of pools with a free-gas cap are shown in the typical chart in Figure 10-15. The gas-oil ratio increases to its maximum as the rate of oil production reaches its minimum.

Pools that would normally have gas caps at shallow depths are quite different at the high pressures and temperatures that come with depth, until finally the oil and gas become indistinguishable. Such pools are in the area of B in Figure 10-4. The density of the gas increases with pressure (or depth) because it is being compressed into a smaller space. On the other hand, the density of the oil decreases with an increase in pressure, or depth, because more and more gas is being forced into solution. Finally, at some pressure between 5,000 and 6,000 psia, which is the critical point, the densities of the gas and the oil are the same, and the viscosities, surface tensions, and compressibilities likewise are so nearly alike that it is difficult or impossible to distinguish between gas and oil; they are in a single-phase state. These are the retrograde condensate pools in which the condensation of the liquid is in a reverse direction from that in a single-component system in that the condensation occurs on a reduction of pressure and vaporization on an increase of pressure.

*Water (Edge-water) Drive**

Water-drive pools[24] are those in which the reservoir pressure is transmitted from the surrounding aquifer to the contact between water and oil or gas at the edge and bottom of the pool. The energy in the water-drive pool comes chiefly from outside the limits of the pool and is transmitted along the pressure gradient toward the pool that is formed when fluids are withdrawn; water moves in, replaces the volume of oil, gas, and water withdrawn, and thereby maintains the reservoir pressure. The pressure on the water-oil or water-gas contact of the pool may be hydrostatic or hydrodynamic or may be caused by the elastic compression of the water, by the elastically compressed, dissolved, and entrained gases contained in the water, by the elastically compressed reservoir rock, or by several or all of these together. A free-gas cap may be present, but in a water-drive pool it does not add energy as in pools with a free-gas drive.

If the pressure in a reservoir remains at or near the original reservoir pressure during production, new water under reservoir pressure is entering the reservoir, in equal volume, and as fast as fluids are being withdrawn. Such a pool is said to have an *active water drive*. When the reservoir pressure of a pool is maintained near the original pressure at the time the pool is produced, the cause may be: (1) an active recharge, or inflow of water, as at the outcrop of the confined fluid system, that balances the production and maintains fluid pressure; (2) a small expansion per unit volume, under diminishing pressure, of a large volume of water surrounding the pool; (3) a small reduction in pore space per unit volume of a large volume of reservoir rock outside the pool limits, due to the increased influence of the geostatic pressure on the reservoir fluid pressure, which may occur as the reservoir pressure is reduced in the area of the oil pool by the withdrawal of fluids; (4) any combination of these causes. There are all gradations from a pool with an active water drive to one in which the water enters the reservoir so slowly that reservoir pressure declines rapidly, even at a low rate of oil production. A very inactive or partial water drive is difficult to distinguish in its operation from gas expansion or gravity drainage, and, in fact, all three agencies may operate in the same pool without one's knowing the relative importance of each.

Geologically, a pool with an active water drive must be in a permeable reservoir rock extending over a large surrounding area or to some recharge area capable of supplying enough water to replace the fluids produced. If the reservoir rock is lenticular, or is cut into fault blocks, or contains facies causing low permeabilities and other obstructions, the chances of an active water drive are greatly diminished. In some regions the conditions affecting the drive can be anticipated before or shortly after discovery. More frequently, however, the character of the water drive cannot be determined until enough petroleum has been removed to reveal trends in the changing physical behavior of the reservoir fluids, in quantitative measurements of their changing volumes, and in the rate of decline in reservoir pressure per barrel produced.

As the water advances from the surrounding region to replace the oil and gas that have been produced, it tends to accomplish three things: (1) to lessen a fluid potential difference between the pool and the well and between the surrounding region and the pool; (2) to flush oil out of the pores of the reservoir; (3) to drive the oil and gas to the well. If production is more rapid than the inflow of the water, reservoir pressure as well as production declines, and the flushing action is decreased. The reservoir energy in a pool with an active water drive, unlike that in a gas-drive pool, *may be replenished* by shutting the wells in and allowing the reservoir pressure to build up again.

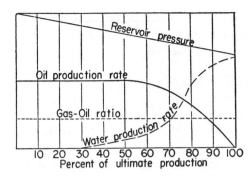

FIGURE 10-16

The production history characteristic of pools having a water drive as source of energy. [*Redrawn from Murphy, Petrol. Engr., August 1952, p. B-94.*]

If the pressure builds up quickly to what it was originally, it indicates that there is an abundance of available energy, together with a permeable reservoir rock. The energy of an active water drive is large or even unlimited, whereas the energy of a dissolved-gas drive or a free-gas drive comes chiefly from within the pool, and is therefore limited. For these reasons the recoveries from pools with an active water drive are often higher than those from pools in which the drive is supplied by any other single mechanism, recoveries being sometimes as much as 70 or 80 percent of the original oil in place, although most such pools recover less than 60 percent.

The production history of a typical pool having an active water drive is shown in Figure 10-16. The reservoir pressure has a slow decline, depending upon the rate of production, and also on whether a uniform gas-oil ratio can be maintained throughout the producing life of the pool. During the typical producing history of a water-drive pool, the oil production does not begin to decline appreciably until the water appears in the wells. Ultimately the oil production declines to zero, and the wells produce only water.

Some evidence is available regarding the distance to which the water pressures are affected by the withdrawal of oil and water from a pool. The Ain Dar pool in Saudi Arabia is twenty-seven miles west of the Abqaiq pool, and both are within a common aquifer. A measurable drop of the pressure in the Ain Dar pool was caused by the drop in pressure in the Abqaiq pool, showing that they were interconnected.[25] Another example is in the pressure drop extending out several miles from the Schuler pool in Arkansas, which has already been noted. (See pp. 400 and 562–563.) Studies of the East Texas Basin show that the drop in pressure in the Woodbine sand (Cretaceous) of the East Texas pool caused a fluid potential gradient to form throughout an area extending seventy miles to the west and including nearly all of the basin.[26] Isobars showing the extent of the pressure decline are given in Figure 10-17.

The source of the energy in most water-drive pools has generally been taken to be the weight of a column of water to the potentiometric surface. In recent years, however, it has become more and more evident that often, where a water drive exists, there is no free connection between the petroleum

RESERVOIR MECHANICS [CHAPTER 10]   467

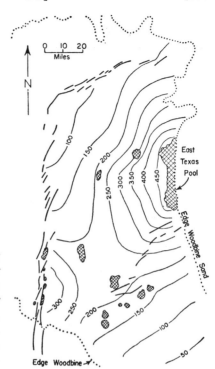

FIGURE 10-17

Map showing the decline in pressure by contours (isobars) in the East Texas–Tyler basin as a result of the production of oil in the East Texas and Mexia pools. Contour interval is 50 psi. [Redrawn from Rumble, Spain, and Stamm, Trans. Amer. Inst. Min. Met. Engrs., Vol. 192 (1951), p. 336, Fig. 9.]

reservoir and a source of new water entering the reservoir formation along its outcrop; so other sources of energy that could operate in water-drive pools have been considered. Among these are, (1) the compressed water within the reservoir rock extending outward into the surrounding region or basin, (2) elastically compressed reservoir rock, both within the reservoir and in the surrounding region, and (3) osmotic forces resulting from the semipermeable membrane nature of the average shale formation.

Compressed Water. The expansion of the compressed fluids in the surrounding region is a possible source of the energy in some water-drive pools. Although the compressibility of water is low, as shown in Table 10-3, the volumes of water involved are often so large that the supply of reservoir energy from the expansion of compressed water in the surrounding region may be large enough to maintain the reservoir pressure. The expansion of dissolved gas, and of micro-gas pockets entrained within the water throughout the surrounding region, would be difficult to distinguish from the expansion of the water in many cases. Some of the reservoir energy is also stored in the compressed oil and interstitial water in the pool, but the volume of liquids in the pool is generally infinitesimal compared with that of the water in the region tributary to a water-drive pool.

TABLE 10-3   Compressibilities of Water, Oil, and Rock

COMPRESSIBILITY OF WATER

Woodbine sand water, East Texas basin (35,000 ppm)
  1.11×10^{-7} cu ft/cu ft/lb/sq ft [Ref. 1]
  1.85×10^{-8} cu ft/cu ft/lb/sq ft (measured on samples) [Ref. 2]
  5.3×10^{-8} cu ft/cu ft/lb/sq ft (computed) [Ref. 2]
Artesian aquifers
  0.015 of 1% per 100 ft of head [Ref. 3]
Spraberry field, West Texas
  3.2×10^{-6} (equivalent to 11,650 bbl/acre, or 0.037 bbl/acre/psi) [Ref. 4]

COMPRESSIBILITY OF CRUDE OIL

Average reservoir fluid
  10×10^{-6} [Ref. 5]
Redwater, Alberta, Canada
  6.73×10^{-6} bbl/bbl/psi [Ref. 6]
East Texas field
  Original reservoir pressure 1,625 psi, saturation pressure 740 psig. 0.1% per 100 psi [Ref. 2]
Crude oils
  Of order of 10^{-4} per atm, or $68 \times 10^{-4}/100$ psi [Ref. 7]
Spraberry field, West Texas
  12.2×10^{-6} (equivalent to 10,060 bbl/acre, or 0.124 bbl/acre/psi) [Ref. 4]

COMPRESSIBILITY OF ROCK

Spraberry siltstone and fine sandstone
  1.88×10^{-7} (equivalent to 240,000 bbl/acre, or 0.045 bbl/acre/psi) (pore-volume change/bulk vol/psi) [Ref. 4]
Woodbine (Cretaceous), Strawn and Bartlesville sands (Pennsylvanian)
  3.0×10^{-6} (pore-volume change/unit pore volume/psi) [Ref. 5]

---

1. Stuart E. Buckley, "The Pressure-Production Relationship in the East Texas Field," in *Production Practice*, Amer. Petrol. Inst. (1938), p. 141.
2. R. C. Rumble, H. H. Spain, and H. E. Stamm III, "A Reservoir Analyzer Study of the Woodbine Basin," Tech. Pub. 3219, Petrol. Technol. (December 1951), Trans. Amer. Inst. Min. Met. Engrs., Vol. 192 (1951), pp. 335–336.
3. Oscar Edward Meinzer, "Compressibility and Elasticity of Artesian Aquifers," Geol., Vol. 23 (May–June 1928), p. 285.
4. Lincoln F. Elkins, "Reservoir Performance and Well Spacing, Spraberry Trend Area Field of West Texas," Tech. Pub. 3622, Trans. Amer. Inst. Min. Met. Engrs., Vol. 198 (1953), p. 186.
5. Howard N. Hall, "Compressibility of Reservoir Rock" (Tech. Note 149), Jour. Petrol. Technol. (January 1953), p. 18; Trans., Vol. 198, p. 310.
6. Irene Haskett, "Reservoir Analysis of Redwater Pool," O. & G. Jour., October 18, 1951, p. 146, Table 3.
7. Morris Muskat, *Physical Principles of Oil Production*, McGraw-Hill Book Co., New York (1949), p. 271.

Laboratory measurements of the compressibility of a number of reservoir rocks have been made.[27] The total effective compressibility of a rock formation may be divided into (1) the compression of the contained fluids and (2) the compression of the rock and rock particles. The total compressibility of the reservoir rocks tested is shown in Figure 10-18. When the compressibility of the fluids (taken as 10×10^{-6} change in volume per unit volume per psi) is subtracted from the total compressibility, the result is the component due to the compaction of the rock. Figure 10-19 shows the rock component of compressibility, and Figure 10-20 shows that the effect of compressibility of the oil in place is enough to cause important differences between the calculated and the actual oil in place in a reservoir.

Even in the East Texas pool and the Tyler basin, where some of the most intensive studies have been made of reservoir conditions,[28] especially reservoir pressures, it is difficult to separate the different causes of the water drive. The decline in reservoir pressure extends outward long distances (see Fig. 10-17); so presumably there is not an active recharge of sufficient water to replace the fluids produced.

Donahue[29] has suggested, for example, that for the Woodbine sand of the East Texas pool the expansion in volume of the Woodbine water is 28.6×10^{-5} percent for each 100 psi of pressure decline, thereby expelling 0.57 barrel of water for

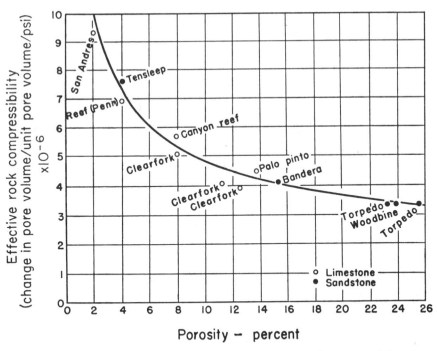

FIGURE 10-18 *The relation of porosity to effective or total compressibility for a group of reservoirs. It can be seen that there is a good correlation between rock compressibility and rock porosity. All tests were made at 95°F. [Redrawn from Hall, Tech. Note 149, Jour. Petrol. Technol., January 1953, p. 18, Fig. 2.]*

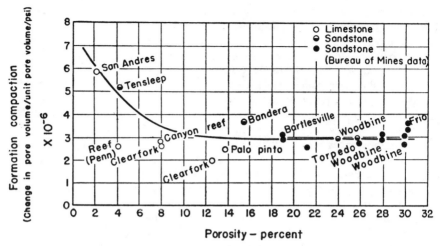

FIGURE 10-19 *The part of the total compressibility (Fig. 10–18) that is due to formation compaction. The difference between the compressibility values shown in the two figures represents the amount of rock expansion that can be attributed to the expansion of the individual rock particles as the pressure in the surrounding fluids declines.* [Redrawn from Hall, Tech. Note 149, Jour. Petrol. Technol., January 1953, p. 18, Fig. 3.]

each acre-foot of sand, and that the compaction, or decrease in porosity, of the Woodbine sand is on the order of 45×10^{-5} percent per 100 psi of pressure decline, or that an acre-foot of sand expels 0.9 barrel of water as a result of this compaction.* The total amount of water that would move out of the water-saturated region around the pool, therefore, would be 1.47 barrels for each acre-foot of sand. Since there are 300 feet of sand thickness, one acre would yield 440 barrels of

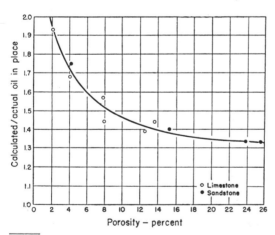

FIGURE 10-20

The effect of rock compressibility on the oil in place as calculated for undersaturated reservoirs. It shows that rock compressibility is of a magnitude to cause the calculated values for oil in place to be from 30 to 100 percent higher than the actual values. [Redrawn from Hall, Tech. Note 149, Jour. Petrol. Technol., January 1953, p. 19, Fig. 4.]

---

* Rumble, Spain, and Stamm (*op. cit.* in reference note 26, p. 336) arrive at a figure of 4.7×10^{-8} vol./vol./lb./sq.ft. as the compressibility of the Woodbine aquifer by adding the compressibility of the Woodbine water (1.85×10^{-8} vol./vol./lb./sq.ft.) to the reduction in pore volume of the Woodbine sand (2.85×10^{-8} vol./vol./lb./sq.ft.).

water, and one square mile would yield 280,000 barrels of water under a pressure drop of 100 psi. Compaction of the reservoir rock surrounding the pool becomes effective, however, only as the reservoir pressure declines. If the sand compacts, it loses porosity, which in turn tends to maintain the pressure. We should thus expect a tendency for maintenance or partial maintenance of pressure if compaction of the sand is the sole cause of reservoir pressure. It appears that the water drive may be due to a combination of several pressure sources, such as partial recharge at the outcrop, expansion of compressed water in the East Texas basin, expansion of compressed rock, expulsion of water from deformed and compressed clay particles within the reservoir as the reservoir pressure declines and the geostatic pressure increases, and expansion of dissolved gas and of micro-gas pockets entrained in the Woodbine water of the region.

Compressed Reservoir Rock. The rate of water encroachment on a pool with an active water drive varies greatly, but is generally within the range of 100–1,000 feet per year.[30] Pools in which only a partial water drive is present show much slower rates of water advance, and many of them, especially some of the stratigraphic type, show a total water encroachment of only a few hundred feet after many years of production. The reservoir energy originally available in such pools is largely supplied by expansion of dissolved gas and by gravity drainage, and only partly by a water drive. A slower rate of production permits a slower water encroachment, and is generally more efficient, because it gives time for the water to displace the oil from the less permeable areas of the reservoir, and consequently less oil is left trapped behind as nonrecoverable.

Osmotic Forces. Clays that make up most shales act as effective semipermeable membranes and permit osmotic phenomena to operate. Thus it is believed that where a shale separates two aquifers of widely differing salt concentrations, the water in the aquifer of lower concentration will pass through the clay layers and enter the aquifer of higher concentration. This process will continue until the fluid pressure in the aquifer of higher concentration is enough to resist the fluid movement and equilibrium is reached; the shale is permeable to the water but less so to the salt in the water. The water moves through the shale to dilute the more salty aquifer until stopped by the back pressure. The movement of additional water into an aquifer raises the pressure and thus forms an added source of reservoir pressure.

While these effects probably go on in all areas where there are shales separating reservoir rocks of differing salt concentrations, the pressure effect is lost in the areas where permeability is high and uniform and the reservoir rocks widespread. However, where there are "dirty" sands and lenticular and variable-permeability formations, the effects of osmotic phenomena are probably important.

Gravity

Gravity is active as a reservoir energy force, to some extent at least, throughout the life of all producing pools. The pressure exerted by the column of petroleum will contribute to the fluid potential gradient in gas-drive and other pools, and thereby aid in the flow of oil to the well. This force may be significant in traps of high relief. Gravity may also be said to be the fundamental source of the reservoir energy in water-drive pools, for the weight of the water column provides hydrostatic pressure, and, together with the weight of the overburden, stores energy in the compressed water, in its dissolved gas, in the entrained micro-gas caps, and in the compressed reservoir rock. Separation of the gas, oil, and water, on the basis of their specific gravity, is also an effect of gravitational energy in the reservoir.

Gravity becomes the dominant active source of reservoir energy in the later stages of the life of some pools with dissolved-gas drive. This occurs after the gas has been exhausted and there is no further encroachment of water. At such a time the reservoir pressure is at a minimum, and the weight of the remaining oil causes it to seep or drain through the pores and down the dip to lower levels in the reservoir, where it may be pumped up to the surface. Gravitational energy prolongs the life of many oil pools that would otherwise be abandoned. Pools that produce as a result of the gravity drainage of oil from higher to lower levels are called *gravity-drainage pools*. Under favorable conditions the recoveries by gravity drainage are important, and may include almost all of the recoverable oil.[31] It has been found, in general, that oil production by gravity drainage is greater where the oil is of low viscosity, in reservoir rocks with a high permeability to oil, and where the reservoir pressure is sufficient to provide a displacement pressure that will overcome the capillary pressure of the oil-water interface and push the oil out of the pores.

Oil has been mined in Alsace, Romania, Burma, Japan, the USSR, and Germany.[32] The oil recovered by mining flows almost wholly by gravity drainage, as it would be impossible to do much mining where gas existed under pressure.

Combined Sources of Reservoir Energy

Most pools utilize several sources of reservoir energy during their producing life, although it may be difficult, if not impossible, to distinguish the dominant form of reservoir energy at any one time. For example, even in a pool with an active water drive, the very first reservoir energy utilized when production begins is probably the expansion of compressed liquids, the water and oil next to the well; this is followed by the expansion of gas that comes out of solution in the oil. Only after a fluid potential gradient has been established by production from the well to the contact between oil and edge or bottom water does water start moving in to replace the oil and gas as they are removed. In a partial water drive, some water moves in and helps to maintain the reservoir pressure or at least to prevent a more rapid decline. Potential reservoir energy from some application of gravity acts throughout the life of every pool; but in many pools gravity, as a separate force, is the last source of energy to be utilized; only after all of the other sources have been exhausted does it become the dominant reservoir force.

Because of the differing efficiencies of the various sources of reservoir energy, it becomes important to determine what kind of reservoir energy is present as soon as possible after production starts in a new pool. The production of the oil must be so planned as to utilize the existing reservoir energy to the greatest advantage in bringing the most oil into the well bores. The earlier the dominant source of energy present in the reservoir is known, the more effective can the necessary planning be. How great the water drive is may be learned by shutting in the wells, permitting the pressure to stabilize itself, and then comparing the shut-in reservoir pressure with the original reservoir pressure. The sooner and the more nearly the shut-in pressure approaches the original reservoir pressure, the more active the water drive. Where an early drop in reservoir pressue per unit produced (barrel of oil or Mcf of gas) is great, and the pressure is not being replenished, a gas-expansion drive or a too rapid rate of production for other types of drive is indicated. The rate of withdrawal of the fluids from a reservoir is probably the most important control the petroleum engineer has over the ultimate production of a pool.

The most efficient reservoir energy is found in oil pools that have both an active water drive and a free-gas cap, a combination most likely to be found in a blanket reservoir rock and a structural trap. When such a pool is most efficiently produced, the gas expanding downward from the top eventually meets the water encroaching upward from the bottom. All of the recoverable oil in such a pool may be produced by means of the natural reservoir energy, for the reservoir rocks are swept clean of their recoverable oil content by the gas and the water, and a fluid potential gradient is maintained toward the wells throughout the life of the pool by the encroaching water from below.

MOVEMENT OF OIL AND GAS IN A POOL

A petroleum geologist should know as much as possible about how oil and gas move through the connected rock pores. No one has ever been down in a producing pool to observe at first hand what goes on there. Our ideas, then, of what happens when a pool is produced are based on deductions from phenomena that can be observed and measured at the surface, from data given by recording instruments let down into the well, and from laboratory experiments. The most effective line of attack is a study of the phenomena associated with the production of an oil and gas pool—the movement of oil and gas from the pool into the well.

Two fundamental conditions are necessary in order that oil may move through the reservoir rock into the well:[33]

1. A fluid potential gradient or, for horizontal flow, a pressure gradient must be established and maintained between the reservoir and the well. The most efficient recoveries can be achieved by maintaining the reservoir pressure as long and at as high a level as possible. The local fluid potential gradient is controlled by the rate of production, the resistance of the reservoir rock to the flow of the fluid, and by the resistance concentrated around the well bore because of the drilling and well-completion practices, sometimes known as the *skin effect*.[34]

2. The oil saturation in the vicinity of the well must be maintained at as high a point as possible, so that the oil will move from the reservoir rock into the well more readily than gas or water. This condition can be met by producing at a rate low enough to prevent such a rapid release of gas from solution that the viscosity of the oil is lowered or that the gas expands to a point where its high saturation permits it to move more readily than oil. The rate should be sufficiently low to prevent preferential "channeling" or "coning" of water instead of oil into the well.

Probably the simplest way to think of the movement of oil and gas from a pool into a well is to consider it as the movement of a unit particle of oil and its dissolved gas, enclosed within the pattern of connected pores. For our example we may take the most common case, that of a water-wet sand in which the oil content is undersaturated with gas. The water, because of the lower interfacial tension between the rock particle and the water than between the rock particle and the oil, may be expected to fill the finer pores and the attenuated ends of the larger pores (pendular) and to be adsorbed as a thin film, possibly only a few molecules thick, on the rock surfaces; the water wets the rock. The individual pore, provided it is large enough, is thus found to be lined with a film of water, inside which there is a globule of oil. (See Fig. 10-7, p. 448.) According to the relative saturation of oil and water, the oil globules may connect with one another in adjacent pores, through narrow and tortuous openings (funicular), or they may be dispersed as separate globules (insular). As long as the temperature and pressure remain unchanged, the position of the oil and water is presumed to remain stationary in a state of static equilibrium.

As soon as a well taps a pool and fluids are withdrawn from the reservoir, the reservoir pressure around the well bore is lowered. A man-made fluid potential gradient is thus formed that ranges from the original fluid potential in the reservoir at some distance from the well bore, where the fluid pressure has not been disturbed by the withdrawal of reservoir fluid, to a minimum potential (atmospheric pressure) at the surface or, in some cases, below the surface in the well. At some elevation within this fluid system the fluid pressure drops below the bubble point and the dissolved gas then begins to come out of solution and expands. One means of efficient production is so to control the reservoir pressure that the bubble point will be reached as near the stock tank as possible. Commonly, however, the bubble point is reached within the reservoir rock, either before or as the fluid enters the well. The fluid potential gradient within the reservoir, however, really controls the reservoir performance in moving oil into the well.

As a bubble of gas comes out of solution and expands, it joins with other gas bubbles, forming in adjacent pores, until the gas occupies so large a proportion of the pore space that a continuous, connecting web of gas extends through the connecting, larger pores of the rock. The lower the oil saturation, the higher the gas saturation, and the lower viscosity of the gas permits the gas to move more readily than the oil toward the area of lower fluid potential—in this case the well. Whatever oil clings to the gas-oil interface is carried with the gas into the well, where it is produced. By the coalescing of gas bubbles within the larger pores, phase continuity of the gas is established, and, if the gas saturation is high enough, the gas moves toward the area of low fluid potential. If the reservoir pressure stays above or only slightly below the bubble point, most of the fluid produced will be oil, and the free gas will be dispersed through it as minute bubbles. But, if the reservoir pressure drops considerably below the bubble point, the gas coming out of solution expands so much that it occupies most of the pore space near the well, and then the fluid produced is chiefly gas. Moreover, the amount of oil that clings to the surface of the gas will be small when gas is the dominant fluid being produced. As a result, a high gas-oil ratio will develop. In other words, the relative permeability of the rock near the hole will be high for gas and low for oil because of the high gas saturation and the low oil saturation, and the resulting efficiency will be low.

As the reservoir pressure declines in the pore space nearest the well, the pressure gradient extends to the next outward pore space. The slope of the extended pressure gradient is less steep than in the pores near the well because of the greater area for the same net decline in pressure. The reservoir pressure gradient extends radially outward, as shown in Figure 10-21, in successive increments from one pore to the next for a variable distance, depending on the viscosity of the oil, the permeability of the rock, and the distance to the next well.

So far we have been considering what happens to an oil in which gas is dissolved, or what might be termed a typical pool with gas-expansion drive. An effective reservoir pressure is maintained for a much shorter time in such a pool than in one with either water drive or free-gas drive; the higher recoveries are achieved with water drives and free-gas drives because these best satisfy the condition of maintaining the reservoir pressure for a given volume of withdrawn fluid. When the zone of pressure draw-down in a dissolved-gas pool finally reaches to the edge of the reservoir, the reservoir pressure continues to decline without any renewal of the energy. In a free-gas pool, however, the original energy supply is higher, and a higher reservoir pressure is maintained longer. When the oil has all been produced from a

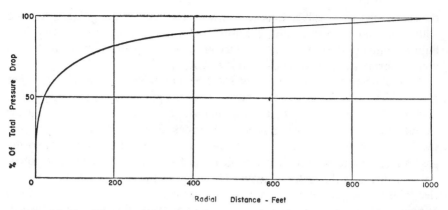

FIGURE 10-21 *The increasing rate of pressure drop between the pool and the well as the well is approached.*

pool of this kind, the overlying gas has expanded until it meets the underlying water, and only gas is produced. The reservoir pressure in a water-drive pool is maintained throughout the life of the pool, because of the relatively large volume of water moving in from the surrounding area as compared with the volume of fluid produced from the reservoir. In the pool with a perfect, active water drive, the rate of production of water when the oil has been exhausted tends to equal the maximum rate for the oil previously produced, or, because of the greater mobility (lower viscosity) of water, even exceeds it. The accumulation of oil into a pool is aided by a fluid potential gradient that provides a force that acts in a direction opposed to the buoyancy force and prevents the further migration of the oil and/or gas phase. During the production of oil and gas from a pool, the reverse is true—the movement of the oil and gas into the well bore is aided by the development of a local fluid potential gradient around the well bore.

Too rapid a rate of production in reservoirs of variable porosity and permeability may permit a channeling of the flow of the less viscous fluids, such as gas or water, into the well in preference to oil.[35] Once a channeling pattern has formed, it is difficult to change, and it may isolate substantial quantities of nonrecoverable oil in patches or islands within the reservoir rock where permeability is lower. This result is due in part to the disruption of the fluid equilibrium within the capillary pores. Under equilibrium conditions, when production begins, there is a higher percentage of water in the finer and less permeable sands than in the coarser sands. At a rapid rate of production, however, the oil-water contact moves up the dip more rapidly than the finer sands can expel their oil. The fluid equilibrium is disrupted, and the water that is taken into the fine sands as the oil is expelled (a phenomenon called *imbibition*) moves so slowly that the oil-water contact has moved up the dip, leaving the oil trapped in the low-permeability lens.[36] The phenomenon is illustrated in Figure 10-22 and may be observed in the laboratory if a fragment or core of oil-saturated shale or siltstone, as from the Spraberry formation of western Texas, is placed in a beaker of water. Within a few hours droplets of oil ooze out over the surface of the rock as the water enters it through the same openings to displace the oil.[87]

PRODUCTION PHENOMENA

A few of the phenomena involved in the production of a pool that the petroleum geologist may have some interest in are discussed in the following section The list is by no means complete but is intended to give some idea of the scope of production problems and a few of the more common concepts that are of general interest.

Maximum Efficient Rate

The *maximum efficient rate,* or the rate at which a well or a pool produces the most efficiently, is called the MER. The abbreviation also means *maximum economic recovery,* which is generally considered to result from the highest rate of production without waste, waste being the unreasonable use of reservoir energy to produce oil. The MER of a water-drive pool is such that the water does not come in unevenly and thereby entrap oil in the less permeable

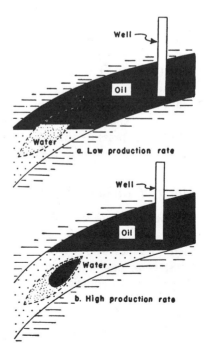

FIGURE 10-22

Sections showing the effect of water imbibition. A fine-grained sand lens of low permeability occurs within a porous and permeable sand reservoir. Under a low production rate the water advances faster in the sand of low permeability, driving the oil out ahead of it. The sand is said to imbibe water, taking the water in at the bottom and expelling oil at the top, and thereby maintaining a higher water saturation in the fine sand than in the coarse sand. At a high rate of production the water front moves so fast that the tight lens has no opportunity to imbibe water; hence the oil originally in the lens is only partly produced, and some remains behind, trapped and lost to production. [Redrawn from Buckley and Leverett, Trans. Amer. Inst. Min. Met. Engrs., Vol. 146 (1942), p. 158, Fig. 8, and O. & G. Jour., August 18, 1952, p. 157.]

areas, that the pool output is limited to the rate of water encroachment, whether natural or artificial, and that the gas-oil ratio is held to a minimum.[38] The MER for an oil pool or an oil field has been officially defined as "the highest daily rate of production that can be sustained by a field or pool for a period of 6 months without jeopardizing the maximum practicable ultimate recovery from the reservoir." [39] The MER of a pool may be determined only after enough oil and gas have been removed to show the kind of reservoir energy involved and the rates of production that cause the lowest decline per unit of pressure drop.

The MER of most oil pools ranges between 3 and 8 percent of the recoverable reserve per year and thus assumes for the pool an ultimate life of 12–33 years. The recoverable reserves of the United States, for example, have been for a number of years between 12 and 15 times the annual production, which thus takes place at annual overall rates of $6\frac{1}{2}$–8 percent. Many pools are produced at an annual rate between 5 and 6 percent of their ultimate reserve as the MER for the pool. Government regulative bodies sometimes use the term *maximum permissible rate* (MPR) of production. The MPR is a combination of the MER and the market or pipe-line demand for crude oil.

Productivity Index

The reservoir performance is frequently indicated by stating its *productivity index* (PI).[40] The PI is the barrels per day of stock-tank oil produced per pound per square inch difference between reservoir and producing bottom-hole pressure. The PI, in other words, is a measure of the ability of a reservoir to yield oil at the reservoir face. The *specific productivity index* (SPI) is the same except that it is calculated per foot of reservoir rock. The PI of a well is subject to wide fluctuations and is dependent upon such factors as reservoir permeabilities, fluid volumes and saturations, production rates, and the past production history, especially the state of depletion of the reservoir. Generally the PI decreases with time, largely because of the decline in reservoir pressure, but also because of the increase in the viscosity of the oil, and in its resistance to flow, after its solution gas is gone.

Material Balance Equation

Reservoir analyses and predictions of what will happen as a pool is produced may be made by use of the *material balance equation,* which relates such variables as reservoir fluid volumes, reservoir pressures and temperatures, compressibilities, stock-tank volumes, and water encroachment. This rather complex equation[41] is used by petroleum engineers to calculate the volumes of oil, gas, and encroaching water in the reservoir, and to predict changes of volume during the future, and its discussion is beyond the scope of this book. The point to remember here is that the petroleum reservoir is characterized by many interdependent variables, and that changes occurring in one variable can cause predictable changes in others; the results depend on the accuracy of the values used in the equation for the different variables involved. After a pool has had some producing history as a basis, quantitative and semiquantitative predictions of its future behavior may be made.

A knowledge of the physical laws behind the material balance equation gives clues as to extensions of pools already producing. Thus, where the pressure drops less during production than the original pool measurements indicate that it should, a feeder source is suggested. For example, the granite reservoir in the Mara field in western Venezuela was found by searching for the feeder to the original pool in the Cretaceous sediments when it was learned that the production was not following the material balance equation. (See p. 125 and Fig. 6-31.) The side of an oil pool where the reservoir pressures and the production remain at higher levels after production has declined throughout the remainder of the pool suggest an area of extension to the pool, and this kind of reasoning has discovered much oil.[42]

Exceptional Wells

Many individual wells of exceptionally high productivity have been described in the literature. Such wells are of interest because they show how much a single well can produce and give some idea of the forces underground. The number of these outstanding wells, however, is not as large now, with modern methods of producing under strict engineering controls, as it was formerly, when wells were permitted to flow wide open at full capacity for as long as they were able to produce oil.* A few unusually large wells, some of which are rather shallow, are briefly described below.

The Petrólea well No. 200 was put down near a large oil seepage on the Petrólea North Dome, Barco Concession, Colombia, South America (see map, p. 649), and came in with an initial production of over 5,000 barrels in ten hours from a depth of only 30 meters (approximately 100 feet) before it could be plugged and the production stopped.[43]

The famous Potrero del Llano well No. 4 of the Mexican Eagle Oil Company, in the "Golden Lane," Mexico, which has been flowing continuously since 1910, produced up to May 1, 1949, a total of 115,160,000 barrels of oil.[44] It was brought in with an initial flow close to 100,000 barrels a day, then got out of control and flowed a total of approximately 2,000,000 barrels of oil, most of which was wasted before the well could be controlled. Well No. 4 of the Aguila Company, also in the "Golden Lane" of Mexico, began

---
* A *flowing well* is a well from which the oil flows out through the tubing or casing at the surface. The flow may be a continuous, steady, solid stream of oil; it may be a frothy mixture of oil and gas, flowing intermittently, or *by heads;* the oil may be accompanied by varying amounts of water, part or all of which may be in an oil-water emulsion. Large flowing wells are called *gushers* in the United States, and *spouters* or *fountain wells* in the USSR.

flowing on the morning of December 27, 1910, and flowed wild for sixty days. When it was finally capped and under a 21-day gauge, it showed a production varying from 100,000 to 110,000 barrels per day. At the end of 1933 its total production had reached 93,000,000 barrels, excluding the production lost during the first sixty days.

In the Pincher Creek pool, in Alberta, Canada, the Gulf No. 1 Bruder was completed during the summer of 1953, in the Mississippian limestone at from 11,971 to 12,415 feet, at the rate of 168,000,000 cubic feet of gas a day, and each million cubic feet of gas contained about 34 barrels of liquid condensate petroleum, or a total of over 5,700 barrels of liquid.

Well No. 54 in the Staro (Old) Grozny oil field of the Chechenian-Ingshetian ASSR, in the USSR, flowed oil for eight years and during this time produced a total of 1.5 million tons of oil [45] (about 10,720,000 barrels). The Droojba well at Balakhany (Baku area) produced at the rate of 400,000 poods* per day (50,000 barrels) in 1875, from a depth of 574 feet, and produced a total of 3,750,000 barrels before it quit flowing. Another well in the Baku area produced a total of 3,750,000 barrels before it quit flowing, and still another produced from a depth of 700 feet at the rate of 87,500 barrels per day (700,000 poods) in 1886.[46] In fact, thousands of flowing wells were found in the Baku district at depths of 1,000–2,500 feet. Fifty million barrels was produced from the great well No. 7-7 in the Masjid-i-Sulaiman field of Iran before it was abandoned;[47] production was from porous, fractured Asmari limestone (Oligocene-Miocene), found at depths of 1,000–3,500 feet.

During the spring and summer of 1862 more than five million barrels of oil floated down the surface of Black Creek, Ontario, and formed a film of oil that covered Lake Erie. It came from the newly discovered pool in Enniskellen Township, southwestern Ontario, and because of the low price of ten cents per barrel could not be marketed. Individual wells ranged from 1,000 to 7,500 barrels of oil per day from depths of 108–237 feet.[48]

The Ohio Oil Co. Yates No. 30 well, in the Yates pool of western Texas, flowed 8,528 barrels of oil of 30° API gravity in one hour from a depth of 1,070 feet, or at a rate of 204,681 barrels for twenty-four hours.[49] This well was completed in July, 1929, and is still flowing its allowable production. The producing formation is Permian limestone, and much of the pool's porosity is in vugs or is even cavernous.

Stripper

A *stripper well* or a *stripper pool* † is any oil well or oil pool in which the cost of operation is nearly as great as the revenue from the sale of the oil

---

* A pood is a pre-1917 measurement approximately equivalent to ⅓ U.S. barrel.

† For statistical purposes, nonprorated wells that produce less than 10 barrels per day are classified as stripper wells.

produced. A stripper well is a marginal operation in the sense that the profit is small or about to end. The depth or size of the well does not necessarily determine its classification. A stripper well may produce half a barrel of oil per day from a depth of 500 feet, or it may produce fifty barrels per day from a depth of 10,000 feet, but make hundreds of barrels of water. The principal factor in defining a stripper well is the cost of production as compared with the amount of money received for the oil. Approximately two-thirds (392,535) of the total number of producing wells in the United States are classified as stripper wells, but they produce only about a fifth of the total production.[50] When a well, a property, or a pool finally produces only enough oil or gas to pay the cost of operation, it is said to have reached its *economic limit*. Estimates of oil and gas reserves assume that a well, a property, or a pool will be abandoned when it has reached its economic limit.

Life of Wells and Pools

The producing life of a well, a property, or a pool begins when the first barrel of oil or cubic foot of gas is brought to the surface, and it ends when the economic limit is reached—when the well, the property, or the pool is abandoned as uneconomic because the cost of producing the oil is greater than the price received for it. This period may cover only a few years in a small pool or one that is difficult and expensive to operate, or it may extend over many decades. Some oil pools in Pennsylvania and West Virginia have been producing continuously for seventy-five years or more. Most pools developed in the past reached their peak production, either by flowing or by pumping, during the first few years, when drilling was most active and initial production high, and thereafter declined until finally abandoned at an age of 15–25 years. The *primary production* of an oil well or pool is the amount of oil that can be produced at a profit by using the natural energy within the pool, with the assistance of pumping and other methods of raising the oil to the surface. Modern practice tends to produce the wells in a pool at a lower and more uniform rate during the early years, reserving a part of the reservoir energy to produce the remaining oil during the later years. Such conservation of reservoir energy makes for longer flowing life and more efficient operation. Many pools now being developed will have economic lives of fifty years or more.

SECONDARY RECOVERY

The depletion of the reservoir energy before the depletion of the recoverable oil leaves a portion of the oil in the ground without a natural propulsive energy to move it. After the reservoir energy is about exhausted, and the production is approaching its economic limit, much of the remaining oil may be recovered by supplying a new energy from the outside, through some such medium as water, air, gas, miscible fluid, underground combustion, or steam injection, and a second "crop" of oil harvested. The reserves of oil in the United States that may be recovered by secondary methods have been estimated as more than seven billion barrels.[51] The secondary reserves of Arkansas, for example, have been estimated to be almost equivalent to the primary reserves. In the Bradford field of Pennsylvania, where the primary natural production amounted to 250 million barrels (3,000 barrels per acre), the estimated secondary production from water flooding has been 320 million barrels (4,000 barrels per acre) and will add 170 million barrels more by 1980, and it is estimated that there will still be 800 million barrels (10,000 barrels per acre) left in the sand for a possible tertiary recovery.[52] The progress being made continually in secondary recovery technology is good reason to believe that much of the remaining oil in Bradford, as well as in many pools that are in the stripper class, will eventually be recovered.

Basically, water flooding depends on the ability of water to displace the remaining oil in the reservoir in the same way it displaces oil in the primary production of a water-drive pool. Water is injected into the reservoir through intake wells at regularly spaced intervals, the spacing depending upon the ultimate recovery expected, the price of oil, and the cost of the wells. As the water enters the reservoir, it moves toward the area of lower fluid potential and, as it moves, drives or flushes the funicular and insular oil left behind during the primary recovery phase. An increased oil saturation, called a *bank of oil*, develops ahead of the moving water and finally reaches the production wells. As the property or the pool is produced, the ratio of water to oil produced increases, until only water is produced. A graph of a typical performance history of a water-flooded pool, the Pettit lime (Cretaceous) pool in the Haynesville field, northern Louisiana, is shown in Figure 10-23.

Injection of air or gas does not drive the oil ahead of it as a bank, for the low viscosity of the gas causes it to bypass the oil by traveling through the more permeable layers. Rather, the gas, which is injected under pressure, causes the decline in reservoir pressure to be minimal, and the energy required to sustain production is kept at a maximum. In addition, the gas, by dissolving in the oil, lowers its viscosity, permitting it to move more freely through the reservoir. Natural gas is more desirable than air as a repressuring agent because of the corrosive action of air on the equipment and also because of the danger of explosion when air and gas are mixed. As the property or the pool is produced, the gas-oil ratio rises, and the rate of oil production declines, until finally only gas is produced. A gas drive has generally been found to be more effective in a reservoir with a high interstitial water saturation, whereas a water flood is more effective in a reservoir with a high oil saturation and a low water saturation. Where gas is introduced into a de-

RESERVOIR MECHANICS [CHAPTER 10] 483

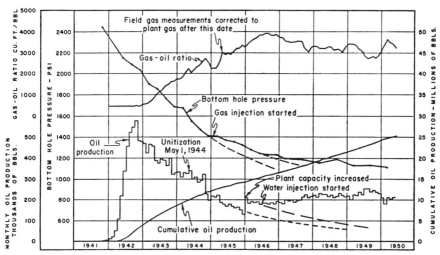

FIGURE 10-23 *The performance history of a typical pool, the Pettit lime (Cretaceous) pool in the Haynesville field, northern Louisiana. Gas repressuring began in January 1945, and water flooding in January 1946. The effects on gas-oil ratios, reservoir pressure, and oil production are readily seen in the curves.* [Redrawn from Akins, Trans. Amer. Inst. Min. Met. Engrs., Vol. 192 (1951), p. 242, Fig. 4.]

pleted reservoir of undersaturated residual oil, either before or accompanying the water flood, increased recoveries result. The addition of carbon dioxide to water used in laboratory flooding experiments also increases the amount of oil recovered, the cause of the increase being both physical and chemical.[53]

Oil recovery by in-situ combustion[54] is rapidly developing into a commercial method of extracting most of the remaining oil from many "exhausted" reservoirs. It consists of injecting air into the reservoir and then igniting the residual oil at the injection well site. Igniting is done by down-hole electrical heaters, down-hole gas burners, and chemical methods that give high temperatures. Oxygen may be added to step up the rate of combustion. Once the formation is ignited, the aerated region becomes heated and moves outward in a slowly burning front at the advancing edge of the air. The crude oil burned at the front is carbonized and becomes the chief fuel for the advancing front. The heat and pressure that develop from the combustion set up a bank of oil that is produced through the production wells.

The injection of hot water and steam into a reservoir is a variation of the thermal methods that are also rapidly coming into favor.[55] The problem is to get the water and steam to the reservoir at an elevated temperature. The injection temperatures are around 400°F, with temperatures up to 700°F, reported. The recovery of the residual oil is due largely to the improved mobility at the higher temperatures and to the thermal expansion of the oil. The hot water and steam are generally injected cyclically and the oil may be produced through the injection well when the steam is shut off or it may be produced through production wells continuously.

A large number of shallow oil fields in the United States are currently being produced by secondary recovery methods. The most commonly used injection medium is natural gas, water being the next most common. Air is used only where gas is not readily available and where the operation is not suitable for water flooding. Water flooding has given the best results.

Where conditions are favorable and the most efficient engineering methods are followed, either all or most of the recoverable oil may be produced by primary production operations, and there will be no need for a secondary recovery. Oil produced by secondary recovery methods is expensive. The objective in programming the development of a new pool, therefore, should be to produce as much of the oil as possible by primary methods. The best prospects for secondary recovery operations are in pools where little attention was given to the conservation of the reservoir energy during the primary production period, and where a high percentage of residual oil is left behind.

GAS PRODUCTION

Natural gas may be produced from the operation of oil fields, as a by-product (associated), or from pools producing only gas (nonassociated).

Natural Gas from Oil-field Operation

Large volumes of gas are commonly produced during the normal operation of oil fields. Even if no free gas accompanies the oil, the gas in solution expands into free gas as the reservoir pressure declines. (See Fig. 5-26, p. 200.) Upon reaching the surface, the excess gas is separated from the oil in a separator and then processed for recovery of its natural gasoline content or for production of liquefied gas products, known also as liquefied petroleum gas (LPG), such as propane, butane, and pentane. After processing, the residue gas and the dry natural gas produced from gas pools are mixed and make the natural gas of commerce that is supplied to the gas pipelines.

Casing-head gas, or the natural gas produced with the oil, has two important functions: (1) to supply the energy that brings the oil from the reservoir rock into the well bore and sometimes up to the surface; (2) by dissolving in the oil, to reduce the viscosity of the oil, thus rendering it more mobile and easier to recover.

The casing-head gas produced with the oil is often utilized as fuel for operating the property. Where large amounts of casing-head gas are available, natural gasoline plants are established, and the casing-head gas is gathered and passed through these plants for recovery of the gasoline and other hydrocarbons. If these plants are so located geographically that the residue dry gas has no commercial value, and the amounts are in excess of the needs for field operations, the dry gas is vented into the air or burned in flares. Field operations require dry gas as fuel for the steam and gas engines that are used for pumping wells, heating, and lighting. A major use of excess gas is in repressuring depleted oil pools, where gas returned to the reservoir may again carry its load of oil into the well bore.

Gas from Gas Pools

Most of the gas used in commerce and transported in gas pipelines comes from fields and pools that produce only gas. Commonly the gas so produced is dry gas and goes directly from the well to the pipelines. If the pressures have declined so much that the gas cannot be put into the line against the higher pressure in the pipes, the gas is compressed in compressor stations until its pressure exceeds the line pressure.

The maximum efficient rate (MER) of production of natural gas pools is generally much higher, in relation to reserves, than the maximum efficient rate of production of oil fields. No gas is wasted by flaring or venting into the air in natural gas fields.

The petroleum in a condensate pool is dissolved in the gas in the reservoir. (See Fig. 10-14, B.) Upon a reduction of pressure, but not of temperature, a portion of the gas condenses into hydrocarbon liquids. In order to obtain the maximum recovery from this type of accumulation, it is necessary to maintain the reservoir pressure above the level at which important condensation takes place. Failure to do so causes a large part of the liquid to condense within the reservoir and become so firmly attached to the sand grains that it is unrecoverable. The method developed for producing retrograde condensate pools is to produce the reservoir fluid at a high pressure in the gaseous phase. The liquid constituents are recovered, with a drop in pressure, after leaving the reservoir. The dry gas residue is repressured and returned to the producing formation to pick up another load of condensate; it also serves to maintain the reservoir pressure, thus preventing condensation within the reservoir. Substantial increases in the ultimate recovery of hydrocarbon liquids from reservoirs of this type are attained by recycling operations.

In many provinces a complete gradation exists between pools that contain only gas and pools that contain oil with a very low gas content. Some fields contain dry gas in the shallow sands, undersaturated oil in the median reservoirs, and condensate in the deep reservoirs. Or the same reservoir rock may contain dry gas in one pool and undersaturated oil in another pool nearby. The distribution of oil and gas pools in an area typical of many is shown in Figure 10-24.

Several possible reasons may help to explain the variations in gas content of pools: (1) traps form at different times, even in the same province, and

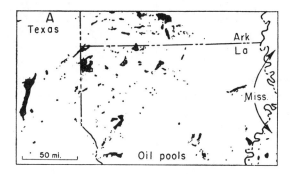

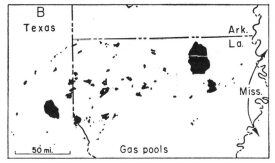

FIGURE 10-24

Two maps of the same area, showing oil pools in A and gas pools in B. The area is associated with the Sabine uplift, which centers in northwestern Louisiana, and whose influence extends into the neighboring states of Arkansas and Texas.

gas and oil may migrate in different concentrations at different times, thereby permitting one trap to catch the moving gas whereas another catches the moving oil; (2) the vagaries of petroleum origin may make one formation or one area rich in gas or hydrocarbons of low boiling point, and another rich in oil; (3) hydrodynamic conditions at some time in the past may have caused the oil to move out from the trap and leave the gas behind; (4) selective trapping of oil and gas may have occurred (see also pp. 573–574); (5) the reservoir rock through which a mixture of oil and gas migrated may have contained oil-wet particles, to which the oil would have become attached while the gas moved on.

ECONOMIC FACTORS AND REGULATIONS

The production engineer is charged with producing oil and gas at the lowest cost and with the highest efficiency. He is governed in his approach to the engineering part of the problem by various economic factors and also by many governmental and industry-approved regulations.

Among the many economic factors that enter into the manner in which an oil pool is produced are the depth of the wells, the cost of drilling, the cost of producing, the price of oil, and the lease restrictions. Ultimately a profit must be obtained: the cost of getting the oil to the surface of the ground and ready to put into the pipeline must be less than the price received for it. The anticipated income from the property is determined by the capacity of the rock to produce and the reserve of recoverable oil, and these factors limit the amount and kind of equipment that may be installed if a profit is to be assured. With a low-profit property, only the least amount of the less expensive equipment can be used, and such equipment does not generally do as good a job of production as better equipment. The objective in some special-purpose properties, such as the United States Naval Reserves and the nationally owned and operated fields of some countries, is not profit in the money sense, and as a consequence their development and operation are quite different from what they would be if a financial profit were the objective.

Practically all oil and gas are produced under some governmental or industry-approved regulations or restrictions. Complete control is exercised by countries that both own the mineral rights and operate the wells. Even countries where the individual landowner generally owns the mineral rights, such as the United States, have some control measures to ensure the most efficient production and the least waste of a national resource. Such regulations are generally adopted as a part of the conservation laws of a nation, state, or province, and they usually pertain to such problems as rate of production, casing and cementing methods, disposal of salt or mineralized water, and the most efficient use of the reservoir energy. The petroleum industry has learned through long experience that efficient operation is profitable operation; in fact, many of the governmental regulations directed toward more efficient operation were inspired by the more forward-looking producers. Large government-granted concessions to oil-producing companies are generally operated with the utmost efficiency, even though they may be subject to few if any governmental regulations. The reason is simple—it is the most profitable way to produce oil and gas.

Petroleum engineering has caused a steady increase in the percentage of recoverable oil through the years, but the greatest increases came with the understanding by both the engineers and the industry of the importance of making the fullest use of the reservoir energy. Because the oil left behind is largely nonrecoverable when the reservoir energy is exhausted, wasting energy was found to be the equivalent of wasting oil.

CONCLUSION

We may summarize some of the chief contributions of reservoir mechanics to the geology of petroleum as follows:

1. A large mass of quantitative and semiquantitative data is continually being added to our knowledge, by which many old concepts of what goes on in the reservoir are being modified, and out of which many new concepts are formed.

2. We see the reservoir and its fluid content as a complex combination

of many sensitive and interdependent variables. When one variable is changed, equilibria are upset; things begin to happen. The same situation has undoubtedly prevailed through long geologic time, and we come to realize that almost every past change in geologic conditions has had a bearing on the history of the oil and gas content of the rocks. From this we see the importance of knowing as much of the total geologic history of a region as possible if we are to translate the story of the rocks into petroleum discovery. We also see that a great many sciences and nearly all phases of geology have a part in our understanding of the geology and the geologic history of a region.

3. Reservoir mechanics has demonstrated that a large amount of oil is left behind as nonrecoverable; but—what is probably even more important—it has shown why such oil has been nonrecoverable. Everything we learn about the reservoir and its fluids is a step toward the solution of the enormously large problem: "How do we recover the nonrecoverable oil?" The solution would probably give the United States, in new oil reserves, as much oil as the country has produced in the past. The solution is, in part, geological.

Selected General Readings

THEORETICAL RESERVOIR MECHANICS

John E. Sherborne, "Fundamental Phase Behavior of Hydrocarbons," Tech. Pub. 1152, Trans. Amer. Inst. Min. Met. Engrs., Vol. 136 (1940), pp. 119–133. Bibliog. 44 items. A concise statement of the phase relationships encountered in the petroleum reservoir as they are used by the production engineer.

Floyd E. Bartell (director), Project No. 27, Amer. Petrol. Inst., "Function of Water in the Production of Oil from Reservoirs." A number of important articles were published between 1943 and 1949 in the progress reports on Fundamental Research on the Occurrence and Recovery of Petroleum.

Anthony M. Schwartz and James W. Perry, *Surface Active Agents,* Interscience Publishers, New York (1949). The theory of boundary relations clearly stated with many examples.

Emil J. Burcik, *Properties of Petroleum Reservoir Fluids,* John Wiley & Sons, Inc., New York 1957), 190 pages. In addition to the properties of the reservoir fluids, this book considers their behavior as a function of temperature and pressure.

S. J. Gregg, *The Surface Chemistry of Solids,* Reinhold Publishing Corp., New York, 2nd ed. (1961), 393 pages. A discussion of the phenomena and the principles associated with the solid-gas, solid-liquid, solid-solid, liquid-liquid, and liquid-gas interfaces.

J. T. Davies and E. K. Rideal, *Interfacial Phenomena*, 2nd ed., Academic Press, New York (1963), 480 pages. Discusses many of the physical relations that exist between liquids, solids, and gases in petroleum reservoirs.

Gilman A. Hill, William A. Colburn, and Jack W. Knight, "Reducing Oil-Finding Costs by Use of Hydrodynamic Evaluations," Southwestern Legal Foundation, Dallas, Texas, Prentice-Hall, Inc., Englewood Cliffs (1961). This is one of the few published statements of many of the applications of hydrodynamics to exploration problems.

J. J. Arps, "Engineering Concepts Useful in Oil Finding," Bull. Amer. Assoc. Petrol. Geol., Vol. 48 (February, 1964), pp. 157–165. A concise statement of the principles involved in capillary pressure, relative permeability, and the material balance as applied to the problem of finding oil.

OIL AND GAS PRODUCTION

M. L. Haider and Carl E. Reistle (chairmen), *Joint Progress Report on Reservoir Efficiency and Well Spacing*, Standard Oil Development Company, New York. A concise statement of the fundamentals of reservoir behavior, recovery mechanisms, and well spacing, with a number of specific examples.

E. N. Kemler, "Physical Properties of a Reservoir and of Reservoir Fluids," Oil Weekly: Part I, April 10, 1944, pp. 26 ff.; Part II, April 17, 1944, pp. 17 ff.

Sylvain J. Pirson, *Elements of Oil Reservoir Engineering*, McGraw-Hill Book Co., New York, 2nd ed. (1958). A standard reference book on petroleum engineering practice.

Engineering Committee, Interstate Oil Compact Commission, *Oil and Gas Production*, University of Oklahoma Press, Norman, Okla. (1951), 128 pages. A simple and concise statement of the fundamentals of oil and gas production, reservoir mechanics, and reservoir engineering.

Morris Muskat, *Physical Principles of Oil Production*, McGraw-Hill Book Co., New York (1949). A standard reference book.

Norman J. Clark, "Elements of Petroleum Reservoirs," Soc. Petrol. Engrs., of Amer. Inst. Min., Met. and Petrol. Engineers (April 1960), 243 pages. Elements and engineering processes of the science of petroleum recovery are discussed.

Frank W. Cole, "Basic Principles of Reservoir Engineering," Univ. of Oklahoma, Dept. Business and Industrial Services," Norman, Oklahoma (1964), 84 pages.

Ralph W. Edie, "Distribution and Behavior of Oil and Water in Mississippian Limestone Reservoirs, Southeastern Saskatchewan," Jour. Can. Petrol. Technol., Vol. 2 (Fall 1963), pp. 122–132. Discusses movement of oil and water in limestone reservoirs.

Stuart E. Buckley (ed.) and fifteen contributors, *Petroleum Conservation*. Amer. Inst. Min. Met. Engrs., New York (1951), 304 pages. An authoritative and excellent discussion of the petroleum reservoir, reservoir mechanics, and efficient recovery principles and practices. About half of the book is devoted to secondary recovery, pressure maintenance, conservation, and unit operation.

Lester Charles Uren, *Petroleum Production Engineering—Oil Field Development and Oil Field Exploitation,* 4th ed., 2 vols., McGraw-Hill Book Co., New York (1953). A standard reference work.

M. C. Brunner (chairman) and eleven others, *Primer of Oil and Gas Production,* Amer. Petrol. Inst., New York (1954), 73 pages. Descriptions, with drawings and photographs, of modern drilling and production equipment and methods, and a glossary of terms used.

SECONDARY RECOVERY

Secondary Recovery of Oil in the United States, Amer. Petrol. Inst., New York; 1st ed. (1942), 259 pages; 2nd ed. (1950), 838 pages. Technical articles on all phases of secondary recovery and over 2,000 references listed.

George H. Fancher with Donald K. Mackay, *Secondary Recovery of Petroleum in Arkansas—A Survey,* Ark. Oil and Gas Commission, El Dorado, Ark. (1946), 264 pages. Contains a summary of the principles of secondary recovery, with a description of every pool in the state and an analysis of its possibilities for secondary recovery operations.

Reference Notes

1. L. E. Elkins, "Petroleum Reservoir Mechanics," Scientific Monthly, Vol. 70, No. 2 (February 1950), pp. 122–126.

2. B. H. Sage and W. N. Lacey, *Volumetric and Phase Behavior of Hydrocarbons,* Stanford University Press, Stanford, California (1939), 299 pages.

John E. Sherborne, "Fundamental Phase Behavior of Hydrocarbons," Tech. Pub. 1152, Trans. Amer. Inst. Min. Met. Engrs., Vol. 136 (1940), pp. 119–133. Discussion to p. 135. Bibliog. 44 items.

Norman J. Clark, "It Pays to Know Your Petroleum," World Oil: Part I, March 1953, pp. 165–172; Part II, April 1953, pp. 208–213. 10 references. A clear statement of the phase behavior of hydrocarbons under varying reservoir and producing conditions.

3. John E. Sherborne, *op. cit.* (note 2), p. 2.

4. A. S. C. Lawrence, "Interfacial Tension," in *The Science of Petroleum,* Oxford University Press, London and New York, Vol. 2 (1938), pp. 1369–1374.

Neil Kensington Adam, *The Physics and Chemistry of Surfaces,* 3rd ed., Oxford University Press, London (1941), 436 pages. A standard reference work.

Project No. 27 of the American Petroleum Institute, Function of Water in the Production of Oil from Reservoirs (Floyd E. Bartell, director), resulted in a number of important articles, published between 1943 and 1949 in the Progress Reports on Fundamental Research on the Occurrence and Recovery of Petroleum.

Norman J. Clark, "Fundamentals of Reservoir Fluids," Jour. Petrol. Technol. (January 1962), pp. 11–16; (May 1962), pp. 491–501.

5. H. K. Livingston, "Surface and Interfacial Tensions of Oil-Water Systems in Texas Oil Sands," Tech. Pub. 1001, Petrol. Technol., Amer. Inst. Min. Met. Engrs., November 1938. 27 references.

6. F. E. Bartell and D. O. Niederhauser. *Film-Forming Constituents of Crude Petroleum Oils,* Research on Occurrence and Recovery of Petroleum, Amer. Petrol. Inst. (1946-1947), pp. 57-80. 18 references.

7. Gerald L. Farrar, "Metals in Petroleum," O. & G. Jour., April 7, 1952, p. 79.

8. F. C. Benner and F. E. Bartell, *The Effect of Polar Impurities upon Capillarity and Surface Phenomena in Petroleum Production,* Research on Occurrence and Recovery of Petroleum, Amer. Petrol. Inst. (1943), pp. 79-93. 15 references.

9. C. M. Blair, Jr., "Prevention and Treatment of Petroleum Emulsions," O. & G. Jour., Vol. 44 (1945), pp. 116-127.

10. F. E. Bartell and H. J. Osterhof, "Determination of the Wettability of a Solid by a Liquid," Ind. & Engr. Chem., Vol. 19 (November 1927), pp. 1277-1280.

F. E. Bartell and F. L. Miller, "Degree of Wetting of Silica by Crude Petroleum Oils," Ind. & Engr. Chem., Vol. 20 (1928), pp. 738-742, and O. & G. Jour., Vol. 27, No. 9 (1928), pp. 102-104.

P. T. Kinney and R. F. Nielsen, "Wettability in Oil Recovery," World Oil, March 1951, pp. 145-154. Bibliog. 8 items.

"Surface Phenomena," Science, Vol. 145 (August 1964), p. 955.

S. S. Marsden, "Wettability; The Elusive Key to Waterflooding," Petrol. Engr. (April 1965), pp. 82-87.

11. M. C. Leverett, "Capillary Behavior in Porous Solids," Tech. Pub. 1223, Trans. Amer. Inst. Min. Met. Engrs., Vol. 142 (1941), pp. 152-169; also in reprinted Vol. 136 & 142, pp. 341-358. 24 references.

12. F. E. Bartell and F. L. Miller, *op. cit.* (note 10).

13. Irving Fatt, "Effect of Fractional Wettability on Multiphase Flow Through Porous Media," Tech. Note 2043, Petrol. Trans. AIME (October 1959), pp. 71-76.

J. E. Bobek, C. C. Mattax, and M. O. Denekas, "Reservoir Rock Wettability—Its Significance and Evaluation," Tech. Paper 8021, Petrol. Trans. AIME, Vol. 213 (1958), pp. 155-160.

14. Foster Dee Snell and Cornelia T. Snell, "Surface Activity as Applied to Secondary Petroleum Production," World Oil, November 1951, pp. 184-196. 20 references cited.

15. P. G. Nutting, "Some Physical and Chemical Properties of Reservoir Rocks Bearing on the Accumulation and Discharge of Oil," in *Problems of Petroleum Geology,* Amer. Assoc. Petrol. Geol., Tulsa, Okla. (1934), pp. 830-831.

16. Gilman A. Hill, Panel discussion, "What's the Secret of Successful Gas Search," O. & G. Jour., April 24, 1961, p. 95.

17. G. L. Hassler, E. Brunner, and T. J. Deahl, "The Role of Capillarity in Oil Production," Tech. Pub. 1623. Petrol. Technol., September 1944, and Trans. Amer. Inst. Min. Met. Engrs., Vol. 155 (1944), pp. 155-174.

18. Sylvain J. Pirson, *Elements of Oil Reservoir Engineering,* McGraw-Hill Book Co., New York (1950): see "Capillary Forces," pp. 245-269.

19. Ionel I. Gardescu, "Behavior of Gas Bubbles in Capillary Spaces," Tech. Pub. 306, Trans. Amer. Inst. Min. Met. Engrs. (1930), pp. 351-370. A clear statement of the mechanism of forcing a bubble from one pore into the next.

20. Sylvain J. Pirson, *op. cit.* (note 18), pp. 258-266.

21. Walter Rose and W. A. Bruce, "Evaluation of Capillary Character in Petroleum Reservoir Rock," Tech. Pub. 2594, Trans. Amer. Inst. Min. Met. Engrs., Vol. 186 (1949), pp. 127-142. 39 references.

22. M Muskat and M. O. Taylor, "Effect of Reservoir Fluid and Rock Characteristics on Production Histories of Gas-Drive Reservoirs," Trans. Amer. Inst. Min. Met. Engrs., Vol. 165 (1946), pp. 78-93.

23. John F. Carll, "Geology of the Oil Regions," in *2nd Geol. Surv. of Pennsylvania, 1875-1879,* Vol. 3 (1880), p. 262.

24. H. C. Miller, *Oil-Reservoir Behavior Based upon Pressure-Production Data,* RI 3634, U.S. Bur. Mines (1942), 36 pages.

George R. Elliott, "Behavior and Control of Natural Water-Drive Reservoirs," Tech. Pub. 1880, Trans. Amer. Inst. Min. Met. Engrs., Vol. 165 (1946), pp. 201–218. Bibliog. 65 items. Reprinted in Oil Weekly, August 6, 1945, pp. 54–60.

25. Dahl M. Duff, "New Life for Abqaiq," O. & G. Jour., December 27, 1951, p. 46.

26. R. C. Rumble, H. H. Spain, and H. E. Stamm III, "A Reservoir Analyzer Study of the Woodbine Basin," Tech. Pub. 3231, Trans. Amer. Inst. Min. Met. Engrs., Vol. 192 (1951), pp. 331–346.

27. Howard N. Hall, "Compressibility of Reservoir Rocks," Tech. Note 149, Jour. Petrol. Technol., Amer. Inst. Min. Met. Engrs., January 1953, pp. 17–19; Trans., Vol. 198, pp. 309–311.

28. Ralph J. Schilthuis and William Hurst, "Variations in Reservoir Pressure in the East Texas Field," Trans. Amer. Inst. Min. Met. Engrs., Vol. 114 (1935), pp. 164–176.

Morris Muskat, *Physical Principles of Oil Production*, McGraw-Hill Book Co., New York (1949), pp. 372–376.

John S. Bell and J. M. Shepherd, "Pressure Behavior in the Woodbine Sand," Tech. Paper 3000, Trans. Amer. Inst. Min. Met. Engrs., Vol. 192 (1951), pp. 19–27. Discussion to p. 28.

29. David Donahue, "Elasticity of Reservoir Rocks and Fluids, with Special Reference to East Texas Oil Fields" (Geol. Note), Bull. Amer. Assoc. Petrol. Geol., Vol. 28 (July 1944), pp. 1032–1035. 18 references cited.

30. F. G. Miller and H. C. Miller, *Résumé of Problems Relating to Edgewater Encroachment in Oil Sands*, RI 3392, U.S. Bur. Mines (March 1938), pp. 12–15.

31. James O. Lewis, "Gravity Drainage in Oil Fields," Tech. Paper 1611, Petrol. Technol. (September 1943), Trans. Amer. Inst. Min. Met. Engrs., Vol. 155 (1944), pp. 133–154.

E. P. Burtschaell, "Reservoir Performance of a High Relief Pool," Tech. Paper 2645, Petrol. Technol. (July 1949), Trans. Amer. Inst. Min. Met. Engrs., Vol. 186 (1949), pp. 171–179.

32. George S. Rice, *Mining Petroleum by Underground Methods*, Bull. 351, U.S. Bur. Mines (1932), 159 pages.

Gerald B. Shea, "Mining for Oil in Japan," Prod. Monthly, December 1947, pp. 19–28.

A. Beeby Thompson, *Oil Exploration and Development*, 2nd ed., Technical Press, London, Vol. 2 (1950), pp. 855–861.

F. Hoffmann and W. Ruehl, "Oil Mining in Germany," Prod. Monthly: Part I, July 1953, pp. 20–32; Part II, August 1953, pp. 25–33; Part III, September 1953, pp. 25–38.

33. T. V. Moore, "Behavior of Fluids in Oil Reservoirs," Bull. Amer. Assoc. Petrol. Geol., Vol. 22 (September 1938), pp. 1237–1249.

34. A. F. Van Everdingen, "The Skin Effect and Its Influence on the Productive Capacity of a Well," Trans. Amer. Inst. Min. Met. Engrs., Vol. 198 (1953), pp. 171–176. 8 references cited.

35. H. D. Wilde, "The Value of Gas Conservation and Efficient Use of a Natural Water-Drive as Demonstrated by Laboratory Models," Prod. Bull. 210, Amer. Petrol. Inst. (December 1932), p. 4.

36. S. E. Buckley and M. C. Leverett, "Mechanism of Fluid Displacement in Sands," Trans. Amer. Inst. Min. Met. Engrs., Vol. 146 (1942), pp. 107–116. 7 references.

37. E. R. Brownscombe and A. B. Dyes, "Water-imbibition Displacement—a Possibility for the Spraberry," in *Drilling and Production Practice*, Amer. Petrol. Inst. (1952), pp. 383–390.

38. Edgar Kraus, "MER—A History," in *Drilling and Production Practice*, Amer. Petrol. Inst. (1947), pp. 108–112.

Stuart Buckley (ed.), *Petroleum Conservation*, Amer. Inst. Min. Met. Engrs., New York (1951), pp. 151–163.

39. Conservation Committee of California Oil Producers (June 26, 1946).

40. James A. Lewis, William L. Horner, and Marion Stekoll, *Productivity Index and Measurable Reservoir Characteristics,* Tech. Pub. 1467, Amer. Inst. Min. Met. Engrs. (March 1942), 9 pages. 15 references.
Joseph A. Kornfeld, "Applications of the Productivity Index to Oil Field Problems," O. & G. Jour., July 26, 1951, pp. 236 ff.

41. Alton B. Cook, "Derivation and Application of Material Balance Equations to Magnolia Field, Arkansas," RI 3720, U.S. Bur. Mines. Also see Petrol. Engr., Part I, February 1944, pp. 91–96; Part II, March 1944, pp. 222–230.

42. John Arps, Address at Dallas, October 5, 1962.

43. Frank B. Notestein, Carl W. Hubman, and James W. Bowler, "Geology of the Barco Concession, Republic of Colombia, South America," Bull. Geol. Soc. Amer., Vol. 55 (October 1944), pp. 1165–1216.

44. Independent Petroleum Association of America, Monthly, August 1949, p. 49.

45. I. Brod, N. Elin, E. Starobinetz, and V. Tilupo, "Oil Fields of the Chechenian-Ingshetian (The Grozny Region) A.S.S.R.," XVIIth Int. Geol. Cong., Fascicle 3 (1937), p. 47.

46. A. Beeby Thompson, *The Oil Fields of Russia,* Crosby Lockwood & Son, London (1904), p. 303.

47. Julius Fohs, Sci. Amer., Vol. 179, No. 3 (September 1948), p. 12.

48. Alexander Winchell, *Sketches of Creation,* Harper & Brothers, New York (1871), pp. 287 and 443–444.

49. Ray V. Hennen and R. J. Metcalfe, "Yates Pool, Pecos County, Texas," Bull. Amer. Assoc. Petrol. Geol., Vol. 13 (December 1929), p. 1556.

50. World Oil, June 1961, p. 140.

51. O. & G. Jour., September 14, 1953, p. 83.

52. Charles R. Fettke, "Bradford Oil Field, Pennsylvania and New York," Bull. M 21, Penn. Geol. Surv., 4th series (1938), pp. 290–293.
John F. Buckwalter, "How Much Oil Will Bradford Produce after January 1, 1953?" Prod. Monthly, October 1953, pp. 22–23.

53. Pennsylvania Grade Crude Oil Association, "Laboratory Experiments with Carbonated Water and Liquid Carbon Dioxide as Oil Recovery Agents," Prod. Monthly, November 1952, pp. 15–45. 6 references.

54. James S. McNiel, Jr., and Jon T. Moss, "Recent Progress in Oil Recovery by In-Situ Combustion," Prod. Monthly (December 1958), pp. 25–32.

55. C. F. Gates and H. J. Ramey, Jr., "Better Technology Opens Way for More Thermal Projects," O. & G. Jour., July 13, 1944. Includes many references.

PART FOUR

The Geologic History of Petroleum

11. *The Origin of Petroleum*
12. *Migration and Accumulation of Petroleum*

INTRODUCTION
TO PART FOUR

THE TWIN PROBLEMS of the origin of petroleum and the manner of its migration and accumulation into pools have puzzled geologists for a long time and the questions they raise have yet to be completely answered.[1] The two problems are intimately related, and the solution of either would throw much light on the other. Part of the difficulty has been the inadequacy of the analytical methods. This is fast being remedied by modern extraction techniques and chromatography, which make possible rapid analysis for individual hydrocarbon molecules through C_{10} and individual group species through about C_{40}; this should permit tracing hydrocarbons from their original source material to the petroleum pool. We find the oil pool, for example, as the end product of a series of events that we have so far been unable to determine precisely. Much of this lack of precise and accurate data should be remedied during the next decade, thus permitting the full history of petroleum to become known.

Many theories have been proposed to explain the origin, migration, and accumulation of petroleum, but none has proved to be without some fundamental flaw or some missing link in the chain of events. In general, the early theories were based on laboratory experiments that attempted to simulate field conditions and on geologic reasoning applied to the various data uncovered during the exploration for and production of oil and gas. The more recent theories are based on modern analytical methods, especially carbon-isotope geochemistry. Any theory, if it is to be acceptable, must be based on common natural processes that will account not only for the extensive production of petroleum but for its widespread geologic and geographic distribution.[2] Petroleum hydrocarbons are common even though minor constituents of the earth's crust, and their origin is not to be sought in the extraordinary or unusual circumstance.

At present, it might be said that most geologists believe in a three-stage sequence of origin, migration, and accumulation into pools:

1. A disseminated mixture of soluble hydrocarbons, petroleum-like hydrocarbon compounds, and petroleum, together with insoluble organic matter, is deposited with the sediments that form shales and other nonreservoir rocks. Some of these hydrocarbons and petroleum-like hydrocarbon compounds are similar or closely related to those found in living organisms, and presumably they too were deposited along with the sediments. (Some believe that they constitute the chief primary source material for petroleum.) The more complex of the hydrocarbon compounds may have been formed with the help of one or more of the available energy sources or through slight chemical changes.

2. As the water is expelled from the shales during their compaction by loading, the petroleum is carried along and moves into the nearby porous and permeable sediments, such as sandstones. These become the reservoir rocks. This process, called *primary migration,* probably occurs chiefly during diagenesis and serves to make the initial separation of the petroleum from its water environment. The petroleum that is moved during the primary migration consists of petroleum hydrocarbons and asphalts.

3. Once the petroleum is in the reservoir rock, a subsequent, *secondary migration,* due to the buoyancy of the oil and to the movement of water through the permeable rocks, carries the oil to places where further movement is barred: thus petroleum pools are formed.

CHAPTER 11

The Origin of Petroleum

> Framework of limiting conditions. Inorganic origin. Organic origin: nature of organic source material – modern organic matter. Transformation of organic matter into petroleum: bacterial action – heat and pressure – alteration of petroleum.

THEORIES of the origin of petroleum* may be divided into two groups according to their view of the primary source material as organic or as inorganic. Early ideas leaned toward the inorganic sources, whereas the modern theories, with few exceptions, assume that the primary source material was organic. The change was brought about by an increasing number of objections to the inorganic ideas; but, since these objections have not completely eliminated the possibility that inorganic substances—especially hydrogen—played some part in the origin of petroleum, inorganic theories still find occasional favor.

Although agreement on the organic origin of petroleum is nearly complete, there are many differences of opinion about the details of the processes by which it was formed and about the relative importance of the different source materials. Were they primarily marine or terrestrial? How much petroleum was derived from hydrocarbons that were part of living organisms and how much was derived from the transformation of hydrocarbon compounds into petroleum? What was the nature of the energy involved in the transformation? Bacterial action, heat and pressure, radioactive bombardment, and catalytic phenomena have all been suggested as energy sources that may have made conversion possible, and each may have participated either separately or in combination with one or more of the others.

Theories on the accumulation of oil and gas have a direct bearing on origin.

---

* The reader is especially referred to the thorough analysis by Hollis D. Hedberg, "Geologic Aspects of Origin of Petroleum," Bull. Amer. Assoc. Petrol. Geol., Vol. 48, No. 11 (November 1964), pp. 1755–1803. Includes an extensive bibliography.

Some geologists hold that all petroleum was formed in place, either at or adjacent to the position of the present pools; others hold that petroleum has migrated from areas of origin to trap areas, and that the source area does not necessarily coincide with the accumulation area. Some theories assume that oil was transported along with circulating water, whereas others assume that it migrated independently of water movement. Some geologists believe that the source material was deposited in the shale formations or transformed into petroleum within them, and migrated from there into the reservoir rocks, while others believe that the source material, possibly in the form of colloidal or water-soluble organic matter, was concentrated in the reservoir rocks, or even in the places where the traps are now found, and was there changed to petroleum. Each theory has some evidence and reasoning in its support, and each has some against it.

The theory one adopts—and it must always be a theory, since no one was present when the origin and accumulation took place—has a bearing on the method of exploration one pursues. If one believes, for example, that petroleum originated *in situ,* one looks for areas favorable to origin; if one believes that petroleum has migrated into traps at some distance from the area of origin, the critical question becomes the location of suitable traps and barriers to migration. Belief in a marine environment of origin suggests that the best place to explore is in marine sediments, whereas belief in the possibility of fresh-water sources would encourage exploration in areas underlain by continental and fresh-water sediments. If migration is permissible, however, the question of marine or nonmarine environment of origin is not always decisively significant, for migrating petroleum could concentrate wherever there was a trap within the limits of the barriers to migration.

FRAMEWORK OF LIMITING CONDITIONS

Hydrocarbon compounds* similar to those found in petroleum may be formed in the laboratory from various source materials. Laboratory conditions, however, are quite different from those in oil and gas pools. It is helpful, therefore, to list some of the natural environmental conditions known to prevail in present pools, in order to ensure that any theories or laboratory results proposed will be applied within a framework of conditions that can reasonably be compared to the known field conditions. These limiting conditions have been called a *geologic fence*.[3] Some of the limiting factors are summarized as follows:

1. Nearly all petroleum occurs in sediments. These sediments are chiefly of marine origin, and it follows that the contained petroleum is also most likely

---

* A hydrocarbon compound is a complex hydrocarbon that contains, in addition to hydrogen and carbon, small amounts of other elements, such as sulfur, nitrogen, and oxygen.

THE ORIGIN OF PETROLEUM [CHAPTER 11] 501

marine, or related to marine conditions. Substantial amounts of petroleum are also found in sediments of continental, or nonmarine, origin. This petroleum, therefore, might conceivably have originated within the continental formations; but, since continental deposits are generally oxidized and grade into or are in unconformable contact with marine sediments, the petroleum in them is more likely to have been of marine origin and to have migrated into the nonmarine rocks.

2. Petroleums are extremely complex mixtures of many hydrocarbons occurring in homologous series, and no two petroleums are exactly alike in composition. This variation in composition is probably chiefly due to variations in the primary source material, or it may in part be the result of subsequent environments and of such vicissitudes as migration, catalysis, polymerization, pressure and temperature changes, and metamorphism. Although the components of petroleum unite to form an extremely complex mixture, the elemental chemical analyses of most petroleums are remarkably similar, even of those that vary greatly in physical properties. Most petroleum is chiefly composed of 11–15 percent hydrogen and 82–87 percent carbon by weight.

3. Petroleum is found in rocks from the Precambrian to the Pleistocene, although the occurrences in Precambrian and Pleistocene rocks are rare and anomalous. Organic carbon that is possibly of petroleum or petroleum-like origin, however, has been identified in rocks of Precambrian age.[4] Some petroleum existed as such in Ordovician and Pennsylvanian time, for certain conglomerates of these two ages contain oil-saturated pebbles embedded in a barren matrix. Inspissated (?) petroleum deposits in the unconformity zone below Pennsylvanian rocks in Oklahoma would appear to be at least as old as Pennsylvanian. Asphaltic sandstone and grahamite in an unconformity zone within the Ordovician in the Lucien field, Oklahoma, appear to have formed in Ordovician time.[5] These occurrences show that petroleum, once formed, may have been preserved against the forces of destruction and decay over long periods of geologic time.[6]

4. Until the advent of chromatographic and similar tools, no soluble (in organic solvents) liquid petroleum hydrocarbons had been found in the shales and carbonates that make up such a high percentage of the sediments of the world. Insoluble organic matter was found to be almost universal in the sediments, but no soluble petroleum hydrocarbons had been detected. Such soluble petroleum hydrocarbons as well as petroleum have now been found by a number of investigators,[7] the amounts varying up to 50 or more barrels per acre-foot but commonly less than 10 barrels per acre-foot. Many of the hydrocarbons are also found in living organisms.

5. The temperatures of petroleum reservoirs rarely exceed 225°F (100.7°C), but temperatures as high as 300°F (141°C) have been measured in some of the deeper reservoirs. Minimum temperatures approaching the mean atmospheric temperature occur in some shallow pools. The presence of porphyrins in some petroleums indicates that the temperature of such petro-

leums has never exceeded 392°F (200°C), for porphyrins are destroyed at slightly lower temperatures. In other words, it suggests that the origin of petroleum is a low-temperature phenomenon.

6. The origin of petroleum is within an anaerobic and reducing environment. The presence of porphyrins in some petroleums means that anaerobic conditions developed early in the life of such petroleums, for chlorophyll derivatives, such as the porphyrins, are easily and rapidly oxidized and decomposed under aerobic conditions. The low oxygen content of petroleums, generally under 2 percent by weight, also indicates that they were formed in a reducing environment.

7. The fluctuations in pressure and temperature within a petroleum reservoir that result from the erosion, deformation, and deposition accompanying uplift, truncation, and burial in many basins are known to have been extensive. Pressure variations on petroleum have ranged from atmospheric pressure up to 8,000 or 10,000 psi. Temperatures also have fluctuated by as much as 250 degrees Fahrenheit. Petroleum, therefore, can undergo considerable change in pressure and temperature without being appreciably changed in physical character; changes in chemical composition have been observed, however, which may be attributed to either the environment of deposition or to depth of burial.[8]

8. The geologic history of oil pools indicates that in some of them neither lateral nor vertical oil migration of any consequence has occurred, whereas in others there has been extensive lateral and vertical migration. No consistent differences have been observed between oils that are known to have migrated and those that are known to have been formed where they are now found. Any valid theory of the origin of petroleum, therefore, must be independent of migration effects.

9. The time required to form petroleum and concentrate it into pools is probably less than one million years. The highest ratio of oil pool occurrence to volume of sediments occurs in the Pliocene series, which ended about one million (?) years ago.[9] Late Pliocene sandstones and rocks such as the Plio-Pleistocene rocks in the Quirequire field of eastern Venezuela contain commercial oil. Small oil pools at Baku and in the Turkmen SSR, USSR, have been found in the "Baku stage" and in the "Apsheron stage" of the Pleistocene. Oil has also been produced from Plio-Pleistocene rocks in the Summerland field of California. Pleistocene oil in pools such as these may have moved in from older rocks, yet could equally well have formed from equivalent marine rocks, and thus could measure the minimum time for forming petroleum and concentrating it into a major oil pool. The time required for hydrocarbons to form, however, may be much less, as shown by the hydrocarbons found in shallow cores (3–103 feet) of Recent sediments in the Gulf of Mexico,[10] where measurable amounts of paraffinic, naphthenic, and aromatic hydrocarbons were found to range between 9 and 11,700 parts per million of dried sediments. The measurement of age by carbon 14 indicates a recent

origin, some 11,800–14,600 ± 1,400 years ago, of both the source material and the enclosing sediments.

At Pedernales in eastern Venezuela,[11] a 20-foot sand enclosed within the 200-foot Paria clay formation contains an appreciable concentration of hydrocarbons—about four times that of the surrounding clays or the sands opening to the surface. Carbon-14 dating indicates that the entire Paria formation was deposited in less than 10,000 years and that the enclosed sand was deposited in about 5,000 years. The average hydrocarbon concentration in the sand is about 150 ppm; subtracting the average—25 ppm of hydrocarbons for the entire formation—leaves 125 ppm hydrocarbons as having accumulated since the sand was deposited. This is an average of 0.025 ppm per year, which if projected into the future would require only 1,000,000 years to accumulate 500 barrels per acre-foot of oil—a rich field.

INORGANIC ORIGIN

The chief interest of the "inorganic theories" on the origin of petroleum is historical, for most of them have long since been abandoned. It was natural that they should develop, however, in view of what was known about the universe during the eighteenth century. The chief support for theories of inorganic origin lies in the fact that in the laboratory the hydrocarbons methane, ethane, acetylene, and benzene have repeatedly been made from inorganic sources. There has not been, however, any field evidence that the processes have occurred in nature, while there is an expanding mass of evidence of organic origin.

Theories that uphold the inorganic origin of petroleum have few supporters[12] today for several good reasons. In the first place, optical rotary power is a characteristic of petroleums, and especially of the intermediate boiling fractions (250–300°F). As far as is known, this phenomenon is almost entirely confined to organic matter and is observed only where biological agencies have prevailed. Another serious objection to any inorganic origin is that several homologous series of hydrocarbon compounds, containing great numbers of individual members, are found in all petroleums. All known compounds of this kind are of organic origin and could hardly be formed by inorganic agencies.

The lack of association of petroleum with vulcanism or its products, except in rare and anomalous occurrences, is another reason for doubting that there is any important relation between volcanic action and the origin of petroleum. As has been pointed out by White* and by DeGolyer,[13] areas where oil or gas is found associated with igneous material, hot springs, or other evidence of vulcanism are underlain by sediments; there is no record of the occurrence of petroleum in volcanic areas where the underlying material is igneous. Most

---
* Don E. White in a personal communication.

occurrences of hydrocarbons associated with volcanic rock appear to be better explained as having emanated from associated sediments than as being genetically related to the igneous material.

If petroleum were of cosmic origin, we would expect to find it more uniformly distributed over the earth than it is, and to find it abundant in the older rocks. Petroleum of cosmic origin would be no respecter of the age of the rock, and should be uniformly distributed throughout the geologic column wherever there is permeability. We find, however, that the Precambrian, Cambrian, Triassic, and Pleistocene rocks are relatively low in hydrocarbons, even though they all contain large volumes of porous and permeable rocks.

ORGANIC ORIGIN

Three compelling reasons favor the belief that the chief primary source material of petroleum—the "protopetroleum"—was organic:

1. The vast amounts of organic matter and hydrocarbons now found in the sediments of the earth. Carbon and hydrogen predominate in the remains of organic material, both plant and animal. Furthermore, lesser but still important amounts of carbon and hydrogen and hydrocarbons are continually produced by the life processes of plants and animals. An abundant and widely distributed source of the two essential elements of petroleum—carbon and hydrogen—is therefore provided by organic material.

2. The fact that many crude oils have been found to contain porphyrin pigments,* and the fact that nearly all petroleums contain nitrogen, are more or less direct indication of the animal or vegetable origin, or both, of petroleum, because all organic matter contains both porphyrins and nitrogen.[16] The porphyrins occur in the asphalts, and in the medium-to-heavy fractions when they have not been filtered and still contain asphaltic components. The amounts present vary, but in nearly half of the samples examined, which came from nearly all parts of the world, the porphyrin content ranged from 0.004 to 0.02 mg per 100 grams. In others it ranged from 0.4 to 4.0 mg per gram.

Nitrogen is an essential component of the amino acids [$CH_2(NH_2)COOH$] —that is, of the hydrolyzed protein of all living matter. Our noses remind us of that whenever we smell the ammonia (NH_3) given off by rotting refuse. Trask and Patnode[17] found that the amount of organic nitrogen in sediments varied in weight almost in direct proportion with the organic carbon content. Either the nitrogen or the carbon content may be used to give an approximate measure of the organic matter present in a sediment; the organic matter in the

---

* Porphyrins are formed from the red coloring matter of blood (hemin) or from the green coloring matter of plants (chlorophyll). The porphyrins in petroleum occur in the form of complex hydrocarbon compounds that oxidize readily. Treibs found the vegetable porphyrins ($C_{32}H_{36}N_4$ and $C_{32}H_{35}N_4COOH$), derived from chlorophyll, to be far more plentiful than the animal porphyrins ($C_{32}H_{38}N_4$ and $C_{32}H_{36}N_4COOH$), derived from hemin.

ancient sediments is, on the average, 1.1 times as abundant as the carbon and 24 times as abundant as the nitrogen. Nitrogen is present in practically all petroleums, chiefly as a constituent of complex hydrocarbon compounds. The continuous chain of occurrence of nitrogen, from living matter through organic matter in sediments to petroleum, therefore seems to offer a reasonable indication of the organic nature of the source material.

3. Optical activity—the power to rotate the plane of polarization of polarized light—is a property of most petroleums, and is not known to occur in oils of inorganic origin or in inorganic substances or minerals with the exception of cinnabar (HgS) and quartz (SiO_2). The power of optical rotation is not uniform throughout the distillation range, but is usually at a maximum in the fractions having intermediate boiling points (from 250° to 300°C). It is believed that the optical activity in most petroleum is due to the presence of cholesterol ($C_{26}H_{45}OH$), which is found in both vegetable and animal matter.[18]

4. A wide variety of petroleum hydrocarbons, and even crude oil, have been found included in the organic material that is found in nearly all nonreservoir rocks, such as the shales and carbonates. The same types of hydrocarbons occur in both the fine-grained sediments and in crude oil. The intimate relation of the organic material and the petroleum in the sediments leaves no doubt that organic matter was the original source of the petroleum.

Sir William Logan may have been the first to express the view that petroleum is possibly of organic origin. In 1863 Robb[19] credited him with the view that ". . . petroleum owes its origin, in all probability, to the slow decomposition and bituminization of organic matter . . . deposited with the other materials of which the rocks are composed."

Hackford[20] showed in 1922 that the ash content of algae is similar to that of crude oil—both contain iodine, bromine, phosphorus, and ammonium salts. Later he enlarged on this idea[21] and reached the conclusion that oil and bitumen could be produced by the pyrolysis and hydrolysis of algae at low temperatures.

An even more direct approach to a correlation between petroleum and its source material is contained in the work of Sanders.[22] He found a wide variety of microobjects in crude oils from many pools. The long list of materials he found includes calcified or siliceous skeletal tests or frameworks, petrified wood fragments, foraminiferal tests, minute pyrite globules or concretions, vegetable remains encrusted with silica, small crustaceans, insect scales, barbules, spore coats, algae, fungi, cuticles, resins, and fragments of coal and lignite. Some of this material may well have been entrained in the petroleum from a foreign source as it moved through the rocks. The large variety of organic material, however, strongly suggests a genetic relation between it and the petroleum in which it occurs.

The evidence of petroleum hydrocarbons in modern Gulf of Mexico sediments is an additional proof of organic origin.[23] The age of the oil studied, as determined by carbon, indicates that the oil has not migrated and is not seepage oil. It was most likely formed at such shallow depths and in such late sediments by the decomposition of organic matter deposited along with the sediments.

CURRENT THEORY

A recent far-reaching advance in our thinking concerns the discovery that petroleum hydrocarbons and related hydrocarbon compounds occur in many living organisms and are deposited in the sediments with little or no change. Many investigators have participated in these discoveries; a few are listed at the end of the chapter.[24]

Nearly all shales and carbonates contain disseminated organic matter of three general kinds: soluble liquid hydrocarbons, soluble asphalts, and insoluble kerogen. The quantities range from a fraction of a barrel to more than 50 barrels per acre-foot of soluble hydrocarbons, and this represents the material remaining after the removal of an unknown amount during and after diagenesis. A significant observation by Hunt[24] is that with the exception of a few red shales, sandstones, and metamorphosed sediments, the presence of petroleum hydrocarbons is practically universal in the nonreservoir sediments.

In addition to the soluble petroleum hydrocarbons, the organic matter contains numerous insoluble hydrocarbon compounds, asphalts, and complex organic substances, some of which through bacterial action, heat, pressure, or catalytic action, or combinations of these, may be transformed into petroleum hydrocarbons.

These complex insoluble organic substances constitute kerogen—a pyrobitumin. Two types of kerogen are now known, a coaly type and an oily type. The coaly type does not form petroleum but contributes to the cannel coal and lignite deposits. The oily type does form petroleum. These organic substances tend to be oil-wet, and petroleum hydrocarbons may travel far attached to and protected by them.

The same kinds of hydrocarbons are found in both the nonreservoir sediments and in the living organisms. Most living organisms contain hydrocarbons,[25] hydrocarbon compounds, fatty acids, terpenoids, and steroids, all of which may have been deposited directly in the shales and carbonate rocks with little or no change. Thus an abundant and direct source is provided for petroleum hydrocarbons found in the sediments of the world. The migration and accumulation into oil and gas pools will be considered in the next chapter.

The organic matter now found in the nonreservoir sediments has the following general composition in percentages by weight as compared with the composition of petroleum:

	ORGANIC MATERIAL	PETROLEUM (CRUDE OIL)
Carbon	52–71	83–87
Hydrogen	5–10	11–15
Oxygen	5–20	Trace to 4
Nitrogen	4–6	Trace to 4
Sulfur		Trace to 4

This organic matter is, in general, of three different kinds:

1. Hydrocarbons, similar in composition and in form to the heavier fractions of the crude oil found in the reservoir rocks.
2. Asphalts, similar in composition and in form to the asphaltic constituent of crude oil.
3. Kerogen, an insoluble, pyrobituminous organic matter that makes up the bulk of the organic matter of most nonreservoir sediments.

Hydrocarbons. The hydrocarbons found in the nonreservoir sediments presumably have come directly, either wholly or partly, from the hydrocarbons that are found in living plant and animal matter. They may result, in part, from processes whereby organic matter, or petroleum-like hydrocarbon compounds in the organic matter, were converted through minor changes into petroleum hydrocarbons. This conversion must have occurred either before or during diagenesis of the sediments, for the petroleum content of shales and carbonate rocks is rather universally distributed.

Studies have been made by Erdman[26] of many of the common petroleum hydrocarbons that are found in the sediments. He has analyzed the following characteristic fractions that are present both in petroleum and in the sediments in appreciable quantities:

1. The low-molecular-weight aromatic hydrocarbons—benzine, naphthalene, toluene, ethylbenzene, the xylenes, and other derivatives boiling up to 250°C—make up as much as 5 percent of the total hydrocarbon fraction of crude oils and are widely distributed in ancient sediments. They are not found in Recent sediments, however, and hence a chemical mechanism must be provided for their genesis by other than life processes.

Erdman suggests that such compounds as squalene, the carotenes, terpenes, and unsaturated fatty acids may have been precursors of the light aromatic hydrocarbons. All of these are formed in considerable quantities by land and sea plants and animals, and are converted by simple chemical reactions into aromatic hydrocarbons.

2. The light aliphatic hydrocarbons—methane, ethane, propane, the butanes, pentanes, etc.—are a characteristic fraction of crude oils and consist of the members through C_5, or *n*-heptane, which is the lowest hydrocarbon known to be a constituent of living organisms, hence it is necessary to seek a chemical mechanism for the generation of these hydrocarbons.

Proteins are an important, and probably sufficient, source of the aliphatic hydrocarbons. Known reactions of many of the amino acids that make up the proteins will yield all the aliphatic isomers needed. Many amino acids are found in ancient sediments, but many more are found in Recent sediments, and in greater abundance.

3. The intermediate and heavy aliphatic, naphthenic, and aromatic hydrocarbons are found both in sediments and in crude oil. Erdman believes that

some were formed from the lipid fraction of both plants and animals and have survived through geologic time with little change because of their stability; and that others were generated from nonhydrocarbon material such as fatty acids, aldehydes, and alcohols.

Asphalt. The asphaltic constituents, such as resins, maltenes, and asphaltenes, are the dark, nonhydrocarbon fractions of the organic portion of the sediments and of petroleum and consist primarily of carbon and hydrogen together with oxygen, nitrogen, sulfur, and the metals vanadium and nickel. They are complex high-molecular-weight substances, probably ranging in number from a few hundred into the hundred thousands, and make up from a trace to 50 percent of some crude oils.

The asphalt found in the nonreservoir sediments is similar to the asphaltic fraction of crude oil and of many natural asphalts deposited in seepages of crude oil. Asphalts have been found in the sediments in amounts that range from about 1 to 70 barrels per acre-foot, or slightly more than the hydrocarbons. Asphalts are complex compounds that have not been found in living organisms but which may have been derived from such materials as cellulose, lignin, purines, and pyrimidines. The presence of porphyrins (natural pigments related to chlorophyll and hemoglobin) in crude oil is evidence of its biological origin. Porphyrin mixtures are found in appreciable amounts chiefly in the crude oils of high asphaltic content, which further suggests an early origin for the asphaltic content.[27] Porphyrins occur as complicated metal-porphyrin mixtures, which suggests a low-temperature history, and also cause petroleum to cling to oil-wet reservoir rock surfaces.

Kerogen. Most of the organic material (85–95%) found in the nonreservoir rocks consists of kerogen, a solid pyrobitumen that is insoluble in ordinary organic solvents.[28] Heat is required to break it down. Its elemental analysis shows that it is chiefly carbon, hydrogen, and oxygen, with lesser amounts of nitrogen and sulfur. Kerogen found in typical marine nonreservoir rocks, is, when dried, a fine amorphous dark-brown to black powder that frequently shows a marked resemblance to coal dust. When the material is heated in test tubes, little or no oily distillate is formed as is formed when oil shales are heated. Apparently kerogen is of different types, even though the elemental analyses are similar. Some are indistinguishable from coal, whereas others, probably with a high asphaltic content, form the oil shales. Except for the variable asphaltic content, kerogen probably is not a source material for petroleum, but might be likened to the carrier material for the petroleum hydrocarbons and the related hydrocarbon compounds.

The Nature of Organic Source Material

Organic matter that contains both soluble and insoluble hydrocarbons that might be considered as a potential source material for petroleum occurs in a wide variety of both animals and plants. This fact may explain the great variation in petroleums found in nature. The source material of a specific oil may, on the other hand, have consisted predominantly of a single type of organic matter, and the variations in composition of petroleum may have developed later as a result of migration, filtering, bacterial action, metamorphism, catalysis, and so on, after the formation of petroleum. Many of the naturally occurring organic compounds are unsaturated hydrocarbons, whereas petroleum consists chiefly or entirely of saturated hydrocarbons. As yet there is no specific indication whether the primary source material consists of many types of organic matter or predominantly of one type.

Proteins. Proteins are nitrogenous substances that make up a large part of all plants and animals; they contain approximately 16 percent nitrogen by weight, and also contain carbon, hydrogen, oxygen, and often other elements, such as phosphorus, sulfur, iron, and copper. Proteins heated in acidic solutions hydrolize into amino acids, which are carboxyl acids in which one hydrogen atom has been replaced by an amino group (NH_2). The simplest of the amino acids is glycine, $CH_2(NH_2)COOH$. Proteins are composed of complex, long-chain molecules, built up by joining together large numbers of amino acid molecules, and may be converted into water-soluble compounds by decomposition.

Carbohydrates. Carbohydrates are found in both plant and animal matter. They consist principally of the glucose sugars ($C_6H_{12}O_6$), starches ($C_6H_{10}O_5$), and celluloses (also of the formula $C_6H_{10}O_5$). All carbohydrates consist of long-chain molecules; most of them are easily decomposed into water-soluble compounds. The degradation products of carbohydrates are chiefly humus and humic acids.

Fats, Fatty Acids, and Fat-soluble Compounds. Fats and fat-soluble compounds are present in all living animals and plants. They are insoluble in water and are much more resistant to decay than either the carbohydrates or the proteins. The natural fats and oils are esters; that is, they are products of reactions between acids and the alcohol glycerol ($C_6H_5O_6$). The sodium salts of the fatty acids are called soaps. Various types of fatty acids occur in nature, as both straight-chain and branched-chain structures, both saturated and unsaturated, and the fatty acid radicals constitute the largest known group of long-chain molecules. The naturally occurring fats and fatty acids, or lipids, are thought by many to constitute the chief primary source of compounds of petroleum.

Hydrocarbons. Petroleum-like hydrocarbons occur in many living organisms, both plant and animal. Their source is thought to be in the lipid fractions of the organism, and the conversion from "living" hydrocarbons to crude-oil hydrocarbons may require no intermediate step. Bacteria may be the chief source of such hydrocarbons, and bacterial action may be the chief agency for the conversion of closely related hydrocarbons into petroleum.

Modern Organic Matter

One guide to the nature and origin of the organic matter in the sediments formed in the geologic past is a consideration of the organic matter that is being formed, decomposed, and buried in modern sediments at the present time. Modern organic matter and its mode of burial will be considered first, then the organic matter contained in the ancient sediments.

The primary source of the organic matter in the sediments may be either animal or vegetable or both. Some of this material is carried to areas of sedimentation by streams, waves, or currents, and some of it remains in the places where it grows. Organic matter may consist either of animal or vegetable remains or of the waste materials formed by organisms during their life cycles. Since most petroleum deposits are closely associated with sediments deposited under marine conditions, most theories of organic origin hold that petroleum originated in sediments of marine environment. Consequently, the organic matter of the oceans is of the most importance. A part of the organic matter in the oceans, however, is of continental, or nonmarine, origin, having been carried to the sea by rivers; when the story is finally told, these terrestrial hydrocarbons and organic materials may form a very important source of petroleum-related organic matter.

The marine organisms[29] that may provide organic source material for petroleum may be classified broadly as (1) *plant life,* including marine fungi, bacteria, algae, and the dinoflagellates; (2) *animal life,* including many diverse groups such as the foraminifers, radiolarians, and other protozoans, the sponges, corals, worms, bryozoans, brachiopods, crustaceans, mollusks, echinoderms, and finally the vertebrates.

The marine conditions that generally favor high organic content have been listed by Sverdrup et al. as:

(1) An abundant supply of organic matter.

(2) A relatively rapid rate of accumulation of inorganic material, particularly if fine-grained.

(3) A small supply of oxygen to the waters in contact with the sediments. The most extreme development of such conditions may be found in basins and landlocked regions where stagnation exists and results in the exclusion of animal life.[30]

Minute or even microscopic forms of marine plants and animals are probably of the greatest importance as a source of organic matter. Much of this material either floats or swims at or near the surface of the water, where sunlight is available. Floating plant life is termed *phytoplankton* (Greek, *plankton,* wanderer), and floating animal life is termed *zooplankton.* The sessile, creeping, and burrowing organisms of the sea bottom are known as the benthonic (Greek, *benthos,* deep sea).

The organic cycle of the sea begins with an abundance in sea water of the necessary inorganic elements—carbon, hydrogen, oxygen, phosphorus, nitrogen, iron, and many trace elements, which, together with sunlight, make plant life possible. Many of the plants are of minute or even microscopic size, and these constitute the food material for animals that, although minute, are larger than the plants on which they feed or graze. These tiny animals, in turn, become food for larger animals. The death of the plant and animal organisms, followed by their decomposition and decay, permits a partial return of the elements of which they consist to the sea in the form of various chemical compounds; only a small part of the organic material, in various stages of decomposition, is buried and preserved under the sediments that are being deposited. Decomposition and decay of organisms involves bacterial action, and bacterial action may be the chief agency that converts or aids in the conversion of hydrocarbon compounds into more petroleum-like hydrocarbons. During the organic cycle of both plants and animals, in fact, many, if not all, of them also produce hydrocarbons as a normal part of their existence.[31]

Most of the organic content of the ocean is either dissolved or in a colloidal form in the sea water. The rest is contained in the plant and animal life, chiefly in the microscopic and semimicroscopic organisms, the suspended, floating, or free-swimming plankton found in the water penetrated by sunlight. It is difficult to estimate the rate at which organic matter is produced in the sea because of the many variable factors concerned in the process, but several observations have shown that phytoplankton is produced at a rate as high as several hundred grams of carbon per cubic meter of sea water per year.[32] Photosynthesis in the oceans has been estimated to produce 12 million tons (80,000,000 barrels) of hydrocarbon material annually.[33] A minute fraction of this amount, preserved in the sedimentary rocks, would supply all the known petroleum deposits, plus those that we can expect to discover in the future.

Plant Life. Marine plants are able to synthesize complex organic substances by photosynthesis from the inorganic compounds dissolved in the sea water. Photosynthesis is the process whereby the pigments in the plant tissue are able to intercept the radiant energy from the sun, and, in the presence of carbon dioxide, to manufacture carbon compounds from water and carbon dioxide. The equation may be expressed as follows:

$$6CO_2 + 6H_2O + \text{energy} \longrightarrow 6O_2 + C_6H_{12}O_6$$
$$\text{carbon dioxide} \quad \text{water} \quad \text{sunlight} \quad \text{oxygen} \quad \text{carbohydrate (hexose sugar)}$$

Photosynthesis is an endothermic reaction (uses heat). The energy of sunlight, stored in the complex organic product, becomes the source of energy for plant life and indirectly for animal life. The free oxygen released by photosynthesis is used in the respiration of both the plant and the animal life of the sea. The carbohydrates, in part, are presumably reduced in the direction of hydrocarbons in the reducing, oxygen-free environments into which they fall and in which they are buried.

The depth to which sunlight penetrates the water therefore determines the thickness of the layer of sea water that is favorable for the production of plant life and consequently of the life that feeds upon plant life. This depth varies, depending on the turbidity of the water, but in clear water it is about 250 feet or more. It has been estimated that coastal waters[34] produce at least fifty times more plant life than do the open ocean waters, and the coastal water that is richest in organic material is off the mouths of rivers.

Marine algae offer some of the most promising source material for petroleum.[35] The blue and green algae are the most common; in addition to utilizing sunlight to form chlorophyll and other, unidentified substances, such as enzymes, they assimilate their mineral constituents through their entire surface from the waters in which they grow. Because of this faculty, the same algal species may vary in chemical composition, depending upon the composition of the water in which it lives. Algae are probably the chief agents of lime secretion and deposition:[36] one genus, *Halimeda,* deposits aragonite, and another, *Lithothamnium,* deposits calcite. Both genera are especially common in organic reef deposits.

Diatoms are algae characterized by siliceous shells, or frustules. They have formed great deposits known as diatomaceous earth and are probably the most important group of algae in the organic economy of the sea. Some geologists believe that they provide the source material for much of our oil.[37] A significant product of their life process is vegetable oil, droplets of which may be frequently seen within diatom shells.[38] It has been estimated that from 5 to 50 percent of the volume of these diatoms consists of oil globules,[39] which may be freed from the shell if it should encounter fresh water and break because of the osmotic pressure from within the shell.[40] In the colonial algae *Elaeophyton,* even the cell walls have been observed to consist largely of oil or oily substances.[41] This oil may be considered the medium in which such organisms store food energy, serving the same purpose as the oils contained in nuts and seeds.

Marine water normally provides an equilibrium between the various organisms that constitute its population. This equilibrium is not stable, but is changing constantly with changes in marine conditions—currents, waves, upwelling, rainfall, and fluctuations in temperature and the supply of nutrients. From time to time, therefore, the conditions will be especially favorable to one organism or type of life, which will then multiply rapidly and thus upset the normal balance. This rapid multiplication of microorganisms is termed *blooming*. It may sometimes be recognized from a discoloration of the water—which is most commonly tinged with blue, red, or green—or from the decomposition of a deposit of organisms washed up on the shore. Along the Copalis

Beach of Washington,[42] for example, rapid propagation of the diatom *Aulacondiscus* periodically causes great masses of it to accumulate for several days along the tidal flats.

Extensive "oily patches" are formed during the months of August and September along the western coast of Japan because of the swarming of the pelagic diatom *Rhizoslenia,* and the same alga covers the Sea of Azov from September to December, giving it the dark-brown color, smooth surface, and marshy odor of a placid swamp. A different effect of blooming is the occasional development of a toxic substance that causes the death of great numbers of fish, whose bodies sink to the bottom or pile up on shore.[43] Twice a year in Whale Bay, on the west coast of Africa, a blooming of plankton secretes a poisonous substance that kills all the fish in the area. The dead organic tissue falls to the bottom and forms a *sapropel* * in the anaerobic environment.[44] A gelatinous agglomeration of diatoms, buoyed up by oxygen bubbles resulting from photosynthesis, forms the *mare sporco* of the Adriatic. The floating masses become large enough to break fishermen's nets by their weight.[45]

We have, then, in marine algae and the modern phenomenon of algal and diatom blooming, a mechanism whereby material with a high carbon and hydrogen content, including petroleum hydrocarbons, may form in large quantities. Whether some form of algae or algal blooming has provided a source material for crude oil is unknown, but they form what is probably the most promising and widespread potential source material now known.

Animal Life. Partly as a result of Engler's distillation of oil from menhaden,† and also because of the common association of petroleum with fossiliferous marine sediments, fossils, especially microfossils, have long been considered as evidence of a possible petroleum source material. The association of oil pools with fossiliferous limestone, shale, or sandstone seems, on the face of it, a good reason for believing that at least a portion of the soft parts of the organisms were decomposed and changed into petroleum, leaving the hard skeletons behind as fossils. Support is given to this idea by the fossil casts filled with oil or with petroleum-like liquids[46] that have been found in many places and by the hydrocarbons that are found in nearly all living organisms. (See also pp. 514 and 516.) The animal population of the modern seas is both enormous and complex, and it alone produces soluble organic matter and hydrocarbons in quantities adequate for all organic source requirements.

The sorting and transporting action of ocean waves and currents may cause large volumes of shells to accumulate far from the areas in which the shell-bearing animals lived. Therefore, a rock that consists largely of shells may not

---
* From Greek *sapros,* "rotten"; organic debris that consists largely of the remains of recently deposited marine plants and animals and accumulates on the ocean floor.

† A fish found along the Atlantic coast of the United States and used for the making of oil and fertilizer.

be as likely a source of organic matter as might at first be thought, since the shells may have been concentrated after the living tissues had left them. In the modern seas the naked masses of protoplasm leave the tests to drift around freely, so that a large proportion of the tests preserved in the rocks were probably empty of animal tissue at the time of burial.[47]

Silliman,[48] in a significant article published in 1846 on the living calcareous corals, stated that the organic content of corals ranged between 4 and 8 percent of the total mass. Silliman also noted that a fatty, wax-like residue, soluble in ether, but insoluble in alcohol, could be produced by dissolving the corals. Later Clarke and Wheeler[49] verified his observations, finding the percentage of organic matter in the shells of marine organisms to vary between 2 and 10 percent for common invertebrates. Many less-common invertebrate shells contained as much as 40–50 percent organic matter. Bergmann and Lester[50] investigated these chemically and found that they contained sterols, cetyl alcohol, low-melting hydrocarbons, and small amounts of ketones. The yield of ether-soluble waxes runs from 0.3 to 0.5 percent by weight of the total mass of the coral, or about 10 percent by weight of the organic content of the stony corals. The organic material was found to penetrate deep into the coral structure, some sea fans (Gorgonaceae) containing up to 3.0 percent of the total weight of nonsaponifiable material. Reef-builders are thus seen to be vast storehouses of potential source material for petroleum.

Nonmarine Organic Matter

Much organic matter reaches areas of marine sedimentation by way of streams. The organic matter found in modern streams may constitute more than 50 percent by weight of the total solids,[51] as shown in Table 11-1.

Probably the most important potential source materials for the biochemical

TABLE 11-1 Percentage of Organic Matter in Dissolved Solids of River Waters

River	Percentage	River	Percentage
Danube	3.25	Amazon	15.03
James	4.14	Mohawk	15.34
Maumee	4.55	Delaware	16.00
Nile	10.36	Lough Neagh, Ireland	16.40
Hudson	11.42	Xingú	20.63
Rhine	11.93	Tapajos	24.16
Cumberland	12.08	Plata	49.59
Thames	12.10	Negro	53.83
Genesee	12.80	Uruguay	59.90

Source: From Clarke, Bull. 770, U.S. Geol. Surv. (1924), p. 110.

organic compounds that form on the land and are carried into the oceans by the rivers are the humus substances, which include humic acid ($C_{20}H_{10}O_6$), geic acid ($C_{20}H_{12}O_7$), and ulmic acid ($C_{20}H_{14}O_6$). These humus substances are formed by the slow decomposition of the lignins in peat and are found in soils highly charged with decaying vegetation.[52] Vast quantities of humic acid are forming constantly in swampy regions and, especially in the tropics, are carried into the oceans, either in solution or in colloidal dispersion. Mingling of fresh and salt water might cause the precipitation of the organic material.[53] One part of humic acid dissolves in 8,333 parts of water at 6°C and in 625 parts of water at 100°C. Temperature changes alone, therefore, would be sufficient cause for extensive precipitation of some of the organic matter upon reaching the oceans.

Waxes and resins are commonly produced by many kinds of vegetation. They are generally more resistant to decay than other organic residual products, and probably are the source of the benzene-soluble montan waxes that are obtained commercially from lignite coals in both Europe and the United States. A similar wax is also found in some peat deposits. These substances are organic, but because of their close association with the rocks they are generally grouped with the "mineral" waxes, such as ozokerite and the paraffin hydrocarbons. The chemical constituents that have been identified in these waxes include a number of esters, acids, alcohols, ketones, hydrocarbons, and resins.[54] An analysis (in percent) of five benzene-soluble waxes from Europe and the United States gives the following average composition:[55]

Hydrogen	11.9
Carbon	79.4
Nitrogen	0.1
Sulfur	0.4
Ash	0.2
Oxygen (by difference)	7.8

A possible important source of hydrocarbon is apparently forming at the uppermost surface of the ground, where it might readily be washed away and into the rivers. McDermott* has had the material occurring in the upper half inch of the soil and at depths of 5–7 feet below the surface analyzed, and finds the following average composition (in percent) of the wax obtained from samples covering the whole of southern Taylor County, Texas:

	UPPER HALF INCH	5–7 FEET BELOW SURFACE
Hydrogen	10.6	11.9
Carbon	71.5	75.2
Oxygen	10.8	12.4
Nitrogen	0.0	0.5
Ash	7.1	0.0
Molecular weight	170	475
Melting point	32–36°C	33–38°C

---

* Eugene McDermott in a personal communication.

The waxes are insoluble in water and appear to be attached to the soil particles. The amounts vary widely at the surface, but average about 300 ppm, whereas they vary little in the deeper samples, where the average is about 75 ppm. McDermott is of the opinion that these hydrocarbon waxes are formed from hydrocarbon gases that migrate from below and become concentrated at the surface of the ground, possibly in part as a function of photosynthesis; an alternative possibility is that they are formed at the upper surface of the ground, where the sunlight can reach the miscroscopic plants. The changed character and lower concentration of the deeper waxes may be explained as being due to their having been carried downward by surface waters. If the subaerial surfaces of unconformity of the geologic column were all the loci of hydrocarbon formation such as this, there has been a continuous source of organic matter to be carried to the oceans for as long as there have been sunlight and microscopic plant life on the earth.

Decomposed grass and root hairs in the upper soil contain unsaturated hydrocarbons and all of the saturated hydrocarbons into the C_6 range. The spectrograms from soil samples correspond closely with those of the vegetation, and this suggests that the soil hydrocarbons are at least partly if not wholly due to vegetation, and not to the upward migration of hydrocarbons from buried petroleum accumulations—the basis for geochemical prospecting. It further suggests that the organic matter of the streams that empty into the oceans may contain large amounts of hydrocarbons that require little if any further treatment to become a part of the petroleum hydrocarbons of the sediments.[56]

Cate[57] suggests that *podzolization* or the downward movement of organic matter, metals, and clay through the weathering profile of soils may involve catalytic reactions that transform some of the organic matter into bituminous substances, including petroleum and coal. Although such a process is complex, it appears to be possible, and certainly should be considered as providing a possible source for some petroleum hydrocarbons and hydrocarbon compounds.

Recycled Petroleum. Petroleum hydrocarbons and related hydrocarbon compounds eroded from the land may conceivably be carried by the rivers to the ocean and redeposited along with recycled sedimentary particles in both reservoir and nonreservoir sediments. Eroded petroleum may come from two sources: (1) the oil disseminated through the nonreservoir shales and carbonates and (2) the oil contained in any fields that have undergone erosion since the beginning of Paleozoic time. Such petroleum is in addition to the hydrocarbons that form at the surface of the ground, probably with the aid of photosynthesis, and are later eroded off. Presumably the more stable, heavier hydrocarbon fractions and the asphaltic fractions eroded from the nonreservoir rocks and from the oil pools would be the most susceptible to preservation and redeposition; this might help explain why the oil in the younger Tertiary rocks is generally heavier (of lower gravity). The oil-wet properties of some clays and of insoluble organic matter may have provided a protective carrier mechanism by which these hydrocarbons were transported from areas of erosion to the oceans. The volume of such petroleum source material is vast beyond comprehension, and it seems reasonable to speculate that at least some of it escaped the various destructive forces that prevail between the areas of erosion and the sea and became reincorporated in sediments.

TRANSFORMATION OF ORGANIC MATTER INTO PETROLEUM

In addition to the hydrocarbons in living organisms, there are hydrocarbon compounds that are petroleum-like in their composition and could readily and easily be transformed into petroleum hydrocarbons. Transformation of some of these petroleum-like hydrocarbons probably accounts for a part of the petroleum supply, and such transformation most likely occurs early, or by the time diagenesis of the sediments is complete, for we find petroleum widely disseminated through the nonreservoir sediments. These various hydrocarbons and hydrocarbon compounds occur in minute quantities per acre-foot of nonreservoir sediments, but the volume of such sediments is so vast that the escape of but a fraction of them would be adequate to supply the petroleum of the world many times over.

The transformation of a primary petroleum-like organic material into petroleum begins with its deposition in the sediments in a reducing environment. The transformation process requires energy, and several possible sources of energy are available. These include: (1) bacterial action, (2) heat and pressure, (3) catalytic reactions, and (4) radioactive bombardment.

Bacterial Action

The well-known, everyday decomposition of organic matter of all kinds by bacterial fermentation, and the evolution of methane as one of the products of such decay, are good reasons for considering bacterial action as an agency in the formation of other petroleum hydrocarbons as well. Many organic chemical reactions are brought about as readily by bacteria in nature as by chemists in the laboratory. Some bacteria require free oxygen (aerobic bacteria); others use combined oxygen and cannot live in the presence of free oxygen (anaerobic bacteria); and still others can live either in the presence or in the absence of free oxygen (facultative bacteria). Bacteria grow readily under a wide variety of temperatures and pressures, in fresh water and in brines, in soils, streams, lakes, and swamps. The rapid decay that takes place in an oxidizing environment, as in contact with the atmosphere, is largely due to the rapid growth of aerobic bacteria. The slow decay or absence of decay in a reducing environment is largely due to the low bacterial population resulting from the low oxygen supply. Available oxygen is supplied through air and carbonic acid dissolved in water. The supply of free oxygen diminishes rapidly below the surface of the sea floor, and anaerobic bacteria become important reducing agents during the diagenesis of the sediments, and possibly long after burial. The presence of sulfate-reducing anaerobic bacteria in wells[59] is indicative of continuing reduction after burial.

Bacteria are thought to function in several ways in aiding the final transformation of organic decay products into petroleum. The evidence for this view is derived in part from results in the laboratory and in part from natural phenomena observed in the field. Investigators differ in regard to the efficiency of bacterial action; some believe that it can completely transform organic matter into petroleum, while others give it credit only for slight changes that merely make the original organic material more petroleum-like. Bacteria, for example, are able to deoxygenate certain kinds of organic matter (decarboxylation), to increase its nitrogen content (ammonification), and to remove the sulfur in producing hydrogen sulfide. Laboratory experiments to date have shown that bacteria are able to produce only methane from organic matter; various investigations are still in progress to determine if bacteria can produce higher hydrocarbons.

Most of the recent work on the relation of bacteria to the origin of petroleum has been done by ZoBell and his associates at the Scripps Institution of Oceanography, at La Jolla, California.[60] ZoBell finds large populations of living bacteria in ocean bottom deposits. Thousands of viable bacteria per gram of bottom sediments were recovered from material more than twenty feet below the sea floor. These bacteria or their enzymes (which act as organic catalysts) are capable of producing many chemical changes in organic material.

Geographic and geologic conditions have an effect on the distribution and type of microorganism[61] and determine the kind of bacteria found, the kind of organic matter that becomes the food for the organisms, and the kind of microbiological activity that occurs. At 3 to 4 feet below the surface, for example, deltaic clayey muds have a much higher bacterial population than carbonate material.

Numerous bacteria have also become recovered from ancient sediments and from petroleum found at depths of several thousands of feet. There is always a question, however, whether these bacteria may have been introduced either during the drilling of the well or afterward. Bacteria have been found, however, in petroleum produced from a well many years after it was drilled, which indicates that they are growing, at present at least, within oil-bearing formations. ZoBell concludes from the observations he has made on these bacteria that neither salinity, hydrostatic pressures of 150,000 pounds per square inch, nor temperatures of 85°C can prevent bacterial activity in a petroliferous environment. Most of these bacteria are facultative.

In a highly reducing environment bacteria tend to convert the organic remains of plants and animals into substances that are more petroleum-like. They do this by splitting oxygen, nitrogen, sulfur, and phosphorus from various organic compounds. The trend of these reactions[62] is illustrated in Table 11-2. The table shows the effect of the oxidizing zone in the high oxygen content of marine sapropel, which supports the aerobic bacteria, and a loss of oxygen with burial, which suggests that, as the source of air and free oxygen is shut off, there is a change from aerobic to anaerobic bacteria.

TABLE 11-2 Effects of Oxidation in Various Zones (in percent)

Type of Material	Carbon	Hydrogen	Oxygen	Nitrogen	Phosphorus
Marine sapropel	52	6	30	11	0.8
Recent sediments	58	7	24	9	0.6
Ancient sediments	73	9	14	0.3	0.3
Crude oil	85	13	0.5	0.4	0.1

Bacteria not only may have a direct part in the origin of organic matter and its transformation into petroleum hydrocarbons, but may promote its development in many other ways. Some of these are as follows:

Evolution of Hydrogen. Fermentation of organic matter in the absence of free oxygen may result in the liberation of appreciable quantities of hydrogen, according to ZoBell.[63] The liberation of hydrogen has been accomplished experimentally. In anaerobic bacterial fermentation that produces no free hydrogen, the reason for the absence of free hydrogen may be (1) that hydrogen is activated and united with carbon dioxide to produce methane, as in the equation

$$CO_2 + 4H_2 \longrightarrow CH_4 + 2H_2O$$

or (2) that certain bacteria cause hydrogen to reduce some sulfate to hydrogen sulfide, as in the equation

$$SO_4 + 5H_2 \longrightarrow H_2S + 4H_2O$$

or (3) that bacteria activate the hydrogenation of unsaturated organic compounds.

All these reactions are known to be carried out by anaerobic bacteria that occur in marine sediments. In marine sediments, therefore, one would expect to find the reduction products (methane, hydrogen sulfide, or saturated hydrocarbons) rather than free hydrogen. It may be significant in this regard that unsaturated compounds are not found in petroleum.

Release of Oil from Sedimentary Rocks. There are several mechanisms by which bacteria might liberate the oil held in oil-bearing sediments. (1) One mechanism is the solution of carbonates by H_2CO_3 and organic acids produced by bacteria; the oil contained in limestones and dolomites is liberated when channels and pores are formed by solution of the carbonates. (2) Carbon dioxide produced by bacteria tends to promote the movement of oil by decreasing its viscosity. (3) The carbon dioxide produced by bacteria *in situ* develops an internal gas pressure, which tends to drive oil from dead-end pockets and through interstitial spaces. (4) Some bacteria are thought to have a greater affinity for solid surfaces than oil does, so that they crowd it off. They are called *thigmotactic bacteria.* Some bacteria produce detergents, or surface-active substances, that liberate oil from solid surfaces, presumably by changing the interfacial tensions.

Bacterial Release of Oil from Plants. Some plants synthesize appreciable amounts of hydrocarbons, and selective decomposition of the plant material by bacteria may release these hydrocarbons for eventual accumulation in the sediments.

Bacterial Oxidation. Most kinds of petroleum hydrocarbons are susceptible to oxidation by bacteria under certain conditions. Many microorganisms that are known to utilize hydrocarbons are commonly found in marine sediments, and they are particularly abundant in the soil around storage tanks, oil wells, and oil and gas seepages. Oil spread on the ground may be completely destroyed in a few months, whereas under water it may retain its physical characteristics for years. Long-chain aliphatic and paraffinic compounds are oxidized more readily than corresponding aromatic and naphthenic compounds.

Heat and Pressure

Heat combined with pressure, or pressure alone, has been advanced as a means of transforming organic matter into petroleum. The reaction may or may not be aided by the presence of catalysts. (See pp. 524–526.) Heat and pressure occur together in rocks in varying degrees; if it could be shown that they were able to accomplish the transformation, they would offer a simple and ready agency. Rich early decided that "the oil was distilled from the carbonaceous rocks in what was, virtually, a giant high-pressure cracking still, and that it found easiest escape laterally along the bedding, being driven out to cooler zones by the gases generated during the distillation." [64] Experiments by Seyer led him to conclude: "The transformation of the waxy substance can then be regarded as a sort of low-temperature, high-pressure cracking process in which production favors cyclic hydrocarbons." [65]

Solid organic matter, such as that found in kerogen shales and "source rocks," is a pyrobitumen; that is, heat is required—a temperature of 350–400°C (662–752°F)—to break it down into gaseous and liquid substances. Doubt has been raised about the generally accepted conclusion that the presence of porphyrins in petroleum is evidence that the temperature of petroleum has never exceeded 200°C (392°F).[66] Time may replace temperature; but even so, if there is any validity to the porphyrin thermometer, 200°C is the maximum allowable temperature. Reservoir temperatures up to 235°F

(113°C) have been measured in oil pools, and a high recorded underground temperature—found in the Frontier sand (Cretaceous) of Wyoming, at a depth of 20,521 feet, in a test well * that was not productive of oil or gas—is 325°F (163°C). Most oil-producing reservoirs have temperatures less than 200°F (93.2°C), and in many reservoirs this temperature has probably never been exceeded.

The explanation offered by Treibs[67] and others,[68] for example, is that time may replace temperature—that some reactions, if given a geologically long period of time, will occur at temperatures lower than those that are necessary in the laboratory. Maier and Zimmerly found that the conversion of the kerogen in a kerogen shale into a bitumen soluble in carbon tetrachloride behaved as a chemical reaction of the first order, which means that it was dependent, not solely on temperature, but also on time. In other words, the reaction would take place at any temperature, but the lower the temperature the longer the time required; an extremely slow but finite reaction may have produced a lot of product in the course of geologic time. If time has not accounted for a complete transformation at the low temperatures of the sediments, the transformation of only a minute percentage of the organic matter available could add a vast amount of petroleum hydrocarbons to those already available; time is the imponderable, as it is in many geologic problems.

Although kerogen may not be broken down into petroleum hydrocarbons, there are many liquid hydrocarbons and petroleum-like hydrocarbon compounds mixed with and attached to the solid matter that may be transformed into petroleum, and heat and pressure may either make or help make such transformation possible. The kerogen, then, is chiefly a nonpetroleum material left behind as a residue.

When folding or rock deformation occurs in sedimentary rocks, the constituent particles of the rock transmit the pressures through their contacts with one another. According to the character of the rock (its clay content, its soluble material, the character of its particles, etc.), its porosity may be increased (dilatancy) or decreased (compaction and cementation). At the points of particle contact, however, the pressures are momentarily great, and the particles glide over one another, so that shearing may occur and the temperature may rise. Fash[69] has called this last effect "skin frictional heat" and suggests it as the source of heat that may bring about the transformation of minute amounts of organic matter into more petroleum-like substances or even into petroleums.

Alteration of Petroleum by Heat and Pressure

There is empirical evidence suggesting that once crude oil has been formed, various changes in its composition and gravity occur along with changes in

---
* Superior Oil Company, Pacific Creek No. 1. Sec. 27 T. 27 N. R. 10W Sublette County, Wyoming.

pressure and temperature. These observable alterations suggest that at least part of the original transformation of organic matter was into asphaltic and naphthenic compounds, which were later further altered by pressure and heat. How the original transformation may have occurred, however, is not known. The evidence for these changes in petroleum character is of two kinds: (1) the changes in composition occurring with increasing depth of burial, presumably as a result of increasing pressure and rising temperature, and first studied by Barton[70] and later by Hunt; (2) changes in gravity and character as a result of regional metamorphism, presumably also a result of increased pressure and higher temperature, and first proposed by David White as the carbon-ratio theory.[71] (See also pp. 630–634.)

Changes in Composition. The progressive change in the character of petroleum that occurs with the increase in pressure and rise in temperature accompanying burial has been studied by Barton,[70] Brooks,[72] Pratt,[14] and Hlauscheck,[73] who all believed that the first crude oil formed—the protopetroleum —was naphthenic and asphaltic in character and heavier than the final oil evolved from it. A chart showing the changes in composition with depth of a group of Gulf Coast Tertiary crude oils is given in Figure 11-1; it suggests that the shallow crudes are closer to the composition of organic matter than are the deeper crudes with their increasing paraffin content. The chart also suggests that, if progressive changes such as these could occur with an increase in pressure and a rise in temperature, and if we extrapolated upward to near the surface of deposition, we might find that even the lower temperatures and pressures prevailing there might have initiated the transformation of at least some of the fresh organic material into petroleum or petroleum-like substances. The change in composition that accompanies the rise in pressure and temperature resulting from increasing depth of burial has been termed "maturing" of the oil.[74] While this chart and others like it suggest that the changes in the character of oil are a function of the increasing depth of burial, the same changes may also be correlated with facies, the higher gravities accompanying a change from continental, shallow-water, and sandy facies to marine, deeper-water, and shaly facies.[75]

There are many exceptions to the progressive and regular increase in lower boiling-point fractions with depth (see also p. 197) but in most of those cases the geologic section is broken by unconformities, hiatuses, and other disturbances, or by abrupt changes in facies with depth, that could account for the difference. For the studied examples of Gulf Coast oils, the geologic column is nearly complete and represents almost continuous deposition down to the Recent. For that reason the over-all changes are significant, for they indicate the normal undisturbed effects of increasing pressure and temperature, even though facies changes may cause some irregularities in the orderly arrangement. Even the Recent sediments near the surface of the water-sediment contact in the Gulf of Mexico suggest an early change, with depth or possibly

THE ORIGIN OF PETROLEUM [CHAPTER 11] 523

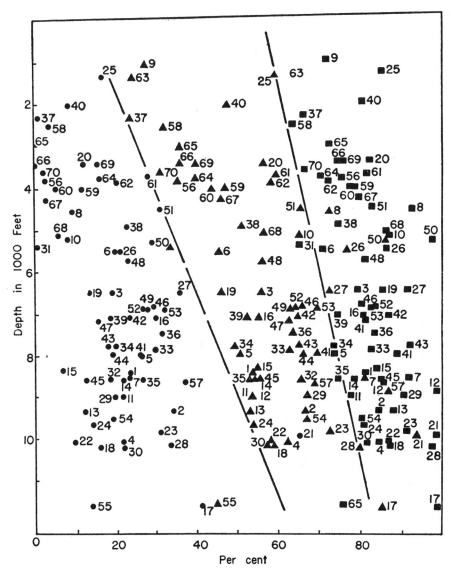

FIGURE 11-1 *Changes that occur with depth in seventy different crude oils of Miocene age from Louisiana. The small circles represent gasoline and naphtha, the distance from the left line being their percentage; the distance from the circles to the triangles represents the percentage of kerosene and gas oil; the distance from the triangles to the squares represents the percentage of lubricating oil distillates; the distance from the squares to the right line represents the percentage of heavy residue. Note the progressive increase in naphthas and gasolines and the decrease in heavier fractions as the oils become deeper.* [Redrawn from Brooks, Bull. Amer. Assoc. Petrol. Geol., Vol. 33 (1949), p. 1604, Fig. 3.]

with age, in the direction of more petroleum-like material.[10] In the top 3–4 feet of sediment, for example, hydrocarbons made up only 7.5 percent of the organic matter extracted, of which 92.5 percent was in the form of asphaltic and complex organic matter, whereas in the samples at 102–103 feet the proportion of hydrocarbons had increased to 30.9 percent and that of asphaltic and complex organic matter had decreased to 69.1 percent.

Hunt[76] has shown that in Wyoming, where only Mesozoic and Paleozoic oils were analyzed, the API gravity increased with depth. In the Tensleep oils, for example, where all of the geologic factors have been comparatively constant except the depth of burial, the deeper oils contained more gasoline and the shallower oils contained more of the heavy residuum.

Changes in Gravity. If the carbon-ratio theory, as originally proposed by David White and later further amplified by him,[77] is correct, and only the application is at fault, then there is reason to believe that low degrees of heat and pressure, as evidenced by the carbon ratios of the affected coals, determine the state of decomposition of the petroleums of any area. The carbon-ratio theory points to the progressive transformation of organic matter into petroleum by low temperatures and pressures, the end point being the formation of the low-boiling hydrocarbons found in natural gas. A more complete discussion of the carbon-ratio theory is deferred to Chapter 14, for its chief importance is its influence on exploration.

Catalytic Reactions. Catalysts are substances that aid or accelerate a chemical reaction but take no part in the reaction; they remain the same in composition at the completion of the reaction as at the beginning. Organic catalysts are called *enzymes*. Catalysis is a complex and essentially a free-surface-energy phenomenon. One method of increasing catalytic action, therefore, is to increase the surface area, either by using a more finely divided catalyst or by increasing the pore surface of the material catalyzed. The atoms that are adsorbed on the surface layers of a catalyst are stimulated into unusual activity by the atoms of the catalyst, and, when atoms of two or more substances are adsorbed by the catalyst, they interact to form new combinations. This is why, the greater the catalytic surface exposed, the more rapidly the reaction goes forward. The large surface areas of fine particles in reservoir rocks, together with the complex chemical substances found there, suggest catalytic effects as an aid in the transformation of organic matter into petroleum, or at least into petroleum-like materials. A stronger reason, however, is that catalysts, at the low temperatures found in the reservoir rocks, may promote reactions that would not otherwise go forward except at high temperatures. In refinery operations, catalysts are exceedingly useful and work in many ways. They should be considered as possible tools in the natural transformations and alterations of organic matter into petroleum.

Thermal cracking in refinery operations—without catalysts—involves the treatment of petroleum stocks at temperatures in the range 850–1,150°F (454–620°C) and at pressures from atmospheric pressure up to 1,600 psi. But with catalysts the reactions are accelerated, the temperatures are reduced to 750–1,000°F (399–538°C), and pressures are reduced to 100 psi and less. Reactions that take place to various degrees in refinery catalytic processing include cracking, polymerization, alkylation, aromatization, isomerization, hydrogenation, dehydrogenation, and cyclization.[78] Such reactions occur quite rapidly and are complex. The catalysts used are clays, synthetic pellets or beads, and synthetic "fluid" catalysts.

Some catalysts may occur as compounds within the petroleum. For example, vanadium, molybdenum, and nickel are common elements in the ash from crude oils, and certain salts of these elements are effective catalysts for the laboratory synthesis of hydrocarbons. Presumably these elements were extracted from sea water by organisms, and would be available for catalytic action during the entire decomposition period of the organic matter.

The organic matter associated with clay sediments may occur as discrete particles mechanically mixed with the clay particles, or it may occur as molecules adsorbed on the lattice planes of the clay minerals, especially the basal cleavage surfaces.[79] The effect of the clay minerals on the organic matter is thought to be catalytic and due to the close molecular structural similarity between some of the clay minerals and some of the organic compounds. Adsorption of organic matter by the clays, and replacements within the lattice structure, provide a possible means of changing organic matter in the direction of petroleum. This phenomenon may have wide significance as a mechanism whereby buried organic material is transformed into petroleum or petroleum-like substances.

A common application of catalytic adsorption is found in the use of Fuller's earth to change the character of many oils. Fuller's earth is a highly siliceous clay consisting essentially of hydrous aluminum silicates. It has the property of selectively adsorbing certain organic coloring matters from vegetable and mineral oils, and is used in many petroleum refinery operations. Its selective adsorptive action is thought to be largely due to the surface activity of its particles, which are of colloidal size.[80] Other commercial adsorptive materials are acid-treated bentonites, dehydrated silica gels, and bauxite. Gayer[81] has pointed out that laboratory experiments with rock samples from many geologic formations, such as the Sylvan shale (Silurian) of Oklahoma, the "green sands" (Cretaceous) of New Jersey and Texas, the Stanley shale (Mississippian) of Oklahoma, the Reagan sandstone (Cambrian) of Texas, and the Calvin sandstone (Pennsylvanian) of Oklahoma, showed that they all aided materially in the polymerization of propylene (C_3H_6) at a temperature of 35°C, which is well below the ordinary cracking temperatures. A number of naturally occurring substances, therefore, are known to have catalytic adsorptive properties.

Two points especially suggest that catalytic action is an important agent at some point in the transformation of organic matter into petroleum, or at least into petroleum-like substances: the general absence of olefins from crude oil, and the general presence of aromatics (benzenes) in crude oil.[82] The olefins that are present in organic matter form paraffins in the presence of catalysts[83] at temperatures below those used in cracking. Because of the absence of olefins from petroleum, it is assumed that they were once present but were hydrogenated to paraffins. Thus, if any hydrogenation occurred, it was catalytic hydrogenation. Clays are usually present and could act as catalysts in reservoir rocks. The benzenes, on the other hand, which do not occur in marine organic matter but are present in humic acid, are formed experimentally from paraffins by active catalysts at temperatures as low as 80°C. Francis[84] has pointed out that, at temperatures below 550°C, reactions without catalysts are of an entirely different kind than reactions with catalysts and would never produce aromatics from paraffins even in geologic time. Erdman[95] concludes that there were many precursors of the aromatic hydrocarbons[85] and that dehydrogenation and a slight change in the arrangement of the bonds, for example, would be sufficient to change terpene into one of the low-molecular-weight aromatic hydrocarbons.

A way in which water-soluble organic matter might be transformed into petroleum hydrocarbons at temperatures below 135°C is suggested by the work of Whitehead and Breger.[86] When a water-soluble fraction of organic matter from Recent muds was collected along the coast of Cuba and was heated up to 135°C, it gave off a complex mixture of gases composed of C_4, C_5, and C_6 hydrocarbons. The mechanism by which this low-temperature pyrolysis of marine organic matter yielded hydrocarbons is unknown, but the presence of catalysts in the mud is suggested as a possibility.

Nutting[87] has described another catalytic effect that may have a bearing on the transformation processes. When vegetable and animal tissues are in intimate contact with finely divided silica and silicates and with some oxides, such as those of aluminum and iron, they are subject to strong disintegrating forces because of the tendency of some of the hydrogen and hydrocarbon radicals to be removed. This makes them oil-wet instead of water-wet, and the products of the reaction, being soluble, tend to hydrolyze. Pressure and shearing increase the rate of disintegration. This process is believed to be active in the breaking down of organic matter for the preparation of plant food by the soils, and it may explain reactions that have sometimes been attributed to bacterial action.

Radioactive Bombardment. The widespread occurrence of radioactive minerals in the earth, together with the known chemical reactions that result from radioactive bombardment, makes it appear possible that some radioactive phenomena, in addition to being a possible source of heat, may also aid or cause the transformation of organic matter into petroleum. Lind and Bardwell[88] first suggested that the small amounts of radioactive material distributed throughout the earth might have a direct influence on the formation of petroleum, and their idea has received increasing attention from a number of investigators. Much of the material that follows is taken from articles by Bell, Goodman, and Whitehead,[89] Beers,[90] Lind,[91] Russell,[92] Shepard,[93] and Beers and Goodman.[94]

The atomic nuclei of naturally radioactive elements, such as uranium, thorium, and potassium, are unstable; they decay by spontaneous disintegration through several transformations until the final, stable end-product is reached. During the decay, some atoms always emit high-velocity electrons known as *beta particles,* while other atoms emit high-velocity helium nuclei known as *alpha particles.* Alpha particles account for more than 75 percent of the energy released by the terrestrial radioactive elements.[95] *Gamma rays* are emitted after the emission of alpha and beta particles; they are high-penetration, electromagnetic waves similar to but shorter than x-rays. Gamma-ray well logging measures the natural gamma-ray emissions of the formations drilled. (See also pp. 81–83.)

Within the disintegration series of uranium 238 (atomic weight), which is probably the best-known of the naturally radioactive elements, uranium 234, ionium 230, and radium 226 are produced. The disintegration of radium 226 produces radon 222, one of the inert, stable, noble gases, which is also radioactive, with a half-life of approximately four days. Radon emits alpha rays during its disintegration and is an excellent source of alpha particles for laboratory purposes. Radium 218, radium 214, radium 210, polonium 210, and finally lead 206 are produced. Radon and uranium reach equilibrium after about thirty days, when the rate of decay of uranium is the same as the rate of decay of radon. The amount of radon remains constant because as many radium atoms are breaking down to form radon as are radon atoms breaking down to form radium 218. Radon is more stable in organic liquids, such as petroleum, than it is in water, and crude oils commonly contain radon in excess of the amount to be expected from the disintegration of the quantity of uranium present in the crude oils. The kinetic energy of the alpha particles emitted in the disintegration of radium accounts for approximately 90 percent of the energy liberated. When these particles penetrate matter until they come to rest, the energy appears as heat. The distance an alpha particle penetrates matter, or its range, is a few centimeters in gases and around one-thousandth of that distance in solids and liquids. As the alpha particle passes through matter, ionization of the atoms and chemical changes in the material take place.[96]

The important radioactive elements found in the sedimentary rocks are uranium, thorium, and potassium.[90,94] They occur as follows: (1) associated with the heavy minerals of the sands and sandstones; (2) as the active isotope (K_{40}) of potassium, which is found in evaporites, oil-field brines, clays, shales, and potassium-bearing minerals; (3) in the uranium and thorium content of clays, shales, impure limestones, and organic substances. Beers found that pure limestones and pure quartzites exhibit practically no radioactivity, while black organic shales contain high concentrations of the three principal radioactive elements. Russell,[92] who examined the radioactive content of 510 samples of sedimentary rocks, found that marine shales are highly radioactive as compared with other sediments, that the Paleozoic shales average higher in radioactivity than the Cenozoic shales, and that there is more contrast between the different strata in the older formations than in the younger.

Bombardment of saturated fatty acids having the formula RCOOH [an example is palmitic acid ($C_{15}H_{31}COOH$)] by alpha particles produced paraffin hydrocarbons.[97] Subsequently bombardment of a naphthenic acid (cyclohexanecarboxylic acid) by alpha particles produced a cyclic hydrocarbon (cyclohexane).[98] These results are significant, for certain fatty acids have been recognized in the organic matter of sediments. The efficiency of the process is low, however, and the rate of conversion is small, which indicates that it would require an even geologically long time to produce important amounts of hydrocarbon in this way. Alpha-particle bombardment of methane and other gaseous hydrocarbons in the laboratory has yielded large percentages of hydrogen and unsaturated hydrocarbons.[99] Bombardment of liquid hydrocarbons gives similar high percentages of hydrogen and perhaps a higher yield of unsaturated hydrocarbons, but these are not present in crude oil except possibly in very minor quantities.

Lind [100] points out that, in general, the thermodynamic relations between the saturated hydrocarbons involve only small amounts of free chemical energy. This means that under ordinary conditions these hydrocarbons are quite inactive toward one another because there is no driving force to cause interaction. It likewise means that there is no large opposing force to overcome in any reaction, and that heats of reaction are low. If, therefore, suitable energies are available, one may expect these reactions to proceed by successive steps in all directions. Lind reasons that a complex product may thus be formed from a relatively simple one, and that, at the pressures and temperatures of the upper part of the earth's crust, any member of the paraffin series might be transformed into the complex hydrocarbons found in petroleums. He suggests that other energies might operate also, such as electrical discharge, alpha radiation, and ultraviolet radiation. All of these have been found effective in causing hydrocarbons to interact. Lind points out the further interesting fact that all of these energies exhibit the striking ability to condense hydrocarbons of low boiling point into liquids and solids, with only such elimination of gaseous compounds as is necessary to avoid chemical supersaturation. Since electrical discharge and ultraviolet radiation do not occur in the earth's crust, they may be eliminated as possible sources of energy in the transformation of simple hydrocarbons to complex hydrocarbons. Alpha radiation, however, is everywhere present, even though of low intensity.

Objections to Radioactive Transformation Processes. The chief objection to ascribing the transformation of organic matter into petroleum to alpha radiation is that, in the laboratory experiments at least, hydrogen atoms are split off during the reaction.[101] This would cause, in geologic time, the formation of progressively heavier oils with a high ratio of carbon to hydrogen, whereas the change from organic matter to petroleum, in general, calls for a progressive increase in the ratio of hydrogen to carbon. The free hydrogen formed, however, might be utilized in other transformation processes that require additional hydrogen.

Another objection to extensive transformation is found in the occurrence of the highly radioactive black organic shales, a well-known example of which in the United States is the Antrim-Chattanooga-Woodford shales (Lower Mississippian and Upper Devonian).[94] If radioactive transformation processes have been operating since Devonian time, we should expect shales such as these to contain free petroleum in the fractures and minute openings, and little or no organic matter. The organic content, however, is high and is altogether pyrobituminous except in a few areas where natural gas has accumulated, and the oil content is low or missing. The natural gas may, indeed, have been formed by radioactive processes, but the irregularity of its occurrence, contrasted with the uniformly high radioactivity and high organic content of the shales over many states, indicates that it probably originated from some other cause.

CONCLUSION

The problem of the origin of petroleum is losing some of its importance as a prerequisite to exploration for petroleum. The reason is that petroleum and petroleum-like hydrocarbons are found to be nearly universal in the nonreservoir sediments; even a fraction of the residual amounts that still exist in these sediments would be adequate to supply the petroleum of the world. A specific source rock is consequently not necessary; practically any or all of the fine-grained sediments provide source material.

We might summarize our ideas on the origin of petroleum as follows:

1. The hydrocarbons and petroleum-like hydrocarbon compounds that are associated with the kerogen-type organic matter in the nonreservoir shales and carbonates are the chief source of petroleum.

2. These hydrocarbons are similar to those that form within living plants and animals, both marine and terrestrial. In ordinary decay processes, little or no change would be required for them to be converted into petroleum.

3. In addition there are many petroleum-like hydrocarbon compounds that may be transformed into petroleum hydrocarbons and asphaltic fractions by rather common chemical and biochemical agencies.

4. There are several sources of energy available to transform soluble, complex organic materials into petroleum. These include such forms of energy as bacterial action, heat and pressure, catalytic reactions, and radioactivity. Thus,

if needed, there are many potential sources for the more complex petroleum compounds.

5. All of these materials together with petroleum are found disseminated throughout most of the fine-grained nonreservoir sediments. Hence, reactions necessary to form petroleum may have occurred before or during diagenesis of the sediments.

6. The insoluble organic matter, kerogen, is not to be considered as a source material for petroleum. It is more closely related to coal than to petroleum.

7. Petroleum forms in a reducing environment.

8. Petroleum hydrocarbons are currently being deposited, for they are found in the clays and muds of the Gulf of Mexico and of San Francisco Bay.[102]

9. Some petroleum hydrocarbons might be recycled; the erosion of the fine-grained nonreservoir sediments and of oil pools furnishes a vast and continuous supply of petroleum hydrocarbons, some of which may be carried to the sea to be reincorporated in sediments.

Selected General Readings

Benjamin T. Brooks, "The Chemical and Geochemical Aspects of the Origin of Petroleum," in *The Science of Petroleum,* Oxford University Press, London and New York, Vol. 1 (1938), pp. 46–53. 51 references cited. A good discussion of many of the chemical and geochemical conditions surrounding the origin of petroleum.

Parker D. Trask, "Organic Content of Recent Marine Sediments," in *Recent Marine Sediments: A Symposium,* Amer. Assoc. Petrol. Geol., Tulsa, Okla. (1939), pp. 428–453. 67 references cited.

Parker D. Trask and H. Whitman Patnode, *Source Beds of Petroleum,* Amer. Assoc. Petrol. Geol., Tulsa, Okla. (1942), 566 pages. Contains a selected bibliography on the origin of oil, pp. 2–3. Contains the results of examination of 32,000 well samples and 3,000 outcrop samples through seven provinces of the United States. Modern analytical methods have outdated parts of this discussion.

American Petroleum Institute, Dallas, Texas, in its Fundamental Research on the Occurrence and Recovery of Petroleum, publishes a series of annual volumes covering research on various projects sponsored by the Institute. The series begins with 1943. Project 43 has to do with the transformation of organic matter into petroleum, and is divided into:

43A, Bacteriological and Sedimentation Phases
43B, Chemical and Biochemical Phases
43C, Studies of the Effect of Radioactivity on the Transformation of Marine Organic Materials into Petroleum Hydrocarbons.

The reader is especially referred to these publications for the results of much original research on the problem of the origin of petroleum.

J. Gordon Erdman, "Some Chemical Aspects of Petroleum Genesis as Related to the Problem of Source Bed Recognition," in Geochimica et Cosmochimica Acta (1961), pp. 16–36, Pergamon Press, Ltd. This is one of the best of the modern discussions of the origin of petroleum by one of the workers in one of the great research organizations, The Mellon Institute in Pittsburgh, Pennsylvania.

W. G. Meinschein, "Origin of Petroleum," Bull. Amer. Assoc. Petrol. Geol., Vol. 43 (May 1959), pp. 925–943. Discusses the similarities and differences between the naturally occurring hydrocarbons in the shales and other sediments and the crude oils that occur in pools. Water serves as the accumulating agent.

H. M. Smith, H. N. Dunning, H. T. Rall, and J. S. Ball, "Keys to the Mystery of Crude Oil," Amer. Petrol. Inst., New York, 24th Midyear Meeting (May 29th, 1959), 32 pages. Extensive bibliography. This is an authoritative statement, by members of the United States Bureau of Mines, of our knowledge concerning the origin and chemistry of crude oil.

Hollis D. Hedberg, "Geologic Aspects of Origin of Petroleum," Bull. Amer. Assoc. Petrol. Geol., Vol. 48 (November 1964), pp. 1755–1803. An authoritative analysis of the current thinking on the problem of the origin of petroleum. Extensive bibliography.

Reference Notes

1. H. Hofer, *Das Erdöl*, 2nd ed. (1906), pp. 160–229. This, as well as the third edition (1912), contains a complete review of the early theories of the origin of petroleum.
Frank Wigglesworth Clarke, "Inorganic and Organic Theories of the Origin of Oil," in *Data of Geochemistry*, Bull. 770, U.S. Geol. Surv., 5th ed. (1924), pp. 731–755. Contains a review of early theories of the origin of petroleum.

2. A. Beeby Thompson, *Oil-Field Exploration and Development*, 2nd ed., Technical Press, London (1950), Vol. 1, p. 19.

3. Ben B. Cox, "Transformation of Organic Material into Petroleum under Geological Conditions—the Geological Fence," Bull. Amer. Assoc. Petrol. Geol., Vol. 30 (May 1946), pp. 645–659. 68 references listed.

4. Kalervo Rankema, "New Evidence of the Origin of Pre-Cambrian Carbon," Bull. Geol. Soc. Amer., Vol. 59 (1948), pp. 389–416.

5. R. B. Whiteside, "Migration in Lucien Oil Field During Ordovician," Bull. Amer. Assoc. Petrol. Geol., Vol. 20 (1936), pp. 617–619.

6. Geo. Edwin Dorsey, "Preservation of Oil During Erosion of Reservoir Rock," Bull. Amer. Assoc. Petrol. Geol., Vol. 17 (July 1933), pp. 827–842. See Discussion, p. 1273.

7. John M. Hunt and George W. Jamieson, "Oil and Organic Matter in Source Rocks of Petroleum," in *Habitat of Oil*, Lewis G. Weeks (ed.), Amer. Assoc. Petrol. Geol., Tulsa, Okla. (1958), pp. 735–746.
J. Gordon Erdman, "Some Chemical Aspects of Petroleum Genesis as Related to the Problem of Source Bed Recognition," Geochimica et Cosmochimica, Pergamon Press (1961), Vol. 22, pp. 16–36.

Sol R. Silverman and Samuel Epstein, "Carbon Isotopic Composition of Petroleums and Other Sedimentary Organic Materials," Bull. Amer. Assoc. Petrol. Geol., Vol. 42 (May 1958), pp. 998–1012.

W. G. Meinschein, "Origin of Petroleum," Bull. Amer. Assoc. Petrol. Geol., Vol. 43 (May 1959), pp. 937–38.

P. V. Smith, Jr., "Studies on the Origin of Petroleum: Occurrence of Hydrocarbons in Recent Sediments," Bull. Amer. Assoc. Petrol. Geol., Vol. 38 (March 1954), pp. 377–404. Bibliog. 29 items.

8. John M. Hunt and J. P. Forsman, "Relation of Crude Oil Composition to Stratigraphy in the Wind River Basin," Wyoming Geol. Assoc. Guidebook (1957), pp. 105–112.

9. L. G. Weeks, "Factors of Sedimentary Basin Development that Control Oil Occurrences," Bull. Amer. Assoc. Petrol. Geol., Vol. 36 (November 1952), p. 2103.

10. P. V. Smith, Jr., *op. cit.* (note 7), pp. 377 and 383.

11. Albert L. Kidwell and John M. Hunt, "Migration of Oil in Recent Sediments of Pedernales, Venezuela," in *Habitat of Oil,* Lewis G. Weeks (ed.), Amer. Assoc. Petrol. Geol., Tulsa, Okla. (1958), pp. 790–817.

12. M. Florkin, "Aspects of the Origin of Life," *Saturday Review of Literature* (July 6, 1963). Discussion by John M. Haun in Saturday Review of Literature (August 3, 1963), pp. 44–45 and The Mines Magazine, Colo. Sch. Mines, Golden, Colorado (May 1964). Florkin believes that life originates from petroleum and that petroleum has an inorganic origin.

13. E. DeGolyer, "The Effect of Igneous Intrusions on the Accumulation of Oil in the Tampico-Tuxpam Region, Mexico," Econ. Geol., Vol. 10 (1915), p. 651.

14. Wallace E. Pratt, "Hydrogenation and the Origin of Oil," in *Problems of Petroleum Geology,* Amer. Assoc. Petrol. Geol. (1934), pp. 235–245.

Frank Wigglesworth Clarke, *op. cit.* (note 1), p. 48.

15. Kalervo Rankèma and Th. G. Sahama, *Geochemistry,* University of Chicago Press (1950), pp. 185–186.

16. A. Treibs, "Porphyrin in bituminösen Gesteinen und Erdöl-Kohlen: Zur Entstehung des Erdöls," Angew. Chem., Vol. 44 (1936), p. 551.

A. Treibs, "Chlorophyll und Häminderivata in bituminösen Gesteinen, Erdöl, Erdwaschen, und Asphalten," Angew. Chem., Vol. 44 (1936), pp. 683–686, and Annalen, 410, Vol. 42 (1934), p. 517, Vol. 43 (1935), pp. 172–196.

H. N. Dunning and J. W. Moore, "Porphyrin Research and Origin of Petroleum," Bull. Amer. Assoc. Petrol. Geol., Vol. 41 (November 1957), pp. 2403–2412.

17. Parker D. Trask and H. Whitman Patnode, *Source Beds of Petroleum,* Amer. Assoc. Petrol. Geol., Tulsa, Okla. (1942), 566 pages, pp. 32–61.

18. C. Engler, "Die Entstehung des Erdöls," Zeit. für angew. Chemie, Vol. 21 (1908), pp. 1585–1597.

19. Robb, Chas., "Mineral Resources of North America," in *Eighty Years' Progress of British North America,* Stebbins, Toronto (1863), pp. 308–372.

20. J. E. Hackford, "The Significance of the Interpretation of the Chemical Analyses of Springs," Jour. Inst. Petrol. Technol., Vol. 8 (1922), p. 197.

21. J. E. Hackford, "The Chemistry of the Conversion of Algae into Bitumen and Petroleum and of the Fucosite-Petroleum Cycle," Jour. Inst. Petrol. Technol., Vol. 18 (1932), pp. 74–114. Discussion on pp. 115–123.

22. J. McConnell Sanders, "The Microscopical Examination of Crude Petroleum," Jour. Inst. Petrol. Technol., Vol. 23 (1937), pp. 525–573. 30 references listed.

23. P. V. Smith, Jr., *op. cit.* (note 7), p. 382.

24. John M. Hunt and George W. Jamieson, "Oil and Organic Matter in Source Rocks of Petroleum," *Habitat of Oil,* Amer. Assoc. Petrol. Geol., Tulsa, Okla. (June 1958), pp. 735–746.

W. G. Meinschein, "Origin of Petroleum," Bull. Amer. Assoc. Petrol. Geol., Vol. 43 (May 1959), pp. 925–943.

Sol R. Silverman and Samuel Epstein, "Carbon Isotope Compositions of Petroleums and Other Sedimentary Organic Materials," Bull. Amer. Assoc. Petrol. Geol., Vol. 42 (May, 1958), pp. 998–1012.

J. Gordon Erdman, "Some Chemical Aspects of Petroleum Genesis as Related to the Problem of Source Bed Recognition," Geochimica et Cosmochimica, Pergamon Press, Ltd., Vol. 22 (1961), pp. 16–36.

P. V. Smith, "The Occurrence of Hydrocarbons in Recent Sediments from the Gulf of Mexico," Science, Vol. 116 (1952), pp. 437–439.

P. V. Smith, "Studies on Origin of Petroleum: Occurrence of Hydrocarbons in Recent Sediments," Bull. Amer. Assoc. Petrol. Geol., Vol. 38 (1954), pp. 337–404.

N. P. Stevens, E. E. Bray, and E. D. Evans, "Hydrocarbons in Sediments of Gulf of Mexico," Amer. Assoc. Petrol. Geol., Vol. 40 (1956), pp. 975–983. Points out the differences between hydrocarbons deposited in the ocean sediments and those found in petroleum.

Keith A. Kvenvolden, "Hydrocarbons in Modern Sediments and the Origin of Petroleum," The Mines Magazine, Colo. Sch. Mines (February 1964), pp. 24–25.

S. R. Silverman, "Migration and Segregation of Oil and Gas," in *Fluids in Subsurface Environments*, Amer. Assoc. Petrol. Geol. (1965).

Ellis E. Bray and Ernest D. Evans, "Hydrocarbons in Non-Reservoir-Rock Source Beds," Bull. Amer. Assoc. Petrol. Geol., Vol. 49 (March 1965), pp. 248–257.

25. Beveridge J. Mair, "Terpenoids, Fatty Acids and Alcohols as Source Materials for Petroleum Hydrocarbons," Geochimica et Cosmochimica, Vol. 28, Pergamon Press, Ltd. (1964), pp. 1303–1321.

26. J. Gordon Erdman, "Some Chemical Aspects of Petroleum Genesis as Related to the Problem of Source Bed Recognition," Geochimica et Cosmochimica, Vol. 22, Pergamon Press, Ltd. (1961), pp. 16–36.

27. H. N. Dunning and J. W. Moore, "Porphyrin Research and Origin of Petroleum," Bull. Amer. Assoc. Petrol. Geol., Vol. 41 (November 1957), pp. 2403–2412.

28. J. P. Forsman and John M. Hunt, "Insoluble Organic Matter (Kerogen) in Sedimentary Rocks," Geochimica et Cosmochimica Acta, Vol. 15, Pergamon Press, Ltd. (1958), pp. 170–182.

29. H. U. Sverdrup, M. W. Johnson, and R. H. Fleming, *The Oceans, Their Physics, Chemistry and General Biology*, Prentice-Hall, New York (1942), 1087 pages. Extensive bibliographies.

30. *Ibid.,* p. 1013.

31. Bibliography on "Isolation of Hydrocarbons from Natural Plant Sources," Research on the Occurrence and Recovery of Petroleum, Amer. Petrol. Inst. (1944–1945), Appendix D, pp. 154–155.

Keith A. Kvenvolden, "Hydrocarbons in Modern Sediments and the Origin of Petroleum," The Mines Magazine, Colo. Sch. Mines (February 1964), pp. 24–25.

32. H. U. Sverdrup, M. W. Johnson, and R. H. Fleming, *op. cit.* (note 29), pp. 937–939.

33. Gordon A. Riley, "The Carbon Metabolism and Photosynthetic Efficiency of the Earth as a Whole," Amer. Sci., Vol. 32, No. 2 (April 1944), p. 134.

34. H. U. Sverdrup, M. W. Johnson, and R. H. Fleming, *op. cit.* (note 29), p. 763, quoting Hans Lohman.

35. J. E. Hackford, *op. cit.* (note 21).

Fred B. Phleger, Jr., and Claude C. Albritton, Jr., "Diatoms as a Source for California Petroleum: A Summary Review," Field and Laboratory, Vol. 6 (November 1937), pp. 25–32. Bibliog. 35 items.

36. F. J. Pettijohn, *Sedimentary Rocks,* Harper & Brothers, New York (1949), p. 163.

37. F. M. Anderson, "Origin of California Petroleum," Bull. Geol. Soc. Amer., Vol. 37 (1926), pp. 585–614.

C. F. Tolman, "Biogenesis of Hydrocarbons by Diatoms," Econ. Geol., Vol. 22 (1927), pp. 454–474.

G. Dallas Hanna, "An Early Reference to the Theory that Diatoms Are the Source of Bituminous Substances," Bull. Amer. Assoc. Petrol. Geol., Vol. 12 (1928), pp. 555–556.

38. H. U. Sverdrup, M. W. Johnson, and R. T. Fleming, *op. cit.* (note 29), pp. 297 and 765.

39. A. Mann, "The Economic Importance of the Diatom," Ann. Rept. Smithsonian Inst. (1916), pp. 386–397.

40. L. B. Becking, C. F. Tolman, H. C. McMillin, John Field, and Tadaechai Hashimoto, "Preliminary Statement Regarding the Diatom 'Epidemics' at Copalis Beach, Washington, and an Analysis of Diatom Oil," Econ. Geol., Vol. 22 (1927), p. 366.

41. Reinhardt Thiessen, *Origin of the Boghead Coals,* Prof. Paper 132, U.S. Geol. Surv. (1925), pp. 121–135.

42. C. F. Tolman, *op. cit.* (note 37), pp. 360–361.

43. Paul S. Galstoff, "The Mystery of the Red Tide," Scientific Monthly, Vol. 68 (February 1949), pp. 109–117.

44. Margaretha Brongersma-Sanders, "The Importance of Upwelling Water to Vertebrate Paleontology and Oil Geology," Verh. Kon. Nederlandsch Akad. van Wetensch. Amsterdam, Afd. Nat., Sec. 2, D1 45, 4 (1948), 112 pages.

45. Nature, Vol. 129 (1932), p. 660.

46. F. M. Van Tuyl, Ben H. Parker, and W. W. Skeeters, "The Migration and Accumulation of Petroleum and Natural Gas," Quart. Colo. Sch. Mines, Vol. 40 (January 1945), pp. 15–16, quoting W. H. Curry. Describes oil-saturated interiors of *Gryphaes* enclosed in a dense white limestone found in cores.

47. Thomas F. Stipp, "The Relation of Foraminifera to the Origin of California Petroleum," Bull. Amer. Assoc. Petrol. Geol., Vol. 10 (1926), pp. 697–702.

48. B. Silliman, Jr., "On the Chemical Composition of the Calcareous Corals," Amer. Jour. Sci., Series 2, Vol. 1 (1846), pp. 189–199.

49. Frank Wigglesworth Clarke and Walter Calhoun Wheeler, *The Inorganic Constituents of Marine Invertebrates,* Prof. Paper 124, U.S. Geol. Surv. (1922), 62 pages. 82 references listed.

50. W. Bergmann and D. Lester, "Coral Reefs and the Formation of Petroleum," Science, Vol. 92 (1940), pp. 452–453.

51. Frank Wigglesworth Clarke, *op. cit.* (note 1), p. 110.

52. Amadeus W. Grabau, *Principles of Stratigraphy,* 3rd ed., A. G. Seiler, New York (1932), p. 173.

53. Chester W. Washburne, "Discussion of Colin C. Rae's 'Organic Material of Carbonaceous Shales'," Bull. Amer. Assoc. Petrol. Geol., Vol. 7 (1923), pp. 440–442.

54. W. A. Selvig, W. H. Ode, B. C. Parks, and H. J. O'Donnell, *American Lignites: Geological Occurrence, Petrographic Composition, and Extractable Waxes,* Bull. 482, U.S. Bur. Mines (1950), 63 pages.

55. *Ibid.,* p. 36.

56. Gerould H. Smith and Max M. Ellis, "Chromatographic Analysis of Gases from Soils and Vegetation, Related to Geochemical Prospecting for Petroleum," Research Center, Union Oil Company of California.

57. Robert B. Cate, Jr., "Can Petroleum Be of Pedogenic Origin?" Bull. Amer. Assoc. Petrol. Geol., Vol. 44 (April 1960), pp. 423–432. Contains nearly 100 references.

58. F. W. Went, "Organic Matter in the Atmosphere, and Its Possible Relation to Petroleum Formation," Proceed. Nat. Acad. Sci., Vol. 46 (February 1960), pp. 212–221.

F. W. Went, "Thunderstorms as Related to Organic Matter in the Atmosphere," Proceed. Nat. Acad. Sci., Vol. 48 (March 1962), pp. 309–316.

59. Edson S. Bastin, "The Problem of the Natural Reduction of Sulphates," Bull. Amer. Assoc. Petrol. Geol., Vol. 10 (1926), pp. 1270–1299.

Claude E. ZoBell and Sydney A. Rittenberg, "Sulphate-Reducing Bacteria in Marine

Sediments," Research on Occurrence and Recovery of Petroleum, Amer. Petrol. Inst., Dallas, Texas (1948–1949), pp. 161–176. 71 references.

60. Claude E. ZoBell, "The Role of Bacteria in the Formation and Transformation of Petroleum Hydrocarbons," Science, Vol. 102 (October 12, 1945), pp. 364–369.

61. Gordon P. Lindblom and Marcia D. Lupton, "Microbiological Aspects of Organic Geochemistry," Developments in Industrial Microbiology, Vol. 2, Plenum Press, Inc., New York (1961), pp. 9–22.

62. Claude E. ZoBell, "Influence of Bacterial Activity on Source Sediments," Research on Occurrence and Recovery of Petroleum, Amer. Petrol. Inst., p. 109 (1943) and p. 69 (1944–1945).

63. Claude E. ZoBell, "Microbial Transformation of Molecular Hydrogen in Marine Sediments, with Particular Reference to Petroleum," Bull. Amer. Assoc. Petrol. Geol., Vol. 31 (October 1937), pp. 1709–1751. Bibliog. 200 items.

64. John L. Rich, "Generation of Oil by Geologic Distillation During Mountain-Building," Bull. Amer. Assoc. Petrol. Geol., Vol. 11 (November 1927), p. 1139.

65. W. F. Seyer, "Conversion of Fatty and Waxy Substances into Petroleum Hydrocarbons," Bull. Amer. Assoc. Petrol. Geol., Vol. 17 (October 1933), p. 1251; Jour. Inst. Petrol. Technol., Vol. 19 (1943), pp. 773–783.

66. J. Gordon Erdman, Virginia G. Ramsey, and William E. Hanson, "Volatility of Metallo-porphyrin complexes," Science, Vol. 123, No. 3195 (1956), p. 502.

67. A. Treibs, *op. cit.*

68. C. G. Maier and F. R. Zimmerly, "The Chemical Dynamics of the Transformation of the Organic Matter to Bitumen in Oil Shale," Bull. Univ. Utah, Vol. 14, No. 7 (1924), pp. 62–81.

David White, "Exchange of Time for Temperature in Petroleum Generation," Bull. Amer. Assoc. Petrol. Geol., Vol. 14 (September 1930), pp. 1227–1229.

P. H. Abelson, "Organic Geochemistry and the Formation of Petroleum," 6th World Petroleum Congress, Frankfort, Germany (June 1963), Sec. I, Paper 41, 9 pages. Preprint.

69. Ralph H. Fash, "Theory of Origin and Accumulation of Petroleum," Bull. Amer. Assoc. Petrol. Geol., Vol. 28 (October 1944), pp. 1510–1518.

70. Donald C. Barton, "Natural History of the Gulf Coast Crude Oil," in *Problems of Petroleum Geology*, Amer. Assoc. Petrol. Geol., Tulsa, Okla. (1934), pp. 109–155.

71. David White, "Some Relations in Origin Between Coal and Petroleum," Jour. Wash. Acad. Sci., Vol. 5 (1915), pp. 189–212.

72. Benjamin T. Brooks, "Active Surface Catalysts in Formation of Petroleum," Bull. Amer. Assoc. Petrol. Geol.: Part I, Vol. 32 (December 1948), pp. 2269–2286, with 33 references; Part II, Vol. 33 (September 1949), pp. 1600–1612, with 15 references.

73. Hans Hlauscheck, *Naphthene and Methane Oils, Their Geological Occurrence and Origin*, Ferdinand Enke, Stuttgart, Germany (1936); reviewed in Bull. Amer. Assoc. Petrol. Geol., Vol. 20 (1936), pp. 1499–1501.

74. Walter K. Link, "Approach to the Origin of Oil," O. & G. Jour., March 16, 1950, pp. 88–92.

75. Fred R. Haeberle, "Relationship of Hydrocarbon Gravities to Facies in Gulf Coast," Bull. Amer. Assoc. Petrol. Geol., Vol. 35 (October 1951), pp. 2238–2248.

76. John M. Hunt, "Composition of Crude Oil and Its Relation to Stratigraphy in Wyoming," Bull. Amer. Assoc. Petrol. Geol., Vol. 37 (August 1953), pp. 1837–1872.

77. David White, "Metamorphism of Organic Sediments and Derived Oils," Bull. Amer. Assoc. Petrol. Geol., Vol. 19 (May 1935), pp. 589–617.

78. H. S. Bell, *American Petroleum Refining*, 3rd ed., D. Van Nostrand Co., New York (1945), pp. 259–261.

79. J. E. Gieseking, "Mechanism of Cation Exchange in Montmorillonites—Beidellite-Nontronite Types of Clay Minerals," Soil Science, Vol. 47 (1939), pp. 1–14.

Ralph E. Grim, "Relation of Clay Mineralogy to Origin and Recovery of Petroleum," Bull. Amer. Assoc. Petrol. Geol., Vol. 31 (August 1947), pp. 1491-1499.

80. A. Rauch, "Fuller's Earth and Its Uses in the Petroleum Industry," Jour. Inst. Petrol. Technol., Vol. 13 (1927), pp. 325-350.

William A. Gruse and Donald R. Stevens, *The Chemical Technology of Petroleum,* 2nd ed., McGraw-Hill Book Co., New York (1942), pp. 327-330.

81. F. H. Gayer, Ind. & Engr. Chem., Vol. 25 (1933), p. 1122.

82. Benjamin T. Brooks, "The Chemical and Geochemical Aspects of the Origin of Petroleum," in *The Science of Petroleum,* Oxford University Press, London and New York, Vol. 1 (1938), pp. 46-53. Bibliog. 51 items.

Benjamin T. Brooks, *op. cit.* (note 83), Part I, pp. 2282-2283.

83. F. H. Gayer, *op. cit.* (note 81).

84. A. W. Francis, "The Free Energies of Some Hydrocarbons," Ind. & Engr. Chem., Vol. 20 (1928), pp. 277-282.

85. J. Gordon Erdman, *op. cit.* (note 26), p. 28.

86. Walter L. Whitehead and Irving A. Breger, "The Origin of Petroleum: Effects of Low Temperature Pyrolysis on the Organic Extract of a Recent Marine Sediment," Science, Vol. 111 (March 1950), pp. 335-337.

87. P. G. Nutting, "Some Geological Consequences of the Selective Adsorption of Water and Hydrocarbons by Silica and Silicates," Econ. Geol., Vol. 23 (November 1928), pp. 773-777.

88. S. C. Lind and D. C. Bardwell, "Chemical Action of Gaseous Ions Produced by Alpha-Particles: Part IX—Saturated Hydrocarbons," Jour. Amer. Chem. Soc., Vol. 48 (1926), pp. 2335-2351.

89. K. G. Bell, Clark Goodman, and W. L. Whitehead, "Radioactivity of Sedimentary Rocks and Associated Petroleum," Bull. Amer. Assoc. Petrol. Geol., Vol. 24 (September 1940), pp. 1529-1547. 45 references.

90. Roland F. Beers, "Radioactivity and Organic Content of Some Paleozoic Shales," Bull. Amer. Assoc. Petrol. Geol., Vol. 29 (January 1945), pp. 1-22. Bibliog. 38 items.

91. Samuel C. Lind, *Chemical Effects of Alpha Particles and Electrons,* Chemical Catalog Co., New York (1928), 177 pages.

Samuel C. Lind, "On the Origin of Petroleum," in *The Science of Petroleum,* Oxford University Press, London and New York, Vol. 1 (1938), pp. 39-41. Bibliog. 14 items.

92. William L. Russell, "Relation of Radioactivity, Organic Content, and Sedimentation," Bull. Amer. Assoc. Petrol. Geol., Vol. 29 (October 1945), pp. 1470-1494.

93. C. W. Sheppard, "Radioactivity and Petroleum Genesis," Bull. Amer. Assoc. Petrol. Geol., Vol. 28 (July 1944), pp. 924-952. Bibliog. 57 items.

94. Roland F. Beers and Clark Goodman, "Distribution of Radioactivity in Ancient Sediments," Bull. Geol. Soc. Amer., Vol. 55 (October 1944), pp. 1229-1254. Bibliog. 35 items.

95. Irving A. Breger, "Transformation of Organic Substances by Alpha Particles and Deuterons," Research on Occurrence and Recovery of Petroleum, Amer. Petrol. Inst., Dallas, Texas (1948-1949), pp. 236-248. 39 references.

96. For further discussion of radioactivity the reader is referred to Konrad Bates Krauskopf, *Fundamentals of Physical Science,* 2nd ed., McGraw-Hill Book Co., New York, Chap. 20, "X Rays and Radioactivity," pp. 272-303, and Linus Pauling, *College Chemistry,* W. H. Freeman & Co., San Francisco, Chap. 33, "Nuclear Chemistry," pp. 663-676.

97. C. W. Sheppard and W. L. Whitehead, "Formation of Hydrocarbons from Fatty Acids by Alpha-particle Bombardment," Bull. Amer. Assoc. Petrol. Geol., Vol. 30 (January 1946), pp. 32-51. Bibliog. 31 items.

Walter L. Whitehead, Clark Goodman, and Irving A. Breger, "The Decomposition of Fatty Acids by Alpha Particles," Research on Occurrence and Recovery of Petroleum, Amer. Petrol. Inst., Dallas, Texas (1950-1951), pp. 208-213. Bibliog. 14 items.

98. Irving A. Breger and Walter L. Whitehead, "Radioactivity and the Origin of Petroleum," Research on Occurrence and Recovery of Petroleum, Amer. Petrol. Inst., Dallas, Texas (1950–1951), p. 217.

99. G. M. Knebel, "Progress Report on API Research Project 43," "The Transformation of Organic Material into Petroleum," Bull. Amer. Assoc. Petrol. Geol. Vol. 30 (November 1946), pp. 1935–1954.

100. Samuel C. Lind, "On the Origin of Petroleum," *op. cit.* (note 91), p. 39.

101. Samuel C. Lind, "On the Origin of Petroleum," Science, Vol. 73 (1931), p. 19. Benjamin T. Brooks, *op. cit.* (note 72), Part I, p. 2285.

102. Keith A. Kvenvolden, "Normal Paraffin Hydrocarbons in Sediments from San Francisco Bay, California," Bull. Amer. Assoc. Petrol. Geol., Vol. 46 (September 1962), pp. 1643–1652.

CHAPTER 12

Migration and Accumulation of Petroleum

Geologic framework. Short or long migration. Primary migration: water squeezed out of clays – normal water circulation – sedimentary oil – recycled oil. Secondary migration: entrained particles – capillary-pressure – displacement-pressure phenomena – buoyancy – dissolved gas effects – accumulation – tilted oil-water contacts – stratigraphic barriers – vertical migration – time of accumulation – petroleum supply.

ONCE PETROLEUM has reached the reservoir rock, it must be concentrated into pools if it is to be commercially available. As we have seen, it was deposited in the nonreservoir shales and carbonates as disseminated soluble hydrocarbon particles associated with the nonsoluble organic matter. By the time diagenesis was complete most of the petroleum hydrocarbons were probably in the form of petroleum. During and after diagenesis, water was squeezed out of the nonreservoir rocks into the reservoir rocks, and a fraction of the petroleum and petroleum hydrocarbons was entrained in the water. Petroleum and petroleum hydrocarbons probably continued to come out of the nonreservoir rocks after diagenesis and some may have been deposited directly in the reservoir rocks. The movement from the nonreservoir rocks to the reservoir rocks is called the *primary migration* to distinguish it from the concentration and accumulation into pools of oil and gas, the *secondary migration*.

There are many facets to the problem of the migration and accumulation of oil and gas into pools. Because of the apparent complexities of the problem and the many opportunities for speculative and imaginative analysis, many theories have been devised to explain the phenomena that have been observed. We must admit that we do not have a completely satisfactory answer to the problem of the migration and accumulation of petroleum, and in this connection, we should again remind ourselves that no one has ever seen an oil or gas pool forming, nor has anyone ever seen an oil pool in place. All of our information is from well and production records. As drilling continues, a vast amount of new data is continually being added, and this eventually should clarify many ideas that are now in doubt.

Presumably the petroleum entrained in the water squeezed out of the water-saturated nonreservoir rocks was dispersed in minute particles, possibly of colloidal or microscopic size, and some might even have been in solution in the water. In the absence of any unbalanced forces, such petroleum might remain more or less stationary for a long time and might become deeply buried. Local fluid potential or temperature gradients might cause local movements, but it would require some regional disturbance or change to cause regional movement. This might be regional folding, tilting, mountain building, or warming—possibly due to igneous activity or to various changes in the hydrodynamic gradients. The normal geologic history of most sedimentary regions includes many such episodes that could upset the equilibrium of the reservoir fluids, and each may have caused some increment of fluid movement.

Before we discuss the theories and many facets of the problem, it would be useful to point out two of the fundamental aspects of the problem that should be kept in mind: (1) the geologic framework within which petroleum migration and accumulation must have occurred, and (2) the distances through which petroleum has migrated. Are we limited to short distances of migration—say less than a mile?

GEOLOGIC FRAMEWORK OF MIGRATION AND ACCUMULATION

From a knowledge of the geologic conditions that prevail in the oil-producing regions, it is possible to set up a framework of limiting geologic factors within which any theory of petroleum migration and accumulation must fit. Such a frame can at best be only general, for it includes many variables, both known and unknown. The sides of the frame might be summarized as follows:

1. Nearly every petroleum pool exists within an environment of water—free, interstitial, edge, and bottom water. This means that the problem of migration is intimately related to hydrology, fluid pressures, and water movement. The interstitial water content of the reservoir rocks generally shows a hydrodynamic fluid pressure gradient, as between wells, and this indicates that the water, which is confined and in phase continuity throughout the reservoir rock, is in motion. The water moves in the direction of the lower fluid potential, and the rate varies with the magnitude of the difference in fluid

potential and the transmissibility of the aquifer. The rate may be small and measured in inches or a few feet per year, but the effect of the hydrodynamic conditions may be extremely important to the movement of the petroleum.

2. The gas and oil are chiefly immiscible in the water, and both are of lower density than the surrounding water.

3. Reservoir rocks that contain petroleum differ from one another in various ways. They range in geologic age from Precambrian to Pliocene, in composition from siliceous to carbonate, in origin from sedimentary to igneous, in porosity from 1 to 40 percent, and in permeability from one millidarcy to many darcies.

4. There is a wide variation also in the character of the trap or barrier that retains the pool. The trap may have been chiefly due to structural causes, to stratigraphic causes, or to combinations of these causes. Where there is a fluid potential gradient in the reservoir rock unit, it may form a barrier to the movement of petroleum and become an important trapping agency; it is a trapping element that combines readily with the structural and stratigraphic elements.

5. The microscopic shapes and sizes of the porosity, the tortuous paths of permeability, and the chemical character of the reservoir rock may vary widely in complexity. It is within these pore spaces and chemical environments that the migration and accumulation must take place.

6. The minimum time for oil and gas to originate, migrate, and accumulate into pools is probably less than one million years. (See p. 50.) The evidence for this lies in the fact that in some pools the trap was not formed until Pleistocene time. An example is the Kettleman Hills pool in California; the oil and gas of this pool are in the Miocene Temblor formation, but the fold that forms the trap cannot be earlier than Pleistocene, for the Temblor formation fold is parallel to the Pleistocene rocks at the surface of the ground.[1] This places the accumulation in late Pleistocene or post-Pleistocene time, possibly within the last 100,000 years and certainly within a million years. An illustration of the short time necessary for a pool to adjust itself to a change in conditions may be seen in the tilting of the Cairo pool in Arkansas.[2] (See p. 562.) The tilt occurred within a period of 10–12 years; if it had gone on for a few years more, at the same rate, the oil would probably have moved completely out of the trap. Thus the time it takes for oil to accumulate into pools may be geologically short, the minimum being measured, possibly, in thousands or even hundreds of years.

7. The upper and/or outer boundary, or roof rock, of *every* pool, whether the trap is structural or stratigraphic, or both, or in an area of hydrodynamic pressure gradient, is a relatively impervious surface that is concave as viewed from below. The exceptions to such a statement are not important and might be only apparent if all the facts were known.

8. The temperatures of the reservoir rock may fluctuate, generally within

the range 50–100°C (122–212°F), although extremes up to 163°C (325°F) have been observed in sediments within oil regions.

9. The fluid pressures within the reservoir rock may also fluctuate during the life of the reservoir, depending on the geologic history of the region. They have been observed to range from one atmosphere up to 1,000 atmospheres or greater, and they may have fluctuated up and down many times during the geologic life of the rock.

10. The geologic history of the trap may vary widely—from a single geologic episode to a combination of many phenomena extending over a long period of geologic time. Pools trapped in limestone and dolomite reservoir rocks, moreover, have the same relations that pools trapped in sandstone rocks have to such things as the reservoir fluids, oil-water and oil-gas contacts, and trap boundaries. Yet the chemical relations of the reservoir rock and the effects of solution, cementation, compaction, and recrystallization are quite different in sandstone and carbonate reservoirs.

SHORT OR LONG MIGRATION

One of the fundamental questions involved in the accumulation of petroleum is whether petroleum migrates considerable distances (i.e., one or more miles) to form pools, or whether it was formed essentially in place. Some believe that migration was negligible and that petroleum was formed virtually where it is now found.[3] If this is so, there is no problem of secondary migration, and exploration for petroleum should be confined chiefly to the search for favorable traps in areas of origin.

Probably the best evidence for migration from a nearby source are the pools in isolated porous and permeable lenses formed by facies changes, reefs, and sand patches. Reservoir rocks such as these are commonly formed in a highly organic shore environment and are contemporaneous with the source rocks; they represent the nearest reservoir to the richest source for the longest time. Petroleum particles entrained in the water squeezed out of nearby shales and other nonreservoir rocks would tend to flocculate in the larger pore spaces of the lens but would not develop enough capillary pressure to re-enter the fine-grained sediments. The petroleum, in effect, would be screened or filtered out and retained in the local reservoir rock.

There are several good reasons, on the other hand, for believing that some oil and gas may have migrated for relatively long distances to accumulate into pools. Rich,[4] one of the most vigorous proponents of extremely long-distance migration, believed that the oil was squeezed out of the sediments in the mobile and deformed belts of the earth, and entered the water-bearing sediments, which he termed "carrier beds," where it was carried until trapped.

Several kinds of evidence direct support to the view that oil may migrate through permeable reservoir rocks from a source of supply to become concentrated in a distant area of accumulation, where pools may be formed.

1. The common occurrence of oil and gas seepages and springs is direct evidence that the movement of petroleum is possible. In some of these the petroleum may be observed being carried along with the water; in others it escapes independently of any water movement.

2. The production of oil and gas from pools demonstrates that oil and gas may move through the permeable rocks and into the bore holes. The distance they move depends on the well spacing, which normally ranges from one-eighth to one-half mile and may be as much as a mile, being wider in gas pools than in oil pools. Given a longer time and a wider spacing, the distances through which oil or gas might be shown to move would no doubt be considerably greater. The movement of oil and gas into the well bore may accompany movement of water, or it may be independent of any water movement. Reservoirs with as much as 50 percent interstitial water, for example, are known to produce only clean oil and gas.

3. A structural trap may not begin to form until long after the reservoir rock is in place. Pools that accumulate in late-forming traps such as these are generally in reservoir rocks of regional extent. The oil in a late-forming trap should, then, have traveled a much greater distance than that in an isolated lens within a rich organic shale.

4. Another factor that argues for free movement of petroleum in a reservoir rock is that very few traps have remained unchanged in size, character, and effectiveness since they were first formed. As a result of the repeated folding, faulting, tilting, erosion, uplift, deposition, solution, and cementation that affect the average sedimentary basin or province, they have undergone numerous modifications, in hydraulic gradients, temperatures, and pressures, and in many physical properties. Each change that a trap undergoes either decreases or increases its capacity to hold oil and gas, or modifies the position of the oil and gas within the trap.

We find that the fluids in oil and gas pools today are in density adjustment to the *present* shape and nature of the trap, including the present hydrodynamic gradients. If this adjustment is true now, it must have been true throughout the life of the pool. In other words, the petroleum content has been at all times in gravity adjustment to the changing position of the trap. In order that a pool may maintain itself in a gravity equilibrium when trap conditions change, there must be a movement of the oil and gas within the reservoir rock. Such movements of the petroleum can in some cases extend for miles.

The regional westward tilt of the Paleozoic rocks across Texas, Oklahoma, and Kansas provides an example; many pools in that region have shifted their position within their traps in order to maintain their gravity adjustment with the changing structure. The tilting is illustrated in Figure 12-1, in which A shows the structure across northern Oklahoma at the time the Permian was

MIGRATION AND ACCUMULATION [CHAPTER 12]   543

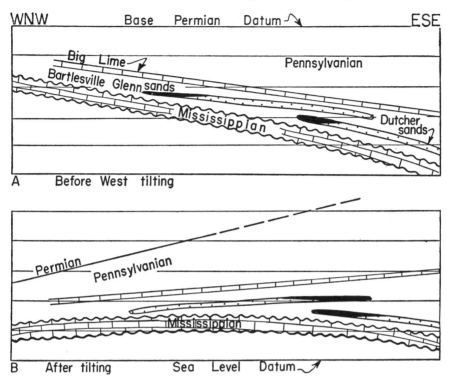

FIGURE 12-1 *Idealized section across northern Oklahoma, showing the dip of the rocks during Permian time (A) and at the present (B). During Permian time, the Pennsylvanian sands all dipped toward the east-southeast because the interval from the Permian to the Mississippian increases in that direction. The oil accumulations would be expected to be along the up-dip western edges. After the west-northwest dipping homocline was formed, probably before Cretaceous time, the dip of some of the Pennsylvanian sands, notably the Bartlesville and Glenn sands, was reversed, and the oil moved from the west end of the sand to the up-dip east end, where it is now found. This adjustment of the oil pools to the changing structure is taken as evidence of the movement of oil through the reservoir rocks, a movement which may extend for many miles. The length of the section above is approximately seventy-five miles. [Redrawn from Levorsen, Bull. Amer. Assoc. Petrol. Geol., Vol. 17 (1933), p. 1122, Fig. 7.]*

being deposited; the base of the Permian is taken as a level reference plane, and it is seen that the bedding of the Mississippian limestone diverges from it toward the east-southeast. When the Permian was being deposited, presumably as a level or nearly level plane, therefore, the intervening sands of the Pennsylvanian were dipping toward the east and southeast. Any oil or gas present in these sands would be expected to move to the highest side of the sand, and consequently would be found along the western, or up-dip, edge. At some time after the deposition of the Permian formations, the region was tilted down toward the west. The resulting structural section is shown at B. This is the present situation, and it can be seen that some of the Pennsylvanian sands (the Burbank-Bartlesville-Glenn sands), which originally dipped toward the east, now dip slightly toward the west; in these sands the westward tilting exceeded earlier tilting toward the east. The oil pools now occur along the eastern, or up-dip, edge of these Pennsylvanian sands; so they presumably migrated from the western edge to the eastern edge between Permian time and the present because of the reversal in the direction of dip.[5] (See also Fig. 7-12.) The lower Pennsylvanian sands, on the other hand, called the Dutcher sands in Oklahoma, were not tilted enough to reverse their dip; so their dip is still slightly toward the east, and the oil pools are generally found along their western edge.

We may conclude, therefore, that in some pools there is good evidence that the petroleum migrated only a short distance, while there is equally good evidence that the oil in other pools migrated for a long distance. Obviously, if petroleum will migrate a distance measured in feet, it can, by increments, multiply the distance into miles and tens or even hundreds of miles during geologic time. The distance it may migrate is determined by the distance from the source area to the nearest trap; if this distance is short, as it usually is, the distance the petroleum has moved is short; but, if there is no obstruction to migration through the reservoir rock, we may expect the petroleum to keep on moving until it reaches a trap or a barrier that can hold it or until it is lost through escape at the surface. It may have to move tens or even hundreds of miles, but the chances of its having to go so far without finding a trap are slight.

PRIMARY MIGRATION

Probably the best explanation of primary migration is that the oil and gas are entrained in the water squeezed out of the shales and clays during diagenesis and, to a lesser extent, in the confined water of the normal hydraulic circulation after diagenesis. Some of the oil and gas is in the form of submicroscopic and colloidal particles and some is in solution, the volumes being on the order of 10 to 50 parts per million. A quite different means by which large quantities of oil may have been moved is by a recycling of oil released during the erosion of oil pools and of nonreservoir rocks carrying petroleum hydrocarbons. (See pp. 516 and 517.) Some of the recycled oil might be deposited directly in the potential reservoir rocks, such as reefs, sand bars, and sand deposits.

Several summaries of theories based primarily on the movement of petroleum along with moving water are available. These include the hydraulic theory of Munn,[6] the hydraulic-buoyancy theories of Rich,[7] Mrazec,[8] and

Daly,[9] the sedimentary compaction theories of King,[10] Monnett,[11] and Lewis,[12] and the compaction-hydraulic theory of Cheney.[13] Their general characteristics are similar in one respect: all involve movements of large quantities of water, which carries the dispersed oil and gas to a trap in which the oil and gas are accumulated and a pool is formed.

Oil and gas in solution in reservoir waters should also be considered in connection with primary migration. The solubility of natural gas in water ranges from 4 cubic feet per barrel of water at 400 psi up to 22 cubic feet between 2,000 and 6,000 psi.[14] The lighter constituents of crude oil also are soluble in water, the amount increasing with pressure. At atmospheric pressure the solubility was found to be 0.014 volume of oil for each 100 volumes of water, and at a pressure of 1,100 psi the amount of oil that went into solution in the water was considerably more.[15] The oil is released from solution as the pressure drops. Some of the petroleum showings seen in water at the surface during a test may actually be oil and gas that have come out of solution in the reservoir waters as the pressure declined from that of the reservoir to that at the well head.

Water Squeezed Out of Clays and Shales

If water containing oil and gas is squeezed out of clay shales and into permeable reservoir rocks in the primary migration, then it must displace water that is already present in the permeable rocks. This would cause a flow of water through the reservoir rocks to some outlet—we may assume to the surface of the ground. A shale formation, for example, may during its deposition have been 100 miles long, 100 miles wide, and 1 mile thick, and thus have had a volume of 10,000 cubic miles. Half of this total volume, or 5,000 cubic miles, might conceivably have been water at some period in the early part of the diagenesis. Now, however, as a result of compaction and diastrophism, the formation as a whole is only 25 percent water by volume, and its total bulk has shrunk to 6,666 cubic miles. Of the original water content, 3,334 cubic miles has been squeezed out, and 1,666 cubic miles remain. Most of the water squeezed out probably escaped upward across the bedding planes that were forming during diagenesis and was thus lost to the overlying ocean. The process might be thought of as a settling of mud through water. A part of the water squeezed out, however, possibly a third, or 1,111 cubic miles, may later, when the clays were buried and were forming shales, have passed out of them laterally and into the porous reservoir rocks in the vicinity; for the lateral permeability would finally become greater than the vertical. A part of the hydrocarbons deposited or formed in the nonreservoir sediments would be carried along with the water entering the reservoir rocks, either as minute, submicroscopic, dispersed particles in colloidal form or in solution.

As the water moves through the reservoir rocks, the entrained petroleum particles would flocculate into discrete patches. These may be carried along in a miscible phase with the water until enough petroleum accumulates in phase continuity to develop buoyancy. Patches several feet in length and a few molecules or even pores in thickness may have to accumulate before buoyancy develops sufficiently to initiate independent movement of petroleum as a separate phase.

During the migration of petroleum as a discrete phase within a water-wet reservoir, the petroleum tends to enter only the larger pores because of the requirement of higher displacement pressure for entry of petroleum into the smaller pore spaces. In a water-wet rock of variable porosity, therefore, if the oil is in continuous phase, it would tend to be displaced from the finer pores by the water and localized in the larger pores, leaving the water to pass through the smaller openings. The net result would be that the oil would be concentrated in the coarser and more permeable rocks, where later movement into traps would be easier, and the water would pass on through the finer openings.

The replacement theory proposed by McCoy[16] is based on the replacement through capillary action of oil in the shales and fine-grained rocks by water from adjacent, more porous, coarser-grained rocks. This process of replacement is based on the fact that in the smaller capillary openings there is greater adhesion between rock and water than between oil and either rock or water. The water is thus able to separate the oil from the rock and to drive it from the smaller capillary openings, such as those in shales, into the larger openings, such as those in sand. McCoy believed that there was little movement of oil except for this cause, and that the places into which the oil had thus been driven became the sites of the present oil fields. This implies that the trap was already formed when the primary migration from the shale into the reservoir rock occurred.

Normal Water Circulation

Presumably at the time diagenesis is complete, the pore spaces of all of the reservoir and nonreservoir sediments are filled with water. Regional circulation patterns develop and are continually changing as the fluid pressure gradients change. Where no fluid potential gradients exist the fluids are static. During the geologic life of the average reservoir rock, the contained water has undoubtedly moved at different times in different directions at different rates; the possible exception would be in the late deposits, where the present fluid potential gradient is the same as the original gradient. That such circulation has occurred right down to the present, long after lithification and diagenesis, is evidenced by the present local and regional fluid potential gradients observed in many reservoir rocks.

There are many possible causes for fluid pressure changes and fluid potential gradient changes in most sedimentary regions. They include such phenomena as diastrophism, mountain building, erosion, deposition, recharge from meteoric waters, and osmosis. Faulting, folding, and cementation affect the permeability and may change the direction of movement. Deep canyons change the outlet area, and new flow paths are developed. Chemical deposits such as salt and anhydrite are practically impermeable and undoubtedly have a strong effect on flow patterns. Flow patterns would be expected to be modified by volcanic activity and other phenomena that change the temperature of a region.

The migration and accumulation of petroleum is intimately associated with the water within which it occurs. Such water is confined, in much the same way that water is confined within the pipes that make up a city's water system, in contrast to water in a lake or a river. As a confined water it may flow either up or down the dip or slope, for the rate and direction of flow of confined water is proportional to the rate of change of head—the hydrodynamic fluid potential gradient—measured in relation to a horizontal datum plane referred to sea level. It is not proportional to the rate of change in hydrostatic pressure along the flow path. For example, water will flow from an area of low fluid pressure to one of high fluid pressure, provided the head is lower in the direction of flow. These relations may be seen in Figure 12-2, a diagrammatic profile along the direction of flow and pressure gradient of confined water. Circulation may be caused by any condition that develops a fluid potential gradient, as evidenced by a sloping potentiometric surface, from one part of the basin to another. See also Chapter 9.

A factor that undoubtedly has an important bearing on the ease of movement of oil and water through reservoir rock is the normal decrease in their viscosity and interfacial tension and also that which comes with an increase in the dissolved gas in oil permitted by the normal increase in temperature and

FIGURE 12-2

Diagrammatic sketch showing the path water takes in flowing from intake area A *to outlet area* B *as it passes across synclines and anticlines. The reservoir pressure at* F *would raise the water to* G; *this fluid pressure is less than at* D, *where the fluid pressure would raise the water to* E, *or at* H, *where the fluid pressure would raise the water to* I. *The surface* AB *is the potentiometric surface; its slope governs the overall flow of water from* A *to* B.

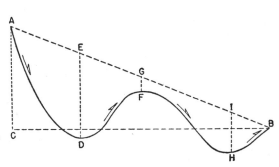

pressure that accompanies burial.[17] An oil saturated with gas at a temperature of 70°F and a pressure of 500 psi, for example, has about half the viscosity of the same oil saturated with gas at surface conditions. By the time the pressure has increased to 1,800 psi, which is the normal pressure for a depth of burial of 4,100 feet, the viscosity of the gas-saturated oil is approximately that of kerosene at atmospheric pressure. The viscosity of the water, too, is reduced by increased temperature; at depths of 10,000 feet water should flow through rocks three times as easily as at the surface, and at a depth of 20,000 feet it should flow six times as easily.[18]

Sedimentary Oil—Recycled Oil

Another way in which oil may travel long distances is to go along with the sediments. Two sources are available: (1) eroding oil pools and (2) eroding nonreservoir shales and carbonates. The source of the oil-bearing sediments in a number of producing provinces can be traced to some nearby topographically high area. Mountains or high topography in sediments generally means that the rocks are structurally high and therefore favorable for the location of oil pools, and also that erosion is rapid and the oil content of the shales would be quickly freed. The association of an oil-producing area with a nearby area of structurally high sediments that have been eroded may be entirely coincidental; but there are so many such areas, and the potential source of petroleum for redeposition is so vast, that the possibilities become worthy of speculation.

Three of the many areas where these conditions appear to prevail are discussed here.

1. The Rocky Mountains of the United States were uplifted into great domes and anticlines in late Cretaceous and early Tertiary time. Erosion began then and continues to the present, the sediments going chiefly to the Gulf of Mexico, where they were deposited and became the Tertiary producing formations of the Gulf Coast province. Presumably these structurally high sedimentary mountains, extending from Mexico to Montana, once contained many great oil fields (certainly many of the folds that are left contain oil pools) and many thousands of cubic miles of oil-bearing shales, and there is a strong possibility that hundreds or even thousands of billions of barrels of oil trapped in these pools and sediments were later eroded along with the sediments. What became of this oil? Did a part of it go along with the sediments and help form the protopetroleum of the Gulf Coast province?

2. The shelf area of southeastern Kansas and northeastern Oklahoma[19] contains a great many oil pools in the Cherokee formation (Pennsylvanian), which was derived from the erosion of the Nemaha Mountains, a great faulted anticline extending across Kansas and into Nebraska, made up of early Paleozoic and Precambrian rocks, now buried beneath the Pennsylvanian. The Devonian, Silurian, and Ordovician formations equivalent to those that

were eroded from the top of the mountain range, during Cherokee time, extend far to the south, into Oklahoma, along the extension of the range, and many prolific oil fields are found in them. (See Figs. 13–13 and 14–6.) The Nemaha Mountains fold was large enough before it was eroded to contain an enormous oil field in any one of several possible reservoir formations.

3. The Wichita Mountains–Amarillo line of uplift, extending for several hundred miles in western Oklahoma and the Texas Panhandle, was eroded down to the Precambrian granite basement in Pennsylvanian time, and the debris was deposited along the northern flank of the range. (See Fig. 3-8.) Many oil fields are now found in the rocks formed from the erosion products of the great faulted anticline that caused the mountain range. Maybe the oil came in with the sediments.

The possibility of oil being deposited as a sediment was first mentioned by Murray,[20] who observed that the iridescent film of waste oil on the waters of the Irrawaddy River was gone within a mile or two of the place where it was dumped. He experimented with the oily, muddy water to see what became of the oil. He found that it settled to the bottom along with the clay. Murray concluded that the settling of oil with clay mixtures is due to the fact that the oil is broken up into separate particles by deflocculation and that it assumes spherical shapes because of the interfacial tension between oil and immiscible water. This mixture corresponds to an oil-water emulsion, except for the effect of the sediment particles, which are water-wet. The oil globules cannot come together because they are separated by the fine mud particles. He believed that the dispersed oil globules were carried down and mechanically mixed with the mud particles.

Later Poirier and Thiel [21] tested the oil-settling capacities of various shales and clays in synthetic salt water. The capacity varied, depending on the nature of the clay; some clays settled their weight in oil, and the volume of oil deposited was inversely proportional to the size of the mineral grains. Table 12-1 shows the results they obtained. Poirier and Thiel concluded from their experiments that the cause of the settling was not mechanical trapping, as suggested by Murray, but was, rather, that the oil globules were carried down by the weight of the adhering clay particles. They found the globules to be in a loose state of packing, with the clay particles actually adhering to them. The presence of organic acids decreased the sediments' settling capacity, possibly through a lowering of the interfacial tension of the oil-water interface.

We can be sure that great volumes of oil were released by erosion in the past. What became of the oil? Was it all destroyed by oxidation and bacterial action? Was a part of it carried by rivers to the oceans? Until we know more about it, the possibility of oil moving great distances, attached to clay particles, cannot be dismissed.

Besides the erosion of oil pools, other sources of petroleum hydrocarbons in the oceans might be (1) submarine seepages and (2) the early transformation of part of the organic matter of the sea bottoms into petroleum, during

TABLE 12-1 Sediments Listed in the Order of Their Oil-settling Capacities
(emulsion with one gram of sediment used in each)

Sediment	Maximum Amount of Oil (cu cm)	Average Amount of Oil (cu cm)	Predominating Minerals
Kaolin	2.50	1.21	Clay minerals
Blue-earth silt	2.10	1.33	Quartz
Marl	1.50	1.11	$CaCl_3$
Diatomaceous earth	1.50	1.00	Silica
Decorah shale (Ordovician)	1.37	0.70	Sericite, clay minerals
Bentonite	1.35	0.68	Montmorillonite
Sand silt	1.25	0.75	Quartz, sericite
Calc. shale	1.15	0.70	$CaCO_3$, sericite
Calc. silt	1.12	0.85	$CaCO_3$, quartz
Kerogen shale	0.70	0.60	Sericite, organic matter
Humus	0.00	0.00	Organic matter

Source: Adapted from Poirier and Thiel, Bull. Amer. Assoc. Petrol. Geol., Vol. 25 (1941), p. 2178.

the diagenesis of the muds into shales, and its escape into the ocean water above. The high concentration of hydrocarbon waxes in the top half inch of the soil (see p. 515) suggests the possibility that they may have formed at many subaerial surfaces of unconformity and, as the surface was being eroded, been carried by the streams into the oceans. Once petroleum has reached the ocean, from whatever source, much of it would presumably rise to the surface of the water and be carried by the waves and currents to where organic muds, off-shore bars and sand patches were forming, or where conditions were right for the development of organic reefs. A small amount of oil caught in a potential reservoir rock might be of more significance as source material than a much larger amount of organic matter caught in muds and shales.

SECONDARY MIGRATION

As with the primary migration from the nonreservoir sediments into the reservoir rocks, there are many facets to secondary migration through the reservoir rocks and accumulation into pools. Some of these will be considered, without particular order, as key elements in the problem of petroleum migration and accumulation. They are (1) entrained particles, (2) capillary pressure—displacement pressure phenomena, (3) buoyancy, (4) dissolved gas effects, (5) accumulation, (6) tilted oil-water contacts, (7) stratigraphic barriers, (8) vertical migration, and (9) time of oil accumulation.

Entrained Particles

The water in any aquifer, which is a potential petroleum reservoir rock, has been and in many cases is now in motion. The rate and the direction of water movement are determined by the magnitude of the fluid potential differences, the transmissibility of the aquifer, and the location of the area of minimum fluid potential. The rate and the direction of water movement have undoubtedly changed, possibly several times, during the geologic life of an aquifer because of continuous changes in the overburden, structure, deformation, erosion, and geochemistry. The movement of fluid would follow any available path of permeability, such as porous and permeable rocks, faults, unconformities, and fracture systems. Microscopic and submicroscopic particles of petroleum and petroleum hydrocarbons entrained in the moving water would be carried along until obstructed by the structure or character of the rock or until separated by pressure, temperature, and volume changes in the mixture, at which time they would be expected to flocculate and accumulate into larger-sized particles and aggregates until buoyancy became effective.

Capillary-Pressure–Displacement-Pressure Phenomena*

The basic requirement for the migration of large patches of oil in a water-wet reservoir is that the capillary pressure of the oil-water interface exceed the displacement pressure of the larger openings or capillaries between the pores. For any specific combination of oil, water, and sand pores, the displacement pressure P_d is a constant value. On the other hand, the capillary pressure is dependent upon buoyancy, pressure gradients, and the length of continuity of the oil phase. Whenever these forces are sufficient to cause the capillary pressure to exceed the displacement pressure, the oil-water interface will enter and pass through the pore-connecting capillaries, and oil migration will result.

In order to evaluate more quantitatively the conditions under which oil migration can occur, it is necessary to estimate the difference between the capillary pressure at the leading end of the migrating oil body and the capillary pressure at the rear. The geometry of an isolated oil globule in a pore when there is little or no effective force tending to cause it to migrate is shown in Figure 12-3. The capillary pressure in part X is

$$P_c = \frac{2\gamma \cos \theta}{r}$$

which is approximately the same at points A, B, and C, as indicated by the same radius of curvature, r, at each of these points. Part Y shows the distorted shape of this globule as it must be just before breaking through the con-

---
* This discussion by courtesy of Gilman A. Hill.

striction at point A and migrating into the next pore to the right. In this condition the capillary pressure at the foremost end is

$$P_c = \frac{2\gamma \cos \theta}{r_c}$$

and the capillary pressure at the rear is

$$P_c = \frac{2\gamma \cos \theta}{r_p}$$

where r_c is the effective radius of the capillary opening between pores and r_p is the effective radius of the pore. The difference in these capillary pressures (ΔP_c) at the fore and aft positions is

$$\Delta P_c = 2\gamma \cos \theta \left(\frac{1}{r_c} - \frac{1}{r_p} \right)$$

If we assume that for an average porous sand the radius, r_c, of the pore-connecting capillaries may be between $\frac{1}{2}$ and $\frac{1}{4}$ of the radius, r_p, of the pores (i.e., that r_c is between $\frac{1}{2} r_p$ and $\frac{1}{4} r_p$), then ΔP_c lies between

$$\frac{2\gamma \cos \theta}{r_p} \quad \text{and} \quad \frac{6\gamma \cos \theta}{r_p}$$

If we take an average value, $r_c = \frac{1}{3} r_p$, then

$$\Delta P_c = \frac{4\gamma \cos \theta}{r_p}$$

If the contact angle, $\theta, = 60°$, $\cos \theta = \frac{1}{2}$, and $\Delta P_c = 2\gamma/r_p$.

This approximate average value of capillary pressure difference required for migration under these assumed conditions can now be computed for a number of possible situations of interfacial tension and size of pore. Table 12-2 gives the results.

Consider now the possibility of creating this required capillary pressure gradient resulting from flowing formation water. If the oil is present in the form of isolated globules in individual pores, as shown in Figure 12-3, and

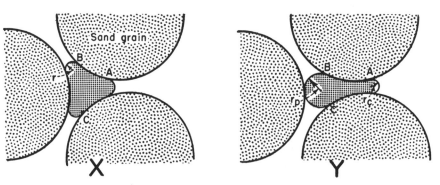

FIGURE 12-3 *Geometry of a pore, showing the distortion that is necessary if an insular particle of oil is to be forced through a pore constriction.*

if the water is flowing horizontally to the right, then the capillary pressure difference is equal to the pressure drop from one pore to the next pore. A normal hydrodynamic gradient of 10 feet per mile would produce a capillary pressure difference of 0.1 dyne/cm² for coarse sand and 0.02 dyne/cm² for fine sand. An extreme hydrodynamic gradient of 100 feet per mile would produce a capillary pressure difference of only 1.0 dyne/cm² for coarse sand and 0.2 dyne/cm² for fine sand. In comparison with the required capillary pressure difference shown in Table 12-2, the ordinary hydrodynamic pressure gradients occurring in nature are far too small to cause migration of oil. Indeed, forces several thousand times greater than those produced by normal hydrodynamic pressure gradients would be needed to cause migration of isolated oil globules.

Buoyancy forces are also insufficient to cause migration of isolated oil globules. The buoyancy pressure, $(\rho_w - \rho_o)gz$, in units of dynes/cm², will

TABLE 12-2 Capillary Pressure Difference Required for Migration, dynes/cm.²

γ Interfacial Tension (dynes/cm)	$\Delta P_c = \dfrac{2\gamma}{r_p}$				
	Very Coarse Sand ($r_p = 0.02$ cm)	Coarse Sand ($r_p = 0.01$ cm)	Average Sand ($r_p = 0.005$ cm)	Fine Sand ($r_p = 0.002$ cm)	Very Fine Sand ($r_p = 0.001$ cm)
30	3,000	6,000	12,000	30,000	60,000
25	2,500	5,000	10,000	25,000	50,000
20	2,000	4,000	8,000	20,000	40,000
10	1,000	2,000	4,000	10,000	20,000
5	500	1,000	2,000	5,000	10,000
1	100	200	400	1,000	2,000

produce for 30° API oil a capillary pressure difference across an isolated pore-size oil globule of only 7.4 dynes/cm² for coarse sand and 1.5 dynes/cm² for fine sand; z equals the vertical interval occupied by the petroleum phase. Therefore the buoyancy force would have to be several thousand times that existing in nature in order to cause migration of isolated oil globules. A simple laboratory experiment illustrating the necessity of oil-phase continuity in obtaining sufficient buoyancy to cause oil movement is shown in Figure 12-4.

Although the normal hydrodynamic forces and buoyancy forces are incapable of causing migration of oil globules isolated in individual pores, these forces can easily cause migration of oil bodies that are continuous through several thousand pores. For instance, a potentiometric surface gradient of 10 ft/mile will produce a capillary pressure difference of 10,000 dynes/cm² if

  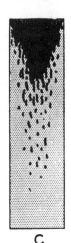

FIGURE 12-4

Successive stages in an experiment devised by Gilman A. Hill to show the effect of buoyancy. The stippled area is the front of a rectangular box, 6 feet high, 4 inches deep, and 1 foot wide, filled with water-soaked sand. The front is of glass, and the effects may be observed by the use of ultraviolet rays. In A three patches of oil have been injected into the water-soaked sand, each a few inches across; they remain in place. More oil has been added in B until the patches coalesce; fingers of oil start moving upward. In C the entire mass of oil has moved to the top of the box, with the exception of some residual oil; this remains in small patches a few pores in diameter. Sufficient buoyancy was achieved in B to overcome the resistance to upward movement that held the separate patches in place in A, and the mass rose as high in the box as it could go. The movement from position B to position C occurs in a few hours.

the oil phase is continuous over a vertical distance of 5.4 meters. This is enough to cause the oil to migrate under the average conditions shown in Table 12-2. Also, the buoyancy force for 30° API oil will cause a capillary pressure difference of 10,000 dynes/cm^2 if the oil phase is continuous over a vertical distance of 68 cm. Continuity of the oil phase over a distance of 1–10 meters, therefore, provides the necessary conditions for oil to migrate under the normal forces existing in nature.

Interesting experiments have been devised and recorded on motion-picture film to show the microscopic nature of oil movement through a container filled with glass and Lucite beads.[22] When water and oil were pumped through the container at rates of 1.5–1,000 feet per day, the flow of each fluid was through its own tortuous network of connecting pores or channels. While these velocities are greater than would be expected in nature, they probably show the details of the movement once the capillary pressure has been overcome.

Buoyancy

A body, whether fluid or solid, that is immersed in a fluid is buoyed by a force equal to the weight of the fluid it displaces. The fluids that petroleum geologists deal with are natural gas with a specific gravity between 0.00073 and 0.000933 (water = 1.0), oil with a specific gravity between 0.7 and 1.0, and water with a specific gravity between 1.0 and 1.2. In a reservoir that contains all three, then, the gas will rise to the top, the oil lie next below, and the water remain at the bottom. Since this density layering is the normal arrangement in the petroleum pools of the world, we naturally conclude that not only the final equilibrium adjustment of the gas, oil, and water in a pool, but also the original concentration of the dispersed particles of petroleum in the water, is in some way a result of the buoyancy of both the oil and the gas in water.

Upward movement due solely to buoyancy may begin where a sufficient local concentration of oil and gas has developed. (See Fig. 12-4.) If a limited portion of the reservoir rock receives enough oil to provide the phase continuity required for developing sufficient buoyancy, this force will overcome the capillary resistance to the entry of the oil into the water-saturated pore spaces. The entire mass or patch of petroleum fluids may be expected to move upward along connected larger pores to the highest point of the reservoir rock, and if there is a dip, or slope, to the reservoir rock, it will move up the dip along the upper surface of the permeable rock and below the impervious cover. Migration of the oil phase will occur and continue so long as the capillary pressure is equal to or greater than the entry pressure of the rock bounding the oil. Once the oil and gas patch started to move, it would be expected to add to its size by picking up dispersed particles of oil and gas en route, thus further increasing the buoyant effect. Its movement would be speeded up where it encountered local areas of steeper dips and decreased displacement pressure. How much oil would be left behind at the rear of the moving patch would depend upon the size of the pores, the wettability of the reservoir rock, the capillary properties of the oil and water, and the rate of movement. Slight pressure changes, such as are caused by earthquakes, for example, may "trigger" the delicate balance between the capillary pressure and the entry pressure at both the fore and the rear end of the moving oil patch, so that the amount left behind may be negligible, but more likely it is appreciable. Some reservoir rocks are left almost entirely free from oil residues, while in others there is an almost continuous thin layer of oil showings along the top of the permeable reservoir rock or a trail of discontinuous patches left behind in the nonproductive areas between pools.

The force of buoyancy acts to trap petroleum in two ways:

1. When the patches of oil and gas entrained in flowing water reach an anticlinal area their buoyancy causes them to tend to resist further movement by the moving water when they reach the crest of the anticline. The net result is that the oil and gas are retained in the highest part of the structure, as shown in Figure 12-5. Water moving across a structural trap thus leaves the oil and gas behind at the point of lowest potential energy, or near the highest point in the reservoir rock. If the magnitude of the hydrodynamic pressure gradient across the structure is great enough, the

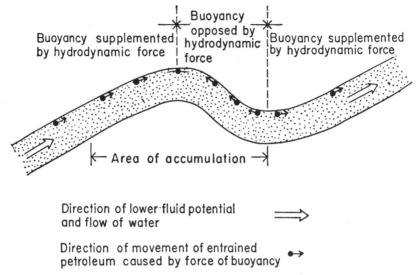

FIGURE 12-5 *The trapping mechanism of an anticline, showing the effect of hydrodynamic force on migration and accumulation of petroleum.*

oil-water contact is tilted in the direction of flow, or it assumes such a high angle that the oil passes out of the trap, as discussed on p. 568. The greater the density of the oil (or the lower the API gravity), the greater the tilt.

2. The situation is somewhat different in a stratigraphic trap, in which the permeability lessens up the dip. The oil and gas migrate up the dip by their buoyancy to a site where the buoyancy or capillary pressure no longer exceeds the displacement pressure of the finer-grained rocks. If the water flows down the dip, it increases the barrier effect. This situation is shown in Figure 12-6. If, however, the water flows up the dip, the combined hydrodynamic force and the buoyancy of the oil and gas is enough to permit the oil and gas to enter smaller and smaller pore spaces, and in many cases migrate through the barrier zone; only a small pool, if any, then results. (See also Fig. 12-7.)

The barrier effect of a down-dip hydrodynamic fluid potential gradient was discussed in Chapter 7. The corks in the analogy given there (see Fig. 7-63) may be likened to particles or patches of petroleum, the constriction to the lessening of rock permeability, and the size of the accumulation of corks to the height of the oil column. If the direction of the fluid flow were reversed, there would, of course, be no accumulation of either corks or petroleum. Moreover, if the fluid system were hydrostatic the accumulation would be much smaller.

This explanation assumes that there is a continuous path allowing the fluids to move beyond the edge of permeability. This concept presents no problem if the reservoir rock is truncated, for it would be normal for connecting channels, even though minute, to extend throughout the surface of the unconformity. This problem is more difficult, however, when there is a facies change, as from sand to shale or from dolomite to limestone. Presumably the shale and the limestone contain water in phase continuity, and this would permit the flow of water, even though possibly through a zone of high resistance to flow. Furthermore, thin sandy patches, sand stringers, and similar minute irregularities in the stratigraphic section at the level of a sand deposit would presumably provide irregular small paths that would allow passage through the water occupying such void spaces of the particles of oil and gas that might overflow or be pushed through the barrier.

The widespread phenomenon of hydrodynamic pressure gradients must be considered as a trapping agency—probably as important and as effective an agency as the rock elements. Basically the situation is that a down-dip flow of water in an aquifer increases the trapping capacity of an area of lessened permeability enough to prevent further up-dip movement of petroleum. When the accumulation is large enough to develop a capillary pressure sufficient to overcome the barrier's displacement pressure, the capacity of the barrier to hold petroleum is reached. If the flow of water is up the dip, it increases the capillary pressure, making it possible for the petroleum to pass through the zone of higher displacement pressure.

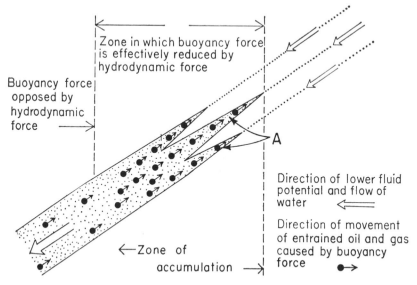

FIGURE 12-6 *The effect of water flow directed down the dip and through a barrier zone of higher displacement pressure and decreased permeability. In this case the influence of the buoyancy force is decreased by the hydrodynamic force, and oil and gas are trapped below the barrier zone.*

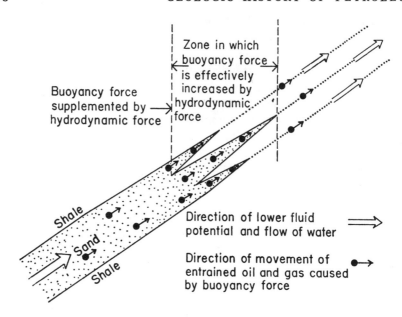

FIGURE 12-7   *The effect of water flow directed up the dip and through a barrier zone of relatively higher displacement pressure and decreased permeability. In this case the hydrodynamic force acts in a direction that is additive to the buoyancy force and makes possible the entry of petroleum into the smaller pores of the barrier zone, thus reducing the trapping capacity of the barrier zone.*

Dissolved Gas Effects

A variable amount of natural gas is dissolved in the oil in practically all oil pools. Completely gas-free oil pools are so seldom found that they must be regarded as anomalous. Natural gas has extremely low viscosity and high buoyancy, as compared with water and oil, and its volume changes in response to changes in pressure and temperature are many times greater. Energy is readily stored in compressed gas, and is just as readily released, so that its volume is in a sensitive adjustment to reservoir conditions. The expansion of compressed gas resulting from a reduction of pressure is an important agent in moving oil from the reservoir rock into the well bore in the normal production of oil pools. The question then naturally arises: does not gas enter actively into the migration process that concentrates oil into pools?

The probable importance of gas in the movement of oil through reservoir rocks was long ago pointed out by Johnson,[23] who noted the ease with which gas is concentrated, and believed that gas was responsible for the facility with which oil evidently moves through pores in the rocks. He believed that the moving oil formed films surrounding the gas bubbles.

The classic experiments of Thiel,[24] followed by the further work of Thiel and Emmons,[25] showed that in the laboratory the presence of small amounts of gas in oil-water-sand mixtures caused the gas and oil to concentrate in the higher portions of the container and that no concentration occurred when gas was absent.*

Further work on the movement of oil in the presence of gas was done by Dodd,[26] who found that when water, with a small amount of gas in solution, was forced through water-saturated sands containing disseminated oil, the oil could be made to migrate rapidly up low dips.

Mills[27] found, while conducting some experiments on gas and oil disseminated through water-saturated sands sealed in glass containers, that several times the containers were cracked.† The result each time was that the gas expanded because of the release of pressure and immediately rushed out through the cracks. The oil and some water were carried along with the gas and escaped until the pressure was equalized. Mills explained what happened by supposing that, when the water, oil, and gas were mixed together, the gas, because it was under pressure, became diffused through the water and the oil. When the pressure was released at the crack in the container, the gas expanded and carried the oil along with it to the point of release. The same thing could occur at a fault in a natural reservoir. Much of the oil entangled with the gas would escape along with the gas, but a part would be left behind at the fault, because of its higher viscosity. Another point made by Mills in support of his

---

* In the Thiel experiments a mixture of crude oil diluted with one-third its volume of kerosene to lower the viscosity, sea water acidized with 0.5 percent acetic acid, and crushed quartzite that passed through a 20-mesh screen was placed in a tube one inch in diameter and four feet in length, which was bent to form a miniature anticline. Four inches of both ends of the tube were then filled with crushed dolomite, and finally the ends were sealed. The carbon dioxide generated by the reaction of the acetic acid with the dolomite moved up the tube, carrying the oil with it. The gas and oil were largely segregated at the crest of the bend in the tube within twenty-four hours, and the separation was even greater at the end of forty-eight hours. Similar experiments in which no gas was generated resulted in no concentration of the oil.

Variations of the above experiment were described by Emmons. Some experiments consisted in using gasoline and heating it to develop gas pressure; others consisted in making additional bends in the tube to simulate terraces, and varying the grain size of the sand to cause permeability variations and other irregularities. In each case the oil was concentrated in the local traps, and oil movement was observed with differential pressures as low as a few ounces per cubic inch and on slopes as low as one-half of one degree.

† Following up this accidental discovery, Mills mixed oil with a fermenting liquid—consisting of water, apple juice, sugar, and yeast—in a fine-grained water-wet sand. The mixture of sand and fluids was tightly packed in ginger-ale bottles, which were then sealed with caps and allowed to stand for three days. Only a slight gravitational segregation occurred. Practically all the gas that was formed went into solution under pressure in the oil and water. A single hole was then punched through the cap of each of six bottles, which had been packed differently. Immediately gas began escaping from each hole, and the oil started moving upward and segregating above the water. At the end of two minutes, oil with a little water was jetting from the holes in all the bottles in practically the same manner. The same result was obtained when the bottles were laid on their sides, the oil moving to the point of gas escape.

idea was that many of the minerals normally associated with faults and found in veins, such as calcite, barite, gypsum, and inclusions of waxy hydrocarbons, are also commonly found in oil as it comes to the well head. In fact, they are so common in some areas as to give trouble in producing the oil.

A reasonable conclusion to all of these experiments is that the first effect of a decline in pressure to the vicinity of the bubble point is an increase in the volume, until phase continuity is reached as the bubbles coalesce, the buoyancy and the mobility of the gas-oil mixture increase and result in a movement toward the area of pressure release. The solubility of gas in oil would cause an expansion of the dispersed oil globules and particles until they formed connected patches large enough to have adequate buoyancy. An oil patch alone may move through or with the water, but the presence of gas certainly makes the movement easier. (See also pp. 607–612.)

If oil and gas move independently of the movement of water, and if the water in a reservoir is moving, the oil and gas are moving either along with the water or against it. We have seen, in Figure 12-2, that the water moves from A to B because of the difference in head; it would presumably carry colloidally dispersed and stabilized oil and gas and also oil and gas in solution along with it. The buoyancy of a patch of oil and gas located to the left of H, for example, would be directed against the flow of water and toward F; a patch of oil or gas to the right of H would move with the flow of water toward B. A high rate of water movement might overcome the buoyant effect, but presumably most water movement is at low rates and would merely retard the rate of oil and gas movement.

Accumulation

The concentration of oil and gas from a disseminated state in the reservoir-rock waters into an accumulation of commercial size is the final step in the formation of a pool. In some reservoirs there appears to be no connecting free water, and presumably there has been little movement of the petroleum; that is, the accumulation is close to the source area. Most traps, however, *are* connected to free water, either moving or stationary. Whether the petroleum is carried along by moving waters, by its buoyancy, or by both, it moves along the upper surface of the permeable reservoir rock, possibly as a film only a few molecules or pores thick. The evidence is in the countless showings of oil and gas found at the top of water-bearing reservoir rocks in oil territory. The traps that are effective in retaining oil and gas, then, are those that bar or obstruct oil and gas moving along the upper surface of the reservoir rock.

The size of the accumulation may be limited by (1) the amount of available petroleum source material, (2) the physical elements present at the place of accumulation, or (3) any combination of the two. This assumes that every trap is as full of petroleum as is possible under the existing conditions—pressure, temperature, fluid potential gradient, relative densities of petroleum and water, slope of the rocks, permeability and permeability variations.

Where the reservoir water is in hydrostatic equilibrium, the oil and gas come to rest at the highest part of the trap (where the potential energy is lowest), and the oil-water contact assumes an approximately level position, although in some the contact is tilted (these are discussed later).

In general, a trap is an area of low potential energy[28] toward which the buoyant oil and gas move from areas of higher energy. The oil and gas separate in accord with their density and remain in the trap as long as the conditions do not change—that is, until the trap is connected to some point of still lower potential energy, as by faulting, by regional tilting, by a change in the rate at which water flows through the rocks, or by a well, at which time they again move, and the effectiveness of the trap decreases.

When large quantities of water in which minute quantities of oil and gas are disseminated enter a trap, the trap retains the oil and gas and lets the water go on. The traps through which water passes most easily are domes and anticlines; fault traps and most stratigraphic traps are partly sealed along one or more edges, and consequently deflect or stop most of the fluid that reaches them. The volume of petroleum-bearing water that passes through a trap is seldom great enough to supply the amount of oil now found in them. It seems reasonable to believe, therefore, that some of the oil and gas that has accumulated into pools has done so independently of the movement of water. The chief transporting force for this oil and gas is its buoyancy, which causes it to move to the highest places in the reservoir rocks, either aided by or hindered by the movement of water in the aquifer.

Folding not only forms high, local areas of lower fluid pressure, but also further decreases the pressure, for when a clastic rock is folded, the bending of the formation causes a fracturing and a rearrangement of any loose grains, so that locally the volume of the rock is increased. The increase in volume, called dilatancy, is due to an increase in the pore space, which in turn results in a lowering of the pressure.[29] It is conceivable that this pressure loss toward the tops of anticlines, if it should occur rapidly, might in some cases further add to the effect of buoyancy.

Tilted Oil-Water Contacts

In many pools the oil-water contact is tilted. Commonly the tilt is at a rate of only a few feet per mile, but occasionally it reaches 800 feet or more per mile, roughly equivalent to a slope of eight degrees. A tilt displaces the oil and gas pool down one flank of the trap, and this displacement may be important during the development, because, if it is recognized early, the pool may be completely developed with fewer dry holes. Where the tilt is high, the pool may be so far displaced that the highest part of the structure is barren or contains only gas, and the entire oil pool is far down the flank. Such a pool may well be missed if wildcat wells are located in the conventional way on the highest part of the structure. Or the tilt may be so high in relation to the local fold that the oil or the oil and the gas have been completely displaced from the structure; the pool is said to have been *flushed out,* and the trap is now barren of petroleum. A few examples showing the nature of the tilt follow.

The Cairo pool, in Union County, Arkansas, has a tilted oil-water contact that is of especial interest because it is probably the result of a man-made fluid potential gradient formed by the withdrawal of reservoir fluids in the

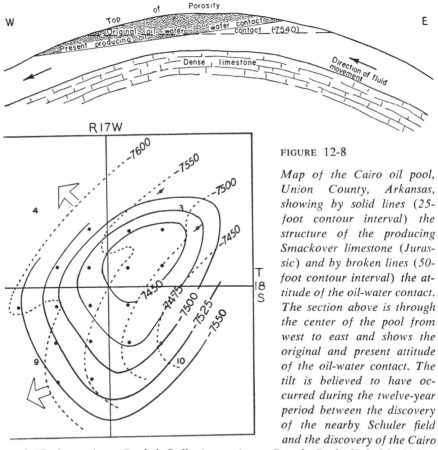

FIGURE 12-8

Map of the Cairo oil pool, Union County, Arkansas, showing by solid lines (25-foot contour interval) the structure of the producing Smackover limestone (Jurassic) and by broken lines (50-foot contour interval) the attitude of the oil-water contact. The section above is through the center of the pool from west to east and shows the original and present attitude of the oil-water contact. The tilt is believed to have occurred during the twelve-year period between the discovery of the nearby Schuler field and the discovery of the Cairo pool. [Redrawn from Goebel, Bull. Amer. Assoc. Petrol. Geol., Vol. 34 (1950), pp. 1956, 1966, and 1972.]

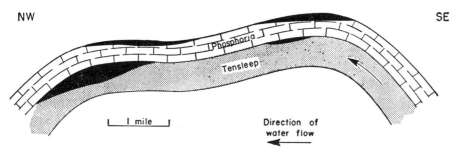

FIGURE 12-9 Section along the axis of the Northwest Lake Creek oil field, showing the location of the oil pools (black). [Redrawn from Green and Ziemer, O. & G. Jour., July 13, 1953, p. 178.]

near-by Schuler field within a period of ten or twelve years.[2] The pool is located three miles up the regional dip from the much larger Schuler field, although it is structurally lower than the top of the Schuler fold. Production is obtained in both pools from the Reynolds oolitic member of the Smackover formation (Jurassic), and the Schuler field was discovered twelve years before the discovery of the Cairo pool. During this time the reservoir pressure dropped about 500 psi and the field produced over 7 million barrels of oil and 10 million barrels of water. It is believed that the decline in reservoir pressure at Schuler created a hydrodynamic fluid potential gradient toward the pool, and that this gradient extended across what was later to become the Cairo pool, for the oil-water contact in the Cairo pool was found inclined toward the Schuler field at a dip of approximately 100 feet in a distance of one mile. A structural map of the Cairo pool, the inclination of the inclined oil-water contact, and a section through the pool are shown in Figure 12-8.

The Northwest Lake Creek pool, in the Big Horn basin of Wyoming, is in the Phosphoria limestone and the Tensleep sandstone, which are folded into a long, narrow anticline extending for a distance of seven miles with a width of about one-half mile. The oil in the Tensleep sand is displaced down the plunge of the fold toward the northwest,[30] as shown in the section along the axis in Figure 12-9. Another displaced oil pool may be seen in the sands in the Stevens zone (Upper Miocene) in both the North Coles Levee and the South Coles Levee fields, Kern County, California. (See Fig. 12-10 for map and section.) Other examples of tilted oil-water contacts have been described at Frannie and Sage Creek, Wyoming[31] (see Figs. 12-11 and 12-12 for maps), at the Wheat pool, in Loving County, Texas,[32] and at the Cushing field, Creek County, Oklahoma.[33]

Tilted oil-water contacts have been ascribed to several possible causes.[34] A lag in the adjustment of the oil-water contact to a late regional tilting of the enclosing rocks has been suggested. This is not a very satisfactory explanation, since in most regions the latest folding of the reservoir rock took place millions

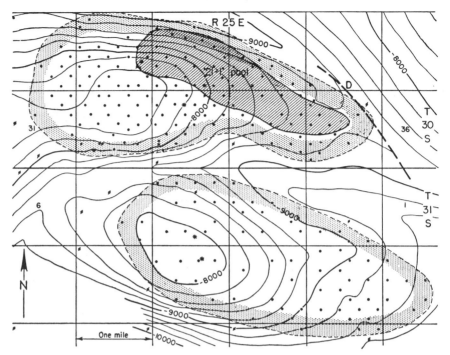

FIGURE 12-10 *Structural map and section of the Coles Levee field, California. The map shows the North and South Coles Levee field; contours on top of the N marker are at a contour interval of 200 feet. The 21-1 pool is trapped in the up-dip pinch-out of one of the Stevens sands (Miocene). The section through the pools shows the eastward tilt of the oil-water contact. The shaded outline of the pools, when applied to the structural contours, also shows that tilt. The section is from west to east along the axis of the North Coles Levee anticline. It shows a tilted oil-water contact in the Lower Western Zone and a sand pinch-out trap in the 21-1 Zone pool, both in the Stevens sand formation (Miocene).* [Redrawn from Davis, Jour. Petrol. Technol., August 1952, pp. 12 and 13, Figs. 1 and 2.]

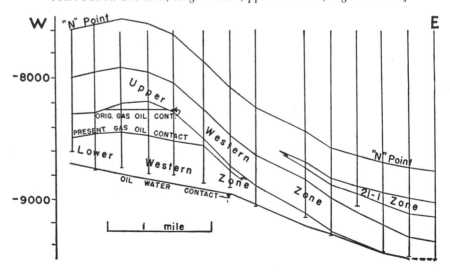

MIGRATION AND ACCUMULATION [CHAPTER 12]   565

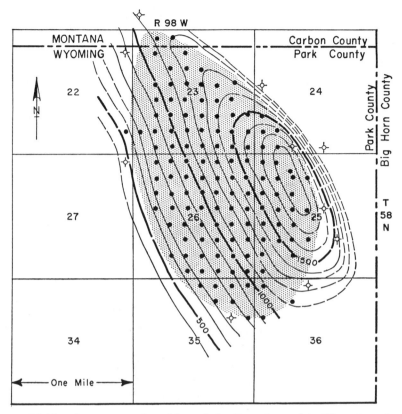

FIGURE 12-11. *Structure and position of the Frannie pool in Wyoming. An ex ample of a tilted water table.* [Redrawn from *Wyoming Geological Association, Symposium, 1957.*]

of years ago, which should be sufficient time for a complete adjustment, especially in view of the apparently short time it took for the tilt in the Cairo pool of Arkansas to develop.

In places the tilting is only apparent; irregularities in the contact resulting from various other causes are mistakenly interpreted as a uniform slope of the contact. Water may appear at higher levels in some wells than in others; for example, because of conditions such as facies changes, minor faulting and fracturing, irregular rates of production, leaking well casing, edge-water encroachment, and water coning. Or a water saturation that is locally variable because of irregularities of porosity and permeability may cause some wells to produce water at a slightly higher level than that at which other wells in the same pool produce only oil.

Any tilting due to capillarity is minor, as the following analysis indicates.*

---

* Gilman A. Hill, unpublished report, Stanford University (March 1951).

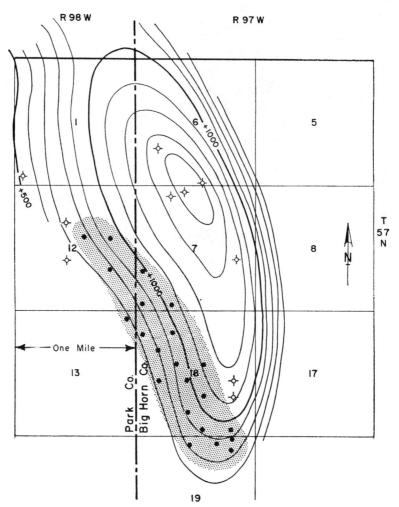

FIGURE 12-12 *Structure and position of the Sage Creek oil pool, Wyoming. An example of a tilted or inclined water table. [Redrawn from Wyoming Geological Association, Symposium, 1957.]*

In the absence of capillary pore openings, the oil-water contact under hydrostatic conditions is a smooth, level, plane surface. It may be called the *free-water table* or the *zero capillary pressure plane*. When capillary pore openings are present, however, the water rises to various heights by capillary pressure. The height to which it will rise depends on the size of the openings above the free-water table and the densities of the water and the oil, according to the equation

$$h = \frac{2\gamma \cos \theta}{(\rho_w - \rho_o)rg}$$

MIGRATION AND ACCUMULATION [CHAPTER 12]

where h is the height of rise of water above the free-water table (in centimeters), θ is the contact angle measured through the wetting phase (water), ρ_w and ρ_o are the specific gravities of water and oil respectively, γ is the interfacial tension between the oil and water, g is the acceleration of gravity (980 cm/sec^2), and r is the radius of the capillary opening (in cm). It may be seen from this equation that the rise of the water against the oil is inversely proportional to the radius of the capillary opening, other conditions being the same. (See Fig. 12-13.)

The tilting of an oil-water contact or of a gas-water contact in a pool is frequently best explained as the result of a hydrodynamic pressure gradient extending through the pool, the evidence being in a sloping potentiometric surface. Under such conditions the oil-water and gas-water contacts are inclined in the direction of the water flow, the amount depending on the gradient and on the differences in specific gravity of the fluids. The brief explanation that follows of the phenomena of tilted oil-water contacts is based on several articles and reports, to which the reader is referred for more complete discussion.[34]

Where the potentiometric surface is horizontal, the fluid pressure at all points of equal elevation in the reservoir rock is the same, and the oil-water contact is horizontal. Where the potentiometric surface is sloping, however, it indicates that there is a hydrodynamic pressure gradient at right angles to the contours of the potentiometric surface, and that water is flowing through the rock down the slope of the potentiometric surface. (See Fig. 12-2, p. 547.) In Figure 12-14, for example, the potentiometric surface slopes at the rate of h feet in distance $AC(l)$, or dh/dl. This means that the water flows from F toward G in the reservoir rock because the potentiometric surface at G is lower than at F. Small differences in the density of the oil and water cause the oil-water contact to slope at a higher rate than the potentiometric

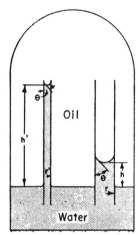

FIGURE 12-13

This diagram shows the higher rise, under hydrostatic conditions, of water against oil in the capillary opening with the small radius r' than in the larger opening with the radius r. As long as the capillary openings are of the same size and hydrostatic conditions prevail, the oil-water contact is a level, smooth plane. Where a coarse sand grades into a fine sand along the oil-water contact, however, the contact is raised, a few feet at the most, in the direction of the fine sand. [Redrawn from Yuster, Trans. Amer. Inst. Min. Met. Engrs., Vol. 198 (1953), p. 150.]

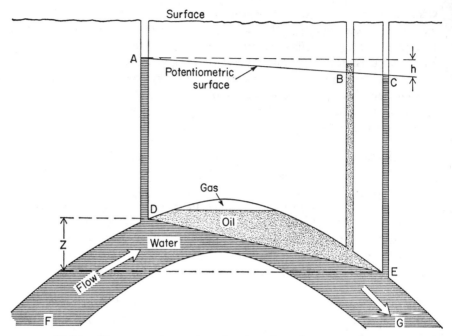

FIGURE 12-14 *Diagrammatic section showing the relation of the slope of the oil-water contact to the slope of the potentiometric surface.*

surface, but in the same direction. The water continues to flow, but the oil is in static equilibrium. The slope of the oil-water contact may be expressed as

$$\frac{dZ}{dl} = \frac{\rho_w}{\rho_w - \rho_o} \times \frac{dh}{dl}$$

where ρ_w and ρ_o are the densities of water and oil respectively, dZ/dl the slope of the oil-water contact, and dh/dl the slope of the potentiometric surface. If the slope of the potentiometric surface is known, the slope of the oil-water contact can be calculated. Figure 12-15 shows the oil-water tilts for varying slopes of the potentiometric surface and API gravities of oil. The lower chart shows the order of magnitude of the velocities of water flow for several cases of sand permeability.

It will be noted in the chart that the slope is much greater for oils of low API gravity (high density) than for oils of high API gravity and for gases. The explanation may be shown in terms of a vector diagram, Figure 12-16. The heavier oil is seen to have the higher inclination for the same hydrodynamic pressure gradient. Under hydrostatic conditions, the oil-water contact is horizontal—or normal to the vertical direction of the buoyancy force.

Stratigraphic Barriers

Stratigraphic barriers to petroleum migration are those geologic phenomena that lessen permeability laterally up the dip. They are common, either as the

MIGRATION AND ACCUMULATION [CHAPTER 12]

chief trapping agency or as an auxilliary agency. Facies changes, truncation and overlap, cementation, solution, and fracturing are the common causes of permeability variations that aid in or result in petroleum pools. It is not necessary that the permeability be completely eliminated, for a mere lessening of permeability may increase the entry pressure enough to bar further movement of oil and gas.

Vertical Migration

Fluids, such as water and petroleum, may move vertically across bedding planes and into formations either above or below, depending on the location of the zone of lower fluid potential. Thus if a high-fluid-potential sand is higher in the section than a low-pressure sand, movement will take place from the high-potential sand downward into the low-potential sand whenever a pathway is available. That such vertical differences in fluid potential occur

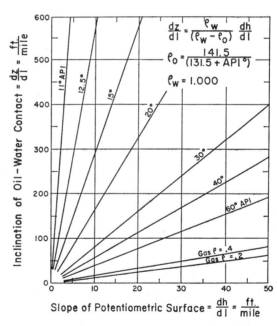

FIGURE 12-15

The inclination of the oil-water contact for varying slopes of the potentiometric surface and for varying densities of the oil, and the velocity of water flow for several cases of sand permeability. [Data from Gilman A. Hill.]

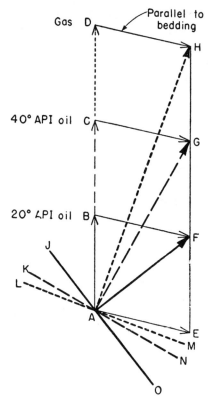

FIGURE 12-16

Vector diagram showing the effect of the same hydrodynamic force on different petroleums. The hydrodynamic force is shown as vectors parallel to the bedding (BF, CG, and DH) for oils of 20° and 40° API gravity and for gas, respectively. The buoyancy force is represented by a vector directed vertically upward; its magnitude varies with the magnitude of the difference between the densities of the given petroleum and water. The greater the difference in density, the greater the buoyancy force. Thus the buoyancy vector for gas (80° API) in a water system is represented by AD, the vector for 40° API oil as AC, and the vector for 20° API oil as AB. When these buoyancy force vectors are combined with the hydrodynamic force vectors, BF, CG, and DH, the vectoral sums become AF, AG, and AH, respectively, for the 20° and 40° oils and the 80° gas. The oil-water contacts are normal to the vector sums AF, AG, and AH and are shown as the slopes JO, KN, and LM, respectively.

is known from several observed phenomena and from comparison of pressure measurements with normal hydrostatic pressure gradients. (See also pp. 405–411.)

During drilling, evidence of fluid pressures that are either higher or lower than the expected hydrostatic pressure is found when zones of lost circulation are encountered, when pressure tests (D.S.T., etc.) are made, and when blowouts occur in a hole filled with drilling mud. Water springs at the surface indicate that the potentiometric surface is above the ground surface.

Three of the several pathways available for vertical migration of fluids are:

1. During diagenesis, the early fluids rise vertically out of the muds that are being compacted by added deposits and enter the overlying water. As diagenesis advances, the upward vertical permeability lessens until finally the lateral permeability becomes greater and the fluids move out laterally along the bedding planes as compaction continues.

2. Faults may produce zones having higher or lower permeability than the rocks that are faulted. The frequent association of springs at the surface outcrop of faults is evidence that fault zones are often permeable and that fluids may pass along them. Thus high-fluid-potential aquifers would be connected to low-fluid-potential aquifers by a permeable path when both are cut by

faults, and fluids would move vertically, either up or down, depending on the local fluid potential gradient. See Figure 9-6.

3. Where planes of unconformity truncate tilted aquifers, the overlying formations, together with the surface of the unconformity, offer a channel along which higher-potential formations are in connection with lower-potential formations. As shown in Figure 9-6 the direction of the fluid flow may be changed or even reversed.

Time of Accumulation

In some pools there is some basis for judging the earliest time at which the oil and gas could have accumulated, and this in turn throws light on the conditions under which the migration took place. The reasoning is based on two simple but entirely different approaches:[35]

The Law of Gases. This law tells us that the volume of a gas varies directly as the absolute temperature and inversely as the pressure. The temperature effects may be neglected in the general discussion of the problem because, with the average pressure-temperature gradients encountered in the oil regions, they are considerably less than the pressure effects. The law of gases can be applied to the problem in this way: the capacity of a trap to hold gas is a function of the pressure, which means that in a trap that is completely full of gas the present accumulation of gas could not have occurred until the present pressure, at least, was reached. Because reservoir pressures in most areas are roughly relative to depth of burial, this also means that the trap could not have been filled until the present depth of overburden was deposited.

If, for example, we start with a given volume of gas, at the temperature (60°F) and the pressure that existed when the reservoir rock was being deposited (one atmosphere, or 14.7 psi, if at sea level), then the increase in reservoir pressure with burial will cause the volume of the gas to decrease, and the increase in temperature will cause the volume to increase. The net volume change is a decrease to a fraction of the original volume. The volume, for example, of a given amount of gas at a pressure of 153 atmospheres (2,250 psi) and a temperature of 160°F, which are the average pressure and temperature found at depths of 5,000 feet, is roughly $\frac{1}{143}$ as great as under surface conditions.* If, however, we find a trap at a depth of 5,000 feet that is full of gas, we may know that the gas could not all have entered the trap until its pressure had reached 143 atmospheres, for at a lower pressure less gas would have been required to fill the trap. The capacity of a trap to con-

---

* The calculations are based on an increase in pressure of 45 psi per 100 feet and an increase in temperature of 2°F per 100 feet. A unit volume at the surface (60°F and 14.7 psi) equals 1. At a constant temperature, the volume at 2,250 psi is 0.0065, or roughly $\frac{1}{154}$ of the surface volume, and at the reservoir temperature of 160°F (60° + 100°) the volume is 0.0077, or roughly $\frac{1}{129}$ of the surface volume.

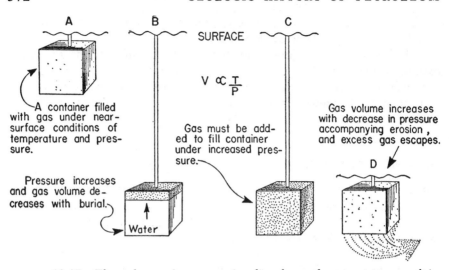

FIGURE 12-17 *The volume of a gas varies directly as the temperature and inversely as the pressure, a physical phenomenon known as the Law of Gasses. For our purpose we may disregard the effect of the temperature changes because they are but a fraction of the pressure changes found underground. The figure shows a unit volume of gas at* A *under surface conditions of fluid pressure as filling the container. If the fluid pressure is increased, as with increasing depth of burial, the same gas would be compressed into the small volume shown in* B. *If, however, we find that the same container is full of gas at the increased pressure, as in* C, *we must conclude that additional gas has entered the container; the amount of gas in* C *is far greater than in* A. *We may say that the capacity of the trap to hold gas at the increased fluid pressure in* C *is much greater than in* A. *A trap full of gas at high pressure, such as illustrated in* C, *is therefore taken as evidence that the gas migrated into the trap as the fluid pressure provided greater capacity. The cycle is completed in* D, *where erosion decreases the fluid pressure and the gas expands and escapes until the original amount of gas remains.*

tain gas increases with pressure; in other words, it may be said to be a direct function of the reservoir pressure. Consequently, the gas that now fills the trap could not have all entered the trap until after burial had reached the depth of 5,000 feet. Or, conversely, in order to keep the trap full of gas, there must have been a movement of gas into the trap as the pressure increased along with depth of burial. The principle, applied to a unit volume of gas, is illustrated in the diagrams in Figure 12-17.

A possible alternative explanation for a trap that is full of gas is that the gas was formed in place from oil or organic matter, being evolved at the expense of the heavier hydrocarbons by biochemical, thermal, or catalytic cracking as the temperature, pressure, and bacterial and chemical conditions of the reservoir changed during geologic time. This explanation is not considered nearly as likely as later movement of gas into the trap from an outside source, for the temperatures and pressures in reservoirs are far lower than those required to carry such cracking reactions forward in our laboratories or refineries. If gas were continually being evolved from oil, moreover, not only should all the older oils be transformed to gas by now, but it would be reasonable to expect most, if not all, pools, especially those in the older rocks, to be saturated with gas and to contain free-gas caps.

The fact that all oil pools contain at least some gas, and that many either approach gas saturation or are saturated, may give added evidence on which to speculate as to the possible time at which the oil accumulated. Oil has its greatest mobility when its viscosity is the lowest and its buoyancy is the highest—that is, at its saturation pressure. Where we find a saturated oil pool with a free-gas cap, and the trap is full to the spill point (meaning that any additional volume of oil or gas would force the excess oil out through the highest bounding syncline), the accumulation must have been completed at the time the present reservoir pressure was attained, which may generally be taken as the time when the present depth of burial was reached, either by deposition or by erosion. If, however, the oil is undersaturated, we may conclude that the accumulation was probably completed at some time before the depth equivalent of the saturation pressure was reached.

In the East Texas pool, by way of example, the pressure at which gas begins to come out of solution in the Woodbine sandstone (Upper Cretaceous) reservoir is 755 psi, whereas the reservoir pressure when the pool was discovered was 1,420 psi. The reservoir saturation pressure of 755 psi corresponds generally to a depth of burial of 1,700 feet, which is about the thickness of the overlying rocks up to the Tertiary-Cretaceous unconformity. We know that, when the erosion period represented by that unconformity began, a trap had been formed that could hold all the oil and gas now in the pool.[36] We might conclude, then, that since migration is easier at the approximate bubble-point pressure, it may be possible that the accumulation in the East Texas pool was largely completed by the beginning of the Tertiary deposition, because thereafter no more free gas was available. Subsequent movement merely resulted in a shifting of the pool as it adjusted itself to the changing relations that accompanied a regional southward tilting of the Woodbine sand reservoir.

Another explanation of the difference in the oil and gas content of nearby traps has been advanced by Gussow.[37] It is, in brief, that when a fold trap, for example, is filled with oil and gas, the oil is eventually forced out at the bottom of the fold, either because additional oil and gas enter the trap or because a loss of pressure brings about an increase in gas volume. The successive steps are shown in Figure 12-18. As the oil is forced out at the bottom, it moves up the dip and is caught in the next trap. The successive steps are shown in Figure 12-19, where trap 1 of A is equivalent to stage 3 of Figure 12-18. A sequence of pools, as in C, may be such that the down-dip trap, 1, is full of

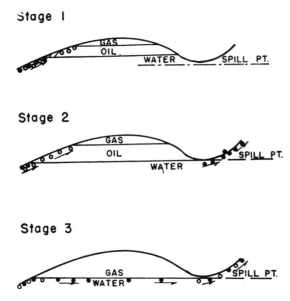

FIGURE 12-18

The selective trapping of oil and gas. Stage 1: *The oil and gas are above the spill point, and both will continue to be trapped until the free water is all displaced and the oil-water contact reaches the spill point.* Stage 2: *Gas continues to be trapped while oil spills out and goes on up the dip.* Stage 3: *Gas has filled the trap and spills out up the dip. Oil bypasses the trap and goes on up the dip.* [Redrawn from Gussow, Proc. Alberta Soc. Petrol. Geol., June 1953, p. 4.]

gas, the next, up-dip trap, 2, contains both oil and gas, the next, up-dip trap, 3, contains only oil, and the next, up-dip trap, 4, still contains water. Trap 4 will eventually be filled with oil from the excess spilled out from the lower traps, 1, 2, and 3.

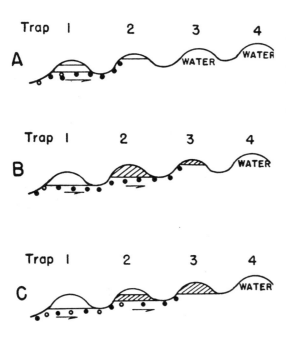

FIGURE 12-19

In the series of traps at A, trap 1 is in the state of Stage 2 (Fig. 12-18): oil is being spilled out into trap 2. Traps 3 and 4 are full of salt water. In B, trap 1 has spilled all its oil into trap 2. Trap 2 is full of oil, and the excess is passing up the dip into trap 3. Trap 1 is full of gas. In C, trap 1 is unchanged, trap 2 has a gas cap (the excess gas coming from trap 2), trap 3 is full of oil, and trap 4 is still full of water but will be the next to be filled with oil spilling out of trap 3. [Redrawn from Gussow. Proc. Alberta Soc. Petrol. Geol., June 1953, p. 4.]

MIGRATION AND ACCUMULATION [CHAPTER 12]   575

Time Trap Formed. Another line of reasoning that may be applied to the time of accumulation is indicated by the almost axiomatic statement that an accumulation cannot take place until *after* the trap is formed. If, then, we know when the trap was formed, we at least know the time *before* which there was no accumulation. The trap may be completely formed in one geologic episode, which may have occurred immediately after the reservoir rock was deposited or a long time later. Or the trap may have grown intermittently and with repeated modifications throughout the life of the reservoir rock. Thus the first accumulation of oil within the trap may have occurred either early or late with respect to the age of the reservoir rock. The accumulation at any time may fill either a fraction or all of the ultimate capacity of the trap to hold oil and gas, depending upon what fraction of the trap was available first, or on whether the trap was formed by one or several repeated geologic episodes. It is difficult to conceive of oil and gas as forming at the exact spot where, during some later geologic period, a trap will be formed. For example, why should oil and gas have accumulated in an Ordovician sandstone at the exact spot where the first sign of folding to form a trap occurred in Pennsylvanian time? Petroleum is not so foreseeing! We must conclude, in such a case, that the oil and gas moved into the trap from some outside area after the trap began to form (which was, in the example, after the first Pennsylvanian folding), and that the full capacity of the trap to hold petroleum was not reached until the last folding and tilting occurred.

The kind of evidence that may be used for dating a trap is shown diagrammatically in Figure 12-20. For example, the up-dip edge of the sand formation is shown as forming a trap at 1. The trap was formed when the overlying shale formation was deposited, or at "a" time. Trap 1 in the sketch is thus the earliest trap to form. Trap 2 was formed after the faulting occurred, during the interval recorded by the unconformity, or at "b" time. The porosity in trap

FIGURE 12-20

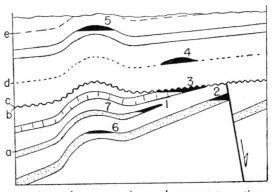

Diagrammatic section illustrating the method of determining the relative time when a trap was formed. The traps are numbered from 1 to 7, and those containing oil pools are black. Stratigraphic time planes are labeled from a to e. The oldest trap to form is 1, which was formed at a time. The last traps to form were 5, 6, and 7, all of which were formed at some time subsequent to e time. Accumulation could not occur until after the trap was formed; hence the date when a trap was formed gives a time before which there could be no accumulation into a pool.

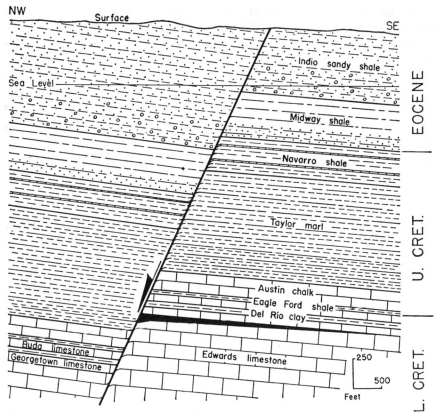

FIGURE 12-21 *Section across the Darst Creek pool, Guadalupe County, Texas. This is typical of many pools in this region in which the trap is formed by a slight arching crossed by a normal fault. Since the displacement along the fault is the same at the surface as it is on the producing Edwards limestone, the date of the faulting must be post-surface. Consequently the trap was formed after the Indio formation, now at the surface, was deposited, and the accumulation of the oil must have been some time after the trap was formed. Both the trap and the accumulation are thus post-Indio in age.* [Redrawn from McCallum, Bull. Amer. Assoc. Petrol. Geol., Vol. 17 (1933), p. 26, Fig. 3.]

3 was formed during the erosion period that caused the unconformity, and the trap was completed when the overlying impervious shale was deposited, or at "c" time. Trap 4 was in a sand lens or sand bar formed at "d" time. Traps 5 and 6 were both formed when the folding occurred. This was after bed "e" was deposited, for it was folded parallel to both 5 and 6. Trap 7 was formed at the same time, but the oil may have gone past it to accumulate in trap 1 during the time interval represented by "e," so that there was no oil

MIGRATION AND ACCUMULATION [CHAPTER 12]   577

or gas left to accumulate during the short period from "e" to the present when the trap was in existence.

A simple example of the method of judging when a trap such as trap 2 was formed is illustrated in Figure 12-21, a section through the Darst Creek pool in Texas. As the faulting that formed the trap has the same displacement on the surface Indio formation as on the producing Edwards limestone, there could have been no trap, and therefore no accumulation, prior to Indio time. Another example is in the typical oil field of the Gulf of Suez graben of Miocene age shown in Figure 12-22. Pre-Miocene faulting caused traps to form in Cretaceous and Carboniferous limestones and sandstones. Later these formations were breached by erosion, and most or all of the oil must have escaped. The trap was again sealed when the reservoir was unconformably overstepped by Miocene shales and anhydrites, and only then could the present oil begin to accumulate.

A somewhat more complex structural history is illustrated by the folding of the Voshell field of McPherson County, Kansas, shown in Figure 12-23. Part A shows the structure at the time the Kinderhook shale was being deposited on the eroded pre-Kinderhook surface. Apparently there was no local folding, but there was a wedge-out of the Hunton (Devonian-Silurian) limestone, a feature that now forms a trap at the south side of the Voshell structure, and there was a Misener sand lens (Mississippian). Part B shows the structure at the beginning of Pennsylvanian time to have been an anticline in all of the pre-Pennsylvanian rocks, which were truncated and overstepped by the Pennsylvanian formations. Part C shows the structure in which the Pennsylvanian rocks up to the Lansing formation were faulted and folded. This folding was added to the pre-Pennsylvanian folding, so that by Lansing time there was a

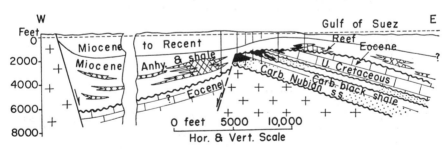

FIGURE 12-22 *Section through a typical oil field in the Miocene graben of the Gulf of Suez. Pools are trapped above the pre-Miocene unconformity plane in organic reefs and below the unconformity plane in overlapped porous limestone and sandstone reservoirs. The pre-unconformity traps were not formed until the Miocene anhydrites and shales were deposited to form the impervious seal; consequently the present oil accumulation could not have formed until after the trap was sealed.* [Redrawn from Weeks, Bull. Amer. Assoc. Petrol. Geol., Vol. 36 (1952), p. 2118, Fig. 25.]

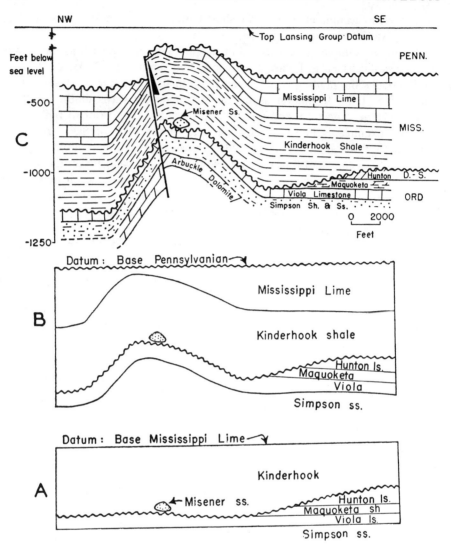

FIGURE 12-23 *Structural episodes that resulted in the Voshell trap, McPherson County, Kansas, in which the Voshell pool is located. A shows the structure at the beginning of Mississippian time, B at the beginning of Pennsylvanian time, and C at the beginning of Lansing time. The Lansing is also folded, which increment is added to the pre-Lansing folding to make the present structure.* [Part C redrawn from Bunte and Fortier, in Stratigraphic Type Oil Fields (*1941*), *p. 111*.]

considerable amount of folding in which to trap oil and gas. The subsequent folding of the Lansing formation was added to all of the pre-Lansing folding to make the folding now observed. Thus we may recognize in this trap three separate periods of folding: pre-Pennsylvanian, pre-Lansing, and pre-surface;

when all are added together, they make the fold now producing oil from Ordovician rocks. Each increment of folding added an equivalent increment of capacity of the trap to contain oil and gas.

Two periods of oil movement may be seen in pools such as the Hastings and the Van, in Texas. (See p. 150.) A section through the Hastings field, seen in Figure 12-24, shows a free-gas cap on the downfaulted side of the field. This suggests that the first accumulation came after the fold had formed and that faulting occurred later, dropping the free-gas cap down to a lower position than the oil in the other fault block. The common oil-water contact suggests free intercommunication across the faults below the *Marginulina* roof rock, which permits much of the oil but none of the gas to cross the central graben area.

A method of graphically showing the time a trap was formed, and, consequently, the time before which there could be no accumulation, is shown in Figure 12-25. The vertical distance represents the interval between the time the reservoir rock was formed, whether that was Ordovician or Pliocene, and the present time. The width of the black represents the percentage of the trap that formed at any time and was available for the accumulation of a pool of oil or gas. Part A represents traps such as lenses, reefs, and sand patches; part B represents traps formed by successive periods of folding, as in many reservoirs of the Mid-Continent region in Ordovician rocks that were first folded during Pennsylvanian time; part C represents traps formed by one period of folding, as in the Pennsylvanian sands of the Rocky Mountain region that were folded during early Tertiary time; and part D represents very late

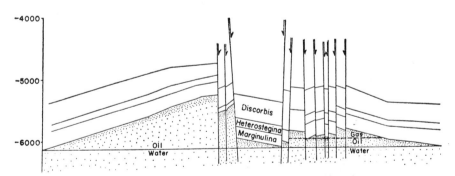

FIGURE 12-24 *Section through the Hastings field, south of Houston, Texas. The common oil-water contact suggests a permeable connection between the various producing members of the Cockfield sand (Claiborne, Eocene). The lower level of the free-gas cap suggests that the faulting occurred after the oil and gas had accumulated in the fold and that the position of the shale roof rock in the central graben kept part of the oil and all of the gas from crossing into the opposite and higher side of the trap. (See Fig. 8-17, p. 367, for a structural map of the field.)* [Redrawn from Halbouty, Houston Geological Society Guidebook (1953).]

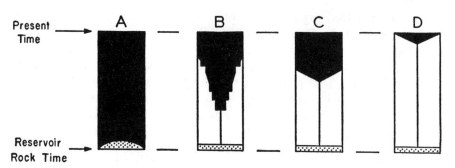

FIGURE 12-25 *A method of graphically showing the time when a trap was formed and consequently the time before which there could be no accumulation. The width of the black shows the percentage of the trap at any time between the deposition of the reservoir rock and the present.*

folding, such as that of the Tertiary reservoirs in California, which occurred during Pleistocene time. There was no trap—and consequently no pool could accumulate—during the period represented by blankness on the chart. A natural question is: Where was the oil then?

It must be concluded, from the evidence bearing on time of accumulation, that little or no migration of oil and gas was necessary to form some pools—in short, that the source was in the trap or very close to it; and that the petroleum of other pools came in from outside source areas that may have been far away. This migration may have occurred early or late, all at once or in many installments during the interval between the formation of the reservoir rock and the present time.

Petroleum Supply

As we have seen (pp. 505–508) petroleum hydrocarbons and petroleum are practically universal in the shales and the carbonates—the nonreservoir rocks. Source rocks should not be lacking except in evaporite and red-bed sequences. Therefore, every anticline and structural trap should be explored thoroughly to find where an accumulation may be located if it is barren at the normally anticipated location. Some anticlines, however, are barren on all sides and in all trap possibilities. The question, then, is, "Why?"

We should differentiate between source material and supply. Although the source material is practically universal, the supply available to any particular trap may be negligible. There are several possible reasons for this: (1) The nonreservoir rocks may contain a high proportion of oil-wet minerals, which would hold the petroleum and not permit it to pass out with the water. (2) The reservoir rocks may contain oil-wet minerals, to which the petroleum would adhere, allowing only the gas to continue on. This may explain why little or no liquid petroleum is found in some of the gas reservoirs. (3) A fault, a facies change, or some similar elongated feature may exist that would serve to deflect the moving petroleum around the trap. (4) The trap may have formed late—after the main circulation had slowed down or after the petroleum content of the aquifer had been exhausted.

CONCLUSIONS

In summary we might list some of the more likely factors that help explain the manner in which oil and gas migrated through reservoir rocks and accumulated into petroleum pools.

1. Chiefly during diagenesis, but also during later continuing compaction of the sediments, water was squeezed out of the nonreservoir sediments and into the reservoir sediments, and along with it a part of the petroleum hydrocarbons—on the order of a few parts per million.

2. The distance that oil and gas move from the source area to the trap area depends on (a) the proximity of a trap or barrier to the source area, and (b) the presence of oil-wetting minerals in the reservoir rock or at the contact between the reservoir rock and the overlying roof rock. These would tend to capture the oil and permit the gas to pass on through.

3. Some traps were formed early, by stratigraphic variations such as facies changes, reefs, channels, bars, and sand and carbonate lenses, all of which were present during the diagenesis of the surrounding rocks and therefore provided trap and reservoir space within the enclosing nonreservoir rocks.

4. The petroleum particles carried along in the water flocculated into larger particles until eventually phase continuity was achieved. Initially, buoyancy permitted the petroleum particles to rise to the upper edge of the reservoir rock; later, phase continuity permitted buoyancy to move the larger patches of petroleum laterally up the dip toward areas of lower potential.

5. Any barrier to the further movement of the petroleum—whether due to structure, stratigraphy, hydrodynamic pressure gradient, or any combination of these—would cause the petroleum to accumulate into a pool.

6. A hydrodynamic pressure gradient, if present, would either increase or decrease the size of the pool, depending on its effect on capillary pressure— that is, whether the flow of water were directed down the dip and added to the buoyancy.

7. The effect of a hydrodynamic pressure gradient in a structural trap containing a pool of oil is to displace the pool down the flank of the structure in the direction of the flow of water. The displacement varies from imperceptible to complete removal of the oil, depending on the direction of flow, the magnitude of the fluid potential gradient, and the relative densities of the oil and water. An irreducible minimum saturation of petroleum would be left behind within the rocks, from which petroleum has been displaced by hydrodynamic conditions.

Selected General Readings

V. C. Illing, "The Migration of Oil," in *The Science of Petroleum*, Oxford University Press, London and New York, Vol. 1 (1938), pp. 209–215. 38 references listed. Summarizes many of the theories concerning the migration of petroleum.

Alfred Chatenever and John C. Calhoun, Jr., "Visual Examinations of Fluid Behavior in Porous Media," Tech. Paper 3310, Petrol. Trans., Trans. Amer. Inst. Min. Met. Engrs., Vol. 195 (1952), pp. 149–156. Describes experiments on migration of oil with photographs of microscopic details.

M. King Hubbert, "Entrapment of Petroleum under Hydrodynamic Conditions," Bull. Amer. Assoc. Petrol. Geol., Vol. 37 (August 1953), pp. 1954–2026. 31 references. A classic article on many phases of oil and gas migration and accumulation. Many laboratory and field examples check the theoretical demonstration.

Gilman A. Hill, William A. Colburn, and Jack W. Knight, "Reducing Oil-Finding Costs by Use of Hydrodynamic Evaluation," in *Economics of Petroleum, Exploration, Development, and Property Evaluation*, Southwestern Legal Foundation, Dallas, Texas, Prentice-Hall, Inc., Englewood Cliffs (1961).

William Carruthers Gussow, "Differential Entrapment of Oil and Gas: A Fundamental Principle," Bull. Amer. Assoc. Petrol. Geol., May 1954, pp. 816–853. 34 references. A discussion of factors controlling selective trapping of oil and gas in pools.

Reference Notes

1. John Galloway, "The Kettleman Hills Oil Fields," Bull. 118, Calif. Div. Mines (April 1943), pp. 491–493.

2. Lawrence A. Goebel, "Cairo Field, Union County, Arkansas," Bull. Amer. Assoc. Petrol. Geol., Vol. 34 (October 1950), pp. 1954–1980.

3. A. W. McCoy, "On the Migration of Petroleum Through Sedimentary Rocks," Bull. Amer. Assoc. Petrol. Geol., Vol. 2 (1918), pp. 168–171.
Frank R. Clark, "Origin and Migration of Oil," in *Problems of Petroleum Geology*, Amer. Assoc. Petrol. Geol., Tulsa, Okla. (1934), pp. 309–335.
Alex W. McCoy and W. Ross Keyte, "Present Interpretations of the Structural Theory for Oil and Gas Migration and Accumulation," *ibid.*, pp. 253–307. 58 references.

4. John L. Rich, "Function of Carrier Beds in Long-Distance Migration of Oil," Bull. Amer. Assoc. Petrol. Geol., Vol. 15 (1931), pp. 911–924.

5. A. I. Levorsen, "Studies in Paleogeology," Bull. Amer. Assoc. Petrol. Geol., Vol. 17 (September 1933), pp. 1107–1132.

6. Malcolm J. Munn, "The Anticlinal and Hydraulic Theories of Oil and Gas Accumulation," Econ. Geol., Vol. 4, No. 6 (September–October 1909), pp. 509–529.

7. John L. Rich, "Moving Underground Water as a Primary Cause of the Migration and Accumulation of Oil and Gas," Econ. Geol., Vol. 16, No. 6 (September–October 1921), pp. 347–371.

MIGRATION AND ACCUMULATION [CHAPTER 12]

8. L. Mrazec, *L'Industrie du pétrole en Romaine: Les Gisements de pétrole,* Imprimeries Indépendantes, Bucharest (1910), 82 pages.

9. Marcel R. Daly, "The Diastrophic Theory," Trans. Amer. Inst. Min. Met. Engrs., Vol. 56 (1917), pp. 733–753. Discussion to p. 781.

10. F. H. King, "Principles and Conditions of the Movements of Ground Water," 19th Ann. Rept. U.S. Geol. Surv., Part II (1899), pp. 59–294.

11. V. E. Monnett, "Possible Origin of Some of the Structures of the Mid-Continent Oil Field," Econ. Geol., Vol. 17, No. 3 (May 1922), p. 200.

12. J. Volney Lewis, "Fissility of Shale and Its Relations to Petroleum," Bull. Geol. Soc. Amer., Vol. 35 (September 1924), pp. 557–590.

13. M. G. Cheney, "Geology of North-Central Texas," Bull. Amer. Assoc. Petrol. Geol., Vol. 24 (January 1940), pp. 65–118.

14. C. R. Dodson and M. B. Standing, "Pressure-Volume-Temperature and Solubility Relations for Natural Gas-Water Mixtures," in *Drilling and Production Practice,* Amer Petrol. Inst. (1944), pp. 173–178. Discussion to p. 179. 14 references.

John J. McKetta, Jr., and Donald L. Katz, "Phase Relations of Hydrocarbon-Water Systems," Trans. Amer. Inst. Min. Met. Engrs., Vol. 170 (1947), pp. 34–41.

Stuart E. Buckley, C. R. Hocott, and M. S. Taggart, Jr., "Distribution of Dissolved Hydrocarbons in Subsurface Waters," in *Habitat of Oil,* Lewis G. Weeks (ed.), Amer. Assoc. Petrol. Geol., Tulsa, Oklahoma (1958), pp. 850–882.

15. "Production of Crude Oil by Solution in High-Pressure Water," World Oil, September 1948, pp. 136–140.

16. Alex W. McCoy, "Notes on Principles of Oil Accumulation," Jour. Geol., Vol. 27 (1919), pp. 252–262.

Alex W. McCoy, "A Brief Outline of Some Oil Accumulation Problems," Bull. Amer. Assoc. Petrol. Geol., Vol. 10 (November 1926), pp. 1015–1035.

Alex W. McCoy and W. Ross Keyte, *op. cit.,* p. 258.

17. C. E. Beecher and I. P. Parkhurst, "Effect of Dissolved Gas upon the Viscosity and Surface Tension of Crude Oils," Petrol. Dev. & Technol., Amer. Inst. Min. Met. Engrs. (1926), pp. 51–69.

18. J. L. Rich, *op. cit.* (note 4), p. 914.

19. Thomas Eugene Weirich, "Shelf Principle of Oil Origin, Migration, and Accumulation," Bull. Amer. Assoc. Petrol. Geol., Vol. 37 (August 1953), pp. 2027–2045.

20. Stuart Murray, *The Geology of Oil, Oil Shale, and Coal,* Mining Publications, London (1926), 104 pages.

21. O. A. Poirier and George A. Thiel, "Deposition of Free Oil by Sediments Settling in Water," Bull. Amer. Assoc. Petrol. Geol., Vol. 25 (December 1941), pp. 2170–2180.

22. Alfred Chatenever and John C. Calhoun, Jr., "Visual Examination of Fluid Behavior in Porous Media—Part I," Tech. Pub. 3310, Petrol. Technol. Trans. Amer. Inst. Min. Met. Engrs., Vol. 195 (1952), pp. 149–156. 10 references.

23. R. H. Johnson, "The Accumulation of Oil and Gas in Sandstone," Science, new series, Vol. 35, No. 899 (March 22, 1912), pp. 458–459.

24. G. A. Thiel, "Gas as an Important Factor in Oil Accumulation," Engr. & Min. Jour., Vol. 109 (April 19, 1920), pp. 888–889.

25. W. H. Emmons, "Experiments on Accumulation of Oil in Sands," Bull. Amer. Assoc. Petrol. Geol., Vol. 5 (January–February 1921), pp. 103–104.

William Harvey Emmons, *Geology of Petroleum,* 2nd ed., McGraw-Hill Book Co., New York (1931), pp. 76–79.

26. Harold V. Dodd, "Some Preliminary Experiments on the Migration of Oil up Low-angle Dips," Econ. Geol., Vol. 17 (June–July 1922), pp. 274–291.

27. R. Van A. Mills, "Natural Gas as a Factor in Oil Migration and Accumulation in the Vicinity of Faults," Bull. Amer. Assoc. Petrol. Geol., Vol. 7 (January–February 1923), pp. 14–24.

28. M. King Hubbert, "Entrapment of Petroleum under Hydrodynamic Conditions," Bull. Amer. Assoc. Petrol. Geol., Vol. 37 (August 1953).

29. Warren J. Mead, "The Geologic Rôle of Dilatancy," Jour. Geol., Vol. 33 (1925), pp. 685–698.

Duncan A. McNaughton, "Dilatancy in Migration and Accumulation of Oil in Metamorphic Rocks," Bull. Amer. Assoc. Petrol. Geol., Vol. 37 (February 1953), pp. 217–231.

30. Thom H. Green and C. W. Ziemer, "Tilted Water Table at Northwest Lake Creek Field, Wyoming," O. & G. Jour., July 13, 1953, p. 178.

31. Philip C. Ingals, O. & G. Jour., July 13, 1953, p. 177.

M. King Hubbert, op. cit. (note 28), p. 2019.

Wyoming Geol. Assn., "Wyoming Oil and Gas Field Symposium" (1957).

32. John Emory Adams, "Oil Pool of Open Reservoir Type," Bull. Amer. Assoc. Petrol. Geol., Vol. 20 (June 1936), pp. 780–796.

33. Carl H. Beal, *Geologic Structure in the Cushing Oil and Gas Field, Oklahoma, and Its Relation to the Oil, Gas, and Water,* Bull. 658, U.S. Geol. Surv. (1917), 64 pages.

34. M. King Hubbert, "The Theory of Ground Water Motion," Jour. Geol., Vol. 48 (November–December 1940), pp. 785–944.

William L. Russell, *Principles of Petroleum Geology,* McGraw-Hill Book Co., New York (1951), pp. 207–209.

Gilman A. Hill, unpublished manuscript, Stanford University, California (March 1951).

M. King Hubbert, op. cit. in note 28, pp. 1954–2020.

S. T. Yuster, "Some Theoretical Considerations of Tilted Water Tables," Tech. Paper 3564, Trans. Amer. Inst. Min. Engrs., Vol. 198 (1953), pp. 149–153.

Roger L. Hoeger, "The Effect of Hydrodynamics on Production in the Miller Creek Oil Field, Crook County, Wyoming," 17th Annual Field Conference Report (1962), Rocky Mtn. Geol. Association, Denver, Colo.

35. A. I. Levorsen, "Time of Oil Accumulation," Bull. Amer. Assoc. Petrol. Geol., Vol. 29 (August 1945), pp. 1189–1194.

36. A. I. Levorsen, op. cit. (note 5), p. 1130, Fig. 11.

37. W. C. Gussow, "Differential Trapping of Hydrocarbons," Alta. Soc. Petrol. Geol. Vol. 1 (June 1953), pp. 4–5; Oil in Canada, May 24, 1954, pp. 19–33.

PART FIVE

Applications

13. *Subsurface Geology*
14. *The Petroleum Province*
15. *The Petroleum Prospect*

INTRODUCTION TO PART FIVE

THE CHIEF OBJECTIVE in acquiring knowledge of the geology of petroleum is to obtain a guide that will aid in the discovery of new oil and gas deposits. The discovery is ultimately made by the drill, but the location of the drill hole is determined by coordinating many diverse elements, some geologic, some economic, and some personal.

This final part of the book is concerned with the manner in which we choose a drill site; how and why we decide to drill "here" instead of "there." To make such a decision we must first gather data from all available sources—for example, other wells, geophysical surveys, pressure and temperature surveys—and then put them into a form that will reveal the underground stratigraphy, structures, and conditions: maps, tables, diagrams, and so on. The early identification of both the regional surface and subsurface geologic features of an area in which a discovery is made can be of great economic value, not only because it will indicate the best method of exploitation of that area, but because it can also lead to the identification of new regions throughout which oil and gas occur in economic quantities.

Quite commonly all the pools in a given area are found to be associated with similar geologic features. In one they may be associated with intrusive salt plugs; in another, with subsurface sand patches, bars, channels, and similar reservoirs; in still another, they may all underlie a regional surface of unconformity. A region of which this is true is called a *petroleum province*.

When study of geologic conditions gives reason to believe that oil and gas lie beneath the surface, the decision to drill (or not to drill) a test well is based on economic considerations. If the decision is favorable, the exact location of the well is, in the final analysis, determined by the personal opinion of the petroleum geologist. As the drill bites into the surface at the selected spot the *petroleum prospect* has begun.

CHAPTER 13

Subsurface Geology

Subsurface maps: structural – isopach – facies – paleogeologic and subcrop – geophysical – geochemical – miscellaneous. Dry holes.

THE TOTAL FOOTAGE of holes drilled each year by the petroleum industry of the world amounts to some hundreds of millions of feet. Of the total footage drilled, approximately one-fifth consists of wildcat wells drilled in the search for new pools, and the other four-fifths consists of wells drilled to develop previously discovered pools and fields. This great footage does not merely represent holes in the ground; because of the information it discloses, it actually represents so many feet of geologic column investigated. Over the years this huge drilled footage forms a vast store of geologic data with which to work; it is the chief basis for our understanding of the combination of underground stratigraphy, structure, and geologic history that is called *subsurface geology*[1] to distinguish it from surface geology.

The purpose of subsurface mapping in the geology of petroleum is to find traps that contain oil and gas pools, and, once a pool is found, to bring to bear the geologic evidence and concepts that aid most in its efficient development and production. The more of the geology of an area a geologist can know, the better job he should be able to do; so every bit of evidence becomes grist for his mill. While the information obtained from wells forms the heart of the data upon which subsurface geology depends, other information comes from geophysical surveys, pressure and temperature surveys, and the production history of producing oil and gas pools. All these diverse kinds of information fit together when correctly interpreted, and their coordination becomes the particular job of the petroleum geologist who is searching for new pools or extensions of known pools.

We are dealing with a great number of variables in subsurface geology, but it is doubtful if we can ever evaluate more than a fraction of them. Each pool may be thought of as a unique situation, resulting from a combination of many geological, chemical, and physical phenomena, most of which cannot be understood until after the pool has been discovered and developed. To one standing at the surface of the ground and attempting to locate a pool miles underground, even with all of the tools that have been developed, the chances of missing the target far outweigh the chances of hitting it. It is as though one were searching for a needle in a haystack with the aid of a small magnet. The magnet helps enormously when one gets near the needle, but to get near it one needs to use all of the clues there are before opening up the stack of hay, for from without all sides look alike. This helps explain the old adage that "oil, like gold, is where you find it."

One thing to bear in mind with regard to subsurface maps is that they are never finished. They may be thought of as progress maps or contemporary maps, only as complete as the data that are available when they are made. Where new wells are being drilled or old holes being re-examined, new information becomes available, and the maps are added to and corrected. The gross, or regional, geologic conditions may be deduced in some areas from only a few well records, and for that reason subsurface maps of all kinds are valuable, even those based on only a few scattered control points. Clues to regional arching, regional facies changes, or regional unconformities may give valuable advance notice of the favorable areas for further exploration. The new information that comes along from year to year from added wells merely fills in the details of the maps, and progressively closer detail is obtained with addition of more and more well data.

Regional subsurface geologic mapping is geologic reconnaissance mapping, and it is approached in the same way as surface geologic reconnaissance. This means that in the early stages of any subsurface work a careful and detailed study should be made of the local area for which the most complete geological data are available, such as an oil field or an area for which a concentration of good well logs is available. In such areas one can learn the habits of the rocks, their stratigraphic relations, the potential reservoir rocks, the position and nature of the unconformities, the time of folding and faulting, the changes in folding with depth, and the facies changes in the section of rocks being mapped. The nature of the petroleum content should also be studied; we want to learn what type of reservoir energy exists; what the direction and rate of the water movement might be; what effect the reservoir fluids have on the electric and radioactive logs, what order of petroleum reserves to expect, in barrels of oil and cubic feet of gas per acre-foot; and what the recovery factors are. Having made such a detailed case-history study, the geologist goes from the known to the unknown with much greater assurance that he will make the best possible application of the many scattered bits of data that make a regional frame of reference useful, even necessary, to the most successful exploration. Careful spot studies thus result in more accurate regional and local predictions.

Most of the basic data of the geology of petroleum and the work of the petroleum geologist are obtained from drilling both wildcat wells and development wells. It is worth while to consider the kinds of information that it is possible to obtain from a single hole or test well. Largely because of the added expense, much of the information listed below often is not generally obtained; and in development wells, especially, no attempt is made to secure all of these data except from a few wells.

INFORMATION	HOW DETERMINED
Formation boundaries, correlations	Well samples; cores; electric logs; radioactive logs; caliper logs; drillers' logs; geophysical surveys; drilling-time logs.
Formation lithology	Well samples; cores; electric logs; caliper logs; drillers' logs; drilling-time logs.
Formation age	Paleontological determinations from well samples and cores.
Elevation of formation boundaries	Subtract depth to formation boundary from surface elevation, either at surface of ground (Gr) or at derrick floor (DF).
Porosity	Laboratory examination of cores and well cuttings; drilling-time logs; electric logs; neutron logs.
Permeability	Laboratory examination of cores and well cuttings; rate of fluid fill-up in drill-stem tests or in open hole, as in cable-tool drilling; electric pilot; production tests.
Fluid content of porous formations	Laboratory analyses of well cores and well samples; drill-stem tests; electric logs; radioactive logs; fill-up in cable-tool holes; drilling-mud analyzers; production tests.
Fluid pressure	Pressure bomb; height of fluid fill-up in open hole, such as cable-tool well tubing and casing; lost circulation; well blowout; drill-stem test.
Formation temperature	Lower self-recording thermometer in hole.

INFORMATION	HOW DETERMINED
Unconformities	Well samples; cores; paleontological hiatus; basal conglomerate; some thin red formations, evidence of weathering; stratigraphic hiatus.
Faults	Cores; well samples; hiatus in section. Electric log correlation with other wells in vicinity.
Fractures	Cores; electric fracture logs (amplitude, sonic).
Velocity of seismic-wave transmission	Geophone placed in hole opposite formations to be measured.
Magnetic susceptibility	Well samples and cores in laboratory.
Dip and strike of formations	Dip-meter readings; oriented cores.

KINDS OF SUBSURFACE MAPS

Subsurface mapping is practiced chiefly in working out the geology of petroleum deposits. Before well data became available, most geologic observations were limited to surface outcrops and mine and quarry openings. The information from the well data, however, makes geologic mapping in three dimensions possible over wide areas in which there are no outcrops and whose underground geology could not be deciphered by the earlier methods. Geologic maps of many kinds are prepared from these data to show various conditions underground, and these maps are interpreted to show the geologic history of the region and to predict the location of petroleum pools yet undiscovered. Active drilling calls for a continuous reinterpretation of the new geologic data uncovered.

Some of the kinds of subsurface geologic maps that are commonly prepared and used are (1) structural maps and sections, (2) isopach maps, (3) facies maps, (4) subcrop and paleogeologic maps, (5) geophysical maps, and (6) geochemical maps. These maps vary widely in scale and amount of detail, depending upon the amount of information available and the purpose for which they are constructed. They are not used competitively; each type contributes an essential facet for a complete understanding of the total geology of an area.[2]

Structural Maps and Sections

Subsurface structure may be mapped on any formation boundary, unconformity, or producing formation that can be identified and correlated by well data. Structure may be shown by contour elevation maps or by cross sections. Since most formations in the oil regions are below sea level, the contoured elevations are generally minus figures, meaning below sea level datum, and subsurface structural maps are sometimes called *sub-sea maps.* Examples may be seen in Figures 6-4, B, 6-10, and 6-36. While it is customary to show the structural relation to sea level, in many cases any arbitrary level datum serves as well. The essential thing is to show the dip of the rocks, faults, folds, etc. in their present attitudes. A *structural cross section* is a vertical section showing the present attitude of the rocks with respect to sea level. Many structural sections have been shown, as in Figures 6-12, 6-16, and 8-12. A *stratigraphic cross section* is one in which some formation boundary or unconformity surface is represented as level and is used as a datum or reference horizon. (See Figs. 7-3 and 7-45.) The basic assumption is made that, at the time the datum or reference horizon was formed, it was essentially a plane surface, possibly level, and that the attitudes of all the rocks below the surface shown in the section are as they were at the time the reference surface was being formed. A stratigraphic structural section is therefore a paleostructural section of the time when the reference plane was actually level. An example may be seen in Figure 12-1. Both structural sections and stratigraphic sections can be made on isometric projection, and are then called *panel,* or *fence,* sections; an example is seen in Figure 13-1.

Subsurface contours should be on an interval consistent with the accuracy and detail available in the well logs. Contour intervals of less than ten feet are generally within the limits of logging errors, and maps having so small a contour interval may reflect anomalies that are not structural, but rather due to errors in measurement or in picking the formation boundaries, either by sampling or by electric logs. Contour intervals of twenty or twenty-five feet are generally about the smallest that are practical, even in regions of low dip.

Most local folds show some faulting, and many traps are defined by faults. The question whether a missing formation or a hiatus in the paleontological section is the result of a fault or of an unconformity comes up frequently in subsurface mapping and is often impossible to answer from a single well. The best evidence for an unconformity is in a knowledge of the regional geological history of the area, and for a fault it is fracturing and veining seen in the cores and samples. The identification of either a fault or an unconformity[3] generally depends on the evidence from several wells. Unconformities especially require several kinds of supporting evidence, sedimentary, stratigraphic, paleontologic, and structural, before their presence in the subsurface can be assured. The common stratigraphic anomalies found in a well that crosses a fault are shown in Figure 13-2. In A and B the well crosses a normal fault, in C and D it crosses a reverse fault, and in E and F the well bore is deflected from the vertical. If the deflection is not known, the interpretation may be just the reverse of the true one.

Many holes were drilled, especially in the early days of rotary drilling,

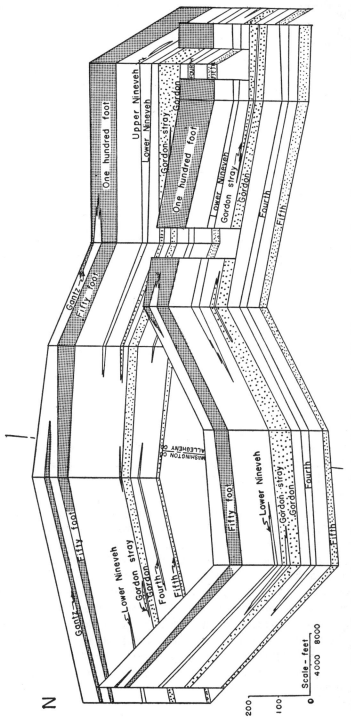

FIGURE 13-1 Panel, or fence, diagram of a productive area in Washington and Allengheny Counties, Pennsylvania. [Redrawn from Ingham, Prod. Monthly, February 1949, Fig. 4, p. 54.]

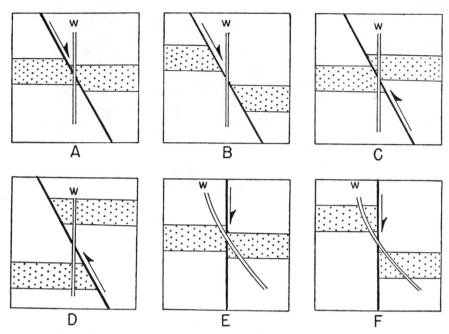

FIGURE 13-2 *The manner in which faulted formations show on the well log. Wells* A *and* B *cut normal-faulted beds;* C *and* D *cut reverse-faulted beds;* E *and* F *are deflected and cut vertical faults. Note the similarity in sequence in logs* C *and* E *and in* D *and* F. *In each diagram* W *shows the position of the well bore.*

that were not vertical but were deflected as much as 40° or 50° from the vertical. The deflections were unknown in some of these wells, but in others they were measured, and all unintentionally deflected holes are known as *crooked holes*. Later many wells were purposely deflected, and surveyed and controlled from top to bottom, so as to give the exact position of the hole at any level. An extreme in directional drilling, but one that shows what may be accomplished, was a well in California that was drilled under the Pacific Ocean. The total footage of the well was 11,440 feet, but the vertical depth was only 4,240 feet. The bottom of the hole was 10,244 feet horizontally from the top, and the maximum angle from the vertical was 75 degrees.[4] One area where wells are purposely deflected is the Huntington Beach field in the Los Angeles Basin, California; there wells are started on the land and are intentionally deflected to penetrate the Tideland pool, which occurs in a dome fold under the ocean. This is illustrated in Figure 13-3, a section through the Tideland pool, and in Figure 13-4, a map of part of the Wilmington field of California. Bore holes are also deflected from marine drilling platforms, as in the Creole field, in the Gulf of Mexico, off the coast of Lousiana. (See Fig. 6-33, p. 265.)

SUBSURFACE GEOLOGY [CHAPTER 13]   595

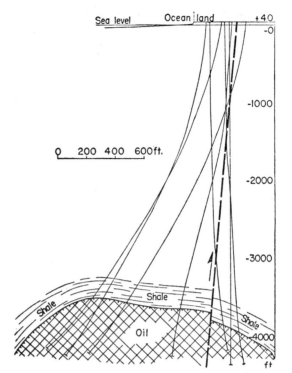

FIGURE 13-3

Section through the Tideland pool in the Huntington Beach field in Orange County, Los Angeles basin, California. It shows a faulted dome fold reached by wells that commence on the land and are deflected under the ocean in order to enter the producing formation (Repetto formation of Pliocene age) on the top of the fold. [Redrawn from Weaver and Wilhelm, Bull. 118, Calif. Div. Mines (1943), p. 330, Fig. 138.]

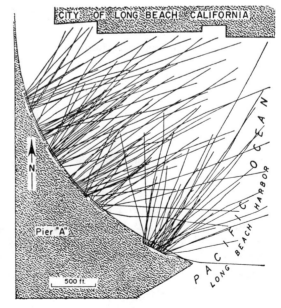

FIGURE 13-4

Map of a part of the Wilmington oil field, California, showing the surface traces of wells drilled under the ocean from locations on a pier. [Redrawn from Stormant, O. & G. Jour., March 1, 1954, p. 42.]

Data from a deflected hole, if the deflection is not known and allowed for, give excessive thicknesses of formations and anomalies in the structural contours, such as to suggest thrust faulting where gravity faults occur. (See Fig. 13-2, E and F.) Holes intended to be drilled vertically generally have an allowed deviation of 2–3°, which is usually not enough to make any appreciable difference in geologic interpretations. An anomalous well log, therefore, arouses a suspicion that the hole may be crooked or deflected.

Isopach Maps

Isopach maps show by means of contours the varying thickness of the rocks intervening between two reference planes, commonly bedding planes or surfaces of unconformity. Isopach maps offer a simple method of showing the distribution of a geologic unit in three dimensions. Thicknesses of individual formations, of reservoirs, of reservoir rocks, of groups of formations, of intervals between unconformities, or of intervals between a surface of unconformity and a normal stratigraphic contact or formation boundary, may be mapped in this manner. The more sharply the reference planes may be defined and the depths to them measured, the more accurate and useful the isopach map will be. Gradational and uncertain contacts are generally not satisfactory for most purposes. Isopach maps may be made to show minor details of areas such as a single pool or field (see Fig. 13-5), or they may be made for regional studies covering thousands of square miles.[5] The number of mappable reference planes generally diminishes as the size of the area to be mapped increases. This is because over wide areas correlations become less certain and fewer formations can be traced. Reference planes for regional isopach mapping should be chosen with these considerations in mind before the map is begun.

Isopach maps are particularly useful in determining the time of faulting and folding. (See Fig. 13-6.) The basic assumption is the same as for a stratigraphic section; the upper reference surface is assumed to have been an approximately level plane when it was formed. Some of the thin limestone formations (from two to ten feet thick) of the Permian and Pennsylvanian, for example, extend for hundreds of miles across Nebraska and Kansas and into Oklahoma in all directions without appreciable variation. Each of them must have been deposited under similar conditions throughout its extent, and this means under a uniform depth of water. At the time they were deposited, therefore, they formed approximately level surfaces. When they are used as the upper reference surface of an isopach map, the isopach contours show the structural attitude of the lower reference plane *at the time the upper reference formation was deposited*. Unconformity surfaces, likewise, may offer good planes of reference. In detail, an unconformity surface may be irregular and far from smooth; over large regions, however, the irregularities are inconsequential, and the overall effect is that of a smooth plane surface. Where the overlying

SUBSURFACE GEOLOGY [CHAPTER 13]

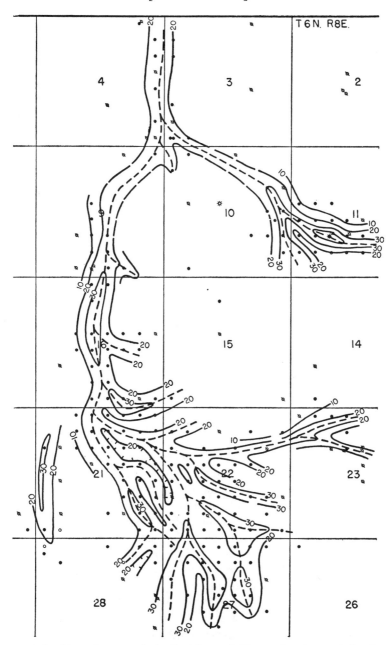

FIGURE 13-5 *Isopach map of the Booch sand (Pennsylvanian) of the Hawkins field, Hughes County, Oklahoma. The contour interval is 10 feet, and the dashed line shows the center of the sand-filled channels. Compare with Figure 3-5, an isopotential map of the same area.* [*By the courtesy of Daniel A. Busch.*]

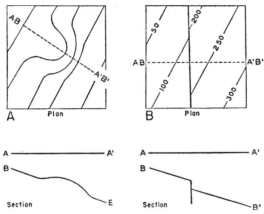

FIGURE 13-6

Sketches showing how thinning on an isopach map may be interpreted as old folding (A) or as old faulting (B). Where isopach maps are interpreted in terms of ancient structure, the upper datum horizon, AA' in both the stratigraphic secitons above, is considered as a horizontal or approximately horizontal reference surface. It may also be thought of as a time plane of reference, and the isopach contours in sketch A show the folding in bed BB' at the time AA' was being formed, and the contours in sketch B show the fault that was present in bed BB' at the time AA' was being formed.

formation is of nearly uniform thickness, as is the Chattanooga shale (Mississippian) and its equivalent formations, which vary between 50 and 150 feet in thickness for several million square miles through central North America, the unconformity surface must be thought of as a plane and essentially level surface when it was formed. When isopach maps are made with this unconformity surface as an upper plane of reference, for example, the reconstruction eliminates all subsequent deformation, and the map may be interpreted as the structure of the lower reference plane when the surface of unconformity was being formed. Where the lower reference plane is an unconformity surface, the isopach map may also be interpreted as showing, at least in part, the topography of the unconformity surface at the time it was overstepped.

Isopach maps are useful in regional studies where the geologic history is desired. The time of trap formation is important because the earlier traps have had a longer time in which to accumulate oil, and also because in different provinces oil and gas accumulation is often related to certain specific times of deformation. Sets of isopach maps of the same region give many clues to favorable areas. The presence of a locally thin interval between reference planes in an isopach map may mean that folding or deformation took place during the time interval represented by the map, and is frequently the first evidence of folding capable of trapping oil and gas. Regional isopach studies of the Mississippian limestone of eastern Kansas, for example, show that the structurally high areas generally coincide with the thin intervals.[6] Isopach maps are especially useful in locating the shelf areas of deposition.[7] Shelf areas are likely to have patchy sands, organic reefs, and distributary channels, all of which form traps in a near-shore environment favorable to the origin of oil and gas.

Isopach maps are also used in the development of a pool, especially in showing the thickness of the pay formation. (See Fig. 13-5.) After the upper reference surface has been penetrated by a well, estimates of the depth to the pay formation—and consequently of its structural position—may be made more accurately than without such maps. Frequently depositional features, such as sand pinch-outs, sand bars, reefs, and lenses, are related to some particular thickness contour pattern shown on the isopach maps. The same isopach contour interval, for example, between the nearest overlying and the nearest underlying reference surface to a sand deposit could be thought of as a measure of a like depositional environment, and as such could be projected into new areas. Isopach maps of sand formations, and also of porous limestone or dolomite formations, are sometimes called "sand maps." The areas where the porosity and permeability pinch out up-dip are possible trap areas and are consequently of special interest. Where the pinch-out is due to regional causes, such as regional truncation, regional shore-line conditions, or regional lithofacies changes, conditions represented by the isopach map may be projected and used to predict the position of the up-dip wedged-out edge of permeability in unknown areas ahead of development.

Isopach maps of formations or groups of formations that thin or converge more or less uniformly in one direction are sometimes called *convergence maps*.[8] The thinning may be due to any one of several causes, such as offlap, onlap,[9] erosion and truncation, bypassing,[10] turbidity currents, and a lesser supply of sediments in the direction of thinning. The various kinds of convergence are graphically shown in the stratigraphic sections of Figure 13-7. The convergence may be gentle, on the order of a few feet per mile, or it may be rapid and measured in hundreds of feet per mile. Section 1 of Figure 13-7 shows thinning due to bypassing or less deposition in the direction of thinning; section 2 shows onlap deposition in the direction of thinning; section 3 shows the result of folding or tilting, erosional truncation, and later overstepping; and section 4 is a combination of sections 2 and 3. Some of the effects of converg-

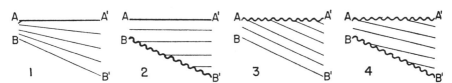

FIGURE 13-7 *Diagrammatic sections showing the manner in which sedimentary units commonly converge or wedge out. The upper surface AA', in each, when considered as an approximate time plane, may also be thought of as a horizontal reference surface. The attitude of the surface BB' may in this manner be considered as it existed during AA' time. This attitude is structural in examples 1 and 3 and either topographic or structural, or both, in examples 2 and 4. Sections such as these are called stratigraphic sections and may be thought of as paleostructural sections—that is, as showing the structure or attitude of surface BB' at AA' time.*

ing strata in shifting the position of the folding with depth and of either increasing or decreasing its intensity have been discussed earlier. (See p. 246.)

Minor changes in formation thickness become significant in an isopach map. Some apparent variations, however, are due to errors in cable-tool drillers' logs. Frequently such errors in the drilling measurements are accumulative, and are first known at the bottom of the hole, immediately above the prospective producing sand. The reason is that a steel-line measurement is taken to the top of the pay formation and the depth correction (the difference in depth between the steel-line measurement and the drillers' logs) may then be distributed arbitrarily through the recorded depths of the overlying formations. Any single anomalous point, especially on a cable-tool log, should be looked at with suspicion. Where two or three wells show the same anomaly, the chances are that the anomaly is actually present and not a result of error. Comparison of a surface measurement of the thickness of a formation with the measurement made in a well may not be of much use, for the errors in surface measurements may be considerable, depending upon the terrain and the character of the exposures. Here, too, a single anomalous point should be viewed with suspicion before interpretations are made that depend upon it. Single anomalous well measurements that do not give the true formation thickness may also be caused by unforeseen local steep dips, by minor drag folding, by local facies changes, or by faults intersected by the well.

Facies Maps

Facies[11] maps are of several kinds,[12] but those most used in the geology of petroleum are lithofacies maps—maps that distinguish the various lithologic types rather than formations.[13] Nearly every formation or group of formations lies within definite stratigraphic boundaries, but within these boundaries one rock type may grade laterally into another, and a lithofacies map is designed to show the nature and the direction of the gradations.

Several devices have been developed to represent changes in rock facies throughout a region on maps. One of the simplest, especially where there are not many data with which to work, is a circle around each well location; the composition percentages of the various rocks in the producing formation are represented by "pie slices." An example of such a map is shown in Figure 13-8. The slices that represent rock types may have either distinctive colors or different black-and-white patterns. The diameter of each circle is proportional to the thickness of the formation or group of formations mapped in the well or outcrop around which it is placed. A glance at the map immediately shows the general changes in thickness as well as the changes in character and relative amounts of the different main rock types. Another type of facies map is seen in Figure 13-9, which shows the distribution of coarse clastics in the Pennsylvanian rocks of southeastern Colorado.

Comparison, with respect to amount and distribution, of formations of clastic origin with those of chemical origin is frequently of use in regional

SUBSURFACE GEOLOGY [CHAPTER 13]   601

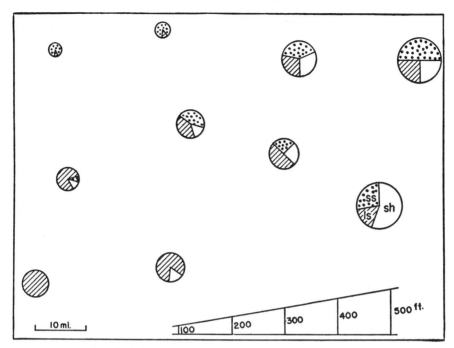

FIGURE 13-8 *Simple lithofacies map showing only the evidence of facies changes in a formation or group of formations. The diameter of a circle represents the thickness of the interval mapped, the pie slices the percentage of each kind of rock encountered.*

studies. The maps that show the results of such comparisons are called *clastic ratio maps*.[14] The clastic ratio at a given point is the ratio of the total thickness of the clastics to that of the nonclastics at that point, and is expressed as

$$\text{clastic ratio} = \frac{\text{conglomerate} + \text{sandstone} + \text{shale}}{\text{limestone} + \text{dolomite} + \text{evaporite}}$$

Variations of the clastic ratio maps are *sand-shale ratio maps* and *clastic-shale ratio maps*. These ratios are expressed as follows:

$$\text{sand-shale ratio} = \frac{\text{conglomerate} + \text{sandstone}}{\text{shale}}$$

$$\text{clastic-shale ratio} = \frac{\text{sandstone} + \text{clastic limestone}}{\text{shale}}$$

A clastic ratio map shows the variations in the clastic ratio from place to place, and other maps show corresponding variations in other ratios. The ratios may be plotted according to the pattern shown in Figure 13-10, or contour lines may be drawn to connect points of equal ratio. An example of a

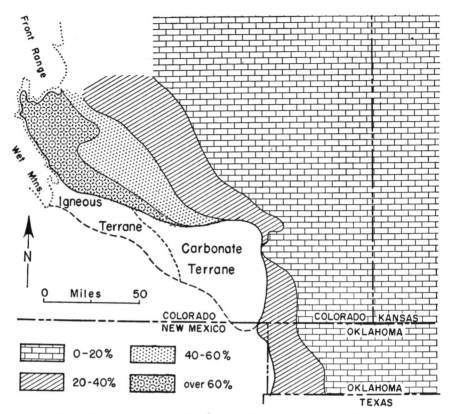

FIGURE 13-9 *Map showing the distribution of coarse clastics in the Pennsylvanian rocks of southeastern Colorado. This facies map shows the coarsening of the sediments toward the west, presumably toward the land mass that furnished the sediments.* [Redrawn from Maher, *Bull. Amer. Assoc. Petrol. Geol., Vol. 37* (1953), p. 2485, Fig. 10.]

clastic ratio map is shown in Figure 13-11 for the Spraberry formation in western Texas.

Paleogeologic and Subcrop Maps

Paleogeology may be defined as the science that treats of the geology as it was during various geologic periods in the past.[15] Paleogeology bears the same relation to paleogeography that geology bears to geography. It requires that one imagine oneself to be alive at some time in the geologic past and then consider the geology of some particular area *as it existed at that time*.

A map that shows the paleogeology of an ancient surface, such as that on a surface of unconformity, is called a *paleogeologic map*. A *subcrop map* is a

SUBSURFACE GEOLOGY [CHAPTER 13] 603

paleogeologic map in which the overlying formation is still present, whereas a paleogeologic map shows the formation boundaries projected, in part, into the areas from which the overlying formation has been eroded. Paleogeologic maps are constructed by plotting the formations that underly and are in contact with the base of the key or datum formation overlying the erosion surface at the unconformity. One paleogeologic map of an area might thus represent the pre-Cretaceous areal geology and another the pre-Mississippian areal geology. In making such a map, the data from well records are combined with geologic sections measured on surface outcrops. When sufficient control has been accumulated, formation boundaries may be drawn as in ordinary geologic mapping. Each surface of unconformity offers an opportunity for making such a map. Unconformities of local extent are suitable reference planes for maps of local areas, but only the widespread unconformities can be used for regional maps. A paleogeologic map may cover the area of a single oil pool, of a sedimentary basin, or of a continent. The general principles of its construction and interpretation are the same in all of these cases.

A local paleogeologic map is seen in Figure 13-12. This shows the areal geology surrounding the Apache pool in Oklahoma[16] as it existed at the time the first Permian sediments were deposited. This is also a subcrop map, for the overlying Permian formations are still present over the area mapped. A fold that had formed in pre-Permian time permitted any oil and gas present to become trapped in the Ordovician sands. Much of the uplifted rocks had been eroded away (though, if a pool had accumulated, it was left intact), leaving a nearly level surface upon which the Permian rocks were laid down. The structure of the pre-Permian rocks was essentially the same then as now, as shown by the lack of deformation in the overlying Permian sediments. (See Fig. 6-5, p. 243, and Fig. 6-19, p. 253.) To judge from the position of the

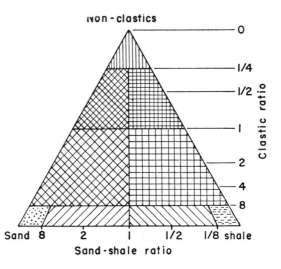

FIGURE 13-10

Triangle diagram illustrating the application of the clastic ratio and the sand-shale ratio to mapping the composite lithologic aspect. The clastic ratios are used as contour intervals, and the patterns are used for areal maps. (See Fig. 13-11 also.) [Redrawn from Sloss, Krumbein, and Dapples, Memoir 39, Geol. Soc. Amer. (1949), p. 101, Fig. 1.]

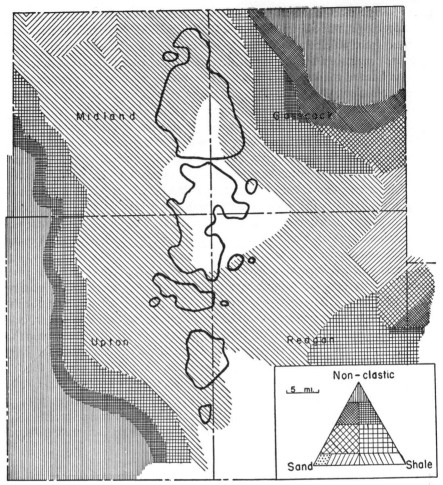

FIGURE 13-11 *Example of a clastic ratio map: the Spraberry formation (Permian) of western Texas. The Spraberry field is outlined in black. (See also Figs. 13-10 and 13-21.)* [Redrawn from Wilkinson, Bull. Amer. Assoc. Petrol. Geol., Vol. 37 (1953), p. 258, Fig. 4.]

wells, the crest of the present dome has shifted slightly to the northwest along the axis of the fold. This slight displacement may be due to slight post-Permian downward tilting to the southeast.

A geologist mapping in northeastern Kansas and southeastern Nebraska, for example, at the time the first Pennsylvanian sediments covered the region, would have made such a map as is shown in Figure 13-13. The formation boundaries shown are all the eroded edges of the truncated formations below the pre-Pennsylvanian surface of unconformity, and therefore do not, as is frequently interpreted, represent shore-line phenomena. The topography was

SUBSURFACE GEOLOGY [CHAPTER 13]   605

nearly flat, except for the low hills along the area of the Nemaha fault, the evidence being in the nearly uniform thickness of the overlying rocks, except for the sudden thinning along the escarpment at the east side of the Nemaha Mountains.[17] The dotted line shows the buried, truncated edges of the Hunton formation (Devonian-Silurian), and the broken lines show the truncated edge of the Viola-Simpson formations (Ordovician), below the pre-Mississippian unconformity, and the truncated edges of the Arbuckle formation (Cambro-Ordovician), below the pre-Simpson unconformity. Because a continuous

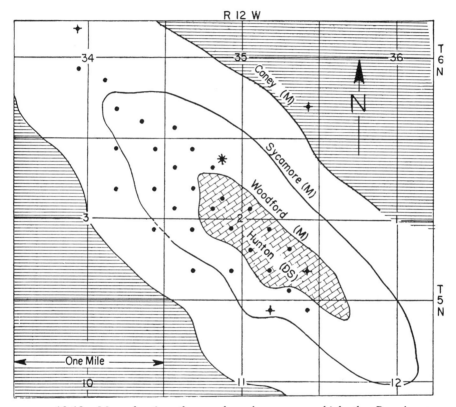

FIGURE 13-12 *Map showing the areal geology upon which the Permian was deposited in the Apache pool and vicinity, Caddo County, Oklahoma. It may be thought of as a paleogeologic map, since it represents the areal geology as it was in pre-Permian time, and it is also a subcrop map, for the Permian formations still overlie the area. If the formation contacts are considered as strike lines of that time, a comparison with the present structure of the underlying Ordovician Bromide sand (Fig. 6-5, p. 243) shows a shift in the position of the high point of the fold toward the northwest. This may be the result of post-Permian regional downward tilting toward the southeast.* [Redrawn from Scott, Bull. Amer. Assoc. Petrol. Geol., Vol. 29 (1945), p. 104, Fig. 4.]

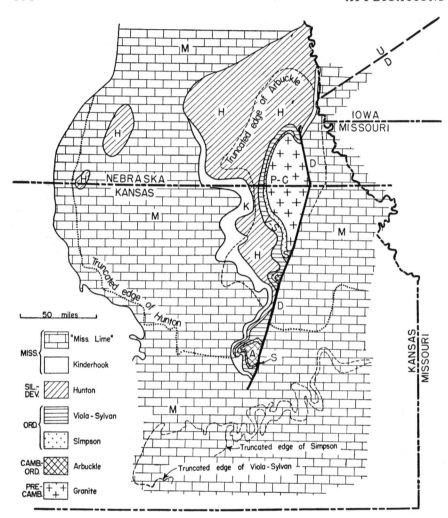

FIGURE 13-13 *Pre-Pennsylvanian paleogeologic map of eastern Kansas and southeastern Nebraska. This is the geologic map of the region as it existed at the beginning of Pennsylvanian time. The formation boundaries shown are all truncated and eroded edges. This is also a subcrop map for the Pennsylvanian sediments that overlie the entire area. The folding that occurred during the post-Mississippian and pre-Pennsylvanian time interval is shown by the outcrop pattern around the Precambrian inlier now known as the Nemeha buried mountains. The volumes of sediments eroded at that time may be calculated from a map such as this.* [Redrawn from Tulsa Geol. Soc., Bull. Amer. Assoc. Petrol. Geol., Vol. 35 (1951), p. 324, Fig. 78.]

succession of uniformly dipping formations are truncated between the Precambrian inlier and the top of the Mississippian rocks, the faulting and folding must have occurred during the time interval between the Pennsylvanian and the Mississippian. The uplift along the fault fold, now known as the "Nemaha buried mountain range," may be traced far to the south, into Oklahoma. The volume of sediments eroded from the uplift at this time may be calculated, and much of it was deposited to the southeast in the Cherokee seas of middle Pennsylvanian age, which were beginning to transgress the region toward the northwest from Oklahoma and Arkansas. The material in many of the shoestrings, offshore sand bars, and sandstone lenses and patches that characterize the Cherokee embayment and the shelf region southeast of the Nemaha range is believed to have been chiefly derived from the uplift during this period of erosion.[18]

Some of the questions that a paleogeologic map such as this raise in the mind of the petroleum geologist are: Where was the pre-Pennsylvanian oil at this time? If a trap is necessary now to contain the pool, what traps existed in pre-Pennsylvanian time? What was the nature of the water circulation and what were the hydrodynamic gradients before the overlying Pennsylvanian sediments covered the region? What was the regional dip? If an early Pennsylvanian geologist were exploring for oil in eastern Kansas, for example, where would he drill? We would like to know because, if no reason for escape of the oil could be found, the sites of early Pennsylvanian pools might be the sites of modern oil pools. A vast amount of oil and gas must have escaped when the Nemaha anticlinal mountain range was uplifted and eroded. Did some of it go along with the debris washed off the mountains and become the source of the Cherokee sand pools to the southeast? Questions such as these are answerable to varying degrees for different regions through the study of facies, isopach, and paleogeologic maps, which thus add immeasurably to our understanding of the geologic history of petroleum provinces and its implications with regard to the occurrence of oil and gas in these regions.

A paleogeologic map of a continental area is shown in Figure 13-14, which represents the areal geology of western North America as it was at the beginning of Cretaceous time. The progressive overlap by the Cretaceous formations across the truncated and eroded edges of all pre-Cretaceous formations, from the Jurassic on the west to the Precambrian on the east, was one of the great transgressive advances of the sea in North American geologic history. It shows that westward tilting occurred throughout a vast area between Jurassic and Cretaceous time. The formation boundaries shown on the map are truncated edges and do not in any sense represent shore lines (a frequent interpretation from paleogeographic maps). The shores may have been close to these contacts, or they may have been hundreds of miles to the east; the answer in most instances will be determined only by additional studies of the facies along the wedged-out edges.

A structural section, AA′, across the southern part of Figure 13-14, from

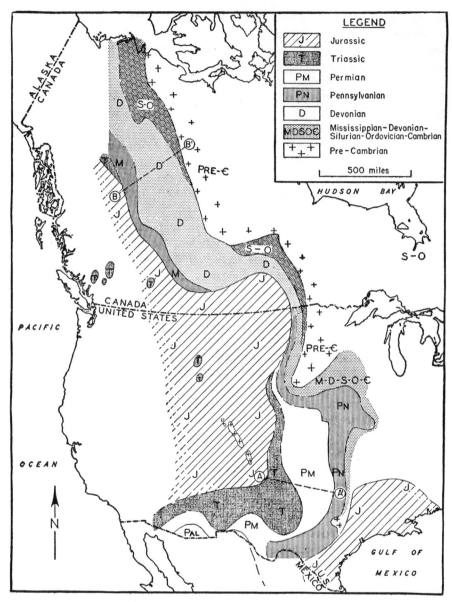

FIGURE 13-14 *Paleogeologic map of western North America showing the areal geology as it was before the deposition of the Cretaceous. It is upon this areal distribution of formations that the Cretaceous was deposited.*

New Mexico to central Texas, is shown in Figure 13-15. It shows the progressive overlap of the Cretaceous onto the tilted and truncated pre-Cretaceous formations. The fluid pressure changes that occurred during the uplift, erosion, and Cretaceous overstep of the region are further illustrated in the sections shown in Figure 13-16. The upper section shows the present relationships of the Cretaceous formations progressively overstepping successively older formations from Jurassic to Pennsylvanian and from west to east across southeastern New Mexico into central Texas (a distance of 700 miles), where the Cretaceous rests on Precambrian rocks in the Llano region of Texas. The truncation of the pre-Cretaceous formations occurred after Jurassic time, for all of the older formations are virtually parallel to one another, and the overstep is uniformly progressive, formation by formation, from Jurassic to Precambrian from west to east across the entire area. The lower section represents the same region restored to the conditions preceding the uplift truncation—that is, prior to the Cretaceous overlap. For example, at that time an oil deposit, located at the black dot, was under a reservoir pressure of about 3,600 psi, corresponding to a burial of 8,000 feet. After the regional uplift at the east and ensuing the erosion, the reservoir pressure in the oil deposit dropped, until, as in the upper section, it amounted to approximately 45 psi, corresponding to a burial of about 100 feet. The decline in pressure that accompanied the uplift and truncation was progressively greater toward the east and created an east-sloping potentiometric surface. The equilibrium of the fluids, together with the pressure and the temperature of every reservoir formation involved, had been changed. A readjustment to accommodate the reservoir fluids to the new conditions might be expected to occur during such a period of regional uplift, truncation, and overlap. The buoyant oil and gas might be expected to expand and move at such times and to become trapped in any traps that are present. Traps that formed later would be barren. The effect of regional tilting is shown diagrammatically in Figure 13-17. The importance of determining the time a

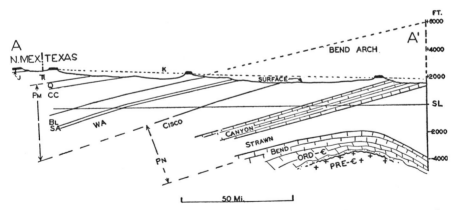

FIGURE 13-15 *Structural section AA' (Fig. 13-14) from New Mexico to the Bend Arch area of Texas, showing the progressive overlap of the Cretaceous (K) on the truncated edges of all earlier formations, beginning with the Jurassic (J) at the west and onto the Pennsylvanian (Pn) at the east. The westward tilting of the pre-Cretaceous formations appears to have occurred after Jurassic time.*

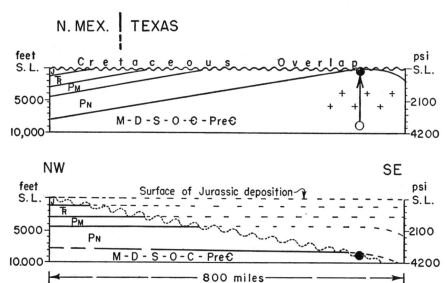

FIGURE 13-16 *Diagrammatic sections AA' across southeastern New Mexico and western Texas to show conditions as they are now (upper section), and as they were before the uplift to the east and the subsequent truncation and overstep of the region by the Cretaceous formations (lower section). The upward movement of the rocks containing the patch of oil, shown in black, from a depth of 8,000 feet in the lower section, to a depth of 100± feet in the upper section, means a reduction in temperature and pressure, with resulting horizontal east–west temperature and fluid pressure gradients, and a readjustment of the fluid content to meet the changing conditions. It is during such a period that migration is probably most active, and traps present then become the loci of oil and gas pools.*

trap was formed with respect to regional disruptions of pressure equilibria becomes evident. Nearly every sedimentary basin contains one or more periods of pressure readjustment resulting from uplift, truncation, and overlap, and it is during these periods that much of the movement of oil and gas may be expected to have occurred as they adjusted their position to the new conditions.

The energy released during a decline in pressure such as accompanied the tilt followed by the Cretaceous transgressive overlap of western North America must be tremendous. The conditions shown in Figure 13-15 for New Mexico and Texas extend far north into Canada, a distance of more than 2,500 miles. (See Fig. 13-18.) The width throughout this distance is about 750 miles. Throughout this area, a total of 1,625,000 square miles, every accumulation of oil and gas present was modified by a reduction in pressure and by a change in the hydrodynamic gradient, and consequently every pool changed its position to adjust itself to the changed conditions. And the conditions mapped may actually have extended far beyond the area shown. Hence

SUBSURFACE GEOLOGY [CHAPTER 13]   611

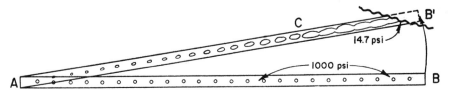

FIGURE 13-17 *The forces involved in a regional tilt of the reservoir rock such as that shown in Figures 13-15 and 13-16. Consider AB as a sealed tube containing sand, oil, gas globules, and water under a fluid pressure of 1,000 psi, equivalent to the hydrostatic pressure from B' to B. If B is raised to B', buoyancy will move the globules of gas and the oil particles from A toward B, provided sufficient phase continuity can be attained to overcome the displacement pressures. However, if the tube is broken, as at B' (equivalent to erosion in nature), the fluid pressure drops to atmospheric pressure, and a hydrodynamic component is added to the buoyancy effect, phase continuity of the gas develops from C to B', and movement up the dip is greatly accelerated.*

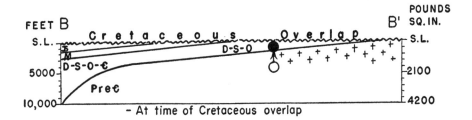

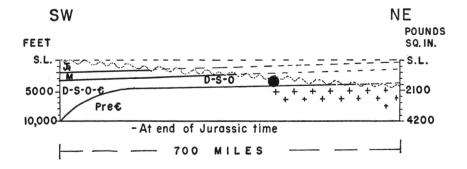

FIGURE 13-18 *Sections across western Canada (BB' of Fig. 13-14) as it was at the end of Jurassic time (lower) and at the time of the Cretaceous overlap (upper). The similarity of the relations here to those in Texas (Fig. 13-16), 2,000 miles to the south, is apparent; it shows the continental extent of the tilting, truncation, and overlap that occurred during the Jurassic-Cretaceous interval.*

every regional unconformity may be looked upon as representing a time of widespread adjustment, migration, and return to equilibrium conditions for every accumulation of oil, large or small, that existed in the pre-unconformity rocks.

Many useful interpretations of the geologic history of an area may be made from a paleogeologic map. Since the pre-unconformity distribution of formations is generally quite unlike that on surface-areal geologic maps, a paleogeologic map is a wholly new areal geologic map, only at a different erosion surface, lower than we see today. It shows the formations that were being eroded on that surface and the areas the sediments came from, and it may aid in locating the positions of shore lines. Through interpretation of the relations of the older to the younger rocks, the structure of the pre-unconformity rocks, as it existed at that time, may be inferred. When several paleogeologic maps are made for a single area, each representing one of several different layers of geology, separated by unconformities, we are able to visualize the chief episodes in the changing geologic history of the area. Such a series of maps, together with isopach and facies maps, enables one to break the geologic history of the area down into its component parts, and to study each geologic episode quantitatively. And, in doing this, it not only provides a method of obtaining a better understanding of the geologic history of a region, but helps us to judge its petroleum prospects and furnishes hints as to where to look for pools.

Geophysical Maps

Geophysical mapping is essentially a kind of subsurface mapping in advance of drilling, although it may require that enough drilling be done beforehand to calibrate the instruments. The mapping consists in measuring various physical properties of the rocks, such as the time of wave transmission for reflection and refraction seismic mapping, magnetic susceptibility for magnetic mapping, rock density for gravity mapping, and radioactivity for radioactivity mapping. The measurements are then translated into geologic data, such as structure, stratigraphy, depth, and position. These data may be combined with those obtained by other kinds of subsurface mapping and become a part of the total geological picture. A local variation or irregularity in the normal pattern is called a *geophysical anomaly*—or, more specifically, a *seismic anomaly,* a *magnetic anomaly,* or a *gravity anomaly.* After all pertinent corrections have been applied to the measurement of any property, the residual local anomaly is considered to be due to some geologic phenomenon, the correct interpretation of which depends on the regional geology, the local geology, and the experience and ability of the geologist and the geophysicist.

It is not the purpose here to describe the various geophysical methods and equipment. Since 1915, when they were first applied in the geology of petroleum, a number of articles and books[19] have been written about them, and

SUBSURFACE GEOLOGY [CHAPTER 13]   613

the current developments are available in *Geophysics,* the monthly journal of the geophysical profession.[20] A petroleum geologist should understand geophysical surveying as well as he understands surface geologic mapping, the use of the Brunton compass, and aerial photographs. Many geophysical electrical and radioactive phenomena are also utilized in obtaining the basic well-log data used in subsurface mapping. Geophysical surveying has been instrumental, either partly or wholly, in mapping the structures that have trapped a great many oil pools. The case histories of a few of these have been told.[21] Two maps showing how reflection seismic surveys permit the structure to be mapped in advance of drilling are presented in Figures 13-19 and 13-20. (See also Fig. 6-4, p. 242.) In each of these cases discoveries of oil pools resulted from drilling on the structure found by the seismic mapping.

The practical value of geophysical methods in the geology of petroleum consists in their ability to measure the physical properties of rocks that are related to potential traps in reservoir rocks. A trap in a reservoir rock is the only kind of trap that counts. Seismic mapping is the most precise of the geophysical

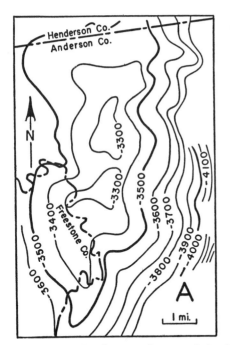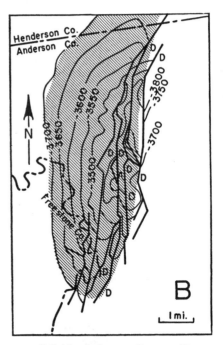

FIGURE 13-19   *Structural maps of the Cayuga oil field, Anderson County, Texas. A, The reflection seismic map on the Austin chalk made in January 1934. The discovery well was completed on March 2, 1934. B, The structure (contour interval 50 feet) on top of the producing Woodbine sand (Upper Cretaceous) after the field (stippled) was developed. [Map B is by Wayne V. Jones. Redrawn from Peacock, in* Geophysical Case Histories, Soc. Explor. Geophys., *Vol. 1 (1948), p. 126, Fig. 4, and p. 131, Fig. 7.]*

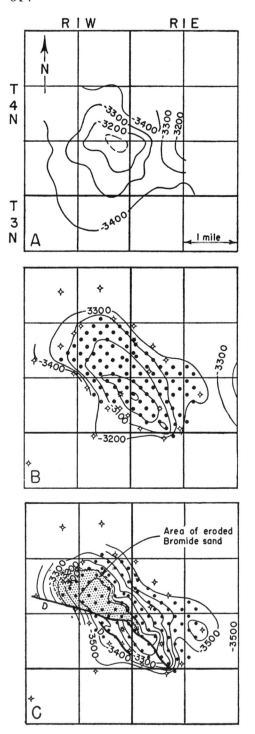

FIGURE 13-20

Maps of the Pauls Valley field, Garvin County, Oklahoma. A is a reconnaissance reflection seismic structural map, made in August 1930; B shows the structure of the base of the Pennsylvanian of the same area after the field was drilled up following the discovery of oil in April 1942; C shows the structure of the producing Bromide sand (Ordovician) with the top restored where it was eroded (dotted) before the deposition of the Pennsylvanian. Note that the Ordovician faulting does not carry above the basal Pennsylvanian unconformity. Eroded folds such as this are sometimes called "baldheaded" structures. [Redrawn from Cram, in Geophysical Case Histories, Soc. Explor. Geophy., Vol. 1 (1948), Figs. 1, 6, and 7.]

methods and applies, moreover, to the widest variety of geological conditions; it leaves us free to choose what formation boundaries are to be surveyed, so that we can map the surface of the potential reservoir rock or other formation boundaries close to it. The dependability of a seismic survey, however, is determined largely by an accurate knowledge of the velocity of the reflecting or refracting waves through the intervening rocks; unforeseen variations in these rocks, such as facies changes, faults, and unconsolidated surface formations, may cause variable velocities and thus introduce errors into the structural interpretation. Special geologic conditions relating to traps in the reservoir rock may be mapped, and useful reconnaissance surveys may be made, by gravity, magnetic and radioactive methods; proper interpretation of the data thus obtained, however, requires care and experience, largely because of the inability of such methods to determine the precise position of the anomaly being mapped.

Geochemical Maps

Geochemical maps are used for mapping various kinds of chemical analyses of rocks and their fluid contents.[22] Such maps may show the surface distribution of the hydrocarbons, of the hydrocarbon waxes, or of the bacteria that utilize hydrocarbons. Where one or more of these is found at the surface in larger amounts than normal, the inference is that there is a leaking oil or gas pool below. The upward migration, if any, of the hydrocarbons is complex and not understood; it probably takes place by some process of diffusion. Hydrocarbon concentration maps, sometimes called *soil analysis maps,* if they overlie oil pools, are, in effect, maps of micro-seepages. In some cases the hydrocarbons, or their residues, are thought to be concentrated directly over the pool; in other cases the center of the pool is overlain by a barren area, and there is a halo or doughnut-shaped concentration of hydrocarbons at the surface above the edges of the pool. Most of the difficulties encountered in making geochemical maps at the surface are due to the extremely low hydrocarbon content of the normal rocks, which is generally measured in parts per million or parts per billion. Errors in the measurements on which the maps are based are likely to be greater than the anomalous variations in hydrocarbon content. Moreover, some or maybe all of the irregularities in hydrocarbon content may be due to purely local surface causes: photosynthesis, differences in the original hydrocarbon content of the formations that crop out, differential erosion processes, or different weathering effects.

Still other geochemical maps and sections are made from data supplied by cores and drill cuttings taken from wells.[23] Ethane, propane, butane, pentane, and higher hydrocarbon fractions are measured. Some oil pools show significant increases in hydrocarbon content in the shales immediately overlying the reservoir rock, and discoveries of oil pools have resulted from continuing drilling after encountering shales with high hydrocarbon values.

Miscellaneous Subsurface Maps

Other kinds of subsurface maps are prepared to illustrate and help visualize specific phenomena. An *isopotential map* shows by contours the initial or calculated daily rate of production of wells in a pool. An example is given in Figure 13-21. (See also Fig. 13-11, p. 604.) *Isoconcentration maps* show the concentration of salts in oil-field waters by contours. (See Fig. 5-14, p. 162.) *Isobar maps* show by contours the reservoir pressure of a pool or field, and these are especially useful in showing the areas of decline in pressure. Sets of isobar maps at regular intervals show the progressive changes in pressure. *Water encroachment maps* show the position of wells from which water is produced along with the oil, and these also are frequently made at regular intervals to show the progressive changes in the advancing water front. *Peg* and *relief models* are sometimes made of pools, especially where it is desired to show subsurface conditions in three dimensions for laymen, such as juries and stockholders. They may be made with pegs having subsurface formations drawn on them to scale, or they may be made by clear plastic or glass layers on top of one another, each layer contoured to represent a different level. An *isochore* map shows by contours the thickness of the pay section of a pool between the oil-water contact and the roof rock; it is used for making calculations of reservoir volumes.

Computers

As might be expected, the use of computers has entered petroleum exploration over a wide front, in both geophysical and geologic applications. It is a rapidly growing field, and a few references to the applications to geologic problems are listed at the end of the chapter.[24] There is a vast amount of detailed data, both past and current, in any actively explored region, and one of the principal uses is in the fast retrieval of these data. A specific kind of data can often be plotted directly on maps, saving the geologist much time in assembling and recording the information. Anomalies that do not show up in ordinary contouring may show up in these maps and give clues to areas favorable for exploration. It will probably be some time before the ultimate application of computers to exploration is achieved and the student is encouraged to keep up on developments in the geologic and industrial publications.

Offshore Exploration. During recent years the offshore continental-shelf areas of the world have become the sites of many major petroleum pools. Wells can now be completed in depths of water up to 500 feet, and the technical problems of drilling and producing petroleum under water have been largely solved. The geology under the water, however, is merely an extension of the geology on the land, or is similar to it; the water is just a different kind of superficial cover. As on land, the reservoirs may be associated with salt domes,

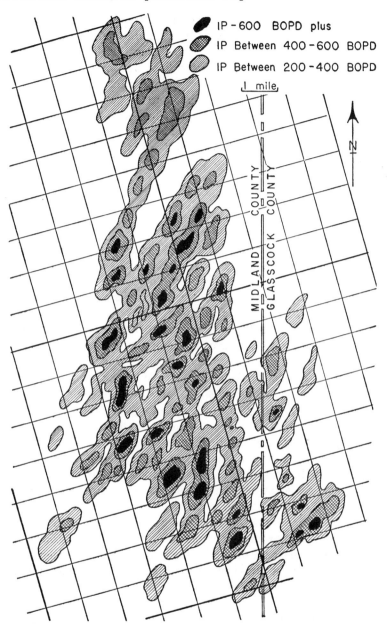

FIGURE 13-21 *Isopotential map of the Tex-Harvey pool in the Spraberry field of western Texas. (See also Fig. 13-11.) The linear pattern is interpreted as a fracture system causing an increase in permeability and consequently in productivity.* [*Redrawn from Wilkinson, Bull. Amer. Assoc. Petrol. Geol., Vol. 37 (1953), p. 264, Fig. 10.*]

anticlines, or faults; they may be in stratigraphic traps or in combination traps. Geophysical surveys may be made as they are on land; in fact they are the chief means by which anomalies are located. Occasionally bathymetric charts suggest topographic features on the ocean floor that in turn suggest anticlinal or other structural features in the underlying sediments. Oil and gas seepages at the surface of the ocean may call attention to the petroleum potential of the area. Costs of offshore production are generally several times higher than production costs on land, a factor that must be considered in the appraisal of any prospect. The subsurface maps that are used are essentially the same as those used on land.

DRY HOLES

Test wells that fail to find commercial oil or gas are called *dry holes*. Of the wildcat wells that are annually drilled in the United States, on the average eight are abandoned as dry holes for every one that is completed. The ratio differs in different provinces, and is generally highest in those where more pools are trapped by stratigraphic variations than by structural anomalies; in such provinces it may reach 20:1. Whether a well is abandoned as a dry hole is in part a function of the economics of the area; a well that could produce only five or ten barrels a day, or that produced excessive volumes of water along with the oil, might be considered noncommercial and be abandoned where the drilling was deep and the lifting costs high, or where a large dry-hole contribution was involved.* On the other hand, a shallow well producing only a fraction of a barrel of high-quality oil per day might be commercial.

A dry hole is not necessarily to be thought of as a total failure. Each hole is capable of supplying a large amount of new data, and if these data are intelligently studied, they may sooner or later become the basis for a new wildcat location. Countless wildcat wells have been located on prospects where one or more structurally high or stratigraphically interesting dry holes had already been drilled. The hope is that the next well will be still higher in the structure or better located—and productive! Some wildcat wells are put down in locations chosen chiefly because they were close to holes that, though classed as dry holes, had oil or gas showings or small and noncommercial production; others are located up the dip from dry holes that encountered only water, in the hope of finding oil trapped against a pinched-out sand. Still others are located close to structurally low dry holes that show evidence of having crossed a buried fault. This kind of geologic evidence may be expensive to obtain, but this very expense is all the more reason why every possible bit of information

---

* An operator, before drilling a wildcat well, may obtain a contract with the owner of the neighboring lease whereby the neighbor pays a stipulated amount of money, a *dry hole contribution*, if the prospective test well is dry, the reason being that the test well helps determine the prospects of the neighbor's lease also.

should be squeezed out of every hole. In the absence of some direct method of locating petroleum pools, the information that may be extracted from dry holes is very often the best guide to discovery.

Holes may be dry for any one of many causes, some avoidable and some unavoidable. In extracting every possible bit of information from the well records, it is always worth while to try to find an explanation when a test well turns out to be dry. This effort may lead to a better location for the next well, or it may result in avoiding similar mistakes in judgment in another area. Some of the common causes of dry holes are listed below. There are others—in fact, one might regard every dry hole as unique—but these are the causes that seem to recur most frequently.

1. *There is no trap in the reservoir rock at the well location.* This is one of the commonest causes of dry holes. It may be due, in turn, to several causes, some of which can be determined only after the hole has been drilled. (a) Miscorrelation of stratigraphy may occur either in the surface mapping or in the subsurface mapping. It may result from poor or insufficient data, from carelessness, or from a lack of understanding of the geologic problems involved. These errors may cause structural highs, or up-dip reservoir-rock pinch-outs, to be mapped where they do not occur. (b) Careless mistakes in surveying the elevation of the surface of the derrick floor, or in subtracting the formation depth from the elevation, may cause a high structure to appear where none actually occurs. Such "structures" disappear when the test well is drilled. (c) Unforeseen facies changes in the reservoir rock may cause either the presence or the absence of a prospective trap. Facies changes also mean changes in velocities of seismic waves, and the resulting seismic anomalies may be translated into structural anomalies. Where unpredicted facies changes occur between the surface and the reservoir rock, therefore, structural anomalies may be mapped without being actually present. Where drilling has been close and the subsurface geology is well understood, facies changes may be anticipated, but in new territory they cannot be known until after many test wells have been drilled. (d) Topographic relief associated with buried unconformities may be misinterpreted as buried faulting or buried structure. Seismic mapping may be on a surface of unconformity and have no direct structural significance, as where either high-velocity rocks or low-velocity rocks project up into overlying rocks of opposite velocity characteristics. It is only after wells have been drilled and subsurface mapping experience has been gained that such misinterpretations may be distinguished.

2. *There is no reservoir rock.* Even when favorable structures have been recognized, and extend to the predicted position of the reservoir rock, it may be found that the reservoir rock is not there. This may be due to such reasons as the following: (a) The reservoir rock may have shaled out or lost its permeability, through a local facies change or a local cementation at the position of the test well. Or, where the permeability is connected with fractures to caverous openings, the test well may not happen to hit any of these fractures or

openings, and thereby miss discovering a pool. (b) The reservoir rock has been faulted out, and the test well may have passed through the normal position of the reservoir rock in a faulted-out section. (See Figs. 6-34, 13-2, B.) (c) The reservoir rock may have been eroded off at the position of the test well; the evidence for this would be a hiatus in the stratigraphic column and the presence of an unconformity. Where the producing formation is eroded from the top of a dome, but occurs as a ring round the flanks, the structure is said to be "bald-headed" or "scalped." (See Fig. 13-20, C, p. 614.) (d) The reservoir rock has not been reached; it may be the "farmer's sand," below the bottom of the hole. This name is given to a sand that the owner of the land feels was present but not reached. Sometimes he is right! Through failure to drill deep enough, because of lack of money, inadequate equipment, or mis-correlation of stratigraphy, a test well may be abandoned above the position of the reservoir rock. (e) If exploring a trap associated with an up-dip pinch-out of permeability, the test well may be beyond the pinch-out and thereby miss the reservoir.

3. *The trap has shifted its position with depth.* A number of causes of lateral shifting or change in character of a structural trap with depth have been considered (pp. 245–259). These may be summarized as regional convergence of the intervening stratigraphic column, recurrent folding or faulting so that structural deformation is greater with depth, squeezing out of evaporites and other incompetent rocks, parallel folding, buried hills, unexpected faulting at depth, shallow weathering phenomena, including slumping and collapse, pre-unconformity folding or faulting, and a displacement of the oil in the direction of water flow in the reservoir rock.

4. *The test well is deflected.* While modern drilling may be accurately directed to reach the reservoir rock at a specific position, there has been much drilling in which there is a deviation from vertical whose amount or direction is unknown. (See p. 594.) The point where the drill penetrates the reservoir may then be at so great a distance, horizontally, from the well head that it will be outside the projected position of the trap and miss the pool. Traps formed by faults, and those around the flanks of salt domes, require especially accurate aiming of the drill; pools can easily be missed by even a slight deviation from the vertical, and the reason for missing them may not be known if directional surveys are not made.

5. *There is no oil or gas in the trap.* The expected trap may exist yet neither oil nor gas be present, or there may be only minor showings of one or both. In fact, most potential reservoir rocks in a structure are usually found to be barren; texturally favorable sands that contain only water are usually penetrated by the drill between the surface and the productive reservoir rock. Why should not every permeable formation, on a fold that extends vertically from the surface to the deepest reservoir formation, be found productive, especially a formation that is productive at other localities? Several causes may account for these permeable yet barren formations. a. No supply of oil or gas was available. Although it is generally believed that nearly all nonreservoir rocks

contain petroleum hydrocarbons, or should be considered as source rocks, the supply of petroleum that has reached a specific trap may vary considerably. (See pp. 541–544.) Probably of greater importance, however, is the possibility that the highest part of the structure is not the location of the trap; each anticline penetrated by a dry hole at its crest should be investigated further for accumulations down the flank. Bald-headed structures especially should not be summarily abandoned because there is "no source rock." b. *The local trap formed too late.* Where the conditions that favor the movement of gas and oil through the reservoir rock existed before the local trap was formed, the gas or oil may have moved past the position of the trap before the trap came into existence. (See pp. 573–579.) Or the trap may have been eroded, and the pool may have escaped, as in Figure 13-22. The time at which the local trap was formed, relative to the time of regional movement of petroleum, thus becomes of importance in exploration. (c) *The oil may have been flushed out of the trap.* Where hydrodynamic conditions now exist or have existed in the geologic past, the gas and oil may have been displaced down a side of the trap or even completely flushed out. The trap, in short, is ineffective for the particular fluid conditions that exist. (See pp. 561–562.) Oil showings and oil-stained reservoir rock throughout a trap suggest that the oil has once been there but has since moved out.

6. *There has been a failure to recognize the pool during drilling.* Many pools capable of producing large quantities of gas or oil have been drilled through without being recognized. Some of these pools have been found by subsequent drilling, but there are undoubtedly many others that are still undiscovered. The reason for such mishaps is that, as drilling becomes deeper and deeper, the drilling-mud pressures increase, the best drilling muds become complex chemical mixtures, the speed of drilling increases, the equipment becomes heavier, and larger power units are employed. Some oil-bearing formations are therefore

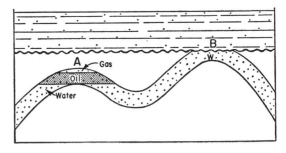

FIGURE 13-22

Diagrammatic section showing how a surface of unconformity may not only conceal an oil pool, as at A, but also explain a barren trap, as at B. Originally there were probably two pools, one at A and the other at B, but the pool at B was breached and exposed to the surface where it was eroded, whereas pool A was preserved. The section also suggests that the accumulation was pre-unconformity in age, and that no additional oil was moving after the post-unconformity shales were deposited. Contrast with the Oklahoma City field (Fig. 14-6), where a large accumulation is found in a breached trap, suggesting a post-unconformity age for the accumulation.

sealed off, either by the pressure of the drilling mud forming an impermeable mud-cake on the walls of the hole or by the expansion of clay minerals in the oil- or gas-bearing formation as the fresh water of the drilling fluid is forced into the pores. The small volume of gas and oil contained in the pores of the rocks exposed in a five-inch hole, at a depth of perhaps two miles or more, may well be difficult to detect when it has been mixed with the drilling mud and raised to the surface of the ground. Modern "mud trucks" with much electronic and analytical equipment have done much to eliminate the uncertainties of evaluating small showings of oil and gas, but there still remains a judgment factor, especially in borderline cases. The pay formation may be missed by the electric log because of low resistance, due to abundant highly saline interstitial formation water that may obscure the high resistance due to gas and oil. Many pools have been discovered by reopening old holes, or drilling beside abandoned test wells, when the original operator did not examine the electric log with sufficient care to identify a pay formation. Or the walls of the hole may be caving, and in that case samples are useless for determining the presence of petroleum. Or the oil may be so high in API gravity and so light in color that it evaporates quickly and is imperceptible when the samples are examined some time after they come out of the hole. Not to be overlooked as a cause of missing the pay formation is inexperience or carelessness on the part of the person charged with watching the well cuttings and cores, or to his being absent from the well at a critical time. Woe unto him!

Every wildcat hole with even a slight showing of gas or oil, recognized by any of the several devices and methods used to indicate the presence of petroleum, presents a question for judgment. Is it worth while to spend money —perhaps a lot of money—in further testing of the productive possibilities of the formation from which the showings have come? Failure to test may be one reason for failure to discover a pool. Practice varies in this regard: some operators make drill-stem tests, or even set a production string of casing, on every oil or gas showing; others spend the extra money only when they think the showing is strong enough to assure a commercial producer. The many pools that have been found as a result of testing what might be called insignificant showings of oil or gas lead to the conclusion that in most cases the extra money spent for testing is well spent.

CONCLUSION

Subsurface geology is a synthesis of all the various geologic and engineering data from producing wells and dry holes and of the results of geophysical and geochemical surveys. These data are placed on maps and sections of many kinds, the sole purpose of which is to help visualize and understand the geologic conditions underground. Locations for wildcat wells, pool wells, and extension wells are based on this understanding.

Probably the most useful information comes from well logs. The accuracy and the detail of the information obtained from wells increase every year, as new techniques and methods are developed; and this is in spite of the continuing increase in depth of drilling. In the light of the newer and more accurate data, maps and sections are being continually revised, many of the older concepts are being changed, and predictions maintain their ratio of success in spite of the deeper drilling and the more complex geology of the older formations.

It cannot be emphasized too strongly that the different kinds of maps and surveys are not competitive—they are tools to be used cooperatively. Each makes its individual contribution to an understanding of the whole, and it is this understanding that results in the finding of unexplored traps, wildcat and extension well locations, and new pools.

Selected General Readings

L. W. LeRoy, "Graphic Representations," in *Subsurface Geologic Methods,* 2nd ed., Colo. Sch. Mines (1951), Chap. 10, pp. 856–893. Shows the common methods of representing subsurface data.

Julian W. Low, "Subsurface Maps and Illustrations," *ibid.,* Chap. 11, pp. 894–968. Most subsurface maps are described, and many illustrations give examples of their preparation and application to subsurface problems.

W. C. Krumbein and L. L. Sloss, *Stratigraphy and Sedimentation,* 2nd ed., W. H. Freeman and Company, San Francisco (1963).

John C. Maher, "Permian and Pennsylvanian Rocks of Southeastern Colorado," Bull. Amer. Assoc. Petrol. Geol., Vol. 37 (May 1953), pp. 913–939. This article is an excellent example of the kind of maps that may be prepared from well data to form the basis of an understanding of a region's geologic history.

A. L. Morgan, "Structural Analysis of Delta Farms Field, LaFourche Parish, Louisiana," Bull. Amer. Assoc. Petrol. Geol., Vol. 37 (December 1953), pp. 2649–2676. A good example of the use of structural and isopach maps to show the growth and geologic history of a complex dome fold.

Society of Exploration Geophysicists, "Geophysical Case Histories," Vol. I (1948) and Vol. II (1956). A total of 113 papers describing case histories of geophysical surveys, chiefly in North America.

Margaret S. Bishop, "Subsurface Mapping," John Wiley & Sons (1960), 198 pages. Discusses the ways in which subsurface data may be represented on maps and sections.

James M. Forgotson, Jr., "Review and Classification of Quantitative Mapping Techniques," Bull. Amer. Assoc. Petrol. Geol., Vol. 44 (1960), pp. 83–100. Describes preparation of different kinds of facies maps.

United States Geological Survey, Oil and Gas Investigations, Preliminary Maps and Preliminary Charts. This series of maps and charts contains many excellent examples of regional subsurface studies, ways of presenting subsurface data, and analyses of geologic history.

Reference Notes

1. Ira H. Cram, "Definitions of Geology—Subsurface Geology," Bull. Amer. Assoc. Petrol. Geol., Vol. 29 (April 1945), p. 470.

Daniel A. Busch, "Subsurface Techniques," in Parker D. Trask (ed.), *Applied Sedimentation,* John Wiley & Sons, New York, and Chapman & Hall, London (1950), pp. 559–578. 30 references.

L. W. LeRoy (ed.), "Subsurface Geologic Methods—A Symposium," Quart. Colo. Sch. Mines, Vol. 44 (July 1949), 826 pages; 2nd ed. (1951), 1,156 pages. A comprehensive discussion of subsurface methods.

2. Marshall Kay, "Paleogeographic and Palinspastic Maps," Bull. Amer. Assoc. Petrol. Geol., Vol. 29 (April 1945), pp. 426–450. Describes many types of maps used in understanding the geologic history of a region.

R. Clare Coffin, "Recent Trends in Geological-Geophysical Exploration and Methods of Improving Use of Geophysical Data," Bull. Amer. Assoc. Petrol. Geol., Vol. 30 (December 1946), pp. 2013–2033.

Geophysical Case Histories, Vol. I. (1948), L. L. Nettleton (ed.), and Vol. II (1956), Paul L. Lyons (ed.). Soc. Explor. Geophys., Shell Bldg., Tulsa, Okla. Numerous maps of surveys that resulted in petroleum discoveries.

W. C. Krumbein and L. L. Sloss, *Stratigraphy and Sedimentation,* 2nd ed., W. H. Freeman and Company, San Francisco (1963), 660 pages. Extensive bibliographies. This contains the most comprehensive discussion of subsurface mapping, especially Chap. 10 (Principles of Correlation), Chap. 13 (Stratigraphic Maps), and Chap. 14 (Paleogeography).

3. W. C. Krumbein, "Criteria for Subsurface Recognition of Unconformities," Bull. Amer. Assoc. Petrol. Geol., Vol. 26 (1942), pp. 36–62.

4. Gordon Jackson, "Directional Drilling Today," Jour. Petrol. Technol., September 1953, Sec. 1, pp. 27–31.

5. Wallace Lee, *The Stratigraphy and Structural Development of the Forest City Basin in Kansas,* Bull. 51, State Geol. Surv., University of Kansas (December 1943), 142 pages. Contains many regional isopach maps.

6. Wallace Lee, *Relation of Thickness of Mississippian Limestones in Central and Eastern Kansas to Oil Deposits,* Bull. 26, State Geol. Surv., Lawrence, Kansas (June 15, 1939), 42 pages. Includes discussion of isopach maps.

7. Thomas Eugene Weirich, "Shelf Principle of Oil Origin, Migration, and Accumulation," Bull. Amer. Assoc. Petrol. Geol., Vol. 37 (August 1953), pp. 2027–2045.

8. A. I. Levorsen, "Convergence Studies in the Mid-Continent Region," Bull. Amer. Assoc. Petrol. Geol., Vol. 11 (July 1927), pp. 657–682.

Robert R. Wheeler and Robert M. Swesnik, "Stratigraphic Convergence Problems," World Oil, April 1950, pp. 57–62.

9. Frederic M. Swain, "Onlay, Offlap, Overstep, and Overlap," Bull. Amer. Assoc. Petrol. Geol., Vol. 33 (April 1949), pp. 634–636. Contains 15 references to earlier articles on this subject.

10. J. E. Eaton, "The By-Passing and Discontinuous Deposition of Sedimentary Materials," Bull. Amer. Assoc. Petrol. Geol., Vol. 13 (July 1929), pp. 713–762.

11. Raymond C. Moore, "Meaning of Facies," Memoir 39, Geol. Soc. Amer. (June 1949), pp. 1–34. 45 references cited. The entire memoir of 171 pages consists of papers and discussion on the problems of facies. Many examples are illustrated.

12. W. C. Krumbein, "Recent Sedimentation and the Search for Petroleum," Bull. Amer. Assoc. Petrol. Geol., Vol. 29 (September 1945), pp. 1233–1261.

W. C. Krumbein, "Principles of Facies Map Interpretation," Jour. Sed. Petrol., December 1952, pp. 200–211.

13. W. C. Krumbein, "Lithofacies Maps and Regional Sedimentary-Stratigraphic Analysis," Bull. Amer. Assoc. Petrol. Geol., Vol. 32 (October 1948), pp. 1909–1923. 17 references.

L. L. Sloss, E. C. Dapples, and W. C. Krumbein, "Lithofacies Maps, An Atlas of the United States and Southern Canada," John Wiley & Sons (1960), 108 pages. Numerous lithofacies, isopach, and paleogeologic maps with a chapter on the preparation and construction of these maps.

14. W. C. Krumbein, op. cit. (note 13), pp. 1910–1916.

L. L. Sloss, W. C. Krumbein, and E. C. Dapples, "Integrated Facies Analysis," Memoir 39, Geol. Soc. Amer. (June 1949), pp. 100–102.

15. A. I. Levorsen, "Pennsylvanian Overlap in United States," Bull. Amer. Assoc. Petrol. Geol., Vol. 15 (February 1931), pp. 113–148.

A. I. Levorsen, "Studies in Paleogeology," Bull. Amer. Assoc. Petrol. Geol., Vol. 17 (September 1933), pp. 1107–1132.

A. I. Levorsen, *Paleogeologic Maps,* W. H. Freeman and Company (1960), 174 pages.

16. V. C. Scott, "Apache Oil Pool, Caddo County, Oklahoma," Bull. Amer. Assoc. Petrol. Geol., Vol. 29 (January 1945), pp. 100–105.

17. Wallace Lee, op. cit. (note 5), p. 126, Fig. 17, A and B.

18. N. W. Bass, "Origin of Bartlesville Shoestring Sands, Greenwood and Butler Counties, Kansas," Bull. Amer. Assoc. Petrol. Geol., Vol. 18 (October 1934), pp. 1333–1342.

Thomas Eugene Weirich, op. cit. (note 7).

19. Donald C. Barton, "Petroleum Geophysics," in *The Science of Petroleum,* Oxford University Press, London and New York, Vol. 1, pp. 319–327. Contains a good historical account of the development of geophysical methods.

C. A. Heiland, *Geophysical Exploration,* Prentice-Hall, New York (1940), 1,013 pages.

L. L. Nettleton, *Geophysical Prospecting for Oil,* McGraw-Hill Book Co., New York (1940), 444 pages.

J. J. Jakosky, *Exploration Geophysics,* Times-Mirror Press, Los Angeles (1940), 786 pages; 2nd ed., Trija Publishing Co., Los Angeles (1950), 1,195 pages.

Milton B. Dobrin, *Introduction to Geophysical Prospecting,* McGraw-Hill Book Co., New York (1952), 435 pages.

C. Hewitt Dix, *Seismic Prospecting for Oil,* Harper & Brothers, New York (1952), 414 pages.

20. Published by the Society of Exploration Geophysicists, Tulsa, Okla., since 1936.

21. *Geophysical Case Histories,* Vol. I (1948), L. L. Nettleton (ed.), 61 authors, and Vol. II (1956), Paul L. Lyons (ed.). Soc. Explor. Geophys., Shell Bldg., Tulsa, Okla. Many geophysical maps and surveys that led to petroleum discoveries are described.

22. Eugene McDermott, "Geochemical Exploration (Soil Analysis)," Bull. Amer. Assoc. Petrol. Geol., Vol. 24 (May 1940), pp. 859–881. 36 references listed.

Esme Eugene Rosarie, "Geochemical Prospecting for Petroleum," Bull. Amer. Assoc. Petrol. Geol., Vol. 24 (August 1940), pp. 1400–1433. 32 references listed.

A discussion of these two articles is contained in the Bull. Amer. Assoc. Petrol. Geol., Vol. 24 (August 1924), pp. 1424–1463.

A. A. Kartsev, Z. A. Tobassaonsii, M. I. Subbota, and G. A. Mogilevskii, "Geochemical Methods of Prospecting and Exploration for Petroleum and Natural Gas," Moscow (1954). Translated by Paul A. Witherspoon and William D. Romey, Univ. of California Press, Berkeley (1959).

23. Leo Horvitz, "Geochemical Well Logging," in Carl A. Moore (ed.), *A Symposium on Subsurface Logging Techniques,* University Book Exchange, Norman, Okla. (1949), pp. 89–94.

24. J. M. Forgotson Jr., "How Computers Help Find Oil," O. & G. Jour., Vol. 61, No. 11 (1963), pp. 100–109.

D. F. Merriam and J. W. Harbaugh, "Computer Helps Map Oil Structures," O. &. G. Jour., Vol. 61, No. 47 (1963), pp. 158–159.

See also Kansas Geological Survey, Computer Contributions, Daniel F. Merriam (ed.), Univ. Kansas, Lawrence, Kansas. A continuing series of reports and publications on the use of computers in geology.

CHAPTER 14

The Petroleum Province

Sediments: carbon-ratio theory – sedimentary basins. Evidences of oil and gas. Unconformities. Wedge belts of permeability. Regional arching. Local traps.

ONE OF THE MOST pressing and acute problems of every petroleum exploration unit—whether an individual, a major oil company, or a nation—is to decide which unexplored or partially explored province or region offers the greatest promise of petroleum discovery for money and effort spent. Few if any organizations can afford to operate on a full scale in every possible area. Each has to choose among a number of areas that have varying characteristics and to hope that it will choose what is best for its particular situation. A large company must allocate its exploration effort according to its judgment of the conditions that prevail. These conditions are not solely geologic; in each area that it wishes to consider, the political climate, the land and lease situation, the distance from markets and pipe lines, and the cost of operating, all enter the picture. (See also Chap. 15.) But the choice must always be based primarily on geology; geology is fundamental to the whole evaluation.

The supreme achievement of the petroleum geologist is to discover an oil or gas pool that opens a new petroleum province. Upon what criteria should the geologic choice of the best area to explore be based? What evidence is significant? How may one compare the oil possibilities of Flordia, for example, with those of Utah, or of Australia with those of North Africa, or of France with those of England? In the business of maintaining and increasing the world's productive capacity, neither the petroleum industry nor the individual organization can afford to depend on the chance of "stumbling into" a new province. New provinces must be anticipated. What kinds of geologic evidence can be used in predicting where a petroleum province will eventually be found? And what weight should be given to each kind of evidence? These are some of the questions that will be considered in this chapter.

If we compare the geologic conditions that are found in the known petroleum provinces, we see that they are extremely diverse. Each province has its own geologic history, its own characteristic deformation, its own stratigraphy, and its own peculiar types of petroleum accumulation. It is doubtful whether many of these geologic conditions could be accurately anticipated for an unknown province in advance of drilling. Yet there are certain empirical characteristics that seem to be present in most productive areas, and that seem to carry more weight than others in a pre-discovery evaluation of the prospects of a region. These characteristics might be classified as (1) sediments, (2) evidences of oil and gas, (3) unconformities, (4) wedge belts of permeability, (5) regional arching, (6) nature of local traps.

Evidence for identifying these characteristics may be found, in part, in the surface geology of the region, especially along the upturned edges of the strata at the borders of a region that lies in a basin. In nearly every potential region, moreover, there has been some exploratory drilling. The stratigraphy and the nature of the different fluids encountered in the holes that have been drilled, considered in relation to the surface data, deserve the most careful study because of the help they can give one who is predicting the underground geologic conditions. Even one deep test well, far out in a basin and at a long distance from the nearest outcrops of the subsurface rocks, can become of tremendous value in preparation of the many types of three-dimensional maps that are the basis for understanding the geologic history of the region. But, in order to make such a well yield the greatest possible amount of useful evidence, the geologic data that it reveals must be fully logged, and good well samples or cores must be taken and intelligently studied.

SEDIMENTS

The presence of sediments may be said to be the one essential element of a potential petroleum province. Sediments provide the source of the petroleum, the reservoir rock and the cover for the individual trap. Sedimentary rocks are the host rock of practically all oil and gas pools, and for that reason merit first consideration in any evaluation of the possibilities of oil and gas production within any hitherto unexplored or unproductive region. In general, we may say that the chances of eventually finding commercial oil are roughly in proportion to the volume of the sediments—the greater it is, the better the chances. The quantity of the sediments in cubic miles offers some basis for comparing the potentialities of different regions. The past and future production of the United States has been estimated to range from 6,000 to 200,000 barrels per cubic mile of sediments in the various petroleum provinces, with an average of 50,000 barrels per cubic mile for the 2 million cubic miles of sediments considered petroliferous.[1] Because the United States, in large areas, has reached a high state of oil development, it might be taken as a basis of comparison, and the figure of 50,000 barrels per cubic mile of sediments applied to the other sedimentary regions of the earth that are yet unexplored and undeveloped. There is no known reason why the sediments of the United States should contain more petroleum than unexplored similar sediments in other parts of the world. That this figure of 50,000 barrels per cubic mile is merely an average, however, to be considered only as a general guide and not as a local guide, is evident when we are reminded of the rich deposits discovered in such small subprovinces as the Los Angeles basin of California and the Baku subprovince of the USSR. Most estimates of the volume of potentially productive sediments eliminate all areas whose total sedimentary section is less than 1,000 feet, all sediments of Pleistocene, Cambrian, and Precambrian ages, and sediments that show even low-grade metamorphism.

The character of the sediments also must be considered. Since most oil and gas found heretofore have occurred in marine sediments, an area in which the strata consist chiefly of marine sediments is considered to possess greater potential value for petroleum than one in which the sediments are mostly nonmarine. Varied lithology, also, is favorable: if the sediments are all shales or sandstones, the chances of finding a number of pools are far less than if they consist of interbedded and variable shales, sandstones, and limestones. Blanket sands call for more local folding to form traps than do laterally variable sands. A rock section, however, consisting wholly of limestones and dolomites would not be nearly so unfavorable in potential value as one that consists wholly of shale or of sandstone, for the carbonate rocks may contain both the reservoirs and the impervious trap covers. In many areas, as in western Texas, for example, oil and gas are produced from limestone and dolomite reservoirs in limestone sections that are thousands of feet thick. The facies of most sediments change laterally, and we must never be too ready to assume that the entire section of a region is shale or limestone or sandstone; we certainly should not deliver such a verdict before several holes have been drilled in different parts of the sedimentary basin.

Metamorphism of the sediments is another factor that is thought by many geologists to have a bearing on the possibilities of their containing commercial petroleum. Metamorphism (a much-abused term) is here considered only as it is related to changes that come from heat and pressure. The presence of secondary minerals such as chlorites, sericites, and oriented mica flakes, and of stretched or deformed particles, such as are formed under higher-than-average heat and pressure, is an expression of mild metamorphism. Orthoquartzites are formed by cementation with silica from either primary or secondary sources and are not classed as metamorphic rocks. One effect of the metamorphism of sediments is to reduce permeability and hence to lower the chances that oil or gas have become concentrated into pools. Another possible effect of low-grade metamorphism is to change petroleum into more volatile forms, and this has led to the development of the carbon-ratio theory. The carbon-ratio theory will be considered in some detail because of the wide influence it has had on petroleum exploration.

Carbon-ratio Theory

The degree of low-grade metamorphism of certain sedimentary rocks, as measured by the carbonization of the contained coals, has been used as an index of the nature of any oil and gas that these rocks may contain. The view that this method would give valid results has been called the *carbon-ratio theory*. It was most effectively stated by David White[2] in 1915, although the idea had been under discussion since the time of the Drake well.[3]

Briefly, the carbon-ratio theory, as applied to regional exploration for petroleum, states that in those areas where there has been little metamorphism and where the coals are brown lignitic coals, the oil is of high (low API) gravity. As the temperatures and pressures increase, the percentage of fixed carbon increases, the grade of the coals rises, and the oils become lighter. In regions where the coal is bituminous, only light oil accompanied by gas may be expected, and with increasing metamorphism the point is reached where only gas will be found. Finally, when the coals are of the anthracite grade, neither oil nor gas in commercial quantities is to be expected. In general, past experience supports the idea of an absence of commercial oil and gas accumulation in the areas of high coal metamorphism. There are exceptions, however, to this generalization; moreover, there is a general lack of drilling for oil or gas in areas thought to be unfavorable because of the high metamorphism of the contained coals.

The degree of metamorphism, or the carbon ratio, of the sediments is measured by the fixed-carbon percentage of the dried coal on a moisture-free and ash-free basis. The carbon ratio is computed by dividing the percentage of fixed carbon of the proximate analysis percentages (by weight) of the coal by the sum of the percentage of fixed carbon plus the volatile matter of the same analysis. Equal carbon ratios are joined by lines called *isocarbs* to form maps. Isocarb maps have been published for many regions.[4] They are based on "as-received and ash-free" analysis of the coal; in other words, the moisture content is included with the volatile matter as a part of the coal content.[5]

The relation between the fixed-carbon content of the coals (on a moisture-free basis) and the oil and gas in the associated sediments has been listed by Fuller[6] as follows:

CARBON RATIOS (surface)	PETROLEUM PRODUCTION
70+	No oil or gas, with rare exceptions.
65–70	Usually only "shows" or small pockets. No commercial production.

CARBON RATIOS (surface)	PETROLEUM PRODUCTION
60–65	Commercial pools rare, but oil exceptionally high grade when found. Gas wells common but usually in isolated pockets rather than in pools.
55–60	Principally light oils and gas of the Appalachian fields.
50–55	Principally medium oils of Ohio-Indiana and Mid-Continent fields.
Under 50	Heavy oils of coastal plain and of unconsolidated Tertiary or other formations.

Because of the commercial implications of the carbon-ratio theory, as applied to the exploration for petroleum, it has received a great amount of thought and study. As a result, the original idea has been considerably modified, and many objections to it have been raised, relating chiefly to the accuracy of the carbon-ratio measurements, but also to the interpretation of these measurements, both as an index of the metamorphism and as a guide to petroleum exploration.

Some of the objections stem from the manner in which the coal samples are obtained and analyzed. These are, briefly:

1. Generally coal samples are not representative of the coal. These samples are taken either from coal seams in various stages of weathering or from single laminae or selected layers within the seam. Furthermore, unless the sample is placed in air-tight containers immediately, weathering begins, and the sample finally analyzed in the laboratory may be quite different from the field sample. The analyses by various coal company laboratories and commercial laboratories vary, and it is only through comparing analyses made with the same or comparable methods that a correct interpretation can be expected.

2. The rigorous statement of the carbon-ratio theory involves comparison of the coals on a moisture-free and ash-free basis. Any moisture in the sample was originally considered as accidentally present and not a component part of the coal. It is now thought, however, that some of the moisture is an actual constituent of the coal,[7] and that moisture is as important as the volatile content as a measure of the fixed-carbon percentage. Tertiary coals, especially, are notably higher in fixed carbon when considered on the "dry and ash-free" basis than when considered on the "as-received and ash-free" basis. The carbon ratios of some of the low-ranking coals, for example, are twice as high when analyzed on a moisture-free basis as when analyzed on the "as-received" basis.

3. Many of the early isocarb maps did not show the carbon ratios of a single coal, but rather were maps of the carbon ratios of whatever coals were found in the region. This practice resulted in maps in which older and deeply buried coals were compared with younger coals. Allowances were not made for the edges of the basin, where the older coals came to the surface and would normally be expected to have the higher carbon ratios. The fixed carbon of the Appalachian region increases at an average rate of 0.69 unit per 100 feet of stratigraphic depth,[8] which means that the carbon ratio is considerably higher in the older and once deeply buried coals found along the eastern boundary of the region than in the younger coals farther west. Yet both coals have presumably undergone approximately the same tectonic deformation. Comparison between these coals,

therefore, may represent, not a measure of tectonic metamorphism, but rather a measure of primary differences in the coals and of differences in environment and depth of burial.

That variations in the method of measuring the carbon ratio might be significant and lead to erroneous conclusions is shown in the following list of the probable maximum variations of the carbon ratio from the norm (percentage of fixed carbon): because of errors in coal sampling, weathering, etc.: 15 percent; because of errors in nonstandard methods of coal analysis, 20 percent; because of change from moisture-free to as-received basis, 50 percent.[9]

The main weakness of the carbon-ratio theory lies in assuming that variations in the percentage of fixed carbon reflect differences in the metamorphism of the area considered. If this assumption, which is fundamental to the carbon-ratio theory, fails, then the explanation of the lack of oil and gas in the regions of high carbon ratio also fails. Some of the objections to this basic assumption are:

1. The principal alternative to metamorphism suggested as the cause of higher carbon ratios is that these variations in ratios are primary—that is, that they are due to differences in the organic matter that was deposited in different places or to differences in the depositional environment. Some coals originally contained higher percentages of algal material, of pollens and spores, or of woody materials, than other coals, and the original content may vary from region to region within the same coal formation, depending on the environment. Such differences in the original organic material presumably would result in differences in the degree of carbonization of various coals at the same temperature and pressure. Similar reasoning applies to the potentially oil-and-gas-bearing sediments: it is generally true that porosities and permeabilities are low where the carbon ratio is high, but this difference also may sometimes be primary rather than secondary.

2. The metamorphic effects of the same heat and pressure will be expressed differently in different sediments. Those containing even small amounts of the more unstable minerals, such as some of the clays and carbonates, may deform with more plasticity and resultant loss of porosity and permeability than sediments that consist almost wholly of the more stable and resistant minerals such as quartz. The competence of coal-bearing formations and that of related petroleum-bearing formations may thus be quite different, so that the metamorphic effects on the coal may mean little or nothing in terms of the metamorphism of a nearby reservoir rock.

3. If a low-grade petroleum or a primary organic material is metamorphosed into a higher-grade petroleum by heat and pressure, a residue of asphalt or coke should be left behind.[10] This expected relationship seems almost axiomatic, since in distillation processes the heavier compounds are left behind at all stages. The fact that residues are not found implies that a slow distillation process is not the answer—except for the possibility that only microscopic amounts of residual hydrocarbon material are present and might be hard to detect. The normal expectation is that, if an area of high pressure and temperature did exist in a reservoir rock, the lighter oil, and especially the gas, would move out of it, in the direction of the water movement, into the areas of lower potential energy, leaving the heavier residues behind. The carbon-ratio theory, however, assumes that the gas will be found in the area of highest metamorphism and presumably of highest pressure and temperature.

4. Hilt,[11] in 1873, first announced the principle that, in a series of coals, the degree of carbonization normally increases with the stratigraphic depth, and the general verification of this principle in the coal fields throughout the world has led to its being called *Hilt's law*. The average rate of increase is thought to be about 0.7 units for each 100 feet of depth. This means that the carbon ratio of many coals measured at the surface does not apply when projected downward, across thousands of feet of stratigraphic section, to a point where the reservoir rocks may be expected to be productive. In fact, as Russell points out,[12] the limestone oils of Kentucky—and probably many other deep reservoirs—occur in reservoir rocks that are as much metamorphosed as the country rock of coals whose carbon ratio is extremely high and, under the carbon-ratio theory, presumably should contain no hydrocarbon unless it were gas. Any coal formation used as a guide to the metamorphism of a reservoir rock should therefore be stratigraphically close to the reservoir.[8]

5. Another possible source of some unaccountable variation in the carbon ratios is varying permeability in the rocks enclosing the coal beds. The more permeable such rocks are, the more easily the volatile gases within the coal can escape. Differences in carbon ratio, therefore, may indicate different permeabilities in the enclosing rocks rather than metamorphism.

6. While White was of the opinion that the pressures that caused the metamorphism were tangential to the earth's surface and chiefly the result of folding and diastrophism, later investigators generally hold that the variations in carbon ratios are due to variations in the depth of burial.[13] This belief would be in accord with Hilt's law. Hendricks, however, after a careful re-examination of the carbon ratios of the coals of the Arkansas-Oklahoma region,[14] concludes that the fixed-carbon content is directly related to the structural deformation of the region by pressure from the south. He finds no increase in carbonization in going from the younger to the older coals of the region, nor does he believe that the original composition had any bearing on the present carbon ratios. Such an interpretation of the carbon ratios in this area is not entirely conclusive, however, for the isopach contours connecting the thicknesses of the coal-bearing formations are more nearly parallel to the isocarbs than are the thrust faults.

We must conclude that at present the evidence for the metamorphism of oil along with coal is inconclusive. There probably is some regional relation between carbon ratios, especially the "as-received ash-free" ratios, and the degree of metamorphism, as a result of either loading or diastrophism or of a combination of both. On many carbon-ratio maps the low carbon ratios are certainly found in the regions of low deformation and low overburden, yet the relations between low carbon ratio and petroleum have not been demonstrated conclusively. It is also true that few pools of either oil or gas have hitherto been found in regions of high carbon ratio, but this relation may be based on nothing more than the general lack of drilling in such areas, which is due to the discouraging influence of the high carbon ratio.

In conclusion, there are a number of exceptions to the original idea that no oil but only gas can exist in areas of high carbon ratio, especially when Hilt's law is applied to the overburden found over many of the modern deep oil pools. If coal was present near these deep reservoirs, it would be expected, in many instances, to have a carbon ratio exceeding the limits regarded as the upper limits for the accumulation of oil or gas. As a guide to exploration, then, the presence of high carbon ratios in the coals of an area does not mean that the area should be condemned for that reason alone, for high carbon ratios, as we have seen, do not necessarily imply temperatures and pressures incompatible with the accumulation of petroleum. They may, in fact, be due to causes that have no unfavorable implications.

Sedimentary Basins

The areas of the world where substantial amounts of unmetamorphosed sediments occur are shown on the inside covers of this book;[15] in each of these areas the sediments are thickest in the interior and thin out at the edges, and each is therefore called a *sedimentary basin*. These basins include all the areas known to contain large volumes of sediments. They contain not only all the petroliferous provinces discovered hitherto, but all those that are likely to be discovered in the future. Other maps showing the sedimentary basins of the world include those by Weeks[16] and Gester.[17]

Sedimentary basins are not as simple as the name might imply. They have the common characteristic of being geologically depressed areas, with thick sediments in the interior and thinner sediments at the edges, but otherwise they may be quite different in origin and character. Some sedimentary basins are *depositional basins*, as shown by a succession of strand-line phenomena around the periphery. Other basins are *structural;* actually they are regionally closed synclines. In these structural basins the depositional sedimentary patterns show no relation to the periphery, and the individual formations are more continuous and more uniform in thickness than those of the depositional basins. Still other areas are called "basins" merely because of the present topographic form of the surface of the ground and may have little or no relation to the attitude of the underlying rocks. Petroleum is equally likely to be found, it seems, in both depositional and structural basins, and along the edges as well as near the centers.

Many sedimentary basins are of composite origin. In some of these the basin structure of the upper strata, as above a surface of unconformity, is of depositional origin, while the underlying sediments form a structural basin. Where the areas covered by two or more basins of separate origin coincide, the resulting basin may be thought of as a *composite basin*. An idealized section through such a composite basin, which may be thought of as *composite vertically*, is shown in Figure 14-1. Two or more depositional and structural basins may be superimposed, offset, or entirely independent of each other. Many examples of composite basins exist in the Rocky Mountain region of the United States; the Tertiary rocks at the surface often form a depositional basin, and the underlying rocks are folded into a structural basin, usually closed. In most of these the folding preceded and localized the deposition of the later Tertiary formations, so that the two basins coincide in position.

THE PETROLEUM PROVINCE [CHAPTER 14]   635

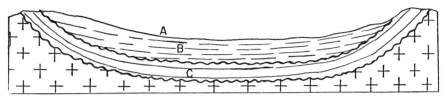

FIGURE 14-1   *Diagrammatic section showing a vertically composite basin. The surface A is in part structural but largely topographic; the sediments B form a depositional basin; and the sediments C are in a structural basin, the depositional limits of which are beyond the limits of the section. The scale is on the order of several miles vertically and several hundreds of miles horizontally.*

In others, a late depositional basin was divided by later folding into several structural basins, all of which were underlain by an extensive early structural basin. Thus there may be basins within basins, forming complex patterns that can be unraveled only by careful stratigraphic and structural analysis.

Another composite type of sedimentary basin, which seems to be especially associated with petroliferous provinces, was formed by two independent episodes, one stratigraphic and one structural, which came at different times and left the intervening rocks in the form of a basin. The sequence of events, which is graphically shown in Figure 14-2, seems to be as follows: One side, B, formed as the result of a gradual transgressive overlap of the sediments over a gently dipping shore surface, called the shield, the platform, or the craton,[18] and characterized by shore and strand-line phenomena. This deposition was followed by sharp linear folding and faulting on the opposite side, A, the "geanticlinal welt" of Weeks,[19] leaving a basin-shaped depression between A' and B'. Such a basin, which may be thought of as *composite laterally,* has two separate periods of origin, and oil and gas pools may accumulate along either side. A few examples of this type, each of which forms a productive

FIGURE 14-2

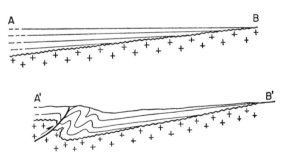

Diagrammatic section to show the development of a laterally composite basin. First to form is side B, which is characterized by shore, shelf, and strand-line phenomena. Later diastrophism along side A leaves the area between A and B as a basin-shaped depression, A'B'. Basins such as this, formed by two separate episodes, one stratigraphic and one structural, are associated with many petroliferous provinces. A few examples are shown in Figure 14-3. The scale is approximately several miles vertically and several hundred miles horizontally.

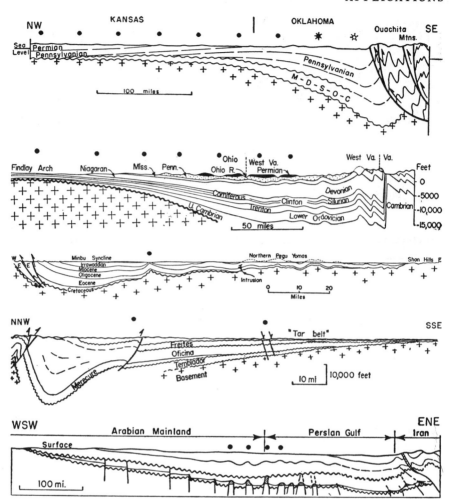

FIGURE 14-3 *Sections showing five productive basins that are composite laterally. [The third, fourth, and fifth sections are redrawn from figures published in Bull. Amer. Assoc. Petrol. Geol., respectively, by Tainsh (Vol. 34, p. 832, Fig. 2), Funkhouser, Sass, and Hedberg (Vol. 32, p. 1864, Fig. 3), and Link (Vol. 36, p. 1515, Fig. 19).]*

province, are illustrated in Figure 14-3. This shows idealized sections through the McAlester basin of eastern Oklahoma, the Appalachian basin of the eastern United States, the central basin of Burma, the Maturin basin of eastern Venezuela, and the Persian Gulf–Arabian basin of the Middle East.

Three implications of a laterally composite basin are as follows:

1. The petroleum had a much longer time in which to accumulate along the onlap side, B, of the basin because traps were available there from the time the reservoir rocks were overlapped by impervious rocks.

2. The equilibrium of the fluids, as well as the pressures and temperatures of all of the reservoir rocks, were upset when the folded and faulted side, A, formed. This deformation was therefore a time of much readjustment of all of the oil and gas occurrences involved in the basin; and, as the reservoir fluids returned to equilibrium conditions, it was a time when migration and accumulation into new pools occurred. Hydrostatic conditions may change into active hydrodynamic conditions, for example, and still later the hydrodynamic conditions may change in such a manner that the original direction of water flow is reversed.

3. Where were the oil and gas along the folded side of the basin, A, before the folding took place? Obviously the reservoir rocks were deposited outward, to the left in Figure 14-2, for some unknown distance until they were terminated, either by a succession of shore lines on the opposite side of a depositional basin, or more probably by a thinning out seaward due to lack of sediments, as off the edge of the continental shelf. Oil and gas now found along the side of the folded sediments probably came from rocks formed along that side, for the pressures were generally higher there than on the onlap side, B. Presumably the oil and gas were distributed as dispersed particles through either the reservoir rocks or the shales, or through both, until the folding occurred, at which time they were concentrated until sufficient buoyancy was developed to move them to the low-potential-energy areas that coincide with high structure. Or the patch of oil and gas was carried along with the water in the direction of the slope of the potentiometric surface until caught in a trap.

We may conclude, then, in analyzing the sediments of a potential oil- and gas-producing sedimentary basin, that they are most likely to be productive if they are present in great volume, if they are marine, if they are variable, and if they are not metamorphosed. Sedimentary basins are commonly complex, and the present basin-shaped sediments may have little or no relation to previous basins or to the original basin of deposition. The origin of oil is probably more closely tied to depositional basins, whereas migration and accumulation into pools are more closely related to structural basins and to the structural changes occurring during geologic time.

EVIDENCES OF OIL AND GAS

Geologists are uncertain, at best, about the origin of oil and gas. We have not yet completely bridged the gap between the organic material and the crude oil and natural gas. If the sediments in the region being considered contain evidences of oil and gas, then there must have been a source at some time, whatever it was. While it would be helpful to be able to say that any rock that is highly organic or dark in color is probably a source rock (and many have said it), yet one seepage of oil or a showing of "live" oil in a well is much more direct and useful information. Surface evidence of petroleum has led to the discovery of nearly every petroliferous province in the world.[20] (See pp. 15–19 also.) Direct evidence of the presence of petroleum may consist of surface seepages, springs, or vein fillings at the surface, chromatographic analysis, or subsurface showings in wells, as oil in cores, cuttings, or the drill mud. Indirect evidence may be given by electric logs, by fluorescence under ultraviolet radiation, or by the electrical resistance of material extracted from well cuttings and drilling mud when heated in a vacuum.

If the oil is "live" and contains dissolved gas, it is of much more importance than a "dead" petroleum such as an asphalt or tar. A "bleeding core" * generally does not mean commercial production in the formation from which it comes, but it is significant as an indication that liquid hydrocarbon is present in the region.

Closely related to the presence of oil or gas is the presence of water in the permeable formations. Reservoir water—interstitial, edge, or bottom—occurs in all commercial oil and gas pools. While some pools are completely enclosed by rock boundaries, most of them float upon an underlying water surface. If the water is fresh or of low salt concentration, it means either that the sediment forming the reservoir rock was deposited under continental conditions or that conditions of hydrodynamic flow, under present or past geologic conditions, have permitted fresh water to enter the formation and displace the connate salt water. Water that freely enters the well hole under pressure shows that the formation containing the water has a connected and permeable pore system—a system of water-lined "highways" along and through which oil and gas, if present, may migrate into traps. An abundance of formation water under normal pressure, therefore, is a favorable condition, because it is evidence that a proper environment for migration and accumulation exists.

Several regions that contain many occurrences of petroleum in inspissated deposits, seepages, springs, and vein fillings have not yet been found to be of commercial importance as petroleum provinces. An outstanding example is the island of Cuba; great seepages and asphalt deposits have been found there in many localities, yet the production is negligible, compared with the apparent prospects, in spite of the extensive exploration programs that have been going on there for many years.[21] In Australia, New Zealand, and Brazil, also, there are many surface evidences of oil and gas, but as yet few commercial deposits have been discovered.[22] Several explanations might be given to account for this lack of oil pools, even though there is evidence of petro-

---

* A bleeding core (sometimes called a "weeping core") is one in which little or no oil is visible when it reaches the well-head but from which oil begins to drain out shortly afterward. One explanation is that the oil is in disconnected particles within the rock, and there is not enough gas to drive the oil out by expansion as the pressure is decreased from that in the reservoir rock to that at the surface. Another explanation is that the permeability of the rock is so low that it takes time for the oil to drain out under surface conditions of temperature and pressure.

leum: (1) Most or all of any pools that accumulated may have been exposed by erosion to surface weathering, thus leaving behind the large inspissated deposits. (2) Metamorphism may have driven the lighter hydrocarbons and the gases off, leaving the heavier fractions behind. Or more likely (3) not enough drilling has been done to test properly the petroleum possibilities, and the lack of commercial oil pools may be merely apparent.

Some phenomena that have been taken as evidence of source rock do not really have much significance. The presence of bugs and fossil casts filled with oil is of geological interest, but probably not of much help in evaluating the possibilities of a region, for the oil may be entirely derived from the organic remains that originally occupied only the cavities. The presence of a dark or black kerogen shale or limestone is not nearly so important as the evidence of actual oil or gas in the sediments. Whether dark rocks and fossiliferous rocks are commonly source rocks is still in the realm of personal opinion, since these rocks contain organic material that was not transformable into petroleum under reservoir conditions.

Complete absence of any evidence of petroleum in the early stages of exploration of a region does not necessarily mean that the region is not potentially productive. The evidence may have been overlooked because of insufficient search, geological conditions that permit seepages may be lacking, or there may not be any oil and gas in the particular formations that happen to be exposed or were tested in the wells. Highly volatile petroleums may weather rapidly and leave no trace. Mere absence of visible oil and gas showings need not condemn the region, but the chances that the region will eventually prove productive are greatly increased if such evidences are found.

We may conclude, then, that the actual presence of oil and gas in the region being considered is of great importance in evaluating its petroleum possibilities. The greater the number and the larger the size of the occurrences, the greater their value as indicating the presence of commercial petroleum in the geologic section. And if small accumulations of petroleum occur at the surface, or showings are found in wells, then larger traps might be expected to contain proportionately larger accumulations.

UNCONFORMITIES

As we have seen (Chap. 7), many oil and gas pools are intimately associated with unconformities. A surface of unconformity may mark the boundaries of a trap; it may be associated with permeable rock; or it may either separate or connect formations that were formed under widely different geological conditions. In any of these circumstances, it may determine the position of a pool; and, if unconformities are of such importance locally, they must be important regionally, since many unconformities are known to extend over a wide area. Some of the implications of the presence of an unconformity in the geologic section of a potential petroleum province are these:

1. A surface of unconformity marks the position of permeable rocks. Since an unconformity marks an erosion surface, it commonly overlies a surface of weathering, with its related solution and recementation phenomena. Solution phenomena are especially evident in carbonate rocks but also occur in any rocks that contain soluble minerals. The rocks immediately below an unconformity are therefore likely to be either locally or regionally porous and permeable—to be, in other words, potential reservoir rocks.

2. The debris deposited on the surface of unconformity may provide a permeable connection between possible source rocks and reservoir rock, or between one reservoir rock and another, even when they are widely separated both vertically and laterally. A regional unconformity generally marks a plane of overlap, onlap, offlap, erosion, and angular change in strike and dip. It commonly separates rocks of distinctly different character and may mark a position of great geologic change, both stratigraphic and structural. Many of the overlying formations are deposited directly on many of the underlying formations at some place in the region. Along the porous and permeable unconformity zone, therefore, source rocks, or rocks favorable to either the early or the late origin of oil and gas, have access to, and may drain into, rocks that provide avenues for further migration and contain traps in which petroleum may accumulate.

3. Many unconformities separate rocks formed under quite different geologic conditions. In most of the large sedimentary regions, sedimentary basins, and geosynclines, the data obtained by drilling show that the geologic history was much more complex than could be supposed from surface information alone. An existing basin or syncline may or may not coincide with any former basin of deposition; early faulting and folding may or may not be expressed in the formations above the unconformity surface; nonmarine beds near the surface may be unconformably underlain by marine sediments; and shallow high-carbon-ratio rocks may be underlain below an unconformity by rocks showing no metamorphism. Thus each of the regional surfaces of unconformity in a geologic section frequently marks the boundary between major geologic episodes in the geologic history of the region. Regional unconformities may be thought of as surfaces separating groups of different combinations of geological phenomena somewhat as the filling separates the layers of a layer cake. The term "layer-cake geology" expresses crudely the nature of the relationship.

Just as, in a layer cake, the nature of each underlying layer is unknown until one cuts into it, so, in the geologic history of a region, even the general nature of the geology underlying an unconformity may not be known until the unconformity is penetrated by the drill. Some regions may have two layers while others may have three, five, seven, or more layers. The number

THE PETROLEUM PROVINCE [CHAPTER 14]   641

and nature of the separate geologic layers are not known until many wells have penetrated to the basement rocks of the region.

Each layer of geology in a producing region has its own exploration characteristics, its own individual structure and stratigraphy, its own type of oil and gas trap, its own peculiar richness or leanness of oil and gas production. Each layer differs both from the layer above and from the layer below in many characteristics. It is impossible, in advance of drilling through the overlying surface of unconformity, to predict what the conditions will be in the underlying layer. Crossing an unconformity with a drill for the first time means discovering a whole new geologic environment; in effect, it often means the discovery of a new potential subprovince of oil and gas production —a whole set of new conditions. Thus the first clues to many of the subsurface geologic conditions in a region may not be found until several wells have penetrated to depths of one, two, or three miles below the surface of the ground and the data obtained have been put on maps. A sedimentary region is not completely explored until each of the separate layers of geology has been explored.

An example of layers of geology separated by surfaces of unconformity is seen in the several layers of north-central Texas, shown in Figure 14-4,

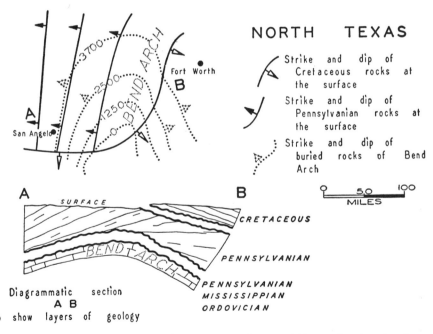

FIGURE 14-4 *Diagrammatic section across north-central Texas showing the three layers of geology superimposed one above another. Each has its own characteristics of oil and gas occurrence. [Redrawn from Levorsen, Bull. Amer. Assoc. Petrol. Geol., Vol. 27 (1943), p. 910, Fig. 19.]*

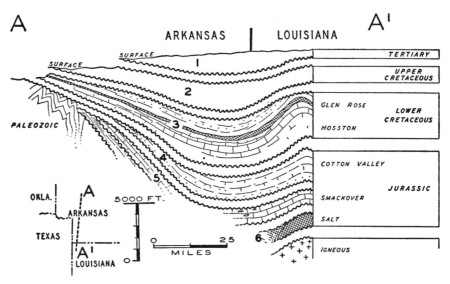

FIGURE 14-5 *Diagrammatic "exploded" section from Arkansas into Louisiana showing the many superimposed layers of geology. Each layer has its own characteristics of oil and gas occurrence. [Redrawn from Imlay, Ark. Geol. Surv. (1940), Cir. No. 12.]*

where the Cretaceous, Pennsylvanian, and pre-Pennsylvanian sequences are separated from one another by unconformities. Each layer has its own characteristic oil and gas accumulations. One reason why this region has had a long period of active development is that the geology underlying each of the various surfaces of unconformity could not be foretold, but had to be discovered by progressively deeper drilling. The Arkansas-Louisiana region likewise is underlain by several layers of related rocks separated by unconformities, each independently productive and the geology of the petroleum of each unknown until the drill penetrated the overlying unconformity. This region is shown diagrammatically in Figure 14-5. A section across a part of western Texas is presented in Figure 6-16, page 251, and it too shows the several layers of rocks, each with its own characteristics of oil and gas accumulation.

4. Some unconformity surfaces seem to mark the upper time boundary of periods peculiarly favorable to oil and gas accumulation. The great unconformity that separates the Pennsylvanian from the pre-Pennsylvanian rocks in the United States, for example, also marks the upper boundary of considerable local and regional folding, faulting, tilting, and arching. This unconformity seems to have a peculiar and widespread geologic significance in petroleum history, for a great many of the domes, anticlines, and faulted folds that were formed in pre-Pennsylvanian and post-Mississippian time are now found to contain oil and gas. The importance of this pre-Pennsylvanian

THE PETROLEUM PROVINCE [CHAPTER 14]   643

deformation is shown in the Oklahoma City region of central Oklahoma, located at the southern end of the buried Nemaha uplift of Kansas and Oklahoma. The pre-Pennsylvanian surface is shown in Figure 14-6. This subcrop-paleogeologic map represents the areal geology of this region at the beginning of Pennsylvanian time. It was upon this distribution of sediments (this layer of geology), in which many oil and gas pools have accumulated, as shown in Figure 14-7, that the Pennsylvanian shales and sandstones were deposited. The structure of the area is also indicated. Of particular interest is the intersection of the pre-Pennsylvanian unconformity plane with the pre-Mississippian unconformity plane to form the trap in the West Edmond pool producing from the Bois d'Arc formation member of the Hunton group (Devonian-Silurian). Similarly, in Illinois, Kansas, Oklahoma, Texas, and New Mexico, pre-Pennsylvanian folding is characteristic of nearly all the pools found in the pre-Pennsylvanian rocks of these regions, and the accumulation of oil in those rocks appears to have been closely associated with this particular geologic episode. Crossing such a surface of unconformity as this by the drill thus renews the opportunity for discovery.

5. Unconformities mark erosion surfaces in the geologic column. Not only were the rocks eroded, but also any oil pools trapped in them. The

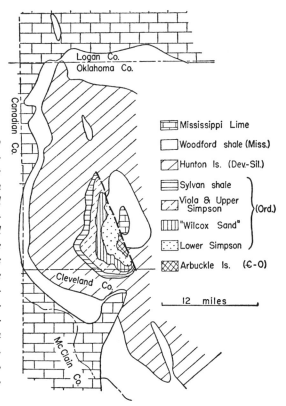

FIGURE 14-6

Paleogeologic (subcrop) map showing the pre-Pennsylvanian areal geology of the West Edmond–Oklahoma City–Moore region in central Oklahoma. The Hunton, Viola, Simpson, "Wilcox" and Arbuckle formations produce in a number of pools in this region, shown also in Figure 14-7. The total ultimate production of this area is estimated to be one billion barrels of oil. [Redrawn from McGee and Jenkins, Bull. Amer. Assoc. Petrol. Geol., Vol. 30 (1946), p. 1806, Fig. 5.]

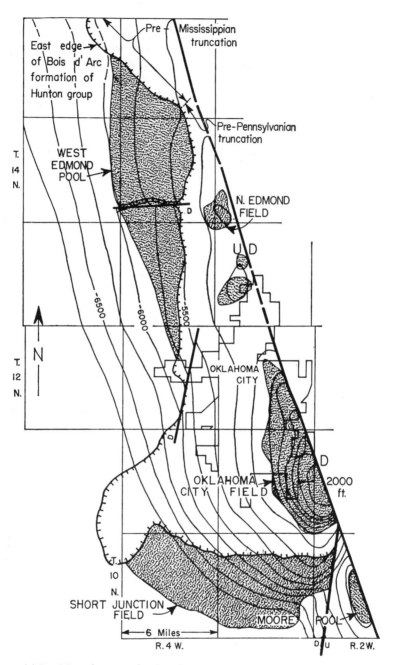

FIGURE 14-7 Map showing the distribution of pools below the Pennsylvanian and the structure of the Haragan-Henryhouse member of the Hunton formation (Devonian-Silurian) of the central Oklahoma region shown in Figure 14-6. The contour interval is 250 feet and is reconstructed where dashed. It was upon this structure that the Pennsylvanian was deposited. Later folding accentuated these early folds. [Redrawn from McGee and Jenkins, Bull. Amer. Assoc. Petrol. Geol., Vol. 30 (1946), p. 1805, Fig. 4.]

topographically high sedimentary regions, which would be the first to erode, are generally structurally high also, and, if structurally high, probably contained oil and gas pools. The number of unconformities found in many sedimentary sections (twenty or more, for example, in the Paleozoic of Kansas[23]) means that a great many oil and gas pools must have been eroded and lost during geologic time. One of the problems is: what became of this eroded petroleum? Was it destroyed or was it redeposited? That so many pools should be preserved through all of the deformational and erosional experience of the average sedimentary basin is one of the remarkable features of the geology of petroleum.

Where formations involved in pre-unconformity folding and tilting intersect the plane of unconformity, they may wedge out and thus form local obstructions that could trap moving oil and gas. The rocks immediately overlying the unconformity surface, being, in general, associated with onlapping shore or shoaling conditions, often exhibit facies changes, coarsening, sand bars, sand deposits, and even conglomeratic formations, all of which may become favorable places for accumulation of oil and gas pools. Shore environments are especially productive of organic matter, and the overlapping beds may also be source beds from which petroleum hydrocarbons drain into the permeable rocks below the unconformity surface.

Surfaces of unconformity within a geologic section, therefore, are places of particular interest. Other factors being equal, the more unconformities present in a region, the better the chances that it will be productive of petroleum.

WEDGE BELTS OF PERMEABILITY

The local up-dip edge of reservoir permeability frequently traps an oil or gas pool, and the same phenomenon on a regional scale may keep oil and gas from moving out of the region. We have seen how local variations in the stratigraphy, porosity, and permeability, either alone or in combination with folding, faulting, and hydrodynamic gradients, may form traps in which oil and gas accumulate into pools. May we not expand this concept of the local up-dip wedging out of permeability, possibly combined with a favorable hydrodynamic gradient, into a regional wedging-out of permeability, again possibly combined with a regionally favorable hydrodynamic gradient, to form a petroleum province with many oil and gas pools? The Nashville Dome in Tennessee[24] is typical of the manner in which unconformities occur in large growing dome folds. Figure 14-8 is a paleogeologic map of the pre-Chattanooga (pre-Mississippian) surface, and each line represents an up-dip wedge edge. Figure 14-9 is a section across the dome and shows the many unconformities in the lower Paleozoic, all of which converge over the crest of the fold and, through truncation, form many thousands of miles of wedge edges.

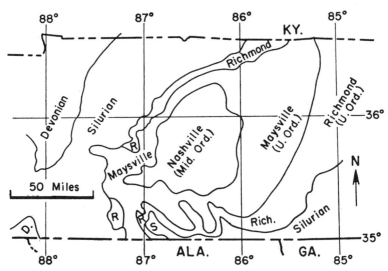

FIGURE 14-8 *Paleogeologic map of pre-Chattanooga (Mississippian) surface in Nashville dome, central Tennessee. See Fig. 14-9 for NW-SE section. [Redrawn from Wilson and Stearns, Bull. Amer. Assoc. Petrol. Geol., Vol. 47 (1963), p. 828.]*

Regional wedge-outs may in some ways be similar to local wedge-outs, such as the truncation in the Nashville dome and facies changes from dolomite to limestone or sand to shale. The evidence for regional hydrodynamic gradients is generally lacking unless drilling has been done relatively recently. Certainly

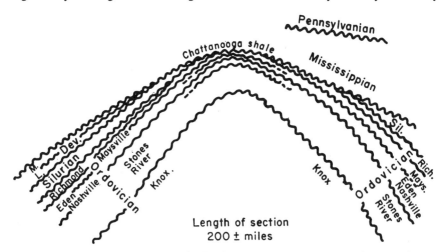

FIGURE 14-9 *Section across Nashville dome, central Tennessee, showing the unconformities in the pre-Chattanooga formations. Each marks the position of many wedge belts of truncated and overlapping sediments. [Redrawn from Wilson and Stearns, Bull. Amer. Assoc. Petrol. Geol., Vol. 47 (1963), p. 825.]*

THE PETROLEUM PROVINCE [CHAPTER 14]   647

the combination of regional down-dip fluid flow and regional up-dip wedges of permeability would attract exploration.

Two belts of oil and gas pools associated with regional up-dip wedge-outs of permeability may be seen in Figure 14-10. They occur on opposite sides of the Cincinnati arch; the Clinton sand (Silurian) wedging out up the dip toward the west on the east side of the arch, and the Trenton permeability wedging out up the dip to the southeast along the northwest side of the arch. (See also p. 307.) The Clinton sand gives way to shale and limestone, and the

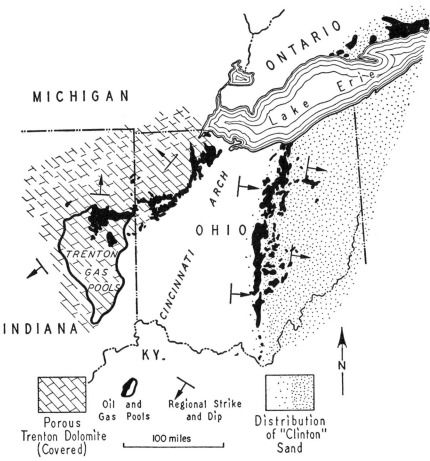

FIGURE 14-10 *Map showing the buried up-dip wedge edges of permeability of the Trenton dolomite (Ordovician) and the Clinton sand (Silurian) on the flanks of the Cincinnati arch in Indiana and Ohio. The association of the oil and gas pools with the edge zone is striking in both examples. The Trenton edge is chiefly the loss of porosity as porous dolomite changes to impervious limestone, and the Clinton edge is the result of strand-line phenomena in which the sands give way to shales. [Redrawn from Bull. Amer. Assoc. Petrol. Geol., Vol. 27 (1943), p. 891, Fig. 3.]*

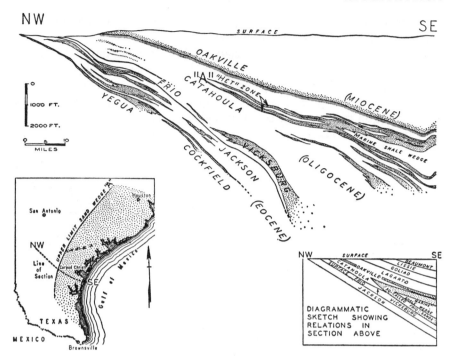

FIGURE 14-11 *Section normal to strand and shore lines of Gulf Coast Tertiary formations, southern Texas, showing the initial up-dip wedging out of permeable sands.* [Redrawn from Deussen and Owen, Bull. Amer. Assoc. Petrol. Geol., Vol. 23 (1939), p. 1626, Fig. 3, and p. 1630, Fig. 5.]

Trenton permeable dolomite changes to impermeable limestone to form the wedge-out. Many examples of up-dip wedge-out of sands occur along the Gulf Coast province of Texas and Louisiana. (See also Fig. 7-6, p. 292.) One is shown in Figure 14-11, a section across the trend showing the wedge-out of the "Het" and Frio sands of the Oligocene. It should be remembered that these wedge-outs occurred before the folds and salt domes formed traps in these formations, and that these up-dip edges early formed regional traps able to catch and hold the early (pre-fold) migration. A highly productive trend, consisting of large sand patches, facies changes, and lenses of permeable rocks along a regional up-dip facies-change barrier associated with folding and faulting, occurs in what are known as the Bolivar Coast fields, which are scattered along the eastern side of Lake Maracaibo in Venezuela.[25] (See Fig. 14-12.) The Tertiary sands in these fields contain many pools, which in the aggregate make up one of the great oil accumulations of the world, the ultimate recoverable reserve being 7 or 8 billion barrels.

A possible reason for the number of petroleum provinces located in wedge-out belts may be that such belts are areas favorable to the origin of oil and

THE PETROLEUM PROVINCE [CHAPTER 14]   649

gas. The fact that many of these belts do not appear to be associated with near-shore depositional environments, where most organic matter occurs, however, suggests that this may apply only to some provinces. A better reason seems to be that they form areas of early accumulation—before the local deformation that causes traps—and thereby catch any moving oil and gas

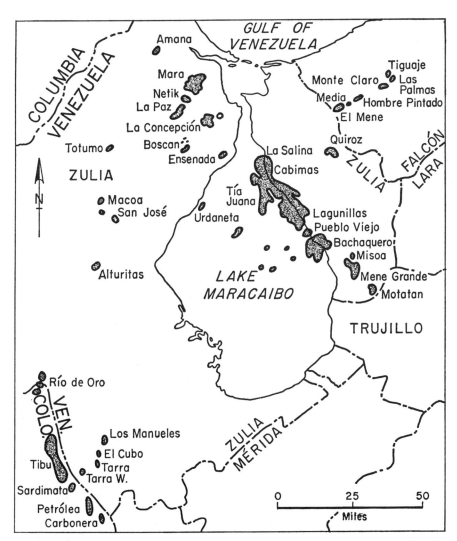

FIGURE 14-12   *Map of western Venezuela showing the location of the oil fields. The Bolivar Coast fields lie along the eastern side of Lake Maracaibo and produce from folds and faulted traps occurring near the eastern, up-dip change in facies from sands to shales. A section across the trend through the Lagunillas field is shown in Figure 7-13, page 297.*

along the wedge edge. Later folding then merely localizes oil and gas that has already had some measure of regional concentration, and folds formed down the dip may be barren because they were formed too late. The up-dip wedge-belt side of the reservoir rock, for example, is where the oil and gas are being held during the period between the sealing of the reservoir rock and the first folding—the blank area in the graphs of Figure 12-25, page 580. The sequence of events is diagrammatically shown in Figure 14-13, where the early accumulation is in A, and the later local folding of the areas in B merely relocalizes into local traps the oil and gas already concentrated along the wedge belt, the folds farther away being barren. An explanation such as this has been used by Link[26] to show the development of the pool in the Turner Valley fold. (See Fig. 14-14.)

Some regional up-dip wedge-outs of permeability are due to facies changes, some to strand-line phenomena, and some to the unconformable overlap of impermeable rocks on truncated edges of permeable rocks. The cause of the wedging out is not the significant element; the shape of the permeable formation, with its up-dip edge to obstruct migration of oil and gas, seems to be the important feature. We may say, then, that a permeable formation, with an up-dip wedge edge of permeability, from any cause, is a potentially favor-

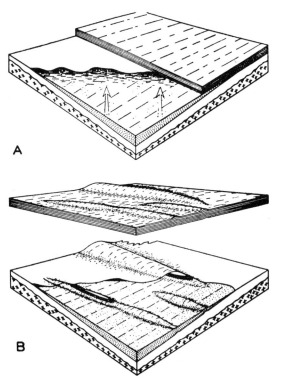

FIGURE 14-13

A, *the early up-dip accumulation of petroleum along the wedge of the permeable reservoir rock and,* B, *the later localization in the traps that form near the wedge edge. The traps down the dip and away from the area of early accumulation are barren because they were formed too late—after the oil had passed by, as in* A. [Redrawn from Bull. Amer. Assoc. Petrol. Geol., Vol. 27 (1943), p. 901, Fig. 13.]

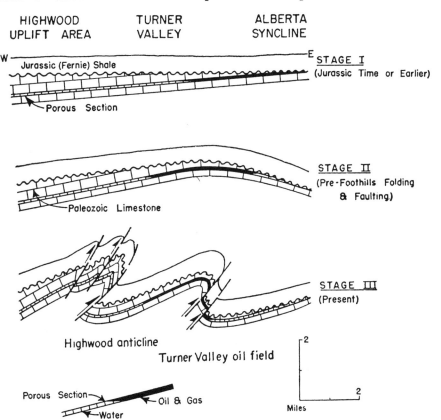

FIGURE 14-14 Sections showing an explanation of the geologic history that resulted in the water-bearing Highwood anticline and the oil- and gas-bearing Turner Valley anticline in Alberta, Canada. During Stage I the porous upper part of the Paleozoic limestone wedged out toward the east by truncation and was transgressively overlapped by the Jurassic shales. Oil and gas moved to the up-dip wedge edge, where they were in a favorable position to move into the nearby anticline shown in Stage II. Still later the oil and gas were in a favorable position to relocalize in the Turner Valley overthrust fold shown in Stage III. The Highwood fold was too far west to receive any of the oil or gas accumulated in Stage II, hence is now water-bearing. [Redrawn from Link, Bull. Amer. Assoc. Petrol. Geol., Vol. 33 (1949), p. 1481, Fig. 7.]

able geologic condition, both for the concentration of oil and gas and for the prevention of their escape. The more up-dip wedge-outs of permeable rocks that can be mapped in a prospective province, the better its chances of becoming productive. Hydrostatic fluid conditions, or better yet, an environment of down-dip water flow, would improve the prospects greatly.

REGIONAL ARCHING

Some provinces are characterized by broad folds or arches of low relief and by a small total thickness of sediments. Not only do such features form centers, or axes, away from which all formations dip, and consequently toward which any oil and gas would converge, but, because they are large and broad, with many recurrent periods of folding, many have been intermittently active for long intervals of geologic time. The stratigraphic sequence is frequently interrupted by truncation and unconformable overlap, and by faulting, local folding, and facies changes. While most producing provinces seem to be closely associated with sedimentary basins and with thick sediments, some provinces appear to be more closely related to the intervening uplifts. Among the better-known such regional folds, on and along which many pools are found, are the Sabine Arch of northwestern Louisiana,[27] the buried Central Basin Platform of western Texas (see Fig. 3-10, p. 73), the buried Central

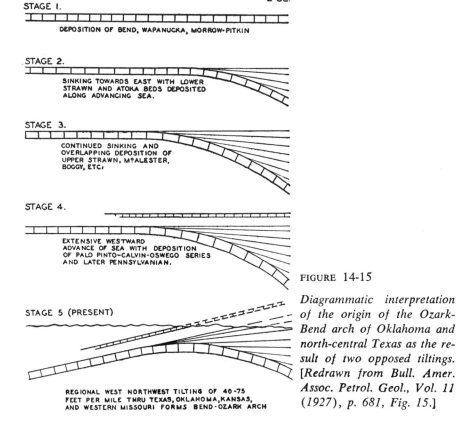

FIGURE 14-15

Diagrammatic interpretation of the origin of the Ozark-Bend arch of Oklahoma and north-central Texas as the result of two opposed tiltings. [Redrawn from Bull. Amer. Assoc. Petrol. Geol., Vol. 11 (1927), p. 681, Fig. 15.]

Kansas Arch (or Barton Arch) of western Kansas,[28] the buried Nehama Mountains of Kansas and Oklahoma (see Fig. 13-13, p. 606, and Fig. 14-6, p. 643), the Cincinnati Arch of Ohio and Indiana (see Fig. 14-10, p. 647), and the Nashville Dome in Tennessee. (See Figs. 14-8 and 14-9, p. 646.)

A special form of regional folding is the broad arching formed by two opposed tiltings, each as a different geologic episode. This cannot be proved to have occurred except where there have been two separate episodes of movement. An example is the Bend Arch of north-central Texas,[29] in which the tilt down to the east during early Pennsylvanian time was followed in post-Permian time by a tilt down to the west, as shown in Figure 14-15. The resultant arch became the site of many oil pools, probably because any regional movement of oil and gas would be up-dip toward the central, high, axial portion of the arch. Its counterpart in Oklahoma is the Hunton-Seminole-Ozark Arch, along which many rich pools have also been found in nearly all the rocks involved in the arching.

LOCAL TRAPS

Unless there are sufficient oil and gas to completely fill a regional trap that exists throughout a province, whatever oil and gas there is in the reservoir formation may be expected to occur within local traps. For example, if only a tenth of the oil found in the East Texas pool were originally present, it probably would be divided into several separate pools, localized by local variations in permeability and folding, instead of in the present one great pool that fills a large part of the regional trap formed by the arching of the wedged-out Woodbine sand up the dip toward the east. It becomes necessary, therefore, to judge what kinds of local traps may be expected within a prospective province and whether they may localize the individual pools.

Some idea of the type of local traps that exist in a potential province can be gained from the nature of the sedimentary section involved. The evidence regarding the type of trap must come in part from the outcrops that usually occur around the edge of the sedimentary basin or region, together with stratigraphic data from such wells as have been drilled. The more wells there are, of course, the more abundant these data will be if they have been properly recorded. Where the sediments are variable in character, laterally and vertically, and where unconformities are present, the traps may be expected to be largely of the stratigraphic and combination types; but where blanket sands prevail, structural traps are necessary. Where the surface relief is high, hydrodynamic gradients and tilted oil-water contacts may be expected; where the surface relief is low, hydrostatic conditions may be more common.

The analysis of a region has as its objective the discovery or the expansion of a producing province. Regional analysis is a step preliminary to pin-pointing the local traps, in which the objective is the discovery of separate pools or fields. Experience has amply demonstrated that in every province the local traps are far more numerous, and more varied in character, than could possibly be predicted from any analysis prior to drilling. Many regions give no surface indications of the underground geologic conditions. But facies changes, lenticular sands, organic reefs, pre-unconformity folding and faulting, and deep-seated folds and uplifts of varying intensity can reasonably be expected to occur in every large volume of sediments of regional size, whether or not there is any evidence of them to be obtained at the surface. Some of the traps may be located by surface mapping, but many are found by subsurface mapping based on the results of drilling, by various geophysical methods, and by combinations of these approaches.

Outside North America, much of the early exploration has been closely related to local features, especially to seepages and to folds exposed at the surface. Where mapping located potential traps, drilling followed, but little or no drilling has been done where there are no surface or geophysical anomalies to suggest buried traps. The result is that pools in stratigraphic traps may be found in many undeveloped regions when enough wells have been drilled to map the subsurface geology and to furnish a more accurate basis for an interpretation of geophysical measurements.

CONCLUSION

We may ask ourselves, in conclusion, whether there are new provinces left to be discovered and explored. The answer is, obviously, "Yes! And many of them." Even the United States, which has received the most aggressive exploration on earth for the longest period of time, still contains many large unexplored and partially explored regions that have good petroleum possibilities.[30] From the time when the first oil well was drilled by Drake in 1859 to the present day, some new petroleum province has continually been in the process of being developed. And for each such province in the limelight of exploration, there are always two or three new provinces coming up over the horizon and waiting to be explored. Each developing province, moreover, has many revivals of interest; new and deeper pays are discovered, or long extensions are made into territory that was once considered unfavorable, and what appeared as a simple province, at first, becomes complex and contains many subprovinces. The result is that a province, once discovered, seems never to finally die out as long as exploration is active; it seems certain to keep alive for at least a few decades. Discoveries are still being made in the eastern Oklahoma province, for example, where the first commercial discovery was made in 1903, and some of the biggest discoveries made in the West Texas–New Mexico province (see Figs. 3-8, 5-16) were made a quarter of a century after the province itself was discovered. If there are still so many favorable areas for exploration in the United States, with its half a million or more producing oil wells and even more dry holes and abandoned wells, then the outlook for new petroleum provinces in the sedimentary regions of the world outside the United States must be practically limitless.

Selected General Readings

E. G. Woodruff, "Petroliferous Provinces," Bull. 150 (1919), pp. 907–912, Trans. Vol. 65 (1921), pp. 199–204 (discussion by Charles Schuchert and others, pp. 204–216), Amer. Inst. Min. Met. Engrs. This is the earliest use of the idea of a petroliferous province.

K. C. Heald, "Essentials for Oil Pools," in *Elements of the Petroleum Industry,* Seely W. Mudd Series, Amer. Inst. Min. Met. Engrs. (1940), pp. 26–62. 33 references. An analysis of many factors surrounding the regional occurrence of oil pools.

Max W. Ball (ed.), "Possible Future Oil Provinces of North America," Bull. Amer. Assoc. Petrol. Geol., Vol. 35 (February 1951), pp. 141–485. The most authoritative analysis of the future possibilities of a continent by over 100 authors.

L. G. Weeks, "Factors of Sedimentary Basin Development that Control Oil Occurrence," Bull. Amer. Assoc. Petrol. Geol., Vol. 36 (November 1952), pp. 2071–2124. Importance of depositional environment stressed. A world-wide analysis of oil occurrence in basin areas.

L. G. Weeks (ed.), *Habitat of Oil,* Amer. Assoc. Petrol. Geol. (1958), 1384 pages. This volume contains a wealth of information on the geology of oil fields and oil provinces throughout the world.

Reference Notes

1. L. G. Weeks, "Concerning Estimates of Potential Oil Reserves," Bull. Amer. Assoc. Petrol. Geol., Vol. 34 (October 1950), p. 1951.

2. David White, "Some Relations in Origin Between Coal and Petroleum," Jour. Wash. Acad. Sci., Vol. 5 (1915), pp. 189–212.

3. W. T. Thom, Jr., "Present Status of the Carbon-Ratio Theory," in *Problems of Petroleum Geology,* Amer. Assoc. Petrol. Geol., Tulsa, Okla. (1934), pp. 69–95. Bibliog. 120 items.

4. *Ibid.,* pp. 71–75. Gives maps, carbon ratio, published up to 1934.

Olive C. Postley, "Natural Gas Developments and Possibilities East of the Main Oil and Gas Fields of Appalachian Region," Bull. Amer. Assoc. Petrol. Geol., Vol. 19 (June 1935), p. 857, Fig. 1, and p. 860, Fig. 2.

Thomas A. Hendricks, "Carbon Ratios in Part of Arkansas-Oklahoma Coal Field," Bull. Amer. Assoc. Petrol. Geol., Vol. 19 (July 1935), p. 938, Fig. 1.

5. W. T. Thom, Jr., *op. cit.* (note 3), pp. 86 and 87, Figs. 1 and 2, carbon ratios of United States.

6. Myron L. Fuller, "Carbon Ratios in Carboniferous Coals of Oklahoma, and Their Relation to Petroleum," Econ. Geol., Vol. 15 (1920), p. 226.

7. W. T. Thom, Jr., *op. cit.* (note 3), pp. 84–85.

8. Frank Reeves, "The Carbon-ratio Theory in the Light of Hilt's Law," Bull. Amer. Assoc. Petrol. Geol., Vol. 12 (August 1928), pp. 795–823.

9. W. T. Thom, Jr., *op. cit.* (note 3), pp. 82 and 84.

10. George Edwin Dorsey, "The Present Status of the Carbon-ratio Theory," Bull. Amer. Assoc. Petrol. Geol. (May 1927), pp. 455–463. Discussion by Russell S. Tarr, pp. 463–465.

11. Carl Hilt, "Die Beziehungen zwischen der Zusammensetzung und den technischen Eigenschaften der Steinkohlen," Zeitschr. des Ver. Deutscher Ingn., Band 17 (1873), Heft 4, pp. 194–202.

12. William L. Russell, "Relation Between Isocarbs and Oil and Gas Production in Kentucky," Econ. Geol., Vol. 20 (May 1925), pp. 254–255.

13. E. T. Heck, "Regional Metamorphism of Coal in Southeastern West Virginia," Bull. Amer. Assoc. Petrol. Geol., Vol. 27 (September 1943), pp. 1194–1227. Bibliog. 80 items.

14. Thomas A. Hendricks, *op. cit.* (note 4), pp. 937–947.

15. L. G. Weeks, "Highlights on 1948 Developments in Foreign Petroleum Fields," Bull. Amer. Assoc. Petrol. Geol., Vol. 33 (June 1949), pp. 1029–1124. The maps are reproductions of Fig. 1, p. 1036, and Fig. 9, p. 1071.

16. L. G. Weeks, *op. cit.* in note 15, pp. 1029–1124.

17. G. C. Gester, "World Petroleum Resources," World Oil, November 1948, p. 253.

18. Marshall Kay, "Geosynclinal Nomenclature and the Craton," Bull. Amer. Assoc. Petrol. Geol., Vol. 31 (July 1947), pp. 1289–1291.

19. L. G. Weeks, "Factors of Sedimentary Basin Development that Control Oil Occurrence," Bull. Amer. Assoc. Petrol. Geol., Vol. 36 (November 1952), p. 2076, Fig. 2.

20. Walter K. Link, "Significance of Oil and Gas Seeps in World Oil Exploration," Bull. Amer. Assoc. Petrol. Geol., Vol. 36 (August 1952), pp. 1505–1540.

21. E. DeGolyer, "Geology of Cuban Petroleum Deposits," Bull. Amer. Assoc. Petrol. Geol., Vol. 2 (1918), pp. 133–165.

J. Whitney Lewis, "Occurrence of Oil in Igneous Rocks of Cuba," Bull. Amer. Assoc. Petrol. Geol., Vol. 16 (August 1932), pp. 809–818.

J. Whitney Lewis, "Geology of Cuba," Bull. Amer. Assoc. Petrol. Geol., Vol. 16 (June 1932), pp. 533–553. 22 references. Discussion to p. 555.

22. Frederick G. Clapp, "Oil and Gas Prospects of New Zealand," Bull. Amer. Assoc. Petrol. Geol., Vol. 10 (1926), pp. 1227–1260. Bibliog. 29 items.

S. Frôes Abreu, "Brazilian Oil Fields and Oil-shale Reserves," Bull. Amer. Assoc. Petrol. Geol., Vol. 33 (September 1949), pp. 1590–1599.

Frank Reeves, "Australian Oil Possibilities," Bull. Amer. Assoc. Petrol. Geol., Vol. 35 (December 1951), pp. 2479–2525. Bibliog. 70 items.

23. Fanny Carter Edson, Subsurface Geologic Cross-sections, Kansas, No. 1 (1945) and No. 3 (1947), State Geol. Surv. of Kansas.

24. Charles W. Wilson, Jr., and Richard G. Stearns, "Quantitative Analysis of Ordovician and Younger Structural Development of Nashville Dome, Tennessee," Bull. Amer. Assoc. Petrol. Geol., Vol. 47 (May 1963), pp. 823–832.

25. F. A. Sutton, "Geology of the Maracaibo Basin, Venezuela," Bull. Amer. Assoc. Petrol. Geol., Vol. 30 (October 1946), pp. 1621–1741. 43 references.

Staff of Caribbean Petroleum Company, "Oil Fields of Royal Dutch-Shell Group in Western Venezuela," Bull. Amer. Assoc. Petrol. Geol., Vol. 32 (April 1948), pp. 517–628. Numerous maps, sections, and oil-field descriptions.

H. D. Borger and E. F. Lenert, "The Geology and Development of the Bolivar Coastal Field at Maracaibo, Venezuela," Fifth World Petrol. Congr., Proc. Sect. I (1959), pp. 481–496.

26. Theodore A. Link, "Interpretation of Foothills Structure, Alberta, Canada," Bull. Amer. Assoc. Petrol. Geol., Vol. 33 (September 1949), p. 1481.

27. C. L. Moody, "Tertiary History of Sabine Uplift, Louisiana," Bull. Amer. Assoc. Petrol. Geol., Vol. 15 (May 1931), pp. 531–551.

28. Hugh W. McClellan, "Subsurface Distribution of Pre-Mississippian Rocks of Kansas and Oklahoma," Bull. Amer. Assoc. Petrol. Geol., Vol. 14 (December 1930), pp. 1535–1556.

Edward A. Koester, "Geology of Central Kansas Uplift," Bull. Amer. Assoc. Petrol. Geol., Vol. 19 (October 1935), pp. 1405–1426.

29. A. I. Levorsen, "Convergence Studies in the Mid-Continent Region," Bull. Amer. Assoc. Petrol. Geol., Vol. 11 (July 1927), pp. 657–682.

30. Max W. Ball (ed.), and 100 or more contributors, "Possible Future Petroleum Provinces of North America: A Symposium," Bull. Amer. Assoc. Petrol. Geol., Vol. 35 (February 1951), pp. 141–498. A presentation of the evidence and reasoning upon which new petroleum provinces may be expected to be discovered in North America.

CHAPTER 15

The Petroleum Prospect

Discovery. Geologic factors. Economic factors. Personal factors.

THE PETROLEUM GEOLOGIST is the center pier in the foundation of the petroleum industry. The industry is completely dependent upon a continuing discovery of new pools, or extension of old pools, for maintenance of its supply of crude oil and thereby of itself; and most certainly "the discovery and production of oil is a geological enterprise." [1]

Because of its dependence upon geologists to guide it in its search for petroleum, the industry has entrusted many geologists with much greater business and management responsibilities than fall to the lot of the average scientifically trained person. This is in part because the geologist is accustomed to thinking in terms of many variables, and the management of the oil industry, possibly more than that of most others, must deal in variables, many of which are not known. It is also because the geologist understands better than anyone else the hazards upon which much of the exploration money and effort is spent, and can therefore better appraise the chances of its bringing in a fair return. The result is that the petroleum geologist is now the key in a complex interlocking of many specialized abilities, ranging from those of persons doing wholly scientific work to those of persons engaged wholly in management. The final chapter of this book is therefore devoted to an analysis of the place of the geologist in the petroleum industry and to some of the factors involved in the end point of his effort—the discovery of new pools or the extension of old ones. These factors might be thought of as centering on the petroleum prospect, which is that combination of geology, economics, and personal opinion that justifies the drilling of a test well.

This chapter is essentially a reprint, with some changes, of A. I. Levorsen, "The Petroleum Prospect," Mines Magazine, Colo. Sch. Mines, Denver, Colorado (October 1951), pp. 24–29.

DISCOVERY

The discovery of petroleum generally follows a more or less similar pattern. First, a prospect is located that is thought to be worth drilling; second, a test well is drilled into the potential reservoir rocks. The answers to the questions "What is a prospect?" and "What makes a prospect worth drilling?" combine not only geology but also economics and many personal attitudes in an almost infinite number of combinations and variations. Each prospect is unique. In giving thought to the various factors that go to make up a petroleum prospect, it should be remembered, first of all, that a favorable prospect is the immediate objective of the petroleum geologist and is a prerequisite for every wildcat well that is drilled. But the value of a prospect is partly determined by the varying needs of operators. Each prospect has a different value and appeal to each operator who might consider spending time and money in testing it. What seems like a good drilling prospect to X Company, for example, may hold no interest for Y Company with its different experience and needs. And what makes a first-class drillable prospect to an individual may have no appeal to a major company.

It should also be remembered that many prospects are found, and that many pools have been discovered, without the benefit of formal geology. By the very nature of the occurrence of oil and gas, especially in areas of stratigraphic traps, if enough holes are poked into the ground, discoveries will be made—even as the plum will be found if enough cuts are made into the pudding. Some of these discoveries come from what might be termed "magic geology": the area "looks like Pennsylvania," the creek "makes a bend like the one at Walnut Bend," or "any prospect near a town named Eldorado is good enough for me—remember the rich oil fields at Eldorado, Kansas, El Dorado, Arkansas, and Eldorado, Texas?" Others come from the thinking of the "lay geologist," who knows he should drill on a "high" or "get up against a show of oil" and remembers that "twenty years ago ole Joe Doakes said there was an oil field under the pile of rocks in the far forty"; and other discoveries come from the advice of "doodle-buggers"—those persons of mystery who continually seem to have some new method of locating oil and gas pools(?). The efficiencies of these methods are low; that is, the ratio of dry holes to discoveries is three or four times higher than with the orthodox methods of geology and geophysics; but in the aggregate they find many pools and furnish many needed well records for more scientific work. The discovery of oil is very much like fishing: you have to keep a baited hook in the water to catch fish, and you have to drill wells to find oil.

Whenever a wildcat well* is drilled, it means that someone, for some reason

---

* Wildcats have been classified by Lahee (*loc. cit.* in reference note 2, p. 1125) into *new-field* wildcats, which are drilled on a structure or in a geologic environment never before productive, and *new-pool* wildcats, which are drilled in the hope of finding a or other, has located a prospect that he believes worth testing. The fact that most wildcats result in dry holes is not a deterrent to the effort, just as the real fisherman is not discouraged if he doesn't get a fish every time he goes fishing.

For many years, the record in the United States has been that one out of every nine new-field wildcat wells produced some oil or gas. Of more significance, however, is that during the past 18 years it has required 38 new-field wildcat wells for the discovery of one oil field containing one million barrels or more.[2] A field containing a million barrels would supply only about one-tenth of the daily demand for the United States and for the operator may be either nonprofitable or barely profitable. The continuing need for locating many new prospects is readily seen.

While the petroleum prospect is first and fundamentally a geologic feature, the ultimate decision whether it justifies drilling depends upon many economic and personal factors along with the geologic factors. These three kinds of factors are useful headings for examining the question of what constitutes a prospect.

GEOLOGIC FACTORS

The geologic factors that enter into a petroleum prospect are essentially those pertaining to a reservoir—the reservoir rock, its pore space and permeability, and the trap. Of these, the trap is the most important, for it is the factor that localizes both the depth and the areal location of the prospect. The trap is a result of various combinations of deformation, such as folding, tilting, and faulting, together with various stratigraphic variations, such as lenticular porosity, facies changes, erosional truncation and depositional overlap, and solution cementation. The structural elements are generally the most readily determined before drilling because in many places they extend through a wide vertical distance and can be mapped by geological or geophysical surveying from the surface of the ground. After holes have been drilled and data concerning the potential reservoir rocks have become available, the stratigraphic information becomes increasingly important as a means of predicting traps, and many new combinations of structure with stratigraphy are revealed.

Probably two out of every three traps thought to be present are found to be nonexistent after drilling. The nonexistence of predicted traps may be due to the unreliable nature of the data that were used or to a misinterpretation of reliable but inconclusive data. Thus poor outcrops, facies changes, unconformities, alluvium cover, and miscorrelations all add hazards to the presence of the surface fold. Where holes drilled in the past have been widely scattered, the well logs and samples poor, and the correlations difficult, the subsurface maps are correspondingly unreliable. An actual showing of live oil and gas

---

new reservoir or an extension of a producing reservoir, but which are located on a structure or in a geologic environment already productive.

in a potential reservoir rock in the vicinity is significant, but if the evidence is obscure and the information has passed through several interested persons in the telling, it becomes of little importance. Seismic measurements may be thrown off by facies changes, buried faults, discordances associated with unconformities, and surface weathering conditions. Finally, when the trap is drilled for, it may not be there at all! Thus the hazards of drilling for a trap are considerably increased by the unreliable nature of the information that often must be used to predict whether a trap exists and, if so, just where it is.

Even after a trap has been located in a potential reservoir rock, an average of probably only one out of three or four traps will prove to be productive of oil or gas. The ratio will vary widely, however, for different regions. Some regions may produce from only one out of a great many porous and permeable formations that are found to occur as traps and are tested. Other regions contain only one or a few potential reservoir formations in the geologic column, but these may all produce wherever a trap occurs. The drilling of dry holes—holes that find only barren traps or no traps at all—is a calculated risk of petroleum exploration.

A geologist attempts to show his opinion of the *validity of the geologic data* on his maps. If he considers the information sufficient and is confident of its validity, he draws continuous contours and formation contact lines. As he becomes more and more doubtful of the data and their interpretation, he draws long-dashed lines, then short-dashed lines, and finally widely spaced dots. So, the shorter the continuous lines, the greater the risk, in his judgment. Frequently he has no clear idea what is happening underground but knows that the condition is not normal—that something anomalous is present. He can express such an idea on a map not only by using dotted and dashed contours, but by using special geologic symbols and by putting in short written descriptions on the map. The geophysicist, also, uses solid and dashed lines to indicate his opinion of the data, and, in addition, he frequently marks the seismic record G for good, F for fair, P for poor, and VP for very poor. Thus both the geologist and the geophysicist endeavor to give, on their maps, not only their idea about what kinds of traps are present but also their estimate of the validity of the evidence upon which their opinion is based.

Another variable besides the evaluation of the geologic data enters into the problem of intelligently predicting the position of a trap. It might be called a *confirmation of the data*. This means that if belief in the presence of a trap at a certain point is based on but one kind of evidence, the trap is not as likely to be found, and therefore not as attractive a prospect, as it would be if two kinds of evidence confirmed each other. It is still better if three independent kinds of evidence all point the same way. A surface terrace fold alone may indicate a trap at depth, but the probability of a trap below that terrace may be doubled if the surface evidence is supported by a seismic anomaly underlying the surface structure. And it might be more than trebled if, in addition to the surface and seismic evidence, a subsurface fold were indicated by good well logs. Three independent kinds of evidence indicating the presence of a trap at a certain location in a known reservoir rock would provide strong assurance to the operator who expected to drill the test well—provided, of course, the evidence for each was reliable. The problem finally resolves itself into a matter of personal judgment as to what weight should be attached to each bit of evidence and what confirmation one kind of evidence gives to the other kinds. In general, the more varied the kinds of evidence and the more certain the data that indicate a trap at a certain locality, the better the prospect geologically.

There is a wide range in the attractiveness of prospects. Suppose, on the one hand, that we have a well-defined surface dome, accompanied by ample seismic and subsurface evidence of deep-seated folding in harmony with the surface fold, all in an area containing well-authenticated showings of oil and gas in a thick potential reservoir rock. Such a prospect would represent a minimum risk. On the other hand, we may have the same trap potential except that all of the data are sketchy, hazy, and inconclusive. In such a case many new hazards are introduced. We find every gradation between an A prospect, which might be considered ideal, down to an F prospect, of such small interest as to be only slightly better than any random location in the surrounding region. The geologic considerations boil down to an estimate of the chances of finding oil or gas in the prospect if a test well is drilled. This estimate may be, and frequently is, expressed as one chance out of three, five, or ten, depending on the geologist's estimate of the number of similar prospects that would have to be drilled in order to get production in one of them.

The petroleum geologist is in many ways like a detective—he is forever following up and evaluating clues that might lead to the discovery of the pool of oil or gas. The location of traps is a deductive process. Some clues give direct evidence of the location of a pool at some specific spot, while others merely suggest the presence of a pool in some general area. In the latter case additional evidence is needed in order to localize the search. Sometimes this can be obtained by reworking the geology with greater detail, by closer spacing of the seismic shot points, by re-examining the geophysical measurements, by shallow core-drilling, or by employing more experienced men. At other times there is enough evidence to justify the drilling of a test well, which, even if it is not productive, will give valuable added information so that the next well location will have a better chance of being productive. Such holes are called "strat holes" or "stratigraphic tests," which implies that their chief purpose is to secure additional information rather than to find a new pool of oil or gas. Many operators expect that to make a discovery on a prospect may require two, three, or even more test wells, and in their planning they take this possibility into consideration. Many prospects have been abandoned by an operator after one test well has been drilled dry, only to be drilled successfully later by some other who interprets differently the information found in the first test well and thereby makes a better location Pools have even been discovered by opening up and testing "dry holes" when the log and data were reinterpreted by another geologist. The ingenuity of the geologist in finding clues and putting them together in their proper relation has much to do with the solution of the mystery—the location of the pool.

ECONOMIC FACTORS

The mere discovery of one small oil or gas pool is not enough; there have to be enough recoverable oil and gas discovered to return a profit above and beyond all the costs of exploration, drilling dry holes, and operation. The petroleum geologist must realize and understand the economic elements involved in exploring for oil, and he is more and more required to show the chances for a potential profit along with his recommendation of a favorable drilling site.[3] Because most pools discovered are not profitable, most operators down-grade a prospect if it does not appear to contain enough petroleum to repay the costs plus a profit and thereby to justify the time and effort spent. Discounting of a potentially small pool is especially frequent among the larger oil companies, whose crude oil requirements run into many millions of barrels per year and whose operating expenses are proportionately large. They are not much interested unless the prospect has some chance of being a substantial help in their supply problems. The amount of oil or gas in a reservoir cannot be predicted in advance of drilling, it is true, but where the trap is small and where the sands are known to be thin and deep, the chances of a profit being realized become progressively smaller. Such prospects may be attractive to the individual operator, who has lower overhead and can operate more cheaply than the major company, or to the operator whose oil and gas requirements are not as large as those of the large, integrated company.

The desirability of a prospect is generally in direct ratio to the potential profit, but profits for the large company are on a totally different scale than profits for the individual. One operator may drill a prospect without hoping to make any profit beyond what would come from the refinery operations after the oil is produced—with no hope, that is, of direct profit from production. The requirements of another operator can be satisfied only by large and directly profitable production. Every geologic prospect, then, must face the profit requirements of the operator when the drilling of a test well is being considered. The greater the potential profit, the less vital it is that all known geologic conditions be favorable, and the greater the sum that can justifiably be risked in testing the possibilities.

The economic evaluation of a prospect also depends on the total money outlay required to acquire the land and then test it. If the probable outlay for a single prospect is large in terms of the budget or in comparison with the resources of the operator, he considers the reports on its geology and geophysics much more critically than if it is small. He may be less critical if he plans to sell a part of the prospect and share the costs as well as the risk and thereby to have funds to participate in the testing of several prospects.

The economic factors that apply to the anticipated profit from geologic prospects and that thus influence their evaluation before they are drilled include the following:

Reserve Estimates. The objective in drilling a petroleum prospect is to discover a pool of oil and/or gas that will return a profit on the investment. The first consideration is to form some idea of the size of the pool—if it should exist within the prospect. Is its potential content one, fifty, or one-hundred million barrels of productive oil? Usually the experience gained from the discovery of pools in areas of similar geology serves as the best guide, both to the amount of oil recoverable by primary methods of production and to the amount recoverable by secondary methods—fire-flooding, steam treatment, water-flooding, and others. The relative importance of the factors that go into locating the prospect is more or less determined by the size of the pool that might be expected.

The Price of Petroleum. The price paid for oil and gas at the well constitutes the entire income of the producer; there is ordinarily no scrap, no by-product, nor any accessory product to sell.* In determining the profit, all of the costs up to the moment the oil or gas leaves the well and enters the pipe line must be deducted from the sale price.

The price of oil and gas at the well varies with the demand, with the character of the oil and gas, with the distance to refineries, and with the kind of transportation available—truck, boat, railroad, or pipe line. Each prospect is therefore evaluated with respect to these factors, which in turn control the price. A prospect close to a company's own pipe line, for example, is more desirable than one a long distance away, and consequently is up-graded and requires less favorable geological reasons to justify drilling. A prospect in a heavy-oil area, where truck transportation is the only method of carrying the oil to the refinery, and where the chief demand is for gasoline rather than for the lower-priced heavy oils, requires much more favorable geological reasons to justify the risk than a prospect located where the oil is of high API gravity and meets the prevailing demand, and where transportation costs are low, all favoring a high price for the oil and potentially larger profits.

As the price of oil and gas at the well fluctuates with the demand, wildcatting activity also fluctuates. Even the prospect of a drop in price will cause a lessening of interest in exploration. When the price drops, drilling tools are stacked, leases are canceled, and operating costs are cut as much as possible. This means that during such a period many operators drill only the best prospects or those that must be drilled in order not to lose an attractive lease. An increase in price, on the other hand, means a renewed demand for prospects, and those of smaller size, with fewer data to go on, and of more uncertain characteristics, become attractive again. The price of oil and gas has an immediate and important effect on

---

* This condition is gradually changing, however, with respect to some gas pools of a high sulfur content. The sulfur may be extracted at a profit and thereby add to the producers' income from the pool.

the rating of every prospect, and it is watched with the greatest interest by all operators.

The net result of the fluctuating price and exploration activity is that for many years the recoverable oil reserve in the United States has ranged between twelve and fifteen times the annual production. This seems to be an optimum equilibrium relationship, possibly based on the cost of discovery and the carrying charges involved in developing the oil fields and keeping the oil in the ground for the ten or fifteen years until it is produced. At any rate, when the ratio of reserves to the annual production increases, the price tends to go down; the wildcat effort consequently declines, and less oil is discovered. Then the ratio of reserves to annual production decreases, the price advances, more wildcat drilling follows, and so the cycle keeps repeating itself through the years.

Geological and Geophysical Costs. The geological and geophysical costs—sometimes called the G & G costs—that enter into the location of a prospect are variable. They may range from a nominal sum to more than the cost of drilling the test well. Sometimes all the geologic evidence is found in reports published either by geological surveys or by individuals in the technical journals, and in such cases the cost is negligible. At the other extreme, we find areas that have been worked over by every known geological and geophysical method and have required years of effort to secure the necessary evidence for a prospect. For example, a seismograph crew may work in an area for months or even years at $20,000, $30,000, or even $50,000 a month, and find only a few prospects. The seismic work may be preceded by reconnaissance and detailed gravity and magnetic surveys; core holes may have been drilled, and the surface mapped geologically by surface and aerial crews. The geological and geophysical investment in such a prospect becomes enormous. The chances are that even if the potential of such a high-cost prospect were rather doubtful, it would nevertheless be drilled, since the cost of drilling would probably be only a minor addition to the prior costs. It might be looked upon in one of two ways—either as "good money following bad" or as "protecting the prior investment"—depending on the operator and the geologic conditions in the surrounding region.

Lease Costs. The cost of the oil and gas lease,* whether of a small tract containing only a few acres or of a large concession, is divided into three parts: the initial cost, or bonus payment; the royalty, or amount of oil reserved to the landowner or the owner of the mineral rights; and the rental, or cost of maintaining the lease in force through its specified lifetime. In marginal regions where the geologic prospects for discovery are indefinite or uncertain, the initial lease costs per acre are generally low, and leases of large tracts may be obtained at small cost. As the chances of discovery increase, either as a result of discoveries in the region or in similar geologic situations elsewhere, or because the geology of the region has become better known, the initial lease costs increase, and so do the rental payments and sometimes the royalty. The leases that bring the highest

---

* Ownership of the oil and gas by the operator is sometimes by fee title; generally it is by a lease that sets out the division of ownership of any oil or gas produced during the term of the lease. The proportion of oil and gas the operator (the *lessee*) owns is called the *working interest* and that of the landowner (the *lessor*) the *royalty interest*.

prices are those close to large discovery wells, where the probability of obtaining a similar large production is correspondingly high.

In evaluating a prospect it is generally expected that the unknown geologic variables will probably outnumber the known variables. For that reason many of the actual underground conditions can only be surmised, and may turn out, when the test wells are drilled, to be different in many essential respects from what was predicted at the surface of the ground. Experience has often shown that pools are larger or smaller than predicted, that facies changes place the oil and gas on one side of the anomalous geologic area instead of on top, or that the subsurface conditions can be determined only after several test wells have been drilled. In fact, the geologic reasons for drilling discovery wells are often not the reasons that are later found to cause the pool to be where it is. Embarrassing to the geologist—but nevertheless frequently true. It is therefore generally desirable to obtain a large block of leases, covering the geologic anomaly and extending out on all sides to include every conceivable possibility of production. Large holdings permit the operator to drill additional test wells if the first should prove to be dry, and if production is obtained the chances of profit are greater with a single than with a divided ownership.

The imminence of expiration of leases frequently causes an upgrading of all prospects in the same region. An operator feels that it is better to drill some test wells than to spend the same money for bonuses for lease renewals, which will still require drilling if he is to learn whether they are productive. Large-scale drilling campaigns, in which a large number of wildcat wells are suddenly started throughout a region where no important discovery has recently been made, are often due to the fact that a large number of leases will expire within a year or two. Such a situation is especially common in regions where the stratigraphy is known to be favorable for petroleum production and many leases have been taken, even though it is difficult to obtain good localized geological and geophysical data. In these circumstances, a prospect that would normally be regarded as of quite low grade may become drillable.

Large tracts of land are frequently called *concessions,* especially when the lessor is a government. Concessions, as well as large leases from individuals or land company lessors, frequently call for a continuous drilling or exploration program, which may be considered as a part of the lease cost.

In evaluating every geologic prospect, then, the cost of obtaining a lease has to be taken into account. Where that cost is high, the geologic conditions must be more definitely known and more favorable than where it is low. Where the desirable and available area is large—where, in other words, there is a chance of high profits—the operator may be ready to take a heavier risk, and may not demand that the geologic conditions be as fully known. Low lease costs also permit a greater percentage of the available money to be put into drilling—which, after all, is the only way of actually discovering the petroleum pool.

Drilling Costs. The cost of drilling is chiefly a function of depth, but other factors may be of considerable importance in the overall cost. The character of the formations to be penetrated, the number of production tests to be made, or the cost of preparing the well site may greatly increase the normal expense of drilling. The distance from supplies, the cost of the road into the location, the manner of

THE PETROLEUM PROSPECT [CHAPTER 15] 667

transportation (train, truck, or air), and the distance from which water has to be brought or fuel obtained, may all become important in certain locations. And when drilling is at maximum depths for the equipment, the costs rise rapidly for the lower part of the hole.

Since the cost of drilling is an essential element in the amount of profit that can be obtained, the necessary depth of drilling is an important factor in the value of every geologic prospect. The depth factor applies not only to the first test well but also to the additional drilling expense that will be necessary if production is discovered. The cost of producing the oil after the pool has been drilled is likewise influenced by the depth of the pay formation. A prospect at 2,000 feet, for example, does not require nearly as much potential petroleum reserves in order to be attractive as one at 15,000 feet. The risk involved in drilling one test well to a deep reservoir of larger potential reserve, moreover, may equal the total risk of drilling many shallow prospects with a smaller potential reserve, and this difference has a varying appeal to different operators. A shallow prospect is ordinarily more desirable, however, than a deep prospect of equal geologic merit. In other words, the shallow prospect is upgraded because the drilling and producing costs involved are generally less than in the deep prospect.

Production Costs. The anticipated cost of producing the oil or gas after it has been discovered also has a bearing on the evaluation of a prospect. Some reservoirs may flow throughout their productive life, whereas other reservoirs require expensive pumping soon after they have been discovered. Oil from some reservoirs contains much loose fine sand, which damages equipment and causes a high percentage of shut-down time on the wells. In some provinces, much trouble is caused by corrosive water that eats holes in the pipe, necessitating costly repairs, pulling of tubing and casing, and new equipment. The nature of the expected reservoir energy—dissolved gas, gas cap, water drive, or any combination of them—has a great bearing on the cost of production as well as on the amount of recoverable oil reserve. And finally the royalty oil, which must be produced and furnished free of cost to the owner of the mineral rights, has a bearing on the cost of production.

While the standard primary royalty rate in the United States is 12.5 percent, the rate varies widely with special situations, especially where brokers' commissions are taken as additional royalty. In foreign countries, royalty rates plus special taxes or profit-sharing arrangements often add up to a high rate of payment to the lessor, at times reaching 50 percent of the net profit or even more. Royalty oil production, plus any added payments to the lessor, must be added to the cost of producing the oil belonging to the operator. Water is produced along with the oil at the expense of the operator, and when large amounts are involved the profits are materially reduced. Royalty payments may then become disproportionately large compared with the operating expense and, if not reduced, may result in early abandonment of the property.

Depletion. The United States government recognizes that every time the producer pumps up a barrel of oil, or produces a cubic foot of gas, he produces not ordinary income alone but also a portion of his capital. And when capital consists of petroleum underground, it is, like other mineral substances, irreplaceable once it is produced. Because it is illegal to tax capital as income, an arbitrary allowance is made that enables the petroleum producer in the United States to maintain his capital as it is depleted by production. It is somewhat analogous to the depreciation allowance on equipment, which enables the operator to replace equipment as it is worn out with use, although depreciation is an allowance to replace something that can be rebuilt, whereas depletion is an allowance to replace something that must be discovered. Depreciation and depletion, which oil-field accountants often speak of together as D and D, are both allowances to protect the business against exhaustion of capital. The depletion allowance is deducted from income before taxes are calculated. As a result of the depletion allowance, the producer who wishes to restore his capital reserve may either buy oil and gas to replace the oil and gas produced annually or risk the allowance to drill test wells on prospects, in the hope that these will discover enough new petroleum to replace the oil and gas he has produced and thus keep his capital reserve of petroleum intact.

As we have seen, eight out of nine wildcats are dry—wildcatting is a risky business. Yet it is absolutely essential to the national interest of the United States that new oil and gas be continually discovered to meet the increasing petroleum demands, and the only known way to discover new petroleum reserves is to drill wildcat wells to test the geologic prospects. Experience has amply demonstrated that, as more prospects are drilled, more petroleum is discovered. Many discoveries, moreover, are made from prospects that originally did not appear to be especially attractive—high-risk prospects, they might be called. The money provided by the depletion concept and ear-marked for exploration encourages the drilling of more of these apparently lower-grade prospects. Thus depletion enters into the consideration of most prospects, and the result is that many wells are drilled that would not otherwise be drilled.

Tax Position. If an operator has high profits and is subject to high income and excess-profits taxes, both federal and state, the greater part of his income annually may have to be paid as taxes. Instead of paying out such a large proportion of his income in taxes, he may decide, paradoxically, to spend some more money to drill additional wildcat test wells. In doing so he reduces his net income and profits, but at the same time he also reduces his taxes. If, for an oversimplified example, he has to pay out 75 percent of his income annually in taxes, he may take a part of this money and drill a prospect that offers large returns if the drilling leads to a discovery. If the wildcat well is a failure, he charges off the cost—known as the *intangible drilling cost*—as expense, and the actual cost to him is only 25 percent of the total, since the other 75 percent is money that would otherwise have been paid out in taxes. He can afford to take a greater risk because he assumes only a small part of the cost of the venture. And, from the standpoint of the taxing agency, such as a state or the federal government, if he is successful in discovering oil or gas and profits result, the ultimate tax payments will be so much the greater; in other words, the government assumes a large part of the risk of the wildcat venture, and in return gains a large part of the increased income and profits. The government, because of its complex income-tax regulations, is, in effect, in the oil business, as indeed it is in all business.

Exploration money resulting from high tax brackets comes both from without and from within the petroleum industry. The money that comes from without the industry is obtained from individuals and companies in other lines of business who have high taxes and who are looking for a place to invest money where, though the chances of success may be small, the profits would be exceptionally large if the venture should prove to be successful. Drilling a wildcat well on a prospect that has possibilities for discovering a large reserve of oil or gas is often an ideal investment for such money.

The tax money that comes from within the petroleum industry generally shows up after the first half of the taxable year has passed and after it is learned that large taxes will be due at the end of the year. An operator in such a position may decide to drill more wildcat wells with the money that he would otherwise pay out in taxes, and in doing so he will upgrade many of his prospects—prospects that he may not have considered for drilling earlier. The net result is that many of the lower-grade prospects are drilled because of such a tax position, and consequently more petroleum is discovered. His tax position has some bearing on the decision of every operator to drill or not to drill a particular prospect, and, for the operator with a large taxable income, it may become a dominant factor.

PERSONAL FACTORS

In addition to the objective elements of a prospect—the geology and economics—there is also the personal, or subjective, element, which frequently dominates the entire evaluation. Past experiences, whether good or bad, whether geological or financial, whether successful or unsuccessful, all permeate and influence the operator's judgment regarding every prospect considered for drilling. One operator may be conservative and cautious, while another may be inclined to take longer chances; one operator may like to work with heavy equipment, big power units, and deep holes, whereas another may be better satisfied with lighter equipment and shallow holes; and one operator may be more scientific, more technical, or more curious than another who has a more practical or more down-to-earth background. Personal factors such as these may be multiplied indefinitely, and each has a bearing on the final and overall judgment as to the class or rank a prospect is assigned.

Probably the most important personal element that applies to a particular prospect is empirical—the past experience of the operator either with the same kind of geology or in the same general region. A few dry holes or an unusually expensive and difficult drilling operation often cools the ardor of an operator for another prospect near by. After such an experience, a frequent reaction is that, no matter how good the prospect appears geologically, it is in the "wrong country" or the "wrong township" or is just "no good." The effect is to require better geological salesmanship. In every new province there are some operators, both large and small, who because of their previous experiences became discouraged too soon, and failed to admit or visualize the petroleum possibilities of the region until these were finally developed by other men.

Sales, mergers, and loans sometimes have a bearing on the evaluation of a particular prospect. An operator trying to make such a deal wants to put his best foot forward, and he may think that his chances will be improved if wildcat wells are being drilled on his property. A low-grade prospect may then suddenly become desirable for drilling—there is nothing quite like a wildcat well drilling on a good block of leases to excite one's imagination! Hope, ambition, profits, and success are all fused into one idea—"Let me in!"

Some of the personal factors are geological. Many deeply embedded geologic opinions have very little real evidence to support them; they are really prejudices, yet they have a tremendous effect on the geologic evaluation of many prospects.[4] The personal opinion of a geologist about petroleum source beds, about continental, or nonmarine, rocks, about the effect of faulting, about the presence of fresh, or low-concentration, waters in the reservoir rocks, about long-distance as against short-distance migration, all have a direct bearing on his evaluation. Geologists kept certain operators out of eastern Texas because "there was no Woodbine sand west of the Louisiana line," yet the absence of sand there and its presence still farther west are largely responsible for the great East Texas pool. "No source rocks" kept other operators out of the West Texas province in the early days, and "oil and red beds do not mix" was a common geologic belief when central and western Oklahoma were first being explored.

Petroleum is discovered by drilling favorable prospects. A positive approach is therefore essential on the part of the geologist. Oil and gas pools are not found by searching in a pessimistic spirit for conditions that seem unfavorable and then recommending against drilling because of them, or in merely keeping geologic maps up to date. It is more to the purpose, although it is far more difficult and requires a lot more thought, to find something favorable and worth drilling about an area in which there are some conditions that geologists may consider unfavorable. Many geologic prejudices are extremely difficult to overcome, and it is surprising how often and how much they color geologic thought about an area—most frequently to its disadvantage.

Many of the factors depending on personality and experience are difficult to set down in black and white, and difficult to apply systematically to a new prospect about to be classified. Some operators attempt to set up objective check or rating sheets, by which a prospect may be evaluated and the result given in numerical terms. Various factors may be graded by the letters A, B, C, and D, or in percentages, or in chances of success, as one out of three, one out of five, and so forth. These standards of comparison are based on estimates of such things as geologic factors, expected profit, potential reserve, and capital required. Both similar and dissimilar prospects may thus be compared, and below-standard prospects automatically discarded or "farmed out" to some other operator who has different standards. First-class prospects are difficult and expensive to locate, and some operators feel that by drilling many prospects of a lesser grade they will eventually make discoveries and average out a profit. This is called depending on the "law of averages." No scheme has yet been devised, however, by which prospects could be classified for all operators, for each operator has his own blend of requirements, his own background of geologic, economic, and exploration experience, and his own ideas concerning what *to him* constitutes a good prospect and what a poor one.

The essence of the work of the petroleum geologist is expressed in the three words: "Find a trap!" Although every petroleum pool occurs in a trap, it does not follow that every trap contains a pool. Why are many traps barren of oil and gas? We do not know. Our approach to exploration has been almost entirely empirical. The petroleum geologist is forced to choose among many theories in trying to find an explanation for a dry trap. Consequently, the safe procedure is to test *every* known or suspected trap—at least until we have developed a successful method for predicting the presence or absence of petroleum in a particular trap.

Freedom to explore is greater in the United States than it is anywhere else. Here the mineral rights are generally owned by the surface owner while the exploration is done by a highly competitive industry. In other countries freedom to explore is limited, in greater or less degree, and in some countries all the petroleum is government property, and exploration also is wholly in the hands of the government. Diversity of approach—also called multiple effort— operates only where the economic and political climate permits the widest freedom to test any personal, economic, or geologic idea. This system, as worked out in the United States, has found and will continue to find the highest percentage of the petroleum that exists underground, because it gives the most encouragement to the greatest number of individuals in testing their personal ideas about how to discover petroleum. Once test wells have to be approved and dry holes explained to a Congressional committee or to a commissar representing the political power of a country, there will soon be few if any dry holes because few holes of any kind will be drilled. Every geologist knows that there are no sure prospects and that the majority of wildcat holes must be dry. The inevitable result of freely testing geologic ideas would bring the geologist continually "on the carpet." Politicians find it difficult to understand that the information from dry holes repeatedly furnishes the basis for the discovery of a pool and that dry holes are therefore an essential part of the exploration program. Successful exploration requires that many dry holes be drilled, even though the aim is to drill as few as possible.

A rather significant measure of the ability of the petroleum industry to maintain its ratio of successful to unsuccessful prediction of oil under the ground is contained in the figures that relate the new crude oil reserves found in the United States to the number of feet of hole drilled.[5] During the period 1935–1941 there were 25.1 barrels of oil found per foot drilled; in the period 1942–1945 the rate was 24.6 barrels of new reserves per foot drilled; and in the period 1946–1951 the rate was 26.5 barrels of new oil found for each foot drilled. The footage of exploratory wells drilled each year that result in dry holes bears about the same relation to the footage that results in discoveries,[6] ranging between 3.2 and 3.9 feet in dry holes to each foot in a petroleum discovery well. This suggests, in other words, that the industry is maintaining its ability to find new oil in spite of the increasing difficulties, that it is developing techniques and new ideas that meet the new problems of discovering and producing oil at continually greater depths and of finding new areas for exploration as old areas decline. The petroleum geologist may well be proud of the large part he has contributed to this record of achievement, and it shows that both he and the industry are keeping pace with the petroleum needs of the nation.

CONCLUSION

Behind the rating of a petroleum prospect, then, we see an infinite diversity of fact, reason, experience, and prejudice. Some factors are quantitative, some qualitative, and some are merely sensed. It is this very diversity of approach, however, that has made the petroleum industry strong in the United States and wherever else it is permitted to operate. If exploratory effort—and that means first of all the selection of the prospect that is to be tested by the drill—were to be guided by any one exploration philosophy, only a fraction of the ultimate petroleum reserve would ever be found. Many, many prospects that were originally found dry at one location have later turned out to contain pools because some other operator, with a different approach, drilled them at a slightly different location and found oil or gas. Even minimum geologic prospects, such as drilling where the only known geologic reason for doing so is that a sedimentary section will be tested, have produced pools such as the East Texas and Glenn pools and countless other stratigraphic accumulations that gave no advance clue to their presence underground. The blending together of all shades of geologic, economic, and experience factors to find a drillable prospect is thus a singularly personal and individualistic process. Only by such a process, however, can we hope to find the thousands of drillable prospects needed every year to guide the wildcat effort necessary to meet our rising world-wide demand for oil and gas.

Selected General Readings

John M. Campbell, *Oil Property Evaluation*, Prentice-Hall, Inc., Englewood Cliffs (1959), 523 pages. An authoritative statement of the economic factors involved in evaluating an oil property. Contains extensive section on interpretations of logs and cores.

Reference Notes

1. Wallace E. Pratt, *Oil in the Earth,* University of Kansas Press, Lawrence, Kansas (1942), p. 9.

2. J. Ben Carsey and Marion S. Roberts, "Exploratory Drilling in 1962," Bull. Amer. Assoc. Petrol. Geol., Vol. 47 (June 1963), pp. 889–934. Detailed results of exploratory drilling in North America are presented in the June issue of the bulletin each year. A great volume of detailed information is assembled and analyzed by a large number of geologists.

3. Ben F. Rumfield and Norman S. Morrisey, "How to Evaluate Exploration Prospects," Geophysics, Vol. 29 (June 1964), pp. 434–444. Includes bibliography.

4. Earl B. Noble, "Geological Masks and Prejudices," Bull. Amer. Assoc. Petrol. Geol., Vol. 31 (July 1947), pp. 1109–1117.

5. Western Hemisphere Oil Study Committee, Indep. Petrol. Assoc. Amer., Washington, D.C., *Report on Petroleum in the Western Hemisphere* (October 1952), pp. 61 and 75.

6. J. Ben Carsey and Marion S. Roberts, *op. cit.* (note 2), p. 915.

APPENDIX

Many useful tables and data are published by, and may be obtained from, the trade journals, especially *The Petroleum Engineer* (Dallas, Texas), *The Oil and Gas Journal* (Tulsa, Oklahoma), and *World Oil* (Houston, Texas), and may be found in various handbooks, such as Joseph Zaba and W. T. Doherty, *Practical Petroleum Engineers Handbook,* 3rd ed., Gulf Publishing Co. (Houston, Texas), and in brochures and handbooks of the logging and cementing service companies.

GLOSSARY OF PETROLEUM EXPLORATION

Albertite. One of the asphaltic pyrobitumens commonly found in veins at Albert Mines, New Brunswick, Canada, where it was first called Albert coal.

Asphalt (Greek, bitumen). A brown-to-black, solid or semisolid bituminous substance that occurs naturally but is also obtained as a residue from the refining of certain petroleums and then known as artificial asphalt. Asphalt melts between 150° and 200°F. The substance *asphaltite,* which imparts the hardness to asphalts, is believed to belong to a colloidal system. Asphaltite resins are responsible for the ductility of asphalt, or its capacity to elongate under tension without breaking. Natural asphalts generally include inert mineral matter of specific gravity 1.0–1.15, are from 10 to 70 percent soluble in petroleum naphtha, and are wholly soluble in carbon disulfide.

Asphaltic pyrobitumens. Harder and more infusible than asphaltites. Their specific gravity does not range above 1.25, and they do not melt; rather they swell and decompose before a melting temperature is reached. They are insoluble in petroleum naphthas and only slightly soluble in carbon disulfide. Wurtzilite, elaterite, and albertite are examples.

Asphaltites. The harder solid hydrocarbons, with melting points between 250° and 600°F. The specific gravity is less than 1.20; the solubility in petroleum naphtha ranges between 0 and 60 percent and in carbon disulfide between 60 and 90 percent. Gilsonite and glance pitch are asphaltites.

Bailer. A long cylindrical container with a valve at the bottom, which is used for removing the water, cuttings, and mud from the bottom of a cable-tool well.

Basic sediment. The residue that settles out of oil kept in a tank and is generally found covering the tank bottom when it is cleaned.

Bit. The drilling tool that actually cuts the hole in the rock. Both cable-tool and rotary-tool bits are of various designs, depending on the kind of rock being drilled, and many are patented.

Bitumen. Originally, native mineral pitch, tar, or asphalt. Now the term generally means any of the inflammable, viscid, liquid or solid hydrocarbon mixtures soluble in carbon disulfide; often used interchangeably with "hydrocarbons."

Bituminous. Containing oil, or yielding oil on distillation. A kerogen shale and an asphaltic sand are examples of bituminous rocks.

Bottom settlings. Basic sediment, *q.v.*

Brea. A viscous asphalt formed from the evaporation of oil seepages.

Cable-tool drilling. A method of drilling whereby a steel bit of various design is fastened to a drill stem and jars, and the whole fastened on a wire line. When given an up-and-down motion, the bit cuts the rock by impact. The drill cuttings are removed by bailing them out after the tools have been removed from the hole. Formation waters and caving walls are controlled by setting steel pipe, called "casing," in the hole and sealing it in with cement. Drilling is then resumed with a smaller bit size. Holes range from three to thirty inches, the larger sizes being near the surface and used for deep holes.

Casing. Steel pipe of varying diameter and weight that comes in joints from sixteen to thirty-four feet in length, which are joined together by threads and couplings at the well. Casing is "run" into the well hole for the purpose of supporting the walls of the well and preventing them from caving, for shutting off water either above or below the pay formation, and for shutting off portions of the pay formation such as shales, loose sands, or gas. The pay formations are opened up by perforating.

Cement. Placed in the hole and forced in behind the casing in order to seal the casing to the walls of the hole and prevent any unwanted leakage of formation fluids into the well.

Chapopoteras. In Mexico, seepages, chiefly asphaltic.

Christmas tree. The assemblage of valves, pipes, and fittings assembled at the top of the well to control the flow of oil and gas.

Churn drilling. Cable-tool drilling, *q.v.*

Coorongite. An algal accumulation of the waxy or gelatinous cell walls of the alga *Eleaophyton* forming on lagoonal waters in South Australia.

Crude (oil). The liquid petroleum as it comes out of wells.

Directional drilling. The controlled drilling of deflected holes by special orientation surveys.

Drill column. Drill pipe, *q.v.*

Drilling mud. The watery mud pumped down the drill pipe, out through the bit, and back up in the annular space between the drill pipe and the walls of the well, in rotary drilling. The mud carries the cuttings up from the bottom, sheathes the walls of the holes with a mud cake that prevents the walls from caving and keeps out formation water. The weight of the mud is regulated by adding various substances so that it is greater than the formation pressures expected. See Table A-5.

Drill pipe. A string of steel pipe, screwed together and extending from the rig floor to the drill collar and bit at the bottom of the hole. The drill pipe transmits the rotating motion from the derrick to the bit and conducts the drilling mud ($q.v.$) from the surface to the bottom of the hole.

Drill stem. A cylindrical bar of steel or iron screwed onto the cable-tool bit to give it weight. The jars are attached either above or below the stem, depending on the action wanted.

Drill-stem test. A test of the productive capacity of a well when it is still full of drilling mud. The testing tool is lowered into the hole attached to the drill pipe and placed opposite the formation to be tested. Packers are set to shut off the weight of the drilling mud, and the tool is opened to permit the flow of any formation fluid into the drill pipe, where the flow can be measured.

Elaterite. An asphaltic pyrobitumen, originally found in the lead mines of Derbyshire, England, in Carboniferous limestone, and known, because of its elasticity, as "mineral rubber."

Fishing. Searching for a piece of drill pipe or drilling equipment broken off from the drilling tools and left in the hole. Many special tools have been designed for catching hold of the missing tool in the well.

Gilsonite. A solid pyrobitumen, an asphaltite, named after S. H. Gilson, the owner of the deposit where it was first observed, and found in the Uinta Mountains of eastern Utah and in western Colorado. It occurs in veins in Tertiary shales. Also called uintaite.

Glance pitch. An asphaltite, also known as manjak. It occurs on Barbados Island, BWI, and in Colombia.

Grahamite. An asphaltite, named after J. A. and J. C. Graham, owners of the mine where it was first observed. Found in West Virginia, Oklahoma, Cuba, and Mexico.

Gravel packing. The placing of gravel (coarse sand, mesh 8–15) opposite a producing sand to prevent or retard the movement of loose sand grains into the well along with the oil. Usually forced through perforations under pressure.

Gun perforating. A method of completing producing wells by shooting steel bullets through the casing into the producing formation. The mechanism for firing the bullets is operated electrically, and the effect is to open up the reservoir rock to the well bore at the exact depth for the greatest production, as indicated by various logging and sample data.

Hatchettite. A yellow-white mineral wax, somewhat similar to paraffin wax, found in cavities in certain nodules of ironstone that occur in the coal-bearing strata of South Wales. Also called mineral tallow.

Hydrocarbon. Any compound of hydrogen and carbon, whether gas, liquid, or solid. The term is used interchangeably with "bitumen" for liquid and solid petroleums.

Impsonite. An asphaltic pyrobitumen, often found in Impson Valley, Oklahoma. High in fixed carbon and only slightly fusible.

Jars. Two links, resembling chain links, hooked loosely together, that are fastened to the stem above the bit in a cable-tool drill for the purpose of giving a sharp jerk to the bit.

Kerogen. A mineraloid of indefinite composition, consisting of a complex of macerated organic debris and forming the hydrocarbon content of kerogen shales. It consists chiefly of low forms of plant life, such as algae, of pollen, spores, and spore coats, and it may also contain the remains of insect adults and larvae that lived in the enclosed basins where it formed. Chemically it is a complex mixture of hydrocarbon compounds of large molecules, containing hydrogen, carbon, oxygen, nitrogen, and sulfur. The term "kerogen" was first used to describe the material in the Scottish shales, which on destructive distillation yield oil.

Maltha. A black, viscid, mineral tar, or asphalt; a variety of ozokerite.
Manjak. A solid pyrobitumen found on Barbados Island, BWI.

Naphtha. Originally, liquid petroleum; now, the hydrocarbons that boil below 250°C and are liquid at standard conditions. Used as a cleaning fluid and solvent.
Natural gas. A petroleum that is a gaseous mixture under surface conditions of temperature and pressure but some of which becomes liquid under ground with higher temperature and pressure. Natural gas consists predominantly of paraffin hydrocarbons, chiefly methane and varying but generally small amounts of the heavier paraffins.

"Oil shale." Properly called kerogen shale because the hydrocarbon content is pyrobituminous and insoluble in ordinary petroleum solvents. "Oil shales" are organic shales that yield petroleum hydrocarbons upon destructive distillation.

Ojos de chapopote. In Mexico, seepages, chiefly asphaltic.

Ozokerite (Greek, "odoriferous wax"). A naturally occurring mineral wax, usually dark brown, believed to result from the drying out of a paraffin-base oil. Veins occur in the Boryslav area, USSR (formerly Poland), where it has been mined. There it extrudes from holes when the pick is driven into the vein, and the mine has been called an "asparagus mine." Ozokerite is also found at Soldier Summit, Utah, on the Cheleken Peninsula, Turkmen SSR, in the Fergana Valley, Uzbek SSR, and in India. Fully refined ozokerite is a white-to-yellow microcrystalline wax, also known as ceresin, and is used in Vaseline. Ozokerite has the same relation to paraffinic oils that natural asphalt bears to the naphthenic oils. It is essentially a paraffin derivative ranging from C_{22} to C_{29}.

Packing. See Gravel packing.

"Paraffin dirt." A clayey soil of rubbery or curdy appearance and texture, which resembles Art Gum in appearance. It occurs in the upper few inches of the soil and is probably formed by a mass of low-rank organisms that live on the gas that seeps out along the Gulf Coast of Texas and Louisiana. There is probably no true paraffin in it.

Perforation. See Gun perforating.

Petroleum. Rock oil; a gaseous, liquid, or solid mixture of a great many hydrocarbons and hydrocarbon compounds occurring naturally in the rocks. The term was first used by Agricola in 1546.

Pitch. Asphalt, or mineral tar.

Plugging. Sealing up a hole that was dry and is to be abandoned. It is generally done under governmental regulations, so that deep saline waters do not enter surface-water wells or contaminate other reservoirs in the vicinity.

Pyrobitumen. The solid hydrocarbons and kerogen shales that give up liquid and gaseous hydrocarbons only when heated.

Rig. The derrick and other surface equipment used in drilling.

Rotary drilling. The commonest method of drilling wells. A cutting bit, of varying design for different kinds of rock, is screwed to one or more hollow drill collars to give added weight, and these are screwed onto the bottom of the hollow drill-pipe, which is rotated from the surface. A stream of drilling mud passes down the drill pipe and drill collar and out through the bit, from where it is forced back up the hole outside the drill pipe and into pits, where it is again picked up by the pumps and forced back down. Holes more than 21,000 feet deep have been drilled in this manner.

Separator. An apparatus for separating oil from the gas that is dissolved in it or is produced with it as it reaches the surface.

Shale-shaker. A device on the rig that separates the coarser well cuttings from the drilling mud as it comes out of the well.

Side-wall coring (side-wall sampling). A device whereby a gun shoots from six to thirty hollow cylindrical bullets into the formations exposed in the well hole. Each bullet is fired separately by electrical control, and the bullets, with their rock cores, are retrieved by wires attached to the gun. Cores are three-quarters of an inch in diameter and may be up to two inches in length.

Spud (in). To commence actual drilling operations on a well.

Surface pipe. The first string of casing set in a well, generally used to shut off and protect shallow, fresh-water sands from contamination by deeper saline waters.

Tar. A thick, black or dark-brown, viscous liquid obtained by distilling coal, wood, peat, or other organic matter. Called coal tar, wood tar, etc. Loosely, the same as pitch, or asphalt.

Torbanite. An organic mineral substance intermediate between kerogen shale and coal. It differs from coal in that it produces paraffin and olefin upon destructive distillation, whereas coals produce the benzine hydrocarbons. Named from Torbane Hill, Bathgate, Scotland. It also occurs in France, New South Wales, South Africa, and the eastern United States.

Tubing. The pipe through which oil or gas or both are brought from the reservoir to the surface. It is generally placed ("run") inside the casing, and is 2–4 inches in diameter.

Uintaite. Gilsonite, *q.v.*

Whipstock. A long, slender, steel wedge, with a groove along one side, that is placed at the bottom of the hole and is used to deflect the bit during drilling.

Wurtzilite. An asphaltic pyrobitumen, elastic like elaterite, found in veins in the Uinta region of northeastern Utah. It was named in honor of Dr. H. Wurtz for his work on hydrocarbons.

ABBREVIATIONS USED ON LOGS AND SCOUT REPORTS*

A/	acidized with	bbl	barrel
abd, abnd	abandoned	b/d	barrels per day
		BCPD	barrels condensate per day
ac	acres	BCPH	barrels condensate per hour
AS	after shot	BFPH	barrels fluid per hour

* Adapted from IPAA Monthly, October 1952, and Julian W. Low, "Examination of Well Cuttings," Quart. Colo. Sch. Mines, October 1951, p. 46.

ABBREVIATIONS USED ON LOGS AND SCOUT REPORTS

BHC	bottom-hole choke	CP	casing pressure
BHP	bottom-hole pressure	CPSI	casing pressure shut in
BHPF	bottom-hole pressure flowing	crd	cored
		crg	coring
BHPSI	bottom-hole pressure shut in	crse	coarse
		csg	casing
bl	black	CSL	Center Section Line
bld	bailed	CT	cable tools
bldg drk	building derrick	ctg	cutting
bldg rds	building roads		
blk	black	D&A	dry and abandoned
blr	bailer	DC	drill collar
B/H	bailers per hour	DD	drilling (drilled) deeper
BO	barrels oil	DF	derrick floor
BOPD	barrels oil per day	dk	dark
BP	back pressure	DO	drilled out
BPD	barrels per day	dol, dolo	dolomite
BPH	barrels per hour	DP	drill pipe
BPWPD	barrels per well per day	D/P	drill plug
brkn	broken	drk	derrick
BS	basic sediment	drld	drilled
BS&W	basic sediment and water	drlg	drilling
btm	bottom	DST	drill-stem test
B.T.U.	British thermal unit		
BW	barrels water	EL	east line
BWPD	barrels water per day	elec log	electric log
BWPH	barrels water per hour	elev	elevation
		E log	electric log
C	center	E of W/L	east of west line
C&P	cellar and pits	est	estimate or estimated
CD	contract depth		
CFG	cubic feet gas	f	fine
CFGPD	cubic feet gas per day	FIH	fluid in hole
CGS	centimeter-gram-second system	fl/	flowed or flowing
		fld	field
chk	choke	fluor, flur	fluorescence
circ	circulate or circulation		
clng	cleaning	fm	formation
CO	clean out	fos	fossils, fossiliferous
comp	completed, completion	FP	flowing pressure
cond	condensate	fr E/L	from east line
congl	conglomerate		
contr	contractor	GA	gallons acid
cor	corner	gal(s)	gallon, gallons
corr	corrected	G&O	gas and oil

G&OCM	gas and oil cut mud	MIR	moving in rig
GC	gas cut	MIRT	moving in rotary tools
GCM	gas cut mud	MIST	moving in standard tools
GCR	gas condensate ratio	MIT	moving in tools
ggd	gauged	MO	moving out
gge	gauge		
GO	gas odor	nat	natural
G/O	gas and oil	NL	north line
GOR	gas-oil ratio	NS	no show
gr	gray, ground		
Gran W	granite wash	O&G	oil and gas
grav	gravity	O&GCM	oil and gas cut mud
grd	ground	O&SW	oil and salt water
grn	green	OAW	old abandoned well
		OC	oil cut
hd	hard	OCM	oil cut mud
HFO	hole full oil	OF	open flow
HFW	hole full water	OH	open hole
HGOR	high gas-oil ratio	OIH	oil in hole
HO	heavy oil	OO	odor oil
hr(s)	hour(s)	Ool	oolitic
hvy	heavy	op	opaque
		O sd	oil sand
ig	igneous	OT	own tools
incl	inclusion, including	OTD	old total depth
interst	interstitial	OWDD	oil well drilled deeper
IP	initial production	OWPB	old well plugged back
		OWWO	oil well worked over
KO	kicked off	ox	oxidized
li	lime	p	pump
loc	located or location	P&A	plugged and abandoned
LS, ls	limestone	PB	plugged back
lse	lease	PBTD	plugged back total depth
		PD	per day
mass	massive	per	permeability
MCF	thousand cubic feet	perf	perforated
MCFG-PD	thousand cubic feet gas per day	perf csg	perforated casing
		perm	permeability
md	millidarcys	pk	pink
mi	miles	Pkr	packer
MI	moving in	PL	pipeline
MICT	moving in cable tools	pld	pulled
mil	million	PLO	pipe-line oil
MIM	moving in materials	POL	petroleum-oils-lube

POP	putting on pump	SI	shut in
por	porosity, porous	SIBHP	shut-in bottom-hole pressure
psi	pounds per square inch		
psia	pounds per square inch, absolute	SICP	shut-in casing pressure
		SIP	shut-in pressure
psig	pounds per square inch, gauge	SIS	stopped in sand
		SITP	shut-in tubing pressure
pt	part	Sl	slight
PVT	pressure-volume-temperature	SL	south line
		SLM	steel-line measurement
		SO, S/O	show oil
qtz	quartz	SO&G	show oil and gas
qtze	quartzite	SO&W	show oil and water
		SP	self-potential (electric log)
		spd	spudded
R	range	squ	squeeze or squeezed
rec	recovered	ss	sand
refl	reflection	SSO	slight show oil
refr	refraction	S/T	sample tops
Rge	range	stds	stands
rmg	reaming	stn	stain, stained
rng	running	strks	streaks
RP	rock pressure	sul	sulphur
RT	rotary table	sul wtr	sulphur water
RUCT	rigging up cable tools	sur	survey
RUM	rigging up machine	surf	surface
RUP	rigging up pump	SW	southwest, salt water
RUR	rigging up rotary	S/W	salt water
RUST	rigging up standard tools	swbd	swabbed
		swbg	swabbing
sat	saturated or saturation	SWS	sidewall samples
SC	show condensate	sx	sacks
Sd	sand		
SD	shut down	T	township
SDO	shut down for orders	T/	top
SD rep	shut down for repairs	tbg	tubing
Sd SO	sand showing oil	tbg chk	tubing choke
sdy	sandy	TD	total depth
SE	southeast	temp	temporary
Sec	section	TP	tubing pressure
sed	sediment	T/Pay	top pay
seis	seismograph	TPSI	tubing pressure shut in
SG, S/G	show gas	T/sd	top sand
SG&C	show gas and condensate	tstg	testing
sh	shale	Twp	township

Twst	town site	WO/O	waiting on orders
		WOR	waiting on rig
Unconf	unconformity	WORT	waiting on rotary tools
UR	underreaming	WOST	waiting on standard tools
		WOT	waiting on tools
W	water	WP	working pressure
WC	wildcat	Wpstk	whipstock
W/C	water cushion	wtg	waiting
wh	white	wtr	water
WI	washing in	WW	wash water
WL	west line		
W/L	water load	Xin, Xln	crystalline
WO	waiting on		
WOC, WOCS	waiting on cement to set	Y	yellow
WOCT	waiting on cable tools	Z	zone

LIST OF BIBLIOGRAPHIES OF PETROLEUM

Everett DeGolyer and Harold Vance, *Bibliography of the Petroleum Industry,* Bull. 83, A. & M. College of Texas, College Station, Texas (September 1, 1944), 725 pages.

Clarence P. Dunbar and Lucille Dunbar, *A Selected List of Periodicals, Serials, and Books Dealing with Petroleum and Allied Subjects,* Department of Conservation, New Orleans, Louisiana (1939), mimeographed, 218 pages.

Economic Geology Publishing Company, Urbana, Illinois, *Annotated Bibliography of Economic Geology,* published annually since 1928. Includes a section on petroleum.

Geological Society of America, *Bibliography and Index of Geology Exclusive of North America,* published annually since 1933. Annotated. Includes a section on petroleum.

Robert E. Hardwicks, *Petroleum and Natural Gas Bibliography,* University of Texas, Austin (1937), 167 pages. Has 1,397 entries.

Donald L. Katz and Michael J. Rzasa, *Bibliography for Physical Behavior of Hydrocarbons under Pressure and Related Phenomena,* J. W. Edwards, Ann Arbor, Michigan (1946), 306 pages. Has 1,951 references from 1860 to 1945.

S. F. Peckham, *Report on the Production, Technology and Uses of Petroleum and Its Products,* Tenth Census of the United States, Government Printing Office, Washington, D.C. (1884).

Sir Boverton Redwood, *Petroleum,* 5th ed., 3 vols., C. Griffin & Co., London, and J. B. Lippincott Co., Philadelphia, 1,353 pages. Includes the Dalton bibliography of 8,804 entries.

Parker D. Trask, "Bibliography Relating to Organic Content of Sediments," Report of Committee on Sedimentation, Division of Geology and Geography, National Research Council (1937–1938), pp. 37–43. Has 83 references.

U.S. Geological Survey, *Bibliography of North American Geology* [since 1785], published at intervals. Includes a section on petroleum.

TABLE A-1 Density of Oil-field Waters

Specific Gravity of Water at 60°F	Approximate Total Solids in Parts per Million (mg/liter)	Hydrostatic Pressure Gradient (psi/ft[1])
1.000	none	0.433
1.010	13,500	0.437
1.020	27,500	0.441
1.030	41,400	0.445
1.040	55,400	0.450
1.050	69,400	0.454
1.060	83,700	0.459
1.070	98,400	0.463
1.080	113,200	0.467
1.090	128,300	0.471
1.100	143,500	0.476
1.110	159,500	0.480
1.120	175,800	0.485
1.130	192,400	0.489
1.140	210,000	0.493

Source: From C. E. Reistle and C. E. Lane, Tech. Paper 432, U.S. Bur. Mines (1928). See also O. & G. Jour., February 3, 1945, p. 69, and December 23, 1948, p. 75.

[1] Water compressibility and the presence of gas in solution have been ignored.

TABLE A-2 Viscosity Conversion Factors

Saybolt Universal Seconds	Engler Degrees	Absolute Viscosity, Centipoises (also Relative Viscosity)[1]	Kinematic Viscosity, Poises per g/cc n/d	Redwood No. 1 Seconds	Saybolt Furol Seconds	Redwood Admiralty Seconds
32	1.08	1.41	0.0141			
40	1.31	4.30	0.0430			
50	1.58	7.40	0.0740			
60	1.88	10.20	0.1020			
70	2.17	12.83	0.1283			
80	2.46	15.35	0.1535			
90	2.74	17.80	0.1780			
100	3.02	20.20	0.2020	10	15	86
110	3.31	22.56	0.2256	11	16	94
120	3.60	24.90	0.2490	12	17	102
130	3.69	27.21	0.2721	13	18	111
140	4.19	29.51	0.2951	13	18	119
160	4.71	34.07	0.3407	15	20	136
180	5.35	38.60	0.3860	17	22	153
200	5.92	43.10	0.4310	19	23	170
250	7.35	54.28	0.5428	23	28	212
300	8.79	65.40	0.6540	28	32	254
400	11.68	87.55	0.8755	37	42	338
500	14.00	109.6	1.096	46	52	422
600	17.00	131.7	1.317	55	61	507
1,000	29.00	219.8	2.198	92	101	845
1,500	43.00	329.9	3.299	138	150	1,261
2,000	58.00	439.9	4.399	184	200	1,690
3,000	87.00	659.9	6.599	276	300	2,535

Source: Adapted from The Petroleum Engineer, October 1952, p. E-23.

[1] Values in this column must be multiplied by the specific gravity of the fluid at the temperature of the measurement to complete the conversion.

TABLE A-3 Densities and Specific Volumes of Petroleum
(all measurements at 60°F)

Degrees API Gravity	Specific Gravity	Gallons per Pound	Pounds per Gallon	Pounds per Barrel	Barrels per Short Ton	Barrels per Metric Ton	Barrels per Long Ton
0	1.076	0.1116	8.962	376.40	5.31	5.86	5.95
10	1.000	0.1201	8.328	349.78	5.71	6.30	6.40
15	0.9659	0.1243	8.044	337.85	5.92	6.53	6.63
18	0.9465	0.1269	7.882	331.04	6.04	6.66	6.77
20	0.9340	0.1286	7.778	326.68	6.12	6.75	6.86
22	0.9218	0.1303	7.676	322.39	6.20	6.84	6.95
24	0.9100	0.1320	7.578	318.28	6.28	6.93	7.04
26	0.8984	0.1337	7.481	314.20	6.37	7.02	7.13
28	0.8871	0.1354	7.387	310.25	6.45	7.11	7.22
30	0.8762	0.1371	7.296	306.43	6.53	7.19	7.31
32	0.8654	0.1380	7.206	302.65	6.61	7.28	7.40
34	0.8550	0.1405	7.119	299.00	6.69	7.37	7.49
36	0.8448	0.1422	7.034	295.43	6.77	7.46	7.58
38	0.8348	0.1439	6.951	291.94	6.85	7.55	7.67
40	0.8251	0.1456	6.870	288.54	6.93	7.64	7.76
42	0.8155	0.1473	6.790	285.18	7.01	7.73	7.85
44	0.8063	0.1490	6.713	281.95	7.09	7.82	7.94
46	0.7972	0.1507	6.637	278.75	7.17	7.91	8.04
48	0.7883	0.1524	6.563	275.65	7.26	8.00	8.13
50	0.7796	0.1541	6.490	272.58	7.34	8.09	8.23
55	0.7587	0.1583	6.316	365.27	7.54	8.31	8.44
60	0.7389	0.1626	6.151	258.34	7.74	8.53	8.67
65	0.7201	0.1668	5.994	251.75	7.94	8.76	8.90
70	0.7022	0.1711	5.845	245.49	8.15	8.98	9.12
75	0.6852	0.1753	5.703	239.53	8.35	9.20	9.35
80	0.6690	0.1796	5.568	233.86	8.55	9.43	9.58
85	0.6536	0.1838	5.440	228.48	8.75	9.65	9.80
90	0.6388	0.1881	5.316	223.27	8.96	9.87	10.03
95	0.6247	0.1924	5.199	219.36	9.16	10.10	10.26
100	0.6112	0.1966	5.086	213.61	9.36	10.32	10.49

Source: Adapted from *Petroleum Facts and Figures*, 9th ed., American Petroleum Institute.

TABLE A-4　API Pipe Capacities

Pipe Size in Inches	Weight in Pounds per Foot	Inside Diameter	Capacity per 1,000 Feet	
			Barrels	Gallons
2 API	4.6	1.995	3.869	162.498
2 EUE	4.7	1.995	3.870	162.540
2½ Reg.	6.4	2.441	5.792	243.264
2½ EUE	6.5	2.441	5.794	243.348
3 Reg.	9.2	2.992	8.701	365.442
4½ API	9.5	4.090	16.25	682.50
5½ API	14.0	5.012	24.40	1,024.80
5½ API	17.0	4.892	23.26	976.92
5½ API	20.0	4.778	22.19	931.98
6 API	20.0	5.352	27.84	1,169.28
6⅝ API	24.2	5.921	34.07	1,430.94
7 API	17.0	6.538	41.52	1,743.84
7 API	20.0	6.456	40.50	1,701.00
7 API	22.0	6.398	39.79	1,671.18
7⅝ API	30.0	6.875	45.93	1,929.06
8⅝ API	36.7	7.825	59.50	2,499.00
9⅝ API	40.6	8.835	75.82	3,184.44
10¾ API	54.6	9.784	93.02	3,906.84
11¾ API	54.0	10.880	115.0	4,830.00
13¾ API	54.5	12.615	154.6	6,493.20

TABLE A-5 Conversion Table for Mud Gradient

Gradient, PSI per 1000 Ft of Depth	Density Lb per Gal	Density Lb per Cu Ft	Specific Gravity	Gradient, PSI per 1000 Ft of Depth	Density Lb per Gal	Density Lb per Cu Ft	Specific Gravity
433	8.3	62.4	1.00	800	15.4	115.2	1.85
440	8.5	63.4	1.02	810	15.6	116.6	1.87
450	8.7	64.8	1.04	820	15.8	118.1	1.89
460	8.9	66.2	1.06	830	16.0	119.5	1.92
470	9.1	67.7	1.09	840	16.2	121.0	1.94
480	9.2	69.1	1.11	850	16.4	122.4	1.96
490	9.4	70.6	1.13	860	16.6	123.8	1.99
500	9.6	72.0	1.15	870	16.8	125.3	2.01
510	9.8	73.4	1.18	880	16.9	126.7	2.03
520	10.0	74.9	1.20	890	17.1	128.2	2.06
530	10.2	76.3	1.22	900	17.3	129.6	2.08
540	10.4	77.8	1.25	910	17.5	131.0	2.10
550	10.6	79.2	1.27	920	17.7	132.5	2.12
560	10.8	80.6	1.29	930	17.9	133.9	2.15
570	11.0	82.1	1.32	940	18.1	135.4	2.17
580	11.2	83.5	1.34	950	18.3	136.8	2.19
590	11.4	85.0	1.36	960	18.5	138.2	2.22
600	11.6	86.4	1.39	970	18.7	139.7	2.24
610	11.7	87.8	1.41	980	18.9	141.1	2.26
620	11.9	89.3	1.43	990	19.1	142.6	2.29
630	12.1	90.7	1.45	1000	19.2	144.1	2.30
640	12.3	92.2	1.48	1010	19.4	145.5	2.33
650	12.5	93.6	1.50	1020	19.6	146.0	2.35
660	12.7	95.0	1.52	1030	19.8	148.4	2.38
670	12.9	96.5	1.55	1040	20.0	149.9	2.40
680	13.1	97.9	1.57	1050	20.2	151.3	2.42
690	13.3	99.4	1.59	1060	20.4	152.7	2.45
700	13.5	100.8	1.62	1070	20.6	154.2	2.47
710	13.7	102.2	1.64	1080	20.8	155.6	2.50
720	13.9	103.7	1.66	1090	21.0	157.1	2.52
730	14.1	105.1	1.69	1100	21.2	158.5	2.54
740	14.3	106.6	1.71	1110	21.3	160.0	2.57
750	14.4	108.0	1.73	1120	21.5	161.4	2.58
760	14.6	109.4	1.76	1130	21.7	162.8	2.60
770	14.8	110.9	1.78	1140	21.9	164.3	2.63
780	15.0	112.3	1.80	1150	22.1	165.7	2.65
790	15.2	113.8	1.82				

Source: The AAODC-API Joint Committee on the Standard Daily Drilling Report Form, Petroleum Engineer (Jan. 1958).

INDEX

Abbreviations, logs and scout reports, 680–684
Abo-Wichita Albany reef, 320
Abqaiq pool, Saudi Arabia, 119, 244; crude oil properties, 212; pressure drop, 466
Absolute porosity, 101
Absorption, 443
Abu Dhabi, production, reserves, 34
Abyssal environment, 61
Academic geology, 10
Accumulation, 5; after trap formed, 560–561, 575; geologic framework, 539–541; near unconformities, 642; near wedge-belts of permeability, 648–649; and migration, 538–584; relation to secondary migration, 560–561; relation to structural basins, 637; theories, bearing on exploration, 500; time of, 38, 40, 571–580
Acetylene, 180, 503
Achi-Su field, USSR, 275, 279
Acidization, of a reservoir, 135
Acre-foot, 99, 175
Acropora, 315
Active water-drive pools, 465
Adhesion, 66, 443–444, 452, 546
Adsorbed water, squeezed out during compaction, 130
Adsorption, 66, 100, 173, 443; electric logs, 79; in Fuller's earth, 525; of water on clay minerals, 173
Advancing shoreline, reefs, 324
Aerobic bacteria, 517
Africa, production, reserves, 32, 35; salt domes, 359; Whale Bay, blooming, 513
Age, 39; of reservoir rock, 38–39; and rock classification, 37–41; years, geologic time scale, 39
Agglomerates, reservoir rock, 427
Agha Jari field, Iran, fracture permeability, 125; trap obscured by thrust fault, 254, 275
Ain Dar pool, Saudi Arabia, pressure drop, 466
Air, secondary recovery agent, 482
Air permeability, high, 107; measurement, 105–110
Air-water table, 50
Alberta, *see* Canada
Albertite, 3, 25, 674

Algae, 28, 315–316; Precambrian, 40; rock-builders, 71; source material, 512
Algeria, production, reserves, 35
Aliso Canyon field, California, fault traps, 275, 277, 278
Alkalinity, water analysis, 163–164
Alkane, 178; series, 183–185
Alkene series (olefins), 185–186
Alkylation, 525
Alkyl group, 189
Alpha particle, 527–529; bombardment, forms hydrocarbons, 528
Alsace, mining oil, 472
Alteration, of petroleum by heat and pressure, 521–529
Aluminum, 526
Amarillo Arch, Texas, 65
Amarillo-Texas Panhandle field, composition natural gas, 217; granite reservoir, 64, 120–121; low pressure, 411
Amarillo uplift, Texas, erosion of, 549
Amelia field, Texas, complex structure, 270–272
American Institute of Mining & Metallurgical Engineers, New York, 30
American Petroleum Institute (API), definition of darcy, 104; gravity, crude oil, 188, 194–198; pipe capacities, 688
Amino acids, 504
Ammonia, 504
Ammonification, 518
Anacacho limestone, Cretaceous, Texas, asphalt cemented, 25
Anaerobic bacteria, 517
Ancient seepages, 19
Ancón, Ecuador, seepages, 19
Angola, production, reserves, 35
Angular unconformity, 254, 333–340, 599, 621
Anhydrite, in salt plugs, 364, 367–371
Animal life, 510, 513–514
Ankavandra Beds, Madagascar, nonmarine reservoir, 88
Anomaly, geophysical, 612; gravity, 612; magnetic, 612; seismic, 612; subsurface, 600
Antelope Hills field, California, 335, 337
Anticlinal theory, 233, 234–236
Anticline, residual, 378
Antrim radioactive black organic shale, 529

Apache pool, Oklahoma, 243–244, 253, 257, 603, 605
Apco pool, Texas, 127, 338–340
Appalachian region, barium in Paleozoic brines, 165; carbon ratios, 631–632; composite basin, 635–636; petroleum province, 32; prospects, 128
Application, of geology to discovery, 585–673
Apsheron Peninsula, USSR, 32; gravity of oil, 197; map, 21; mud volcanoes, 21–22; reservoirs, 87–88; seepages, 19
Aquitaine, France, diapir folds, 250; salt, 358, 359
Aquifers, 393
Arabia, *see* Saudi Arabia
Arab zone, Jurassic, Saudi Arabia, 71, 119, 212, 244
Aragonite, 71, 118, 371; from *Halimeda*, 512
Arakan Coast, Burma, mud volcanoes, 22
Arbuckle dolomite, Cambro-Ordovician, reservoir, 127
Arbuckle-Ellenburger group, Cambro-Ordovician, reservoirs, 72
Arbuckle Mountains, Oklahoma, asphalt, 24
Arbuckle waters, Kansas, 170–171
Archean algae, 40
Archeozoic, 39
Arching, regional, 652–653
Argentina, production, reserves, 34; San Pedro field, 87
Arkansas, Cairo pool, 540, 562–563, 565; carbon ratios, 633; isogradient map, 419; Magnolia field, 116; natural gas high in hydrogen sulfide, 222; Reynolds oolite pools, 400; Schuler field, 116–117, 466, 562–563; secondary reserves, 482; section showing geologic layers, 642; Smackover limestone, Jurassic, 106, 116–117, 394–395, 400, 562–563; southwestern oil and gas pools, 486
Arkose, reservoir rock, 56–57, 63–65
Armstrong pool, Texas, 292–293
Aromatic series, 186–187, 526
Aromatization, 525
Artesian flow, 80, 132
Artesian water, 394
Artesian well, flowing, 394
Artificial porosity and permeability, 135–136
Ash, in crude oil, 190; in metals, 444; in oil and algae, 505
Ash-free coal analysis, 630
Asia-Pacific, production, reserves, 35
Asmari limestone, Tertiary, Iran, 18, 71, 480; fractured reservoirs, 125; hydrodynamic pressure gradient, 400; Masjid-i-Sulaiman field, 248, 480
"Asparagus" mine, 679
Asphalt, 3, 4, 8, 16, 24, 25, 27, 508, 638, 675; clogging porosity, 336; composition, 177
Asphalt-base crude oil, 186
Asphaltene, in crude oil, 189
Asphaltic pyrobitumens, 675
Asphaltites, 675
Assam, seepages, 88
Associated gas, 211, 484–485
Asymmetric folding, 250, 255–256
Athabaska oil sands, Canada, 25, 128
Atmospheric pressures, cause of subsurface pressure changes, 404
Atoll, 313
Aulacondiscus, 513
Austin chalk, Cretaceous, reservoir rock, 280, 282
Australia, Great Barrier Reef, 313, 316–317; reserves, 35; seepages, 638
Austria, production, reserves, 35
Available pore space, 101
Avery Island, Louisiana, sandstone in salt plugs, 361

Back pressure, 392
Bacteria, 70; formation of gas pockets under pressure, 405; oxidation, 520; reduction of sulfate, 222, 518, 519; release of oil from plants, 520; transformation of organic matter into petroleum, 517–520; at weathering surface, 118
Bahamas, reefs, 312
Bahrein, production, reserves, 34
Bailer, 393, 675
Baku province, USSR, 58, 629; composition of natural gas, 217; exceptional wells, 480; gravity of crude oil, 195; Kala field, 262–263; map, 21; mud volcanoes, 21–22; seepages, 16, 19
"Bald headed" structures, 614, 620
Balk, Robert, 361
Bank of oil, 482
Barbados Island, burnt clay, 27; seepages, 19
Barbers Hill salt dome, Texas, high salinity in surface water, 172–173; overhang, 370
Barco district, Columbia, Petrólea field, 129, 266, 269, 479
Bardsdale pool, California, nonmarine reservoir, 86–87
Bardwell, D. C., 526
Barite, 371
Barium, in Appalachian waters, 165
Barrackpore field, Trinidad, associated

INDEX 693

mud flow, 23; composition of natural gas, 217
Barrel, U.S., 101, 175
Barrier reef, 312; shifting, 324–325
Bars, offshore, traps, 269–301
Bartlesville sand, northeastern Oklahoma, 292, 296, 543–544; garvity of oil, 197
Barton, Donald C., 522
Base-exchange, 66–67
Basement complex, reservoir rocks, 64–65, 74–75, 120–121, 263
Basic sediment, 675
Basins, composite, 634–637; depositional, 634; enclosed, 173–174; sedimentary, 634–637, 652; structural, 634
Bass, N. W., 296
Bathyal environment, 61
Baumé gravity, crude oil, 194
Bay of Batavia, reefs, 314
Bayou St. Denis field, Louisiana, excess pressure, 409
Beaumont, Elie de, 126
Bedding planes, 15, 70, 74, 99, 115
Beers, Roland F., 527
Belcher field, Ontario, Canada, dolomite reservoir, 127, 307–308
Bell, K. G., 527
Bellevue field, Louisiana, porosity and permeability, 133
Bend Arch, Texas, 652–653
Benthonic organisms, 511
Benthos, 511
Benzene, 3, 503; formation from paraffin, 526; series, 186–187
Bergmann, W., 514
Bermudez Pitch Lake, Venezuela, 17
Berry, F. A. F., 403, 406
Beta particles, 527
Bibi-Eibat field, USSR, 22; gravity of oil, 198
Bibliographies, petroleum, 684–685
Bicarbonate, in water, 152
Big Injun sand, Mississippian, Ohio and West Virginia, 60
Bikini Island, coral rock in boring, 314
Bikini reef, blowholes, 317; room-and-pillar structure, 317–318
Biochemical activity in weathering zone, 118
Biochemical causes of pressure changes, 405
Biochemical reservoir rocks, 54, 68–72
Biofacies, 68, 288
Bioherm, 306, 310; cause of buried hill, 252; ioslated reservoir, 411; regressive, 324–325
Biology, 9
Biostrome, 306–308

Bit, 676
Bitkow field, Poland, complex traps, 275, 281
Bitumen, 3, 4, 14, 15, 24, 27, 676
Bituminous, defined, 676
Bituminous dikes, 25
Bituminous limestones, 24
Bituminous sands, 24–25
Black Hand oil shale, Ohio, 27
Black oil, Rocky Mountain region, 168, 170
Black shale, Kentucky, reservoir rock, 280
Blanket sands, 58, 629, 653
Bleeding core, 638
Bloom, 207
Blooming, 512
Blowholes, reefs, 317
Boghead coals, 29
Bois d'Arc limestone, Oklahoma, reservoir, 122, 337
Bolivar Coast fields, Venezuela, production from strandline sands, 293, 297, 648–650
Bolivia, production, reserves, 34
Bombardment of acids to form hydrocarbons, 528
Bordon Group, Indiana, reefs, 322
Boryslaw, Poland, ozokerite, 185
Boscan field, Venezuela, gravity of oil, 198
Botacatu sandstone, Triassic, Brazil, seepages, 427
Bottom sediment, 676
Bottom settlings, 675
Bottom water, 151
Bottom-hole pressure, 390; flowing, 392
Boundary surfaces, 434
Boundary tension, 187, 443
Bowsher, A. L., 322
Boyle, R. W., 219
Bradford field, Pennsylvania, 58–59; CI of crude oil, 194; crude oil composition, 192; crude oil properties, 212; oil analysis, 190, 192; oil-wet sand, 450; porosity and permeability, 106, 133–134; secondary reserves, 482; water analysis, 166
Bradford sand, Devonian, Pennsylvania, 133
Bramlette, M. N., 73
Brazil, production, reserves, 34; seepages, 427, 638
Brea, defined, 676
"Break down" of pressure, 411
Breccia, mud, 20–21
Breger, Irving A., 526
Brine, see Oil-field water
British thermal unit (Btu), 211, 218
Broistedt salt dome, Germany, 361, 365
Broken-bond water, 173
Bromide sand, Ordovician, Oklahoma,

Apache pool, 243, 253, 605; Pauls Valley pool, 614
Brooks, Benjamin T., 522
Bubble point, 199, 213, 437–438, 560; curve, 437–438; migration at, 573; pressure, 205
Buoyancy, 147, 390, 454, 551, 553, 554–558, 560, 561; cause of movement, 546, 581; effect in regional tilting, 609–611
Burbank sands, Pennsylvanian, 544
Bureau of Mines, U.S., Hempel analysis, 191
Burgan field, Kuwait, 58; gravity of oil, 197; water analysis, 166
Buried hills, 251–255
Burma, central basin, 636; mining oil, 472; mud volcanoes, 22–23; production, reserves, 35
Burnt clay, 27
Burnt Hill, Barbados Island, burnt clay, 27
Bush City pool, Kansas, channel deposit, 302, 304
Butane, 178, 191, 216, 217, 484, 615
Buttress sands, 335, 337, 355

Caballos novaculite, Devonian, Texas, 73
Cabin Creek field, West Virginia, absence of free water, 152
Cable-tool drilling, 676; logs, 75, 76; permeability measurements, 109; pressure measurement, 393
Cairo pool, Arkansas, 540, 562–563, 565
Calcarenite, 68
Calcite, 27, 57, 68, 70, 71, 361; from *Lithothamnium,* 512; solubility, 118
Calcium, 173
Caliche, 256
California: age of traps, 579–580; Aliso Canyon field, 275, 277, 278; Antelope Hills field, 335, 337; Bardsdale pool, 86–87; bituminous sand, 24; burnt clay, 27; Capitan field, 86–87; Coalinga field, 336; continental shelf, reservoirs, 36–37; Cymric pool, 27; East Cat Canyon field, 353, 355; East Coalinga pool, 353–354; Edison field, 74, 120; Elk Hills field, 67; El Segundo pool, 120; gravity of crude oil, 195; Huntington Beach field, 594–595; Inglewood field, 263–264; Jurassic reserves, 120; Kettleman Hills field, 199, 540; Long Beach field, 132; Los Angeles basin, 629; Midway field, 335; Montebello field, 353, 355; Monterey formation, Miocene, 73, 121; natural gas high in carbon dioxide, 221; nitrogen in crude oil, 189; nonmarine reservoirs, 86; North Coles Levee field, 563–564; Oxnard field, 198; Playa del Rey field, 60, 120; Pleasant Valley pool, 336, 352–353; production, fractured reservoirs, 120; reservoirs, 58; Round Mountain field, 257, 261; Russell Ranch field, 255, 257; San Joaquin valley, east side, 337; Santa Fe Springs field, 238–239; Santa Maria field, 67, 121, 338, 341; seepages, 16, 17, 19; Shiells Canyon field, 86–87, 174; South Coles Levee field, 563–564; South Mountain field, 86–87, 174, 271–272, 274; sulfate-free brine, San Joaquin valley, 165; Tehachapi-Bakers field earthquake, 404; Ten Section field, 106, 147, 149; Tideland pool, 594–595; Timber Canyon field, 272, 274; Torrance pool, 120; Venice pool, 120; Ventura Basin reservoirs, 60, 86–87; Ventura field, 278, 409–410; West Edison field, 258, 261; Wilmington field, 120, 594–595
Caliper log, 84, 103, 590
Calorific value, crude oil, 211
Calvin sandstone, Pennsylvanian, Oklahoma, 525
Cambrian Period, 39
Cambrina rocks, lack of oil, 38, 40, 504
Canada, Athabaska oil sands, 25, 128; Belcher oil field, 127, 307–308; burnt clay, 27; dolomitic reservoirs, 126–127; exceptional wells, Ontario, 480; Fort Norman pool, 329; gravity of Alberta crude oil, 195; Highwood fold, 276, 651; Leduc field, 117, 126–127, 327–328, 330–331; Norman Wells reef, 328; Pincher Creek pool, 480; production, reserves, 34; Redwater pool, 117, 330, 468; sections across western, 611; seepages at Gaspé, 234; Turner Valley field, 217, 272, 275, 276, 651; western, reservoir rocks, 72
Cannel coal, 25, 29
Capacities, API pipe, 688
Capacity of trap, 241–242, 542, 571–572, 574–575
Capillarity, 66, 100; equation of, 451
Capillary pressure, 147, 451–457, 458, 546, 551–554, 555, 566–567; curves, 110, 455; zero plane of, 566–567
Capitan field, California, 86–87
Capitan reef, New Mexico and Texas, 320–321, 326
Cap rock, 367–371
Carbocyclic series, 186
Carbohydrates, 509
Carbon, in organic matter, 506, 515; in petroleum, 177, 506; isotopes, 38, 40
Carbonate cements, source of, 130
Carbonate reservoir rocks, 68–72, 115–119; porosity, 117–133
Carbon black, 216

Carbon dioxide, 511; in natural gas, 218, 219, 220–221; produced by bacteria, 519–520; in secondary recovery, 483; in water, 152
Carbon disulfide, 3, 25
Carbon-ratio theory, 522, 524, 630–634
Carbon tetrachloride, 28
Carboxyl group, 189
Caribbean region, petroleum province, 32; reservoirs, 61
Carll, John F., 460
Carpathians, diapir folds, 250; excess pressures, 410; oil pools, 250
Carrier beds, 541
Carthage gas field, Texas, composition of natural gas, 217; oolite reservoir, 117, 352
Case, L. C., 160
Casing, defined, 676
Casing pressure, 392, 393
Casing-head gas, 218, 484
Caspian Sea, pressure changes, 404; seepages, 19
Catalyst, defined, 524
Catalytic reactions, 405, 517, 524–526
Cate, R. B., Jr., 516
Caucasus province, USSR, Achi-Su field, 275, 279; excess pressures, 410; Maikop region, 58, 292, 296, 304–305; reservoirs, 58, 60; seepages, 19
Cavernous porosity, 73, 118–119, 317, 480
Caverns, 102
Cayuga field, Texas, structure maps, 613
Cedar Lake field, Texas, variable porosity and permeability, 101–102
Celestite, 371
Cement, 55–56, 68, 99, 103, 114; drilling, 616; silica, 67, 73, 129–130
Cementation, 128–130, 521, 541, 542; cause of pressure, 403–404; effect on permeability, 115; fragmental rocks, 67–68; primary, 128; relation to formation resistivity factor, 159
Cenozoic, 39
Centipoise, 104, 207
Centistoke, 207
Central Basin Platform, Texas, 73, 326, 652
Central Kansas Arch, Kansas, 652
Ceres pool, Oklahoma, shoestring-sand, 300–301
Chalk, reservoir rock, 70
Chanac formation, Miocene(?), California, 128, 258, 337
Channel filling, 301–305; Hawkins field, Oklahoma, 597
Channel sands, distributary, 61–62
Channeling, of water and oil, 474
Chapopote, 188
Chapopoteras, 676
Chattanooga radioactive black organic shale, 67, 529
Chebuda island, Burma, mud volcano, 22
Cheney, M. G., 545
Cherokee shales, Pennsylvanian, Kansas, gas pools, 67, 279–280; source of sediments, 548–549
Chert, 56, 63, 64, 70, 72, 73
Chester, Albert Huntington, 4
Chester strata, Mississippian, Illinois, high chloride in oil-field water, 160; oil in Minerva fluorite mine, 28; section, 290
Chile, Manantiales oil field, 100–101; production, reserves, 34; Springhill sand, Cretaceous, 100–101
China, Laochunmaie field, Tertiary, 88; nonmarine reservoirs, 88; production, reserves, 35; Shensi series, Jura-Triassic, 88; Tsupinkai field, Tertiary, 88
Chlorides, in water, 152
Chloroform, 28
Chlorophyll, 504
Cholesterin, 209
Cholesterol, 209, 505
Christmas tree, defined, 676
Churn-drilling, 675
CI (correlation index), 191, 194
Cincinnati Arch, Ohio and Indiana, 127, 307, 653; wedge belts of permeability, 647
Circle Ridge field, Wyoming, fault traps, 275, 277
Clapp, Frederick G., 236
Clarke, Frank Wigglesworth, 514
Classification, oil-field waters, 152–157; and origin, pore space, 113–136; reservoir rocks, 52–54; rock, time, 37–38; structural traps, 240; traps, 236–238
Clastic reservoir rocks, 53, 54–68
Clastic-ratio maps, 600–602, 603, 604
Clastic-shale ratio maps, 601
Clay, 65–67, 99, 103, 115, 136, 455; burnt, 27; cementing material, 67, 128; compaction and porosity reduction, 130–133, 412–414; in crude oil, 190; effect of interstitial water, 156; effect on permeability measurements, 107; minerals, 66; minerals, catalysts, 525; oil-settling capacities, 549–550; relation to oil-field brines, 173; swelling, 256; waters squeezed out of, 545–546
Clinton sand, Cincinnati Arch, up-dip wedge-out, 647
Clogging, reservoir rock, 336–337
Cloos, Ernst, 126
Closed-in pressure, 390
Closure, Erath field, Louisiana, 247; structural, 241

Coal, cannel, 29; carbon-ratio, 630–634; kerogen, 15, 29; resource, 8
Coalinga field, California, 336
Cockfield sand, Eocene, Texas-Louisiana, 147–148, 363, 368
Coefficient of expansion, crude oil, 200, 210–211; gas, 425
Coefficient of linear expansion, rock, 425–426
Cohesion, 445
Coldwater field, Michigan, pressure-solution-viscosity relationships, 201–202
Coles Levee fields, California, 563–564
Colloidal oil particles, 539, 551
Colloids, compaction of, 130
Colombia, Petrólea field, Barco district, 129–130, 266, 269, 479; petroleum province, 32; production, reserves, 34
Colonial organisms, reef-builders, 315
Color, of petroleum, 209
Colorimeter, 209
Colorado, facies map, 602; florence field, 67, 121, 278–279, 282; McCallum field, 221; natural gas high in carbon dioxide, 221; nonmarine reservoirs, 87; Paradox salt dome region, 358; Rangely field, 67, 106, 153, 196, 201, 203, 212, 419, 421, 423; reefs, Permian, 322–323
Columbia Plateau, Washington and Oregon, 74
Combination traps, 237–238, 286, 349–356
Commercial deposits, 6, 15, 628
Compaction, 117, 127, 130–133, 469, 521, 541, 545–546; clays, 130–133, 412–414; effect on permeability, 115; elastic, 130; expels water, 412–414; kinds, 130; over buried hills, 252–253; over reef, 329; plastic, 130; processes, 412–413; reservoir rock, 130–133
Compaction-hydraulic theory of primary migration, 545
Competent and incompetent beds, 248–251
Competition, in exploration, 36
Complex, reef, 324–325
Component, phase relationships, 434–441, 434–441
Composite basin, 634–637
Composition, kerogen, 28–29, 506; natural gas, 177, 183, 216–218; oil-field water, 163–171; petroleum, 177, 180–181, 191, 501, 506; products from crude oil, 182; reservoir rocks, 540
Compressed crude oil, 468–471
Compressed reservoir rock, 133, 468–471
Compressed water, 467–471
Compressor station, 485
Computers, 616

Concentration, salts in oil-field water, 160–163
Concessions, to oil companies, 487, 666
Concretions, petroleum in, 27
Condensate pool, 462, 485
Conductivity, thermal, 419, 421, 423–424
Confined reservoir, 400, 401, 426
Confirmation of data, 661
Confirmation well, 7
Conglomerate, 56, 59, 591; oil-saturated pebbles, 501
Congo, production, reserves, 35
Coning, of water and oil, 474
Connate water, 80, 111–112, 152–153, 545
Conroe field, Texas, permeable connection between sands, 147–148; water analysis, 166
Conservation laws, 487
Contact angle, 446
Contacts, gas, oil, water, 149
Contemporary maps, 589
Continental shelf, reservoirs, 36–37; Creole field, Louisiana, 263–267, 594
Continuous phase, see Phase continuity
Convergence, combined with asymmetrical folding, 250, 255–256; maps, 599; shifts in structure, 245–246
Conversion, factors, viscosity, 686; table for mud gradient, 689; to ionic form, water analysis, 168
Coorongite, 675
Copalis Bay, Washington, blooming, 512
Copely pool, West Virginia, absence of free water, 152
Coquina, 56, 68, 115
Corals, 70, 329
Core: analysis, 10, 84, 108, 176, 590, 591; drilling, 239; loss indicating porosity, 103–104
Correlation, 10; by electric logs, 78; by gamma-ray logs, 83; by thermal measurements, 422; difficulties in variable sediments, 60; how determined, 590; index (CI), 191, 194
Corrosion, 173, 221–222
Cosmic origin, petroleum, 504
Costs, drilling, 666–667; geological and geophysical, 665; lease, 665–666; petroleum, 9, 664–665; production, 486–487, 667; secondary recovery, 484
Covalent bonds, 178–180
Cracking, 179, 524–525
Craton, 635
Creole field, Louisiana, Gulf of Mexico, 263–267; deflected wells, 265, 594
Cretaceous, 39; asphalt, 24–25; chalk in salt dome, 369; overlap on western North America, 607–611; reef, Mexico,

INDEX 697

329, 331; world oil production and reserves, 38, 40
Cretaceous reservoir rock, Carthage gas field, Texas, 352; Eagle ford sand, Texas, 261; Florence field, Colorado, 67; Germany, 369; Manantiales field, Chile, 100–101; Mara field, Venezuela, 263; Mexico, 115–116, 124–125; Montana, 336; Nacatoch sandstone, Louisiana, 133; Paluxy sand, Texas, 261; Petrólea field, Colombia, 269; Pettet formation, Texas, 117, 352; Pettit lime, Louisiana, 483; Pierre shale, Colorado, 279, 382; Quitman field, Texas, 261; Rocky Mountain region, 60; Rodessa field, Louisiana, 350–351; Tamabra limestone, Mexico, 127, 353, 356; Venezuela, 72; Woodbine sand, Texas, 58, 67, 155, 259, 261, 264, 268, 351, 466, 468, 574
Cretaceous waters, Rocky Mountain region, 165, 168, 170
Cricondentherm, 440
Critical temperature, water, 418, 435–436, 439
Crooked holes, 594
Cross bedding, clastic carbonates, 68
Cross section, structural and stratigraphic, 592
Crude oil, 3, 7, 175, 676; adhesion to sand, 448; ash, 190; characteristics, graphic diagrams, 180–181; chemical properties, 176–180; coefficient of expansion, 200, 210–211; composition, 177; composition, compared to organic matter, 505, 519; compressibility, 468–471; constituents, typical, 181; gravity (density), 194–198; measurement, 175–176; physical properties, 192–211, 212; pools, gas-oil ratio, 461–462; sulfur content, 187–188; surface tension, 442–443; viscosity, 201–207; volume, 198–201; world production and reserves, 33, 34–35, 38, 40; see also Oil and Petroleum
Cuba, gases from muds, 526; Jatibonico pool, 306; seepages, 638
Cubical expansion of solids, 425
Cumarebo field, Venezuela, composition natural gas, 217; porosity and permeability, 106
Cummings, Edgar R., 310
Cushing field, Oklahoma, 563
Cut, oil analysis, 191
Cutbank pool, Montana, buttress sands, 335–336
Cuttings, well, 84, 103, 376, 590
Cyclization, 525
Cyclohexane, 186
Cycloparaffin series, 180, 185–186
Cyclopentane, 186

Cyclopropane, 185
Cymric field, California, 27

Daly, Marcel R., 545
Darcy, 104
Darcy, Henri, 104
Darcy's law, 104, 110
Darst Creek pool, Texas, 576–577
Data, confirmation of, 661; inadequate, 240; sources, 10; validity of, 240, 661
Datum elevation, 391
Davenport pool, Oklahoma, water analysis, 166
David, Max W., 118
Dead oil, occurrences, 15, 26–27, 638
Dead Sea, 19, 359
Decarboxylation, 518
Decay, radioactive, 526–529
Decline-curve, method of measuring oil in reservoir, 176; solution gas drive pools, 461
Decline in reservoir pressure, 399–400, 458
Deep River pool, Michigan, fracture permeability, 122–123, 308
Deep-seated salt dome, 267, 359, 363; faulting above, 366; Van field, Texas, 268
Deep wells, pressures, 420; temperature, 420; temperature gradients, 420
Deflected wells, 592, 594–596, 620
Deformational traps, 236
De Golyer, E., 503
Degree of freedom, 435
Dehydrogenation, 525
Delaware Basin, Texas, 73
Delhi pool, Louisiana, 335
Deltaic deposits, 61
Delta of Mississippi, section across, 360
Dense limestone, 105
Density, brine in Smackover limestone, 394–395; crude oil, 194–198; differences, oil-water, 569; indication of water composition, 160–161; Kansas brines, 160–161; layering, 555; oil-field waters, 160–161, 394–395, 401–402, 569, 685; relation to mineral matter in water, 160–161; salt and sediment, differences, 377; water, relation to temperature, 425–426; see also Gravity of crude oil (API) and Specific gravity
Depletion, 667, 668; of reservoir energy, 481–484
Depletion-drive pool, 459
Depositional basin, 634
Depositional traps, 236, 287
Deposits, commercial, 6, 15, 628
Depth, sunlight penetration in sea water, 512

Descriptive geometry, 11
Detrital chert, reservoirs, 63
Detrital reservoir rocks, 54–68
Development, anticlinal theory, 234–236; organic theories on origin of oil, 505; well, 7
Devonian, 39; Bradford sand, Pennsylvania, 133; Caballos novaculite, Texas, 73; Deep River pool, Michigan, 122–123, 308; Hunton group reservoir, Oklahoma, 122, 644; off-shore sandbars, Pennsylvania, 296–300; Oriskany sand, fractured reservoir, 121–122, 272, 274–275; reefs, Canada, 327–328, 330–331; reservoirs, Canada, 72; Rogers City field, Michigan, 122–123; Third Stray sand, Pennsylvania, 288–289; West Edmond pool, Oklahoma, 337–338, 644; Woodbend formation, Canada, 117
Dew point, 438; curve, 437–438
Diagenesis, 67, 115; clays to shales, 253, 545, 550; compaction during, 130–131; reduction in volume, 120
Diagenetic traps, 236, 287
Diamond core head, 104
Diapir fold, 20, 23, 249–251; diagram, 249; excess pressure, 410
Diapiric folding, 246, 249–251
Diastem, 333
Diastrophism, 173, 423, 545; cause of excess pressures, 407; cause of fractures, 119; formation of emulsions, 445
Diatoms, 73, 512
Dibblee, Tom, 54
Differential pressure, 392
Dikes, petroleum, 24, 26
Dilatancy, 521, 561
Dipmeter, 85, 591
Directional drilling, 263, 265, 594, 675
"Dirty" sand, 58–59, 412
Discharge area, water flow, 396
Disconformity, 333
Discordant folding, 246, 249–251
Discovery, concepts of, 8–11; future, 654–655; of oil by Drake, 234; of organic reefs in subsurface, 332; petroleum prospect, 659–660; petroleum province, 627, 637–639; through re-examination of electric logs, 160; relation to total geologic history, 488; by nongeological methods, 659; by wildcats, 659–660
Discovery well, 7
Displaced pool, 258
Displacement meter, 214
Displacement pressure, 147, 453–457, 551–554, 557
Disseminated occurrences of petroleum, 24–25

Dissolved gas, in oil, 211, 213, 459–463, 558–560
Dissolved-gas pool, 459–463
Dissolved salts, in oil-field waters, 160–163
Distance, oil migrates, 541–544, 581
Distributary channels, 61, 303
Distribution, gas, oil, water in reservoir, 147–150
Ditch samples, 76
Dodd, Harold V., 559
Dollarhide pool, Texas, cavernous porosity, 119
Dolomite, 57, 68, 70, 99, 123, 371; capillary pressure curve, 455; porosity, 126–127; reservoir rocks, 126–127, 353; solubility, 118
Dolomitization, 117, 126–127
Domes, 241
Donahue, D., 469
Doodle-buggers, 659
Dora pool, Oklahoma, 290–291
Dossor field, USSR, fault trap, 260, 262–263
Drake, E. L., 4, 234, 654
Drake well, 288, 630, 654
Drewite, 319
Drill, 9–11
Drill column, 676
Driller's logs, 75–76, 590
Drilling, costs, 666–667; deepest, 428; directed, 263–265; errors in measurement, 600; insufficient, 639; lost circulation, 109; order of, 379
Drilling mud, 79, 80, 81, 84, 121, 145, 393, 622, 677; oil-base, 80, 146; radioactive, permeability, measurement, 109–110; see also Mud
Drilling-time logs, 83–84, 103, 590; relation to porosity and permeability, 103, 109
Drill pipe, 677
Drill stem, 677
Drill-stem test, 590, 622, 677
Dry gas, 216, 485
Dry hole, 7, 151, 235, 618–622
Duster, 7, 151
Dutcher sands, Pennsylvanian, Oklahoma, 544
Dynamic pressure gradients, 396–401
Dyne, 442

Eagle Ford sands, Cretaceous, Texas, 261
Eakring field, England, 254
Earth pressure, 402
Earthquakes, cause of pressure, 404; effect on migration, 555; formation of emulsions, 445; Grand Banks, 63; Tehachapi-Bakersfield, California, 404

INDEX 699

East Cat Canyon field, California, 353, 355
East Coalinga pool, California, 353–354
Eastern Hemisphere, crude oil production, 33
East Indies, modern reefs, 312; reservoirs, 61
East Texas Basin, compressibility of water & rock, 468; pressure decline, 400–401, 466–467, 469
East Texas pool, CI of crude oil, 194; largest in US, 31, 58, 351, 653; porosity and permeability, 106; pressure decline, 466–467; volcanic ash, 67; water analysis, 166; when formed, 573
East Wasson field, Texas, excess pressure, 409
Economic factors, petroleum production, 486–487, 667; petroleum prospect, 663–669
Economic limit, production, 481
Ecuador, production, reserves, 34; seepages, 19
Edge water, 151; Edge water drive, 464–471
Edison field, California, basement reservoir, 74, 120
Edwards Limestone, Darst Creek pool, Texas, 576–577
Effective permeability, 110–112, 136
Effective porosity, 101, 136
Efficiency of oil recovery, 473; free-gas cap drive, 463; gas expansion drive, 460–461; gravity-drainage, 472; water-drive, 466
Efficient rate, maximum, production, 461, 477–478
Egypt, Gulf of Suez, typical trap, 577; production, reserves, 35
El Abra limestone reservoir, Mexico, 115–116
Elaeophyton, 512
Elastic compaction, 130–133
Elaterite, 677
Eldorado, oil towns, 659
Electric conductivity, 80
Electric pilot, permeability measurement, 109, 590
Electric potential logs, 78–80
Electrical properties, rocks, 10
Electric log, 77–81, 103, 176, 590, 591, 638; for fluid determination, 145; historical, 78; measure of porosity, 103; resistivity from water analysis, 171–172; salt dome stratigraphy, 376; uses, 78
Electrofiltration, SP log, 79
Electrolytes, 79, 80; reservoir water, 156
Electromotive forces, 79
Electro-osmosis, 79

Elements, catalysts in synthesis of hydrocarbons, 525; in crude-oil ash, 190
Elevations, subsurface, how obtained, 590
Elk Basin field, Wyoming, geothermal gradient, 416; gravity of oil, 196
Elk Hills field, California, 67
Ellenburger dolomite, Texas, fractures in, 123
Ellenburger limestone, Ordovician, Apco pool, Texas, 338–340; Todd field, Texas, 309; see Arbuckle
El Segundo pool, California, basement reservoir, 120
Emba province, USSR, composition of natural gas, 217; fault traps, 260, 262–263; salt domes, 358, 365, 372
Emmons, W. H., 559
Emulsion, 445, 549
Endothermic reaction, 511
Energy, of adhesion, 443–444; chemical, 528; combined sources in reservoir, 472–473; depletion, 481–482; free chemical, in hydrocarbon transformation, 528; kinds of reservoir energy, 459; potential, 401, 458, 561; release of, 458, 610; replenished, 441; reservoir, 176, 458–473, 484; source in water-drive pools, 466–467; stored in elastic compaction, 131–132; stored in water system, 401; surface, 442–451; surface free, 442; see also Reservoir energy
Engineer, see Petroleum engineer
Engineering, 9, 11; petroleum, 389, 487
England, burnt clay, 27; Eakring field, 254
Engler, C., 513
Eniwetok Island, basement beneath reefs in boring, 314–315
Entrained particles, 551
Entry pressure, 453, 555
Enzymes, organic catalysts, 518, 524
Eocene, 39; Eniwetok Island, 314; nonmarine reservoirs, 87
Eocene reservoir rock, Armstrong pool, Texas, 293; Cockfield sand, Texas-Louisiana, 147–148, 363, 368; Conroe field, Texas, 147–148; Delhi pool, Louisiana, 335; Germany, 369; Government Wells field, Texas, 294; Petrólea field, Colombia, 129; Pettus sand, Texas, 293; trends, south Texas, 291–292; Uinta basin, Utah, 87; Wilcox sandstone, Texas, 133; Yequa-Jackson sands, Texas, 292
Eola Field, Louisiana, 363, 368
Epoch, 38, 39
Equilibrium, hydrodynamic, 394, 398; hydrostatic, 394, 398
Era, 38, 39
Erath field, Louisiana, 247, 249
Erdman, J. G., 507, 526

Erg, 442
Eroded oil pools, 645
Erosion, Amarillo Uplift, Texas, 65, 549; Nemaha Mountains, 548–549, 606; releases oil in pools, 548–550; of reservoir rock, 620; results in fractures, 119; Rocky Mountains, 548; of salt plugs, 364, 376; sandstones, 57–58; unconformities, 643–645; Wichita Mountains, 65, 549
Erratic blocks, boulders, pebbles, 19–20, 23
Errors, drilling measurements, 600; logs, 592–596; soil analysis maps, 615
Eruptions, mud and mud volcanoes, 20–23
Essential elements, pool, 6
Eternal Fires, Baku, USSR, 19
Ethane, 180, 191, 216, 217, 218, 503, 615
Ether, 3, 28
Ethylene, 180
Europe, geologic time scale, 39; production, reserves, 35
Evaporation, cause of high concentration of salt in water, 173–174; relation to oil-field brines, 174
Evidence, for buried hills, 254; of cementation, 128–129; of elastic or plastic, compaction; 131–133; for mapping subsurface structures, 240; of oil and gas, petroleum province, 637–639
Evolution of hydrogen, bacteria, 519
Examination of reservoir fluids, 145–146
Exceptional wells, 479–480
Excess reservoir pressures, 405–414
Exodus, 19
Experiments, by Hill on buoyancy, 554; by Mills on dissemination, 559–560; by Nettleton on salt domes, 376–377; by Thiel and Emmons on oil migration, 559
"Exploded" section, Arkansas-Louisiana, 642
Exploration, competition effect, 36; continental shelves, 37; geologist, 31, 98; geophysics, aim, 10; glossary, 675–680; historical, 233–235; influenced by theories of origin, 500; influence of carbon-ratio theory, 630–631; nonmarine sediments, 88; oil and gas, 379–380; philosophy, 672; political climate, 32–33, 36; reef pools, 332; shoestring sand pools, 305; stages, 379–380; stratigraphic and structural traps, 342–343; unconformities, 119; use of data, 622–623
Extrusive rocks, heat from, 427–428
Exudates, 14

Facies, 68; changes, 557, 619, 629, 654; and lenses, chemical rocks, traps, 306–332; and lenses, clastic rocks, traps, 287–306; and lenses, primary stratigraphic traps, 287–332; maps, 600–602, 603, 604; porous carbonate, traps, 306–308
Facultative bacteria, 517
False cap rock, salt dome, 368
Fanglomerate, reservoir rock, 87
"Farmer's sand," 620
Fash, Ralph N., 521
Fats, fatty acids, and fat-soluble compounds, 509
Faulting, cause of traps, 258, 259, 260–278; normal, cause of traps, 261–270; reverse and thrust, cause of traps, 270–278; *see also* Graben faulting
Faults, cause of dry traps, 581; how determined, 591; intersecting to form traps, 258; overriding underlying structure, 254, 255, 257; over salt domes, 362–367; seepages along, 15–16, 260; subsurface, 592, 594
Fence, diagram, 592–593; geologic, for origin of petroleum, 500–503
Field, oil and gas, 5, 15, 30–31
Findlay Arch, Ohio, 127, 307
Finland, organic carbon, Precambrian, 38, 40
Fishing, 677
Fissures, 102
Fixed carbon, 630, 631, 632
Flares, gas, 485
Florence field, Colorado, 67, 121, 278–279, 282
Florida, reefs, 312
Flowing, artesian well, 394; by heads, 479; pressure, 392; well, 458, 479
Fluid content, how determined, 590; flow, 9; mechanics, 10, 11; potential, 396–397
Fluid potential gradient, 238, 259, 340–342, 394, 396–401, 474, 581; causes of changes, 546–547; man-made, 562–563; *see also* Hydrodynamic pressure gradient
Fluid pressure, 390, 396–397, 541; anomalous, 405–414; causes of changes, 546–547; gradient, 394–401, 428–429, 464, 474, 546, 551; how determined, 390, 590; *see also* Pressure
Fluid traps, 340–342
Fluids, effects of heat, 424–429; reservoir, 144–223; reservoir, compression of, 130; reservoir, examination of, 145–146
Fluorescence, oil, 84, 98, 207–209, 638
Flushed pool, 562
Flysch, reservoir rock, Carpathians, 250
Fold, anticlinal dome, 241; asymmetrical, 250; changes with depth, 245–259; diapir, piercement, 249–250; discordant, 248; localized by buried features, 249; obscured by thrust fault, 254, 255; paral-

lel, 248; pre-unconformity, 253, 254; recumbent, 244–245; repeated, superimposed, 246, 247; superficial, 252
Folding, cause of traps, 240–259; causes, 242; diapiric, 20, 249–251; discordant, 249–251; obscured by thrust fault, 257; over buried hills, 251–254; "pan-of-biscuit," 256; parallel, 248–249; pre-unconformity, 257; repeated, 247–249
Foraminifera, 70
Forefront pressure, 453
Forest Reserve field, Trinidad, excess pressure, 410
Formation, age, how determined, 590; attitude, how determined, 591; boundaries, correlations, how determined, 590; factor, 157–160; gas-oil ratio, 198–199; lithology, how determined, 590; pressure, 390; resistivity factor, 157–160; temperature, how determined, 590; -volume factor, 199–202
Formosa, reserves, 35
Fort Norman pool, Canada, 329
Fossil, casts, petroleum in, 27, 115, 639; mud volcano, 22; oil field, 24; petroleum, 15; reefs, 318–325; seepages, 26; siliceous, 70; slumps, 256; stream channels, 301, 303
Fountain well, 479
Fraction, oil analysis, 191
Fracture porosity, 119–125
Fractured basement rocks, reservoirs, 120–121
Fractured sediments, reservoirs, 121–125
Fractures, 15, 73, 99, 102, 115, 117, 118, 119–125, 591; causes of, 119–120; effect of, 119; man-made in reservoir, 135–136
Fracturing, cause of traps, 278–282
Fragmental reservoir rocks, 53, 54–68
Framework of limiting conditions, migration and accumulation theories, 539–541; theories of origin, 500–503
France, production, reserves, 34; salt plugs, Aquitaine, 358, 359
Francis, A. W., 526
Franciscan, reserves, California, 120
Frannie field, Wyoming, tilted oil-water contact, 563, 565
Free energy, chemical, in hydrocarbon transformation, 528
Free gas, 211
Free-gas cap, 213, 463–464; Hastings field, Texas, 579; original, 460; secondary, 460; structural traps, 280–281
Free-gas drive pool, 463–464
Freeport Coal, Pennsylvanian, Ohio, 27
Free surface energy, 100, 524
Free water, 109, 153, 457, 560; absent in Bush City field, Kansas, 304; absent in Florence field, Colorado, 282; absent in Lytton Springs field, Texas, 306; absent in Mt. Calm pool, Texas, 282; absent in various pools in West Virginia, 152; expelled by compaction, 413; rare in serpentine fields, 305; table, 566–567
Fringing reef, 312
Frio formation, Oligocene, Amelia field, Texas, 270–272; Hastings field, Texas, 367; thickness, southeastern Texas, 365, 371; up-dip wedge-out, 292, 648
Frio-Vicksburg trend, Texas, 291–292
Fuller, Myron L., 630
Fuller's earth, adsorption, 525
Fumaroles, 218, 220
Funafuti Island, coral in boring, 314
Funicular saturation, 448–449, 474
Fusselman limestone, Silurian, Texas, cavernous reservoir, 119

Gaban, production, reserves, 35
Galena, 371
Gamma-ray, 527; log, 81–83, 103; Garber field, Oklahoma, density of rocks, 131; water analysis, 166
Gas, absent at dry hole, 620–621; associated, 211, 484–485; chromatography, 177; column, 242; composition, 177, 183, 216–218; detection, 84; dissolved in oil, 198–199, 205, 211–213, 459–463; dissolved in water, 211, 213–214, 545; distribution in the reservoir, 147–150; effect on migration, 558–560; expansion, effect on brines, 174; formed in place, 572–573; from gas pools, 485–486; from oil-field operation, 484–485; impurities, 218–228; injection medium, 484; measurement, 84, 214–216; movement into trap, 571–572; nonassociated, 211, 484–486; nonwetting phase, 447; pools, New York and Pennsylvania, 272, 274, 275; pools, relation to oil pools, 485–486; production, 484–486; in reservoirs, 211–223; secondary recovery agent, 484; in solution, relation to interfacial tension, 451; specific gravity, 554; *see also* Natural gas and Petroleum
Gas-cap, drive pools, 463; original free, 460; secondary free, 460
Gases, law of, 571–574
Gas-free oil pools, 558
Gas-oil ratio, 475, 483; examples, 462; formation, 198–199
Gasoline, 183, 523
Gasoline age, 4
Gas-water contact, tilted, 567
Gatchell sand, Tertiary, California, 336, 352–354

Gayer, F. H., 525
Geanticlinal welt, 635
Geic acid, 515
Geiger counter, 109
Genesis, 19
Geochemical maps, 615
Geochemistry, 9, 11
Geodes, petroleum in, 27
Geologic location, petroleum provinces, world, 32–37
Geologic age, helium method, 219; reservoir rock, 37–41; world production and reserves, 40
Geologic fence, for origin of petroleum, 500–503
Geologic framework of limiting conditions, theories of migration and accumulation, 539–541; theories of origin, 500–503
Geologic history, 9, 488; Gulf Coast salt domes, 378; of petroleum, 496–584; variations, trap, 541
Geologic time scale, 39
Geologic-time units, 38
Geological and geophysical costs, 665
Geological factors, petroleum prospect, 600–663
Geologist, well-site, 76; see also Petroleum geologist
Geology, academic, 10; "layer cake," 640; "magic," 659; subsurface, 588–626
Geometry, descriptive, 11; of a pore, 552
Geomorphology, 11
Geophysical anomaly, 612
Geophysical and geological costs, 665
Geophysical mapping, 282, 612–614
Geophysical maps, 242, 612–615
Geophysics, 9, 11
Georgia, USSR, nonmarine reservoirs, 88
Geostatic pressure, 402
Geosyncline, 640
Geothermal gradient, 415–422
Germany, Broistedt salt dome, 361, 365; Gretham-Hademstorf salt dome, 374; Hanigsen salt dome, 374; mining oil, 472; production, reserves, 34; Reitbrook salt dome, 363–364, 369; salt domes, 356–374; Weinhausen-Eicklingen salt dome, 369, 373–374
Gester, G. C., 634
Gibbs, Willard, 435
Gilluly, James, 132
Gilsonite, 3, 25, 189, 677
Glance pitch, 189, 677
Glenn pool, Oklahoma, porosity and permeability, 106
Glenn sands, Pennslyvanian, Oklahoma, 106, 543–544
Glossary, petroleum exploration, 675–680
Goatseep reef, New Mexico, 320–321

Golden Lane, Mexico, exceptional wells, 479–480; reef reservoir, 329–331
Goodman, Clark, 527
Gouge, 361, 364
Government Wells field, Texas, up-dip wedge-out, 292, 294
Graben faulting, 362–364; Eola field, Louisiana, 363, 368; Hastings field, Texas, 363, 579; over salt domes, hypothesis, 366
Gradient, see Dynamic pressure, Fluid potential, Fluid pressure, Geothermal, Hydraulic, Hydrodynamic, Hydrodynamic pressure, Hydrostatic, Potentiometric surface, Pressure, Static, and Temperature gradient
Grahamite, 3, 25, 677
Grain, 98; contact points, increase with depth, 130; size analysis, 54–56; size, effect on capillary pressure, 154; size, shape, and packing, effect on permeability and porosity, 114–115
Grand Banks earthquake, 63
Granite, fresh distinguished from weathered by magnetic susceptibility, 63–64; reservoir, Mara field, Venezuela, 125, 263; reservoir, Amarillo field, Texas, 120–121; wash, 57, 63–64, 411
Granny Creek pool, West Virginia, absence of free water, 152
Grant, U. S., 132
Gravel packing, 677
Gravity, anomaly, 612; drainage pools, 472; faulting, 260–270; source of reservoir energy, 459, 471–472
Gravity of crude oil (API), 188, 194–198; relation to calorific value, 211; relation to index of refraction, 207–208; relation to sulfur content, 188; relation to temperature and viscosity, 204; relation to viscosity and dissolved gas, 205; in structural traps, 280; table, 687; see also Density and Specific gravity
Graywacke, 55, 56, 57; cemented by detrital material, 128; thin section, 59, 449
Great Barrier Reef, Australia, 313, 316–317
Great Britain, see England and United Kingdom
Green River formation, Eocene, Utah, nonmarine reservoir, 87
Greensands, Cretaceous, New Jersey and Texas, catalyst, 525
Gretham-Hademstorf, Germany, salt dome, overhang, 374
Griffithsville pool, West Virginia, absence of free water, 152
Grozny district, USSR, 275, 279; composition of natural gas, 217; exceptional

INDEX 703

well, 480; gravity of crude oil, 195; resins in crude oil, 189
Gruner, John W., 40
Guadalupe-Capitan reef, New Mexico section, 321
Gulf Coast, U.S., average hydrostatic pressure gradient, 394; changes in crude oil compositions with depth, 522–523; excess pressures, 408–409; fields, continental shelf, 36–37; gravity of oil, 195, 197; hydrostatic pressure explained, 414; natural gas high in hydrogen sulfide, 222; organic reefs, 323; origin of salt domes, 374–379; petroleum province, 32; salt domes, 356–379; salt domes, cap rock, 367–371; salt plugs, 359–367; source of sediments, 548; Tertiary reservoirs, 58; traps associated with salt domes, 371–374; traps on down-side of fault, 266, 270; up-dip wedge-out, 292, 648; water permeability, 154; *see also* Gulf of Mexico
Gulf of Mexico, Creole field, 263–267, 594; directional drilling, 263, 265, 594; "oil axis," "oil pole," 32; oil and gas seepages, 19; oil in sediments, 505; source of sediments, 548; Tertiary reservoirs, 61; *see also* Gulf Coast
Gulf sand, Miocene, Creole field, 266
Gulf of Suez, typical field in graben, 577
Gumbo, 75, 368
Gun perforating, 677
Gusher, 479
Gussow, W. C., 573
Guyot, reef, 314
Gypsum, in salt plugs, 364, 367–370

Hackford, J. E., 505
Haft Kel pool, Iran, fracture permeability, 125
Halimeda, 512
Halite, Gulf Coast salt plugs, 361
Hanigsen, salt dome, Germany, overhang, 374
Hanway, Jonas, 404
"Hard scale," 173
Hastings field, Texas, permeable connection between producing sands, 579; structure caused by salt intrusions, 363, 367
Hatchettite, 678
Hauerite, 371
Hawkins field, Oklahoma, 61–62, 597
Hawkins field, Texas, gravity of oil, 197; repeated folding, 249
Haynesville field, Louisiana, 59; performance history, 483

Hazards, structural mapping, 240
Head, potentiometric surface, 397; water, 395
Heads, flowing intermittently, 479
Heald, K. C., 237
Heat, alteration of petroleum, 521–529; conductivity, 423; effects, 424–428; from intrusive and extrusive rocks, 427–428; relation to viscosity, 424–425; sources, 422–424; transformation of organic matter into petroleum, 517, 520–521; *see also* Temperature
Heaving shale, 373–374
Helium, method of geologic age determination, 219; in natural gas, 218–220; production, 219
Hematite, 371
Hemin, 504
Hempel method, oil analysis, 191, 193
Hendrick pool, Texas, reef reservoir, 251, 326; slumping, 251, 256
Hendricks, Thomas A., 633
Henry's law, 201
Heptane, 191, 216
Heroy, William B., 236
Heterostegina antillea, 323
"Het" sands, Gulf Coast, 648
Hexane, 191, 216
Hiatus, represented by unconformity, 333
High Island, salt dome, Louisiana, overhang, 374
High permeability, 108
High porosity, 114
High pressure conditions, effects of, 428
High temperature conditions, effects of, 428
Higher fluid pressure above lower fluid pressure, 411
Higher temperature over folds, 421
Highwood fold, Alberta, Canada, 276, 651
Hilbig pool, Texas, 305
Hill, Gilman A., 236, 450, 551–554, 565
Hilt, Carl, 633
Hilt's law, 633
Historical geology, 11
History, geologic, *see* Geologic history; production, free-gas cap drive pools, 464; production, gas-drive pools, 461; production, water-drive pools, 466; theories of organic origin of petroleum, 505
Hit, Iran, seepage, 17–18
Hlauscheck, Hans, 522
Hohlt, Richard B., 126
Holocene, 39
Homolog, 183
Homologous series, 183, 501, 503
Horizontal permeability, 108
Hoskins Mound salt dome, sulfur in cap rock, 371, 375

Host rock, 628
Howard, W. V., 118
Hubbert, M. King, 236, 396, 407, 456
Hugoton pool, Kansas, composition of natural gas, 217
Hull-Silk field, Texas, patchy sand reservoir, 290–291
Humic acid, 515, 526
Humus, 515
Hunt, J., 506, 522, 524
Huntington Beach field, California, deflected wells, 594–595
Hunton formation, Devonian-Silurian, Oklahoma, 122, 643–644
Hunton-Seminole-Ozark Arch, Oklahoma, 653
Hydrafrac, 136
Hydraulic fracturing, 136
Hydraulic gradient, 104–105, 542; effect on permeability, 114
Hydraulic theory of primary migration, 544
Hydraulic-buoyancy theory of primary migration, 544
Hydrocarbons, 3, 15, 497, 678; bearing on theories of petroleum origin, 507–508; compound, 500; concentration maps, 615; saturated, 178, 509, 519; series, 180–187; source material for petroleum, 509–510; unsaturated, 178, 509, 519
Hydrodynamic conditions, 396–401, 454–455, 486, 637
Hydrodynamic equilibrium, 394, 398
Hydrodynamic gradient, 553
Hydrodynamic pressure gradient, 553, 567–568; trapping agent, 340, 555–558, 581; see also Fluid potential gradient
Hydrodynamics, 342
Hydrogen, effect on neutron logs, 83; evolution of, 519; in organic matter, 506, 515; in petroleum, 177, 506; rare in natural gas, 216; source of, 180, 504; split off during radiation, 529
Hydrogenation, 180, 519, 525
Hydrogen sulfide, 168, 187, 209, 519; in natural gas, 218, 221–223
Hydrophilic, 66, 445
Hydrophobic, 445
Hydrostatic conditions, 341, 397, 556, 566–567, 637
Hydrostatic equilibrium 394, 398, 414
Hydrostatic pressure, 79; gradient, 390–391, 394–396, 547
Hysteresis, 446

Igneous intrusions, causes of lack of oil, 357; associated with petroliferous deposits, 427–428
Igneous and metamorphic reservoirs, 64–65, 74–75, 120–121, 263
Illinois, basin, petroleum province, 32; Chester strata, Mississippian, 28, 160, 290; clay in carbonate reservoirs, 66; Mississippian sands, 288–290; Omaha pool, 427; pre-Pennsylvanian unconformity, 643; Silurian reefs 72, 318–319; Silurian reservoir rock, 72, 319
Illite, 66
Imbibition, 476–477
Impsonite, 678
Impurities in natural gas, 218–223
Incompetence, relation to salt dome formation, 358, 378
Incompetent formations, 248–251
Index of refraction, see Refractive index
India, production, reserves, 35
Indian Ocean, atolls, 313–314
Indiana, Bordon Group, reefs, 322; Cincinnati Arch, 127, 307, 647, 653; Knobstone group, reefs, 322; Lima-Indiana field, 127, 307; Silurian reefs, 72, 318–319; Silurian reservoir rock, 72, 319
Indonesia, gravity of crude oil, 195; production, reserves, 35
Induced porosity, 117
Induction log, 80
Inerts, 218
Information, kinds useful to petroleum geologist, 590–591
Inglewood field, California, 263–264
Initial dip, 254, 322; over reef, 329
Initial production, 176, 305
Injected material, diapir folds, 249–250
Injection pressure, 136
Inorganic material in crude oil, 190
Inorganic origin of petroleum, theories, 503–504
Inspissated deposits, 8, 14, 15, 24–25, 26, 501
Inspissated pool, Kentucky, 303–304
Insular particle, movement of, 552–553
Insular saturation, 448–449, 474
Intangible drilling costs, 668
Interface phenomena, 434, 441–451, 456–457
Interfacial tension, 66, 442–451, 454, 456, 547, 549, 552–553
Intermediary water, 151
Intermediate porosity, 117–133
Intermediate salt domes, 359, 361
Intermittent folding, 246, 247–249
Internal-gas pool, 459
Internal structure, salt plugs, 361–362
Interstitial water, 111–112, 146, 147, 151, 152; effects on reservoir, 153–157; relation to porosity, permeability, and grain

INDEX 705

size, 153–155; *see also* Water saturation
Intrusive rocks, effects of heat, 427–428
Ion exchange, 173
Ionic statement, water analysis, 163
Ionization chamber, neutron logs, 83
Iran, Agha Jari field, 125, 254, 275; fractured reservoirs, 125; Haft Kel pool, 125; Masjid-i-Sulaiman field, 19, 106, 125, 223, 248, 480; Naft-i-Shah field, 19; Naft Khaneh field, 19; natural gas high in hydrogen sulfide, 222–223; petroleum province, 32; production, reserves, 34; Quaiyarah field, 19; salt plugs, unproductive, 359; seepages, 18–19
Iraq, Kirkuk field, 19, 212; petroleum province, 32; production, reserves, 34
Irawaddy River, oil on, 549
Iron, 526
Irreducible water saturation, 453, 457
Ishimbaevo field, USSR, reef reservoir, 323, 331–332
Iskine salt dome, USSR, 365, 372
Isobar maps, 616
Isobaric temperature-composition diagram, 437–438
Isocarbs, 630
Isochore maps, 616
Isoconcentration (Isocon) maps, 161–162, 616
Isogeotherm, 417, 423
Isogeothermal contours, 417
Isogradient contours, 417–419
Isogradient map, Texas and Louisiana area, 417–419
Isolated reservoirs, pools trapped in, 411–414
Isomerization, 525
Isomers, 178–179
Isometric diagram, 438
Isopach maps, 248, 254, 365, 371, 596–600
Isopotential maps, 61–62, 616, 617
Isothermal pressure-composition chart, 438
Isothermal surface, 415–418
Israel, production, reserves, 34
Italy, Vallezza field, complex structure, 244–245
Italy-Sicily, production, reserves, 34

Japan, mining oil, 472; "oily patches" from blooming diatoms, 513; production, reserves, 35
Jars, 678
Jatibonico pool, Cuba, 306
Java Sea, reefs, 313–314
Jefferson Island, Louisiana, sandstone in salt plugs, 361
Jennings, Louisiana, salt dome, overhang, 373, 374, 376
Johnson, J. Harlan, 322
Johnson, R. H., 558
Joints, effect on porosity and permeability, 15, 99, 108, 119–125
Jones, Park J., 99
Joule, 442
Jurassic, 39; Arab zone, Saudi Arabia, 71, 119, 212, 244; Emba province, USSR, 260, 262–263, 358, 372; production, reserves, 40; reserves, California, 120; salt deposition, Gulf Coast, 378; Shensi Series, China, 88; Smackover limestone, Arkansas, 106, 116–117, 394–395, 400, 562–563

Kaibab limestone, Permian, Utah, oil-filled geodes, 27
Kala oil and gas field, USSR, 262–263
Kalmes mud volcano, USSR, 21
Kansas, analysis of water sands, 162–163, 165, 168; Arbuckle waters, 170–171; Bush City pool, 302, 304; Central Kansas (Barton) Arch, 652; Cherokee shales, Pennsylvanian, 67, 279–280, 548–549; fossil channel filling, 302–304; gas fields, shale reservoir, 67; Hugoton pool, composition natural gas, 217; Kansas City-Lansing group, 72; karst topography, 118; Kraft-Prusa field, 127, 338; paleogeologic map of eastern, 606; pre-Pennsylvanian unconformity, 643; relation of water concentration to depth, 162–163; relation of water density to concentration of contained minerals, 160–161; reservoir rocks, 59, 63, 72; Sallyards shoestring-sand trend, 297, 299; shelf area, source of sediments, 548–549; shoestring-sand pools, 297; Sooy conglomerate, Pennsylvanian, 63; tilt of Paleozoic rocks, 542; uncomformities, Paleozoic, 645; viscosity-temperature relations in crude oils, 203–205; Voshell pool, 577–579
Kansas City-Lansing group, Kansas, limestone reservoir, 72
Kaolinite, 66
Karro formation, Madagascar, seepages, 88
Karst topography, Kansas, 118
Kekule structure, 187
Kentucky, channel sands, 303–304; Ohio shale, gas reservoir, 280; Poole field, 290; Silurian reservoir rocks, reefs, 72
Kerogen, 28, 508, 678; coal, 29; composition, 28–29, 506; effects of heat and pressure, 520–521; resources, 8; shale (Oil shale), 14, 15, 28–29
Kerosene age, 4
Kettleman Hills field, California, 540; volume of reservoir oil, 199

Keweenawan, Precambrian, bitumen, 27
Keyte, W. Ross, 235
Khadyzhinskaya Field, USSR, up-dip wedge-outs, 292, 296
Khuar field, Pakistan, excess pressure, 409–410
King, F. H., 545
Kinzi Dag mud volcano, USSR, 21
Kirkuk field, Iraq, crude oil properties, 212; seepages, 19
Kita-Daita-Zima, coral in boring, 314
Klinkenberg permeability factor, 107–108
Knobstone group, Indiana, reefs, 322
Kolm black shale, Cambrian, Sweden, oil and uranium mine, 28
Kraft Prusa field, Kansas, dolomite reservoir, 127; section, 338
Krumbein, W. C., 334
Krynine, Paul D., 55, 128
Kuban-Black Sea area, USSR composition natural gas, 217
Kuenen, Ph. H., 63
Kugler, Hans G., 20
Kuwait, Burgan field, 58, 166, 197; petroleum province, 32; production, reserves, 34

La Concepcion field, Venezuela, composition natural gas, 217
Lagunillas field, Venezuela, section, 297; water analysis, 166
Lake Asphaltites (Dead Sea), 19
Lake Maracaibo district, Venezuela, production from shale, 67; production from strandline sands, 293, 297, 648–650
Lake Valley formation, Mississippian, New Mexico, bioherm, 322
Laminar flow, 104
Lance Creek field, Wyoming, CI of crude oil, 194
Landslides, submarine, 19, 20, 21, 63, 324
Laochunmaie field, China, Tertiary, 88
La Paz field, Venezuela, fractures in, 125; temperature gradients, 419, 422
Las Cruces field, Venezuela, oil-field water, fresh, 161
Late-forming traps, 542
Lateral migration, 502, 545
Lateral permeability, 108, 545
Laterolog, 81, 82
Laudon, L. R., 322
Lava flows, reservoir rocks, 74
"Layer-cake" geology, 640
Layers, gas, oil, water in reservoir, 147; of geology, 640–642
Lead mines, petroleum associated, 27
Lean gas, 216
Lease, costs, 665–666; prices, structural traps, 281

Leduc field, Canada, dolomitized reef reservoirs, 117, 126–127, 327–328, 330–331
Lenses, evidence for short migration, 541; and facies, chemical rocks, traps, 306–332; and facies, clastic rocks, traps, 287–306; and facies, primary stratigraphic traps, 287–332; isolated, anomalous pressures, 400, 411–414; volcanic rocks, traps, 305–306
Leonard-Capitan formation, Texas, reefs, 324
Lessee, 665
Lessor, 665
Lewis, J. Volney, 545
Lias, Madagascar, seepages, 88
Libya, production, reserves, 35
Life, wells and pools, 481
Lima-Indiana field, Ohio and Indiana, 127, 307
Lime, banks, 310; reefs, 310
Limestone, bituminous, 24; capillary pressure curve, 455; massifs, 310; porosity, 117–133; reservoir rocks, 68–72
Limiting conditions, migration and accumulation theories, 539–541; origin theories, 500–503
Lind, Samuel C., 526, 527, 528
Link, Theodore A., 650
Liquefied gas, 211, 214, 484
Liquefied petroleum gas (LPG), 484
Lisbon field, Louisiana, primary porosity in carbonate reservoir, 115; structure, 308, 311
Lithofacies, 68, 288; maps, example, 600–601
Lithologic logs, 76–77
Lithostatic pressure, 402
Lithothamnium, 70, 315, 512
Live oil occurrences, 15, 638
Load pressure, effect on compaction, 130–133, 402
Local traps, petroleum province, 653–654
Logan, Sir William, 505
Logs, 10, 75–85; abbreviations used, 680–684; caliper, 84, 103, 590; cuttings analysis, 84; driller's, 75–76, 590; drilling-time, 83–84, 103, 109, 590; electric, 77–81, 103, 176, 590, 591, 638; electric potential, 78–80; errors, 592–596; gamma-ray, 81–83, 103; induction, 80; Laterolog, 81, 82; lithologic, 76–77; Microlaterolog, 81; Microlog, 80–81, 82, 85, 103; neutron, 81–83, 103, 590; nuclear magnetism, 85; paleontologic, 77; radiation, 81–83, 103; resistivity, 80; sample, 76–77; self-potential, 78–80; sonic, 85, 103; spontaneous-potential, 78–80; temperature, 85; types, 75; uses, 75

Logan, Sir William, 234
Long Beach field, California, subsidence, 132
Long distance migration, 541–544
Los Angeles Basin, California, 629
Loss of core, indicates porosity, 103–104
Lost circulation, 109, 590
Louisiana, Avery Island, 361; Bayou St. Denis field, 409; Bellevue field, 133; changes in crude oil composition with depth, 522–523; continental shelf, reservoirs, 36–37; Creole field, 263–267, 594; Delhi pool, 335; down-dropped fault traps, 266, 270; Eola field, 363, 368; Erath field, 247, 249; excess pressures, 409; geothermal gradients, 417; graywacke reservoirs, 58; Haynesville field, 59, 483; High Island salt dome, 374; isogradient map, 417, 419; Jefferson Island, 361; Jennings salt dome, 373, 374, 376; Lisbon field, 115, 308, 311; Mississippi river delta region, 359–360; Nacatoch sandstone, 133; oil and gas pools, 486; Rodessa field, 166, 212, 350–351; Sabine arch, 486, 652; salt plugs, 359–364; section showing geologic layers, 642; Tertiary reservoirs, 58; *see also* Gulf Coast
Lower fluid pressure below higher fluid pressure, 411
LPG, liquefied petroleum gas, 484
Lucien field, Oklahoma, oil formed in Ordovician, 501
Lytton Springs pool, Texas, serpentine reservoir, 306

McAlester basin, Oklahoma, laterally composite, 636
McCallum field, Colorado, rich in carbon dioxide, 221
McCloskey "sand," oolite reservoir, 117
McCollough, E. H., 236
McCoy, Alex W., 235, 546
McDermott, Eugene, 515–516
McMurray formation, Cretaceous, Canada, cemented by oil, 128
Madagascar, nonmarine seepages, 88, 427
Madison limestone, Mississippian, Wyoming, crude oil properties, 212
Magic geology, 659
Magma, source of heat energy, 423
Magnesium limestone, 70
Magnesium, 170, 190
Magnetic anomaly, 612
Magnetic susceptibility, granite, 63–64; how determined, 591
Magnetism, rocks, 10
Magnolia field, Arkansas, oolitic reservoir, 116
Maier, C. G., 521

Maikop region, USSR, channel-sand pools, 304–305; sandstone reservoir, 58; up-dip wedge-out, 292, 296
Major pool, 30
Malaysia, production, reserves, 35
Maltha, 678
Manantiales oil field, Chile, variable porosity, 100–101
Manjak, 25, 678
Man-made fractures in reservoir, 135–136
Man-made porosity and permeability, 135–136
Manometer, 395
Mapping, geophysical, 282, 612–614; methods, 238–240, 282; structural, 238–240, 282; subsurface, 11, 239, 282, 588–618; surface, 11, 238–239, 282
Maps, 10, 30; clastic-ratio, 600–602, 603, 604; clastic-shale ratio, 601; contemporary, 589; facies, 600–602, 603, 604; geochemical, 615; geophysical, 242, 612–615; hydrocarbon concentration, 615; isobar, 616; isochore, 616; isoconcentration (isocon), 161–162, 616; isopach, 248, 254, 365, 371, 596–600; isopotential, 61–62, 616, 617; miscellaneous subsurface, 616; paleogeologic, 339, 602–612, 643, 646; potentiometric surface, 397; progress, 589; sand, 599; sand-shale ratio, 601, 603; seismic, 242, 613–615; soil analysis, 615; structural, 591–596; subcrop, 602–603, 605, 606, 643; sub-sea, 592; subsurface, 591–618; water encroachment, 616; *see also* Paleogeologic maps
Mara field, Venezuela, fractures in, 72, 125, 263
Mare sporco, Adriatic, 513
Marginulina, 579
Marine organisms, source of petroleum, 510–514; Marine reservoir rocks, 86–89
Marine seepages, 19
Marl, reservoir rock, 70
Marsh gas, 183
Masjid-i-Sulaiman field, Iran, exceptional well, 480; fracture permeability, 125; natural gas high in hydrogen sulfide, 223; porosity and permeability, 106; section, 248; seepages, 19
Mass spectrometry, 177
Materials balance equation, 478
Mathematics, 11
Matrix, 55–56; effect on porosity and permeability, 99, 103, 114, 455
Maturin basin, Venezuela, composite laterally, 636
Maximum economic recovery (MER), 477–478
Maximum efficient rate (MER), 461, 477–478, 485

Maximum permissible rate, (MPR), 478
Measurement, crude oil; 175–176; excess pressures, 405–406; natural gas, 84, 214–216; oil-field water, 157–165; original pressure, 391; permeability, 105–110; porosity, 102–104; pressure, 393–394; viscosity, 206–207
Mechanics, reservoir, 433–493
Meinzer, Oscar Edward, 132
Mene Grande field, Venezuela, complex structure, 353, 355, 357
Menhaden, 513
Meotian, Romania, nonmarine reservoirs, 87
MER, 461, 477–478, 485
Mercury, associated with oil, 27
Mercy field, Texas, porosity and permeability, 133
Mesozoic, 39
Metals, surface-active substances, 444, 450
Metamorphic reservoir rocks, 74
Metamorphism, 9; carbon ratio theory, 630–634; effect on sediments, 629–630, 639
Metaquartzite, 68, 128
Meteoric water, 152
Meter, displacement, 214; orifice, 214
Methane, in hydrocarbon fluids, 191; from inorganic sources, 503; in natural gas, 3, 216–218; from organic decay, 405, 517, 519; series, 183–185
Methyl, 187
Mexico: ash in crude oil, 190; Golden Lane, 329–331, 479–480; gravity of crude oil, 195; high sulfur in heavy-oil seepages, 188; Northern fields, 124–125; Panuco district, 198; petroleum province, 32; pools rich in carbon dioxide, 221; Potrero del Llano well, 479; Poza Rica field, 72, 127, 330, 353, 356; production, reserves, 34; Southern fields, 115–116; Tampico-Tuxpan region, 222, 427
Michigan, bitumen, Precambrian, 27–28; Coldwater field, 201–202; Deep River pool, 122–123, 308; gas pools in offshore sand-bar deposits, 297–299; Salina dolomite, Silurian, 160; Scipio-Albian field, 123–124; Six-Lakes gas pool, 298–299
Micro-gas pockets, 467, 471
Microlaterolog, 81
Microlog, 80–81, 82, 85, 100, 103
Micro-pressures, from earthquakes, 404
Microscopic examination, well cuttings, 103
Microscopist, 103
Mid-Continent, U.S., age of traps, 579–580; Arbuckle-Ellenburger group, 72; characteristic pools in Pennsylvanian rocks, 290–291; gravity of crude oil, 195; high nitrogen in natural gas, 220; petroleum province, 30, 32; reservoirs, 59–60; stage of exploration, 379–380
Middle East, ash in crude oil, 190; exploration, 36; gravity of crude oil, 195; high-sulfur crude oil, 188; location map of seepages and pools, 18; "oil axis," "oil pole," 32; Persian Gulf-Arabian basin, 636; production, reserves, 35; rich reservoirs, 31, 32, 71; seepages, 17–18; stage of exploration, 379–380; Tertiary reservoirs, 61; *see also* Iran, Iraq, Kuwait, and Saudi Arabia
Midland Basin, Texas, 73
Midway field, California, section, 335
Migration, 5; and accumulation, 538–584; at bubble-point pressure, 573; evidence for, 541–544; geologic framework, 539–541; lateral, 502, 504; minimum saturation necessary, 112; physical conditions necessary, 551–554; primary, 538, 544–550; relation to structural basins, 637; secondary, 538, 550–581; short- or long-distance, 541–544; theories, 500, 544–545; vertical, 502, 569–571
Millidarcy, 105
Mills, R. van A., 559–560
Mimbu, Burma, mud volcanoes, 22–23
Mineral, 4; fuel, 4; hydrocarbons, 4; resources, 4; rights, 33, 36; substance, 4
Mineralogy, 11
Mineraloid, 4
Minerals, associated with faults and oil, 560; secondary, evidence of metamorphism, 629
Minerva fluorite mine, oil recovered, 28
Minnesota, algae in Precambrian rocks, 40
Minor occurrences, 31–32
Minor showings, 15, 29–30
Miocene, 39; in Bikini Island boring, 314; graben, Gulf of Suez, 577; Gulf sand (Creole field), 266; in Kita-Daita-Zima boring, 314; Louisiana, changes in crude oil compositions with depth, 522–523; Monterey formation, California, 73, 121; Talang Akar sandstone, Sumatra, 273
Miscellaneous, reservoir rocks, 74–75; substances in crude oil, 190–191; subsurface maps, 616; surface occurrences, 27–28
Mississippi, salt plugs, 359–360
Mississippi River delta region, salt plugs, 359–360
Mississippian, 39; Big Injun sand, Ohio and West Virginia, 60; Bordon or Knobstone group, Indiana, 322; Chester strata, Illinois, 28, 160, 290; gas pools in offshore sand-bar deposits, 297–299;

limestone reservoir, rock, 63; McCloskey "sand," 117; Madison limestone, Wyoming, 212; reef reservoirs, New Mexico and Texas, 322, 326; sands, lateral variations, 288-290; Stanley shale, Oklahoma, 525
Missouri, fossil stream channels, 303
Mixed water, 152-153
Moberly channel, Missouri, 303
Mode of occurrence, petroleum, 15-32
Modern organic matter, 510-514
Modern reefs, 312-318
Moisture, in coal, 630
Molecular replacement, limestone by dolomite, 126
Molybdenum, 525
Monnet, V. E., 545
Montana, Cutbank pool, buttress sands, 335-336; natural gas high in carbon dioxide, 221
Montebello field, California, combination traps, 353, 355
Monterey formation, Miocene, California, fractured siliceous reservoir rock, 73, 121
Montmorillonite, 66, 107
Moreni-Bana field, Romania, complex structure, 275, 280
Morocco, production, reserves, 35
Mount Calm pool, Texas, fracture trap, 280, 282
Movement, oil and gas in a pool, 473-477; see also Migration
Moving water, 560
MPR, maximum permissible rate, 478
Mrazec, L., 544
Mud: bottom reefs, 312; cake, 622; flows, 19-23; to shale, diagenesis, 550; trucks, 622; volcanoes, 14, 19-23; see also Drilling mud
Muddy sands, 58-59, 297
Multicomponent system, 436-437
Multi-pay field, Eakring, England, 254; Midway, California, 335; Montebello, California, 355; Pickett Ridge, Texas, 295; Santa Fe Springs, California, 239; Ten Section, California, 147, 149
Munn, Malcolm J., 544
Murphy, John A., 404
Murray, Stuart, 549
Music Mountain pool, Pennsylvanian, offshore bar reservoir, 298-300; water-free, 151-152

Nacatoch sandstone, Cretaceous, Louisiana, porosity and permeability, 133
Naft-i-Shah field, Iran, seepages, 19
Naft Khaneh field, Iran, seepages, 19
Names, producing formations, 53-54
Naphtha, 3, 28, 678

Naphthene series, 185-186
Nashville Dome, Tennessee, 653; wedge belts of permeability, 645-646
Natural energy of reservoir, 458-473
Natural gas, 3, 7, 678; composition, 177, 183, 216-218; impurities, 218-223; measurement, 84, 214-216; in reservoirs, 211-223; specific gravity, 554; see also Gas and Petroleum
Natural gasoline plant, 484-485
Natural pressure, 391
Near East, see Middle East
Nebraska, southeastern, paleogeologic map, 606
Nemaha Mountains, Kansas, Nebraska, Oklahoma, 548-549, 605-606, 652
Neogene, 39
Neritic environment, 61
Netherlands, production, reserves, 34
Nettleton, L. L., 376
Neutral Zone, production, reserves, 34
Neutron log, 81-83, 590; measure of porosity, 103
New Grozny field, USSR, relation of water concentration to depth, 162
New Guinea, gravity of crude oil, 195; production, reserves, 35
New Jersey, coast compared to Cherokee seas, 297, 299
New Mexico, Abo-Wichita Albany reef, 320; bioherm, Lake Valley limestone, 322; Capitan reef, 320-321, 326; different temperature gradients in different sediments, 418-419; geothermal gradients, 418-419; isogradient map, 417-419; Ouray formation, 219-220; natural gas high in carbon dioxide, 221; natural gas high in hydrogen sulfide, 222; Permian Basin, map, 73; pools in reefs, 326; pre-Pennsylvanian unconformity, 643; Rattlesnake field, 219-220; sections, 608-610
New York, Oriskany sand, 121-122, 272, 274-275; Woodhull-Tuscarora gas pools, 274-275
New Zealand, seepages, 638
Niagaran (Middle Silurian) reefs, 72, 318-320
Nickel, 190, 525
Nigeria, production, reserves, 35
Nimadric Series, Pakistan, seepages, 88
Nitrogen, in crude oil, 177, 189, 506; in natural gas, 177, 189, 218, 220; in organic matter, 506, 515, 518; in petroleum, 177, 504-505, 506
No/B/Hr, Number of bailerfuls per hour, 393
Noble gases, 218, 527
Nomenclature, reservoir rocks, 53-54
Nonane, 216

Nonassociated gas, 211, 484–486
Non-colonial organisms, reef builders, 315
Nonconformity, 333
Nonmarine, organic matter, 514–516
Nonmarine reservoir rocks, 86–89
Nonrecoverable oil, 461, 488
Non wetting fluid, 112, 446
Normal faults, cause of traps, 260–270
Norman Wells reef, Canada, 328
North America, geologic time scale, 39; paleogeologic map of western, pre-Cretaceous, 607–608; production, reserves, 35
North Coles Levee field, California, displaced oil pool, 563–564
North Snyder reef, Texas, 329
Northern fields, Mexico, fracture permeability, 124–125
Northwest Lake Creek field, Wyoming, displaced oil pool, 563
Novaculite, 63
Nuclear Magnetism Logs, 85
Nutting, P. G., 526

Objections to radioactive transformation processes, petroleum origin, 528–529
Occurrence, disseminated, petroleum, 24–25; minor, 31–32; miscellaneous surface, 27–28; mode, 15–32; petroleum, 14–46; solid petroleum, 23–27; subsurface, 15, 29–32; surface, 14, 15–29
Octane, 216
Offlap, 334
Offshore bars, traps, 296–301
Oficina district, Venezuela, geothermal gradient, 416
Ohio, Big Injun sand, Mississippian, 60; Black Hand oil shale, 27; Cincinnati Arch, 127, 307, 647, 653; Clinton sand, up-dip wedge-out, 647; Lima-Indiana field, 127, 307; shale, Kentucky, gas reservoir, 280; Trenton dolomite, up-dip wedge-out, 647–648
Oil, absent at dry hole, 620–621; "axis," 32; bacterial oxidation, 520; -base drilling mud, 146; bank, 482; barrels per foot drilled, 671; cement, 128; column, 242; "crop," 7; detection, 84; difficulties of analysis, 176–177; dissolved in water, 545, 560, 581; erosion of pools, 643, 645; fields, see individual pool and field names; industry, beginning of, 234; "maturing," 522; mining, 472; nonrecoverable, 461, 488; patches, 546, 554–555; "poles," 32; pools, relation to gas pools, 485–486; recoverable, 461; recycled, 548–550; release from plants by bacteria, 520; release from sedimentary rocks, 519–520; in reservoirs, 175–211; saturation maintained near well, 474; sedimentary, 548–550; "seed," 7; -settling capacities, 549–550; "shale," 14, 15, 27–29, 678; specific gravity, 554; see also Crude oil and Petroleum
Oil and Gas Journal, The, 30, 675
Oil-field water, 150–174, 638; analyses, 166, 171–173; "channeling" and "coning," 474; character, 157–173; chemical composition, 163–171; circulation, 546–548; classification, 152–157; compressed, 467–471; concentration, 160–163; connate, 80, 111–112, 152–153, 545; density, 160–161, 394–395, 401–402, 569, 685; distribution in the reservoir, 147–150; effect on electric logs, 80, 156–157; encroachment maps, 616; free, 109, 153, 457, 560; interstitial, 111–112, 146, 147, 151, 152, 153–157; measurement, 157–165; measure of permeability in cable-tool holes, 75; meteoric, 152; mixed, 152–153; origin, 173–174; residual, 453, 457; saturation, 157–160; specific gravity, 555; squeezed out of clay and shales, 412–414, 545–546; see also Water
Oil-water contact, 50; Amarillo Arch, Texas, level, 65; Cairo pool, Arkansas, tilted, 562–563; Coles Levee field, California, tilted, 564; Conroe field, Texas, level, 147–148; Frannie pool, Wyoming, tilted, 565; level, 561; Northwest Lake Creek pool, Wyoming, tilted, 563; Quitman field, Texas, stepped, 261; relation to potentiometric surface, 567–569; Sage Creek pool, Wyoming, tilted, 566; Salt Creek field, Wyoming, tilted, 151; Santa Fe Springs field, California, different levels, 239; structural traps, 281; Ten Section field, California, stepped, 149; tilted, 236, 561–569; Van field, Texas, level, 150
Oil-water table, 50; see also Oil-water contact
Oil-wet capillary, 447
Oil-wet minerals, 580
Oil-wet reservoirs, 450
Ojos de chapopote, 679
Oklahoma, Apache pool, 243–244, 253, 257, 603, 605; asphalt, 24; Bartlesville sand, Pennsylvanian, 197, 292, 296, 543–544; carbon ratio of coals, 633; Ceres pool, 300–301; Cushing field, 563; Davenport pool, 166; Dora pool, 290–291; Garbor field, 131, 166; Glenn pool, 106; Glenn sands, Pennsylvanian, 106, 543–544; Hawkins field, 61–62, 597; Hunton formation, 122, 643–644; Hunton-Seminole-Ozark Arch, 653; isogradient map, 419; isothermal surfaces, 415, 417–418; Lucien field, 501; McAlester Basin, 636; northeastern, distri-

INDEX 711

bution of oil pools, 296; northeastern, gravity of oils, 197; Oklahoma City field, 106, 114, 166, 212, 249, 336, 415, 450, 644; Pauls Valley field, 614; petroleum province since 1903, 654; pre-Pennsylvanian paleogeologic map, 643; pre-Pennsylvanian unconformity, 643; Red Fork sand, Pennsylvanian, 197, 301; Reddin district, 336; reservoirs, 59; "siliceous lime," 72; South Ceres pool, 300–301; South Earlsboro pool, 242–243; Tatums pool, 338; tilt of Paleozoic rocks, 542–544; Velma field, 217; viscosity-temperature-gravity relations in crude oils, 204, 206; West Edmond pool, 122, 205–206, 337; Wichita Mountains, 65, 549; "Wilcox" sand, Ordovician, 106, 114, 242
Oklahoma City field, crude oil properties, 212; isothermal surface, 415; oil-wet sands, 450; porosity and permeability, 106, 114, 336; repeated folding, 249, 644; water analysis, 166
Old Grozny region, USSR, 279
Old Testament, 19
Olefins, 185–186; cracked to paraffins, 526
Oleophobic, 445
Oleophylic, 66, 445
Oligocene, 39; Amelia field, Texas, 270–272; Barrackpore field, Trinidad, 23, 217; Frio formation, Texas, 270–272, 292, 365, 367, 371, 648; reservoir rocks, Germany, 369
Omaha pool, Illinois, associated with igneous activity, 427
One-component system, 434
Onlap, 334
Ontario, exceptional wells, 480; Silurian reefs, 72, 318–319; see also Canada
Oolite, 56, 68, 116; deformation of, 126
Open-flow test, permeability measurement, 109
Opman-Box-Dag mud volcano, USSR, 21
Optical activity, petroleum, 209, 503, 505
Optical rotary power, petroleum, 209, 503, 505
Order of solubility, carbonates, 118
Ordovician, 39; asphalt, 24–25; Bromide sand, 243, 253, 605, 614; Ellenburger limestone, 309, 338–340; limestones, karst topography, 118; Lucien field, Oklahoma, 501; Pulaski channel, Tennessee, 301, 303; Simpson sands, 59; traps in Mid-Continent region, 579–580; Trenton formation, 127, 301, 303, 307, 647–648; water concentrations, Kansas, 162–163; "Wilcox" sand, 59, 106, 114, 242, 243
Organic acids, weathering zone, 118
Organic carbon, Precambrian, 38, 40, 501
Organic chemistry, 178
Organic cycle of sea, 511
Organic matter, composition, nonreservoir sediments, 506; in crude oil, 190; kerogen, 28, 508; kinds, 506–508; modern, 510–514; nonmarine, 514–517; in river waters, 514; source material for petroleum, nature of, 508–510; transformation into petroleum, 517–529
Organic origin of petroleum, 504–530
Organic reefs: see Reefs
Organisms, benthonic, 511; marine, source of petroleum, 510–514
Orifice meter, 214
Origin: cap rock, salt domes, 368–371; and classification, pore space, 113–136; dolomite, 126; helium in natural gas, 218–219; oil-field brines, 173–174; petroleum, 5, 499–537; petroleum, current theory, 506–517; petroleum, inorganic, 503–504; petroleum, organic, 504–505; petroleum, relation to depositional basins, 637; reservoir rocks, 86, 540; salt domes, 374–379; theories, bearing on exploration, 500; theories, framework of limiting conditions, 500–503
Original free-gas cap, 460
Original porosity, 113
Original pressure, 391, 473
Original saline content of water, 174
Oriskany sand, Devonian, fractured reservoir, 121–122, 272, 274–275
Orthoquartzite, 68, 128, 129, 629
Orton, Edward, 126, 235
Osmosis, source of reservoir pressure, 402–403
Osmotic forces, 471
Ouray formation, New Mexico, natural gas high in nitrogen-helium, 219–220
Overburden, effect on compaction, 130–133; 402; pressure, 402
Overhang, salt dome cap rock, 370, 374
Overlap, 334
Overstep, 334
Overturned fold traps, 244–245
Oxidation, bacterial, 520, 549
Oxidizing agencies, 24
Oxnard field, California, gravity of oil, 198
Oxygen, in meteoric water, 152; in organic matter, 506, 515; in petroleum, 177, 189, 506; relation to bacteria, 517–519
Ozokerite (Paraffin), 185, 305, 515, 679

Packing, 113, 679; cubic, 113; effect on porosity and permeability, 108, 113, 114–115, 131; oil globules, 549; rhombohedral, 100, 113
Pakistan, Khuar field, 409–410; nonmarine seepages, 88; production, reserves, 35
Pacific Ocean, atolls, 313–314

Packer, 393
Paleocene, 39
Paleogene, 39
Paleogeologic maps, 339, 602–612, 643, 646
Paleontologic logs, 77
Paleontology, 9, 11
Paleozoic, 39; brines, Appalachian region, 165; dolomites, 70; production, reserves, 38, 40; regional tilt, Texas, Oklahoma, Kansas, 542–544; reservoirs, 59, 72; shales high in radioactivity, 528; unconformities, Kansas, 645
Palmer, Chase, 163
Palmer System, water analysis, 163–165
Paluxy sand, Cretaceous, Texas, 261
Panel diagram, 592–593
"Pan-of-biscuit" folds, 256
Panhandle field, Texas, 64–65; composition of natural gas, 217; natural gas high in hydrogen sulfide, 222; *see also* Amarillo-Texas Panhandle field
Panuco district, Mexico, gravity of oil, 198
Paradox basin, salt domes, Utah and Colorado, 358–359
"Paraffin" dirt, 361, 679
Paraffin series, 178, 183–185; physical properties, 184
Paraffins, effect of catalysts, 526; effect of radioactive bombardment, 528; in natural gas, 216
Parallel folding, 248–249
Parker method, water analysis diagram, 165, 169
Particle size, classification, U.S. Bureau of Soils, 56; clays, 66; range in clastic sediments, 56
Particles, entrained, 551
Parts per million (ppm), 160
Patches, oil, 546, 554–555
Patnode, H. Whitman, 504
Pattern diagrams, crude oil and reservoir rock, 182
Pauls Valley field, Oklahoma, maps of, 614
Pay formations, names, 53–54
Peat, 4
Pechelbronn, France, electric logs, 78
Pendular rings, 449, 474
Pennsylvanian, 39; arkoses, reservoirs, 64–65; Bradford sand, 133; Calvin sandstone, Oklahoma, 525; Cherokee shale, Kansas, 67, 279–280, 548–549; crude oil, low sulfur content, 188; Devonian sands, 59, 288–289, 300; discovery of oil by Drake, 234; facies map, Colorado, 602; formation of traps, 579–580; fossil stream channels, 303–304; gas wells on anticlines, 235; Hawkins field, Oklahoma, 61–62, 597; Music Mountain pool, 151–152, 298–300; Oriskany sand, Devonian, 121–122, 272, 274–275; Paradox formation, 358–359; Red Fork sand, Oklahoma, 197, 301; reefs, Texas and New Mexico, 326–329; shoestring-sand traps, Kansas, 297–299; Silverville sand, Devonian, 298–300; Sooy conglomerate, Kansas, 63; South Ceres pool, Oklahoma, 300–301; Tensleep sands, Wyoming, 196, 450, 563; Third Stray sand, Devonian, 288–289; Todd field, Texas, 115, 308–310; water concentrations, Kansas, 162–163, 165, 168
Pentane, 178–179, 183, 191, 216, 217, 484, 615
Perforation, 679
Period, 38, 39
Perm Basin, USSR, reefs, 323, 330–332
Permanent pressure system, 401
Permeability, 40, 49, 97, 98, 104–112; to air, 107; Anahuac and Tomball fields, Texas, 155; artificial or man-made, 135–136; Bradford sand, Pennsylvania, 133–134; Cedar Lake field, Texas, variable, 101–102; Ceres pool, Oklahoma, 300; conditions affecting, 114–115; Conroe field, Texas, 147–148; East Texas field, 155; effective, 110–112; effects of plastic compaction, 130; fracture, 73, 108, 119–125; to gas, 107–108; horizontal, 108; how determined, 590; in igneous and metamorphic rocks, 74, 629; isolated reservoirs, 412, 414; Klinkenberg, 107–108; lateral, 108; Mara field, Venezuela, 125; Masjid-i-Sulaiman and Haft Kel pools, Iran, 125; measured in cable-tool holes, 75; measurement, 104, 105–110, 146; Nacatoch sandstone, Louisiana, 133; nonmarine sediments, 88; offshore bars, 297; Oriskany sand, 121–122; Petrólea field, Colombia, 129–130; profile, 109; range, 105, 540; Redwater pool, Canada, 117; relation to buoyancy, 556–558; relation to capillary pressure, 454–456; relation to carbon ratio, 632–633; relation to formation factor, 158–159; relation to interstitial water, 153–155; relation to porosity, 133–136; relation to unconformities, 74, 333–339, 639–645; relative, 110–112, 461; representative, 106; rough field appraisal, 105; Smackover limestone, 116–117; Spraberry field, Texas, 121; Springhill sand, Chile, varibale, 101; tilted oil-water contacts, 565, 569; up-dip pinch-outs, 599; Upper Wilcox sandstone, Texas, 133; variable, 60, 101–102, 147, 455, 456, 565; vertical, 108;

INDEX

to water, 107; water-drive pools, 464–467
Permeameter, 105
Permian, 39; arkoses, reservoirs, 64–65; Basin, rich reservoirs, 72–73; Eakring field, England, 254; Guadalupe-Capitan reef, 320–321, 326; nonmarine reservoir, Argentina, 87; oil-filled geodes, 27; Perm Basin, USSR, reefs, 323, 330–332; reefs, Colorado, 322–323; salt deposition, Gulf Coast, 378; salt domes, Emba district, USSR, 358; salt domes, Zechstein, Germany, 358; San Andreas limestone, 101–102; Sand Belt reservoirs, Texas, 73, 292; seepages, Brazil, 427; Spraberry field, Texas, 28, 67, 121, 468, 477, 604
Persian Gulf–Arabian basin, Middle East, composite laterally, 636
Personal factors, petroleum prospect, 669–672
Peru, production, reserves, 34
Petrochemicals, 4
Petrography, 9, 11
Petrólea field, Barco district, Colombia, 129–130, 266, 269; exceptional well, 479
Petroleum, 679; absence of evidence of, 639; alteration by heat and pressure, 521–529; bibliographies, 684–685; catalytic reactions, 524–526; changes in composition, 522–524; changes in gravity, 524; classification, 3–4; composition, 177, 180–181, 191, 501, 506; cost, 9, 664–665; definition, 3–4; density and specific volume table, 687; deposit, classification, 14–15; dikes, 26; engineer, 7, 10, 78, 98, 433, 459; engineering, 389, 487; geologist, 5, 6, 8, 9, 10, 14, 20, 31, 42, 50, 78, 379, 433, 434, 473, 477, 613, 627, 658–659, 662, 671; indirect evidence, 27; pools, fields, provinces, 30–31; price, 480–481, 664–665; prospect, 10, 658–673; province, 15, 31, 627–657; recycled, 516–517; reserves, 7–8; resources, 7–8; solid, occurrence, 23–27; supply, 580–581; undiscovered, 8; uses, 4; *see also* Crude Oil, Gas, Natural gas, and Oil
Petroleum Engineer, The, 675
Petrology, 11
Petrophysics, 98, 136
Pettet formation, Cretaceous, Texas, limestone reservoir, 117, 352
Pettit lime, Cretaceous, Louisiana, performance history of water-flooded pool, 483
Pettus sand, Eocene, Texas, 293
Phase, 434; continuity, 546, 554, 555, 560, 611; diagram, water, 435; relationships, 434–441; rule, 435

Phenols, 189
Philippines, reserves, 35
Phosphoria limestone, Wyoming, 563
Photosynthesis, 511
Physical: chemistry, 11; geology, 11; laws, unknown variables in, 457; properties, common paraffins, 184; properties, crude oils, 192–211
Physically recoverable oil, 461
Physics, 9, 11
Phytoplankton, 510
PI, productivity index, 478
Pickett Ridge field, Texas, multiple-pool, 292, 295
Piercement fold, 250
Piercement salt dome, 359
Pierre shale, Cretaceous, Colorado, 279, 282
Piezometric surface, 395; intersecting trap, 407; low position, 132, 411; Salt Creek field, Wyoming, 151; sloping, 160, 567; *see also* Potentiometric surface
Pincher Creek pool, Canada, exceptional well, 480
Pinnacles, in reefs, 315, 332
Pinpoint porosity, 98, 116
Pipe capacities, API, 688
Pisolite, 116
Pitch, 3, 17, 24, 679
Pitch Lake, Trinidad, 17
Plankton, 510
Plant life, 510, 511–513
Plants, bacterial release of oil from, 520
Plastic, compaction, 130–133; flow theory, origin salt domes, 377; salt, 361
Platform, 635
Playa del Rey field, California, basement reservoir, 60, 120
Pleasant Valley Pool, California, 336, 352–353
Pleistocene, 39; formation of traps, California, 540, 579–580; lack of oil, 38, 501, 504; relation of sea level changes to reefs, 313
Pliocene, 39; Bibi-Eibat field, USSR, 22; bituminous sand, 24; East Cat Canyon field, California, 353, 355; nonmarine reservoirs, 87–88; Productive series, USSR, 87–88, 198, 262; rich in oil pools, 502
Ploesti region, Romanian fields, 275, 280
Plugging, 679
Plugs: *see* Salt domes
Podzolization, 516
Poirier, O. A., 549
Poise, 207
Poland, Bitkow field, 275, 281; Boryslaw, ozokerite, 185; fields, discordant folding, 250; salt, 358

Polar hydrocarbon compounds, 444, 450
Political climate, 32–33, 36, 627
Polymerization, 179–180, 525
Pood, 480
Pool, 5–6, 15, 30; depletion-drive, 459; displaced, 258; dissolved gas, 459–463; essential elements, 6; gas, 485–486; internal-gas, 459; life, 481; major, 30; not recognized, 621–622; producing history, 146–147; saturated, 213; solution-gas, 459; solution-gas expansion, 459; stripper, 480–481; undersaturated, 213; water-drive, 464–471; water environment, 539–540
Poole field, Kentucky, 290
Porcellanite, 21, 27, 63
Pore: capacity per acre-foot, 99, 101; pattern, 98–99; sizes, 552–553
Pore space, 49; classification and origin, 113–136; geometry of, 552; primary, 97; reservoir, 97–143; secondary, 97; *see also* Porosity and Permeability
Porosity, 49, 97–98, 100–104; absolute, 101; artificial or man-made, 135–136; available pore space, 101; Bradford sand, Pennsylvania, 133–134; carbonate, 115–117; causes of low values, 103; cavernous, 119, 480; Cedar Lake field, Texas, variable, 101–102; Ceres pool, Oklahoma, 300; connected (*see* Permeability); cubically packed, 113; decrease, expells water, 412–414; effective, 101; effects of folding, 521; effects of plastic compaction, 130; how determined, 590; induced, 117; intercrystalline, 115; intergranular, 113–117; intermediate, 117–133; Ishimbaevo field, USSR, 331; isolated reservoirs, 412–414; marginal, 102; Masjid-i-Sulaiman and Haft Kel pools, Iran, 125; measurement, 100, 102–104, 146; modern reefs, 315–318; Nacatoch sandstone, Louisiana, 133; nonmarine sediments, 88; Oriskany sand, 121–122; Petrólea field, Colombia, 129–130; pinpoint, 98, 116; primary, 113–117; range, 102, 540; Redwater pool, Canada, 117; relation to carbon ratio, 632; relation to formation factor, 157–160; relation to interstitial water, 153–154; relation to neutron logs, 83; relation to permeability, 133–136; relation to unconformities, 118–119, 333–339, 639–645; representative, 106; rhombohedrally packed, 113; rough field appraisal, 102; secondary, 117–133; secondary, due to cementation and compaction, 127–133; secondary, due to fractures and joints, 119–125, 338; secondary, due to recrystallization-dolomitization, 126–127; secondary, due to solution, 117–119; Smackover limestone, 116–117; Spraberry field, Texas, 121; Springhill sand, Chile, variable, 100–101; titled oil-water contacts, 565; total, 101; up-dip pinch-out, 599; Upper Wilcox sandstone, Texas, 133; variable, 100–102, 147, 456, 565; "Wilcox" sand, Oklahoma City field, 114
Porous carbonate facies, stratigraphic traps, 306–308
Porphyrin, 501–502, 504, 520
Port Neches salt dome, Texas, 364–365, 370
Potassium, radioactive element, 527
Potential energy, 401, 458, 561
Potentiometric surface, 282, 391, 394–403, 406, 407; gradient, 553; maps, 397; Point Lookout sandstone, 403, 406; relation to ground surface, 396; relation to oil-water contact, 567–570; *see also* Piezometric surface
Potrero del Llano well, Mexico, 479
Powell field, Texas, crude oil properties, 212
Poza Rica field, Mexico, combination trap, 72, 353, 356; gravity of crude oil, 195; reef reservoir, 127, 330
ppm (parts per million), 160
Practical Petroleum Engineers Handbook, 675
Pratt, Wallace, 522
Precambrian, 39; algae, 40; bitumen, 27–28; lack of oil, 38, 40, 501, 504; organic carbon, 38, 40, 501
Precipitated cement, 128–130
Precipitation, secondary, porosity changes, 403–404
Pressure, alteration of petroleum, 521–529; anomalous, 405–414; back, 392; bomb, 393, 590; bottom-hole, 390; bottom-hole flowing, 392; "break down," 411; casing, 392, 393; changes, 10, 391–392, 542; closed-in, 390; decline, 146, 392, 399–401, 458–459, 466–467, 474–477, 562–563; differential, 392; displacement, 147, 453–457, 551–554, 557; earth, 402; effect of load, 130; effect on solubility of oil in water, 545; effect on volume of oil, 146–147, 198–201; flowing, 392; fluctuation, 541; fluid, *see* Fluid pressure; forefront, 453; formation, 390; geodynamic, 402; geostatic, 402; gradient, fluid, 389, 394–401, 428–429, 464, 474, 546, 551; injection, 136; interpreted through carbon ratios, 630–634; kinds of, 390–393, 402; lithostatic, 402; maintenance, 460; measurement, 393–394; micro-, from earthquakes, 404; migration at bubble-point, 573; natural,

391; original, 391, 473; osmotic, 402–403; overburden, 402; phase relationships, 434–441; rate of recovery, 109; relation to density of oil, 464; relation to depth, 394–396; relation to gas in solution, 201–202, 213–216, 464, 484–486; relation to interfacial tension, 451; relation to power of adsorption, 173; relation to solubility of salts, 173–174; relation to viscosity, gas and oil, 204, 389, 547–548; relation to viscosity of water, 425; relation to volume, natural gas, 215–216, 571–574; relation to volume and temperature in oil, 203; reservoir, 10, 65, 145, 176, 214, 389, 390–414, 464–471; reservoir, calculation of, 392; reservoir, relation to saturation pressure, 213; reservoir, sources of, 401–405; "rock," 390, 402; saturation, 213, 460, 573; shearing, 521; shut-in, 392, 473; static bottom-hole, 391–392; static formation, 392; in structural trap reservoirs, 281; surface, 392; -temperature diagram, oil and gas system, 439; transformation of organic matter, into petroleum, 517, 520–521; transmission of, 132; triple point, water, 435; tubing, 392; vapor, 213; virgin, 391; -volume method, gas measurement, 215–216

Price of petroleum, 480–481, 664–665

Primary: disseminated deposits, 25; migration, 538, 544–550; porosity, 113–117; production, 481; recoverable oil, 461; stratigraphic traps, 286, 287–332

Producing history, pool, 146–147

Production, Carthage gas field, Texas, ultimate, 352; costs, 486–487, 667; Cutbank pool, Montana, ultimate, 336; Delhi pool, Louisiana, ultimate, 335; Dora pool, Oklahoma, ultimate, 290; economic limit, 481; engineer, 136, 392; gas, 484–486; Golden Lane, Mexico, 329; Hastings field, Texas, ultimate, 367; history, 461, 464, 466; Hull-silk field, Texas, ultimate, 291; individual well, Dossor field, USSR, 260; individual wells, Ishimbaevo field, USSR, 331; individual wells, Maikop field, USSR, 305; individual wells, Music Mountain pool, Pennsylvania, 300; individual wells, Santa Maria field, California, 121, 341; initial, 176, 305; Lima-Indiana field, 127; major pool, 30; Masjid-i-Sulaiman field, Iran, 248; Mene Grande field, Venezuela, 357; Northern Fields, Mexico, 125; Oklahoma City region, Oklahoma, ultimate, 643; per cubic mile of sediments, U.S., 628–629; Permian Basin, 72; phenomena, 477–481; Pitch Lake, Trinidad, ultimate, 17; Poza Rica field, Mexico, ultimate, 330; primary, 481; Quirequire field, Venezuela, 335; rate, 175–176; Santa Maria field, California, 121; "serpentine" reservoirs, Texas, 305; stripper wells, 480–481; Talang Akar field, Sumatra, 273; test, 590; Van field, Texas, ultimate, 268; world, by geographic location, 32–35; world, by geologic age, 40

Productive reefs, 325–332

Productive series, Pliocene, USSR, nonmarine reservoir, 87–88; oil density changes with depth, 198; structure, 262

Productivity index (PI), 478

Profit, 481, 486–487, 663–669

Progress maps, 589

Propane, 191, 216, 217, 484, 615

Prophylene, 185

Prospect, petroleum, 10, 658–673; discovery, 659–660; economic factors, 663–669; geological factors, 660–663; personal factors, 669–672

Prospecting, 10–11

Proteins, 509

Proterozoic, 39

Protopetroleum, 504

Province, petroleum, 15, 31, 627–657; carbon-ratio theory, 630–634; evidences of oil and gas, 637–639; local traps, 653–654; regional arching, 652–653; sedimentary basins, 634–637; sediments, 628–636; unconformities, 639–645; wedge belts of permeability, 645–651

Psi, 390

Psig, 390

Pulaski fossil stream channel, Tennessee, 301, 303

Pumping well, 175

PVT, 389

Pye, Willard D., 27

Pyrite, 371

Pyrobitumen, 15, 23, 28, 506, 675, 679

Pyrolysis, of muds at low temperature, 526

Qatar, production, reserves, 34

Quaiyarah field, Iran, seepages, 19

Quartz, as cementing material, 129; compressibility, 133

Quartzite, 68

Quaternary, 39

Quicksilver mines, petroleum associated, 27

Quirequire field, Venezuela, buttress sands, 335, 337; nonmarine reservoir, 87, 335; variable concentrations, oil-field water, 161, 335

Quirequire formation, Venezuela, nonmarine reservoir, 87

Quitman field, Texas, faulting, 261, 263

Radiation logs, 81–83, 103

Radioactive: bombardment, transformation of organic matter into petroleum, 517, 526–529; decay products, in natural gas and brines, 190; disintegration, source of heat, 423; drilling mud, permeability measurement, 109–110; elements in sedimentary rocks, 527–528; sedimentary rocks, 82–83, 527–528; source of helium, 218–219
Radium, disintegration, 527; source of helium, 218–219
Radon, disintegration, 527
Ramparts, reef, 314
Rangely field, Colorado, 67; crude oil properties, 212; gravity of oil, 196; interstitial water, 153; pressure-volume-temperature relations, crude oil, 201, 203; porosity and permeability, 106; temperature gradients, 419, 421, 423
Rankama, Kalervo, 38
Rattlesnake field, New Mexico, helium-nitrogen gas content, 219–220
Reaction value, water analysis, 163, 170
Reagan sandstone, Cambrian, Texas, 525
Realgar, 371
Recent, geologic time, 39
Recharge, aid to maintaining original pressure, 465; area, determined from water analyses, 172
Recoverable reserves, estimating, 176
Recovery factor, 176
Recrystallization, 56, 57, 69, 70, 99, 117, 126–127, 129, 173, 541
Recumbent fold trap, 244–245
Recycling, 440, 460, 516–517
Reddin district, Oklahoma, 336
Redeposition, 57, 118
Red Fork sand, Oklahoma, gravity of oil, 197; offshore sand reservoir, 301
Red Sea, fringing reef, 312; region, salt domes, 359
Reducing environment, 517–518
Reduction of sulfates to form hydrogen sulfide, 222
Redwater pool, Canada, compressibility of crude oil, 468; limestone reservoir, 117; section, 330
Reefs, atoll, 313; barrier, 312; bioherm, 306, 310; biostrome, 306; "blowholes," 317; builders, kinds of organism, 315; cause of buried hills, 252; complex, 324; exploration for, 332; fossil, 318–325; fringing, 312; guyot, 314; lime, 310; lime banks, 310; limestone massifs, 310; masses, 310; modern, 312–318; organic, 306, 308–310; pools in, evidence of migration, 541; primary stratigraphic traps, 306, 308–332; productive, 325–332; ramparts, 314; reef-like complex, 324; reef-like deposits, 310; reefy complex, 324; reservoirs, 70–72; "room and pillar" structure, 317–318; sea mounts, 314; shingle ramparts, 314; table, 314
Refiner, 177
Reflecting power, rocks, 10
Refractive index (RI) of crude oils, 207–208
Regional: arching, petroleum province, 652–653; mapping, 589
Regressive overlap, 334
Regressive shore lines, effect on reefs, 324–325
Regulations, petroleum production, 486–487
Reitbrook salt dome, Germany, faulting, 363–364, 369
Relative densities, 10
Relative permeability, 110–112, 461
Release of oil, from plants, bacterial, 520; from sedimentary rocks, 519–520
Relief model, 616
Repeated folding, 246, 247–249
Replacement theory, primary migration, 546
Repressuring, 460
Reserves, 7–8; estimates, 176, 664; producible, 7; proved, 7; recoverable, 176, 481–484; secondary, 482; source, 8; world, by geographic location, 32–35; world, by geologic age, 40
Reservoir, 6, 49–384; conditions, 389–432; confined, 400, 401, 426; dynamics, 385–493; energy, 176, 458–473, 484; essential elements, 49–50; fluids, 144–231; isolated, 411–414; mechanics, 433–493; performance, 478; pore space, 97–143; sealed, 400, 401, 403, 411, 414, 426; temperature, 414–428; traps, 50, 232–384; unsealed, 400, 405
Reservoir rock, 6, 49, 52–96, 97; absent at dry hole, 619–620; acidization, 135–136; associated with unconformities, 333; carbonate, 68–72, 115–119; chemical, 68–73; chemically precipitated carbonate, 70–72; classification, 52–54; compaction, 130–133; differences, 540; dolomitic, 126–127, 353; fractured, 119–125; fragmental, 54–68; geologic age, 15, 37–41; geologic life, 247; igneous and metamorphic, 74–75, 120–121, 305–306; marine and nonmarine, 86–89; nomenclature, 53–54; pores per acre-foot, 99–100; shooting, 135; siliceous, 72–73; trap must occur in, 239; *see also* individual pools, fields, formations, lithologies, and ages
Residual: anticline, 378; dome, 378; oil, 112, 461; water, 453, 457
Residue gas, 216, 484–485
Resin, in crude oil, 189

INDEX 717

Resistivity, calculated from water analysis, 171–172; electric logs, 77, 80; formation factor, 157–160
Resources, 7; transformed to reserves, 8
Retrograde condensate pool, 439–440, 464, 485
Retrograde condensation, 439
Reverse fault, traps, 270–278
Reynolds oolite pools, Arkansas, zone of reduced pressure, 400
Rhizoslenig, blooming, 513
RI (Refractive index) of crude oil, 207–208
Rich, John L., 520, 541, 544
Richland pool, Texas, fault trap, 259, 262
Riecke principal, 129
Rig, 679
Rim syncline, 378
River water, organic matter in, 514–515
Robinson Syncline pool, West Virginia, no free water, 152
Rock, barrier, 456; compressibility of, 468–471; geophysical properties, 10; pressure, 390, 402
Rocky Mountain region, age of traps, 579–580; composite basins, 634–635; correlation by water analysis, 172; erosion of, 548; low-concentration oil-field water, 161; petroleum province, 32; reservoirs, 59; sulfate-free water, 165, 168, 170
Rodessa field, Louisiana and Texas, crude oil properties, 212; structural map, 350–351; water analysis, 166
Rogers, G. Sherburne, 219
Romania, electric logs, 78; mining oil, 472; Moreni-Bana field, 275, 280; nonmarine reservoirs, 87; Ploesti region fields, 275, 280; salt anticlines, 250, 358; Tertiary reservoirs, 60
Roof rock, impervious, 6, 50, 232, 540, 579
"Room and pillar" structure, reefs, 317–318
Rotary drillers' logs, 75–76
Rotary drilling, 679
Rotary holes, permeability measurements, 109; pressure measurement, 393; sample logs, 75–76
Round Mountain field, California, normal-fault trap, 257, 261
Royalty, 665
"Rubber structures," 240
Rubey, W. W., 407
Russell Ranch field, California, trap obscured by thrust fault, 255, 257
Russell, William L., 236, 527, 528, 633
Russia: *see* USSR

Sabine arch, Louisiana, regional fold, 652; relation of oil to gas pods, 486

Sage Creek pool, Wyoming, tilted oil-water contact, 563, 566
St. Peter sand waters, isocon map, 161–162
Ste. Genevieve formation, Illinois, oolite reservoir, 117
Salina dolomite, Silurian, Michigan, high mineral concentration in oil-field water, 160
Salinity, water analysis, 163
Sallyards shoestring-sand trend, Kansas, 297, 299
Salt: anticlines, 250, 358; domes, *see* Salt domes; effect on electric logs, 79; "glacier," 358; in oil, 156, 190
Salt Creek field, Wyoming, CI of crude oil, 194; separate oil-water contacts, 149, 151; shale reservoir, 67
Salt domes, 356–379; cap rock, 367–371; characteristics, petroleum accumulation and production, 378–379; classification, 359, 361; deep, 359; Gulf Coast, 359–367; intermediate, 359; internal structure, 361–362; occurrence, 358–359; origin, 374–379; overlying faults, 362–369; piercement, 359; pools, 356–357; reservoir traps, 356–379; seepages associated, 16; surface expression, 361; traps associated, 371–374
Salt plug, *see* Salt domes
Salt-plug pools, 356–357
Sample logs, 76–77
San Andreas limestone, Permian, variable porosity and permeability, 101–102
Sand Belt pools, Texas, 292
Sand, bituminous, 24; produced with oil in high pressure wells, 410; -shale ratio maps, 601, 603
Sanders, J. McConnell, 505
Sandfrac, 136
Sandstone reservoir rock, 56–65
San Joaquin Valley, California, sulfate-free water, 165
San Pedro field, Argentina, nonmarine reservoir, glacial, 87
Santa Fe field, California, dome forming many traps, 238–239
Santa Maria field, California, fractured reservoir, 121; siliceous shale reservoir, 67; unconformity trap, 338, 341
Sapropel, 513; composition, 519
Saturated hydrocarbon, 178, 509, 519
Saturated pool, 213, 463
Saturation, 110–112, 146–147, 546; funicular, 448–449, 474; insular 448–449, 474; irreducible water, 453, 455, 457; kinds in a pore, 449; method, gas measurement, 214–215; method, oil measurement, 176; oil, maintained near well,

474; pressure, 213, 460, 573; relation to capillary pressure, 453–457; relation of oil to gas, 475; water, 157–160
Saudi Arabia, Abqaiq field, 119, 212, 244, 466; Ain Dar pool, 466; Arab zone, Jurassic, 71, 119, 212, 244; petroleum province, 32; production, reserves, 34; salt domes, 359
"Scale," hard and soft, 173
"Scalped" structure, 620
Schuler field, Arkansas, oolite reservoir, 116–117, 563; pressure drop, 466, 562–563
Scipio-Albion field, Michigan, fracture permeability, 123–124
Scout reports, abbreviations used, 680–684
Scripps Institution of Oceanography, work on relation of bacteria to origin of petroleum, 518
Scurry reef, Texas, 327
Scurry-Snyder field, Texas, reef reservoir, 326–329
Scurry-Snyder reef, Texas, 326–327, 329
Sea mounts, reefs, 314
Sea of Azov, "oily patches," 513
Sea water, dissolved mineral matter, 160, 166, 173
Sealed reservoir, 400, 401, 403, 411, 414, 426
Secondary: free-gas cap, 460; migration, 538, 550–581; minerals, evidence of metamorphism, 629; porosity, 117–133; recoverable oil, 461; recovery, 456, 460, 481–484; stratigraphic traps, 286, 333–339
Sections, 591–596
Sedimentary: basins, petroleum provinces, 634–637, 652; compaction theory of primary migration, 545; volcanism, 20
Sedimentation, 9, 11
Sediments, abyssal, 61; bathyl, 61; carbon-ratio theory, 630–634; compaction of, 130–133; composition of organic matter, 506, 519; fractured reservoirs, 121–125; neritic, 61; oil-settling capacities, 550; in petroleum province, 628–637; in petroleum province, character, 629; in petroleum province, metamorphism, 629–630
Seepages, 14, 15–19, 26–27, 88, 234, 260, 269, 427, 638
Seismic: anomaly, 612; mapping, discoveries, 613; maps, 242, 613–615; surveys, determination of structure at depth, 255; wave velocity, how determined, 591
Selective trapping, 486, 574
Self-potential log, 78–80
Separator, 175, 484, 679
Series, time-stratigraphic unit, 38

"Serpentine" reservoirs, Texas, 305–306
Seyer, W. F., 520
Shale: heaving, 373–374; reservoir rocks, 67
Shale-shaker, 680
Shallow and surface weathering phenomena, changes in structure with depth, 251, 252, 256
Shape, clastic rock particles, 114–115
Shearing, 521, 526
Sheathing, along salt plugs, 361, 364, 365, 369, 373
Shensi series, China, nonmarine reservoir, 88
Sheppard, C. W., 527
Shield, 635; areas, no petroleum prospects, 40
Shiells Canyon field, California, calcium chloride brine, 174; nonmarine reservoir, 86–87
Shikhan, 323, 330
Shingle ramparts, reef, 314
Shoestring-sand traps, 293–305
Short distance migration, 541–544
Showings, minor, subsurface, 15, 29–30
Shrinkage factor, 176, 199, 200, 205, 206
Shut-in pressure, 392, 473
Siderite, 371
Side-wall coring, 680
Silica cement, 67, 73, 129–130
"Siliceous lime," Oklahoma, 72
Siliceous reservoir rocks, 72–73
Silliman, B., Jr., 17, 514
Silurian, 39; Fusselman limestone, Texas, 119; Niagaran reefs, 72, 318–320; Salina dolomite, Michigan, 160; Sylvan shale, Oklahoma, 525
Simpson sands, Ordovician, clean sandstone reservoir, 59
Six-Lakes gas pool, Michigan, in offshore sand-bar deposits, 298–299
"Skin effect," 474
"Skin frictional heat," 521
Sleeves, channel sands, 304–305
Slick, Tom, 54
Slimepits, 19
Silverville sand, Devonian, Pennsylvania, offshore sand-bar reservoir, 298–300
Slumping, superficial folds, 252, 256
Smackover limestone, Jurassic, Arkansas, oolite reservoir, 116–117; porosity and permeability, 106; static pressure gradient, 394–395; tilted oil-water contact, 562–563; zone of reduced pressure, 400
Smithsonite, 371
Sodium, 173
Sodium chloride: see Salt
"Soft scale," 173
Soil analysis maps, 615

Solid petroleum, occurrence, 23–27
Solubility, order of carbonates, 118; relation to pressure, 173–174, 201–202, 213–216, 464, 484–486, 545; relation to temperature, 173, 201, 203, 213–214, 427, 464
Solution, cause of anomalous structures, 251, 252, 256; effect on trap, 541, 542; -gas expansion pool, 459; -gas pool, 459; relation to cementation, 130; relation to fractures, 119–120; relation to porosity and permeability, 57, 99, 117–119, 327; relation to unconformities, 333, 335
Sonde, 77–78, 80
Sonic logs, 85, 103
Sooy conglomerate, Pennsylvanian, Kansas, chert reservoir, 63
Sorption, 443
Sour gas, 218
Sour water springs, 361
Source, carbonate cements, 130; heat energy, 422–424; hydrogen, 180; oil-bearing sediments, 548–550; reservoir energy, combined, 472–473; reservoir pressure, 401–405; rock, 31–32; siliceous cement, 129; *see also* Migration, Origin, and Reservoir
South America, ash in crude oil, 190; production, reserves, 35
South Ceres pool, Oklahoma, shoestring-sand reservoir, 300–301
South Coles Levee field, California, displaced oil pool, 563–564
South Earlsboro pool, Oklahoma, maps, 242–243
South Liberty salt dome, Texas, 369
South Mountain field, California, calcium-chloride brine, 174; nonmarine reservoir, 86–87
South Texas petroleum province, trends, 292
Southern fields, Mexico, primary porosity in carbonate reservoirs, 115–116
S. P. log, 78–80
"Sparkling" sandstone, Petrólea oil field, Colombia, 129
Specific gravity, differences with depth between salt and sediment, Gulf Coast, 377; gas, 216, 554; oil, 554; relation to interfacial tension, 451; various sediments, 402; water, 160, 554–555; *see also* Density and Gravity of crude oil
Specific productivity index (SPI), 478
Specific surface, 135
Sphalerite, 371
SPI, specific productivity index, 478
Spill point, 573–574
Spindletop field, Texas, CI of crude oil, 194

Spontaneous-potential log, 78–80
Spouter, 479
Spraberry field, Texas, clastic ratio map, 604; compressibility of water, oil, and rock, 468; fracture permeability, 121; imbibition, 477; shale-siltstone reservoir, 28, 67, 121
Spraberry formation, Texas, shale-siltstone reservoir, 28
Spring Creek field, Wyoming, crude oil properties, 212
Springhill sand, Cretaceous, Chile, variable porosity and permeability, 100–101
Springs, 4, 14, 15–19, 570
Spud in, 680
Stage, time-stratigraphic unit, 38
Stanley shale, Mississippian, Oklahoma, 525
Static: bottom-hole pressure, 391–392; formation pressure, 392; gradients, 394–396; pressure gradient, 397; water level, 393
Stevens zone, Coles Levee fields, California, 563–564
Stiff method, water analysis diagram, 165, 169
Stock-tank oil, 199, 461
Stoke, 207
Straight-run gasoline, 185
Strand-line trend, 297
Stratafrac, 136
Strat hole, 662
Stratigraphic: barriers to migration, 568–569; cross section, 592; tests, 662; traps, 232, 237, 286–339, 556–558, 653–654; traps, exploration for, 342–343; traps, primary, 286, 287–332; traps, secondary, 286, 333–339
Stratigraphy, 9, 11
Stripper pool, 480–481
Stripper well, 480–481
Strontianite, 371
Structural: basins, 634; closure, 241; cross section, 592; geology, 9, 11; mapping, 238–240, 282; maps, 591–596; maps, *see also* individual pools, fields, and areas; relief, 241; trap reservoirs, comparison with others, 280–281; traps, 232, 237, 238–285, 555–556, 653–654; traps, exploration for, 342–343; traps, important features, 282
"Structure," 238
Structure, internal, salt plugs, 361–362
"Structures, rubber," 240
Styolites, 66
Subcrop maps, 602–603, 605, 606, 643; *see also* Paleogeologic maps
Submarine: earthquakes, 404; landslides, 19, 20, 21, 63, 324; seepages, 19, 549

Subnormal pressure, 396, 405, 407, 411
Sub-sea maps, 592; *see also* Structural maps
Subsidence, Long Beach oil field, California, 132
Subsurface: geology, 587–626; mapping, 11, 239, 282, 588–618; maps, kinds, 591–618; occurrence, 14, 15, 29–32; occurrence, *see also* Occurrence
Sulfate: -free waters, 165, 168; in oil-field water, 152; reduction to sulfide, 168, 222, 518, 519
Sulfide, oxidized to sulfate, 152; *see also* Hydrogen sulfide
Sulfur, in cap rock of salt domes, 367, 370–371, 375; in crude oil, 177, 187–188, 506; in natural gas, 177, 664; in organic matter, 506, 515; water springs, 361
Sumatra, Talang Akar pool, 271, 273; Talang Akar sandstone, Miocene, 273
Sunlight penetration, sea water, 512
Surakhany field, USSR, gravity of oil, 197
Surface-active agents, relation to interfacial tension, 444–445, 449, 450, 451, 520
Surface: energy, 442–451; expression, salt domes, 361; free energy, 442; mapping, 11, 238–239, 282; occurrence, 14, 15–29; pipe, 680; pressure, 392; tension, 442–451; weathering phenomena, 252, 256
Sverdrup, H. U., 510
Sweden, Kolm black shale, Cambrian, oil and uranium mine, 28
Sweet gas, 218
Sylvan shale, Silurian, Oklahoma, 525
Syncline, rim, 378
System: phase relationships, 434–439; time-stratigraphic unit, 38

Table reefs, 314
Talang Akar pool, Sumatra, overthrust anticline, 271, 273
Talang Akar sandstone, Miocene, Sumatra, 273
Talus, in reef complex, 312
Tamabra limestone, Cretaceous, Mexico, dolomitized reservoir, 127; up-dip wedge-out, 353, 356
Tampico-Tuxman region, Mexico, natural gas high in hydrogen sulfide, 222; seepages associated with igneous intrusions, 427
Tar, 3, 4, 8, 23, 25, 680
Tatums pool, Oklahoma, complex trap, 338
Tax position, 668
Tehachapi-Bakersfield earthquake, California, 404

Temperature: bomb, 415; constant during production, 213, 415, 421; effect on permeability, 114, 427; effect on pressure, 403, 426; effect on volume, 203, 214–215, 424–426, 571–574; fluctuation, 540–541, 542; formation, how determined, 590; gradient, 389, 415–421, 428–429; interpreted through carbon ratios, 630–634; logs, 85; measurement, 415; measurements, uses, 422; phase relationships, 434–441; -pressure diagram, oil-gas system, 439; relation to density of water, 426; relation to gas in solution, 201, 203, 213–214, 427, 464, 547; relation to gravity of crude oil, 195–196, 206; relation to interfacial tension, 442, 451; relation to power of adsorption, 173; relation to pressure and volume in oil, 203; relation to solubility of gas in oil, 427; relation to solubility of salts, 173, 427; relation to structural position, 421; relation to viscosity, 114, 204, 206, 389–390, 424–425, 547–548; reservoir, 145, 176, 214, 414–428, 501–502, 540–541; triple point, water, 435; *see also* Heat
Tennessee: Nashville Dome, 645–646, 653; Pulaski fossil stream channel, 301, 303
Ten Section field, California, multi-pay, 147, 149; porosity and permeability, 106
Tension, boundary, 187, 443; interfacial, 66, 442–451, 454, 456, 547, 549, 552–553; surface, 442–451
Tensleep sands, Pennsylvanian, Wyoming, displaced oil pool, 563; oil-wet, 450; range in specific gravity, 196
Tertiary, 39; Asmari limestone, Iran, 18, 71, 125, 248, 400, 480; formation of traps, 579–580; mud volcanoes, 19–23; nonmarine reservoirs, 86–88; production, reserves, 38, 40–41; reservoir rocks, 38, 40–41, 58, 60, 61, 71, 155, 304, 580; seepages, Iran, 18–19; up-dip wedge-outs, Gulf Coast, 648; water analyses, California, 171
Test well, 7, 590, 618, 620; *see also* Wildcat well
Texas, Abo-Wichita Albany reef, 320; Amarillo Arch, 65; Amarillo uplift, 549; Amarillo-Texas Panhandle field, 64, 120–121, 217, 411; Amelia field, 270–272; Apco pool, 127, 338–340; Armstrong pool, 292–293; asphaltic limestone, 25; Barbers Hill salt dome, 172–173, 370; Bend Arch, 652–653; Capitan reef, 320–321, 326; Carthage gas field, 117, 217, 352; Cayuga field, 613; Cedar Lake field,

101–102; Central Basin Platform, 73, 326, 652; Conroe field, 147–148, 166; continental shelf, 37; Darst Creek pool, 576–577; different characteristics in different geologic layers, 641–642; Dollarhide pool, 119; down-fault traps, 266, 270–272; East Texas Basin, 400–401, 466–467, 468, 469; East Texas pool, 31, 58, 67, 106, 166, 194, 351, 466–467, 573, 653; East Wasson field, 409; geothermal gradients, 417; Government Wells field, 292, 294; Hastings field, 363, 367, 579; Hawkins field, 197, 249; Henrick pool, 251, 256, 326; Hoskins Mound salt dome, 371, 375; Hull-Silk field, 290–291; interfacial tension between reservoir water and crude oil, 442; isogradient map, 417, 419; Lytton Springs pool, 306; Mercy field, 133; Mount Calm pool, 280, 282; natural gas high in hydrogen sulfide, 222; North Snyder reef, 329; novaculite, 73; Panhandle field, 64–65, 217, 222; Permian Basin, 72–73; Pickett Ridge field, 292, 295; pools in reefs, 326–327; Port Neches salt dome, 364–365, 370; Powell field, 212; pre-Pennsylvanian unconformity, 643; Quitman field, 261, 263; relation of oil pools to gas pools, 486; Richland pool, 259, 262; Rodessa field, 166, 212, 350–351; Sand Belt pools, 292; Scurry-Snyder reef and field, 326–329; South Liberty salt dome, 369; Spindletop field, 194; Spraberry field, 28, 67, 121, 468, 477, 604; sulfur content of crude oil, 188; sulfur content of gasolines, 188; Tertiary reservoirs, 58; Tex-Harvey pool, 121; tilt of Paleozoic rocks, 542; Todd field, 115, 308–310; Tyler Basin, 400–401, 467, 469; Van field, 149–150, 264, 268, 363, 579; Westbrook field, 220; Wheat pool, 563; Yates pool, 480; *see also* Gulf Coast

Tex-Harvey pool, Texas, reservoir, 121
Texture, 54; chemically precipitated carbonates, 70; chert, 63; clastic reservoir rocks, 58
Thermal conductivity, 419, 421, 423–424
Thermal cracking, 524
Thermodynamics, 11
Thiel, G. A., 549, 559
Thigmotactic bacteria, 520
Thinning, determined from isopach maps, 599–600; toward salt domes, 364–365, 370, 371, 372
Third Stray sand, Devonian, Pennsylvania, patchy reservoir, 288–289
Thorium, radioactive element, 218–219, 527; source of helium, 218–219

Three component systems, 435
Thrust fault, traps, 270–278, 279, 280, 281
Tickell, method of water analysis diagram, 165, 169
Tidal-flat channels, 61
Tideland pool, California, deflected wells, 594–595
Tides, cause of pressure changes, 404
Tierra del Fuego, Chile, variable porosity and permeability, 100–101
Tight sand, 105
Tilted oil-water contacts, 50, 236, 547, 561–568, 569, 570, 653
Tilting, opposed, formation of trap, 652–653; regional, forces involved, 611
Timber Canyon field, California, overthrust fault trap, 272, 274
Time, mud return, rotary drilling, 76–77; necessary for origin, migration, and accumulation of petroleum, 540; of petroleum accumulation, 571–580; substitute for temperature in some geologic reactions, 520–521; trap formed, 542, 575–580, 581
Time-stratigraphic units, 38
Todd field, Texas, reef reservoir, 115, 308–310
Top water, 151
Torbanites, 8, 29, 680
Torrance pool, California, basement reservoir, 120
Tortuosity, 98; effect on permeability, 115
Total porosity, 101
Touragai mud volcano, USSR, 21
Tow Creek field, Wyoming, shale reservoir, 67
Transformation of organic matter into petroleum, 517–529
Transgressive overlap, 334
Transgressive shore lines, effect on reefs, 324–325
Traps, 6, 50, 236, 660–661; absence, cause of dry holes, 619; anticlinal theory, 233, 234–236; areas of low potential energy, 561; associated with unconformities, 252, 257, 310, 333–339, 639–645; caused by faulting, 258, 259, 260–278; caused by folding, 240–259; caused by fracturing, 278–282; changes with depth, 245–259; changes with time, 542; classification, 236–238; combination, 237–238, 286, 349–356; deformational, 236; depositional, 236, 287; diagenetic, 236, 287; different kinds in different kinds of sediments, 653; different position at depth, cause of dry holes, 620; effect of intrusives, 427–428; fluid, 340–342; lenses and facies in chemical rocks, 306–332; lenses and facies in clastic rocks, 287–

305; lenses of volcanic rock, 305–306; local, 653–654; porous carbonate facies, 306–308; primary stratigraphic, 286, 287–332; reefs, 306, 308–332; reservoir, 50, 232–384; salt domes, 356–379; secondary stratigraphic, 286, 333–339; in serpentine, 305–306; shoestring-sand, 293–305; stages of accumulation, 574; stratigraphic, 232, 237, 286–339, 342–343, 556–558, 653–654; structural, 232, 237, 238–285, 342–343, 555–556, 653–654; time formed, 542, 575–580, 581; variation in character, 540

Trask, Parker D., 504

Tra-tau, USSR, reef, 323

Treibs, A., 521

Trenton formation, Ordovician, carbonate reservoir, 127, 307; fossil channel, 301, 303; up-dip wedge-outs, 647–648

Triassic, 39; low in hydrocarbons, 38, 504; production, reserves, 38, 40; reservoirs, 38, 40, 41; seepages, Brazil, 427; Shensi Series, China, 88; Barrackpore field, 23, 217; burnt clay, 27; Forest Reserve field, 410; gravity of crude oil, 195; graywacke reservoirs, 55; mud flow, 23; petroleum province, 32; production, reserves, 34; seepages, 16, 17, 19

Trinity Forest Reserve, Trinidad, seepages, 17

Tri-State mining district, oil-filled geodes, 27

Triumph Streak, Pennsylvania, 289

Tsunamis, cause of pressure changes, 404

Tsupinkai field, Tertiary, China, nonmarine reservoir, 88

Tubing, 680; pressure, 392

Tuffs, reservoir rock, 74–75, 427

Tupambi formation, Argentina, glacial reservoir rock, 87

Turbidites, 63

Turbidity currents, 63

Turbulent flow, 104

Turkey, production, reserves, 34

Turkman, USSR, nonmarine reservoirs, 88

Turner Valley field, Canada, composition of natural gas, 217; geologic history, 651; overthrust field, 272, 275, 276

Tuscarora-Woodhill gas pools, New York, traps associated with reverse faults, 274–275

Two-component system, 434–435

Tyler Basin, Texas, pressure decline, 400–401, 467, 469

Tzuliuching limestone, Cretaceous, China, nonmarine, seepages, 88

Uinta Basin, Utah, nonmarine reservoirs, 87

Uintaite, 25, 680

Ulmic acid, 515

Ultraviolet radiation, 98, 208–209

Unconformities, 15, 333, 639–645; angular, 254, 333–340, 599, 621; associated with arkose, 64; with associated traps, 252, 257, 310, 333–339, 639–645; "bald-headed" structures, 614, 620; cause of barren traps, 621; cause of salinity variations, in oil-field brines, 152, 172, 174; effect on permeability, 74, 333–339, 639–645; effect on porosity, 118–119, 333–339, 639–645; fracturing below, 119; how determined, 334, 591; topographic relief on, 249, 254

Undersaturated pool, 213

United Kingdom, production, reserves, 34

United States, amount of petroleum exploration, 654–655; barrels per cubic mile of sediments, 628–629; consumption of crude oil, 660; gravity of crude oil, 195; MER, 478; number of producing wells, 481, 654–655; number of stripper wells, 481; oil found per foot drilled, 671; production regulations, 487; production, reserves, 33, 34, 478, 482; secondary recovery, 484; secondary reserves, 482

Unsaturated hydrocarbon, 178, 509, 519

Unsealed reservoir, 400, 405

Uranium, in black shale, 28, 527–528; disintegration, 218–219, 527; in petroleum, 190; source of helium, 218–219

USSR, Achi-Su field, 275, 279; Apsheron Peninsula, 19, 21–22, 32, 87–88, 197; Baku province, 16, 19, 21–22, 58, 195, 217, 262–263, 480, 629; Bibi-Eibat field, 22, 198; diapir folds, 250; Dossor field, 260, 262–263; Emba province, 217, 260, 262–263, 358, 365, 372; Grozny district, 189, 195, 217, 275, 279, 480; Ishimbaevo field, 323, 331–332; Iskine salt dome, 365, 372; Kala oil and gas field, 262–263; Khadyzhinskaya field, 292, 296; Kuban-Black Sea area, 217; Maikop region, 58, 292, 296, 304–305; mining oil, 472; mud volcanoes, 21–22; natural gas measurements, 214; New Grozny field, 162; nonmarine reservoirs, 87–88; petroleum province, 32; sandstone reservoirs, 58, 60, 61; seepages, 16–17, 19; Surakhany field, 197; Tra-tau, reef, 323; use of electric logs, 78

USSR-China, production, reserves, 35

Utah: bituminous sands, 24; Green River formation, Eocene, 87; Kaibab limestone, Permian, 27; natural gas high in carbon dioxide, 221; nonmarine reservoirs, 87; oil-filled geodes, 27; ozokerite,

185; Paradox salt dome region, 358–359; Uinta Basin, 87

Validity of data, 240, 661
Vallezza field, Italy, complex structure, 244–245
Vanadium, 190, 525; mines, petroleum associated, 27
Van der Waals forces, 441–444
Van field, Texas, fold caused by salt intrusion, 149–150, 363; structure, 149–150, 264, 268; two periods of oil movement, 579
Van Tuyl, F. M., 377
Vapor pressure, 213
Variability, lateral and vertical, clastic reservoir rocks, 60–65
Variables, unknown in geology of petroleum, 457
Variations from expected pressures, 405–414
Vedder, Dwight D., 54
Vedder sand, Round Mountain field, California, 257, 261
Vein fillings, solid petroleum, 14, 15, 24, 25–27
Velma field, Oklahoma, composition of natural gas, 217
Venango group, Devonian, Pennsylvania, patchy reservoir, 288–289
Venezuela: Bolivar Coast fields, 293, 297, 648–650; Boscan field, 198; Cumarebo field, 106, 217; exploration, 33, 36, 379–380; gravity of crude oil, 195; La Concepcion field, 217; Lagunillas field, 166, 297; Lake Maracaibo district, 67, 293, 297, 648–650; La Paz field, 125, 419, 422; Las Cruces field, 161; location of oil fields, western, 649; Mara field, 72, 125, 263; Maturin basin, 636; Mene Grande field, 353, 355, 357; Oficina district, 416; petroleum province, 32; production, reserves, 34; Quirequire field, 87, 161, 335, 337; refractive index of crude oils, 207–208; sandstone reservoirs, 58; seepages, 16–17; use of electric logs, 78
Venice pool, California, basement reservoir, 120
Ventura Basin, California, petroleum province, 60, 86–87
Ventura field, California, complex thrust faulting, 278, 409–410; excess pressures, 409–410
Vertical migration, 502, 569–571
Vertical permeability, 108
Virgin pressure, 391
Viscosimeter, 206–207
Viscosity, 201–207; chagnes in, 547–548; conversion factors, 686; crude oil, 201–207; crude oil, measurement, 206–207; relation to dissolved gas, 205–206, 462–463, 484, 547–548, 558–560; relation to formation-volume factor, 202; relation to gravity of crude oil, 204–206; relation to interfacial tension, 451, 547–548; relation to pressure, 204, 389, 425, 547–548; relation to shrinkage and gas solubility, 206; relation to temperature, 114, 204, 206, 389–390, 424–425, 547–548; relation to temperature and gravity, 204, 206; of water, 424–425, 548

Viscous flow, 104
Voids: *see* Pore space
Volcanic: ash, reservoir rock, 60, 67; gases rich in carbon dioxide, 221; gases rich in hydrogen sulfide, 222; rock lenses, reservoirs, 305–306; theory, origin of salt domes, 374–375
Volcanism: effect on salinity of oil-field water, 174; not associated with petroleum occurrences, 503–504; sedimentary compared to igneous, 20
Volcanoes, mud, 14, 19–23
Volume: of crude oil, 198–201; plastically deformed rock, 130–133; relation to pressure, natural gas, 215–216, 571–574; relation to pressure, oil, 146–147, 198–201; relation to pressure and temperature in oil, 203; relation to temperature, 203, 214–215, 424–426, 571–574; of reservoir, folded trap, 242
Volume-for-volume replacement, dolomitization, 126
Volume-pressure method, gas measurement, 215–216
Volumetric method, gas measurement, 214–215; oil measurement, 176
Voshell pool, Kansas, structural history, 577–579
Vugs, 14, 15, 27, 102; porosity, 127, 480

Warrensburg channel, Missouri, 303
Wasatch formation, Eocene, Colorado, nonmarine reservoir, 87
Washington, Copalis Bay, blooming, 512
Waste, reservoir energy, 477, 487
Water: capillary pressure, 451–457; density, 426; flooding, 107, 460, 482–484; interface phenomena, 441–451; specific gravity, 160, 554; table, 50; viscosity, 424–425, 548; *see also* Oil-field water
Water-drive pools, 464–471
Water pressure, 390
Water-wet, 445; capillary opening, 447; reservoir rocks, 107, 156, 446–451, 546; sand grains, diagram, 156

Wax, 23; composition, 515; paraffin, 183, 185
Weathering, effect on porosity, 118; phenomena, surface, 252, 256
Wedge belts of permeability, 349–356, 645–651
Weeks, L. G., 634, 635
Weeping core, 638
Well: confirmation, 7; cuttings, 84, 103, 376, 590; deflected, 592, 594–596, 620; development, 7; discovery, 7; exceptional, 479–480; first drilled by Drake, 288, 630, 654; flowing, 458, 479; flowing artesian, 394; footage drilled annually, 588; fountain, 479; information available from, 590–591; life, 481; manometer, 395; number producing in U.S., 481, 654–655; pressure, 390; pumping, 175; samples, 590, 591; sitting on, 76; stripper, 480–481; test, 7, 590, 618, 620; wet, 7, 151; wildcat, 7, 10, 84, 562, 618, 659–660
Well-site geologist, 76
Wentworth particle size classification, 55
Wertz Dome field, Wyoming, natural gas high in carbon dioxide, analysis, 221
West Edison field, California, fault trap, 258, 261
West Edmond pool, Oklahoma, fracture permeability, 122; paleogeographic map, 643; up-dip wedge-out, 337; viscosity changes, 205–206
Westbrook field, Texas, natural gas high in nitrogen, 220
Western Hemisphere, crude oil production, 33
West Indies, reefs, 323
West Virginia: ash in crude oil, 190; Big Injun sand, Mississippian, 60; Cabin Creek field, 152; Copley pool, 152; gas wells on anticlines, 235; Granny Creek pool, 152; Griffithsville pool, 152; Robinson Syncline pool, 152
Wet gas, 216
Wet well, 7, 151
Wettability, 66, 100, 445–451
Wetting fluid, 111–112
Whale Bay, Africa, blooming, 513
Wheat pool, Texas, tilted oil-water contact, 563
Wheeler, Walter Calhoun, 514
Whipstock, 680
White, David, 522, 524, 630
White, Don E., 404, 503
White, I. C., 234
Whitehead, Walter L., 526, 527
Whitehorse formation, Permian, Texas, reservoir, 292

Wichita Mountains, Oklahoma-Texas, erosion of, 65, 549
Wienhausen-Eichlingen salt dome, Germany, overhang, 369, 373–374
Wilcox, Homer F., 54
"Wilcox" sand, Ordovician, 59, 242, 243; porosity and permeability, 106, 114
Wilcox sandstone, Eocene, Texas, porosity and permeability, 133
Wildcat well, 7, 10, 84, 562, 618, 659–660
Wildcatter, 7
Wild flysch, 20
Wilhelm, O., 237
Wilmington field, California, basement reservoir, 120; deflected wells, 594–595
Wilson, W. B., 236
Wisconsin, Silurian reefs, 318
Woodbend formation, Devonian, Canada, 117
Woodbine sand, Cretaceous, Texas, 58, 67, 155, 259, 261, 264, 268, 351, 466, 468, 574
Woodford radioactive black organic shale, 529
Woodhill-Tuscarora gas pools, New York, traps associated wtih reverse faults, 274–275
Working interest, 665
World, production, reserves, 33, 34–35, 38, 40
World Oil, 30, 675
Wurtzilite, 25, 680
Wyoming: Circle Ridge field, 275, 277; Elk Basin field, 196, 516; Frannie field, 563, 565; Lance Creek field, 194; natural gas high in hydrogen sulfide, 222; Northwest Lake Creek field, 563; Sage Creek pool, 563, 566; Salt Creek field, 67, 149, 151, 194; Spring Creek field, 212; sulfur content of crude oil, 188; Tensleep sands, Pennsylvanian, 196, 450, 563; Tow Creek field, 67; Wertz Dome field, 221

Yates pool, Texas, exceptional well, 480
Yegua-Jackson trend, Eocene, Texas, up-dip wedge-outs, 291–292
Yenang pool, China, nonmarine reservoir, 88
Yucatan area, Gulf of Mexico, marine seepages of petroleum, 19
Yugoslavia, production, reserves, 34

Zechstein area, Germany, salt domes, 358
Zero capillary-pressure plane, 566
Zimmerly, F. R., 521
Zinc mines, petroleum associated, 27
Zobell, Claude E., 518, 519
Zooplankton, 510